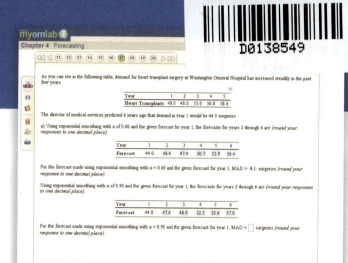

Built-in Student Help ▶

The exercises correspond to the exercises in this textbook, and they regenerate algorithmically to give students unlimited opportunity for practice and mastery.

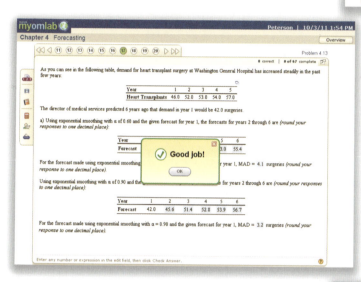

◀ Student Learning Aids

Break the problem into steps for a more active learning process.

Review a problem similar to the one assigned.

Link to the section in the textbook where this problem is covered.

See an instructor explain this concept.

Study Plan for Self-paced Learning ▶

MyOMLab generates a personalized Study Plan for each student based on his or her test results. The Study Plan links directly to interactive tutorial exercises for topics the student has not mastered.

MyOMLab®

Student Purchasing Options

- Purchase an access kit bundled with a new textbook
- Purchase access online at **www.myomlab.com**

TENTH EDITION

Operations Management

PROCESSES AND SUPPLY CHAINS

LEE J. KRAJEWSKI

Professor Emeritus at
The Ohio State University
and the University of Notre Dame

LARRY P. RITZMAN

Professor Emeritus at
The Ohio State University
and Boston College

MANOJ K. MALHOTRA

University of South Carolina

PEARSON

Boston   Columbus   Indianapolis   New York   San Francisco   Upper Saddle River
Amsterdam   Cape Town   Dubai   London   Madrid   Milan   Munich   Paris   Montreal   Toronto
Delhi   Mexico City   Sao Paulo   Sydney   Hong Kong   Seoul   Singapore   Taipei   Tokyo

Editorial Director: Sally Yagan
Editor in Chief: Donna Battista
Senior Acquisitions Editor: Chuck Synovec
Editorial Project Manager: Mary Kate Murray
Editorial Assistant: Ashlee Bradbury
Director of Marketing: Maggie Moylan
Executive Marketing Manager: Anne Fahlgren
Senior Managing Editor: Judy Leale
Production Project Manager: Mary Kate Murray
Senior Operations Supervisor: Arnold Vila
Operations Specialist: Cathleen Peterson
Creative Director: Blair Brown
Senior Art Director, Interior: Kenny Beck
Art Director, Cover: Steve Frim
Text and Cover Designer: Tamara Newnam

Manager, Visual Research: Karen Sanatar
Photo Researcher: PreMediaGlobal
Manager, Rights and Permissions: Hessa Albader
Cover Art: Shutterstock
Associate Media Project Manager: Sarah Petersen
Media Project Manager: John Cassar
Full-Service Project Management: Haylee Schwenk/
PreMediaGlobal
Composition: PreMediaGlobal
Printer/Binder: Courier/Kendallville
Cover Printer: Courier/Kendallville
Text Font: Utopia 9/11

Credits and acknowledgments borrowed from other sources and reproduced, with permission, in this textbook appear on appropriate page within text.

Microsoft® and Windows® are registered trademarks of the Microsoft Corporation in the U.S.A. and other countries. Screen shots and icons reprinted with permission from the Microsoft Corporation. This book is not sponsored or endorsed by or affiliated with the Microsoft Corporation.

Library of Congress Cataloging-in-Publication Data

Krajewski, Lee J.
Operations management : processes and supply chains / Lee J. Krajewski,
Larry P. Ritzman, Manoj K. Malhotra. — 10th ed.
    p. cm.
    Includes index.
    ISBN 978-0-13-280739-5 (alk. paper)
1. Production management. I. Ritzman, Larry P. II. Malhotra, Manoj K. (Manoj Kumar), 1960- III. Title.
    TS155.K785 2013
    658.5—dc23   2011043411

10 9 8 7 6 5 4 3 2
V011
www.pearsonhighered.com

ISBN 10:   0-13-280739-4
ISBN 13: 978-0-13-280739-5

Dedicated with love to our families.

■

Judie Krajewski

Gary and Christine; Gabrielle
Lori and Dan; Aubrey, Madeline, Amelia, and Marianna
Carrie and Jon; Jordanne, Alaina, and Bradley
Selena and Jeff; Alex
Virginia and Jerry
Virginia and Larry

■

Barbara Ritzman

Karen and Matt; Kristin and Alayna
Todd; Cody, Cole, Taylor, and Clayton
Kathryn and Paul
Mildred and Ray

■

Maya Malhotra

Vivek, Pooja, and Neha
Santosh and Ramesh Malhotra
Indra and Prem Malhotra; Neeti and Neil Ardeshna, and Deeksha
Sadhana Malhotra
Leela and Mukund Dabholkar
Aruna and Harsha Dabholkar; Aditee
Mangala and Pradeep Gandhi; Priya and Medha

About the Authors

Lee J. Krajewski is Professor Emeritus at The Ohio State University and Professor Emeritus at the University of Notre Dame. While at The Ohio State University, he received the University Alumni Distinguished Teaching Award and the College of Business Outstanding Faculty Research Award. He initiated the Center for Excellence in Manufacturing Management and served as its director for 4 years. In addition, he received the National President's Award and the National Award of Merit of the American Production and Inventory Control Society. He served as president of the Decision Sciences Institute and was elected a fellow of the institute in 1988. He received the Distinguished Service Award in 2003.

Lee received his PhD from the University of Wisconsin. Over the years, he has designed and taught courses at both graduate and undergraduate levels on topics such as operations strategy, introduction to operations management, operations design, project management, and manufacturing planning and control systems.

Lee served as the editor of *Decision Sciences*, was the founding editor of the *Journal of Operations Management*, and has served on several editorial boards. Widely published himself, Lee has contributed numerous articles to such journals as *Decision Sciences, Journal of Operations Management, Management Science, Production and Operations Management, International Journal of Production Research, Harvard Business Review*, and *Interfaces*, to name just a few. He has received five best-paper awards. Lee's areas of specialization include operations strategy, manufacturing planning and control systems, supply chain management, and master production scheduling.

Larry P. Ritzman is Professor Emeritus at The Ohio State University and Professor Emeritus at Boston College. While at The Ohio State University, he served as department chairman and received several awards for both teaching and research, including the Pace Setters' Club Award for Outstanding Research. While at Boston College, he held the Thomas J. Galligan, Jr. chair and received the Distinguished Service Award from the School of Management. He received his doctorate at Michigan State University, having had prior industrial experience at the Babcock and Wilcox Company. Over the years, he has been privileged to teach and learn more about operations management with numerous students at all levels—undergraduate, MBA, executive MBA, and doctorate.

Particularly active in the Decision Sciences Institute, Larry has served as council coordinator, publications committee chair, track chair, vice president, board member, executive committee member, doctoral consortium coordinator, and president. He was elected a fellow of the institute in 1987 and earned the Distinguished Service Award in 1996. He has received three best-paper awards. He has been a frequent reviewer, discussant, and session chair for several other professional organizations.

Larry's areas of particular expertise are service processes, operations strategy, production and inventory systems, forecasting, multistage manufacturing, and layout. An active researcher, Larry's publications have appeared in such journals as *Decision Sciences, Journal of Operations Management, Production and Operations Management, Harvard Business Review*, and *Management Science*. He has served in various editorial capacities for several journals.

Manoj K. Malhotra is the Jeff B. Bates Professor in the Moore School of Business, and has served as the chairman of the Management Science Department at the University of South Carolina (USC), Columbia, since 2000. He is the founding director of the Center for Global Supply Chain and Process Management (GSCPM), which has been in operation since 2005. He earned an engineering undergraduate degree from the Indian Institute of Technology (IIT), Kanpur, India, in 1983, and a PhD in operations management from The Ohio State University in 1990. He is a fellow of the Decision Sciences Institute and is certified as a fellow of the American Production and Inventory Management Society (CFPIM). Manoj has conducted seminars and consulted with firms such as Cummins Turbo Technologies, John Deere, Metso Paper, Palmetto Health Richland, Phelps Dodge, Sonoco, UCB Chemicals, Verizon, Walmart Global Logistics, and Westinghouse Nuclear Fuels Division, among others.

Apart from teaching operations management, supply chain management, and global business issues at USC, Manoj has also taught at the Terry School of Business, University of Georgia; Wirtschaftsuniversität Wien in Austria; and the Graduate School of Management at Macquarie University, Australia. His research has thematically focused on the deployment of flexible resources in manufacturing and service firms, and on the interface between operations and supply chain management and other functional areas of business. His work on these and related issues has been published in refereed journals such as *Decision Sciences, European Journal of Operational Research, IIE Transactions, International Journal of Production Research, Journal of Operations Management, OMEGA,* and *Production and Operations Management Journal.* He is a recipient of the Decision Sciences Institute's Outstanding Achievement Award for the best application paper in 1990, and the Stan Hardy Award in 2002 and 2006 for the best paper published in the field of operations management. In 2007, his co-authored study on the evolution of manufacturing planning systems was a finalist for the best paper award in the *Journal of Operations Management.* In 2007, Manoj won the University of South Carolina Educational Foundation Award for Professional Schools, which is the university's most prestigious annual prize for innovative research, scholarship, and creative achievement. More recently, he received the *Decision Sciences* journal best paper award for the year 2011.

Manoj has won several teaching awards, including the Michael J. Mungo Outstanding Graduate Teaching Award in 2006 from the University of South Carolina and the Alfred G. Smith Jr. Excellence in Teaching Award in 1995 from the Moore School of Business. He was voted by the students as an outstanding professor in the international MBA program by the classes of 1997, 1998, 1999, 2000, 2005, and 2008; and as the outstanding professor in the IMBA-Vienna program by the classes of 1998 and 2004. He was designated as one of the first "Master Teachers" in the Moore School of Business in 1998, and has been listed in "Who's Who among America's Teachers" in 1996 and 2000.

Manoj is an associate editor of *Decision Sciences* and senior editor for the *POMS* journal. He has served as the past area editor for *POMS* journal (2000–2003) and an associate editor for the *Journal of Operations Management* (2001–2010). He is an active referee for several other journals in the field, and has served as the co-editor for special focus issues of *Decision Sciences* (1999) and *Journal of Operations Management* (2002). He was the program chair for the 36th International Meeting of the Decision Sciences Institute (DSI) in San Francisco in 2005, and has also served as an associate program chair for the POMS national meeting. He has been involved in the Mid-Carolina chapter of APICS as its past president, executive board member, and as an instructor of professional level CPIM certification courses. He is a founding board member of Shingo Prize for Lean Excellence in South Carolina.

Brief Contents

PART 1   Creating Value through Operations Management   **1**

  1 USING OPERATIONS TO COMPETE   **1**

  SUPPLEMENT A DECISION MAKING   **31**

  2 PROJECT MANAGEMENT   **49**

PART 2   Managing Processes   **89**

  3 PROCESS STRATEGY   **89**

  4 PROCESS ANALYSIS   **119**

  5 QUALITY AND PERFORMANCE   **157**

  6 CAPACITY PLANNING   **201**

  SUPPLEMENT B WAITING LINES   **225**

  7 CONSTRAINT MANAGEMENT   **243**

  8 LEAN SYSTEMS   **275**

PART 3   Managing Supply Chains   **307**

  9 SUPPLY CHAIN INVENTORY MANAGEMENT   **307**

  SUPPLEMENT C SPECIAL INVENTORY MODELS   **345**

  10 SUPPLY CHAIN DESIGN   **359**

  11 SUPPLY CHAIN LOCATION DECISIONS   **385**

  12 SUPPLY CHAIN INTEGRATION   **411**

  13 SUPPLY CHAIN SUSTAINABILITY AND HUMANITARIAN LOGISTICS   **441**

  14 FORECASTING   **463**

  15 OPERATIONS PLANNING AND SCHEDULING   **507**

  16 RESOURCE PLANNING   **543**

  SUPPLEMENT D LINEAR PROGRAMMING   **587**

  Appendix 1 NORMAL DISTRIBUTION   **617**

  Appendix 2 TABLE OF RANDOM NUMBERS   **618**

  Glossary   **619**

  Name Index   **633**

  Subject Index   **637**

MYOMLAB SUPPLEMENTS

  SUPPLEMENT E SIMULATION   **E-1**

  SUPPLEMENT F FINANCIAL ANALYSIS   **F-1**

  SUPPLEMENT G ACCEPTANCE SAMPLING PLANS   **G-1**

  SUPPLEMENT H MEASURING OUTPUT RATES   **H-1**

  SUPPLEMENT I LEARNING CURVE ANALYSIS   **I-1**

  SUPPLEMENT J OPERATIONS SCHEDULING   **J-1**

Contents

Preface xiii

PART 1 Creating Value through
Operations Management 1

1 USING OPERATIONS TO COMPETE 1

Scholastic and Harry Potter 1

Operations and Supply Chain Management across the Organization 2

Historical Evolution of Operations and Supply Chain Management 3

A Process View 4
How Processes Work 4
Nested Processes 5
Service and Manufacturing Processes 5

The Supply Chain View 6
Core Processes 7
Support Processes 7
Operations Strategy 7
Corporate Strategy 8
Market Analysis 10

Competitive Priorities and Capabilities 11
Order Winners and Qualifiers 11
Using Competitive Priorities: An Airline Example 13

Operations Strategy as a Pattern of Decisions 15

Trends in Operations Management 16
Productivity Improvement 16
Global Competition 17

Managerial Practice 1.1 Japanese Earthquake and its Supply Chain Impact 19
Ethical, Workforce Diversity, and Environmental Issues 19

Operations Management as a Set of Decisions 20
Computerized Decision-Making Tools 21

Addressing the Challenges in Operations Management 21
Part 1: Creating Value through Operations Management 21
Part 2: Managing Processes 21
Part 3: Managing Supply Chains 22
Adding Value with Process Innovation in Supply Chains 22

Managerial Practice 1.2 Operational Innovation Is a Competitive Weapon at Progressive Insurance 22
Learning Goals in Review 23
MyOMLab Resources 23
Key Equation 23
Key Terms 24
Solved Problems 24
Discussion Questions 25
Problems 26
Advanced Problems 27
Active Model Exercise 28

Video Case Operations as a Competitive Weapon at Starwood 29
Case Chad's Creative Concepts 29
Selected References 30

SUPPLEMENT A Decision Making 31

Break-Even Analysis 31
Evaluating Services or Products 32
Evaluating Processes 33

Preference Matrix 35

Decision Theory 36
Decision Making under Certainty 36
Decision Making under Uncertainty 37
Decision Making under Risk 38

Decision Trees 39
Learning Goals in Review 41
MyOMLab Resources 41
Key Equations 42
Key Terms 42
Solved Problems 42
Problems 45
Selected References 48

2 PROJECT MANAGEMENT 49

XBOX 360 49

Project Management across the Organization 51

Defining and Organizing Projects 51
Defining the Scope and Objectives of a Project 51
Selecting the Project Manager and Team 52
Recognizing Organizational Structure 52

Planning Projects 53
Defining the Work Breakdown Structure 53
Diagramming the Network 54
Developing the Schedule 57
Analyzing Cost–Time Trade-Offs 60
Assessing Risks 64

Managerial Practice 2.1 Boston's Big Dig Project Poses Many Challenges 65
Analysis 66

Monitoring and Controlling Projects 69
Monitoring Project Status 69
Monitoring Project Resources 70
Controlling Projects 70
Learning Goals in Review 71
MyOMLab Resources 71
Key Equations 72
Key Terms 72
Solved Problems 72
Discussion Questions 77
Problems 77
Advanced Problems 82
Active Model Exercise 85

Video Case Project Management at the Phoenician 86
Case The Pert Mustang 87
Selected References 88

PART 2 Managing Processes   **89**

3 PROCESS STRATEGY 89

eBay 89

Process Strategy across the Organization 90
  Supply Chains Have Processes 90
  Processes Are Not Just in Operations 91
Process Strategy Decisions 91
Process Structure in Services 92
  Nature of Service Processes: Customer Contact 92
  Customer-Contact Matrix 93
  Service Process Structuring 94
Process Structure in Manufacturing 94
  Product-Process Matrix 94
  Manufacturing Process Structuring 95
  Production and Inventory Strategies 96
Layout 97
  Gather Information 97
  Develop a Block Plan 98
  Applying the Weighted-Distance Method 98
  Design a Detailed Layout 100
Customer Involvement 100
  Possible Disadvantages 101
  Possible Advantages 101
Resource Flexibility 102
  Workforce 102
  Equipment 102
Capital Intensity 102
  Automating Manufacturing Processes 103
  Automating Service Processes 103
  Economies of Scope 104
Strategic Fit 104
Managerial Practice 3.1 Flexible Automation
  at R.R. Donnelley 105
  Decision Patterns for Service Processes 105
  Decision Patterns for Manufacturing Processes 106
  Gaining Focus 107
Strategies for Change 108
  Process Reengineering 108
  Process Improvement 109
  Learning Goals in Review 109
  MyOMLab Resources 109
  Key Equations 110
  Key Terms 110
  Solved Problems 110
  Discussion Questions 111
  Problems 112
  Active Model Exercise 114
Case Custom Molds, Inc. 115
  Selected References 116

4 PROCESS ANALYSIS 119

McDonald's Corporation 119

Process Analysis across the Organization 121
A Systematic Approach 121
  Step 1: Identify Opportunities 121
  Step 2: Define the Scope 122
  Step 3: Document the Process 122
  Step 4: Evaluate Performance 122
  Step 5: Redesign the Process 123
  Step 6: Implement Changes 123

Documenting the Process 123
  Flowcharts 123
  Swim Lane Flowcharts 125
  Service Blueprints 126
  Work Measurement Techniques 127
Evaluating Performance 131
  Data Analysis Tools 132
Redesigning the Process 135
Generating Ideas: Questioning and Brainstorming 135
Benchmarking 137
Managerial Practice 4.1 Baptist Memorial Hospital 137
Managing and Implementing Processes 138
  Learning Goals in Review 140
  MyOMLab Resources 140
  Key Terms 141
  Solved Problems 141
  Discussion Questions 145
  Problems 145
  Advanced Problems 151
  Active Model Exercise 152
Video Case Process Analysis at Starwood 152
Case José's Authentic Mexican Restaurant 154
  Selected References 155

5 QUALITY AND PERFORMANCE 157

Verizon Wireless 157

Quality and Performance across the Organization 158
Costs of Quality 159
  Prevention Costs 159
  Appraisal Costs 159
  Internal Failure Costs 159
  External Failure Costs 159
Ethics and Quality 159
Total Quality Management 160
  Customer Satisfaction 160
  Employee Involvement 161
Managerial Practice 5.1 Quality and Performance
  at Steinway & Sons 162
  Continuous Improvement 163
Six Sigma 164
  Six Sigma Improvement Model 164
Acceptance Sampling 165
Statistical Process Control 166
  Variation of Outputs 166
  Control Charts 169
Statistical Process Control Methods 170
  Control Charts for Variables 170
  Control Charts for Attributes 174
Process Capability 177
  Defining Process Capability 178
  Using Continuous Improvement to Determine
    the Capability of a Process 179
  Quality Engineering 179
International Quality Documentation Standards 181
  The ISO 9001:2008 Documentation Standards 181
  ISO 14000:2004 Environmental Management
    System 181
  ISO 26000:2010 Social Responsibility Guidelines 181
  Benefits of ISO Certification 182
Baldrige Performance Excellence Program 182
  Learning Goals in Review 183
  MyOMLab Resources 183

Key Equations 184
Key Terms 185
Solved Problems 185
Discussion Questions 188
Problems 189
Advanced Problems 194
Active Model Exercise 197
Video Case Process Performance and Quality at Starwood Hotels & Resorts 197
Experiential Learning Statistical Process Control with a Coin Catapult 198
Selected References 199

6 CAPACITY PLANNING 201

Sharp Corporation 201
Planning Capacity across the Organization 203
Planning Long-Term Capacity 203
Measures of Capacity and Utilization 203
Economies of Scale 204
Diseconomies of Scale 204
Capacity Timing and Sizing Strategies 205
Sizing Capacity Cushions 205
Timing and Sizing Expansion 206
Managerial Practice 6.1 Expansionist Capacity Strategy in the Ethanol Industry 207
Linking Capacity and Other Decisions 207
A Systematic Approach to Long-Term Capacity Decisions 208
Step 1: Estimate Capacity Requirements 208
Step 2: Identify Gaps 210
Step 3: Develop Alternatives 210
Step 4: Evaluate the Alternatives 210
Tools for Capacity Planning 212
Waiting-Line Models 212
Simulation 213
Decision Trees 213
Learning Goals in Review 213
MyOMLab Resources 214
Key Equations 214
Key Terms 215
Solved Problems 215
Discussion Questions 217
Problems 217
Advanced Problems 220
Video Case Gate Turnaround at Southwest Airlines 222
Case Fitness Plus, Part A 223
Selected References 224

SUPPLEMENT B Waiting Lines 225

Why Waiting Lines Form 225
Uses of Waiting-Line Theory 226
Structure of Waiting-Line Problems 226
Customer Population 226
The Service System 227
Priority Rule 229
Probability Distributions 229
Arrival Distribution 229
Service Time Distribution 230
Using Waiting-Line Models to Analyze Operations 230
Single-Server Model 231
Multiple-Server Model 233
Little's Law 234
Finite-Source Model 235
Waiting Lines and Simulation 236
Decision Areas for Management 237
Learning Goals in Review 238
MyOMLab Resources 238
Key Equations 239
Key Terms 240
Solved Problem 240
Problems 240
Advanced Problems 242
Selected References 242

7 CONSTRAINT MANAGEMENT 243

British Petroleum Oil Spill in Gulf of Mexico 243
Managing Constraints across the Organization 245
The Theory of Constraints 245
Key Principles of the TOC 246
Identification and Management of Bottlenecks 247
Managing Bottlenecks in Service Processes 247
Managing Bottlenecks in Manufacturing Processes 248
Managerial Practice 7.1 The Drum-Buffer-Rope System at a U.S. Marine Corps Maintenance Center 253
Managing Constraints in a Line Process 254
Line Balancing 254
Managerial Considerations 258
Learning Goals in Review 259
MyOMLab Resources 259
Key Equations 260
Key Terms 260
Solved Problems 260
Discussion Questions 262
Problems 262
Advanced Problems 265
Experiential Learning Min-Yo Garment Company 269
Video Case Constraint Management at Southwest Airlines 272
Selected References 273

8 LEAN SYSTEMS 275

Panasonic Corporation 275
Lean Systems across the Organization 276
Continuous Improvement Using a Lean Systems Approach 277
Supply Chain Considerations in Lean Systems 278
Close Supplier Ties 278
Small Lot Sizes 279
Process Considerations in Lean Systems 279
Pull Method of Work Flow 279
Quality at the Source 280
Uniform Workstation Loads 281
Standardized Components and Work Methods 282
Flexible Workforce 282
Automation 282
Five S Practices 282
Total Preventive Maintenance (TPM) 283
Toyota Production System 284
House of Toyota 284

Designing Lean System Layouts 285
  One Worker, Multiple Machines 285
  Group Technology 286
Value Stream Mapping 287
The *Kanban* System 290
  General Operating Rules 291
  Determining the Number of Containers 291
Other *Kanban* Signals 293
Operational Benefits and Implementation Issues 293
  Organizational Considerations 293
Managerial Practice 8.1 Lean Systems at the University of
  Pittsburgh Medical Center Shadyside 294
  Process Considerations 295
  Inventory and Scheduling 295
  Learning Goals in Review 296
  MyOMLab Resources 296
  Key Equation 296
  Key Terms 296
  Solved Problems 297
  Discussion Questions 299
  Problems 299
  Advanced Problems 301
Video Case Lean Systems at Autoliv 303
Case Copper Kettle Catering 304
  Selected References 305

PART 3 Managing Supply Chains   307

9 **SUPPLY CHAIN INVENTORY
      MANAGEMENT** 307

Inventory Management at Walmart 307
Inventory Management across the Organization 308
Inventory and Supply Chains 309
  Pressures for Small Inventories 309
  Pressures for Large Inventories 310
  Types of Inventory 311
  Inventory Reduction Tactics 313
ABC Analysis 314
Economic Order Quantity 315
  Calculating the EOQ 315
  Managerial Insights from the EOQ 319
Inventory Control Systems 319
  Continuous Review System 320
  Periodic Review System 325
Managerial Practice 9.1 The Supply Chain Implications of
  Periodic Review Inventory Systems at Celanese 326
  Comparative Advantages of the Q and P Systems 328
  Hybrid Systems 329
  Learning Goals in Review 329
  MyOMLab Resources 330
  Key Equations 331
  Key Terms 332
  Solved Problems 332
  Discussion Questions 336
  Problems 336
  Advanced Problems 340
  Active Model Exercise 341
Experiential Learning Swift Electronic Supply, Inc. 341
Case Parts Emporium 343
  Selected References 344

SUPPLEMENT C Special Inventory
Models 345
Noninstantaneous Replishment 345
Quantity Discounts 348
One-Period Decisions 350
  Learning Goals in Review 352
  MyOMLab Resources 353
  Key Equations 353
  Key Term 353
  Solved Problems 354
  Problems 356
  Selected References 357

10 **SUPPLY CHAIN DESIGN** 359

Nikon 359
Supply Chain Design across the Organization 361
Supply Chains for Services and Manufacturing 362
  Services 362
  Manufacturing 362
Measures of Supply Chain Performance 363
  Inventory Measures 364
  Financial Measures 365
Inventory Placement 367
Mass Customization 367
  Competitive Advantages 367
  Supply Chain Design for Mass Customization 368
Outsourcing Processes 369
  Vertical Integration 370
  Outsourcing 370
Managerial Practice 10.1 Building a Supply Chain for the
  Dreamliner 372
Strategic Implications 372
  Efficient Supply Chains 372
  Responsive Supply Chains 373
  The Design of Efficient and Responsive Supply Chains 374
  Learning Goals in Review 375
  MyOMLab Resources 376
  Key Equations 376
  Key Terms 376
  Solved Problem 377
  Discussion Questions 377
  Problems 378
  Advanced Problems 378
Experiential Learning Sonic Distributors 380
Case Brunswick Distribution, Inc. 381
  Selected References 383

11 **SUPPLY CHAIN LOCATION
      DECISIONS** 385

Bavarian Motor Works (BMW) 385
Location Decisions across the Organization 387
Factors Affecting Location Decisions 387
  Dominant Factors in Manufacturing 388
  Dominant Factors in Services 389
**Geographical Information Systems and Location
  Decisions** 390
Managerial Practice 11.1 How Fast-Food Chains Use GIS
  to Select Their Sites 391
Locating a Single Facility 391
  Comparing Several Sites 392

Applying the Load–Distance Method **393**
Using Break-Even Analysis **395**
Locating a Facility Within a Supply Chain Network 396
Managerial Practice 11.2 General Electric's Expansion
in India **397**
The GIS Method for Locating Multiple Facilities **397**
The Transportation Method **397**
Learning Goals in Review 400
MyOMLab Resources **400**
Key Equations **401**
Key Terms **401**
Solved Problems **401**
Discussion Questions **404**
Problems **405**
Advanced Problems **408**
Active Model Exercise **410**
Selected References **410**

12 SUPPLY CHAIN INTEGRATION 411

Eastman Kodak **411**
Supply Chain Integration across the Organization 412
Supply Chain Dynamics 413
External Causes **414**
Internal Causes **414**
Implications for Supply Chain Design **414**
Integrated Supply Chains **415**
New Service or Product Development Process 416
Design **416**
Analysis **417**
Development **417**
Full Launch **417**
Supplier Relationship Process 417
Sourcing **418**
Design Collaboration **421**
Negotiation **421**
Buying **422**
Information Exchange **424**
The Order Fulfillment Process 425
Customer Demand Planning **425**
Supply Planning **425**
Production **425**
Logistics **425**
Managerial Practice 12.1 Order Fulfillment aboard the Coral
Princess **428**
The Customer Relationship Process 428
Marketing **428**
Order Placement **429**
Customer Service **429**
Levers for Improved Supply Chain Performance 430
The Levers **430**
Performance Measures **431**
Learning Goals in Review 431
MyOMLab Resources **431**
Key Equations **432**
Key Terms **432**
Solved Problems **432**
Discussion Questions **434**
Problems **434**
Advanced Problems **436**
Video Case Sourcing Strategy at Starwood **438**
Case Wolf Motors **439**
Selected References **440**

13 SUPPLY CHAIN SUSTAINABILITY AND HUMANITARIAN LOGISTICS 441

FedEx **441**
Sustainability across the Organization 443
Supply Chains and Environmental Responsibility 443
Reverse Logistics **444**
Managerial Practice 13.1 Recycling at Hewlett-Packard and
Walmart **446**
Energy Efficiency **447**
**Supply Chains, Social Responsibility, and Humanitarian
Logistics 452**
Disaster Relief Supply Chains **453**
Supply Chain Ethics **455**
Managing Sustainable Supply Chains 457
Learning Goals in Review 457
MyOMLab Resources **457**
Key Equations **458**
Key Terms **458**
Solved Problems **458**
Discussion Questions **459**
Problems **460**
Video Case Supply Chain Sustainability at Clif Bar &
Company **461**
Selected References **462**

14 FORECASTING 463

Motorola Mobility **463**
Forecasting across the Organization 465
Demand Patterns 465
Key Decisions on Making Forecasts 466
Deciding What to Forecast **466**
Choosing the Type of Forecasting Technique **466**
Forecast Error **467**
Computer Support **470**
Judgment Methods 470
Causal Methods: Linear Regression 470
Time-Series Methods 472
Naïve Forecast **473**
Estimating the Average **473**
Trend Projection with Regression **476**
Seasonal Patterns **479**
Choosing a Quantitative Forecasting Method 481
Criteria for Selecting Time-Series Methods **481**
Tracking Signals **481**
Using Multiple Techniques 482
Managerial Practice 14.1 Combination Forecasts and the
Forecasting Process **483**
Putting It All Together: Forecasting as a Process 483
A Typical Forecasting Process **484**
Adding Collaboration to the System **485**
Forecasting as a Nested Process **485**
Learning Goals in Review 486
MyOMLab Resources **486**
Key Equations **487**
Key Terms **488**
Solved Problems **488**
Discussion Questions **492**
Problems **493**
Advanced Problems **496**

Video Case Forecasting and Supply Chain Management at Deckers Outdoor Corporation **500**

Case Yankee Fork and Hoe Company **502**

Experiential Learning 14.1 Forecasting with Holdout Sample **503**

Experiential Learning 14.2 Forecasting a Vital Energy Statistic **504**

Selected References **505**

15 OPERATIONS PLANNING AND SCHEDULING 507

Air New Zealand **507**

Operations Planning and Scheduling across the Organization 508

Stages in Operations Planning and Scheduling 509

Aggregation **509**

The Relationship of Operations Plans and Schedules to Other Plans **509**

Managing Demand 511

Demand Options **511**

Managerial Practice 15.1 Harrah's Cherokee Casino & Hotel **512**

Sales and Operations Plans 513

Information Inputs **514**

Supply Options **514**

Planning Strategies **515**

Constraints and Costs **516**

Sales and Operations Planning as a Process **516**

Using Spreadsheets **518**

Scheduling 521

Gantt Charts **521**

Scheduling Employees **522**

Sequencing Jobs at a Workstation **525**

Software Support **528**

Learning Goals in Review **528**

MyOMLab Resources **529**

Key Terms **529**

Solved Problems **530**

Discussion Questions **533**

Problems **533**

Active Model Exercise **538**

Video Case Sales and Operations Planning at Starwood **539**

Case Memorial Hospital **540**

Selected References **541**

16 RESOURCE PLANNING 543

Dow Corning **543**

Resource Planning across the Organization 544

Enterprise Resource Planning 544

How ERP Systems Are Designed **545**

Material Requirements Planning 546

Dependent Demand **547**

Bill of Materials **547**

Master Production Scheduling **549**

Inventory Record **553**

Planning Factors **556**

Outputs from MRP **559**

Managerial Practice 16.1 Material Requirements Planning at Winnebago Industries **563**

MRP, Core Processes, and Supply Chain Linkages **563**

MRP and the Environment **564**

Resource Planning for Service Providers 565

Dependent Demand for Services **565**

Bill of Resources **566**

Learning Goals in Review **567**

MyOMLab Resources **567**

Key Terms **568**

Solved Problems **568**

Discussion Questions **572**

Problems **572**

Advanced Problems **577**

Active Model Exercise **582**

Case Flashy Flashers, Inc. **582**

Selected References **585**

SUPPLEMENT D Linear Programming 587

Basic Concepts 587

Formulating a Problem **588**

Graphic Analysis 590

Plot the Constraints **590**

Identify the Feasible Region **592**

Plot an Objective Function Line **593**

Find the Visual Solution **594**

Find the Algebraic Solution **594**

Slack and Surplus Variables **595**

Sensitivity Analysis **596**

Computer Solution 596

Simplex Method **596**

Computer Output **597**

The Transportation Method 599

Transportation Method for Production Planning **600**

Applications 603

Learning Goals in Review **604**

MyOMLab Resources **604**

Key Terms **604**

Solved Problems **605**

Discussion Questions **607**

Problems **607**

Case R.U. Reddie for Location **614**

Selected References **616**

Appendix 1 Normal Distribution 617

Appendix 2 Table of Random Numbers 618

Glossary 619

Name Index 633

Subject Index 637

MyOMLab SUPPLEMENTS

SUPPLEMENT E   Simulation **E-1**

SUPPLEMENT F   Financial Analysis **F-1**

SUPPLEMENT G   Acceptance Sampling Plans **G-1**

SUPPLEMENT H   Measuring Output Rates **H-1**

SUPPLEMENT I   Learning Curve Analysis **I-1**

SUPPLEMENT J   Operations Scheduling **J-1**

Preface

Creating Value through Operations Management

Operations management is a vital topic that every business student needs to understand because it is at the heart of the creation of wealth for businesses and the improvement in the living standard of citizens of all countries. Operations managers are responsible for the production of services and products in an ethical and environmentally responsible way while being responsive to the market. Sound like a challenge? Add to it the need to manage supply chains of materials, information, and funds reaching to all areas of the world. While challenging, there are concepts, tools and methods that managers use to deal with operating problems in a global environment. The mission of this text is to provide you with a comprehensive framework for addressing operational and supply chain issues. We accomplish this mission by using a systemized approach while focusing on issues of current interest to you. It is important to be efficient and capable with respect to internal processes; however, it is critical for organizations to be able to link those processes to those of their customers and their suppliers to provide competitive supply chains. This text is unique in that it builds the concept of a supply chain from the ground up. Starting with the analysis of business processes and how they relate to the overall operational goals of a firm, our text proceeds to show how these processes are integrated to form supply chains and how they can be managed to obtain efficient flows of materials, information, and funds. This approach reinforces the concept that supply chains are only as good as the processes within and across each firm in them.

This text has been thoroughly revised to meet your needs regardless of your major. Any manager needs to know the global implications of supply chains and how to make decisions in a dynamic environment. We address these contemporary issues of interest through opening vignettes and managerial practices in each chapter. We show you the essential tools you will need to improve process performance. Irrespective of the industry in which you are seeking a career, processes and supply chains are analyzed from the perspective of service as well as manufacturing firms. Our philosophy is that you will learn by doing; consequently, the text has ample opportunities for you to experience the role of a manager with challenging problems, cases, a library of videos customized to the individual chapters, simulations, experiential exercises, and tightly integrated online computer resources. With this text, you will develop the capability to analyze problems and support managerial decisions.

What's New in the Tenth Edition?

Since the *ninth* edition, we have been hard at work to make the *tenth* edition even better, based upon the suggestions of adopters and nonadopters. We have carefully monitored for errors in the book and all supplements. We have more figures, photos, company examples, cases, and problems to test your understanding of the material. Here are some of the highlights of the many changes:

1. Major overhaul of MyOMLab as major teaching and learning tool.

2. Five chapters devoted to supply chain management, beginning with "Supply Chain Inventory Management."

3. New Chapter 13, "Supply Chain Sustainability and Humanitarian Logistics," which addresses critical issues such as reverse logistics, energy efficiency, disaster relief, and ethics, and provides new problem-solving exercises.

4. Added "Learning Goals in Review" at the end of each chapter, which highlights where each goal is addressed in the chapter.

5. Supplement B "Simulation" is now MyOMLab Supplement E.

6. Major overhaul of references so almost all are 2005 or later, with emphasis on references that are student friendly rather than research based.

7. Updates of most Managerial Practices, giving current examples of operations management to students.

8. Continual upgrade of extensive set of software (OM Explorer, POM for Windows, Active Models, and SimQuick). One such example is OM Explorer's *Time Series* Solver which replaces Trend-Adjusted Exponential Smoothing with Projection with Regression, and is expanded to support holdout samples which can be used as an experiential exercise by class teams.

Using Operations to Compete
Project Management

Process Strategy
Process Analysis
Quality and Performance
Capacity Planning
Constraint Management
Lean Systems

Supply Chain Inventory
Management

Supply Chain Design

Supply Chain Location
Decisions

Supply Chain Integration

Supply Chain Sustainability
and Humanitarian Logistics

Forecasting

Operations Planning and
Scheduling

Resource Planning

9. Added MyOMLab Resources at the end of each chapter, which cross references a rich set of MyOMLab videos, advanced problems, cases, virtual tours, and internet exercises at the instructor's disposal.

10. Refreshment of about 20 percent of the Problems, all of which are are now fully coded and represented in the MyOMLab.

Chapter-by-Chapter Changes

- **Chapter Count:** Relative to the *ninth* edition, we have added one chapter and moved one supplement to MyOMLab, for a total of only 16 chapters and four supplements in the book and six supplements in MyOMLab. A central figure in the margin of each chapter shows how each chapter fits into our general theme of processes to supply chains.

- **Part 1: Creating Value through Operations Management** – The first part of the text lays the foundation for why operations management is a strategic weapon.

  - Chapter 1, "Using Operations to Compete," defines operations management and supply chain management.
  - Chapter 2, "Project Management," opens with the product development story of Xbox 360 and shows how you can manage the projects needed to achieve efficient processes and supply chains.

- **Part 2: Managing Processes** – The second part of the text shows how you can design and manage the internal processes of a firm.

  - Chapter 3, "Process Strategy," brings out that supply chains also have processes through a revised opening section, explains the importance of the four key process decisions in the revised "Process Strategy Decisions" section.
  - Chapter 4, "Process Analysis," begins with a new opening vignette on McDonalds, and now has a section on swim lanes and a major expansion of the "Service Blueprints" section.
  - Chapter 5, "Quality and Performance," with additional material on ethics and the environment, provides the essential statistical tools for identifying the onset of process performance problems.
  - Chapter 6, "Capacity Planning," focuses on the long-term capacity decisions that define the process capacities of the firm to do business in the future.
  - Chapter 7, "Constraint Management," shows how you can get the best output rates within the process capacities you have to work with.
  - Chapter 8, "Lean Systems," which now presents Value Stream Mapping (VSM) as a major tool for analyzing and improving Lean Systems, reveals the methods you can use to improve the system performance.

- **Part 3: Managing Supply Chains** – The third part of the text provides the tools and perspectives you will need to manage the flow of materials, information, and funds between your suppliers, your firm, and your customers.

  - Chapter 9, "Supply Chain Inventory Management," combines, from the *ninth* edition, the introductory inventory material of Chapter 9 with Chapter 12 to create a consistent and compact chapter on inventory.
  - Chapter 10, "Supply Chain Design," has been completely revised to focus on supply chain design, with new material on the motivation for supply chain design and outsourcing in today's perspective.
  - Chapter 11, "Supply Chain Location Decisions," with reduced GIS coverage and the addition of the transportation method from the *ninth* edition's Supplement D, provides guidelines and tools for finding the best location for single or multiple facilities in a supply chain.
  - Chapter 12, "Supply Chain Integration," by making a better connection to supply chain design and the addition of material on the implications of supply chain dynamics on supply chain design, focuses on the importance of integrating processes along the supply chain, how to choose a supplier, how to determine the capacity of a logistics system, and how to design supply chains that are environmentally responsible.

- Chapter 13, "Supply Chain Sustainability and Humanitarian Logistics," a completely new chapter addressing sustainability, focuses on how supply chains can support environmental and social responsibility and provides quantitative tools to analyze these issues.

- Chapter 14, "Forecasting," now has a new opening vignette about Motorola Mobility, a stronger discussion of the use of the Error Analysis module for POM for Windows, and has replaced the Trend-Adjusted Exponential Smoothing model with the Trend Projection with Regression model.

- Chapter 15, "Operations Planning and Scheduling," shows that operations planning and scheduling provide a link between a firm, its suppliers, and its customers to create a capability that lies at the core of supply chain integration.

- Chapter 16, "Resource Planning," focuses on translating the demands for services and products to requirements for the resources to produce them.

- **Supplements** – The book also offers 4 supplements that dig deeper on technical topics, and another 6 MyOMLab supplements.

Helping You Learn

Key Features

Several new additions and changes have been made to the book to retain and enhance its theme of processes and supply chains and to expand these themes through new content, Managerial Practices, Examples, and End-of-Chapter Problems and Cases. Several key features designed to help aid in the learning process are highlighted next:

Chapter Opening Vignettes engage and stimulate student interest by profiling how real companies apply specific operational issues addressed in each chapter.

PART 2 Managing Processes

3

PROCESS STRATEGY

At any given time eBay has approximately 113 million listings worldwide, and yet its workforce consists of just 15,000 employees. The explanation? Customers do most of the work in eBay's buying and selling processes. Here a customer prepares items for shipping from sales on his Ebay account.

eBay

Most manufacturers do not have to contend with customers waltzing around their shop floors, showing up intermittently and unannounced. Such customer contact can introduce considerable variability, disrupting carefully designed production processes. Costs and quality can be adversely affected. While customer contact is an issue even with manufacturers, (each process does have at least one customer), extensive customer contact and involvement are business as usual for many processes of service providers. Customers at restaurants or rental car agencies are directly involved in performing the processes. The area where the sales person interacts with the customer *is* the shop floor.

How much should customers be involved in a process, so as to provide timely delivery and consistent quality, and at sustainable cost? Various ways are available—some accommodate customer-introduced variability and some reduce it. eBay illustrates one way to accommodate variability. As an online auction house, eBay has high volume and request variability. Its customers do not want service at the same time or at times necessarily convenient to the company. They have request variability, seeking to buy and sell an endless number of items. They also have variability in customer capability, some with considerable Internet experience and some needing more handholding. Such variability would greatly complicate workforce scheduling if eBay required its employees to conduct all of its processes. It connects hundreds of millions of people around

89

Managerial Practices provide current examples of how companies deal—either successfully or unsuccessfully—with process and supply chain issues facing them as they run their operations.

MANAGERIAL PRACTICE 1.1 Japanese Earthquake and its Supply Chain Impact

Northeast Touhoku district of Japan was struck by a set of massive earthquakes on the afternoon of March 11, 2011, which were soon followed by a huge tsunami that sent waves higher than 33 feet in the port city of Sendai 80 miles away and travelling at the speed of a jetliner. At nearly 9.0 on the Richter scale, it was one of the largest recorded earthquakes to hit Japan. It shifted the Earth's axis by 6 inches with an impact that was felt 250 miles inland in Tokyo, and which moved Eastern Japan 13 feet toward North America. Apart from huge loss of life and hazards of nuclear radiation arising from the crippled Daiichi Nuclear Reactors in Fukushima, the damage to the manufacturing plants in Japan exposed the hazards of interconnected global supply chains and their impact on factories located half way around the globe.

The impact of the earthquake was particularly acute on industries that rely on cutting edge electronic parts sourced from Japan. Shin-Etsu Chemical Company is the world's largest producer of silicon wafers and supplies 20 percent of the global capacity. Its centralized plant located 40 miles from the Fukushima nuclear facility was damaged in the earthquake, causing ripple effects at Intel and Toshiba that purchase wafers from Shin-Etsu. Similarly, a shortage of automotive sensors from Hitachi has slowed or halted production of vehicles in Germany, Spain, and France, while Chrysler is reducing overtime at factories in Mexico and Canada to conserve parts from Japan. Even worse, General Motors stopped production altogether at a plant in Louisiana and Ford closed a truck plant in Kentucky due to the quake. The supply of vehicles such as Toyota's Prius and Lexus will be limited in the United States because of production disruptions in its Japanese factories. China has been affected too, where ZTE Corporation is facing shortages of batteries and LCD screens for its cell phones. Similarly, Lenovo in China is looking at reduced

Following the strong earthquakes and tsunami, flames and smoke rise from a petroleum refining plant next to a heating power station in Shiogama, Miyagi Prefecture, northern Japan, about 220 km north of Tokyo.

supplies of components from Japan for assembly of its tablet computers. These disruptions due to reliance on small concentrated network of suppliers in Japan and globally connected production and logistics systems have caused worker layoffs an increase in prices of affected products, and economic losses that have been felt around the world.

Sources: Don Lee and David Pearson, "Disaster in Japan exposes supply chain weakness," *The State* (April 8, 2011), B6-B7; "Chrysler reduces overtime to help Japan," *The Associated Press* (April 8, 2011) printed in *The State* (April 6, 2011), B7; Krishna Dhir, "From the Editor," *Decision Line*, vol. 42, no. 2, 3.

Examples demonstrate how to apply what students have learned and walk them through the solution process modeling good problem-solving techniques. These examples always close with a unique feature called **Decision Point**, which focuses students on the decision implications for managers.

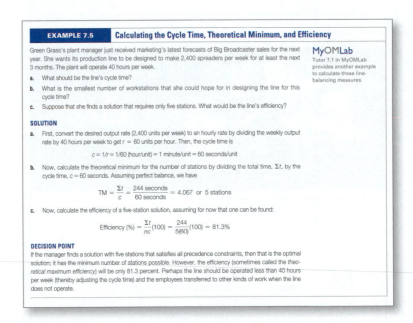

EXAMPLE 7.5   Calculating the Cycle Time, Theoretical Minimum, and Efficiency

Green Grass's plant manager just received marketing's latest forecasts of Big Broadcaster sales for the next year. She wants its production line to be designed to make 2,400 spreaders per week for at least the next 3 months. The plant will operate 40 hours per week.

MyOMLab
Tutor 7.1 in MyOMLab provides another example to calculate these line-balancing measures.

a. What should be the line's cycle time?

b. What is the smallest number of workstations that she could hope for in designing the line for this cycle time?

c. Suppose that she finds a solution that requires only five stations. What would be the line's efficiency?

SOLUTION

a. First, convert the desired output rate (2,400 units per week) to an hourly rate by dividing the weekly output rate by 40 hours per week to get $r = 60$ units per hour. Then, the cycle time is

$$c = 1/r = 1/60 \text{ (hour/unit)} = 1 \text{ minute/unit} = 60 \text{ seconds/unit}$$

b. Now, calculate the theoretical minimum for the number of stations by dividing the total time, Σt, by the cycle time, $c = 60$ seconds. Assuming perfect balance, we have

$$TM = \frac{\Sigma t}{c} = \frac{244 \text{ seconds}}{60 \text{ seconds}} = 4.067 \text{ or } 5 \text{ stations}$$

c. Now, calculate the efficiency of a five-station solution, assuming for now that one can be found:

$$\text{Efficiency (\%)} = \frac{\Sigma t}{nc}(100) = \frac{244}{5(60)}(100) = 81.3\%$$

DECISION POINT

If the manager finds a solution with five stations that satisfies all precedence constraints, then that is the optimal solution; it has the minimum number of stations possible. However, the efficiency (sometimes called the *theoretical maximum efficiency*) will be only 81.3 percent. Perhaps the line should be operated less than 40 hours per week (thereby adjusting the cycle time) and the employees transferred to other kinds of work when the line does not operate.

End of Chapter Resources

- **Learning Goals in Review** for review purposes.
- **MyOMLab Resources** lists the resources found in MyOMLab and how those resources relate back to the topics and discussions in the chapter.
- **Key Equations** for review purposes.
- **Key Terms** for review purposes, the page references highlight where the concept was first discussed.
- **Solved Problems** reinforce and help students prepare their homework assignments by detailing how to solve model problems with the appropriate techniques presented in the chapter.

- **Discussion Questions** test student comprehension of the concepts through the use of short scenarios.

- **Problems** sharpen students' quantitative skills by providing a bridge between chapter materials with a wide selection of homework material. Advanced problems are also included to increase the level of difficulty. Most of the homework problems can be done manually, or students can utilize a variety of software tools through MyOMLab, which is discussed in a later section.

- **Active Model Exercises** enable students to use pre-created spreadsheets to do "what-if" analysis of examples presented in the text to see what would happen if certain parameters were changed.

- **Video Cases** provide a summary of content covered in a series of on-location video profiles of real-world service and manufacturing companies and challenges they face in their operations. Questions are included for classroom discussion or assignment purposes.

- **Cases** challenge students to grapple with a capstone problem that can be used as an in-class exercise or a homework assignment or team project.

- **Experiential Learning** forms students into teams who work both in and out of class on exercises that actively involve them in team-based discussion questions and decisions. The six exercises reinforce student learning. Each exercise has been thoroughly tested in class and proven to be a valuable learning tool.

- A **Video Library** of 23 cases in MyOMLab (including 3 tutorials) offers at least one video customized for each chapter, which make for excellent class discussion and learning.

Teaching and Learning Support

MyOMLab A key capability of MyOMLab is as an online homework and assessment tool designed to help students practice operations management problems and improve their understanding of course concepts, and to give their instructors feedback on their performance. This online product expands the student's learning experience with out-of-class quizzes that are automatically graded and tutorials to guide the problem solving process, keeping students up to date, and freeing instructors for more creative use of class time.

　　MyOMLab lets you teach your course your way. Use MyOMLab as an out-of-the-box resource for students who need extra help, or take full advantage of its advanced customization options.

MyOMLab

For Instructors

Instructor's Resource Center—Reached through a link at **www.pearsonhighered.com/krajewski**, the Instructor's Resource Center contains the electronic files for the complete Instructor's Solutions Manual, PowerPoint lecture presentations, the Image Library, and the Test Item File.

- **Register, redeem, log in** at **www.pearsonhighered.com/irc**, instructors can access a variety of print, media, and presentation resources that are available with this book in downloadable digital format. Resources are also available for course management platforms such as Blackboard, WebCT, and CourseCompass.

- **Need help?** Pearson Education's dedicated technical support team is ready to assist instructors with questions about the media supplements that accompany this text. Visit **http://247pearsoned.com** for answers to frequently asked questions and toll-free user support phone numbers. The supplements are available to adopting instructors. Detailed descriptions are provided at the Instructor's Resource Center.

Instructor's Solutions Manual (0-13-280741-6)—Prepared by John Jensen at The University of South Carolina, this resource begins with the video notes and solutions. They are followed by the instructor notes, and the solutions, and answers to end-of-chapter questions, problems, and cases. This manual is available for download by visiting **www.pearsonhighered.com/krajewski**.

Instructor's Resource Manual—Prepared by John Jensen at The University of South Carolina, this resource begins with sample syllabi for the course suited to various situations: with or without MyOMLab, quarter vs. 7-week course, undergraduate vs. MBA, quantitative vs. qualitative orientation, and process vs. supply chain orientation. It then offers generic (in both Word and PDF versions) Instructor and Student Notes. Both must be revised to reflect the instructor's approach to the course. The Student Notes can be handed out or posted so that the students can have them during class to simplify note taking and concentrate more on what is being said. The Image Library provides possible inserts to the Student Notes. The Instructor Notes offer a course outline, chapter outlines, teaching notes, sample course syllabi, and solutions to the videos. This manual is available for download by visiting **www.pearsonhighered.com/krajewski**.

PowerPoint lecture slides in chapter-by-chapter files for classroom presentation purposes are available for download by visiting **www.pearsonhighered.com/krajewski.** PowerPoints can be customized by the instructor, including inserts from Image Library, just as with the Student Notes.

Image Library—most of the images and illustrations featured in the text are available for download by visiting **www.pearsonhighered.com/krajewski**.

Test Item File—this resource offers an array of questions and problems ranging from easy to difficult. This resource includes true/false and multiple choice questions, which can be accessed by MyOMLab, and short answer, and essay questions. These files are available for download by visiting **www.pearsonhighered.com/krajewski.**

TestGen EQ—Pearson Education's test-generating software is available from **www .pearsonhighered.com/irc**. The software is PC/MAC compatible and preloaded with all of the Test Item File questions. You can manually or randomly view test questions and drag and drop to create a test. You can add or modify test-bank questions as needed.

For Students

Besides having access to study plans and tutorial resources in MyOMLab, students can utilize the following additional course resources within MyOMLab:

OM Explorer a text-specific software tool consisting of Excel worksheets and including tutors, additional exercises, and solvers.

- **Tutors** provide coaching for more than 60 analytical techniques presented in the text. The tutors also provide additional examples for learning and practice.
- **Additional Exercises** pose questions and can be answered with one or more of the tutor applications.
- **Solvers** provide powerful general purpose routines often encountered in practice. These are great for experiential exercises and homework problems.

POM for Windows an easy-to-use software program that covers over 25 common OM techniques.

Active Models include 29 spreadsheets requiring students to evaluate different situations based on problem scenarios.

Download page offers access to software (such as OM Explorer, POM for Windows, SimQuick, and Active Models), and links to free trial of software (such as MS Project, MS MapPoint, and SmartDraw).

CourseSmart

CourseSmart eTextbooks were developed for students looking to save on required or recommended textbooks. Students simply select their eText by title or author and purchase immediate access to the content for the duration of the course using any major credit card. With a CourseSmart eText, students can search for specific keywords or page numbers, take notes online, print out reading assignments that incorporate lecture notes, and bookmark important passages for later review. For more information or to purchase a CourseSmart eTextbook, visit **www.coursesmart.com.**

Acknowledgments

No book is just the work of the authors. We greatly appreciate the assistance and valuable contributions by several people who made this edition possible. Thanks to Beverly Amer of Aspenleaf

Productions for her efforts in filming and putting together the new video segments for this edition; and Annie Puciloski for her diligent work of accuracy checking the book and ancillary materials. Special thanks are due to Howard Weiss, of Temple University, whose expertise in upgrading the software for this edition was greatly appreciated.

Many colleagues at other colleges and universities provided valuable comments and suggestions for this and previous editions. We would also like to thank the following faculty members who gave extensive written feedback and commentary to us:

Harold P. Benson, *University of Florida*

James P. McGuire, *Rhode Island College*

David L. Bakuli, *Westfield State College*

David Levy, *Bellevue University*

Tobin Porterfield, *Towson University*

Anil Gulati, Western *New England College*

Linda C. Rodriguez, *University of South Carolina–Aiken*

Kathryn Marley, *Duquesne University*

Qingyu Zhang, *Arkansas State University*

Ching-Chung Kuo, *University of North Texas*

We would like to thank the people at Pearson Prentice Hall, including Chuck Synovec, Mary Kate Murray, Ashlee Bradbury, Anne Fahlgren, Judy Leale, Sarah Petersen, and Lauren McFalls and Haylee Schwenk at PreMediaGlobal. Without their hard work, dedication, and guidance this book would not have been possible.

At the University of Notre Dame Mendoza College of Business, we want to thank Jerry Wei, Sarv Devaraj, Dave Hartvigsen, Carrie Queenan, Xuying Zhao and Daewon Sun for their constant encouragement and for their willingness to share their teaching secrets. At the University of South Carolina, we thank Sanjay Ahire, Jack Jensen, and Ashley Metcalf for contributing their thoughts and insights on classroom pedagogical issues to this text. In particular, we gratefully acknowledge Jack Jensen for the stellar contributions he has made to the development of ISM and MyOMLab. Thanks go to colleagues at The Ohio State University for their encouragement and ideas on text revision.

Finally, we thank our families for supporting us during this project involving multiple teleconference calls and long periods of seclusion. Our wives, Judie, Barb, and Maya, have provided the love, stability, and encouragement that sustained us while we transformed the ninth edition into the tenth.

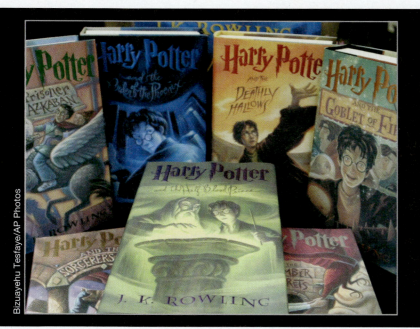

Bizuayehu Testaye/AP Photos

1

USING OPERATIONS TO COMPETE

The seventh novel in the Harry Potter series was released on July 21, 2007 and became an instant best seller around the globe. Because the book had to be delivered in a tight time window to the customers, Scholastic coordinated its publishing and distribution processes in USA months in advance of the release date.

Scholastic and Harry Potter

Scholastic is the world's largest publisher and distributor of children's books and educational materials. Founded in 1920, it had $1.9 billion in revenues in fiscal 2011 with offices in 16 countries including North America, Europe, Southeast Asia, Latin America, the Middle East, Australia, New Zealand, and Africa. Scholastic started planning in early 2007 for the worldwide release of the eagerly awaited seventh book *Harry Potter and the Deathly Hallows* in the acclaimed series by J.K. Rowling on the boy wizard. When the author finished the book in spring 2007, Scholastic's printers R.R. Donnelly & Sons and Quebecor World worked around the clock to make sure that the book would be ready by the release date. To save time in loading and unloading, Scholastic bypassed its own warehouses and required its truckers, Yellow Transportation and JB Hunt Transport Services, to use the same size trailers and pallets to ship books directly from six printing sites to big retailers like Barnes & Noble and Amazon.com. This fleet of trucks, if lined up bumper-to-bumper, would stretch for 15 miles. GPS transponders were used to alert Scholastic by e-mail if the driver or the trailer veered off the designated routes. The timing was particularly tricky for e-tailers, who had to directly ship books in advance for individual orders to arrive simultaneously around the country in order to minimize the risk of someone leaking the book's ending.

Since close to 90 percent of sales of such special books occur in the first week, they get special treatment to save time, money, space, and work. Scholastic had to customize, coordinate, and synchronize its operations and supply chain processes across multiple partners at the printing, warehousing, distribution, and retailing locations to ensure that the last book in the Harry Potter series reached the final customers no more than a few hours before the scheduled July 21, 12:01 A.M. release deadline. Not bad for a bunch of Muggles who transported 12 million copies in a short time window without the magical floo powder, portkeys, and broomsticks!

Source: Dean Foust, "Harry Potter and the Logistical Nightmare," *Business Week* (August 6, 2007), p. 9; Michelle Regenold, "Shipping Harry Potter: How Do They Do That?" **www.go-explore-trans.org/2007/mar-apr/shipping_HP.cfm; www.scholastic.com,** 2011.

LEARNING GOALS *After reading this chapter, you should be able to:*

1 Describe operations and supply chains in terms of inputs, processes, outputs, information flows, suppliers, and customers.

2 Define an operations strategy and its linkage to corporate strategy, as well as the role it plays as a source of competitive advantage in a global marketplace.

3 Identify nine competitive priorities used in operations strategy, and their linkage to marketing strategy.

4 Explain how operations can be used as a competitive weapon.

5 Identify the global trends and challenges facing operations management.

operations management

The systematic design, direction, and control of processes that transform inputs into services and products for internal, as well as external, customers.

process

Any activity or group of activities that takes one or more inputs, transforms them, and provides one or more outputs for its customers.

operation

A group of resources performing all or part of one or more processes.

supply chain

An interrelated series of processes within and across firms that produces a service or product to the satisfaction of customers.

supply chain management

The synchronization of a firm's processes with those of its suppliers and customers to match the flow of materials, services, and information with customer demand.

Operations management refers to the systematic design, direction, and control of processes that transform inputs into services and products for internal, as well as external customers.

This book deals with managing those fundamental activities and processes that organizations use to produce goods and services that people use every day. A process is any activity or group of activities that takes one or more inputs, transforms them, and provides one or more outputs for its customers. For organizational purposes, processes tend to be clustered together into operations. An operation is a group of resources performing all or part of one or more processes. Processes can be linked together to form a supply chain, which is the interrelated series of processes within a firm and across different firms that produce a service or product to the satisfaction of customers.[1] A firm can have multiple supply chains, which vary by the product or service provided. Supply chain management is the synchronization of a firm's processes with those of its suppliers and customers to match the flow of materials, services, and information with customer demand. For example, Scholastic must schedule the printing of a very large quantity of books in a timely fashion, receive orders from its largest customers, directly load and dispatch a fleet of trucks by specific destination while bypassing regular warehouses, keep track of their progress using technology, and finally, bill their customers and collect payment. The operational planning at Scholastic, along with internal and external coordination within its supply chain, provides one example of designing customized processes for competitive operations.

Operations and Supply Chain Management across the Organization

Broadly speaking, operations and supply chain management underlie all departments and functions in a business. Whether you aspire to manage a department or a particular process within it, or you just want to understand how the process you are a part of fits into the overall fabric of the business, you need to understand the principles of operations and supply chain management.

Operations serve as an excellent career path to upper management positions in many organizations. The reason is that operations managers are responsible for key decisions that affect the success of the organization. In manufacturing firms, the head of operations usually holds the

---

[1]The terms *supply chain* and *value chain* are sometimes used interchangeably.

title chief operations officer (COO) or vice president of manufacturing (or of production or operations). The corresponding title in a service organization might be COO or vice president (or director) of operations. Reporting to the head of operations are the managers of departments, such as customer service, production and inventory control, and quality assurance.

Figure 1.1 shows operations as one of the key functions within an organization. The circular relationships in Figure 1.1 highlight the importance of the coordination among the three mainline functions of any business, namely, (1) operations, (2) marketing, and (3) finance. Each function is unique and has its own knowledge and skill areas, primary responsibilities, processes, and decision domains. From an external perspective, finance generates resources, capital, and funds from investors and sales of its goods and services in the marketplace. Based on business strategy, the finance and operations functions then decide how to invest these resources and convert them into physical assets and material inputs. Operations subsequently transforms these material and service inputs into product and service outputs. These outputs must match the characteristics that can be sold in the selected markets by marketing. Marketing is responsible for producing sales revenue of the outputs, which become returns to investors and capital for supporting operations. Functions such as accounting, information systems, human resources, and engineering make the firm complete by providing essential information, services, and other managerial support.

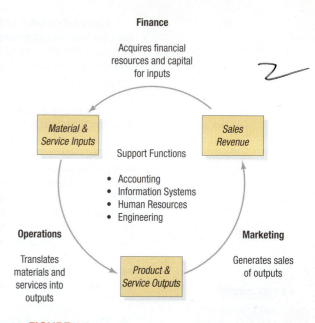

▲ **FIGURE 1.1**
Integration between Different Functional Areas of a Business

These relationships provide direction for the business as a whole, and are aligned to the same strategic intent. It is important to understand the entire circle, and not just the individual functional areas. How well these functions work together determines the effectiveness of the organization. Functions should be integrated and should pursue a common strategy. Success depends on how well they are able to do so. No part of this circle can be dismissed or minimized without loss of effectiveness, and regardless of how departments and functions are individually managed, they are always linked together through processes. Thus, a firm competes not only by offering new services and products, creative marketing, and skillful finance, but also through its unique competencies in operations and sound management of core processes.

Historical Evolution of Operations and Supply Chain Management

The history of modern operations and supply chain management is rich and over two hundred years old, even though its practice has been around in one form or another for centuries. James Watt invented the steam engine in 1785. The subsequent establishment of railroads facilitated efficient movement of goods throughout Europe, and eventually even in distant colonies such as India. With the invention of the cotton gin in 1794, Eli Whitney introduced the concept of interchangeable parts. It revolutionized the art of machine-based manufacturing, and coupled with the invention of the steam engine, lead to the great industrial revolution in England and the rest of Europe. The textile industry was one of the earliest industries to be mechanized. The industrial revolution gradually spread to the United States and the rest of the world in the nineteenth century, and was accompanied by such great innovations as the internal combustion engine, steam-powered ships, metallurgy of iron making, large-scale production of chemicals, and invention of machine tools, among others. The foundations of modern manufacturing and technological breakthroughs were also inspired by the creation of a mechanical computer

Henry Ford with a Model T in Buffalo, New York, in 1921. The Ford Motor Company, founded in 1903, produced about one million Model T's in 1921.

by Charles Babbage in the early part of the nineteenth century. He also pioneered the concept of division of labor, which laid the foundation for scientific management of operations and supply chain management that was further improved upon by Frederick Taylor in 1911.

Three other landmark events from the twentieth century define the history of operations and supply chain management. First is the invention of the assembly line for the Model T car by Henry Ford in 1909. The era of mass production was born, where complex products like automobiles could be manufactured in large numbers at affordable prices through repetitive manufacturing. Second, Alfred Sloan in the 1930s introduced the idea of strategic planning for achieving product proliferation and variety, with the newly founded General Motors Corporation offering "a car for every purse and purpose." Finally, with the publication of the Toyota Production System in 1978, Taiichi Ohno laid the groundwork for removing wasteful activities from an organization, a concept that we explore further in this book while learning about lean systems.

The recent history of operations and supply chains over the past three decades has been steeped in technological advances. The 1980s were characterized by wide availability of computer aided design (CAD), computer aided manufacturing (CAM), and automation. Information technology applications started playing an increasingly important role in 1990s, and started connecting the firm with its extended enterprise through Enterprise Resource Planning Systems and outsourced technology hosting for supply chain solutions. Service organizations like Federal Express, United Parcel Service (UPS), and Walmart also became sophisticated users of information technology in operations, logistics, and management of supply chains. The new millennium has seen an acceleration of this trend, along with an increased focus on sustainability and the natural environment. We cover all these ideas and topical areas in greater detail throughout this book.

A Process View

You might wonder why we begin by looking at processes, rather than at departments or even the firm. The reason is that a process view of the firm provides a much more relevant picture of the way firms actually work. Departments typically have their own set of objectives, a set of resources with capabilities to achieve those objectives, and managers and employees responsible for performance. Some processes, such as billing, may be so specific that they are contained wholly within a single department, such as accounting.

The concept of a process, however, can be much broader. A process can have its own set of objectives, involve a work flow that cuts across departmental boundaries, and require resources from several departments. You will see examples throughout this text of companies that discovered how to use their processes to gain a competitive advantage. You will notice that the key to success in many organizations is a keen understanding of how their processes work, since an organization is only as effective as its processes. Therefore, operations management is relevant and important for all students, regardless of major, because all departments have processes that must be managed effectively to gain a competitive advantage.

How Processes Work

Figure 1.2 shows how processes work in an organization. Any process has inputs and outputs. Inputs can include a combination of human resources (workers and managers), capital (equipment and facilities), purchased materials and services, land, and energy. The numbered circles in Figure 1.2 represent operations through which services, products, or customers pass and where processes are performed. The arrows represent flows, and can cross because one job or customer can have different requirements (and thus a different flow pattern) than the next job or customer.

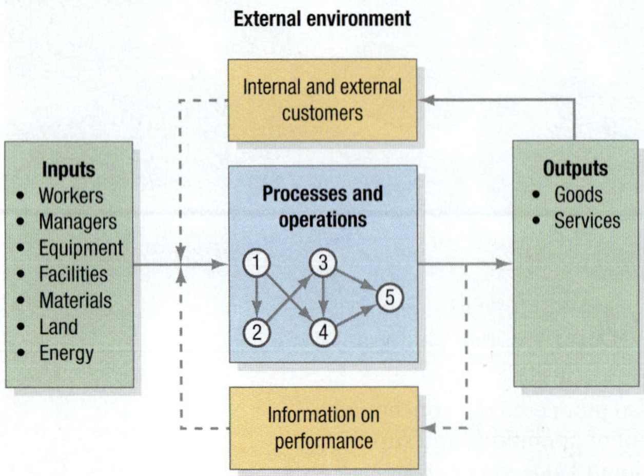

Processes provide outputs to customers. These outputs may often be services (that can take the form of information) or tangible products. Every process and every person in an organization has customers. Some are **external customers,** who may be end users or intermediaries (e.g., manufacturers, financial institutions, or retailers) buying the firm's finished services or products. Others are **internal customers**, who may be employees in the firm whose process inputs are actually the outputs of earlier processes managed within the firm. Either way, processes must be managed with the customer in mind.

In a similar fashion, every process and every person in an organization relies on suppliers. **External suppliers** may be other businesses or individuals who provide the resources, services, products, and materials for the firm's short-term and long-term needs. Processes also have **internal suppliers**, who may be employees or processes that supply important information or materials.

---

external customers

A customer who is either an end user or an intermediary (e.g., manufacturers, financial institutions, or retailers) buying the firm's finished services or products.

internal customers

One or more employees or processes that rely on inputs from other employees or processes in order to perform their work.

external suppliers

The businesses or individuals who provide the resources, services, products, and materials for the firm's short-term and long-term needs.

internal suppliers

The employees or processes that supply important information or materials to a firm's processes.

▼ **FIGURE 1.2**

Processes and Operations

Inputs and outputs vary depending on the service or product provided. For example, inputs at a jewelry store include merchandise, the store building, registers, the jeweler, and customers; outputs to external customers are services and sold merchandise. Inputs to a factory manufacturing blue jeans include denim, machines, the plant, workers, managers, and services provided by outside consultants; outputs are clothing and supporting services. The fundamental role of inputs, processes, and customer outputs holds true for processes at all organizations.

Figure 1.2 can represent a whole firm, a department, a small group, or even a single individual. Each one has inputs and uses processes at various operations to provide outputs. The dashed lines represent two special types of input: participation by customers and information on performance from both internal and external sources. Participation by customers occurs not only when they receive outputs, but also when they take an active part in the processes, such as when students participate in a class discussion. Information on performance includes internal reports on customer service or inventory levels and external information from market research, government reports, or telephone calls from suppliers. Managers need all types of information to manage processes most effectively.

Nested Processes

Processes can be broken down into subprocesses, which in turn can be broken down further into still more subprocesses. We refer to this concept of a process within a process as a **nested process**. It may be helpful to separate one part of a process from another for several reasons. One person or one department may be unable to perform all parts of the process, or different parts of the process may require different skills. Some parts of the process may be designed for routine work while other parts may be geared for customized work. The concept of nested processes is illustrated in greater detail in Chapter 4, "Process Analysis," where we reinforce the need to understand and improve activities within a business and each process's inputs and outputs.

nested process
The concept of a process within a process.

Service and Manufacturing Processes

Two major types of processes are (1) service and (2) manufacturing. Service processes pervade the business world and have a prominent place in our discussion of operations management. Manufacturing processes are also important; without them the products we enjoy as part of our daily lives would not exist. In addition, manufacturing gives rise to service opportunities.

Differences   Why do we distinguish between service and manufacturing processes? The answer lies at the heart of the design of competitive processes. While Figure 1.3 shows several distinctions between service and manufacturing processes along a continuum, the two key differences that we discuss in detail are (1) the nature of their output and (2) the degree of customer contact. In general, manufacturing processes also have longer response times, are more capital intensive, and their quality can be measured more easily than those of service processes.

Manufacturing processes convert materials into goods that have a physical form we call products. For example, an assembly line produces a 350 Z sports car, and a tailor produces an outfit for the rack of an upscale clothing store. The transformation processes change the materials on one or more of the following dimensions:

1. Physical properties
2. Shape
3. Size (e.g., length, breadth, and height of a rectangular block of wood)
4. Surface finish
5. Joining parts and materials

The outputs from manufacturing processes can be produced, stored, and transported in anticipation of future demand.

If a process does not change the properties of materials on at least one of these five dimensions, it is considered a service (or nonmanufacturing) process. Service processes tend to produce intangible, perishable outputs. For example, the output from the auto loan process of a bank would be a car loan, and an output of the order fulfillment process of the U.S. Postal Service is the delivery of your letter. The outputs of service processes typically cannot be held in a finished goods inventory to insulate the process from erratic customer demands.

More like a manufacturing process	More like a service process
• Physical, durable output	• Intangible, perishable output
• Output can be inventoried	• Output cannot be inventoried
• Low customer contact	• High customer contact
• Long response time	• Short response time
• Capital intensive	• Labor intensive
• Quality easily measured	• Quality not easily measured

▲ **FIGURE 1.3**

Continuum of Characteristics of Manufacturing and Service Processes

A second key difference between service processes and manufacturing processes is degree of customer contact. Service processes tend to have a higher degree of customer contact. Customers may take an active role in the process itself, as in the case of shopping in a supermarket, or they may be in close contact with the service provider to communicate specific needs, as in the case of a medical clinic. Manufacturing processes tend to have less customer contact. For example, washing machines are ultimately produced to meet retail forecasts. The process requires little information from the ultimate consumers (you and me), except indirectly through market surveys and market focus groups. Even though the distinction between service and manufacturing processes on the basis of customer contact is not perfect, the important point is that managers must recognize the degree of customer contact required when designing processes.

Similarities   At the level of the firm, service providers do not just offer services and manufacturers do not just offer products. Patrons of a restaurant expect good service and good food. A customer purchasing a new computer expects a good product as well as a good warranty, maintenance, replacement, and financial services.

Further, even though service processes do not keep finished goods inventories, they do inventory their inputs. For example, hospitals keep inventories of medical supplies and materials needed for day-to-day operations. Some manufacturing processes, on the other hand, do not inventory their outputs because they are too costly. Such would be the case with low-volume customized products (e.g., tailored suits) or products with short shelf lives (e.g., daily newspapers).

When you look at what is being done at the process level, it is much easier to see whether the *process* is providing a service or manufacturing a product. However, this clarity is lost when the whole company is classified as either a manufacturer or a service provider because it often performs both types of processes. For example, the process of cooking a hamburger at a McDonald's is a manufacturing process because it changes the material's physical properties (dimension 1), as is the process of assembling the hamburger with the bun (dimension 5). However, most of the other processes visible or invisible to McDonald's customers are service processes. You can debate whether to call the whole McDonald's organization a service provider or a manufacturer, whereas classifications at the process level are much less ambiguous.

The Supply Chain View

Most services or products are produced through a series of interrelated business activities. Each activity in a process should add value to the preceding activities; waste and unnecessary cost should be eliminated. Our process view of a firm is helpful for understanding how services or products are produced and why cross-functional coordination is important, but it does not shed any light on the strategic benefits of the processes. The missing strategic insight is that processes must add value for customers throughout the supply chain. The concept of supply chains reinforces the link between processes and performance, which includes a firm's internal processes as well as those of its external customers and suppliers. It also focuses attention on the two main types of processes in the supply chain, namely (1) core processes and (2) support processes. Figure 1.4 shows the links between the core and support processes in a firm and a firm's external customers and suppliers within its supply chain.

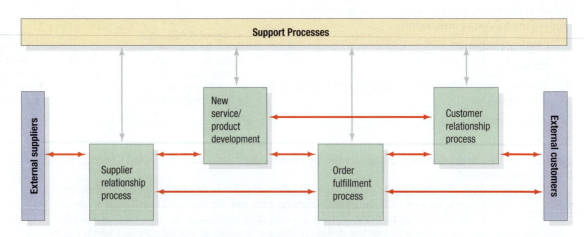

▲ **FIGURE 1.4**
Supply Chain Linkages Showing Work and Information Flows

Core Processes

A **core process** is a set of activities that delivers value to external customers. Managers of these processes and their employees interact with external customers and build relationships with them, develop new services and products, interact with external suppliers, and produce the service or product for the external customer. Examples include a hotel's reservation handling, a new car design for an auto manufacturer, or Web-based purchasing for an online retailer like amazon.com. Of course, each of the core processes has nested processes within it.

In this text we focus on four core processes:

1. *Supplier Relationship Process.* Employees in the **supplier relationship process** select the suppliers of services, materials, and information and facilitate the timely and efficient flow of these items into the firm. Working effectively with suppliers can add significant value to the services or products of the firm. For example, negotiating fair prices, scheduling on-time deliveries, and gaining ideas and insights from critical suppliers are just a few of the ways to create value.

2. *New Service/Product Development Process.* Employees in the **new service/product development process** design and develop new services or products. The services or products may be developed to external customer specifications or conceived from inputs received from the market in general.

3. *Order Fulfillment Process.* The **order fulfillment process** includes the activities required to produce and deliver the service or product to the external customer.

4. *Customer Relationship Process,* sometimes referred to as *customer relationship management.* Employees involved in the **customer relationship process** identify, attract, and build relationships with external customers, and facilitate the placement of orders by customers. Traditional functions, such as marketing and sales, may be a part of this process.

Support Processes

A **support process** provides vital resources and inputs to the core processes and is essential to the management of the business. Firms have many support processes. Examples include budgeting, recruiting, and scheduling. Support processes provide key resources, capabilities, or other inputs that allow the core processes to function.

The Human Resources function in an organization provides many support processes such as recruiting and hiring workers who are needed at different levels of the organization, training the workers for skills and knowledge needed to properly execute their assigned responsibilities, and establishing incentive and compensation plans that reward employees for their performance. The legal department puts in place support processes that ensure that the firm is in compliance with the rules and regulations under which the business operates. The Accounting function supports processes that track how the firm's financial resources are being created and allocated over time, while the Information Systems function is responsible for the movement and processing of data and information needed to make business decisions. Support processes from different functional areas like Accounting, Engineering, Human Resources, and Information Systems are therefore vital to the execution of core processes highlighted in Figure 1.4.

Operations Strategy

Operations strategy specifies the means by which operations implements corporate strategy and helps to build a customer-driven firm. It links long-term and short-term operations decisions to corporate strategy and develops the capabilities the firm needs to be competitive. It is at the heart of managing processes and supply chains. A firm's internal processes are only building blocks: They need to be organized to ultimately be effective in a competitive environment. Operations strategy is the linchpin that brings these processes together to form supply chains that extend beyond the walls of the firm, encompassing suppliers as well as customers. Since customers constantly desire change, the firm's operations strategy must be driven by the needs of its customers.

Developing a customer-driven operations strategy begins with *corporate strategy*, which, as shown in Figure 1.5, coordinates the firm's overall goals with its core processes. It determines the markets the firm will serve and the responses the firm will make to changes in the environment. It provides the resources to develop the firm's core competencies and core processes, and it identifies the strategy the firm will employ in international markets. Based on corporate strategy, a *market analysis* categorizes the firm's customers, identifies their needs, and assesses competitors' strengths. This information is used to develop *competitive priorities*. These priorities help managers develop the services or products and the processes needed to be competitive in the

core process
A set of activities that delivers value to external customers.

supplier relationship process
A process that selects the suppliers of services, materials, and information and facilitates the timely and efficient flow of these items into the firm.

new service/product development process
A process that designs and develops new services or products from inputs received from external customer specifications or from the market in general through the customer relationship process.

order fulfillment process
A process that includes the activities required to produce and deliver the service or product to the external customer.

customer relationship process
A process that identifies, attracts, and builds relationships with external customers, and facilitates the placement of orders by customers, sometimes referred to as *customer relationship management*.

support process
A process that provides vital resources and inputs to the core processes and therefore is essential to the management of the business.

operations strategy
The means by which operations implements the firm's corporate strategy and helps to build a customer-driven firm.

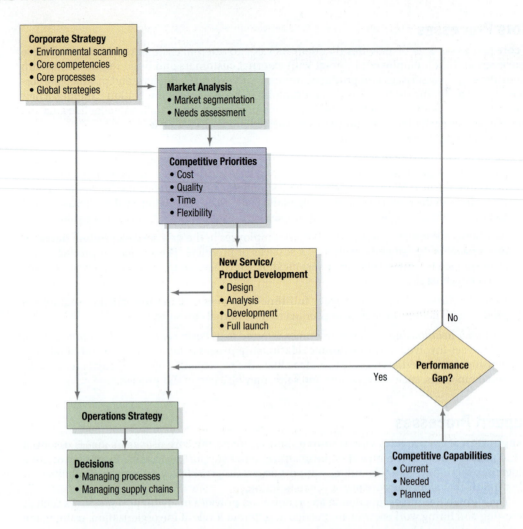

marketplace. Competitive priorities are important to the design of existing as well as new services or products, the processes that will deliver them, and the operations strategy that will develop the firm's capabilities to fulfill them. Developing a firm's operations strategy is a continuous process because the firm's capabilities to meet the competitive priorities must be periodically checked and any gaps in performance must be addressed in the operations strategy.

Corporate Strategy

Corporate strategy provides an overall direction that serves as the framework for carrying out all the organization's functions. It specifies the business or businesses the company will pursue, isolates new opportunities and threats in the environment, and identifies growth objectives.

Developing a corporate strategy involves four considerations: (1) monitoring and adjusting to changes in the business environment, (2) identifying and developing the firm's core competencies, (3) developing the firm's core processes, and (4) developing the firm's global strategies.

Environmental Scanning   The external business environment in which a firm competes changes continually and an organization needs to adapt to those changes. Adaptation begins with *environmental scanning*, the process by which managers monitor trends in the environment (e.g., the industry, the marketplace, and society) for potential opportunities or threats. A crucial reason for environmental scanning is to stay ahead of the competition. Competitors may be gaining an edge by broadening service or product lines, improving quality, or lowering costs. New entrants into the market or competitors that offer substitutes for a firm's service or product may threaten continued profitability. Other important environmental concerns include economic trends, technological changes, political conditions, social changes (i.e., attitudes toward work), and the availability of vital resources. For example, car manufacturers recognize that dwindling oil reserves will eventually require alternative fuels for their cars. Consequently, they have designed prototype cars that use hydrogen or electric power as supplements to gasoline as a fuel.

Developing Core Competencies   Good managerial skill alone cannot overcome environmental changes. Firms succeed by taking advantage of what they do particularly well—that is, the organization's unique strengths. **Core competencies** are the unique resources and strengths that an organization's management considers when formulating strategy. They reflect the collective learning of the organization, especially in how to coordinate processes and integrate technologies. These competencies include the following:

1. *Workforce.* A well-trained and flexible workforce allows organizations to respond to market needs in a timely fashion. This competency is particularly important in service organizations, where customers come in direct contact with employees.

2. *Facilities.* Having well-located facilities (offices, stores, and plants) is a primary advantage because of the long **lead time** needed to build new ones. In addition, flexible facilities that can handle a variety of services or products at different levels of volume provide a competitive advantage.

3. *Market and Financial Know-How.* An organization that can easily attract capital from stock sales, market and distribute its services or products, or differentiate them from similar services or products on the market has a competitive edge.

4. *Systems and Technology.* Organizations with expertise in information systems have an edge in industries that are data intensive, such as banking. Particularly advantageous is expertise in Internet technologies and applications, such as business-to-consumer and business-to-business systems. Having the patents on a new technology is also a big advantage.

core competencies
The unique resources and strengths that an organization's management considers when formulating strategy.

lead time
The elapsed time between the receipt of a customer order and filling it.

Developing Core Processes   A firm's core competencies should drive its core processes: customer relationship, new service/product development, order fulfillment, and supplier relationship. Many companies have all four processes, while others focus on a subset of them to better match their core competencies, since they find it difficult to be good at all four processes and still be competitive. For instance, in the credit card business within the banking industry, some companies primarily specialize in finding customers and maintaining relationships with them. American Airlines's credit card program reaches out and achieves a special affinity to customers through its marketing database. On the other hand, specialized credit card companies, such as CapitalOne, focus on service innovation by creating new features and pricing programs. Finally, many companies are taking over the order fulfillment process by managing the processing of credit card transactions and call centers. The important point is that every firm must evaluate its core competencies and choose to focus on those processes that provide it the greatest competitive strength.

Global Strategies   Identifying opportunities and threats today requires a global perspective. A global strategy may include buying foreign services or parts, combating threats from foreign competitors, or planning ways to enter markets beyond traditional national boundaries. Although warding off threats from global competitors is necessary, firms should also actively seek to penetrate foreign markets. Two effective global strategies are (1) strategic alliances and (2) locating abroad.

One way for a firm to open foreign markets is to create a *strategic alliance.* A strategic alliance is an agreement with another firm that may take one of three forms. One form of strategic alliance is the *collaborative effort,* which often arises when one firm has core competencies that another needs but is unwilling (or unable) to duplicate. Such arrangements commonly arise out of buyer–supplier relationships.

The popular smiling red bee, the mascot of Jollibee, welcomes customers at an outlet in Manila. What began from a two ice cream parlors in Manila in 1975, Jollibee has grown into the biggest Philippines fast-food company employing over 26,000 people in over 1,000 stores in seven countries. By catering to local tastes and preferences, Jollibee took 65 percent of the fiercely competitive Philippine fast-food market, pushing the world giant McDonald's into second place.

Romeo Gacad/AFP/Getty Images/Newscom

Another form of strategic alliance is the *joint venture,* in which two firms agree to produce a service or product jointly. This approach is often used by firms to gain access to foreign markets. For example, to get access to the large Chinese market, General Motors (GM) and Volkswagen (VW) each developed joint ventures with Shanghai Automotive Industry Corporation or SAIC.[2] The Chinese partner is a large manufacturer of automobiles, producing more than 600,000 cars with GM and VW. In 2010, SAIC upped its total share to 51% in Shanghai GM, which is now among the top three passenger vehicle producers in mainland China. Finally, *technology licensing* is a form of strategic alliance in which one company licenses its service or production methods to another. Licenses may be used to gain access to foreign markets.

Another way to enter global markets is to locate operations in a foreign country. However, managers must recognize that what works well in their home country might not work well elsewhere. The economic and political environment or customers' needs may be significantly different. For example, the family-owned chain Jollibee Foods Corporation has become the dominant fast-food chain in the Philippines by catering to a local preference for sweet and spicy flavors, which it incorporates into its fried chicken, spaghetti, and burgers. Jollibee's strength is its creative marketing programs and an understanding of local tastes and claims that its burger is similar to the one a Filipino would cook at home. McDonald's responded by introducing its own Filipino-style spicy burger, but competition is stiff. This example shows that to be successful, corporate strategies must recognize customs, preferences, and economic conditions in other countries.

Locating abroad is a key decision in the design of supply chains because it affects the flow of materials, information, and employees in support of the firm's core processes. Chapter 10, "Supply Chain Design," and Chapter 11, "Supply Chain Location Decisions," offer more in-depth discussion of these other implications.

Market Analysis

One key to successfully formulating a customer-driven operations strategy for both service and manufacturing firms is to understand what the customer wants and how to provide it. A *market analysis* first divides the firm's customers into market segments and then identifies the needs of each segment. In this section, we examine the process of market analysis and we define and discuss the concepts of market segmentation and needs assessment.

Market Segmentation   *Market segmentation* is the process of identifying groups of customers with enough in common to warrant the design and provision of services or products that the group wants and needs. To identify market segments, the analyst must determine the characteristics that clearly differentiate each segment. The company can then develop a sound marketing program and an effective operating strategy to support it. For instance, The Gap, Inc., a major provider of casual clothes, targets teenagers and young adults while the parents or guardians of infants through 12-year-olds are the primary targets for its GapKids stores. At one time, managers thought of customers as a homogeneous mass market, but now realize that two customers may use the same product for different reasons. Identifying the key factors in each market segment is the starting point in devising a customer-driven operations strategy.

Needs Assessment   The second step in market analysis is to make a *needs assessment*, which identifies the needs of each segment and assesses how well competitors are addressing those needs. Each market segment's needs can be related to the service or product and its supply chain. Market needs should include both the tangible and intangible attributes and features of products and services that a customer desires. Market needs may be grouped as follows:

- *Service or Product Needs.* Attributes of the service or product, such as price, quality, and degree of customization.
- *Delivery System Needs.* Attributes of the processes and the supporting systems, and resources needed to deliver the service or product, such as availability, convenience, courtesy, safety, accuracy, reliability, delivery speed, and delivery dependability.
- *Volume Needs.* Attributes of the demand for the service or product, such as high or low volume, degree of variability in volume, and degree of predictability in volume.
- *Other Needs.* Other attributes, such as reputation and number of years in business, after-sale technical support, ability to invest in international financial markets, and competent legal services.

[2]Alex Taylor, "Shanghai Auto Wants to Be the World's Next Great Car Company," *Fortune* (October 4, 2004), pp. 103–110.

Once it makes this assessment, the firm can incorporate the needs of customers into the design of the service or product and the supply chain that must deliver it. We further discuss these new service and product development-related issues in Chapter 12, "Supply Chain Integration."

Competitive Priorities and Capabilities

A customer-driven operations strategy requires a cross-functional effort by all areas of the firm to understand the needs of the firm's external customers and to specify the operating capabilities the firm requires to outperform its competitors. Such a strategy also addresses the needs of internal customers because the overall performance of the firm depends upon the performance of its core and supporting processes, which must be coordinated to provide the overall desirable outcome for the external customer.

Competitive priorities are the critical operational dimensions a process or supply chain must possess to satisfy internal or external customers, both now and in the future. Competitive priorities are planned for processes and the supply chain created from them. They must be present to maintain or build market share or to allow other internal processes to be successful. Not all competitive priorities are critical for a given process; management selects those that are most important. **Competitive capabilities** are the cost, quality, time, and flexibility dimensions that a process or supply chain actually possesses and is able to deliver. When the capability falls short of the priority attached to it, management must find ways to close the gap or else revise the priority.

We focus on nine broad competitive priorities that fall into the four capability groups of cost, quality, time, and flexibility. Table 1.1 provides definitions and examples of these competitive priorities, as well as how firms achieve them at the process level.

At times, management may emphasize a cluster of competitive priorities together. For example, many companies focus on the competitive priorities of delivery speed and development speed for their processes, a strategy called **time-based competition**. To implement the strategy, managers carefully define the steps and time needed to deliver a service or produce a product and then critically analyze each step to determine whether they can save time without hurting quality.

To link to corporate strategy, management assigns selected competitive priorities to each process (and the supply chains created from them) that are consistent with the needs of external as well as internal customers. Competitive priorities may change over time. For example, consider a high-volume standardized product, such as color ink-jet desktop printers. In the early stages of the ramp-up period when the printers had just entered the mass market, the manufacturing processes required consistent quality, delivery speed, and volume flexibility. In the later stages of the ramp-up when demand was high, the competitive priorities became low-cost operations, consistent quality, and on-time delivery. Competitive priorities must change and evolve over time along with changing business conditions and customer preferences.

Order Winners and Qualifiers

Competitive priorities focus on what operations can do to help a firm be more competitive, and are in response to what the market wants. Another useful way to examine a firm's ability to be successful in the marketplace is to identify the order winners and order qualifiers. An **order winner** is a criterion that customers use to differentiate the services or products of one firm from those of another. Order winners can include price (which is supported by low-cost operations) and other dimensions of quality, time, and flexibility. However, order winners also include criteria not directly related to the firm's operations, such as after-sale support (Are maintenance service contracts available? Is there a return policy?); technical support (What help do I get if something goes wrong? How knowledgeable are the technicians?); and reputation (How long has this company been in business? Have other customers been satisfied with the service or product?). It may take good performance on a subset of the order-winner criteria, cutting across operational as well as nonoperational criteria, to make a sale.

Order winners are derived from the considerations customers use when deciding which firm to purchase a service or product from in a given market segment. Sometimes customers demand a certain level of demonstrated performance before even contemplating a service or product.

competitive priorities

The critical dimensions that a process or supply chain must possess to satisfy its internal or external customers, both now and in the future.

competitive capabilities

The cost, quality, time, and flexibility dimensions that a process or supply chain actually possesses and is able to deliver.

time-based competition

A strategy that focuses on the competitive priorities of delivery speed and development speed.

The lavish interior lobby decor of the Ritz Carlton resort in Palm Beach, Florida, USA

America/Alamy

order winner

A criterion customers use to differentiate the services or products of one firm from those of another.

TABLE 1.1 | DEFINITIONS, PROCESS CONSIDERATIONS, AND EXAMPLES OF COMPETITIVE PRIORITIES

Cost	Definition	Processes Considerations	Example
1. **Low-cost operations**	Delivering a service or a product at the lowest possible cost to the satisfaction of external or internal customers of the process or supply chain	To reduce costs, processes must be designed and operated to make them efficient using rigorous process analysis that addresses workforce, methods, scrap or rework, overhead, and other factors, such as investments in new automated facilities or technologies to lower the cost per unit of the service or product.	**Costco** achieves low costs by designing all processes for efficiency, stacking products on pallets in warehouse-type stores, and negotiating aggressively with their suppliers. Costco can provide low prices to its customers because they have designed operations for low cost.
Quality			
2. **Top quality**	Delivering an outstanding service or product	To deliver top quality, a service process may require a high level of customer contact, and high levels of helpfulness, courtesy, and availability of servers. It may require superior product features, close tolerances, and greater durability from a manufacturing process.	**Rolex** is known globally for creating precision timepieces.
3. **Consistent quality**	Producing services or products that meet design specifications on a consistent basis	Processes must be designed and monitored to reduce errors, prevent defects, and achieve similar outcomes over time, regardless of the "level" of quality.	**McDonald's** standardizes work methods, staff training processes, and procurement of raw materials to achieve the same consistent product and process quality from one store to the next.
Time			
4. **Delivery speed**	Quickly filling a customer's order	Design processes to reduce lead time (elapsed time between the receipt of a customer order and filling it) through keeping backup capacity cushions, storing inventory, and using premier transportation options.	**Dell** engineered its customer relationship, order fulfillment, and supplier relationship processes to create an integrated and an agile supply chain that delivers reliable and inexpensive computers to its customers with short lead times.
5. **On-time delivery**	Meeting delivery-time promises	Along with processes that reduce lead time, planning processes (forecasting, appointments, order promising, scheduling, and capacity planning) are used to increase percent of customer orders shipped when promised (95% is often a typical goal).	**United Parcel Services (UPS)** uses its expertise in logistics and warehousing processes to deliver a very large volume of shipments on-time across the globe.
6. **Development speed**	Quickly introducing a new service or a product	Processes aim to achieve cross-functional integration and involvement of critical external suppliers in the service or product development process.	**Zara** is known for its ability to bring fashionable clothing designs from the runway to market quickly.
Flexibility			
7. **Customization**	Satisfying the unique needs of each customer by changing service or product designs	Processes with a customization strategy typically have low volume, close customer contact, and an ability to reconfigure processes to meet diverse types of customer needs.	**Ritz Carlton** customizes services to individual guest preferences.
8. **Variety**	Handling a wide assortment of services or products efficiently	Processes supporting variety must be capable of larger volumes than processes supporting customization. Services or products are not necessarily unique to specific customers and may have repetitive demands.	**Amazon.com** uses information technology and streamlined customer relationship and order fulfillment processes to reliably deliver a vast variety of items to its customers.
9. **Volume flexibility**	Accelerating or decelerating the rate of production of services or products quickly to handle large fluctuations in demand	Processes must be designed for excess capacity and excess inventory to handle demand fluctuations that can vary in cycles from days to months. This priority could also be met with a strategy that adjusts capacity without accumulation of inventory or excess capacity.	**The United States Post Office (USPS)** can have severe demand peak fluctuations at large postal facilities where processes are flexibly designed for receiving, sorting, and dispatching mail to numerous branch locations.

Minimal level required from a set of criteria for a firm to do business in a particular market segment is called an **order qualifier.** Fulfilling the order qualifier will not ensure competitive success; it will only position the firm to compete in the market. From an operations perspective, understanding which competitive priorities are order qualifiers and which ones are order winners is important for the investments made in the design and management of processes and supply chains.

Figure 1.6 shows how order winners and qualifiers are related to achieving the competitive priorities of a firm. If a minimum threshold level is not met for an order-qualifying dimension (consistent quality, for example) by a firm, then it would get disqualified from even being considered further by its customers. For example, there is a level of quality consistency that is minimally tolerable by customers in the auto industry. When the subcompact car Yugo built by Zastava Corporation could not sustain the minimal level of quality, consistency, and reliability expected by customers, it had to exit the U.S. car market in 1991 despite offering very low prices (order winner) of under $4,000. However, once the firm qualifies by attaining consistent quality beyond the threshold, it may only gain additional sales at a very low rate by investing further in improving that order-qualifying dimension. In contrast, for an order-winning dimension (i.e., low price driven by low-cost operations), a firm can reasonably expect to gain appreciably greater sales and market share by continuously lowering its prices as long as the order qualifier (i.e., consistent quality) is being adequately met. Toyota Corolla and Honda Civic have successfully followed this route in the marketplace to become leaders in their target market segment.

Order winners and qualifiers are often used in competitive bidding. For example, before a buyer considers a bid, suppliers may be required to document their ability to provide consistent quality as measured by adherence to the design specifications for the service or component they are supplying (order qualifier). Once qualified, the supplier may eventually be selected by the buyer on the basis of low prices (order winner) and the reputation of the supplier (order winner).

order qualifier

Minimal level required from a set of criteria for a firm to do business in a particular market segment.

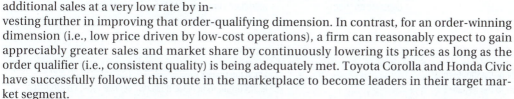

▲ **FIGURE 1.6**
Relationship of Order Winners and Order Qualifiers to Competitive Priorities

Using Competitive Priorities: An Airline Example

To get a better understanding of how companies use competitive priorities, let us look at a major airline. We will consider two market segments: (1) first-class passengers and (2) coach passengers. Core services for both market segments are ticketing and seat selection, baggage handling, and transportation to the customer's destination. The peripheral services are quite different across the two market segments. First-class passengers require separate airport lounges; preferred treatment during check-in, boarding, and deplaning; more comfortable seats; better meals and beverages; more personal attention (cabin attendants who refer to customers by name); more frequent service from attendants; high levels of courtesy; and low volumes of passengers (adding to the feeling of being special). Coach passengers are satisfied with standardized services (no surprises), courteous flight attendants, and low prices. Both market segments expect the airline to hold to its schedule. Consequently, we can say that the competitive priorities for the

One of the competitive priorities of airline companies is on-time delivery of their services. Being able to repair and maintain planes rapidly to avoid delays is a crucial aspect of this.

ROSLAN RAHMAN/AFP/Getty Images

first-class segment are *top quality* and *on-time delivery*, whereas the competitive priorities for the coach segment are *low-cost operations*, *consistent quality*, and *on-time delivery*.

The airline knows what its collective capabilities must be as a firm, but how does that get communicated to each of its core processes? Let us focus on the four core processes: (1) customer relationship, (2) new service/product development, (3) order fulfillment, and (4) supplier relationship. Competitive priorities are assigned to each core process to achieve the service required to provide complete customer satisfaction. Table 1.2 shows some possible assignments, just to give you an idea of how this works.

TABLE 1.2 | COMPETITIVE PRIORITIES ACROSS DIFFERENT CORE PROCESSES FOR AN AIRLINE

	CORE PROCESSES			
Priority	**Supplier Relationship**	**New Service Development**	**Order Fulfillment**	**Customer Relationship**
Low Cost Operations	Costs of acquiring inputs must be kept to a minimum to allow for competitive pricing.		Airlines compete on price and must keep operating costs in check.	
Top Quality		New services must be carefully designed because the future of the airline industry depends on them.	High quality meal and beverage service delivered by experienced cabin attendants ensures that the service provided to first-class passengers is kept top notch.	High levels of customer contact and lounge service for the first-class passengers.
Consistent Quality	Quality of the inputs must adhere to the required specifications. In addition, information provided to suppliers must be accurate.		Once the quality level is set, it is important to achieve it every time.	The information and service must be error free.
Delivery Speed				Customers want immediate information regarding flight schedules and other ticketing information.
On time delivery	Inputs must be delivered to tight schedules.		The airline strives to arrive at destinations on schedule, otherwise passengers might miss connections to other flights.	
Development Speed		It is important to get to the market fast to preempt the competition.		
Customization		The process must be able to create unique services.		
Variety	Many different inputs must be acquired, including maintenance items, meals and beverages.		Maintenance operations are required for a variety of aircraft models.	The process must be capable of handling the service needs of all market segments and promotional programs.
Volume Flexibility	The process must be able to handle variations in supply quantities efficiently.			

Operations Strategy as a Pattern of Decisions

Operations strategy translates service or product plans and competitive priorities for each market segment into decisions affecting the supply chains that support those market segments. Even if it is not formally stated, the current operations strategy for any firm is really the pattern of decisions that have been made for its processes and supply chains. As we have previously seen in Figure 1.5, corporate strategy provides the umbrella for key operations management decisions that contribute to the development of the firm's ability to compete successfully in the marketplace. Once managers determine the competitive priorities for a process, it is necessary to assess the *competitive capabilities* of the process. Any gap between a competitive priority and the capability to achieve that competitive priority must be closed by an effective operations strategy.

Developing capabilities and closing gaps is the thrust of operations strategy. To demonstrate how this works, suppose the management of a bank's credit card division decides to embark on a marketing campaign to significantly increase its business, while keeping costs low. A key process in this division is billing and payments. The division receives credit transactions from the merchants, pays the merchants, assembles and sends the bills to the credit card holders, and processes payments. The new marketing effort is expected to significantly increase the volume of bills and payments. In assessing the capabilities, the process must have to serve the bank's customers and to meet the challenges of the new market campaign; management assigns the following competitive priorities for the billing and payments process:

- *Low-Cost Operations.* It is important to maintain low costs in the processing of the bills because profit margins are tight.
- *Consistent Quality.* The process must consistently produce bills, make payments to the merchants, and record payments from the credit card holders accurately.
- *Delivery Speed.* Merchants want to be paid for the credit purchases quickly.
- *Volume Flexibility.* The marketing campaign is expected to generate many more transactions in a shorter period of time.

Management assumed that customers would avoid doing business with a bank that could not produce accurate bills or payments. Consequently, consistent quality is an order qualifier for this process.

Is the billing and payment process up to the competitive challenge? Table 1.3 shows how to match capabilities to priorities and uncover any gaps in the credit card division's operations strategy. The procedure for assessing an operations strategy begins with identifying good measures for each priority. The more quantitative the measures are, the better. Data are gathered for each measure to determine the current capabilities of the process. Gaps are identified by comparing each capability to management's target values for the measures, and unacceptable gaps are closed by appropriate actions.

The credit card division shows significant gaps in the process's capability for low-cost operations. Management's remedy is to redesign the process in ways that reduce costs but will not impair the other competitive priorities. Likewise, for volume flexibility, management realized that a high level of utilization is not conducive for processing quick surges in volumes while maintaining delivery speed. The recommended actions will help build a capability for meeting more volatile demands.

TABLE 1.3 | **OPERATIONS STRATEGY ASSESSMENT OF THE BILLING AND PAYMENT PROCESS**

Competitive Priority	Measure	Capability	Gap	Action
Low-cost operations	■ Cost per billing statement	■ $0.0813	■ Target is $0.06	■ Eliminate microfilming and storage of billing statements
	■ Weekly postage	■ $17,000	■ Target is $14,000	■ Develop Web-based process for posting bills
Consistent quality	■ Percent errors in bill information	■ 0.90%	■ Acceptable	■ No action
	■ Percent errors in posting payments	■ 0.74%	■ Acceptable	■ No action
Delivery speed	■ Lead time to process merchant payments	■ 48 hours	■ Acceptable	■ No action
Volume flexibility	■ Utilization	■ 98%	■ Too high to support rapid increase in volumes	■ Acquire temporary employees
				■ Improve work methods

Trends in Operations Management

Several trends are currently having a great impact on operations management: productivity improvement; global competition; and ethical, workforce diversity, and environmental issues. Accelerating change in the form of information technology, e-commerce, robotics, and the Internet is dramatically affecting the design of new services and products as well as a firm's sales, order fulfillment, and purchasing processes. In this section, we look at these trends and their challenges for operations managers.

Productivity Improvement

productivity

The value of outputs (services and products) produced divided by the values of input resources (wages, costs of equipment, and so on).

Productivity is a basic measure of performance for economies, industries, firms, and processes. Improving productivity is a major trend in operations management because all firms face pressures to improve their processes and supply chains so as to compete with their domestic and foreign competitors. **Productivity** is the value of outputs (services and products) produced divided by the values of input resources (wages, cost of equipment, and so on) used:

$$\text{Productivity} = \frac{\text{Output}}{\text{Input}}$$

Manufacturing employment peaked at just below 20 million in mid-1979, and shrunk by nearly 8 million from 1979 to 2011.[3] However, the manufacturing productivity in the United States has climbed steadily, as more manufacturing capacity and output has been achieved efficiently with a leaner work force. It is interesting and even surprising to compare productivity improvements in the service and manufacturing sectors. In the United States, employment in the service sector has grown rapidly, outstripping the manufacturing sector. It now employs about 90 percent of the workforce. But service-sector productivity gains have been much lower. If productivity growth in the service sector stagnates, so does the overall standard of living regardless of which part of the world you live in. Other major industrial countries, such as Japan and Germany, are experiencing the same problem. Yet, signs of improvement are appearing. The surge of investment across national boundaries can stimulate productivity gains by exposing firms to greater competition. Increased investment in information technology by service providers also increases productivity.

Measuring Productivity   As a manager, how do you measure the productivity of your processes? Many measures are available. For example, value of output can be measured by what the customer pays or simply by the number of units produced or customers served. The value of inputs can be judged by their cost or simply by the number of hours worked.

Managers usually pick several reasonable measures and monitor trends to spot areas needing improvement. For example, a manager at an insurance firm might measure office productivity as the number of insurance policies processed per employee per week. A manager at a carpet company might measure the productivity of installers as the number of square yards of carpet installed per hour. Both measures reflect *labor productivity*, which is an index of the output per person or per hour worked. Similar measures may be used for *machine productivity*, where the denominator is the number of machines. Accounting for several inputs simultaneously is also possible. *Multifactor productivity* is an index of the output provided by more than one of the resources used in production; it may be the value of the output divided by the sum of labor, materials, and overhead costs. Here is an example:

EXAMPLE 1.1	**Productivity Calculations**

MyOMLab

Tutor 1.1 in MyOMLab provides a new example for calculating productivity.

Calculate the productivity for the following operations:

a. Three employees process 600 insurance policies in a week. They work 8 hours per day, 5 days per week.

b. A team of workers makes 400 units of a product, which is sold in the market for $10 each. The accounting department reports that for this job the actual costs are $400 for labor, $1,000 for materials, and $300 for overhead.

---

[3] Paul Wiseman, "Despite China's Might, US Factories Maintain Edge," *The State* and *The Associated Press* (January 31, 2011).

SOLUTION

a.   Labor productivity $= \dfrac{\text{Policies processed}}{\text{Employee hours}}$

$= \dfrac{600 \text{ policies}}{(3 \text{ employees}) (40 \text{ hours/employee})} = 5 \text{ policies/hour}$

b.   Multifactor productivity $= \dfrac{\text{Value of output}}{\text{Labor cost} + \text{Materials cost} + \text{Overhead cost}}$

$= \dfrac{(400 \text{ units}) (\$10/\text{unit})}{\$400 + \$1{,}000 + \$300} = \dfrac{\$4{,}000}{\$1{,}700} = 2.35$

DECISION POINT

We want multifactor productivity to be as high as possible. These measures must be compared with performance levels in prior periods and with future goals. If they do not live up to expectations, the process should be investigated for improvement opportunities.

The Role of Management   The way processes are managed plays a key role in productivity improvement. Managers must examine productivity from the level of the supply chain because it is the collective performance of individual processes that makes the difference. The challenge is to increase the value of output relative to the cost of input. If processes can generate more output or output of better quality using the same amount of input, productivity increases. If they can maintain the same level of output while reducing the use of resources, productivity also increases.

Global Competition

Most businesses realize that, to prosper, they must view customers, suppliers, facility locations, and competitors in global terms. Firms have found that they can increase their market penetration by locating their production facilities in foreign countries because it gives them a local presence that reduces customer aversion to buying imports. Globalization also allows firms to balance cash flows from other regions of the world when economic conditions are less robust in the home country. Sonoco, a $4-billion-a-year industrial and consumer packaging company in Hartsville, South Carolina, has 335 locations worldwide in Australia, China, Europe, Mexico, New Zealand, and Russia, with 41 industrial product manufacturing facilities and 6 paper mills in Europe alone. These global operations resulted in international sales and income growth even as domestic sales were stumbling during 2007. How did Sonoco do it?[4] Locating operations in countries with favorable tax laws is one reason. Lower tax rates in Italy and Canada helped in padding the earnings margin. Another reason was a weak dollar, whereby a $46 million boost came from turning foreign currencies into dollars as Sonoco exported such items as snack bag packaging, and tubes and cores used to hold tape and textiles, to operations it owned in foreign countries. The exchange

Sonoco is a leading global manufacturer of industrial and consumer packaging goods with more than 300 locations in 35 countries serving 85 nations.

PR Newswire/Associated Press

---

[4] Ben Werner, "Sonoco Holding Its Own," *The State* (February 7, 2008); **www.sonoco.com,** 2008.

rate difference was more than enough to counter the added expense of increased raw materials, shipping, and energy costs in the United States.

Most products today are composites of materials and services from all over the world. Your Gap polo shirt is sewn in Honduras from cloth cut in the United States. Sitting in a Cineplex theater (Canadian), you munch a Nestle's Crunch bar (Swiss) while watching a Columbia Pictures movie (Japanese). Five developments spurred the need for sound global strategies: (1) improved transportation and communications technologies, (2) loosened regulations on financial institutions, (3) increased demand for imported services and goods, (4) reduced import quotas and other international trade barriers due to the formation of regional trading blocks, such as the European Union (EU) and the North American Free Trade Agreement (NAFTA), and (5) comparative cost advantages.

Comparative Cost Advantages   China and India have traditionally been the sources for low-cost, but skilled, labor, even though the cost advantage is diminishing as these countries become economically stronger. In the late 1990s, companies manufactured products in China to grab a foothold in a huge market, or to get cheap labor to produce low-tech products despite doubts about the quality of the workforce and poor roads and rail systems. Today, however, China's new factories, such as those in the Pudong industrial zone in Shanghai, produce a wide variety of products that are sold overseas in the United States and other regions of the world. U.S. manufacturers have increasingly abandoned low profit margin sectors like consumer electronics, shoes, and toys to emerging nations such as China and Indonesia. Instead, they are focusing on making expensive goods like computer chips, advanced machinery, and health care products that are complex and which require specialized labor.

Foreign companies have opened tens of thousands of new facilities in China over the past decade. Many goods the United States imports from China now come from foreign-owned companies with operations there. These companies include telephone makers, such as Nokia and Motorola, and nearly all of the big footwear and clothing brands. Many more major manufacturers are there as well. The implications for competition are enormous. Companies that do not have operations in China are finding it difficult to compete on the basis of low prices with companies that do. Instead, they must focus on speed and small production runs.

What China is to manufacturing, India is to service. As with the manufacturing companies, the cost of labor is a key factor. Indian software companies have grown sophisticated in their applications and offer a big advantage in cost. The computer services industry is also affected. Back-office operations are affected for the same reason. Many firms are using Indian companies for accounting and bookkeeping, preparing tax returns, and processing insurance claims. Many tech companies, such as Intel and Microsoft, are opening significant research and development (R&D) operations in India.

Disadvantages of Globalization   Of course, operations in other countries can have disadvantages. A firm may have to relinquish proprietary technology if it turns over some of its component manufacturing to offshore suppliers or if suppliers need the firm's technology to achieve desired quality and cost goals. Political risks may also be involved. Each nation can exercise its sovereignty over the people and property within its borders. The extreme case is nationalization, in which a government may take over a firm's assets without paying compensation. Exxon and other large multinational oil firms are scaling back operations in Venezuela due to nationalization concerns. Further, a firm may actually alienate customers back home if jobs are lost to offshore operations.

Employee skills may be lower in foreign countries, requiring additional training time. South Korean firms moved much of their sports shoe production to low-wage Indonesia and China, but they still manufacture hiking shoes and in-line roller skates in South Korea because of the greater

Yuriko Nakao/Reuters/Corbis

Shortage of components from suppliers prevented Nintendo from meeting the customer demand for its popular Wii game system.

MANAGERIAL PRACTICE 1.1   Japanese Earthquake and its Supply Chain Impact

Northeast Touhoku district of Japan was struck by a set of massive earthquakes on the afternoon of March 11, 2011, which were soon followed by a huge tsunami that sent waves higher than 33 feet in the port city of Sendai 80 miles away and travelling at the speed of a jetliner. At nearly 9.0 on the Richter scale, it was one of the largest recorded earthquakes to hit Japan. It shifted the Earth's axis by 6 inches with an impact that was felt 250 miles inland in Tokyo, and which moved Eastern Japan 13 feet toward North America. Apart from huge loss of life and hazards of nuclear radiation arising from the crippled Daiichi Nuclear Reactors in Fukushima, the damage to the manufacturing plants in Japan exposed the hazards of interconnected global supply chains and their impact on factories located half way around the globe.

The impact of the earthquake was particularly acute on industries that rely on cutting edge electronic parts sourced from Japan. Shin-Etsu Chemical Company is the world's largest producer of silicon wafers and supplies 20 percent of the global capacity. Its centralized plant located 40 miles from the Fukushima nuclear facility was damaged in the earthquake, causing ripple effects at Intel and Toshiba that purchase wafers from Shin-Etsu. Similarly, a shortage of automotive sensors from Hitachi has slowed or halted production of vehicles in Germany, Spain, and France, while Chrysler is reducing overtime at factories in Mexico and Canada to conserve parts from Japan. Even worse, General Motors stopped production altogether at a plant in Louisiana and Ford closed a truck plant in Kentucky due to the quake. The supply of vehicles such as Toyota's Prius and Lexus will be limited in the United States because of production disruptions in its Japanese factories. China has been affected too, where ZTE Corporation is facing shortages of batteries and LCD screens for its cell phones. Similarly, Lenovo in China is looking at reduced

Following the strong earthquakes and tsunami, flames and smoke rise from a petroleum refining plant next to a heating power station in Shiogama, Miyagi Prefecture, northern Japan, about 220 km north of Tokyo.

Kimimasa Mayama/EPA/Landov

supplies of components from Japan for assembly of its tablet computers. These disruptions due to reliance on small concentrated network of suppliers in Japan and globally connected production and logistics systems have caused worker layoffs an increase in prices of affected products, and economic losses that have been felt around the world.

Sources: Don Lee and David Pearson, "Disaster in Japan exposes supply chain weakness," *The State* (April 8, 2011), B6-B7; "Chrysler reduces overtime to help Japan," *The Associated Press* (April 8, 2011) printed in *The State* (April 6, 2011), B7; Krishna Dhir, "From the Editor," *Decision Line*, vol. 42, no. 2, 3.

skills required. In addition, when a firm's operations are scattered globally, customer response times can be longer. We discuss these issues in more depth in Chapter 10, "Supply Chain Design," because they should be considered when making decisions about outsourcing. Coordinating components from a wide array of suppliers can be challenging, as Nintendo found out in the production and worldwide distribution of its Wii game systems.[5] Despite twice increasing capacity since April 2007 to 1.8 million Wii's a month, Nintendo could only ship the completed units to retailers like Best Buy, Costco, and Circuit City in limited quantities that did not meet the large demand through the 2007 holiday season and beyond. In addition, as Managerial Practice 1.1 shows, catastrophic events such as the Japanese earthquake affect production and operations in Europe and United States because connected supply chains can spread disruptions rapidly and quickly across international borders.

Strong global competition affects industries everywhere. For example, U.S. manufacturers of steel, appliances, household durable goods, machinery, and chemicals have seen their market share decline in both domestic and international markets. With the value of world trade in services now at more than $2 trillion per year, banking, data processing, airlines, and consulting services are beginning to face many of the same international pressures. Regional trading blocs, such as EU and NAFTA, further change the competitive landscape in both services and manufacturing. Regardless of which area of the world you live in, the challenge is to produce services or products that can compete in a global market, and to design the processes that can make it happen.

Ethical, Workforce Diversity, and Environmental Issues

Businesses face more ethical quandaries than ever before, intensified by an increasing global presence and rapid technological change. As companies locate new operations and acquire more

[5] Peter Svensson, "GameStop to Sell Rain Checks for Wii," *The State* (December 18, 2007).

A Chinese consumer looks at Timberland products at a department store in Shanghai, China, November 11, 2010. Timberland seeks to benefit from rising incomes in the worlds fastest-growing major economy, and will also invest in its Hong Kong shops.

suppliers and customers in other countries, potential ethical dilemmas arise when business is conducted by different rules. Some countries are more sensitive than others about conflicts of interest, bribery, discrimination against minorities and women, minimum-wage levels, and unsafe workplaces. Managers must decide whether to design and operate processes that do more than just meet local standards. In addition, technological change brings debates about data protection and customer privacy. In an electronic world, businesses are geographically far from their customers, so a reputation of trust is paramount.

In the past, many people viewed environmental problems, such as toxic waste, poisoned drinking water, poor air quality, and climate change as quality-of-life issues; now, many people and businesses see them as survival issues. The automobile industry has seen innovation in electric and hybrid cars in response to environmental concerns and economic benefits arising from using less expensive fuels. Industrial nations face a particular burden because their combined populations consume proportionally much larger resources. Just seven nations, including the United States and Japan, produce almost half of all greenhouse gases. Now China and India have added to that total carbon footprint because of their vast economic and manufacturing expansion over the past decade.

Apart from government initiatives, large multinational companies have a responsibility as well for creating environmentally conscious practices, and can do so profitably. For instance, Timberland has over 110 stores in China because of strong demand for its boots, shoes, clothes, and outdoor gear in that country. It highlights its environmental credentials and corporate social responsibility through investments such as the reforestation efforts in northern China's Horqin Desert. Timberland hopes to double the number of stores over the next 3 years by environmentally differentiating itself from the competition. We discuss these issues in greater detail in Chapter 13, "Supply Chain Sustainability and Humanitarian Logistics."

The challenge is clear: Issues of ethics, workforce diversity, and the environment are becoming part of every manager's job. When designing and operating processes, managers should consider integrity, respect for the individual, and customer satisfaction along with more conventional performance measures such as productivity, quality, cost, and profit.

Operations Management as a Set of Decisions

In this text, we cover the major decisions operations managers make in practice. At the strategic level, operations managers are involved in the development of new capabilities and the maintenance of existing capabilities to best serve the firm's external customers. Operations managers design new processes that have strategic implications, and they are deeply involved in the development and organization of supply chains that link external suppliers and external customers to the firm's internal processes. Operations managers are often responsible for key performance measures such as cost and quality. These decisions have strategic impact because they affect the processes the firm uses to gain a competitive edge.

The operations manager's decisions should reflect corporate strategy. Plans, policies, and actions should be linked to those in other functional areas to support the firm's overall goals and objectives. These links are facilitated by taking a process view of a firm. Regardless of whether you aspire to be an operations manager, or you just want to use the principles of operations management to become a more effective manager, remember that effective management of people, capital, information, and materials is critical to the success of any process and any supply chain.

As you study operations management, keep two principles in mind:

1. Each part of an organization, not just the operations function, must design and operate processes that are part of a supply chain and deal with quality, technology, and staffing issues.

2. Each function of an organization has its own identity and yet is connected with operations through shared processes.

Great strategic decisions lead nowhere if the tactical decisions that support them are wrong. Operations managers are also involved in tactical decisions, including process improvement and performance measurement, managing and planning projects, generating production and staffing plans, managing inventories, and scheduling resources. You will find numerous examples of these decisions, and the implications of making them, throughout this text. You will also learn about the decision-making tools practicing managers use to recognize and define the problem and then choose the best solution.

Computerized Decision-Making Tools

MyOMLab contains a unique set of decision tools we call OM Explorer. This package contains powerful Excel-based computer routines to solve problems often encountered in practice. OM Explorer also has several tutors that provide coaching for all of the difficult analytical techniques in this text, and can be accessed from the drop-down menu. MyOMLab also contains POM for Windows, which is an extensive set of useful decision-making tools to complete your arsenal for solving operations problems, many Active Models (spreadsheets designed to help you learn more about important decision-making techniques), and a spreadsheet-based simulation package called SimQuick.

MyOMLab

Addressing the Challenges in Operations Management

How can firms meet challenges today and in the future? One way is to recognize challenges as opportunities to improve existing processes and supply chains or to create new, innovative ones. The management of processes and supply chains goes beyond designing them; it requires the ability to ensure they achieve their goals. Firms should manage their processes and supply chains to maximize their competitiveness in the markets they serve. We share this philosophy of operations management, as illustrated in Figure 1.7. We use this figure at the start of each chapter to show how the topic of the chapter fits into our philosophy of operations management. In addition, this text also contains several chapter supplements that are not explicitly shown in Figure 1.7.

The figure shows that all effective operations decisions follow from a sound operations strategy. Consequently, our text has three major parts: "Part 1: Creating Value through Operations Management," Part 2: "Managing Processes," and "Part 3: Managing Supply Chains." The flow of topics reflects our approach of first understanding how a firm's operations can help provide a solid foundation for competitiveness before tackling the essential process design decisions that will support its strategies. Each part begins with a strategy discussion to support the decisions in that part. Once it is clear how firms design and improve processes, and how they implement those designs, we examine the design and operation of supply chains that link processes, whether they are internal or external to the firm. The performance of the supply chains determines the firm's outcomes, which include the services or products the firm produces, the financial results, and feedback from the firm's customers. These outcomes, which are considered in the firm's strategic plan, are discussed throughout this text.

Part 1: Creating Value through Operations Management

The concluding chapter of Part 1 is a discussion of the methods and tools of project management. Project management is an effective approach to implementing operations strategy through the introduction of new services or products as well as any changes to a firm's processes or supply chains. Supplement A, "Decision Making," follows this chapter and covers some basic decision techniques that apply to multiple chapters.

Part 2: Managing Processes

In Part 2, we focus on analyzing processes and how they can be improved to meet the goals of the operations strategy. We begin by addressing the strategic aspects of process design and then present a six-step systematic approach to process analysis. Each chapter in this part deals with some aspect of that approach. We discuss the tools that help managers analyze processes, and we reveal the methods firms use to measure process performance and quality. These methods provide the foundation for programs such as Six Sigma and total quality management.

Determining the best process capacity with effective constraint management and making processes "lean" by eliminating activities that do not add value while improving those that do are also key decisions in the redesign of processes. The activities involved in managing processes are

Creating Value through Operations Management

↓

Using Operations to Compete
Project Management

Managing Processes

↓

Process Strategy
Process Analysis
Quality and Performance
Capacity Planning
Constraint Management
Lean Systems

Managing Supply Chains

↓

Supply Chain Inventory Management
Supply Chain Design
Supply Chain Location Decisions
Supply Chain Integration
Supply Chain Sustainability and Humanitarian Logistics
Forecasting
Operations Planning and Scheduling
Resource Planning

▲ **FIGURE 1.7**
Managing Processes and Supply Chains

essential for providing significant benefits to the firm. Effective management of its processes can allow a firm to reduce its costs and also increase customer satisfaction.

Part 3: Managing Supply Chains

The management of supply chains is based upon process management and operations strategy. In Part 2, we focus on individual processes. The focus of Part 3, however, is on supply chains involving processes both internal and external to the firm and the tools that enhance their execution. We begin this part with a look at managing inventory in supply chains. We follow that with understanding how the design of supply chains and major strategic decisions, such as outsourcing, inventory placement, and locating facilities affect performance. We also look at contemporary issues surrounding supply chain integration and the impact of supply chains on the environment. We follow that with chapters focused on three key planning activities for the effective operation of supply chains: (1) forecasting, (2) operations planning and scheduling, and (3) resource planning.

Adding Value with Process Innovation in Supply Chains

It is important to note that the effective operation of a firm and its supply chain is as important as the design and implementation of its processes. Process innovation can make a big difference even in a low-growth industry. Examining processes from the perspective of the value they add is an important part of a successful manager's agenda, as is gaining an understanding of how core processes and related supply chains are linked to their competitive priorities, markets, and the operations strategy of a firm. As illustrated by Progressive Insurance in Managerial Practice 1.2, who says operations management does not make a difference?

MANAGERIAL PRACTICE 1.2   Operational Innovation Is a Competitive Weapon at Progressive Insurance

Progressive Insurance, an automobile insurer that started business in 1937, had approximately $1.3 billion in sales in 1991. By 2011, it was one of the largest U.S. private passenger auto insurance groups with annual premiums in excess of $14 billion. How did it accomplish this amazing growth rate in a 100-year-old industry that traditionally does not experience that sort of growth?

The answer is simple but the implementation was challenging: offer low prices, better service, and more value to customers through operational innovation. *Operational innovation* means designing entirely new processes by dramatically changing the way work is done. For example, Progressive reinvented claims processing to lower costs and increase customer satisfaction and retention. Progressive's agency-dedicated Web site, ForAgentsOnly.com (FAO), lets agents quickly, easily, and securely access payments; view policy, billing, and claims information; and send quote information directly to customers via e-mail. Customers are encouraged to go online to perform routine tasks such as address changes or simple billing inquiries. In addition, Immediate Response Claims Handling allows a claimant to now reach a Progressive representative by telephone 24 hours a day. The representative immediately sends a claims adjuster to inspect the damaged vehicle. The adjuster drives to the vehicle accident site in a mobile claims van, examines the vehicle, prepares an onsite estimate of damage, and if possible, writes a check on the spot. It now takes only 9 hours to complete the cycle, compared with 7–10 days before the changes were made.

The operational innovations to the processes in the customer relationship–order fulfillment supply chain for claims processing produced several benefits. First, claimants received faster service with less hassle, which helped retain them as customers. Second, the shortened cycle time significantly reduced costs. The costs of storing a damaged vehicle and providing a rental car can often wipe out the expected underwriting profit for

Via operational innovations that add value to its products, and catchy promotional advertisements, Progressive Insurance has been able to achieve amazing growth in a low-growth industry.

a six-month policy. This cost becomes significant when you realize that the company processes more than 10,000 claims a day. Third, the new supply chain design requires fewer people for handling the claim, which reduces operational costs. Finally, the operational innovations improved Progressive's ability to detect fraud by arriving on the accident scene quickly and helped to reduce payouts because claimants often accept less money if the payout is quick and hassle-free. Progressive Insurance found a way to differentiate itself in a low-growth industry without compromising profitability, and it accomplished that feat with operational innovation.

Source: Michael Hammer, "Deep Change: How Operational Innovation Can Transform Your Company," *Harvard Business Review* (April 2004), pp. 85–93; http://www.progressive.com/about-progressive-insurance.aspx, 2011.

The topics in this text will help you meet operations challenges and achieve operational innovation regardless of your chosen career path.

LEARNING GOALS IN REVIEW

① **Describe operations and supply chains in terms of inputs, processes, outputs, information flows, suppliers, and customers.** Review Figure 1.4 for the important supply chain linkage and information flows. The section "Operations and Supply Chain Management Across the Organization," pp. 2–3, shows how different functional areas of business come together to create value for a firm.

② **Define an operations strategy and its linkage to corporate strategy, as well as the role it plays as a source of competitive advantage in a global marketplace.** See the sections on "Operations Strategy" and "Corporate Strategy," pp. 7–10, and review Figure 1.5.

③ **Identify nine competitive priorities used in operations strategy, and their linkage to marketing strategy.** The section "Competitive Priorities and Capabilities," pp. 11–14, discusses the important concept of order winners and qualifiers. Review Table 1.1 for important illustrations and examples of how leading edge firms implemented different competitive priorities to create a unique positioning in the marketplace.

④ **Explain how operations can be used as a competitive weapon.** The section "Operations Strategy as a Pattern of Decisions," p. 15, shows how firms must identify gaps in their competitive priorities and build capabilities through related process and operational changes. Make sure that you review Table 1.3 that provides a nice illustrative example.

⑤ **Identify the global trends and challenges facing operations management.** The section "Trends in Operations Management," pp. 16–20, describes the pressures managers face for achieving productivity improvements, along with managing sustainability and work force diversity related issues in the face of global competition.

MyOMLab helps you develop analytical skills and assesses your progress with multiple problems on labor and multifactor productivity.

MyOMLab Resources	Titles	Link to the Book
Video	*Operations as a Competitive Weapon at Starwood*	A Process View; Operations Strategy as a Pattern of Decisions
Active Model Exercise	1.1 Productivity	Trends in Operations Management; Example 1.1 (pp. 16–17); Solved Problem 1 (p. 24); Solved Problem 2 (pp. 24–25)
OM Explorer Tutors	1.1 Productivity Measures	Trends in Operations Management; Example 1.1 (pp. 16–17); Solved Problem 1 (p. 24); Solved Problem 2 (p. 24–25)
Tutor Exercises	1.1 Ticket sales	Example 1.1 (pp. 16–17); Solved Problem 1 (p. 24); Solved Problem 2 (pp. 24–25)
Virtual Tours	L'Oréal Cosmetics EDS Industries EDS Services	The Supply Chain View; Trends in Operations Management A Process View; The Supply Chain View A Process View; The Supply Chain View
Internet Exercises	1. Coca-Cola and Nestlé 2. Xerox 3. L'Oréal 4. Environment, Health, and Safety at Xerox	The Supply Chain View; Operations Strategy as a Pattern of Decisions The Supply Chain View; Competitive Priorities and Capabilities A Process View; Trends in Operations Management Trends in Operations Management
Additional Cases	BSB, Inc., The Pizza Wars Come to Campus	A Process View; The Supply Chain View; Competitive Priorities and Capabilities
Key Equations		
Image Library		

Key Equation

1. Productivity is the ratio of output to input:

$$\text{Productivity} = \frac{\text{Output}}{\text{Input}}$$

Key Terms

competitive capabilities 11
competitive priorities 11
consistent quality 12
core competencies 9
core process 7
customer relationship process 7
customization 12
delivery speed 12
development speed 12
external customers 4
external suppliers 4
internal customers 4

internal suppliers 4
lead time 9
low-cost operations 12
nested process 5
new service/product development
    process 7
on-time delivery 12
operation 2
operations management 2
operations strategy 7
order fulfillment process 7
order qualifier 13

order winner 11
process 2
productivity 16
supplier relationship process 7
supply chain 2
supply chain management 2
support process 7
time-based competition 11
top quality 12
variety 12
volume flexibility 12

Solved Problem 1

Student tuition at Boehring University is $150 per semester credit hour. The state supplements school revenue by $100 per semester credit hour. Average class size for a typical 3-credit course is 50 students. Labor costs are $4,000 per class, materials costs are $20 per student per class, and overhead costs are $25,000 per class.

a. What is the *multifactor* productivity ratio for this course process?

b. If instructors work an average of 14 hours per week for 16 weeks for each 3-credit class of 50 students, what is the *labor* productivity ratio?

SOLUTION

a. Multifactor productivity is the ratio of the value of output to the value of input resources.

$$\text{Value of output} = \left(\frac{50 \text{ students}}{\text{class}}\right)\left(\frac{3 \text{ credit hours}}{\text{students}}\right)\left(\frac{\$150 \text{ tuition} + \$100 \text{ state support}}{\text{credit hour}}\right)$$

$$= \$37,500/\text{class}$$

$$\text{Value of inputs} = \text{Labor} + \text{Materials} + \text{Overhead}$$

$$= \$4,000 + (\$20/\text{student} \times 50 \text{ students/class}) + \$25,000$$

$$= \$30,000/\text{class}$$

$$\text{Multifactor productivity} = \frac{\text{Output}}{\text{Input}} = \frac{\$37,500/\text{class}}{\$30,000/\text{class}} = 1.25$$

b. Labor productivity is the ratio of the value of output to labor hours. The value of output is the same as in part (a), or $37,500/class, so

$$\text{Labor hours of input} = \left(\frac{14 \text{ hours}}{\text{week}}\right)\left(\frac{16 \text{ weeks}}{\text{class}}\right) = 224 \text{ hours/class}$$

$$\text{Labor productivity} = \frac{\text{Output}}{\text{Input}} = \frac{\$37,500/\text{class}}{224 \text{ hours/class}}$$

$$= \$167.41/\text{hour}$$

Solved Problem 2

Natalie Attire makes fashionable garments. During a particular week, employees worked 360 hours to produce a batch of 132 garments, of which 52 were "seconds" (meaning that they were flawed). Seconds are sold for $90 each at Attire's Factory Outlet Store. The remaining 80 garments are sold to retail distribution at $200 each. What is the *labor* productivity ratio of this manufacturing process?

SOLUTION

$$\text{Value of output} = (52 \text{ defective} \times 90/\text{defective}) + (80 \text{ garments} \times 200/\text{garment})$$

$$= \$20,680$$

$$\text{Labor hours of input} = 360 \text{ hours}$$

$$\text{Labor productivity} = \frac{\text{Output}}{\text{Input}} = \frac{\$20,680}{360 \text{ hours}}$$

$$= \$57.44 \text{ in sales per hour}$$

Discussion Questions

1. Consider your last (or current) job.

   a. What activities did you perform?

   b. Who were your customers (internal and external), and how did you interact with them?

   c. How could you measure the customer value you were adding by performing your activities?

   d. Was your position in accounting, finance, human resources, management information systems, marketing, operations, or other? Explain.

2. Consider amazon.com, whose Web site enjoys millions of "hits" each day and puts customers in touch with millions of services and products. What are amazon.com's competitive priorities and what should its operations strategy focus on?

3. A local hospital declares that it is committed to provide *care* to patients arriving at the emergency unit in less than 15 minutes and that it will never turn away patients who need to be hospitalized for further medical care. What implications does this commitment have for strategic operations management decisions (i.e., decisions relating to capacity and workforce)?

4. FedEx built its business on quick, dependable delivery of items being shipped by air from one business to another. Its early advantages included global tracking of shipments using Web technology. The advancement of Internet technology enabled competitors to become much more sophisticated in order tracking. In addition, the advent of Web-based businesses put pressure on increased ground transportation deliveries. Explain how this change in the environment has affected FedEx's operations strategy, especially relative to UPS, which has a strong hold on the business-to-consumer ground delivery business.

5. Suppose that you were conducting a market analysis for a new textbook about technology management. What would you need to know to identify a market segment? How would you make a needs assessment? What should be the collection of services and products?

6. Although all nine of the competitive priorities discussed in this chapter are relevant to a company's success in the marketplace, explain why a company should not necessarily try to excel in all of them. What determines the choice of the competitive priorities that a company should emphasize for its key processes?

7. Choosing which processes are core to a firm's competitive position is a key strategic decision. For example, Nike, a popular sports shoe company, focuses on the customer relationship, new product development, and supplier relationship processes and leaves the order fulfillment process to others. Allen Edmonds, a top-quality shoe company, considers all four processes to be core processes. What considerations would you make in determining which processes should be core to your manufacturing company?

8. A local fast-food restaurant processes several customer orders at once. Service clerks cross paths, sometimes nearly colliding, while they trace different paths to fill customer orders. If customers order a special combination of toppings on their hamburgers, they must wait quite some time while the special order is cooked. How would you modify the restaurant's operations to achieve competitive advantage? Because demand surges at lunchtime, volume flexibility is a competitive priority in the fast-food business. How would you achieve volume flexibility?

9. Kathryn Shoemaker established Grandmother's Chicken Restaurant in Middlesburg 5 years ago. It features a unique recipe for chicken, "just like grandmother used to make." The facility is homey, with relaxed and friendly service. Business has been good during the past 2 years, for both lunch and dinner. Customers normally wait about 15 minutes to be served, although complaints about service delays have increased recently. Shoemaker is currently considering whether to expand the current facility or open a similar restaurant in neighboring Uniontown, which has been growing rapidly.

   a. What types of strategic plans must Shoemaker make?

   b. What environmental forces could be at work in Middlesburg and Uniontown that Shoemaker should consider?

   c. What are the possible distinctive competencies of Grandmother's?

10. Wild West, Inc., is a regional telephone company that inherited nearly 100,000 employees and 50,000 retirees from AT&T. Wild West has a new mission: to diversify. It calls for a 10-year effort to enter the financial services, real estate, cable TV, home shopping, entertainment, and cellular communication services markets—and to compete with other telephone companies. Wild West plans to provide cellular and fiber-optic communications services in markets with established competitors, such as the United Kingdom, and in markets with essentially no competition, such as Russia and former Eastern Bloc countries.

    a. What types of strategic plans must Wild West make? Is the "do-nothing" option viable? If Wild West's mission appears too broad, which businesses would you trim first?

    b. What environmental forces could be at work that Wild West should consider?

    c. What are the possible core competencies of Wild West? What weaknesses should it avoid or mitigate?

11. You are designing a grocery delivery business. Via the Internet, your company will offer staples and frozen foods in a large metropolitan area and then deliver them within a customer-defined window of time. You plan to partner with two major food stores in the area. What should be your competitive priorities and what capabilities do you want to develop in your core and support processes?

Problems

The OM Explorer and POM for Windows software is available to all students using the 10th edition of this textbook. Go to **www.pearsonhighered.com/krajewski** to download these computer packages. If you purchased MyOMLab, you also have access to Active Models software and significant help in doing the following problems. Check with your instructor on how best to use these resources. In many cases, the instructor wants you to understand how to do the calculations by hand. At the least, the software provides a check on your calculations. When calculations are particularly complex and the goal is interpreting the results in making decision, the software entirely replaces the manual calculations.

1. (Refer to Solved Problem 1.) Coach Bjourn Toulouse led the Big Red Herrings to several disappointing football seasons. Only better recruiting will return the Big Red Herrings to winning form. Because of the current state of the program, Boehring University fans are unlikely to support increases in the $192 season ticket price. Improved recruitment will increase overhead costs to $30,000 per class section from the current $25,000 per class section. The university's budget plan is to cover recruitment costs by increasing the average class size to 75 students. Labor costs will increase to $6,500 per 3-credit course. Material costs will be about $25 per student for each 3-credit course. Tuition will be $200 per semester credit, which is supplemented by state support of $100 per semester credit.

    a. What is the multifactor productivity ratio? Compared to the result obtained in Solved Problem 1, did productivity increase or decrease for the course process?

    b. If instructors work an average of 20 hours per week for 16 weeks for each 3-credit class of 75 students, what is the *labor* productivity ratio?

2. Suds and Duds Laundry washed and pressed the following numbers of dress shirts per week.

Week	Work Crew	Total Hours	Shirts
1	Sud and Dud	24	68
2	Sud and Jud	46	130
3	Sud, Dud, and Jud	62	152
4	Sud, Dud, and Jud	51	125
5	Dud and Jud	45	131

    a. Calculate the *labor* productivity ratio for each week.

    b. Explain the labor productivity pattern exhibited by the data.

3. CD players are produced on an automated assembly line process. The standard cost of CD players is $150 per unit (labor, $30; materials, $70; and overhead, $50). The sales price is $300 per unit.

    a. To achieve a 10 percent multifactor productivity improvement by reducing materials costs only, by what percentage must these costs be reduced?

    b. To achieve a 10 percent multifactor productivity improvement by reducing labor costs only, by what percentage must these costs be reduced?

    c. To achieve a 10 percent multifactor productivity improvement by reducing overhead costs only, by what percentage must these costs be reduced?

4. The output of a process is valued at $100 per unit. The cost of labor is $50 per hour including benefits. The accounting department provided the following information about the process for the past four weeks:

	Week 1	Week 2	Week 3	Week 4
Units Produced	1,124	1,310	1,092	981
Labor ($)	12,735	14,842	10,603	9,526
Material ($)	21,041	24,523	20,442	18,364
Overhead ($)	8,992	10,480	8,736	7,848

    a. Use the multifactor productivity ratio to see whether recent process improvements had any effect and, if so, when the effect was noticeable.

    b. Has labor productivity changed? Use the labor productivity ratio to support your answer.

5. Alyssa's Custom Cakes currently sells 5 birthday, 2 wedding, and 3 specialty cakes each month for $50, $150, and $100 each, respectively. The cost of labor is $50 per hour including benefits. It takes 90 minutes to produce a birthday cake, 240 minutes to produce a wedding cake, and 60 minutes to produce a specialty cake. Alyssa's current multifactor productivity ratio is 1.25.

    a. Use the multifactor productivity ratio provided to calculate the average cost of the cakes produced.

    b. Calculate Alyssa's labor productivity ratio in dollars per hour for each type of cake.

    c. Based solely on the labor productivity ratio, which cake should Alyssa try to sell the most?

    d. Based on your answer in part (a), is there a type of cake Alyssa should stop selling?

Advanced Problems

6. The Big Black Bird Company (BBBC) has a large order for special plastic-lined military uniforms to be used in an urgent military operation. Working the normal two shifts of 40 hours each per week, the BBBC production process usually produces 2,500 uniforms per week at a standard cost of $120 each. Seventy employees work the first shift and 30 employees work the second. The contract price is $200 per uniform. Because of the urgent need, BBBC is authorized to use around-the-clock production, 6 days per week. When each of the two shifts works 72 hours per week, production increases to 4,000 uniforms per week but at a cost of $144 each.

    a. Did the multifactor productivity ratio increase, decrease, or remain the same? If it changed, by what percentage did it change?

    b. Did the labor productivity ratio increase, decrease, or remain the same? If it changed, by what percentage did it change?

    c. Did weekly profits increase, decrease, or remain the same?

7. Mack's guitar fabrication shop produces low-cost, highly durable guitars for beginners. Typically, out of the 100 guitars that begin production each month, only 80 percent are considered good enough to sell. The other 20 percent are scrapped due to quality problems that are identified after they have completed the production process. Each guitar sells for $250. Because some of the production process is automated, each guitar only requires 10 labor hours. Each employee works an average 160 hours per month. Labor is paid at $10/hour, materials cost is $40/guitar, and overhead is $4,000.

    a. Calculate the labor and multifactor productivity ratios.

    b. After some study, the operations manager Darren Funk recommends three options to improve the company's multifactor productivity: (1) increase the sales price by 10 percent, (2) improve quality so that only 10 percent are defective, or (3) reduce labor, material, and overhead costs by 10 percent. Which option has the greatest impact on the multifactor productivity measure?

8. Mariah Enterprises makes a variety of consumer electronic products. Its camera manufacturing plant is considering choosing between two different processes, named Alpha and Beta, which can be used to make a component part. To make the correct decision, the managers would like to compare the labor and multifactor productivity of process Alpha with that of process Beta. The value of process output for Alpha and Beta is $175 and $140 per unit, and the corresponding overhead costs are $6,000 and $5,000, respectively.

Product	PROCESS ALPHA		PROCESS BETA	
	A	**B**	**A**	**B**
Output (units)	50	60	30	80
Labor ($)	$1,200	$1,400	$1,000	$2,000
Material ($)	$2,500	$3,000	$1,400	$3,500

    a. Which process, Alpha or Beta, is more productive?

    b. What conclusions can you draw from your analysis?

9. The Morning Brew Coffee Shop sells Regular, Cappuccino, and Vienna blends of coffee. The shop's current daily labor cost is $320, the equipment cost is $125, and the overhead cost is $225. Daily demands, along with selling price and material costs per beverage, are given below.

	Regular Coffee	Cappuccino	Vienna coffee
Beverages sold	350	100	150
Price per beverage	$2.00	$3.00	$4.00
Material ($)	$0.50	$0.75	$1.25

Harald Luckerbauer, the manager at Morning Brew Coffee Shop, would like to understand how adding Eiskaffee (a German coffee beverage of chilled coffee, milk, sweetener, and vanilla ice cream) will alter the shop's productivity. His market research shows that Eiskaffee will bring in new customers and not cannibalize current demand. Assuming that the new equipment is purchased before Eiskaffee is added to the menu, Harald has developed new average daily demand and cost projections. The new equipment cost is $200, and the overhead cost is $350. Modified daily demands, as well as selling price and material costs per beverage for the new product line, are given below.

	Regular Coffee	Cappuccino	Vienna coffee	Eiskaffee
Beverages sold	350	100	150	75
Price per beverage	$2.00	$3.00	$4.00	$5.00
Material ($)	$0.50	$0.75	$1.25	$1.50

    a. Calculate the change in labor and multifactor productivity if Eiskaffee is added to the menu.

    b. If everything else remains unchanged, how many units of Eiskaffee would have to be sold to ensure that the multifactor productivity increases from its current level?

Active Model Exercise

This Active Model appears in MyOMLab. It allows you to evaluate the important elements of labor productivity.

QUESTIONS

1. If the insurance company can process 60 (10 percent) more policies per week, by what percentage will the productivity measure rise?

2. Suppose the 8-hour day includes a 45-minute lunch. What is the revised productivity measure, excluding lunch?

3. If an employee is hired, what will be the weekly number of policies processed if the productivity of five policies per hour is maintained?

4. Suppose that, during the summer, the company works for only 4 days per week. What will be the weekly number of policies processed if the productivity of five policies per hour is maintained?

▶ **ACTIVE MODEL 1.1**
Labor Productivity Using Data from Example 1.1

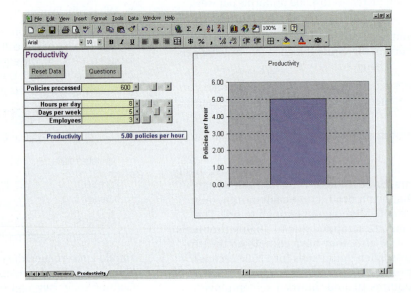

VIDEO CASE | Operations as a Competitive Weapon at Starwood

Starwood is one of the world's largest hotel companies, with more than 750 owned, managed, and franchised properties in more than 80 countries. The company's lodging brands include The Luxury Collection, St. Regis, Sheraton, Westin, Four Points, and W Hotels. Its hotels regularly appear on lists of top hotels around the world. On any given night, guests in the hotels may be individual leisure travelers, independent business guests, or part of a meeting or convention.

In 2002, Starwood standardized its operating processes so that it could measure, improve, and ultimately grow its convention business. Each meeting is assigned a Star Meeting Concierge who works closely with meeting planners.

When guests stay at a Starwood property as part of a meeting or convention, arrangements are typically made by a meeting planner. The meeting planner works with a location to arrange meeting facilities, banquet rooms, lodging, and events for participants. Prior to 2002, the company's individual properties had their own approaches to convention planning, yet no consistent, coordinated program within or across brands made it easy for meeting planners to do business with Starwood. For example, paperwork for confirming program details, rooms, and food and beverage requirements differed between properties and brands. Some hotels had diagrams of meeting space, while others did not. Technology available for meeting rooms varied widely, and a hotel liaison was not always immediately available during the event in case a need arose.

Recognizing that Starwood's future growth and success relied heavily on its relationships with meeting planners, the company held focus groups to gather information about their needs and expectations. One clear priority emerged: consistency in the meeting planning process, whether that meeting was held at the Sheraton in New York, the Westin Kierland in Phoenix, or the W Hotel in Lakeshore, Chicago. Such a program could create consistency across all brands, and generate loyalty and increased revenues from those meeting planners who drive large volumes of business to Starwood properties annually.

As a result of the meetings, Starwood created the Starwood Preferred Planner program. Every hotel property now has the same paperwork for the meeting planning process, and shares that paperwork electronically across properties and brands. Contracts were standardized and new standards created to recognize and reward frequent VIP meeting planners. Each meeting is assigned a "Star Meeting Concierge" whose sole responsibility is to anticipate and fulfill any needs of the meeting planner during the event. Handheld Nextel radio phones are now issued at check-in to the meeting planners at no extra charge so that they have 24-hour access to the concierge.

To measure the performance of the new process, Starwood set high internal targets for scores on the surveys given to meeting planners after their events concluded. For instance, at the Luxury Collection and St. Regis brands, individual meeting scores must be 4.55 on a 5-point scale. At the Westin and W Hotels, scores must be above 4.35 on the 5-point scale. Scores from Sheraton properties must exceed 4.30, and Four Points hotels have a target of 4.25 on the 5-point scale. Because the expectations for an airport location one-day meeting (not held at the St. Regis or Luxury Collection) differ from a multiday resort experience, the targets reflect those expectations.

QUESTIONS

1. What are the key inputs and outputs associated with Starwood's new meeting planning process?
2. How does the meeting planning process at Starwood interact with the following core processes in their hotels?
   a. Customer relationship (internal and external)
   b. New service or product development
   c. Order fulfillment
   d. Supplier relationship

CASE | Chad's Creative Concepts

Chad's Creative Concepts designs and manufactures wood furniture. Founded by Chad Thomas on the banks of Lake Erie in Sandusky, Ohio, the company began by producing custom-made wooden furniture for vacation cabins located along the coast of Lake Erie and on nearby Kelly's Island and Bass Island. Being an "outdoors" type himself, Thomas originally wanted to bring "a bit of the outdoors" inside. Chad's Creative Concepts developed a solid reputation for creative designs and high-quality workmanship. Sales eventually encompassed the entire Great Lakes region. Along with growth came additional opportunities.

Traditionally, the company focused entirely on custom-made furniture, with the customer specifying the kind of wood from which the piece would be

made. As the company's reputation grew and sales increased, the sales force began selling some of the more popular pieces to retail furniture outlets. This move into retail outlets led Chad's Creative Concepts into the production of a more standard line of furniture. Buyers of this line were much more price-sensitive and imposed more stringent delivery requirements than did clients for the custom line. Custom-designed furniture, however, continued to dominate sales, accounting for 60 percent of volume and 75 percent of dollar sales. Currently, the company operates a single manufacturing process in Sandusky, where both custom furniture and standard furniture are manufactured. The equipment is mainly general purpose in nature to provide the flexibility needed for producing custom pieces of furniture. The layout puts together saws in one section of the facility, lathes in another, and so on. The quality of the finished product reflects the quality of the wood chosen and the craftsmanship of individual workers. Both custom and standard furniture compete for processing time on the same equipment by the same craftspeople.

During the past few months, sales of the standard line steadily increased, leading to more regular scheduling of this product line. However, when scheduling trade-offs had to be made, custom furniture was always given priority because of its higher sales and profit margins. Thus, scheduled lots of standard furniture pieces were left sitting around the plant in various stages of completion.

As he reviews the progress of Chad's Creative Concepts, Thomas is pleased to note that the company has grown. Sales of custom furniture remain strong, and sales of standard pieces are steadily increasing. However, finance and accounting indicate that profits are not what they should be. Costs associated with the standard line are rising. Dollars are being tied up in inventory, both in raw materials and work-in-process. Expensive public warehouse space has to be rented to accommodate the inventory volume. Thomas also is concerned with increased lead times for both custom and standard orders, which are causing longer promised delivery times. Capacity is being pushed, and no space is left in the plant for expansion. Thomas begins a careful assessment of the overall impact that the new standard line is having on his manufacturing process.

QUESTIONS

1. What types of decisions must Chad Thomas make daily for his company's operations to run effectively? Over the long run?

2. How did sales and marketing affect operations when they began to sell standard pieces to retail outlets?

3. How has the move to producing standard furniture affected the company's financial structure?

4. What might Chad Thomas have done differently to avoid some of the problems he now faces?

Source: This case was prepared by Dr. Brooke Saladin, Wake Forest University, as a basis for classroom discussion. Copyright © Brooke Saladin. Used with permission.

Selected References

Chase, Richard B., and Uday M. Apte. "A History of Research in Service Operations: What's the Big Idea?" *Journal of Operations Management*, vol. 25, no. 2 (2007), pp. 375–386.

Collis, David J. and Michael G. Rukstad. "Can You Say What Your Strategy Is?" *Harvard Business Review*, vol. 86, no. 4 (2008), pp. 82–90.

Fitzsimmons, James A., and Mona Fitzsimmons. *Service Management*. New York: McGraw-Hill, 2005.

Gaimon, Cheryl. "The Management of Technology: A Production and Operations Management Perspective." *Production and Operations Management*, vol. 17, no. 1 (2008), pp. 1–11.

Hammer, Michael. "Deep Change: How Operational Innovation Can Transform Your Company." *Harvard Business Review* (April 2004), pp. 85–93.

Heineke, Janelle, and Mark Davis. "The Emergence of Service Operations as an Academic Discipline." *Journal of Operations Management*, vol. 25, no. 2 (2007), pp. 364–374.

Hill, Terry. *Manufacturing Strategy: Text and Cases*, 3rd ed. Homewood, IL: Irwin/McGraw-Hill, 2000.

Huckman, Robert S., and Darren E. Zinner. "Does Focus Improve Operational Performance? Lessons from the Management of Clinical Trials." *Strategic Management Journal*, vol. 29 (2008), pp. 173–193.

Karmarkar, Uday. "Will You Survive the Services Revolution?" *Harvard Business Review*, vol. 82 (2004), pp. 100–108.

Kaplan, Robert S., and David P. Norton. *Balanced Scoreboard*. Boston, MA: Harvard Business School Press, 1997.

King Jr., Neil. "A Whole New World." *Wall Street Journal* (September 27, 2004).

Meyer, Christopher and Andre Schwager. "Understanding customer experience." *Harvard Business Review*, vol. 85 (2007), pp. 116–126.

Neilson, Gary L., Karla L. Martin, and Elizabeth Powers. "The secrets to successful strategy execution." *Harvard Business Review*, vol. 86, no. 6 (2008), pp. 60–70.

Pande, Peter S., Robert P. Neuman, and Roland R. Cavanagh. *The Six Sigma Way*. New York: McGraw-Hill, 2000.

Porter, Michael. *Competitive Advantage*. New York: The Free Press, 1987.

Porter, Michael E., and Mark R. Kramer. "Strategy and Society: The Link Between Competitive Advantage and Corporate Social Responsibility." *Harvard Business Review*, vol. 84, no. 12 (2006), pp. 78–92.

Powell, Bill. "It's All Made in China Now." *Fortune* (March 4, 2002), pp. 121–128.

Safizadeh, M. Hossein, Larry P. Ritzman, Deven Sharma, and Craig Wood. "An Empirical Analysis of the Product–Process Matrix." *Management Science*, vol. 42, no. 11 (1996), pp. 1576–1591.

Skinner, Wickham. "Manufacturing—Missing Link in Corporate Strategy." *Harvard Business Review* (May–June 1969), pp. 136–145.

Svensson, Peter. "GameStop to Sell Rain Checks for Wii." *The State* (December 18, 2007).

Voss, Chris, Aleda Roth, and Richard Chase. "Experience, Service Operations Strategy, and Services as Destinations: Foundations and Exploratory Investigation" *Production and Operations Management*, vol. 17, no. 3 (2008), pp. 247–266.

Ward, Peter T., and Rebecca Duray. "Manufacturing Strategy in Context: Environment, Competitive Strategy and Manufacturing Strategy." *Journal of Operations Management*, vol. 18 (2000), pp. 123–138.

Wiseman, Paul. "Despite China's might, US factories maintain edge," *The State* and *The Associated Press* (January 31, 2011).

Womack, James P., Daniel T. Jones, and Daniel Roos. *The Machine That Changed the World*. New York: HarperPerennial, 1991.

DECISION MAKING

Operations managers make many decisions as they manage processes and supply chains. Although the specifics of each situation vary, decision making generally involves the same basic steps: (1) recognize and clearly define the problem, (2) collect the information needed to analyze possible alternatives, and (3) choose and implement the most feasible alternative.

Sometimes, hard thinking in a quiet room is sufficient. At other times, interacting with others or using more formal procedures are needed. Here, we present four such formal procedures: break-even analysis, the preference matrix, decision theory, and the decision tree.

- Break-even analysis helps the manager identify how much change in volume or demand is necessary before a second alternative becomes better than the first alternative.
- The preference matrix helps a manager deal with multiple criteria that cannot be evaluated with a single measure of merit, such as total profit or cost.
- Decision theory helps the manager choose the best alternative when outcomes are uncertain.
- A decision tree helps the manager when decisions are made sequentially—when today's best decision depends on tomorrow's decisions and events.

Break-Even Analysis

To evaluate an idea for a new service or product, or to assess the performance of an existing one, determining the volume of sales at which the service or product breaks even is useful. The **break-even quantity** is the volume at which total revenues equal total costs. Use of this technique is known as **break-even analysis**. Break-even analysis can also be used to compare processes by finding the volume at which two different processes have equal total costs.

break-even quantity

The volume at which total revenues equal total costs.

break-even analysis

The use of the break-even quantity; it can be used to compare processes by finding the volume at which two different processes have equal total costs.

LEARNING GOALS *After reading this supplement, you should be able to:*

1. Explain break-even analysis, using both the graphic and algebraic approaches.

2. Define a preference matrix.

3. Explain how to construct a payoff table.

4. Identify the maximin, maximax, Laplace, minimax regret, and expected value decision rules.

5. Describe how to draw and analyze a decision tree.

A manager is doing some hard thinking and analysis on his computer before reaching a final decision.

Evaluating Services or Products

We begin with the first purpose: to evaluate the profit potential of a new or existing service or product. This technique helps the manager answer questions, such as the following:

- Is the predicted sales volume of the service or product sufficient to break even (neither earning a profit nor sustaining a loss)?
- How low must the variable cost per unit be to break even, based on current prices and sales forecasts?
- How low must the fixed cost be to break even?
- How do price levels affect the break-even volume?

Break-even analysis is based on the assumption that all costs related to the production of a specific service or product can be divided into two categories: (1) variable costs and (2) fixed costs.

variable cost

The portion of the total cost that varies directly with volume of output.

fixed cost

The portion of the total cost that remains constant regardless of changes in levels of output.

The **variable cost**, c, is the portion of the total cost that varies directly with volume of output: costs per unit for materials, labor, and usually some fraction of overhead. If we let Q equal the number of customers served or units produced per year, total variable cost = cQ. The **fixed cost**, F, is the portion of the total cost that remains constant regardless of changes in levels of output: the annual cost of renting or buying new equipment and facilities (including depreciation, interest, taxes, and insurance); salaries; utilities; and portions of the sales or advertising budget. Thus, the total cost of producing a service or good equals fixed costs plus variable costs multiplied by volume, or

$$\text{Total cost} = F + cQ$$

The variable cost per unit is assumed to be the same no matter how small or large Q is, and thus, total cost is linear. If we assume that all units produced are sold, total annual revenues equal revenue per unit sold, p, multiplied by the quantity sold, or

$$\text{Total revenue} = pQ$$

If we set total revenue equal to total cost, we get the break-even quantity point as

$$pQ = F + cQ$$
$$(p - c)Q = F$$
$$Q = \frac{F}{p - c}$$

We can also find this break-even quantity graphically. Because both costs and revenues are linear relationships, the break-even quantity is where the total revenue line crosses the total cost line.

EXAMPLE A.1	**Finding the Break-Even Quantity**

MyOMLab

Active Model A.1 in MyOMLab provides additional insight on this break-even example and its extensions with four "what-if" questions.

A hospital is considering a new procedure to be offered at $200 per patient. The fixed cost per year would be $100,000, with total variable costs of $100 per patient. What is the break-even quantity for this service? Use both algebraic and graphic approaches to get the answer.

SOLUTION

The formula for the break-even quantity yields

$$Q = \frac{F}{p - c} = \frac{100,000}{200 - 100} = 1,000 \text{ patients}$$

MyOMLab

Tutor A.1 in MyOMLab provides a new example to practice break-even analysis.

To solve graphically we plot two lines: one for costs and one for revenues. Two points determine a line, so we begin by calculating costs and revenues for two different output levels. The following table shows the results for $Q = 0$ and $Q = 2,000$. We selected zero as the first point because of the ease of plotting total revenue (0) and total cost (F). However, we could have used any two reasonably spaced output levels.

Quantity (patients) (Q)	Total Annual Cost ($) (100,000 + 100Q)	Total Annual Revenue ($) (200Q)
0	100,000	0
2,000	300,000	400,000

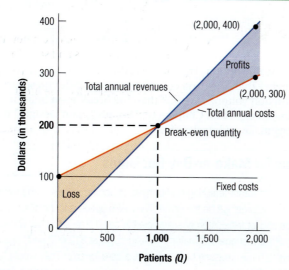

◄ **FIGURE A.1**
Graphic Approach to
Break-Even Analysis

We can now draw the cost line through points (0, 100,000) and (2,000, 300,000). The revenue line goes between (0, 0) and (2,000, 400,000). As Figure A.1 indicates, these two lines intersect at 1,000 patients, the break-even quantity.

DECISION POINT
Management expects the number of patients needing the new procedure will exceed the 1,000-patient break-even quantity but first wants to learn how sensitive the decision is to demand levels before making a final choice.

Break-even analysis cannot tell a manager whether to pursue a new service or product idea or drop an existing line. The technique can only show what is likely to happen for various forecasts of costs and sales volumes. To evaluate a variety of "what-if" questions, we use an approach called **sensitivity analysis**, a technique for systematically changing parameters in a model to determine the effects of such changes. The concept can be applied later to other techniques, such as linear programming. Here we assess the sensitivity of total profit to different pricing strategies, sales volume forecasts, or cost estimates.

sensitivity analysis
A technique for systematically changing parameters in a model to determine the effects of such changes.

EXAMPLE A.2	**Sensitivity Analysis of Sales Forecasts**

If the most pessimistic sales forecast for the proposed service in Figure A.1 were 1,500 patients, what would be the procedure's total contribution to profit and overhead per year?

SOLUTION
The graph shows that even the pessimistic forecast lies above the break-even volume, which is encouraging. The procedure's total contribution, found by subtracting total costs from total revenues, is

$$pQ - (F + cQ) = 200(1,500) - [100,000 + 100(1,500)]$$
$$= \$50,000$$

DECISION POINT
Even with the pessimistic forecast, the new procedure contributes $50,000 per year. After evaluating the proposal with present value method (see MyOMLab Supplement F), management added the new procedure to the hospital's services.

MyOMLab

Evaluating Processes

Often, choices must be made between two processes or between an internal process and buying services or materials on the outside. In such cases, we assume that the decision does not affect revenues. The manager must study all the costs and advantages of each approach. Rather than find the quantity at which total costs equal total revenues, the analyst finds the quantity for which the total costs for two alternatives are equal. For the make-or-buy decision, it is the quantity for which the total "buy" cost equals the total "make" cost. Let F_b equal the fixed cost (per year) of the buy option, F_m equal the fixed cost of the make option, c_b equal the variable cost (per unit) of the buy option, and c_m equal the variable cost of the make option. Thus, the total cost to buy is $F_b + c_b Q$ and the total cost to make is $F_m + c_m Q$. To find the break-even quantity, we set the two cost functions equal and solve for Q:

$$F_b + c_b Q = F_m + c_m Q$$

$$Q = \frac{F_m - F_b}{c_b - c_m}$$

The make option should be considered, ignoring qualitative factors, only if its variable costs are lower than those of the buy option. The reason is that the fixed costs for making the service or product are typically higher than the fixed costs for buying. Under these circumstances, the buy option is better if production volumes are less than the break-even quantity. Beyond that quantity, the make option becomes better. Chapter 10, "Supply Chain Design," brings out other considerations when making make-or-buy decisions.

EXAMPLE A.3	**Break-Even Analysis for Make-or-Buy Decisions**

MyOMLab

Active Model A.2 in MyOMLab provides additional insight on this make-or-buy example and its extensions.

The manager of a fast-food restaurant featuring hamburgers is adding salads to the menu. For each of the two new options, the price to the customer will be the same. The make option is to install a salad bar stocked with vegetables, fruits, and toppings and let the customer assemble the salad. The salad bar would have to be leased and a part-time employee hired. The manager estimates the fixed costs at $12,000 and variable costs totaling $1.50 per salad. The buy option is to have preassembled salads available for sale. They would be purchased from a local supplier at $2.00 per salad. Offering preassembled salads would require installation and operation of additional refrigeration, with an annual fixed cost of $2,400. The manager expects to sell 25,000 salads per year.

What is the make-or-buy quantity?

MyOMLab

Tutor A.2 in MyOMLab provides a new example to practice break-even analysis on make-or-buy decisions.

SOLUTION

The formula for the break-even quantity yields the following:

$$Q = \frac{F_m - F_b}{c_b - c_m}$$

$$= \frac{12,000 - 2,400}{2.0 - 1.5} = 19,200 \text{ salads}$$

FIGURE A.2 ▶

Break-Even Analysis Solver of OM Explorer for Example A.3

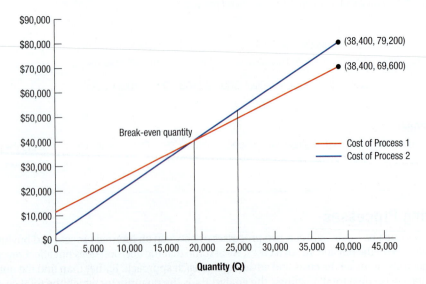

	Process 1	Process 2
Fixed costs (F)	$12,000	$2,400
Variable costs (c)	$1.50	$2.00
Expected demand	25,000	
Break-even quantity	19,200.0	
Decision: Process 1		

Figure A.2 shows the solution from OM Explorer's *Break-Even Analysis* Solver. The break-even quantity is 19,200 salads. As the 25,000-salad sales forecast exceeds this amount, the make option is preferred. Only if the restaurant expected to sell fewer than 19,200 salads would the buy option be better.

DECISION POINT

Management chose the make option after considering other qualitative factors, such as customer preferences and demand uncertainty. A deciding factor was that the 25,000-salad sales forecast is well above the 19,200-salad break-even quantity.

Preference Matrix

Decisions often must be made in situations where multiple criteria cannot be naturally merged into a single measure (such as dollars). For example, a manager deciding in which of two cities to locate a new plant would have to consider such unquantifiable factors as quality of life, worker attitudes toward work, and community reception in the two cities. These important factors cannot be ignored. A **preference matrix** is a table that allows the manager to rate an alternative according to several performance criteria. The criteria can be scored on any scale, such as from 1 (worst possible) to 10 (best possible) or from 0 to 1, as long as the same scale is applied to all the alternatives being compared. Each score is weighted according to its perceived importance, with the total of these weights typically equaling 100. The total score is the sum of the weighted scores (weight × score) for all the criteria. The manager can compare the scores for alternatives against one another or against a predetermined threshold.

preference matrix

A table that allows the manager to rate an alternative according to several performance criteria.

EXAMPLE A.4	**Evaluating an Alternative with a Preference Matrix**

The following table shows the performance criteria, weights, and scores (1 = worst, 10 = best) for a new product: a thermal storage air conditioner. If management wants to introduce just one new product and the highest total score of any of the other product ideas is 800, should the firm pursue making the air conditioner?

MyOMLab

Tutor A.3 in MyOMLab provides a new example to practice with preference matrixes.

Performance Criterion	Weight *(A)*	Score *(B)*	Weighted Score *(A × B)*
Market potential	30	8	240
Unit profit margin	20	10	200
Operations compatibility	20	6	120
Competitive advantage	15	10	150
Investment requirement	10	2	20
Project risk	5	4	20
			Weighted score = 750

SOLUTION

Because the sum of the weighted scores is 750, it falls short of the score of 800 for another product. This result is confirmed by the output from OM Explorer's *Preference Matrix* Solver in Figure A.3.

◀ **FIGURE A.3**
Preference Matrix Solver for Example A.4

Insert a Criterion	Add a Criterion	Remove a Criterion

	Weight (A)	Score (B)	Weighted Score (A x B)
Market potential	30	8	240
Unit profit margin	20	10	200
Operations compatability	20	6	120
Competitive advantage	15	10	150
Investment requirement	10	2	20
Project risk	5	4	20
Final Weighted Score			750

DECISION POINT
Management should drop the thermal storage air-conditioner idea. Another new product idea is better, considering the multiple criteria, and management only wanted to introduce one new product at the time.

Not all managers are comfortable with the preference matrix technique. It requires the manager to state criteria weights before examining the alternatives, although the proper weights may not be readily apparent. Perhaps, only after seeing the scores for several alternatives can the manager decide what is important and what is not. Because a low score on one criterion can be compensated for or overridden by high scores on others, the preference matrix method also may cause managers to ignore important signals. In Example A.4, the investment required for the thermal storage air conditioner might exceed the firm's financial capability. In that case, the manager should not even be considering the alternative, no matter how high its score.

Decision Theory

decision theory

A general approach to decision making when the outcomes associated with alternatives are often in doubt.

Decision theory is a general approach to decision making when the outcomes associated with alternatives are often in doubt. It helps operations managers with decisions on process, capacity, location, and inventory because such decisions are about an uncertain future. Decision theory can also be used by managers in other functional areas. With decision theory, a manager makes choices using the following process:

1. List the feasible *alternatives*. One alternative that should always be considered as a basis for reference is to do nothing. A basic assumption is that the number of alternatives is finite. For example, in deciding where to locate a new retail store in a certain part of the city, a manager could theoretically consider every grid coordinate on the city's map. Realistically, however, the manager must narrow the number of choices to a reasonable number.

2. List the *events* (sometimes called *chance events* or *states of nature*) that have an impact on the outcome of the choice but are not under the manager's control. For example, the demand experienced by the new facility could be low or high, depending not only on whether the location is convenient to many customers, but also on what the competition does and general retail trends. Then, group events into reasonable categories. For example, suppose that the average number of sales per day could be anywhere from 1 to 500. Rather than have 500 events, the manager could represent demand with just 3 events: 100 sales/day, 300 sales/day, or 500 sales/day. The events must be mutually exclusive and collectively exhaustive, meaning that they do not overlap and that they cover all eventualities.

payoff table

A table that shows the amount for each alternative if each possible event occurs.

3. Calculate the *payoff* for each alternative in each event. Typically, the payoff is total profit or total cost. These payoffs can be entered into a **payoff table**, which shows the amount for each alternative if each possible event occurs. For 3 alternatives and 4 events, the table would have 12 payoffs (3 × 4). If significant distortions will occur if the time value of money is not recognized, the payoffs should be expressed as present values or internal rates of return (see MyOMLab Supplement F.) For multiple criteria with important qualitative factors, use the weighted scores of a preference matrix approach as the payoffs.

MyOMLab

4. Estimate the likelihood of each event, using past data, executive opinion, or other forecasting methods. Express it as a *probability*, making sure that the probabilities sum to 1.0. Develop probability estimates from past data if the past is considered a good indicator of the future.

5. Select a *decision rule* to evaluate the alternatives, such as choosing the alternative with the lowest expected cost. The rule chosen depends on the amount of information the manager has on the event probabilities and the manager's attitudes toward risk.

Using this process, we examine decisions under three different situations: certainty, uncertainty, and risk.

Decision Making under Certainty

The simplest situation is when the manager knows which event will occur. Here the decision rule is to pick the alternative with the best payoff for the known event. The best alternative is the highest payoff if the payoffs are expressed as profits. If the payoffs are expressed as costs, the best alternative is the lowest payoff.

EXAMPLE A.5	Decisions under Certainty

A manager is deciding whether to build a small or a large facility. Much depends on the future demand that the facility must serve, and demand may be small or large. The manager knows with certainty the payoffs that will result under each alternative, shown in the following payoff table. The payoffs (in $000) are the present values of future revenues minus costs for each alternative in each event.

Alternative	POSSIBLE FUTURE DEMAND	
	Low	High
Small facility	200	270
Large facility	160	800
Do nothing	0	0

What is the best choice if future demand will be low?

SOLUTION

In this example, the best choice is the one with the highest payoff. If the manager knows that future demand will be low, the company should build a small facility and enjoy a payoff of $200,000. The larger facility has a payoff of only $160,000. The "do nothing" alternative is dominated by the other alternatives; that is, the outcome of one alternative is no better than the outcome of another alternative for each event. Because the "do nothing" alternative is dominated, the manager does not consider it further.

DECISION POINT

If management really knows future demand, it would build the small facility if demand will be low and the large facility if demand will be high. If demand is uncertain, it should consider other decision rules.

Decision Making under Uncertainty

Here, we assume that the manager can list the possible events but cannot estimate their probabilities. Perhaps, a lack of prior experience makes it difficult for the firm to estimate probabilities. In such a situation, the manager can use one of four decision rules:

1. *Maximin.* Choose the alternative that is the "best of the worst." This rule is for the *pessimist,* who anticipates the "worst case" for each alternative.

2. *Maximax.* Choose the alternative that is the "best of the best." This rule is for the *optimist* who has high expectations and prefers to "go for broke."

3. *Laplace.* Choose the alternative with the best *weighted payoff.* To find the weighted payoff, give equal importance (or, alternatively, equal probability) to each event. If there are n events, the importance (or probability) of each is $1/n$, so they add up to 1.0. This rule is for the *realist.*

4. *Minimax Regret.* Choose the alternative with the best "worst regret." Calculate a table of regrets (or opportunity losses), in which the rows represent the alternatives and the columns represent the events. A regret is the difference between a given payoff and the best payoff in the same column. For an event, it shows how much is lost by picking an alternative to the one that is best for this event. The regret can be lost profit or increased cost, depending on the situation.

EXAMPLE A.6	Decisions under Uncertainty

Reconsider the payoff matrix in Example A.5. What is the best alternative for each decision rule?

SOLUTION

a. *Maximin.* An alternative's worst payoff is the *lowest* number in its row of the payoff matrix, because the payoffs are profits. The worst payoffs ($000) are

Alternative	Worst Payoff
Small facility	200
Large facility	160

The best of these worst numbers is $200,000, so the pessimist would build a small facility.

MyOMLab

Tutor A.4 in MyOMLab provides a new example to make decisions under uncertainty.

b. *Maximax.* An alternative's best payoff ($000) is the *highest* number in its row of the payoff matrix, or

Alternative	Best Payoff
Small facility	270
Large facility	800

The best of these best numbers is $800,000, so the optimist would build a large facility.

c. *Laplace.* With two events, we assign each a probability of 0.5. Thus, the weighted payoffs ($000) are

Alternative	Weighted Payoff
Small facility	0.5(200) + 0.5(270) = **235**
Large facility	0.5(160) + 0.5(800) = **480**

The best of these weighted payoffs is $480,000, so the realist would build a large facility.

d. *Minimax Regret.* If demand turns out to be low, the best alternative is a small facility and its regret is 0 (or 200 − 200). If a large facility is built when demand turns out to be low, the regret is 40 (or 200 − 160).

	REGRET		
Alternative	Low Demand	High Demand	Maximum Regret
Small facility	200 − 200 = **0**	800 − 270 = **530**	530
Large facility	200 − 160 = **40**	800 − 800 = **0**	40

The column on the right shows the worst regret for each alternative. To minimize the maximum regret, pick a large facility. The biggest regret is associated with having only a small facility and high demand.

DECISION POINT

The pessimist would choose the small facility. The realist, optimist, and manager choosing to minimize the maximum regret would build the large facility.

Decision Making under Risk

Here we assume that the manager can list the events and estimate their probabilities. The manager has less information than with decision making under certainty, but more information than with decision making under uncertainty. For this intermediate situation, the *expected value* decision rule is widely used (both in practice and in this book). The expected value for an alternative is found by weighting each payoff with its associated probability and then adding the weighted payoff scores. The alternative with the best expected value (highest for profits and lowest for costs) is chosen.

This rule is much like the Laplace decision rule, except that the events are no longer assumed to be equally likely (or equally important). The expected value is what the *average* payoff would be if the decision could be repeated time after time. Of course, the expected value decision rule can result in a bad outcome if the wrong event occurs. However, it gives the best results if applied consistently over a long period of time. The rule should not be used if the manager is inclined to avoid risk.

EXAMPLE A.7	Decisions under Risk

Reconsider the payoff matrix in Example A.5. For the expected value decision rule, which is the best alternative if the probability of small demand is estimated to be 0.4 and the probability of large demand is estimated to be 0.6?

MyOMLab

Tutor A.5 in MyOMLab provides a new example to make decisions under risk.

SOLUTION

The expected value for each alternative is as follows:

Alternative	Expected Value
Small facility	0.4(200) + 0.6(270) = **242**
Large facility	0.4(160) + 0.6(800) = **544**

DECISION POINT

Management would choose a large facility if it used this expected value decision rule, because it provides the best long-term results if consistently applied over time.

Decision Trees

The decision tree method is a general approach to a wide range of processes and supply chain decisions, such as product planning, process analysis, process capacity, and location. It is particularly valuable for evaluating different capacity expansion alternatives when demand is uncertain and sequential decisions are involved. For example, a company may expand a facility in 2013 only to discover in 2016 that demand is much higher than forecasted. In that case, a second decision may be necessary to determine whether to expand again or build a second facility.

A **decision tree** is a schematic model of alternatives available to the decision maker, along with their possible consequences. The name derives from the tree-like appearance of the model. It consists of a number of square *nodes*, representing decision points, which are left by *branches* (which should be read from left to right), representing the alternatives. Branches leaving circular, or chance, nodes represent the events. The probability of each chance event, *P(E)*, is shown above each branch. The probabilities for all branches leaving a chance node must sum to 1.0. The conditional payoff, which is the payoff for each possible alternative-event combination, is shown at the end of each combination. Payoffs are given only at the outset, before the analysis begins, for the end points of each alternative-event combination. In Figure A.4, for example, payoff 1 is the financial outcome the manager expects if alternative 1 is chosen and then chance event 1 occurs.

No payoff can be associated yet with any branches farther to the left, such as alternative 1 as a whole, because it is followed by a chance event and is not an end point. Payoffs often are

decision tree

A schematic model of alternatives available to the decision maker, along with their possible consequences.

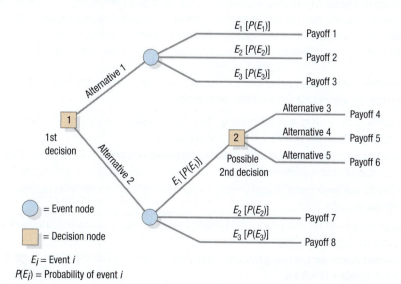

◄ **FIGURE A.4**

A Decision Tree Model

E_i = Event i
$P(E_i)$ = Probability of event i

expressed as the present value of net profits. If revenues are not affected by the decision, the payoff is expressed as net costs.

After drawing a decision tree, we solve it by working from right to left, calculating the *expected payoff* for each node as follows:

1. For an event node, we multiply the payoff of each event branch by the event's probability. We add these products to get the event node's expected payoff.

2. For a decision node, we pick the alternative that has the best expected payoff. If an alternative leads to an event node, its payoff is equal to that node's expected payoff (already calculated). We "saw off," or "prune," the other branches not chosen by marking two short lines through them. The decision node's expected payoff is the one associated with the single remaining unpruned branch. We continue this process until the leftmost decision node is reached. The unpruned branch extending from it is the best alternative to pursue. If multistage decisions are involved, we must await subsequent events before deciding what to do next. If new probability or payoff estimates are obtained, we repeat the process.

Various software is available for drawing decision trees. PowerPoint can be used to draw decision trees, although it does not have the capability to analyze the decision tree. More extensive capabilities, in addition to POM for Windows, are found with SmartDraw (**www.smartdraw.com**), PrecisionTree decision analysis from Palisade Corporation (**www.palisade.com**), and TreePlan (**www.treeplan.com/treeplan.htm**).

EXAMPLE A.8	**Analyzing a Decision Tree**

MyOMLab

Active Model A.3 in MyOMLab provides additional insight on this decision tree example and its extensions.

A retailer must decide whether to build a small or a large facility at a new location. Demand at the location can be either low or high, with probabilities estimated to be 0.4 and 0.6, respectively. If a small facility is built and demand proves to be high, the manager may choose not to expand (payoff = $223,000) or to expand (payoff = $270,000). If a small facility is built and demand is low, there is no reason to expand and the payoff is $200,000. If a large facility is built and demand proves to be low, the choice is to do nothing ($40,000) or to stimulate demand through local advertising. The response to advertising may be either modest or sizable, with their probabilities estimated to be 0.3 and 0.7, respectively. If it is modest, the payoff is estimated to be only $20,000; the payoff grows to $220,000 if the response is sizable. Finally, if a large facility is built and demand turns out to be high, the payoff is $800,000.

Draw a decision tree. Then analyze it to determine the expected payoff for each decision and event node. Which alternative—building a small facility or building a large facility—has the higher expected payoff?

SOLUTION

The decision tree in Figure A.5 shows the event probability and the payoff for each of the seven alternative-event combinations. The first decision is whether to build a small or a large facility. Its node is shown first, to the left, because it is the decision the retailer must make now. The second decision node—whether to expand at a later date—is reached only if a small facility is built and demand turns out to be high. Finally, the third decision point—whether to advertise—is reached only if the retailer builds a large facility and demand turns out to be low.

Analysis of the decision tree begins with calculation of the expected payoffs from right to left, shown on Figure A.5 beneath the appropriate event and decision nodes.

1. For the event node dealing with advertising, the expected payoff is 160, or the sum of each event's payoff weighted by its probability [0.3(20) + 0.7(220)].

2. The expected payoff for decision node 3 is 160 because *Advertise* (160) is better than *Do nothing* (40). Prune the *Do nothing* alternative.

3. The payoff for decision node 2 is 270 because *Expand* (270) is better than *Do not expand* (223). Prune *Do not expand*.

4. The expected payoff for the event node dealing with demand, assuming that a small facility is built, is 242 [or 0.4(200) + 0.6(270)].

5. The expected payoff for the event node dealing with demand, assuming that a large facility is built, is 544 [or 0.4(160) + 0.6(800)].

6. The expected payoff for decision node 1 is 544 because the large facility's expected payoff is largest. Prune *Small facility*.

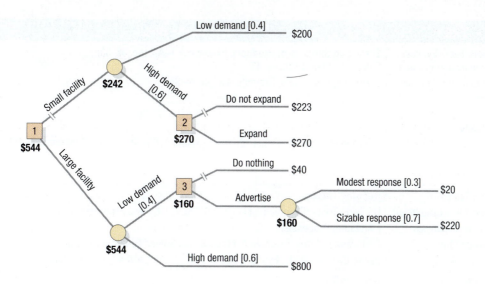

◀ **FIGURE A.5**
Decision Tree for Retailer
(in $000)

DECISION POINT
The retailer should build the large facility. This initial decision is the only one made now. Subsequent decisions are made after learning whether demand actually is low or high.

LEARNING GOALS IN REVIEW

① **Explain break-even analysis, using both the graphic and algebraic approaches.** The section "Evaluating Services and Products," pp. 32–33, covers this analysis. Example A.1 and Solved Problem 1 demonstrate both approaches. Example A.3 shows its use in evaluating different processes.

② **Define a preference matrix.** See the section "Preference Matrix," pp. 35–36, for making decisions involving unquantifiable factors, where some factors are rated more important than others. Example A.4 and Solved Problem 2 demonstrate the calculations.

③ **Explain how to construct a payoff table.** The section "Decision Theory," pp. 36–38, begins with the construction of a payoff table that shows the payoff for each feasible alternative and each event. See the table in Example A.5.

④ **Identify the maximin, maximax, Laplace, minimax regret, and expected value decision rules.** The sections "Decision Making under Uncertainty" and "Decision Making under Risk," pp. 37–38, cover these decision rules for when the outcomes associated with alternatives are in doubt. Examples A.6 and A.7 demonstrate how these rules work, and so does Solved Problem 3.

⑤ **Describe how to draw and analyze a decision tree.** The section "Decision Trees," pp. 39–41, show how to draw and analyze decision trees where several alternatives are available over time. Example A.8 and Solved Problem 4 shows how to work back from right to left, pruning as you go, until the best alternate is found for decision node 1.

MyOMLab helps you develop analytical skills and assess your progress with multiple problems on break-even analysis, sensitivity analysis, make-or-buy, the preference matrix, decisions under uncertainty, decisions under risks, and decision trees.

MyOMLab Resources	Titles	Link to the Book
Active Model Exercise	A.1 Break-Even Analysis A.3 Make-or-Buy Decision A.3 Decision Tree	Evaluating Services or Products; Examples A.1–A.2 (pp. 32–33); Solved Problem 1 (pp. 42–43) Evaluating Processes; Example A.3 (pp. 34–35) Decision Trees; Example A.8 (p. 40)
OM Explorer Solvers	Break-Even Analysis Decision Theory Preference Matrix	Break-Even Analysis; Examples A.1–A.3 (pp. 32–35) Decision Theory; Example A.5–A.7 (pp. 37–39) Preference Matrix; Examples A.4–A.6 (pp. 35–36); Solved Problem 2 (p. 43)

MyOMLab Resources	Titles	Link to the Book
OM Explorer Tutors	A.1 Break-Even, Evaluating Services and Products	Evaluating Services or Products; Example A.1–A.2 (pp. 32–33); Solved Problem 1 (pp. 42–43)
	A.2 Evaluating Processes	Evaluating Processes; Example A.3 (pp. 34–35)
	A.3 Preferences Matrix	Preference Matrix; Example A.4 (pp. 35–36)
	A.4 Decisions under Uncertainty	Decisions under Uncertainty; Example A.6 (pp. 37–38)
	A.5 Decisions under Risk	Decision Making under Risk; Solved Problem A.7 (p. 45); Solved Problem 3 (p. 44)
	A.6 Location Decisions under Uncertainty	Tutor A.6
POM for Windows	Decision Tables	Decision Theory
	Decision Trees (graphical)	Decision Trees (Graphical); Example A.8 (p. 40); Solved Problem 4 (p. 44)
	Cost-Volume Analysis	Evaluating Processes
	Preference Matrix	Preference Matrix; Example A.4 (pp. 35–36); Solved Problem 2 (p. 43)
	Break-Even Analysis	Break-Even Analysis; Example A.1–A.3 (p. 32–35); Solved Problem 1 (pp. 42–43)
Virtual Tours	Flir infrared cameras	Decision Theory
	E* Trade investment services	Decision Theory
MyOMLab Supplements	F. Financial Analysis	MyOMLab Supplement F
Internet Exercises	Florida Small Business	Break–Even Analysis
Key Equations		
Image Library		

Key Equations

1. Break-even quantity: $Q = \dfrac{F}{p - c}$

2. Evaluating processes, make-or-buy indifference quantity: $Q = \dfrac{F_m - F_b}{c_b - c_m}$

Key Terms

break-even analysis 31
break-even quantity 31
decision theory 36

decision tree 39
fixed cost 32
payoff table 36

preference matrix 35
sensitivity analysis 33
variable cost 32

Solved Problem 1

The owner of a small manufacturing business has patented a new device for washing dishes and cleaning dirty kitchen sinks. Before trying to commercialize the device and add it to his or her existing product line, the owner wants reasonable assurance of success. Variable costs are estimated at $7 per unit produced and sold. Fixed costs are about $56,000 per year.

   a. If the selling price is set at $25, how many units must be produced and sold to break even? Use both algebraic and graphic approaches.

   b. Forecasted sales for the first year are 10,000 units if the price is reduced to $15. With this pricing strategy, what would be the product's total contribution to profits in the first year?

SOLUTION

   a. Beginning with the algebraic approach, we get

$$Q = \frac{F}{p - c} = \frac{56,000}{25 - 7}$$
$$= 3,111 \text{ units}$$

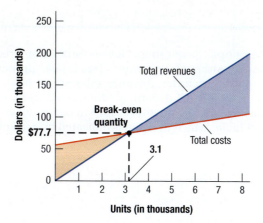

Using the graphic approach, shown in Figure A.6, we first draw two lines:

$$\text{Total revenue} = 25Q$$
$$\text{Total cost} = 56{,}000 + 7Q$$

The two lines intersect at $Q = 3{,}111$ units, the break-even quantity.

b. Total profit contribution = Total revenue – Total cost

$$= pQ - (F + cQ)$$
$$= 15(10{,}000) - [56{,}000 + 7(10{,}000)]$$
$$= \$24{,}000$$

Solved Problem 2

Herron Company is screening three new product ideas: A, B, and C. Resource constraints allow only one of them to be commercialized. The performance criteria and ratings, on a scale of 1 (worst) to 10 (best), are shown in the following table. The Herron managers give equal weights to the performance criteria. Which is the best alternative, as indicated by the preference matrix method?

	RATING		
Performance Criterion	**Product A**	**Product B**	**Product C**
1. Demand uncertainty and project risk	3	9	2
2. Similarity to present products	7	8	6
3. Expected return on investment (ROI)	10	4	8
4. Compatibility with current manufacturing process	4	7	6
5. Competitive advantage	4	6	5

SOLUTION

Each of the five criteria receives a weight of 1/5 or 0.20.

Product	**Calculation**	**Total Score**
A	$(0.20 \times 3) + (0.20 \times 7) + (0.20 \times 10) + (0.20 \times 4) + (0.20 \times 4)$	= 5.6
B	$(0.20 \times 9) + (0.20 \times 8) + (0.20 \times 4) + (0.20 \times 7) + (0.20 \times 6)$	= 6.8
C	$(0.20 \times 2) + (0.20 \times 6) + (0.20 \times 8) + (0.20 \times 6) + (0.20 \times 5)$	= 5.4

The best choice is product B. Products A and C are well behind in terms of total weighted score.

Solved Problem 3

MyOMLab

Tutor A.6 in MyOMLab examines decisions under uncertainty for a location example.

Adele Weiss manages the campus flower shop. Flowers must be ordered three days in advance from her supplier in Mexico. Although Valentine's Day is fast approaching, sales are almost entirely last-minute, impulse purchases. Advance sales are so small that Weiss has no way to estimate the probability of low (25 dozen), medium (60 dozen), or high (130 dozen) demand for red roses on the big day. She buys roses for $15 per dozen and sells them for $40 per dozen. Construct a payoff table. Which decision is indicated by each of the following decision criteria?

a. Maximin

b. Maximax

c. Laplace

d. Minimax regret

SOLUTION

The payoff table for this problem is

Alternative	DEMAND FOR RED ROSES		
	Low (25 dozen)	Medium (60 dozen)	High (130 dozen)
Order 25 dozen	$625	$625	$625
Order 60 dozen	$100	$1,500	$1,500
Order 130 dozen	($950)	$450	$3,250
Do nothing	$0	$0	$0

a. Under the maximin criteria, Weiss should order 25 dozen, because if demand is low, Weiss's profits are $625, the best of the worst payoffs.

b. Under the maximax criteria, Weiss should order 130 dozen. The greatest possible payoff, $3,250, is associated with the largest order.

c. Under the Laplace criteria, Weiss should order 60 dozen. Equally weighted payoffs for ordering 25, 60, and 130 dozen are about $625, $1,033, and $917, respectively.

d. Under the minimax regret criteria, Weiss should order 130 dozen. The maximum regret of ordering 25 dozen occurs if demand is high: $3,250 – $625 = $2,625. The maximum regret of ordering 60 dozen occurs if demand is high: $3,250 – $1,500 = $1,750. The maximum regret of ordering 130 dozen occurs if demand is low: $625 – (–$950) = $1,575.

Solved Problem 4

White Valley Ski Resort is planning the ski lift operation for its new ski resort. Management is trying to determine whether one or two lifts will be necessary; each lift can accommodate 250 people per day. Skiing normally occurs in the 14-week period from December to April, during which the lift will operate 7 days per week. The first lift will operate at 90 percent capacity if economic conditions are bad, the probability of which is believed to be about a 0.3. During normal times the first lift will be utilized at 100 percent capacity, and the excess crowd will provide 50 percent utilization of the second lift. The probability of normal times is 0.5. Finally, if times are really good, the probability of which is 0.2, the utilization of the second lift will increase to 90 percent. The equivalent annual cost of installing a new lift, recognizing the time value of money and the lift's economic life, is $50,000. The annual cost of installing two lifts is only $90,000 if both are purchased at the same time. If used at all, each lift costs $200,000 to operate, no matter how low or high its utilization rate. Lift tickets cost $20 per customer per day.

Should the resort purchase one lift or two?

SOLUTION

The decision tree is shown in Figure A.7. The payoff ($000) for each alternative-event branch is shown in the following table. The total revenues from one lift operating at 100 percent capacity are $490,000 (or 250 customers × 98 days × $20/customer-day).

Alternative	Economic Condition	Payoff Calculation (Revenue – Cost)
One lift	Bad times	0.9(490) – (50 + 200) = 191
	Normal times	1.0(490) – (50 + 200) = 240
	Good times	1.0(490) – (50 + 200) = 240
Two lifts	Bad times	0.9(490) – (90 + 200) = 151
	Normal times	1.5(490) – (90 + 400) = 245
	Good times	1.9(490) – (90 + 400) = 441

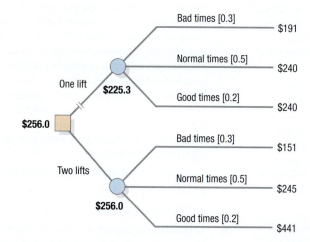

◀ **FIGURE A.7**

Problems

The OM Explorer and POM for Windows software is available to all students using the 10th edition of this textbook. Go to **www.pearsonhighered.com/krajewski** to download these computer packages. If you purchased MyOMLab, you also have access to Active Models software and significant help in doing the following problems. Check with your instructor on how best to use these resources. In many cases, the instructor wants you to understand how to do the calculations by hand. At the least, the software provides a check on your calculations. When calculations are particularly complex and the goal is interpreting the results in making decisions, the software entirely replaces the manual calculations.

BREAK-EVEN ANALYSIS

1. Mary Williams, owner of Williams Products, is evaluating whether to introduce a new product line. After thinking through the production process and the costs of raw materials and new equipment, Williams estimates the variable costs of each unit produced and sold at $6 and the fixed costs per year at $60,000.

   a. If the selling price is set at $18 each, how many units must be produced and sold for Williams to break even? Use both graphic and algebraic approaches to get your answer.

   b. Williams forecasts sales of 10,000 units for the first year if the selling price is set at $14 each. What would be the total contribution to profits from this new product during the first year?

   c. If the selling price is set at $12.50, Williams forecasts that first-year sales would increase to 15,000 units. Which pricing strategy ($14.00 or $12.50) would result in the greater total contribution to profits?

   d. What other considerations would be crucial to the final decision about making and marketing the new product?

2. A product at the Jennings Company enjoyed reasonable sales volumes, but its contributions to profits were disappointing. Last year, 17,500 units were produced and sold. The selling price is $22 per unit, the variable cost is $18 per unit, and the fixed cost is $80,000.

   a. What is the break-even quantity for this product? Use both graphic and algebraic approaches to get your answer.

   b. If sales were not expected to increase, by how much would Jennings have to reduce their variable cost to break even?

c. Jennings believes that a $1 reduction in price will increase sales by 50 percent. Is this enough for Jennings to break even? If not, by how much would sales have to increase?

d. Jennings is considering ways to either stimulate sales volume or decrease variable cost. Management believes that either sales can be increased by 30 percent or that variable cost can be reduced to 85 percent of its current level. Which alternative leads to higher contributions to profits, assuming that each is equally costly to implement? (Hint: Calculate profits for both alternatives and identify the one having the greatest profits.)

e. What is the percent change in the per-unit profit contribution generated by each alternative in part (d)?

3. An interactive television service that costs $10 per *month* to provide can be sold on the information highway for $15 per client per *month*. If a service area includes a potential of 15,000 customers, what is the most a company could spend on *annual* fixed costs to acquire and maintain the equipment?

4. A restaurant is considering adding fresh brook trout to its menu. Customers would have the choice of catching their own trout from a simulated mountain stream or simply asking the waiter to net the trout for them. Operating the stream would require $10,600 in fixed costs per year. Variable costs are estimated to be $6.70 per trout. The firm wants to break even if 800 trout dinners are sold per year. What should be the price of the new item?

5. Spartan Castings must implement a manufacturing process that reduces the amount of particulates emitted into the atmosphere. Two processes have been identified that provide the same level of particulate reduction. The first process is expected to incur $350,000 of fixed cost and add $50 of variable cost to each casting Spartan produces. The second process has fixed costs of $150,000 and adds $90 of variable cost per casting.

a. What is the break-even quantity beyond which the first process is more attractive?

b. What is the difference in total cost if the quantity produced is 10,000?

6. A news clipping service is considering modernization. Rather than manually clipping and photocopying articles of interest and mailing them to its clients, employees electronically input stories from most widely circulated publications into a database. Each new issue is searched for key words, such as a client's company name, competitors' names, type of business, and the company's products, services, and officers. When matches occur, affected clients are instantly notified via an on-line network. If the story is of interest, it is electronically transmitted, so the client often has the story and can prepare comments for follow-up interviews before the publication hits the street. The manual process has fixed costs of $400,000 per year and variable costs of $6.20 per clipping mailed. The price charged the client is $8.00 per clipping. The computerized process has fixed costs of $1,300,000 per year and variable costs of $2.25 per story electronically transmitted to the client.

a. If the same price is charged for either process, what is the annual volume beyond which the automated process is more attractive?

b. The present volume of business is 225,000 clippings per year. Many of the clippings sent with the current process are not of interest to the client or are multiple copies of the same story appearing in several publications. The news clipping service believes that by improving service and by lowering the price to $4.00 per story, modernization will increase volume to 900,000 stories transmitted per year. Should the clipping service modernize?

c. If the forecasted increase in business is too optimistic, at what volume will the new process (with the $4.00 price) break even?

7. Hahn Manufacturing purchases a key component of one of its products from a local supplier. The current purchase price is $1,500 per unit. Efforts to standardize parts succeeded to the point that this same component can now be used in five different products. Annual component usage should increase from 150 to 750 units. Management wonders whether it is time to make the component in-house, rather than to continue buying it from the supplier. Fixed costs would increase by about $40,000 per year for the new equipment and tooling needed. The cost of raw materials and variable overhead would be about $1,100 per unit, and labor costs would be $300 per unit produced.

a. Should Hahn make rather than buy?

b. What is the break-even quantity?

c. What other considerations might be important?

8. Techno Corporation is currently manufacturing an item at variable costs of $5 per unit. Annual fixed costs of manufacturing this item are $140,000. The current selling price of the item is $10 per unit, and the annual sales volume is 30,000 units.

a. Techno can substantially improve the item's quality by installing new equipment at additional annual fixed costs of $60,000. Variable costs per unit would increase by $1, but, as more of the better-quality product could be sold, the annual volume would increase to 50,000 units. Should Techno buy the new equipment and maintain the current price of the item? Why or why not?

b. Alternatively, Techno could increase the selling price to $11 per unit. However, the annual sales volume would be limited to 45,000 units. Should Techno buy the new equipment and raise the price of the item? Why or why not?

9. The Tri-County Generation and Transmission Association is a nonprofit cooperative organization that provides electrical service to rural customers. Based on a faulty long-range demand forecast, Tri-County overbuilt its generation and distribution system. Tri-County now has much more capacity than it needs to serve its customers. Fixed costs, mostly debt service on investment in plant and equipment, are $82.5 million per year. Variable costs, mostly fossil fuel costs, are $25 per megawatt-hour (MWh, or million watts of power used for one hour). The new person in charge of demand forecasting prepared a short-range forecast for use in next year's budgeting process. That forecast calls for Tri-County customers to consume 1 million MWh of energy next year.

 a. How much will Tri-County need to charge its customers per MWh to break even next year?

 b. The Tri-County customers balk at that price and conserve electrical energy. Only 95 percent of forecasted demand materializes. What is the resulting surplus or loss for this nonprofit organization?

10. Earthquake, drought, fire, economic famine, flood, and a pestilence of TV court reporters have caused an exodus from the City of Angels to Boulder, Colorado. The sudden increase in demand is straining the capacity of Boulder's electrical system. Boulder's alternatives have been reduced to buying 150,000 MWh of electric power from Tri-County G&T at a price of $75 per MWh, or refurbishing and recommissioning the abandoned Pearl Street Power Station in downtown Boulder. Fixed costs of that project are $10 million per year, and variable costs would be $35 per MWh. Should Boulder build or buy?

11. Tri-County G&T sells 150,000 MWh per year of electrical power to Boulder at $75 per MWh, has fixed costs of $82.5 million per year, and has variable costs of $25 per MWh. If Tri-County has 1,000,000 MWh of demand from its customers (other than Boulder), what will Tri-County have to charge to break even?

PREFERENCE MATRIX

12. The Forsite Company is screening three ideas for new services. Resource constraints allow only one idea to be commercialized at the present time. The following estimates have been made for the five performance criteria that management believes to be most important:

Performance Criterion	RATING		
	Service A	Service B	Service C
Capital equipment investment required	0.6	0.8	0.3
Expected return on investment (ROI)	0.7	0.3	0.9
Compatibility with current workforce skills	0.4	0.7	0.5
Competitive advantage	1.0	0.4	0.6
Compatibility with EPA requirements	0.2	1.0	0.5

 a. Calculate a total weighted score for each alternative. Use a preference matrix and assume equal weights for each performance criterion. Which alternative is best? Worst?

 b. Suppose that the expected ROI is given twice the weight assigned to each of the remaining criteria. (The sum of weights should remain the same as in part (a).) Does this modification affect the ranking of the three potential services?

13. You are in charge of analyzing five new suppliers of an important raw material and have been given the information shown below (1 = worst, 10 = best). Management has decided that criteria 2 and 3 are equally important and that criteria 1 and 4 are each four times as important as criterion 2. No more than 2 new suppliers are required but each new vendor must exceed a total score of 70 percent of the maximum total points to be considered.

Performance Criterion	RATING				
	Vendor A	Vendor B	Vendor C	Vendor D	Vendor E
Quality of raw material	8	7	3	6	9
Environmental impact	3	8	4	7	7
Responsiveness to order changes	9	5	7	6	5
Cost of raw material	7	6	9	2	7

 a. Which new vendors do you recommend?

 b. Would your decision change if the criteria were considered equally important?

14. Accel Express, Inc., collected the following information on where to locate a warehouse (1 = poor, 10 = excellent):

Location Factor	Factor Weight	LOCATION SCORE	
		A	B
Construction costs	10	8	5
Utilities available	10	7	7
Business services	10	4	7
Real estate cost	20	7	4
Quality of life	20	4	8
Transportation	30	7	6

 a. Which location, A or B, should be chosen on the basis of the total weighted score?

 b. If the factors were weighted equally, would the choice change?

DECISION THEORY AND DECISION TREE

15. Build-Rite Construction has received favorable publicity from guest appearances on a public TV home improvement program. Public TV programming decisions seem to be unpredictable, so Build-Rite cannot estimate the probability of continued benefits from its relationship with the show. Demand for home improvements next year may be either low or high. But Build-Rite must decide now whether to hire more employees, do nothing, or develop subcontracts with other home improvement contractors. Build-Rite has developed the following payoff table:

	DEMAND FOR HOME IMPROVEMENTS		
Alternative	Low	Moderate	High
Hire	($250,000)	$100,000	$625,000
Subcontract	$100,000	$150,000	$415,000
Do nothing	$50,000	$80,000	$300,000

Which alternative is best, according to each of the following decision criteria?

a. Maximin

b. Maximax

c. Laplace

d. Minimax regret

16. Analyze the decision tree in the following figure. What is the expected payoff for the best alternative? First, be sure to infer the missing probabilities.

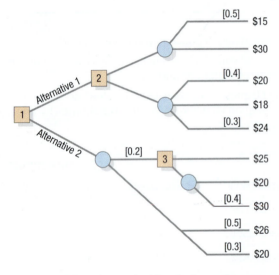

17. A manager is trying to decide whether to buy one machine or two. If only one is purchased and demand proves to be excessive, the second machine can be purchased later. Some sales will be lost, however, because the lead time for producing this type of machine is six months. In addition, the cost per machine will be lower if both are purchased at the same time. The probability of low demand is estimated to be 0.20. The after-tax net present value of the benefits from purchasing the two machines together is $90,000 if demand is low and $180,000 if demand is high.

If one machine is purchased and demand is low, the net present value is $120,000. If demand is high, the manager has three options. Doing nothing has a net present value of $120,000; subcontracting, $160,000; and buying the second machine, $140,000.

a. Draw a decision tree for this problem.

b. How many machines should the company buy initially? What is the expected payoff for this alternative?

18. A manager is trying to decide whether to build a small, medium, or large facility. Demand can be low, average, or high, with the estimated probabilities being 0.25, 0.40, and 0.35, respectively.

A small facility is expected to earn an after-tax net present value of just $18,000 if demand is low. If demand is average, the small facility is expected to earn $75,000; it can be increased to medium size to earn a net present value of $60,000. If demand is high, the small facility is expected to earn $75,000 and can be expanded to medium size to earn $60,000 or to large size to earn $125,000.

A medium-sized facility is expected to lose an estimated $25,000 if demand is low and earn $140,000 if demand is average. If demand is high, the medium-sized facility is expected to earn a net present value of $150,000; it can be expanded to a large size for a net payoff of $145,000.

If a large facility is built and demand is high, earnings are expected to be $220,000. If demand is average for the large facility, the present value is expected to be $125,000; if demand is low, the facility is expected to lose $60,000.

a. Draw a decision tree for this problem.

b. What should management do to achieve the highest expected payoff?

c. Which alternative is best, according to each of the following decision criterion?

Maximin

Maximax

Minimax regret

19. A manufacturing plant has reached full capacity. The company must build a second plant—either small or large—at a nearby location. The demand is likely to be high or low. The probability of low demand is 0.3. If demand is low, the large plant has a present value of $5 million and the small plant, a present value of $8 million. If demand is high, the large plant pays off with a present value of $18 million and the small plant with a present value of only $10 million. However, the small plant can be expanded later if demand proves to be high, for a present value of $14 million.

a. Draw a decision tree for this problem.

b. What should management do to achieve the highest expected payoff?

20 . Benjamin Moses, chief engineer of Offshore Chemicals, Inc., must decide whether to build a new processing facility based on an experimental technology. If the new facility works, the company will realize a net profit of $20 million. If the new facility fails, the company will lose $10 million. Benjamin's best guess is that there is a 40 percent chance that the new facility will work.

What decision should Benjamin Moses make?

Selected References

Clemen, Robert T., and Terence Reilly. *Making Hard Decisions with Decision Tools Suite*. Cincinnati, OH: South-Western, 2004.

Ragsdale, Cliff. *Spreadsheet Modeling & Decision Analysis: A Practical Introduction to Management Science*, 6th ed. Cincinnati, OH: South-Western, 2011.

AFP/Getty Images

Two boys slug it out in an Xbox 360 wrestling game at the 2011 IT Show in Singapore.

XBOX 360

Four years after the introduction of Xbox, Microsoft needed to quickly design, develop, and produce a new product. Sony's PlayStation 2 was dominating the video game market and Microsoft needed a new product to compete with the impending release of PlayStation 3. Developing such a product is a project of massive proportions. The project consisted of four phases: (1) design, (2) analysis, (3) development, and (4) launch. The result was Xbox 360.

Design

The design of the Xbox 360 was a collaborative effort between Microsoft and many other firms, including Astro Studios in San Francisco, which designed the overall console and controller; IBM, which designed the processor chip; ATI, which designed the graphics chip; and a host of game design firms to develop games for the new product. A key element of the new product was the built-in Internet access that allowed gamers to access online games, buy game add-ons, and access multiplayer games developed exclusively for Xbox 360. Microsoft also included its primary manufacturers, Flextronics and Wistron, in the design process to optimize the production and assembly of the more than 1,000 parts contained in an Xbox 360.

Analysis

Getting an estimate of future sales for a new product is always difficult; however, in this case, the historic patterns for PlayStation 1, PlayStation 2, and Xbox were useful. Analysts found that the peak year for a PlayStation product was 4 years after its introduction and that the life cycle for those products is about 11 years. This information provided a basis for estimating the sales potential of Xbox 360, although actual sales may be limited due to supply constraints. Nonetheless, Microsoft realized that the potential was there to open a new generation of game consoles well ahead of the market.

Development

Microsoft worked closely with Flextronics, Wistron, and the various design firms to iron out manufacturing problems in the early phases of Xbox 360 production. Once initial production was underway, Microsoft brought on Celestica to add production capacity. The decision was made to focus manufacturing operations in China. All told, 10,000 workers in China would be involved in Xbox 360 production.

Launch

Microsoft's Xbox 360 gained an early lead in terms of market share due, in part, to its early launch date, which was one year ahead of its rivals PlayStation 3 and Wii. All told, the product was released in 36 countries in the first year of production, a Herculean effort requiring extensive coordination and a high level of project management skill. Sales of the Xbox 360 exceeded expectations with more than 10 million units sold in the first year alone. Nonetheless, Microsoft experienced difficulties in getting the supply chain to meet customer demands in a timely fashion. The lesson to be learned is that projects can be planned and executed properly; however, the underlying infrastructure that delivers the product is equally important in the ultimate success of the venture.

Source: David Holt, Charles Holloway, and Hau Lee, "Evolution of the Xbox Supply Chain," Stanford Graduate School of Business, Case: GS-49, (April 14, 2006); "Xbox 360," Wikipedia, the free encyclopedia, **http://en.wikipedia.org/wiki/Xbox_360.**

LEARNING GOALS *After reading this chapter, you should be able to:*

1 Define the major activities associated with defining, organizing, planning, monitoring, and controlling projects.

2 Diagram the network of interrelated activities in a project.

3 Identify the sequence of critical activities that determines the duration of a project.

4 Explain how to determine a minimum-cost project schedule.

5 Describe the considerations managers make in assessing the risks in a project and calculate the probability of completing a project on time.

6 Define the options available to alleviate resource problems.

---

project

An interrelated set of activities with a definite starting and ending point, which results in a unique outcome for a specific allocation of resources.

Companies such as Microsoft are experts at managing projects such as Xbox 360. They master the ability to schedule activities and monitor progress within strict time, cost, and performance guidelines. A **project** is an interrelated set of activities with a definite starting and ending point, which results in a unique outcome for a specific allocation of resources.

Projects are common in everyday life as well as in business. Planning weddings, remodeling bathrooms, writing term papers, and organizing surprise parties are examples of small projects in everyday life. Conducting company audits, planning mergers, creating advertising campaigns, reengineering processes, developing new services or products, and establishing a strategic alliance are examples of large projects in business.

The three main goals of any project are (1) complete the project on time or earlier, (2) do not exceed the budget, and (3) meet the specifications to the satisfaction of the customer. When we must undertake projects with some uncertainty involved, it does not hurt to have flexibility with respect to resource availability, deadlines, and budgets. Consequently, projects can be complex and challenging to manage. **Project management,** which is a systemized, phased approach to defining, organizing, planning, monitoring, and controlling projects, is one way to overcome that challenge.

Projects often cut across organizational lines because they need the skills of multiple professions and organizations. Furthermore, each project is unique, even if it is routine, requiring new combinations of skills and resources in the project process. For example, projects for adding a new branch office, installing new computers in a department, or developing a sales promotion may be initiated several times a year. Each project may have been done many times before; however, differences arise with each replication. Uncertainties, such as the advent of new technologies or the activities of competitors, can change the character of projects and require responsive countermeasures. Finally, projects are temporary because personnel, materials, and facilities are organized to complete them within a specified time frame and then are disbanded.

Projects, and the application of project management, facilitate the implementation of strategy. However, the power of this approach goes beyond the focus on one project. Operations strategy initiatives often require the coordination of many interdependent projects. Such a collection of projects is called a **program,** which is an interdependent set of projects with a common strategic purpose. As new project proposals come forward, management must assess their fit to the current operations strategy and ongoing initiatives and have a means to prioritize them because funds for projects are often limited. Projects can be also used to implement changes to processes and supply chains. For example, projects involving the implementation of major information technologies may affect all of a firm's core processes and supporting processes as well as some of their suppliers' and customers' processes. As such, projects are a useful tool for improving processes and supply chains.

Project Management across the Organization

Even though a project may be under the overall purview of a single department, other departments likely should be involved in the project. For example, consider an information systems project to develop a corporate customer database at a bank. Many of the bank's customers are large corporations that require services spanning several departments at the bank. Because no department at the bank knows exactly what services a corporate customer is receiving from other departments, the project would consolidate information about corporate customers from many areas of the bank into one database. From this information, corporate banking services could be designed not only to better serve the corporate customers, but also to provide a basis for evaluating the prices that the bank charges. Marketing is interested in knowing all the services a customer is receiving so that it can package and sell other services that the customer may not be aware of. Finance is interested in how profitable a customer is to the bank and whether the provided services are appropriately priced. The project team, led by the information systems department, should consist of representatives from the marketing and finance departments who have a direct interest in corporate clients. All departments in a firm benefit from sound project management practices, even if the projects remain within the purview of a single department.

Defining and Organizing Projects

A clear understanding of a project's organization and how personnel are going to work together to complete the project are keys to success. In this section, we will address (1) defining the scope and objectives, (2) selecting the project manager and team, and (3) recognizing the organizational structure.

Defining the Scope and Objectives of a Project

A thorough statement of a project's scope, time frame, and allocated resources is essential to managing the project. This statement is often referred to as the *project objective statement.* The scope provides a succinct statement of project objectives and captures the essence of the desired project

project management

A systemized, phased approach to defining, organizing, planning, monitoring, and controlling projects.

program

An interdependent set of projects that have a common strategic purpose.

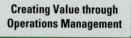

Creating Value through Operations Management

Using Operations to Compete
Project Management

Managing Processes

Process Strategy
Process Analysis
Quality and Performance
Capacity Planning
Constraint Management
Lean Systems

Managing Supply Chains

Supply Chain
Inventory Management
Supply Chain Design
Supply Chain Location
Decisions
Supply Chain Integration
Supply Chain Sustainability
and Humanitarian Logistics
Forecasting
Operations Planning and
Scheduling
Resource Planning

outcomes in the form of major deliverables, which are concrete outcomes of the project. Changes to the scope of a project inevitably increase costs and delay completion. Collectively, changes to scope are called *scope creep* and, in sufficient quantity, are primary causes of failed projects. The time frame for a project should be as specific as possible, as in "the project should be completed by January 1, 2014." Finally, although specifying an allocation of resources to a project may be difficult during the early stages of planning, it is important for managing the project. The allocation should be expressed as a dollar figure or as full-time equivalents of personnel time. A specific statement of allocated resources makes it possible to make adjustments to the scope of the project as it proceeds.

Selecting the Project Manager and Team

Once the project is selected, a project manager must be chosen. The qualities of a good project manager should be well aligned with the roles a project manager must play.

- *Facilitator.* The project manager often must resolve conflicts between individuals or departments to ensure that the project has the appropriate resources for the job to be completed. Successful project managers have good leadership skills and a *systems view*, which encompasses the interaction of the project, its resources, and its deliverables with the firm as a whole.

- *Communicator.* Project progress and requests for additional resources must be clearly communicated to senior management and other stakeholders in a project. The project manager must also frequently communicate with the project team to get the best performance.

- *Decision Maker.* Good project managers will be sensitive to the way the team performs best and be ready to make tough decisions, if necessary. The project manger must organize the team meetings, specify how the team will make decisions, and determine the nature and timing of reports to senior management.

Selecting the project team is just as important as the selection of the project manager. Several characteristics should be considered.

- *Technical Competence.* Team members should have the technical competence required for the tasks to which they will be assigned.

- *Sensitivity.* All team members should be sensitive to interpersonal conflicts that may arise. Senior team members should be politically sensitive to help mitigate problems with upper-level management.

- *Dedication.* Team members should feel comfortable solving project problems that may spill over into areas outside their immediate expertise. They should also be dedicated to getting the project done, as opposed to maintaining a comfortable work schedule.

Recognizing Organizational Structure

The relationship of the project manager to the project team is determined by the firm's organizational structure. Each of the three types of organizational structure described below has its own implications for project management.

- *Functional.* The project is housed in a specific department or functional area, presumably the one with the most interest in the project. Assistance from personnel in other functional areas must be negotiated by the project manager. In such cases, the project manager has less control over project timing than if the entire scope of the project fell within the purview of the department.

- *Pure Project.* The team members work exclusively for the project manager on a particular project. This structure simplifies the lines of authority and is particularly effective for large projects that consist of enough work for each team member to work full time. For small projects, it could result in significant duplication of resources across functional areas.

- *Matrix.* The matrix structure is a compromise between the functional and pure project structures. The project managers of the firm's projects all report to a "program manager" who coordinates resource and technological needs across the functional boundaries. The matrix structure allows each functional area to maintain control over who works on a project and the technology that is used. However, team members, in effect, have two bosses: the project manager and the department manager. Resolving these "line of authority" conflicts requires a strong project manager.

The construction of the 2012 Olympic Stadium in Stratford, East London, required the coordination of materials, equipment, and personnel. Project management techniques played a major role.

Planning Projects

After the project is defined and organized, the team must formulate a plan that identifies the specific work to be accomplished and a schedule for completion. Planning projects involves five steps: (1) defining the work breakdown structure, (2) diagramming the network, (3) developing the schedule, (4) analyzing cost–time trade-offs, and (5) assessing risks.

Defining the Work Breakdown Structure

The **work breakdown structure (WBS)** is a statement of all work that has to be completed. Perhaps, the single most important contributor to delay is the omission of work that is germane to the successful completion of the project. The project manager must work closely with the team to identify all activities. An **activity** is the smallest unit of work effort consuming both time and resources that the project manager can schedule and control. Typically, in the process of accumulating activities, the team generates a hierarchy to the work breakdown. Major work components are broken down to smaller tasks that ultimately are broken down to activities that are assigned to individuals. Figure 2.1 shows a WBS for a major project involving the relocation of a hospital. In the interest of better serving the surrounding community, the board of St. John's Hospital has decided to move to a new location. The project involves constructing a new hospital and making it operational. The work components at level 1 in the WBS can be broken down into smaller units of work in level 2 that could be further divided at level 3, until the project manager gets to activities at a level of detail that can be scheduled and controlled. For example, "Organizing and Site Preparation" has been divided into six activities at level 2 in Figure 2.1. We have kept our example simple so that the concept of the WBS can be easily understood. If our activities in the example are divided into even smaller units of work, it is easy to see that the total WBS for a project of this size may include many more than 100 activities. Regardless of the project, care must be taken to include all important activities in the WBS to avoid project delays. Often overlooked are the activities required to plan the project, get management approval at various stages, run pilot tests of new services or products, and prepare final reports.

Each activity in the WBS must have an "owner" who is responsible for doing the work. *Activity ownership* avoids confusion in the execution of activities and assigns responsibility for timely completion. The team should have a defined procedure for assigning activities to team members, which can be democratic (consensus of the team) or autocratic (assigned by the project manager).

work breakdown structure (WBS)

A statement of all work that has to be completed.

activity

The smallest unit of work effort consuming both time and resources that the project manager can schedule and control.

▶ **FIGURE 2.1**
Work Breakdown Structure for the St. John's Hospital Project

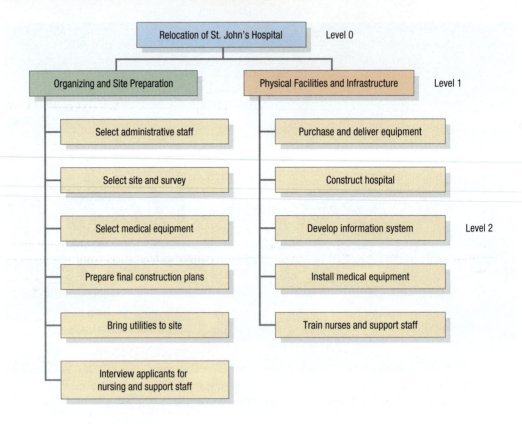

Diagramming the Network

network diagram

A network planning method, designed to depict the relationships between activities, that consists of nodes (circles) and arcs (arrows).

Network planning methods can help managers monitor and control projects. These methods treat a project as a set of interrelated activities that can be visually displayed in a **network diagram**, which consists of nodes (circles) and arcs (arrows) that depict the relationships between activities. Two network planning methods were developed in the 1950s. The **program evaluation and review technique (PERT)** was created for the U.S. Navy's Polaris missile project, which involved 3,000 separate contractors and suppliers. The **critical path method (CPM)** was developed as a means of scheduling maintenance shutdowns at chemical-processing plants. Although early versions of PERT and CPM differed in their treatment of activity time estimates, today the differences are minor. For purposes of our discussion, we refer to them collectively as PERT/CPM. These methods offer several benefits to project managers, including the following:

program evaluation and review technique (PERT)

A network planning method created for the U.S. Navy's Polaris missile project in the 1950s, which involved 3,000 separate contractors and suppliers.

1. Considering projects as networks forces project teams to identify and organize the data required and to identify the interrelationships between activities. This process also provides a forum for managers of different functional areas to discuss the nature of the various activities and their resource requirements.

2. Networks enable project managers to estimate the completion time of projects, an advantage that can be useful in planning other events and in conducting contractual negotiations with customers and suppliers.

critical path method (CPM)

A network planning method developed in the 1950s as a means of scheduling maintenance shutdowns at chemical-processing plants.

3. Reports highlight the activities that are crucial to completing projects on schedule. They also highlight the activities that may be delayed without affecting completion dates, thereby freeing up resources for other, more critical activities.

4. Network methods enable project managers to analyze the time and cost implications of resource trade-offs.

Diagramming the project network involves establishing precedence relationships and estimating activity times.

precedence relationship

A relationship that determines a sequence for undertaking activities; it specifies that one activity cannot start until a preceding activity has been completed.

Establishing Precedence Relationships   A **precedence relationship** determines a sequence for undertaking activities; it specifies that one activity cannot start until a preceding activity has been completed. For example, brochures announcing a conference for executives must first be designed by the program committee (activity A) before they can be printed (activity B). In other words, activity A must *precede* activity B. For large projects, establishing precedence relationships is essential because incorrect or omitted precedence relationships will result in costly delays. The precedence relationships are represented by a network diagram.

Estimating Activity Times   When the same type of activity has been done many times before, time estimates will have a relatively high degree of certainty. Several ways can be used to get time estimates in such an environment. First, statistical methods can be used if the project team has access to data on actual activity times experienced in the past (see MyOMLab Supplement H, "Measuring Output Rates,"). Second, if activity times improve with the number of replications, the times can be estimated using learning curve models (see Supplement I, "Learning Curve Analysis," in MyOMLab). Finally, the times for first-time activities are often estimated using managerial opinions based on similar prior experiences (see Chapter 14, "Forecasting"). If the estimates involve a high degree of uncertainty, probability distributions for activity times can be used. We discuss how to incorporate uncertainty in project networks when we address risk assessment later in this chapter. For now, we assume that the activity times are known with certainty.

Using the Activity-On-Node Approach   The diagramming approach we use in this text is referred to as the **activity-on-node (AON) network,** in which nodes represent activities and arcs represent the precedence relationships between them. Some diagramming conventions must be used for AON networks. In cases of multiple activities with no predecessors, it is usual to show them emanating from a common node called *start*. For multiple activities with no successors, it is usual to show them connected to a node called *finish*. Figure 2.2 shows how to diagram several commonly encountered activity relationships.

MyOMLab

activity-on-node (AON) network

An approach used to create a network diagram, in which nodes represent activities and arcs represent the precedence relationships between them.

◄ **FIGURE 2.2**
Diagramming Activity Relationships

EXAMPLE 2.1 Diagramming the St. John's Hospital Project

Judy Kramer, the project manager for the St. John's Hospital project, divided the project into two major modules. She assigned John Stewart the overall responsibility for the Organizing and Site Preparation module and Sarah Walker the responsibility for the Physical Facilities and Infrastructure module. Using the WBS shown in Figure 2.1, the project team developed the precedence relationships, activity time estimates, and activity responsibilities shown in the following table:

Activity	Immediate Predecessors	Activity Times (wks)	Responsibility
ST. JOHN'S HOSPITAL PROJECT			Kramer
START		0	
ORGANIZING and SITE PREPARATION			Stewart
A. Select administrative staff	Start	12	Johnson
B. Select site and survey	Start	9	Taylor
C. Select medical equipment	A	10	Adams
D. Prepare final construction plans	B	10	Taylor
E. Bring utilities to site	B	24	Burton
F. Interview applicants for nursing and support staff	A	10	Johnson
PHYSICAL FACILITIES and INFRASTRUCTURE			Walker
G. Purchase and deliver equipment	C	35	Sampson
H. Construct hospital	D	40	Casey
I. Develop information system	A	15	Murphy
J. Install medical equipment	E, G, H	4	Pike
K. Train nurses and support staff	F, I, J	6	Ashton
FINISH	K	0	

For purposes of our example, we will assume a work week consists of five work days. Draw the network diagram for the hospital project.

SOLUTION

The network diagram, activities, and activity times for the hospital project are shown in Figure 2.3. The diagram depicts activities as circles, with arrows indicating the sequence in which they are to be performed. Activities A and B emanate from a *start* node because they have no immediate predecessors. The arrows connecting activity A to activities C, F, and I indicate that all three require completion of activity A before they can begin. Similarly, activity B must be completed before activities D and E can begin, and so on. Activity K connects to a *finish* node because no activities follow it. The start and finish nodes do not actually represent activities; they merely provide beginning and ending points for the network.

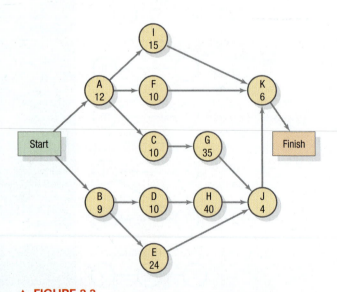

▲ **FIGURE 2.3**
Network Showing Activity Times for the St. John's Hospital Project

Developing the Schedule

A key advantage of network planning methods is the creation of a schedule of project activities that will help managers achieve the objectives of the project. Managers can (1) estimate the completion time of a project by finding the critical path, (2) identify the start and finish times for each activity for a project schedule, and (3) calculate the amount of slack time for each activity.

Critical Path   A crucial aspect of project management is estimating the time of completion of a project. If each activity in relocating the hospital were done in sequence, with work proceeding on only one activity at a time, the time of completion would equal the sum of the times for all the activities, or 175 weeks. However, Figure 2.3 indicates that some activities can be carried on simultaneously given adequate resources. We call each sequence of activities between the project's start and finish a **path**. The network describing the hospital relocation project has five paths: (1) A–I–K, (2) A–F–K, (3) A–C–G–J–K, (4) B–D–H–J–K, and (5) B–E–J–K. The **critical path** is the sequence of activities between a project's start and finish that takes the longest time to complete. Thus, the activities along the critical path determine the completion time of the project; that is, if one of the activities on the critical path is delayed, the entire project will be delayed. The estimated times for the paths in the hospital project network are

Path	Estimated Time (weeks)
A–I–K	33
A–F–K	28
A–C–G–J–K	67
B–D–H–J–K	69
B–E–J–K	43

The activity string B–D–H–J–K is estimated to take 69 weeks to complete. As the longest, it constitutes the critical path. Because the critical path defines the completion time of the project, Judy Kramer and the project team should focus on these activities and any other path that is close in length to the critical path.

Project Schedule   The typical objective is to finish the project as early as possible as determined by the critical path. The project schedule is specified by the start and finish times for each activity. For any activity, managers can use the earliest start and finish times, the latest start and finish times (and still finish the project on time), or times in between these extremes.

- **Earliest Start and Earliest Finish Times** The earliest start and earliest finish times are obtained as follows:

  1. The **earliest finish time (EF)** of an activity equals its earliest start time plus its estimated duration, t, or $EF = ES + t$.
     The **earliest start time (ES)** for an activity is the earliest finish time of the immediately preceding activity. For activities with more than one preceding activity, ES is the latest of the earliest finish times of the preceding activities.
     To calculate the duration of the entire project, we determine the EF for the last activity on the critical path.

- **Latest Start and Latest Finish Times** To obtain the latest start and latest finish times, we must work backward from the finish node. We start by setting the latest finish time of the project equal to the earliest finish time of the last activity on the critical path.

  1. The **latest finish time (LF)** for an activity is the latest start time of the activity that immediately follows. For activities with more than one activity that immediately follow, LF is the earliest of the latest start times of those activities.

  2. The **latest start time (LS)** for an activity equals its latest finish time minus its estimated duration, t, or $LS = LF - t$.

path
The sequence of activities between a project's start and finish.

critical path
The sequence of activities between a project's start and finish that takes the longest time to complete.

earliest finish time (EF)
An activity's earliest start time plus its estimated duration, t, or $EF = ES + t$.

earliest start time (ES)
The earliest finish time of the immediately preceding activity.

latest finish time (LF)
The latest start time of the activity that immediately follows.

latest start time (LS)
The latest finish time minus its estimated duration, t, or $LS = LF - t$.

Aircraft construction is an example of a large project that requires a sound project schedule because of the capital involved. Here several 747s are under construction at Boeing's plant in Everett, Washington.

George Hall/Corbis

| EXAMPLE 2.2 | **Calculating Start and Finish Times for the Activities** |

Calculate the ES, EF, LS, and LF times for each activity in the hospital project. Which activity should Kramer start immediately? Figure 2.3 contains the activity times.

SOLUTION

To compute the early start and early finish times, we begin at the start node at time zero. Because activities A and B have no predecessors, the earliest start times for these activities are also zero. The earliest finish times for these activities are

$$EF_A = 0 + 12 = 12 \text{ and } EF_B = 0 + 9 = 9$$

Because the earliest start time for activities I, F, and C is the earliest finish time of activity A,

$$ES_I = 12, ES_F = 12, \text{ and } ES_C = 12$$

Similarly,

$$ES_D = 9 \text{ and } ES_E = 9$$

After placing these ES values on the network diagram (see Figure 2.4), we determine the EF times for activities I, F, C, D, and E:

$$EF_I = 12 + 15 = 27, EF_F = 12 + 10 = 22, EF_C = 12 + 10 = 22,$$
$$EF_D = 9 + 10 = 19, \text{ and } EF_E = 9 + 24 = 33$$

The earliest start time for activity G is the latest EF time of all immediately preceding activities. Thus,

$$ES_G = EF_C = 22, ES_H = EF_D = 19$$
$$EF_G = ES_G + t = 22 + 35 = 57, EF_H = ES_H + t = 19 + 40 = 59$$

▶ **FIGURE 2.4**

Network Diagram Showing
Start and Finish Times and
Activity Slack

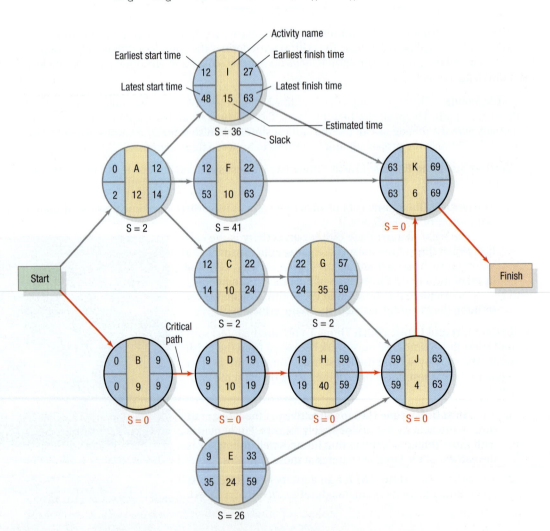

The project team can now determine the earliest time any activity can be started. Because activity J has several predecessors, the earliest time that activity J can begin is the latest of the EF times of any of its preceding activities: EF_G, EF_H, or EF_E. Thus, $EF_J = 59 + 4 = 63$. Similarly, $ES_K = 63$ and $EF_K = 63 + 6 = 69$. Because activity K is the last activity on the critical path, the earliest the project can be completed is week 69. The earliest start and finish times for all activities are shown in Figure 2.4.

To compute the latest start and latest finish times, we begin by setting the latest finish activity time of activity K at week 69, which is its earliest finish time as determined in Figure 2.4. Thus, the latest start time for activity K is

$$LS_K = LF_K - t = 69 - 6 = 63$$

If activity K is to start no later than week 63, all its predecessors must finish no later than that time. Consequently,

$$LF_I = 63, LF_F = 63, \text{ and } LF_J = 63$$

The latest start times for these activities are shown in Figure 2.4 as

$$LS_I = 63 - 15 = 48, LS_F = 63 - 10 = 53, \text{ and } LS_J = 63 - 4 = 59$$

After obtaining LS_J, we can calculate the latest start times for the immediate predecessors of activity J:

$$LS_G = 59 - 35 = 24, LS_H = 59 - 40 = 19, \text{ and } LS_E = 59 - 24 = 35$$

Similarly, we can now calculate the latest start times for activities C and D:

$$LS_C = 24 - 10 = 14 \text{ and } LS_D = 19 - 10 = 9$$

Activity A has more than one immediately following activity: I, F, and C. The earliest of the latest start times is 14 for activity C. Thus,

$$LS_A = 14 - 12 = 2$$

Similarly, activity B has two immediate followers: D and E. Because the earliest of the latest start times of these activities is 9,

$$LS_B = 9 - 9 = 0$$

DECISION POINT

The earliest or latest start times can be used for developing a project schedule. For example, Kramer should start activity B immediately because the latest start time is 0; otherwise, the project will not be completed by week 69. When the LS is greater than the ES for an activity, that activity could be scheduled for any date between ES and LS. Such is the case for activity E, which could be scheduled to start anytime between week 9 and week 35, depending on the availability of resources. The earliest start and earliest finish times and the latest start and latest finish times for all activities are shown in Figure 2.4.

Activity Slack   The maximum length of time that an activity can be delayed without delaying the entire project is called **activity slack**. Consequently, *activities on the critical path have zero slack.* Information on slack can be useful because it highlights activities that need close attention. In this regard, activity slack is the amount of schedule slippage that can be tolerated for an activity before the entire project will be delayed. Slack at an activity is reduced when the estimated time duration of an activity is exceeded or when the scheduled start time for the activity must be delayed because of resource considerations. Activity slack can be calculated in one of two ways for any activity:

$$S = LS - ES \text{ or } S = LF - EF$$

Computers calculate activity slack and prepare periodic reports for large projects, enabling managers to monitor progress. Using these reports, managers can sometimes manipulate slack to overcome scheduling problems. When resources can be used on several different activities in a project, they can be taken from activities with slack and given to activities that are behind schedule until the slack is used up. The slack for each activity in the hospital project is shown in Figure 2.4.

Gantt Chart   The project manager, often with the assistance of computer software, creates the project schedule by superimposing project activities, with their precedence relationships and

activity slack

The maximum length of time that an activity can be delayed without delaying the entire project, calculated as $S = LS - ES$ or $S = LF - EF$.

MyOMLab

Active Model 2.1 in MyOMLab provides additional insight on Gantt charts and their uses for the St. John's Hospital project.

	Task Name	Duration	Start	Finish	Predecessors
1	⊟ St John's Hospital Project	69 wks	Mon 9/12/11	Fri 1/4/13	
2	Start	0 wks	Mon 9/12/11	Mon 9/12/11	
3	⊟ Organizing and Site Prep	33 wks	Mon 9/12/11	Fri 4/27/12	
4	A. Select Staff	12 wks	Mon 9/12/11	Fri 12/2/11	2
5	B. Select Site	9 wks	Mon 9/12/11	Fri 11/11/11	2
6	C. Select Equipment	10 wks	Mon 12/5/11	Fri 2/10/12	4
7	D. Construction Plans	10 wks	Mon 11/14/11	Fri 1/20/12	5
8	E. Utilities	24 wks	Mon 11/14/11	Fri 4/27/12	5
9	F. Interviews	10 wks	Mon 12/5/11	Fri 2/10/12	4
10	⊟ Facilities and Infrastructure	57 wks	Mon 12/5/11	Fri 1/4/13	
11	G. Purchase Equipment	35 wks	Mon 2/13/12	Fri 10/12/12	6
12	H. Construct Hospital	40 wks	Mon 1/23/12	Fri 10/26/12	7
13	I. Information System	15 wks	Mon 12/5/11	Fri 3/16/12	4
14	J. Install Equipment	4 wks	Mon 10/29/12	Fri 11/23/12	8,11,12
15	K. Train Staff	6 wks	Mon 11/26/12	Fri 1/4/13	9,13,14
16	Finish	0 wks	Fri 1/4/13	Fri 1/4/13	15

▲ **FIGURE 2.5**
MS Project Gantt Chart for the St. John's Hospital Project Schedule

Gantt chart

A project schedule, usually created by the project manager using computer software, that superimposes project activities, with their precedence relationships and estimated duration times, on a time line.

estimated duration times, on a time line. The resulting diagram is called a **Gantt chart.** Figure 2.5 shows a Gantt chart for the hospital project created with Microsoft Project, a popular software package for project management. The critical path is shown in red. The chart clearly shows which activities can be undertaken simultaneously and when they should be started. Figure 2.5 also shows the earliest start schedule for the project. Microsoft Project can also be used to show the latest start schedule or to change the definition of the work week to declare Saturday and Sunday as work days, for example. Gantt charts are popular because they are intuitive and easy to construct.

Analyzing Cost–Time Trade-Offs

Keeping costs at acceptable levels is almost always as important as meeting schedule dates. In this section, we discuss the use of PERT/CPM methods to obtain minimum-cost schedules.

The reality of project management is that there are always cost–time trade-offs. For example, a project can often be completed earlier than scheduled by hiring more workers or running extra shifts. Such actions could be advantageous if savings or additional revenues accrue from completing the project early. *Total project costs* are the sum of direct costs, indirect costs, and penalty costs. These costs are dependent either on activity times or on project completion time. *Direct costs* include labor, materials, and any other costs directly related to project activities. *Indirect costs* include administration, depreciation, financial, and other variable overhead costs that can be avoided by reducing total project time: The shorter the duration of the project, the lower the indirect costs will be. Finally, a project may incur *penalty costs* if it extends beyond some specific date, whereas *an incentive* may be provided for early completion. Managers can shorten individual activity times by using additional direct resources, such as overtime, personnel, or equipment. Thus, a project manager may consider *crashing*, or expediting, some activities to reduce overall project completion time and total project costs.

Excavators work on the new Panama Canal project, which has international implications and massive costs.

Cost to Crash   To assess the benefit of crashing certain activities—from either a cost or a schedule perspective—the project manager needs to know the following times and costs:

1. The **normal time (NT)** is the time necessary to complete an activity under normal conditions.

2. The **normal cost (NC)** is the activity cost associated with the normal time.

3. The **crash time (CT)** is the shortest possible time to complete an activity.

4. The **crash cost (CC)** is the activity cost associated with the crash time.

Our cost analysis is based on the assumption that direct costs increase linearly as activity time is reduced from its normal time. This assumption implies that for every week the activity time is reduced, direct costs increase by a proportional amount. For example, suppose that the normal time for activity C in the hospital project is 10 weeks and is associated with a direct cost of $4,000. Also, suppose that we can crash its time to only 5 weeks at a total cost of $7,000; the net time reduction is 5 weeks at a net cost increase of $3,000. We assume that crashing activity C costs $3,000/5 = $600 per week—an assumption of linear marginal costs that is illustrated in Figure 2.6. Thus, if activity C were expedited by 2 weeks (i.e., its time reduced from 10 weeks to 8 weeks), the estimated direct costs would be $4,000 + 2($600) = $5,200. For any activity, the cost to crash an activity by one week is

$$\text{Cost to crash per period} = \frac{CC - NC}{NT - CT}$$

Table 2.1 contains direct cost and time data, as well as the costs of crashing per week for the activities in the hospital project.

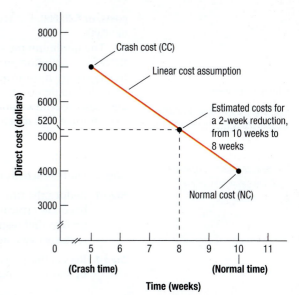

▲ **FIGURE 2.6**
Cost-Time Relationships in Cost Analysis

normal time (NT)

In the context of project management, the time necessary to complete an activity under normal conditions.

normal cost (NC)

The activity cost associated with the normal time.

crash time (CT)

The shortest possible time to complete an activity.

crash cost (CC)

The activity cost associated with the crash time.

TABLE 2.1 | DIRECT COST AND TIME DATA FOR THE ST. JOHN'S HOSPITAL PROJECT

Activity	Normal Time (NT) (weeks)	Normal Cost (NC) ($)	Crash Time (CT) (weeks)	Crash Cost (CC) ($)	Maximum Time Reduction (week)	Cost of Crashing per Week ($)
A	12	$12,000	11	13,000	1	1,000
B	9	50,000	7	64,000	2	7,000
C	10	4,000	5	7,000	5	600
D	10	16,000	8	20,000	2	2,000
E	24	120,000	14	200,000	10	8,000
F	10	10,000	6	16,000	4	1,500
G	35	500,000	25	530,000	10	3,000
H	40	1,200,000	35	1,260,000	5	12,000
I	15	40,000	10	52,500	5	2,500
J	4	10,000	1	13,000	3	1,000
K	6	30,000	5	34,000	1	4,000
	Totals	$1,992,000		$2,209,500		

Minimizing Costs   The objective of cost analysis is to determine the project schedule that minimizes total project costs. Suppose that project indirect costs are $8,000 per week. Suppose also that, after week 65, the Regional Hospital Board imposes on St. John's a penalty cost of $20,000 per week if the hospital is not fully operational. With a critical path completion time of 69 weeks, the hospital faces potentially large penalty costs unless the schedule is changed. For every week that the project is shortened—to week 65—the hospital saves one week of penalty *and* indirect

minimum-cost schedule

A schedule determined by starting with the normal time schedule and crashing activities along the critical path, in such a way that the costs of crashing do not exceed the savings in indirect and penalty costs.

costs, or $28,000. For reductions beyond week 65, the savings are only the weekly indirect costs of $8,000.

The minimum possible project duration can be found by using the crash times of each activity for scheduling purposes. However, the cost of that schedule could be prohibitive. Project managers are most interested in minimizing the costs of their projects so that budgets are not exceeded. In determining the **minimum-cost schedule**, we start with the normal time schedule and crash activities along the critical path, whose length equals the length of the project. We want to determine how much we can add in crash costs without exceeding the savings in indirect and penalty costs. The procedure involves the following steps:

Step 1. Determine the project's critical path(s).

Step 2. Find the activity or activities on the critical path(s) with the lowest cost of crashing per week.

Step 3. Reduce the time for this activity until (a) it cannot be further reduced, (b) another path becomes critical, or (c) the increase in direct costs exceeds the indirect and penalty cost savings that result from shortening the project. If more than one path is critical, the time for an activity on each path may have to be reduced simultaneously.

Step 4. Repeat this procedure until the increase in direct costs is larger than the savings generated by shortening the project.

| EXAMPLE 2.3 | **Find a Minimum-Cost Schedule** |

Determine the minimum-cost schedule for the St. John's Hospital project. Use the information provided in Table 2.1 and Figure 2.4.

MyOMLab

Active Model 2.2 in MyOMLab provides additional insight on cost analysis for the St. John's Hospital project.

SOLUTION

The projected completion time of the project is 69 weeks. The project costs for that schedule are $1,992,000 in direct costs, $69($8,000) = $552,000$ in indirect costs, and $(69 - 65)$ $($20,000)$ $= $80,000$ in penalty costs, for total project costs of $2,624,000. The five paths in the network have the following normal times:

A–I–K:	33 weeks
A–F–K:	28 weeks
A–C–G–J–K:	67 weeks
B–D–H–J–K:	69 weeks
B–E–J–K:	43 weeks

It will simplify our analysis if we can eliminate some paths from further consideration. If all activities on A–C–G–J–K were crashed, the path duration would be 47 weeks. Crashing all activities on B–D–H–J–K results in a project duration of 56 weeks. Because the *normal* times of A–I–K, A–F–K, and B–E–J–K are less than the minimum times of the other two paths, we can disregard those three paths; they will never become critical regardless of the crashing we may do.

STAGE 1

Step 1. The critical path is B–D–H–J–K.

Step 2. The cheapest activity to crash per week is J at $1,000, which is much less than the savings in indirect and penalty costs of $28,000 per week.

Step 3. Crash activity J by its limit of three weeks because the critical path remains unchanged. The new expected path times are

A–C–G–J–K: 64 weeks and B–D–H–J–K: 66 weeks

The net savings are 3 $($28,000) - 3($1,000) = $81,000$. The total project costs are now $2,624,000 - $81,000 = $2,543,000$.

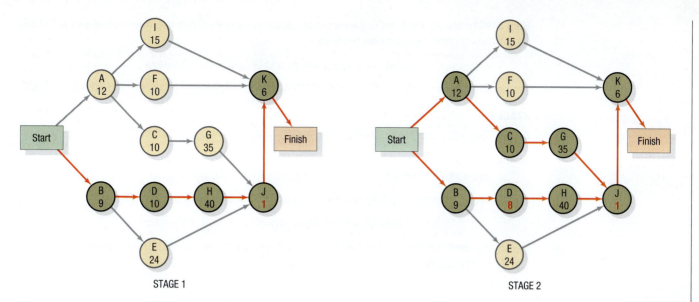

STAGE 1   STAGE 2

STAGE 2

Step 1. The critical path is still B–D–H–J–K.

Step 2. The cheapest activity to crash per week is now D at $2,000.

Step 3. Crash D by two weeks. The first week of reduction in activity D saves $28,000 because it eliminates a week of penalty costs, as well as indirect costs. Crashing D by a second week saves only $8,000 in indirect costs because, after week 65, no more penalty costs are incurred. These savings still exceed the cost of crashing D for a second week. Updated path times are

A–C–G–J–K: 64 weeks and B–D–H–J–K: 64 weeks

The net savings are $28,000 + $8,000 − 2($2,000) = $32,000. Total project costs are now $2,543,000 − $32,000 = $2,511,000.

STAGE 3

Step 1. After crashing D, we now have two critical paths. *Both* critical paths must now be shortened to realize any savings in indirect project costs. If one is shortened and the other is not, the length of the project remains unchanged.

Step 2. Our alternatives are to crash one of the following combinations of activities—(A, B); (A, H); (C, B); (C, H); (G, B); (G, H)—or to crash activity K, which is on both critical paths (J has already been crashed). We consider only those alternatives for which the cost of crashing is less than the potential savings of $8,000 per week. The only viable alternatives are (C, B) at a cost of $7,600 per week and K at $4,000 per week. We choose activity K to crash.

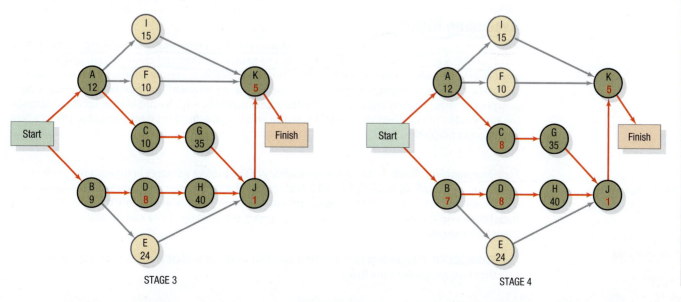

STAGE 3   STAGE 4

Step 3. We crash activity K to the greatest extent possible—a reduction of one week—because it is on both critical paths. Updated path times are

A–C–G–J–K: 63 weeks and B–D–H–J–K: 63 weeks

The net savings are $8,000 − $4,000 = $4,000. Total project costs are $2,511,000 − $4,000 = $2,507,000.

STAGE 4

Step 1. The critical paths are B–D–H–J–K and A–C–G–J–K.

Step 2. The only viable alternative at this stage is to crash activities B and C simultaneously at a cost of $7,600 per week. This amount is still less than the savings of $8,000 per week.

Step 3. Crash activities B and C by two weeks, the limit for activity B. Updated path times are

A–C–G–J–K: 61 weeks and B–D–H–J–K: 61 weeks

Net savings are 2($8,000) − 2($7,600) = $800. Total project costs are $2,507,000 − $800 = $2,506,200. The following table summarizes the analysis:

Stage	Crash Activity	Time Reduction (weeks)	Resulting Critical Path(s)	Project Duration (weeks)	Project Direct Costs, Last Trial ($000)	Crash Cost Added ($000)	Total Indirect Costs ($000)	Total Penalty Costs ($000)	Total Project Costs ($000)
0	—	—	B–D–H–J–K	69	1,992.0	—	552.0	80.0	2,624.0
1	J	3	B–D–H–J–K	66	1,992.0	3.0	528.0	20.0	2,543.0
2	D	2	B–D–H–J–K A–C–G–J–K	64	1,995.0	4.0	512.0	0.0	2,511.0
3	K	1	B–D–H–J–K A–C–G–J–K	63	1,999.0	4.0	504.0	0.0	2,507.0
4	B,C	2	B–D–H–J–K A–C–G–J–K	61	2,003.0	15.2	488.0	0.0	2,506.2

DECISION POINT

Because the crash costs exceed weekly indirect costs, any other combination of activities will result in a net increase in total project costs. The minimum-cost schedule is 61 weeks, with a total cost of $2,506,200. To obtain this schedule, the project team must crash activities B, D, J, and K to their limits and activity C to eight weeks. The other activities remain at their normal times. This schedule costs $117,800 less than the normal-time schedule.

Assessing Risks

Risk is a measure of the probability and consequence of not reaching a defined project goal. Risk involves the notion of uncertainty as it relates to project timing and costs. Often, project teams must deal with uncertainty caused by labor shortages, weather, supply delays, or the outcomes of critical tests. In this section, we discuss risk management plans and the tools managers can use to analyze the risks, such as simulation and statistical analysis, which enable managers to estimate the probability of completing a project on time and the potential for near-critical paths to affect the project completion time.

risk-management plan

A plan that identifies the key risks to a project's success and prescribes ways to circumvent them.

Risk-Management Plans   A major responsibility of the project manager at the start of a project is to develop a **risk-management plan**, which identifies the key risks to a project's success and prescribes ways to circumvent them. A good risk-management plan will quantify the risks, predict their impact on the project, and provide contingency plans. Project risk can be assessed by examining four categories:

- **Strategic Fit** The project may not be a good strategic fit in that it may not be clearly linked to the strategic goals of the firm.

- **Service/Product Attributes** If the project involves the development of a new service or product, there may be market, technological, or legal risks. There is a chance that competitors may offer a superior product, or a technological discovery may render the service or product obsolete before it even hits the market. There may also be a legal risk of potential lawsuits or liability that could force a design change after product development has begun.

- **Project Team Capability** The project team may not have the capability to complete the project successfully because of the size and complexity of the project or the technology involved.

- **Operations** There may be an operations risk because of poor information accuracy, lack of communication, missing precedence relationships, or bad estimates for activity times.

These risks should be identified and the significant ones should have contingency plans in case something goes wrong. The riskier a project is, the more likely the project will experience difficulties as Managerial Practice 2.1 shows.

Simulation   PERT/CPM networks can be used to quantify risks associated with project timing. Often, the uncertainty associated with an activity can be reflected in the activity's time duration. For example, an activity in a new product development project might be developing the enabling

---

MANAGERIAL PRACTICE 2.1   Boston's Big Dig Project Poses Many Challenges

Boston, Massachusetts, has many noteworthy attractions: the world champion Boston Red Sox baseball team, the Freedom Trail, depicting many historic buildings and sights dating back to the 1600s, and the most ambitious road infrastructure project attempted in the United States. The six-lane elevated highway that ran through the center of the city was designed for 75,000 cars per day, but was forced to accommodate close to 200,000 cars per day. The highway was congested for 10 hours a day; congestion was expected to increase to 16 hours a day by 2010. It was costing residents and businesses $500 million a year in accidents, fuel, and late delivery charges.

Solving the traffic problem would take more than adding a few lanes to the existing highway, which was built in 1953 and whose elevated superstructure was rapidly deteriorating. Rather than fixing the old highway, the decision was made to build an 8-to-10-lane underground highway directly beneath the existing road, culminating at the north end of the city in a 14-lane, 2-bridge crossing of the Charles River. On the south end, a four-lane tunnel was built under South Boston and the Boston Harbor to Logan Airport, leaving no doubt how the project got its "Big Dig" nickname; the project spans 7.8 miles of highway with half in tunnels under a major city and harbor! Planning for the project began in 1983, construction began in 1991, and in spring 2007 it was declared 99 percent complete.

Was the project successful? The answer might depend on whom you ask. The residents of Boston have a much more efficient transportation network that allows for growth for many years into the future. However, from a project management perspective, it missed the three goals of every project: (1) on time, (2) under budget, and (3) meet the specifications. The Big Dig was 9 years late (originally scheduled for completion in 1998), more than $10 billion over the budget (originally projected to be about $4 billion in today's dollars), and required significant repairs for leaks shortly after the tunnels were opened. Much negative publicity in summer 2006 resulted from the failure of ceiling panels in the Seaport Access Tunnel, which fell to the roadway and onto a passing vehicle, resulting in the tragic loss of life. Because the project was funded with taxpayer dollars, it is no wonder that the project is the subject of much debate and controversy.

Why did this project experience problems? The Big Dig is an example of a risky project because it was huge and complex. It was called one of the most complex and controversial engineering projects in human history, rivaling the likes of the Panama Canal, the English Channel Tunnel, and the

Michael Dwyer/Alamy

A great deal of controversy surrounded the "Big Dig," a massive highway tunnel built under the city of Boston and Boston Harbor. By the time the Big Dig was completed, it was over budget, late, and did not meet the specifications—in part, because no one had ever undertaken such a complex project before.

Trans-Alaska Pipeline. Project managers held many meetings with environmental and permitting agencies, community groups, businesses, and political leaders to gain consensus on how the project would be built. Because of meetings such as these the project scope was modified over time, thereby causing the project plan to change. From an operational perspective, most of the construction companies involved in the project had never done anything of this size and scope before and had difficulty providing good time estimates for their pieces of the project. Delays and cost overruns were inevitable. Further, quality was difficult to achieve because so many contractors were involved in such a complex project. Projects of this size and complexity are inherently risky; contingency plans should cover the most likely disruptions. Schedule and budget problems are not unusual; however, the job of project managers is to manage the risks and minimize the deviations.

Sources: **http://en.wikipedia.org/wiki/Big_Dig** (2010); **www.massturnpike.com/bigdig/updates** (2007); Seth Stern, "$14.6 Billion Later, Boston's Big Dig Wraps Up," *Christian Science Monitor* (December 19, 2003); "The Big Dig, Boston, MA, USA," **www.roadtraffic-technology.com** (2005); "Big Dig Tunnel Is Riddled with Leaks," *Associated Press,* **http://abcnews.go.com** (November 19, 2004); Michael Roth, "Boston Digs the Big Dig," *Rental Equipment Register* (November 1, 2000), **http://rermag.com/ar** (2005).

MyOMLab

technology to manufacture it, an activity that may take from eight months to a year. To incorporate uncertainty into the network model, probability distributions of activity times can be calculated using two approaches: (1) computer simulation and (2) statistical analysis. With simulation, the time for each activity is randomly chosen from its probability distribution (see MyOMLab Supplement E, "Simulation"). The critical path of the network is determined and the completion date of the project computed. The procedure is repeated many times, which results in a probability distribution for the completion date. We will have more to say about simulation when we discuss near critical paths later in this chapter.

Statistical Analysis   The statistical analysis approach requires that activity times be stated in terms of three reasonable time estimates:

optimistic time (a)

The shortest time in which an activity can be completed, if all goes exceptionally well.

most likely time (m)

The probable time required to perform an activity.

pessimistic time (b)

The longest estimated time required to perform an activity.

1. The **optimistic time** (a) is the shortest time in which an activity can be completed, if all goes exceptionally well.

2. The **most likely time** (m) is the probable time required to perform an activity.

3. The **pessimistic time** (b) is the longest estimated time required to perform an activity.

With three time estimates—the optimistic, the most likely, and the pessimistic—the project manager has enough information to estimate the probability that an activity will be completed on schedule. To do so, the project manager must first calculate the mean and variance of a probability distribution for each activity. In PERT/CPM, each activity time is treated as though it were a random variable derived from a beta probability distribution. This distribution can have various shapes, allowing the most likely time estimate (m) to fall anywhere between the pessimistic (b) and optimistic (a) time estimates. The most likely time estimate is the *mode* of the beta distribution, or the time with the highest probability of occurrence. This condition is not possible with the normal distribution, which is symmetrical, because the normal distribution requires the mode to be equidistant from the end points of the distribution. Figure 2.7 shows the difference between the two distributions.

▶ **FIGURE 2.7**

Differences Between Beta and Normal Distributions for Project Risk Analysis

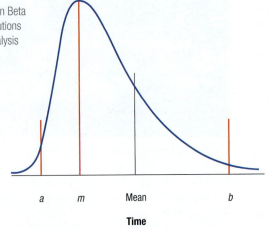

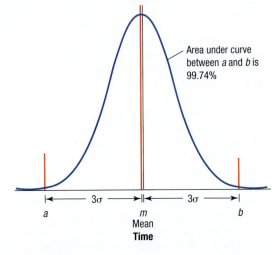

Area under curve between a and b is 99.74%

(a) **Beta distribution:** The most likely time (m) has the highest probability and can be placed anywhere between the optimistic (a) and pessimistic (b) times.

(b) **Normal distribution:** The mean and most likely times must be the same. If a and b are chosen to be 6σ apart, there is a 99.74% chance that the actual activity time will fall between them.

Analysis

Two key assumptions are required. First, we assume that a, m, and b can be estimated accurately. The estimates might best be considered values that define a reasonable time range for the activity duration negotiated between the project manager and the team members responsible for the activities. Second, we assume that the standard deviation, σ, of the activity time is one-sixth the range b − a. Thus, the chance that actual activity times will fall between a and b is high. Why does this assumption make sense? If the activity time followed the normal distribution, six standard deviations would span approximately 99.74 percent of the distribution.

Even with these assumptions, derivation of the mean and variance of each activity's probability distribution is complex. These derivations show that the mean of the beta distribution can be estimated by using the following weighted average of the three time estimates:

$$t_e = \frac{a + 4m + b}{6}$$

Note that the most likely time has four times the weight of the pessimistic and optimistic estimates.

The variance of the beta distribution for each activity is

$$\sigma^2 = \left(\frac{b-a}{6}\right)^2$$

The variance, which is the standard deviation squared, increases as the difference between b and a increases. This result implies that the less certain a person is in estimating the actual time for an activity, the greater will be the variance.

EXAMPLE 2.4   **Calculating Means and Variances**

Suppose that the project team has arrived at the following time estimates for activity B (Select site and survey) of the St. John's Hospital project:

$$a = 7 \text{ weeks}, m = 8 \text{ weeks, and } b = 15 \text{ weeks}$$

a. Calculate the expected time and variance for activity B.
b. Calculate the expected time and variance for the other activities in the project.

SOLUTION

a. The expected time for activity B is

$$t_e = \frac{7 + 4(8) + 15}{6} = \frac{54}{6} = 9 \text{ weeks}$$

Note that the expected time (9 weeks) does not equal the most likely time (8 weeks) for this activity. These times will be the same only when the most likely time is equidistant from the optimistic and pessimistic times. We calculate the variance for activity B as

$$\sigma^2 = \left(\frac{15-7}{6}\right)^2 = \left(\frac{8}{6}\right)^2 = 1.78$$

b. The following table shows expected activity times and variances for the activities listed in the project description.

	TIME ESTIMATES (WEEKS)			ACTIVITY STATISTICS	
Activity	Optimistic (a)	Most Likely (m)	Pessimistic (b)	Expected Time (t_e)	Variance (σ^2)
A	11	12	13	12	0.11
B	7	8	15	9	1.78
C	5	10	15	10	2.78
D	8	9	16	10	1.78
E	14	25	30	24	7.11
F	6	9	18	10	4.00
G	25	36	41	35	7.11
H	35	40	45	40	2.78
I	10	13	28	15	9.00
J	1	2	15	4	5.44
K	5	6	7	6	0.11

DECISION POINT

The project team should notice that the greatest uncertainty lies in the time estimate for activity I, followed by the estimates for activities E and G. These activities should be analyzed for the source of the uncertainties and actions should be taken to reduce the variance in the time estimates.

Analyzing Probabilities Because time estimates for activities involve uncertainty, project managers are interested in determining the probability of meeting project completion deadlines. To develop the probability distribution for project completion time, we assume that the duration time of one activity does not depend on that of any other activity. This assumption enables us to estimate the mean and variance of the probability distribution of the time duration of the entire project by summing the duration times and variances of the activities along the critical path. However, if one work crew is assigned two activities that can be done at the same time, the activity times will be interdependent and the assumption is not valid. In addition, if other paths in the network have small amounts of slack, one of them might become the critical path before the project is completed; we should calculate a probability distribution for those paths as well.

Because of the assumption that the activity duration times are independent random variables, we can make use of the central limit theorem, which states that the sum of a group of independent, identically distributed random variables approaches a normal distribution as the number of random variables increases. The mean of the normal distribution is the sum of the expected activity times on the path. In the case of the critical path, it is the earliest expected finish time for the project:

$$T_E = \Sigma \,(\text{Expected activity times on the critical path}) = \text{Mean of normal distribution}$$

Similarly, because of the assumption of activity time independence, we use the sum of the variances of the activities along the path as the variance of the time distribution for that path. That is, for the critical path,

$$\sigma_P^2 = \Sigma \,(\text{Variances of activities on the critical path})$$

To analyze probabilities of completing a project by a certain date using the normal distribution, we focus on the *critical path* and use the *z*-transformation formula:

$$z = \frac{T - T_E}{\sigma_P}$$

where

$$T = \text{due date for the project}$$

Given the value of *z*, we use the Normal Distribution appendix to find the probability that the project will be completed by time *T*, or sooner. An implicit assumption in this approach is that no other path will become critical during the time span of the project. Example 2.5, part (a), demonstrates this calculation for the St. John's Hospital project.

The procedure for assessing the probability of completing any activity in a project by a specific date is similar to the one just discussed. However, instead of the critical path, we would use the longest time path of activities from the start node to the activity node in question.

Near-Critical Paths A project's duration is a function of its critical path. However, paths that are close to the same duration as the critical path may ultimately become the critical path over the life of the project. In practice, at the start of the project, managers typically do not know the activity times with certainty and may never know which path was the critical path until the actual activity times are known at the end of the project. Nonetheless, this uncertainty does not reduce the usefulness of identifying the probability of one path or another causing a project to exceed its target completion time; it helps to identify the activities that need close management attention. To assess the chances of near-critical paths delaying the project completion, we can focus on the longest paths in the project network keeping in mind that both duration and variance along the path must be considered. Shorter paths with high variances could have just as much a chance to delay the project as longer paths with smaller variances. We can then estimate the probability that a given path will exceed the project target completion time. We demonstrate that approach using statistical analysis in Example 2.5, part (b).

Alternatively, simulation can be used to estimate the probabilities. The advantage of simulation is that you are not restricted to the use of the beta distribution for activity times. Also, activity or path dependencies, such as decision points that could involve different groups of activities to be undertaken, can be incorporated in a simulation model much more easily than with the statistical analysis approach. Fortunately, regardless of the approach used, it is rarely necessary to evaluate every path in the network. In large networks, many paths will have both short durations and low variances, making them unlikely to affect the project duration.

EXAMPLE 2.5	**Calculating the Probability of Completing a Project by a Given Date**

Calculate the probability that St. John's Hospital will become operational in 72 weeks, using (a) the critical path and (b) near-critical path A–C–G–J–K.

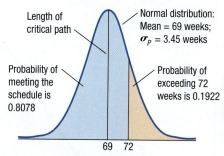

MyOMLab

Active Model 2.3 in MyOMLab provides additional insight on probability analysis for the St. John's Hospital project.

SOLUTION

a. The critical path B–D–H–J–K has a length of 69 weeks. From the table in Example 2.4, we obtain the variance of path B–D–H–J–K: $\sigma_P^2 = 1.78 + 1.78 + 2.78 + 5.44 + 0.11 = 11.89$. Next, we calculate the z-value:

$$z = \frac{72 - 69}{\sqrt{11.89}} = \frac{3}{3.45} = 0.87$$

Using the Normal Distribution appendix, we go down the left-hand column until we arrive at the value 0.8, and then across until we arrive at the 0.07 column, which shows a tabular value of 0.8078. Consequently, we find that the probability is about 0.81 that the length of path B–D–H–J–K will be no greater than 72 weeks. Because this path is the critical path, there is a 19 percent probability that the project will take longer than 72 weeks. This probability is shown graphically in Figure 2.8.

b. From the table in Example 2.4, we determine that the sum of the expected activity times on path A–C–G–J–K is 67 weeks and that $\sigma_P^2 = 0.11 + 2.78 + 7.11 + 5.44 + 0.11 = 15.55$. The z-value is

$$z = \frac{72 - 67}{\sqrt{15.55}} = \frac{5}{3.94} = 1.27$$

The probability is about 0.90 that the length of path A–C–G–J–K will be no greater than 72 weeks.

Length of critical path

Normal distribution: Mean = 69 weeks; σ_P = 3.45 weeks

Probability of meeting the schedule is 0.8078

Probability of exceeding 72 weeks is 0.1922

69   72

Project duration (weeks)

▲ **FIGURE 2.8**

Probability of Completing the St. John's Hospital Project on Schedule

DECISION POINT

The project team should be aware of the 10 percent chance that path A–C–G–J–K will exceed the target completion date of week 72. Although the probability is not high for that path, activities A, C, and G bear watching during the first 57 weeks of the project to make sure no more than 2 weeks of slippage occurs in their schedules. This attention is especially important for activity G, which has a high time variance.

Monitoring and Controlling Projects

Once project planning is over, the challenge becomes keeping the project on schedule within the budget of allocated resources. In this section, we discuss how to monitor project status and resource usage. In addition, we identify the features of project management software useful for monitoring and controlling projects.

Monitoring Project Status

A good tracking system will help the project team accomplish its project goals. Effective tracking systems collect information on three topics: (1) open issues, (2) risks, and (3) schedule status.

Open Issues and Risks One of the duties of the project manager is to make sure that issues that have been raised during the project actually get resolved in a timely fashion. The tracking system should remind the project manager of due dates for open issues and who was responsible for seeing that they are resolved. Likewise, it should provide the status of each risk to project delays specified in the risk management plan so that the team can review them at each meeting. To be effective, the tracking system requires team members to update information periodically regarding their respective responsibilities.

Schedule Status Even the best laid project plans can go awry. A tracking system that provides periodic monitoring of slack time in the project schedule can help the project manager control activities along the critical path. Periodic updating of the status of ongoing activities in the project allows the tracking system to recalculate activity slacks

Monitoring and controlling shipbuilding projects is critical to keeping these complex projects on schedule. Here a propeller is attached to an ocean-going vessel.

and indicate those activities that are behind schedule or are in danger of using up all of their slack. Management can then focus on those activities and reallocate resources as needed.

Monitoring Project Resources

Experience has shown that the resources allocated to a project are consumed at an uneven rate that is a function of the timing of the schedules for the project's activities. Projects have a *life cycle* that consists of four major phases: (1) definition and organization, (2) planning, (3) execution, and (4) close out. Figure 2.9 shows that each of the four phases requires different resource commitments.

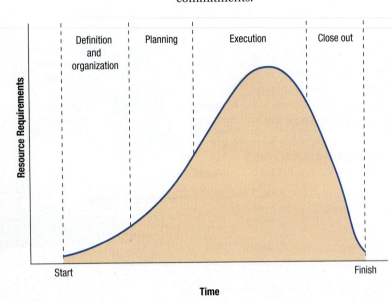

▲ **FIGURE 2.9**
Project Life Cycle

We have already discussed the activities associated with the project definition and organization and project planning phases. The phase that takes the most resources is the *execution phase,* during which managers focus on activities pertaining to deliverables. The project schedule becomes very important because it shows when each resource devoted to a given activity will be required. Monitoring the progress of activities throughout the project is important to avoid potential overloading of resources. Problems arise when a specific resource, such as a construction crew or staff specialist, is required on several activities with overlapping schedules. Project managers have several options to alleviate resource problems, including the following:

- *Resource Leveling.* The attempt to reduce the peaks and valleys in resource needs by shifting the schedules of conflicting activities within their earliest and latest start dates. Software packages such as MS Project have algorithms that move activities to avoid violating resource constraints.

- *Resource Allocation.* The assignment of resources to the most important activities. Most popular project management software packages have a few priority rules that can be used to decide which activity a critical resource should be scheduled to perform when conflicts arise. For example, for all the activities requiring a given resource, assign the resource to the one with the earliest start time. An activity slack report identifies potential candidates for resource shifting—shift resources from high slack activities to those behind schedule.

- *Resource Acquisition.* The addition of more of an overloaded resource to maintain the schedule of an activity. Obviously, this tactic is constrained by the project budget.

Controlling Projects

Project managers have the responsibilities of accounting for the effective use of the firm's resources as well as managing the activities to achieve the time and quality goals of the project. The firm's assets include the physical assets, human resources, and financial resources. Physical assets are controlled by the timely maintenance of machines and equipment so that their failure does not delay the project. Inventories must be received, stored for future use, and replenished. Project managers are also responsible for human resource development. Projects provide a rich environment to develop future leaders; project managers can take advantage of the situation by assigning team members important activities to aid in their managerial development. Last, but not least, project managers must control the expenditures of the firm's financial resources. Most project management software packages contain accounting reports, budget reports, capital investment controls, and cash flow reports. Deviations from the project plan, often referred to as variances, must be periodically reported and analyzed for their causes.

Monitoring and controlling projects are ongoing activities throughout the execution phase of the project life cycle. The project **close out**, however, is an activity that many project managers forget to include in their consideration of resource usage. The purpose of this final phase in the project life cycle is to write final reports and complete remaining deliverables. An important aspect of this phase, however, is compiling the team's recommendations for improving the project process of which they were a part. Many team members will be assigned to other projects where they can apply what they learned.

close out

An activity that includes writing final reports, completing remaining deliverables, and compiling the team's recommendations for improving the project process.

LEARNING GOALS IN REVIEW

1 **Define the major activities associated with defining, organizing, planning, monitoring, and controlling projects.** The entire outline of Chapter 2 revolves around these five very important activities. Nonetheless, be sure to read the opener to the chapter, which shows the four major phases of the project to introduce the new XBOX 360 product, the introduction to the chapter, and the section "Defining and Organizing Projects," pp. 51–52.

2 **Diagram the network of interrelated activities in a project.** See "Defining the Work Breakdown Structure," pp. 53–54, and "Diagramming the Network," pp. 54–56. Figure 2.2 and Example 2.1 are important for achieving this learning goal.

3 **Identify the sequence of critical activities that determines the duration of a project.** Study the section "Developing the Schedule," pp. 57–60, and Example 2.2 for an understanding of the critical path.

4 **Explain how to determine a minimum-cost project schedule.** The section "Analyzing Cost-Time Tradeoffs," pp. 60–64, and Example 2.3 demonstrate how the relevant costs must be considered to minimize costs. Figure 2.6 explains a key assumption in the analysis. Solved Problem 1 contains a detailed solution.

5 **Describe the considerations managers make in assessing the risks in a project and calculate the probability of completing a project on time.** See the section "Assessing Risks," pp. 64–66, which explains the risks faced by project managers. The section "Analysis," pp. 66–69, shows how to compute probabilities. Be sure to understand Examples 2.4 and 2.5 and Solved Problem 2.

6 **Define the options available to alleviate resource problems.** See the section "Monitoring and Controlling Projects," pp. 69–70.

MyOMLab helps you develop analytical skills and assesses your progress with multiple problems on identifying the critical path, calculating an activity's slack, expected time, variance, the project's expected completion time, probability of completing it by a certain date, and minimum-cost schedule.

MyOMLab Resources	Titles	Link to the Book
Video	Project Management at the Phoenician Nantucket Nectars: ERP	Entire chapter. Defining and Organizing Projects
Active Model Exercise	2.1 Gantt Chart 2.2 Cost Analysis 2.3 Probability Analysis	Developing the Schedule ; Active Model Example (p. 85) Analyzing Cost-Time Trade-Offs ; Example 2.3 (pp. 62–64) Assessing Risks ; Exercise 2.5 (p. 69)
OM Explorer Solvers	Single Time Estimates Three Time Estimates Project Budgeting	Developing the Schedule; Example 2.2 (pp. 58–59) Assessing Risks; Example 2.4 (p. 67); Solved Problem 2 (pp. 74–77) Monitoring and Controlling Projects
POM for Windows	Single Time Estimates Triple Time Estimates Crashing Cost Budgeting Mean/Standard Deviation Given	Developing the Schedule; Example 2.2 (pp. 58–59) Assessing Risks; Example 2.4 (p. 67); Solved Problem 2 (p. 74–77) Analyzing Cost-Time Trade-Offs; Example 2.3 (p. 62); Solved Problem 1 (pp. 72–74) Monitoring and Controlling Projects Assessing Risks
SimQuick Simulation Exercises	Software development company	Assessing Risks
Microsoft Project	*Free Trial*	Planning Projects; Figure 2.5 (p. 60)
SmartDraw	*Free Trial*	Diagramming the Network
Virtual Tours	Reiger Orgelbau Pipe Organ Factory and Alaskan Way Viaduct	Planning Projects Assessing Risks; Monitoring and Controlling Projects
MyOMLab Supplements	E. Simulation H. Measuring Output Rates I. Learning Curve Analysis	Assessing Risks Diagramming the Network Diagramming the Network
Internet Exercises	Olympic Movement, London 2012, and Ch2M Hill	Planning Projects Defining and Organizing Projects
Key Equations		
Image Library		

Key Equations

1. Start and finish times:

    t = estimated time duration of the activity

    ES = latest of the EF times of all activities immediately preceding activity

    EF = ES + t

    LF = earliest of the LS times of all activities immediately following activity

    LS = LF − t

2. Activity slack:

    S = LS − ES or S = LF − EF

3. Project costs:

    $$\text{Crash cost per period} = \frac{\text{Crash cost} - \text{Normal cost}}{\text{Normal time} - \text{Crash time}}$$

    $$= \frac{\text{CC} - \text{NC}}{\text{NT} - \text{CT}}$$

4. Activity time statistics:

    t_e = mean of an activity's beta distribution

    $$t_e = \frac{a + 4m + b}{6}$$

    σ^2 = variance of the activity time

    $$\sigma^2 = \left(\frac{b - a}{6}\right)^2$$

5. z-transformation formula:

    $$z = \frac{T - T_E}{\sigma_P}$$

    where

    T = due date for the project

    $T_E = \Sigma$ (expected activity times on the critical path)

    = mean of normal distribution of critical path time

    σ_P = standard deviation of critical path time distribution

Key Terms

activity 53
activity-on-node (AON) network 55
activity slack 59
close out 70
crash cost (CC) 61
crash time (CT) 61
critical path 57
critical path method (CPM) 54
earliest finish time (EF) 57
earliest start time (ES) 57

Gantt chart 60
latest finish time (LF) 57
latest start time (LS) 57
minimum-cost schedule 62
most likely time (*m*) 66
network diagram 54
normal cost (NC) 61
normal time (NT) 61
optimistic time (*a*) 66
path 57

pessimistic time (*b*) 66
precedence relationship 54
program 51
program evaluation and review
    technique (PERT) 54
project 50
project management 51
risk-management plan 64
work breakdown structure
    (WBS) 53

Solved Problem 1

Your company has just received an order from a good customer for a specially designed electric motor. The contract states that, starting on the thirteenth day from now, your firm will experience a penalty of $100 per day until the job is completed. Indirect project costs amount to $200 per day. The data on direct costs and activity precedence relationships are given in Table 2.2.

TABLE 2.2 | ELECTRIC MOTOR PROJECT DATA

Activity	Normal Time (days)	Normal Cost ($)	Crash Time (days)	Crash Cost ($)	Immediate Predecessor(s)
A	4	1,000	3	1,300	None
B	7	1,400	4	2,000	None
C	5	2,000	4	2,700	None
D	6	1,200	5	1,400	A
E	3	900	2	1,100	B
F	11	2,500	6	3,750	C
G	4	800	3	1,450	D, E
H	3	300	1	500	F, G

a. Draw the project network diagram.

b. What completion date would you recommend?

SOLUTION

a. The network diagram, including normal activity times, for this procedure is shown in Figure 2.10. Keep the following points in mind while constructing a network diagram.

1. Always have start and finish nodes.
2. Try to avoid crossing paths to keep the diagram simple.
3. Use only one arrow to directly connect any two nodes.
4. Put the activities with no predecessors at the left and point the arrows from left to right.
5. Be prepared to revise the diagram several times before you come up with a correct and uncluttered diagram.

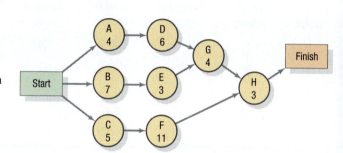

▲ **FIGURE 2.10**
Network Diagram for the
Electric Motor Project

b. With these activity durations, the project will be completed in 19 days and incur a $700 penalty. Determining a good completion date requires the use of the minimum-cost schedule procedure. Using the data provided in Table 2.2, you can determine the maximum crash-time reduction and crash cost per day for each activity. For example, for activity A

$$\text{Maximum crash time} = \text{Normal time} - \text{Crash time} = 4 \text{ days} - 3 \text{ days} = 1 \text{ day}$$

$$\text{Crash cost per day} = \frac{\text{Crash cost} - \text{Normal cost}}{\text{Normal time} - \text{Crash time}} = \frac{CC - NC}{NT - CT} = \frac{\$1,300 - \$1,000}{4 \text{ days} - 3 \text{ days}} = \$300$$

Activity	Crash Cost per Day ($)	Maximum Time Reduction (days)
A	300	1
B	200	3
C	700	1
D	200	1
E	200	1
F	250	5
G	650	1
H	100	2

Table 2.3 summarizes the analysis and the resultant project duration and total cost. The critical path is C–F–H at 19 days, which is the longest path in the network. The cheapest of these acvtivities to crash is H, which costs only an extra $100 per day to crash. Doing so saves $200 + $100 = $300 per day in indirect and penalty costs. If you crash this activity for two days (the maximum), the lengths of the paths are now

A–D–G–H: 15 days, B–E–G–H: 15 days, and C–F–H: 17 days

The critical path is still C–F–H. The next cheapest critical activity to crash is F at $250 per day. You can crash F only two days because at that point you will have three critical paths. Further reductions in project duration will require simultaneous crashing of more than one activity (D, E, and F). The cost to do so, $650, exceeds the savings, $300. Consequently, you should stop. Note that every activity is critical. The project costs are minimized when the completion date is day 15. However, some goodwill costs may be associated with disappointing a customer who wants delivery in 12 days.

TABLE 2.3 | **PROJECT COST ANALYSIS**

Stage	Crash Activity	Time Reduction (days)	Resulting Critical Path(s)	Project Duration (days)	Project Direct Costs, Last Trial ($)	Crash Cost Added ($)	Total Indirect Costs ($)	Total Penalty Costs ($)	Total Project Costs ($)
0	—	—	C–F–H	19	10,100	—	3,800	700	14,600
1	H	2	C–F–H	17	10,100	200	3,400	500	14,200
2	F	2	A–D–G–H B–E–G–H C–F–H	15	10,300	500	3,000	300	14,100

Solved Problem 2

An advertising project manager developed the network diagram shown in Figure 2.11 for a new advertising campaign. In addition, the manager gathered the time information for each activity, as shown in the accompanying table.

	TIME ESTIMATES (WEEKS)			
Activity	Optimistic	Most Likely	Pessimistic	Immediate Predecessor(s)
A	1	4	7	—
B	2	6	7	—
C	3	3	6	B
D	6	13	14	A
E	3	6	12	A, C
F	6	8	16	B
G	1	5	6	E, F

▼ **FIGURE 2.11**
Network Diagram for the Advertising Project

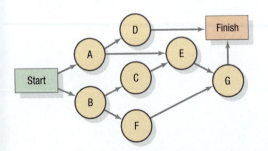

a. Calculate the expected time and variance for each activity.

b. Calculate the activity slacks and determine the critical path, using the expected activity times.

c. What is the probability of completing the project within 23 weeks?

SOLUTION

a. The expected time and variance for each activity are calculated as follows:

$$t_e = \frac{a + 4m + b}{6}$$

Activity	Expected Time (weeks)	Variance (σ^2)
A	4.0	1.00
B	5.5	0.69
C	3.5	0.25
D	12.0	1.78
E	6.5	2.25
F	9.0	2.78
G	4.5	0.69

b. We need to calculate the earliest start, latest start, earliest finish, and latest finish times for each activity. Starting with activities A and B, we proceed from the beginning of the network and move to the end, calculating the earliest start and finish times:

Activity	Earliest Start (weeks)	Earliest Finish (weeks)
A	0	0 + 4.0 = 4.0
B	0	0 + 5.5 = 5.5
C	5.5	5.5 + 3.5 = 9.0
D	4.0	4.0 + 12.0 = 16.0
E	9.0	9.0 + 6.5 = 15.5
F	5.5	5.5 + 9.0 = 14.5
G	15.5	15.5 + 4.5 = 20.0

Based on expected times, the earliest finish for the project is week 20, when activity G has been completed. Using that as a target date, we can work backward through the network, calculating the latest start and finish times (shown graphically in Figure 2.12):

Activity	Latest Start (weeks)	Latest Finish (weeks)
G	15.5	20.0
F	6.5	15.5
E	9.0	15.5
D	8.0	20.0
C	5.5	9.0
B	0.0	5.5
A	4.0	8.0

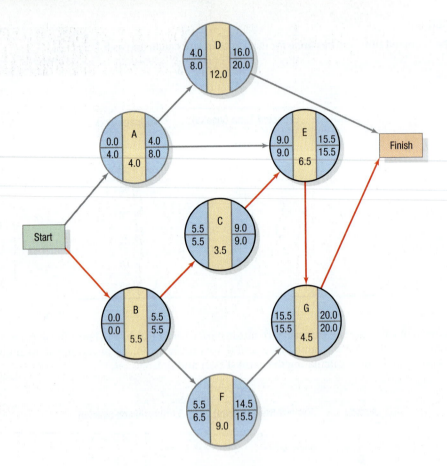

We now calculate the activity slacks and determine which activities are on the critical path:

Activity	START (WEEKS)		FINISH (WEEKS)		Slack	Critical Activity
	Earliest	Latest	Earliest	Latest		
A	0.0	4.0	4.0	8.0	4.0	No
B	0.0	0.0	5.5	5.5	0.0	Yes
C	5.5	5.5	9.0	9.0	0.0	Yes
D	4.0	8.0	16.0	20.0	4.0	No
E	9.0	9.0	15.5	15.5	0.0	Yes
F	5.5	6.5	14.5	15.5	1.0	No
G	15.5	15.5	20.0	20.0	0.0	Yes

The paths, and their total expected times and variances, are

Path	Total Expected Time (weeks)	Total Variance (σ_P^2)
A–D	4 + 12 = 16	1.00 + 1.78 = 2.78
A–E–G	4 + 6.5 + 4.5 = 15	1.00 + 2.25 + 0.69 = 3.94
B–C–E–G	5.5 + 3.5 + 6.5 + 4.5 = 20	0.69 + 0.25 + 2.25 + 0.69 = 3.88
B–F–G	5.5 + 9 + 4.5 = 19	0.69 + 2.78 + 0.69 = 4.16

The critical path is B–C–E–G, with a total expected time of 20 weeks. However, path B–F–G is 19 weeks and has a large variance.

c. We first calculate the z-value:

$$z = \frac{T - T_E}{\sigma_P} = \frac{23 - 20}{\sqrt{3.88}} = 1.52$$

Using the Normal Distribution appendix, we find that the probability of completing the project in 23 weeks or fewer is 0.9357. Because the length of path B–F–G is close to that of the critical path and has a large variance, it might well become the critical path during the project.

Discussion Questions

1. One of your colleagues comments that software is the ultimate key to project management success. How would you respond?

2. Explain how to determine the slack for each activity in a project. Why is it important for managers to know where the slack is in their projects?

3. Define risk as it applies to projects. What are the major sources of risk in a project?

Problems

The OM Explorer and POM for Windows software is available to all students using the 10th edition of this textbook. Go to **www.pearsonhighered.com/krajewski** to download these computer packages. If you purchased MyOMLab, you also have access to Active Models software and significant help in doing the following problems. Check with your instructor on how best to use these resources. In many cases, the instructor wants you to understand how to do the calculations by hand. At the least, the software provides a check on your calculations. When calculations are particularly complex and the goal is interpreting the results in making decisions, the software replaces entirely the manual calculations.

1. Consider the following data for a project:

Activity	Activity Time (days)	Immediate Predecessor(s)
A	2	—
B	4	A
C	5	A
D	2	B
E	1	B
F	8	B, C
G	3	D, E
H	5	F
I	4	F
J	7	G, H, I

a. Draw the network diagram.

b. Calculate the critical path for this project.

c. How much slack is in each of the activities G, H, and I?

2. The following information is known about a project.

Activity	Activity Time (days)	Immediate Predecessor(s)
A	7	—
B	2	A
C	4	A
D	4	B, C
E	4	D
F	3	E
G	5	E

a. Draw the network diagram for this project.

b. Determine the critical path and project duration.

c. Calculate the slack for each activity.

3. A project for improving a billing process has the following precedence relationships and activity times:

Activity	Activity Time (weeks)	Immediate Predecessor(s)
A	3	—
B	11	—
C	7	A
D	13	B, C
E	10	B
F	6	D
G	5	E
H	8	F, G

a. Draw the network diagram.

b. Calculate the slack for each activity. Which activities are on the critical path?

4. The following information is available about a project:

Activity	Activity Time (days)	Immediate Predecessor(s)
A	3	—
B	4	—
C	5	—
D	4	—
E	7	A
F	2	B, C, D
G	4	E, F
H	6	F
I	4	G
J	3	G
K	3	H

a. Draw the network diagram.

b. Find the critical path.

5. The following information has been gathered for a project:

Activity	Activity Time (weeks)	Immediate Predecessor(s)
A	4	—
B	7	A
C	9	B
D	3	B
E	14	D
F	10	C, D
G	11	F, E

a. Draw the network diagram.

b. Calculate the slack for each activity and determine the critical path. How long will the project take?

6. Consider the following information for a project to add a drive-thru window at Crestview Bank.

Activity	Activity Time (weeks)	Immediate Predecessor(s)
A	5	—
B	2	—
C	6	—
D	2	A, B
E	7	B
F	3	D, C
G	9	E, C
H	11	F, G

a. Draw the network diagram for this project.

b. Specify the critical path.

c. Calculate the slack for activities A and D.

7. Barbara Gordon, the project manager for Web Ventures, Inc., compiled a table showing time estimates for each of the activities of a project to upgrade the company's Web page, including optimistic, most likely, and pessimistic.

a. Calculate the expected time, t_e, for each activity.

b. Calculate the variance, σ^2, for each activity.

Activity	Optimistic (days)	Most Likely (days)	Pessimistic (days)
A	3	8	19
B	12	15	18
C	2	6	16
D	4	9	20
E	1	4	7

8. Recently, you were assigned to manage a project for your company. You have constructed a network diagram depicting the various activities in the project (Figure 2.13). In addition, you have asked your team to estimate the amount of time that they would expect each of the activities to take. Their responses are shown in the following table:

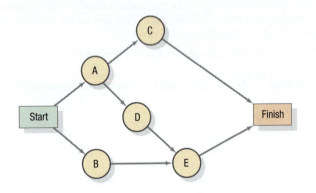

▲ FIGURE 2.13
Network Diagram for Your Company

	TIME ESTIMATES (DAYS)		
Activity	Optimistic	Most Likely	Pessimistic
A	5	8	11
B	4	8	11
C	5	6	7
D	2	4	6
E	4	7	10

a. What is the expected completion time of the project?

b. What is the probability of completing the project in 21 days?

c. What is the probability of completing the project in 17 days?

9. In Solved Problem 2, estimate the probability that the noncritical path B–F–G will take more than 20 weeks. *Hint:* Subtract from 1.0 the probability that B–F–G will take 20 weeks or less.

10. Consider the following data for a project never before attempted by your company:

Activity	Expected Time t_e (weeks)	Immediate Predecessor(s)
A	5	—
B	3	—
C	2	A
D	5	B
E	4	C, D
F	7	D

a. Draw the network diagram for this project.

b. Identify the critical path and estimate the project's duration.

c. Calculate the slack for each activity.

11. The director of continuing education at Bluebird University just approved the planning for a sales training seminar. Her administrative assistant identified the various activities that must be done and their relationships to each other, as shown in Table 2.4.

TABLE 2.4 | ACTIVITIES FOR THE SALES TRAINING SEMINAR

Activity	Description	Immediate Predecessor(s)
A	Design brochure and course announcement	—
B	Identify prospective teachers	—
C	Prepare detailed outline of course	—
D	Send brochure and student applications	A
E	Send teacher applications	B
F	Select teacher for course	C, E
G	Accept students	D
H	Select text for course	F
I	Order and receive texts	G, H
J	Prepare room for class	G

Because of the uncertainty in planning the new course, the assistant also has supplied the following time estimates for each activity:

	TIME ESTIMATES (DAYS)		
Activity	Optimistic	Most Likely	Pessimistic
A	5	7	8
B	6	8	12
C	3	4	5
D	11	17	25
E	8	10	12
F	3	4	5
G	4	8	9
H	5	7	9
I	8	11	17
J	4	4	4

The director wants to conduct the seminar 47 working days from now. What is the probability that everything will be ready in time?

12. Table 2.5 contains information about an environmental clean-up project. Shorten the project three weeks by finding the minimum-cost schedule. Assume that project indirect costs and penalty costs are negligible. Identify activities to crash while minimizing the additional crash costs.

TABLE 2.5 | ENVIRONMENTAL PROJECT DATA

Activity	Normal Time (weeks)	Crash Time (weeks)	Cost to Crash ($ per week)	Immediate Predecessor(s)
A	7	6	200	None
B	12	9	250	None
C	7	6	250	A
D	6	5	300	A
E	1	1	—	B
F	1	1	—	C, D
G	3	1	200	D, E
H	3	2	350	F
I	2	2	—	G

13. The Advanced Tech Company has a project to design an integrated information database for a major bank. Data for the project are given in Table 2.6. Indirect project costs amount to $300 per day. The company will incur a $150 per day penalty for each day the project lasts beyond day 14.

a. What is the project's duration if only normal times are used?

b. What is the minimum-cost schedule?

c. What is the critical path for the minimum-cost schedule?

TABLE 2.6 | DATABASE DESIGN PROJECT DATA

Activity	Normal Time (days)	Normal Cost ($)	Crash Time (days)	Crash Cost ($)	Immediate Predecessor(s)
A	6	1,000	5	1,200	—
B	4	800	2	2,000	—
C	3	600	2	900	A, B
D	2	1,500	1	2,000	B
E	6	900	4	1,200	C, D
F	2	1,300	1	1,400	E
G	4	900	4	900	E
H	4	500	2	900	G

14. You are the manager of a project to improve a billing process at your firm. Table 2.7 contains the data you will need to conduct a cost analysis of the project. Indirect costs are $1,600 per week, and penalty costs are $1,200 per week after week 12.

    a. What is the minimum-cost schedule for this project?

    b. What is the difference in total project costs between the earliest completion time of the project using "normal" times and the minimum-cost schedule you derived in part (a)?

15. Table 2.8 contains data for the installation of new equipment in a manufacturing process at Excello Corporation. Your company is responsible for the installation project. Indirect costs are $15,000 per week, and a penalty cost of $9,000 per week will be incurred by your company for every week the project is delayed beyond week 9.

    a. What is the shortest time duration for this project regardless of cost?

    b. What is the minimum total cost associated with completing the project in 9 weeks?

    c. What is the total time of the minimum-cost schedule?

TABLE 2.7 | DATA FOR THE BILLING PROCESS PROJECT

Activity	Immediate Predecessor(s)	Normal Time (weeks)	Crash Time (weeks)	Normal Cost ($)	Crash Cost ($)
A	—	4	1	5,000	8,000
B	—	5	3	8,000	10,000
C	A	1	1	4,000	4,000
D	B	6	3	6,000	12,000
E	B, C	7	6	4,000	7,000
F	D	7	6	4,000	7,000

TABLE 2.8 | DATA FOR THE EQUIPMENT INSTALLATION PROJECT

Activity	Immediate Predecessor(s)	Normal Time (weeks)	Crash Time (weeks)	Normal Cost ($)	Crash Cost ($)
A	—	2	1	7,000	10,000
B	—	2	2	3,000	3,000
C	A	3	1	12,000	40,000
D	B	3	2	12,000	28,000
E	C	1	1	8,000	8,000
F	D, E	5	3	5,000	15,000
G	E	3	2	9,000	18,000

16. Gabrielle Kramer, owner of Pet Paradise, is opening a new store in Columbus, Ohio. Her major concern is the hiring of a manager and several associates who are animal lovers. She also has to coordinate the renovation of a building that was previously owned by a chic clothing store. Kramer has gathered the data shown in Table 2.9.

a. How long is the project expected to take?

b. Suppose that Kramer has a personal goal of completing the project in 14 weeks. What is the probability that it will happen this quickly?

17. The diagram in Figure 2.14 was developed for a project that you are managing. Suppose that you are interested in finding ways to speed up the project at minimal additional cost. Determine the schedule for completing the project in 25 days at minimum cost. Penalty and project-overhead costs are negligible. Time and cost data for each activity are shown in Table 2.10.

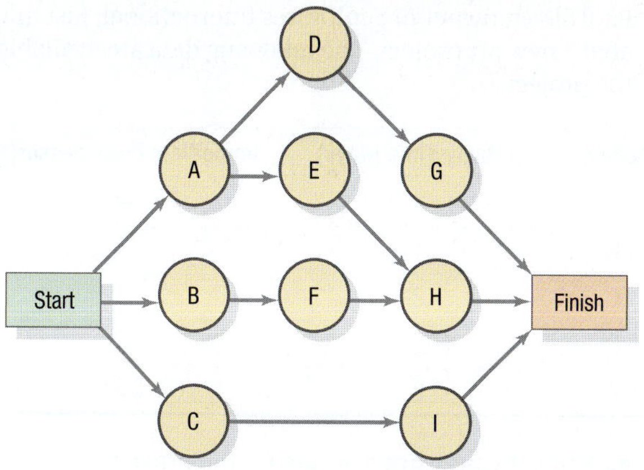

▲ **FIGURE 2.14**
Network Diagram for Problem 17

TABLE 2.9 | **DATA FOR THE PET PARADISE PROJECT**

			TIME (WEEKS)		
Activity	**Description**	**Immediate Predecessor(s)**	***a***	***m***	***b***
A	Interview for new manager	—	1	3	6
B	Renovate building	—	6	9	12
C	Place ad for associates and interview applicants	—	6	8	16
D	Have new manager prospects visit	A	2	3	4
E	Purchase equipment for new store and install	B	1	3	11
F	Check employee applicant references and make final selection	C	5	5	5
G	Check references for new manager and make final selection	D	1	1	1
H	Hold orientation meetings and do payroll paperwork	E, F, G	3	3	3

TABLE 2.10 | **PROJECT ACTIVITY AND COST DATA**

	NORMAL		CRASH	
Activity	**Time (days)**	**Cost ($)**	**Time (days)**	**Cost ($)**
A	12	1,300	11	1,900
B	13	1,050	9	1,500
C	18	3,000	16	4,500
D	9	2,000	5	3,000
E	12	650	10	1,100
F	8	700	7	1,050
G	8	1,550	6	1,950
H	2	600	1	800
I	4	2,200	2	4,000

18. Paul Silver, owner of Sculptures International, just initiated a new art project. The following data are available for the project:

Activity	Activity Time (days)	Immediate Predecessor(s)
A	4	—
B	1	—
C	3	A
D	2	B
E	3	C, D

a. Draw the network diagram for the project.

b. Determine the project's critical path and duration.

c. What is the slack for each activity?

19. Reliable Garage is completing production of the J2000 kit car. The following data are available for the project:

Activity	Activity Time (days)	Immediate Predecessor(s)
A	2	—
B	6	A
C	4	B
D	5	C
E	7	C
F	5	C
G	5	F
H	3	D, E, G

a. Draw the network diagram for the project.

b. Determine the project's critical path and duration.

c. What is the slack for each activity?

20. The following information concerns a new project your company is undertaking:

Activity	Activity Time (days)	Immediate Predecessor(s)
A	10	—
B	11	—
C	9	A, B
D	5	A, B
E	8	A, B
F	13	C, E
G	5	C, D
H	10	G
I	6	F, G
J	9	E, H
K	11	I, J

a. Draw the network diagram for this project.

b. Determine the critical path and project completion time.

Advanced Problems

21. The project manager of Good Public Relations gathered the data shown in Table 2.11 for a new advertising campaign.

a. How long is the project likely to take?

b. What is the probability that the project will take more than 38 weeks?

c. Consider the path A–E–G–H–J. What is the probability that this path will exceed 38 weeks?

TABLE 2.11 | **ACTIVITY DATA FOR ADVERTISING PROJECT**

Activity	TIME ESTIMATES (WEEKS)			Immediate Predecessor(s)
	Optimistic	Most Likely	Pessimistic	
A	8	10	12	START
B	5	8	17	START
C	7	8	9	START
D	1	2	3	B
E	8	10	12	A, C
F	5	6	7	D, E
G	1	3	5	D, E
H	2	5	8	F, G
I	2	4	6	G
J	4	5	8	H
K	2	2	2	H

22. Consider the office renovation project data in Table 2.12. A "zero" time estimate means that the activity could take a very small amount of time and should be treated as a numeric zero in the analysis.

a. Based on the critical path, find the probability of completing the office renovation project by 39 days.

b. Find the date by which you would be 90 percent sure of completing the project.

TABLE 2.12 | DATA FOR THE OFFICE RENOVATION PROJECT

Activity	Optimistic	Most Likely	Pessimistic	Immediate Predecessor(s)
		TIME ESTIMATES (DAYS)		
START	0	0	0	—
A	6	10	14	START
B	0	1	2	A
C	16	20	30	A
D	3	5	7	B
E	2	3	4	D
F	7	10	13	C
G	1	2	3	D
H	0	2	4	G
I	2	2	2	C, G
J	2	3	4	I
K	0	1	2	H
L	1	2	3	J, K
FINISH	0	0	0	E, F, L

23. You are in charge of a project at the local community center. The center needs to remodel one of the rooms in time for the start of a new program. Delays in the project mean that the center must rent other space at a nearby church at additional cost. Time and cost data for your project are contained in Table 2.13. Your interest is in minimizing the cost of the project to the community center.

a. Using the *normal times* for each activity, what is the earliest date you can complete the project?

b. Suppose the variable overhead costs are $50 per day for your project. Also, suppose that the center must pay $40 per day for a temporary room on day 15 or beyond. Find the minimum-cost project schedule.

TABLE 2.13 | DATA FOR THE COMMUNITY CENTER PROJECT

Activity	Normal Time (days)	Normal Cost ($)	Crash Time (days)	Crash Cost ($)	Immediate Predecessor(s)
START	0	0	0	0	—
A	10	50	8	150	START
B	4	40	2	200	START
C	7	70	6	160	B
D	2	20	1	50	A, C
E	3	30	3	30	A, C
F	8	80	5	290	B
G	5	50	4	180	D
H	6	60	3	180	E, F
FINISH	0	0	0	0	G, H

24. The information in Table 2.14 is available for a large fund-raising project.

a. Determine the critical path and the expected completion time of the project.

b. Plot the total project cost, starting from day 1 to the expected completion date of the project, assuming the earliest start times for each activity. Compare that result to a similar plot for the latest start times. What implication does the time differential have for cash flows and project scheduling?

25. You are the project manager of the software installation project in Table 2.15. You would like to find the minimum-cost schedule for your project. There is a $1,000-per-week penalty for each week the project is delayed beyond week 25. In addition, your project team determined that indirect project costs are $2,500 per week.

a. What would be your target completion week?

b. How much would you save in total project costs with your schedule?

TABLE 2.14 | FUND-RAISING PROJECT DATA

Activity	Activity Time (days)	Activity Cost ($)	Immediate Predecessor(s)
A	3	100	—
B	4	150	—
C	2	125	A
D	5	175	B
E	3	150	B
F	4	200	C, D
G	6	75	C
H	2	50	C, D, E
I	1	100	E
J	4	75	D, E
K	3	150	F, G
L	3	150	G, H, I
M	2	100	I, J
N	4	175	K, M
O	1	200	H, M
P	5	150	N, L, O

TABLE 2.15 | DATA FOR SOFTWARE INSTALLATION PROJECT

Activity	Immediate Predecessors	Normal Time (weeks)	Normal Cost ($)	Crash Time (weeks)	Crash Cost ($)
A	—	5	2,000	3	4,000
B	—	8	5,000	7	8,000
C	A	10	10,000	8	12,000
D	A, B	4	3,000	3	7,000
E	B	3	4,000	2	5,000
F	D	9	8,000	6	14,000
G	E, F	2	2,000	2	2,000
H	G	8	6,000	5	9,000
I	C, F	9	7,000	7	15,000

26. Consider the project described in Table 2.16.

   a. If you start the project immediately, when will it be finished?

   b. You are interested in completing your project as soon as possible. You have only one option. Suppose you could assign Employee A, currently assigned to activity G, to help Employee B, currently assigned to activity F. Each week that Employee A helps Employee B will result in activity G increasing its time by one week and activity F reducing its time by one week. How many weeks should Employee A work on activity F?

TABLE 2.16 | PROJECT DATA FOR PROBLEM 26

Activity	Activity Time (weeks)	Immediate Predecessor(s)
START	0	—
A	3	START
B	4	START
C	4	B
D	4	A
E	5	A, B
F	6	D, E
G	2	C, E
FINISH	0	F, G

Active Model Exercise

This Active Model appears in MyOMLab. It allows you to evaluate the sensitivity of the project time to changes in activity times and activity predecessors.

QUESTIONS

1. Activity B and activity K are critical activities. Describe the difference that occurs on the graph when you increase activity B versus when you increase activity K.

2. Activity F is not critical. Use the scroll bar to determine how many weeks you can increase activity F until it becomes critical.

3. Activity A is not critical. How many weeks can you increase activity A until it becomes critical? What happens when activity A becomes critical?

4. What happens when you increase activity A by one week after it becomes critical?

5. Suppose that building codes may change and, as a result, activity C would have to be completed before activity D could be started. How would this affect the project?

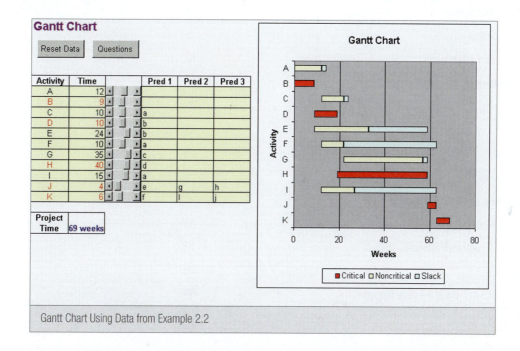

Gantt Chart Using Data from Example 2.2

VIDEO CASE   Project Management at the Phoenician

The Phoenician in Phoenix, Arizona, is part of Starwood's Luxury Collection and its only AAA Five Diamond Award resort in the southwestern United States. Sophistication, elegance, and excellence only begin to describe the guest experience at the hotel. Guests can dine in one of nine restaurants, relax poolside, play tennis, take in 27 holes of golf on three 9-hole courses, or relax with a variety of soothing spa treatments at the 22,000-square-foot Centre for Well-Being.

The Phoenician recently embarked on an ambitious $38 million spa and golf renovation program. The resort's golf and spa programs historically earned high marks from surveys in their industries over the years, but the environment was changing. Evidence of this change was seen in the explosive growth of new golf courses and spas in the Southwest region. Phoenix alone has over 275 golf courses, and the Southwest boasts the largest concentration of new luxury spas anywhere. The Phoenician's facilities, while world-class and highly rated, were more than 15 years old. The hotel's recently awarded Five Diamond status renewed emphasis on bringing every process and service at the property up to Five Diamond level.

The decision to renovate the golf course and existing spa became not a question of *whether* to undertake the projects, but *to what degree* they needed to be pursued. Key considerations centered on (1) whether to build basic facilities or commit to the grandiose luxury level, (2) having a domestic versus international reputation, and (3) developing creative packaging of the new facilities to attract loyal guests, such as a spa and golf "country club-like" membership program. Such a program would be limited to about 600 spa/golf memberships, with a one-time fee of $65,000 each.

The company's senior management considered three options for the Centre for Well-Being spa. First, the existing space in the heart of the resort could be renovated. This option would require relocating the spa to another part of the resort and offering limited treatments during this time, thereby reducing spa revenues significantly. With option 2, hilly terrain directly behind the resort could be carved out to create a new mountainside facility with sweeping vistas. This option meant the closure of one of the hotel's buildings housing 60 guest rooms and suites during the construction period. The existing spa could remain open, however. Under option 3, a parking structure on existing hotel property could be used, having the least impact on revenues. The first option was seen as a short-term fix, while the remaining two were viewed as having longer-term potential.

Additional discussion centered on the type of spa to be built. Recent acquisition of the Bliss spa brand for Starwood's W Hotels was an option, offering day spa amenities and an indulgence atmosphere. The second option was to remain a holistic resort spa with an emphasis on health and restoration. The third option was to become a destination spa with dedicated guest stays and week-long programs. Day spas are the fastest-growing category, with few destination spas.

The Phoenician management team, with assistance from Starwood Field Operations and Corporate offices, prepared an extensive analysis of strengths, weaknesses, opportunities, and threats to better understand the environment. The result of this analysis was used by the team to identify the set of activities necessary for each option. The Corporate Design and Construction group developed architectural and engineering plans, as

Work Breakdown Structure	Activity Time (days)	Activity Precedence Relationships
Project Conception		
A. Kick-off meeting	2	
B. Creation of spa specifications	30	A
Geotechnical Investigation		
C. Preliminary site characterizations	10	B
D. Subsurface investigation	10	C
E. Laboratory testing	5	D
F. Geologic hazard assessments	10	E
Design Development		
G. Initial designs	70	B
H. Preliminary zoning compliance plan	15	C, G
I. Final designs	18	H
J. Owner approval of designs	5	I
Documentation and Cost Estimation		
K. Construction documentation and landscape package	80	F, I
L. Acquisition of contractor estimates and bids	90	J, K
Decision		
M. Owner approval of one of the three projects	60	L

well as the work breakdown structure and diagrams showing the critical path for the possible project options. The work breakdown structure, activity times, and activity precedence relationships are shown in the table on the previous page.

QUESTIONS

1. Coordinating departments in a major project is always a challenge. Which departments within the Starwood organization likely played a role in each of the following project related activities?
   a. Defining and organizing the project
   b. Planning the project
   c. Monitoring and controlling the project
2. Many times, project decision makers do not rely solely on financial hurdles, such as return on investment or internal rates of return, but place a lot of emphasis on intangible factors. Which are the salient intangible factors associated with selecting one of the three options for the spa?
3. Timing is always a challenge in managing projects. Construct a network diagram for the spa selection process. How soon can The Phoenician management make a decision on the spa?

When the Phoenician, a luxury hotel in Phoenix, Arizona, sought to re-design its Center for Well-Being, its management team created a work breakdown structure in order to compare different project options and choose the best one.

CASE | The Pert Mustang

Roberts Auto Sales and Service (RASAS) consists of three car dealerships that sell and service several makes of American and Japanese cars, two auto parts stores, a large body shop and car painting business, and an auto salvage yard. Vicky Roberts, owner of RASAS, went into the car business when she inherited a Ford dealership from her father. She was able to capitalize on her knowledge and experience to build her business into the diversified and successful mini-empire it is today. Her motto, "Sell 'em today, repair 'em tomorrow!" reflects a strategy that she refers to in private as "Get 'em coming and going."

Roberts has always retained a soft spot in her heart for high-performance Mustangs and just acquired a 1965 Shelby Mustang GT 350 that needs a lot of restoration. She also notes the public's growing interest in the restoration of vintage automobiles. Roberts is thinking of expanding into the vintage car restoration business and needs help in assessing the feasibility of such a move. She wants to restore her 1965 Shelby Mustang to mint condition, or as close to mint condition as possible. If she decides to go into the car restoring business, she can use the Mustang as an exhibit in sales and advertising and take it to auto shows to attract business for the new shop.

Roberts believes that many people want the thrill of restoring an old car themselves, but they do not have the time to run down all the old parts. Still, others just want to own a vintage auto because it is different and many of them have plenty of money to pay someone to restore an auto for them.

Roberts wants the new business to appeal to both types of people. For the first group, she envisions serving as a parts broker for NOS ("new old stock"), new parts that were manufactured many years ago and are still packaged in their original cartons. It can be a time-consuming process to find the right part. RASAS could also machine new parts to replicate those that are hard to find or that no longer exist.

In addition, RASAS could assemble a library of parts and body manuals for old cars to serve as an information resource for do-it-yourself restorers. The do-it-yourselfers could come to RASAS for help in compiling parts lists, and RASAS could acquire the parts for them. For others, RASAS would take charge of the entire restoration.

Roberts asked the director of service operations to take a good look at her Mustang and determine what needs to be done to restore it to the condition it was in when it came from the factory more than 40 years ago. She

wants to restore this car in time to exhibit it at the Detroit Auto Show. If the car gets a lot of press, it will be a real public relations coup for RASAS—especially if Roberts decides to enter this new venture. Even if she does not, the car will be a showpiece for the rest of the business.

Roberts asked the director of service operations to prepare a report about what is involved in restoring the car and whether it can be done in time for the Detroit show in 45 working days using PERT/CPM. The parts manager, the body shop manager, and the chief mechanic have provided the following estimates of times and activities that need to be done, as well as cost estimates:

a. Order all needed material and parts (upholstery, windshield, carburetor, and oil pump). Time: 2 days. Cost (telephone calls and labor): $100.

b. Receive upholstery material for seat covers. Cannot be done until order is placed. Time: 30 days. Cost: $2,100.

c. Receive windshield. Cannot be done until order is placed. Time: 10 days. Cost: $800.

d. Receive carburetor and oil pump. Cannot be done until order is placed. Time: 7 days. Cost: $1,750.

e. Remove chrome from body. Can be done immediately. Time: 1 day. Cost: $200.

f. Remove body (doors, hood, trunk, and fenders) from frame. Cannot be done until chrome is removed. Time: 1 day. Cost: $300.

g. Have fenders repaired by body shop. Cannot be done until body is removed from frame. Time: 4 days. Cost: $1,000.

h. Repair doors, trunk, and hood. Cannot be done until body is removed from frame. Time: 6 days. Cost: $1,500.

i. Pull engine from chassis. Do after body is removed from frame. Time: 1 day. Cost: $200.

j. Remove rust from frame. Do after the engine has been pulled from the chassis. Time: 3 days. Cost $900.

k. Regrind engine valves. Do after the engine has been pulled from the chassis. Time: 5 days. Cost: $1,000.

l. Replace carburetor and oil pump. Do after engine has been pulled from chassis and after carburetor and oil pump have been received. Time: 1 day. Cost: $200.

m. Rechrome the chrome parts. Chrome must have been removed from the body first. Time: 3 days. Cost: $210.

n. Reinstall engine. Do after valves are reground and carburetor and oil pump have been installed. Time: 1 day. Cost: $200.

o. Put doors, hood, and trunk back on frame. The doors, hood, and trunk must have been repaired first. The frame must have had its rust removed first. Time: 1 day. Cost: $240.

p. Rebuild transmission and replace brakes. Do so after the engine has been reinstalled and the doors, hood, and trunk are back on the frame. Time: 4 days. Cost: $2,000.

q. Replace windshield. Windshield must have been received. Time: 1 day. Cost: $100.

r. Put fenders back on. The fenders must have been repaired first, the transmission rebuilt, and the brakes replaced. Time: 1 day. Cost: $100.

s. Paint car. Cannot be done until the fenders are back on and windshield replaced. Time: 4 days. Cost: $1,700.

t. Reupholster interior of car. Must have received upholstery material first. Car must have been painted first. Time: 7 days. Cost: $2,400.

u. Put chrome parts back on. Car must have been painted and chrome parts rechromed first. Time: 1 day. Cost: $100.

v. Pull car to the Detroit Auto Show. Must have completed reupholstery of interior and have put the chrome parts back on. Time: 2 days. Cost: $1,000.

Roberts wants to limit expenditures on this project to what could be recovered by selling the restored car. She has already spent $50,000 to acquire the car. In addition, she wants a brief report on some of the aspects of the proposed business, such as how it fits in with RASAS's other businesses and what RASAS's operations task should be with regard to cost, quality, customer service, and flexibility.

In the restoration business there are various categories of restoration. A basic restoration gets the car looking great and running, but a mint condition restoration puts the car back in original condition—as it was "when it rolled off the line." When restored cars are resold, a car in mint condition commands a much higher price than one that is just a basic restoration. As cars are restored, they can also be customized. That is, something is put on the car that could not have been on the original. Roberts wants a mint condition restoration for her Mustang, without customization. (The proposed new business would accept any kind of restoration a customer wanted.)

The total budget cannot exceed $70,000 including the $50,000 Roberts has already spent. In addition, Roberts cannot spend more than $3,600 in any week given her present financial position. Even though much of the work will be done by Roberts's own employees, labor and materials costs must be considered. All relevant costs have been included in the cost estimates.

QUESTIONS

1. Using the information provided, prepare the report that Vicky Roberts requested, assuming that the project will begin immediately. Assume 45 working days are available to complete the project, including transporting the car to Detroit before the auto show begins. Your report should briefly discuss the aspects of the proposed new business, such as the competitive priorities that Roberts asked about.

2. Construct a table containing the project activities using the letter assigned to each activity, the time estimates, and the precedence relationships from which you will assemble the network diagram.

3. Draw a network diagram of the project similar to Figure 2.4. Determine the activities on the critical path and the estimated slack for each activity.

4. Prepare a project budget showing the cost of each activity and the total for the project. Can the project be completed within the budget? Will the project require more than $3,600 in any week? To answer this question, assume that activities B, C, and D must be paid for when the item is received (the earliest finish time for the activity). Assume that the costs of all other activities that span more than one week can be prorated. Each week contains five work days. If problems exist, how might Roberts overcome them?

Source: This case was prepared by and is used by permission of Dr. Sue P. Siferd, Professor Emerita, Arizona State University (Updated September, 2007).

Selected References

Goldratt, E. M. *Critical Chain.* Great Barrington, MA: North River, 1997.

Hartvigsen, David. *SimQuick: Process Simulation with Excel.* 2nd ed. Upper Saddle River, NJ: Prentice Hall, 2004.

Kerzner, Harold. *Advanced Project Management: Best Practices on Implementation,* 2nd ed. New York: John Wiley & Sons, 2004.

Kerzner, Harold. *Project Management: A Systems Approach to Planning, Scheduling, and Controlling,* 10th ed. New York: John Wiley & Sons, 2009.

Lewis, J. P. *Mastering Project Management,* 2nd ed. New York: McGraw-Hill, 2001.

Mantel Jr., Samuel J., Jack R. Meredith, Scott M. Shafer, and Margaret M. Sutton. *Project Management in Practice,* 3rd ed. New York: John Wiley & Sons, 2007.

Meredith, Jack R., and Samuel J. Mantel, *Project Management: A Managerial Approach,* 6th ed. New York: John Wiley & Sons, 2005.

Muir, Nancy C. *Microsoft Project 2007 for Dummies,* New York: John Wiley & Sons, 2006.

Nicholas, John M, and Herman Stein. *Project Management for Business, Engineering, and Technology,* 3rd ed. Burlington, MA: Butterworth-Heinemann, 2008.

"A Guide to Project Management Body of Knowledge," 2008. Available from the Project Management Institute at www.pmi.org.

Srinivasan, Mandyam, Darren Jones, and Alex Miller. "CORPS Capabilities." *APICS Magazine* (March 2005), pp. 46–50.

Don Ryan/AP Photos

3

PROCESS STRATEGY

At any given time eBay has approximately 113 million listings worldwide, and yet its workforce consists of just 15,000 employees. The explanation? Customers do most of the work in eBay's buying and selling processes. Here a customer prepares items for shipping from sales on his Ebay account.

eBay

Most manufacturers do not have to contend with customers waltzing around their shop floors, showing up intermittently and unannounced. Such customer contact can introduce considerable variability, disrupting carefully designed production processes. Costs and quality can be adversely affected. While customer contact is an issue even with manufacturers, (each process does have at least one customer), extensive customer contact and involvement are business as usual for many processes of service providers. Customers at restaurants or rental car agencies are directly involved in performing the processes. The area where the sales person interacts with the customer *is* the shop floor.

How much should customers be involved in a process, so as to provide timely delivery and consistent quality, and at sustainable cost? Various ways are available—some accommodate customer-introduced variability and some reduce it. eBay illustrates one way to accommodate variability. As an online auction house, eBay has high volume and request variability. Its customers do not want service at the same time or at times necessarily convenient to the company. They have request variability, seeking to buy and sell an endless number of items. They also have variability in customer capability, some with considerable Internet experience and some needing more handholding. Such variability would greatly complicate workforce scheduling if eBay required its employees to conduct all of its processes. It connects hundreds of millions of people around the world every day. It has a global presence in 39 markets, with revenue of $9.2 billion in more than 50,000 categories—and only with 15,500 employees. This relatively small workforce is possible in the face of customer-induced variability because its customers perform virtually all of the selling and buying processes through the eBay Web site. When the customer is responsible for much of the work, the right labor is provided at the right moment.

Source: Frances X. Frei, "Breaking the Trade-Off between Efficiency and Service," *Harvard Business Review* (November 2006), pp. 93–101; **http://en.wikipedia.org/wiki/Ebay** (March 19, 2011).

process strategy

The pattern of decisions made in managing processes so that they will achieve their competitive priorities.

LEARNING GOALS *After reading this chapter, you should be able to:*

1. Explain why processes exist everywhere in all organizations.
2. Discuss the four major process decisions.
3. Position a process on the customer-contact matrix or product-process matrix.
4. Configure operations into layouts.
5. Define customer involvement, resource flexibility, capital intensity, and economies of scope.
6. Discuss how process decisions should fit together.
7. Define process reengineering and process improvement.

Creating Value through Operations Management

Using Operations to Compete
Project Management

Managing Processes

Process Strategy
Process Analysis
Quality and Performance
Capacity Planning
Constraint Management
Lean Systems

Managing Supply Chains

Supply Chain
Inventory Management
Supply Chain Design
Supply Chain Location Decisions
Supply Chain Integration
Supply Chain Sustainability and Humanitarian Logistics
Forecasting
Operations Planning and Scheduling
Resource Planning

Process decisions, such as the amount of customer involvement allowed at eBay, are strategic in nature: As we saw in Chapter 1, they should further a company's long-term competitive goals. In making process decisions, managers focus on controlling such competitive priorities as quality, flexibility, time, and cost. Process management is an ongoing activity, with the same principles applying to both first-time and redesign choices.

In this chapter, we focus on **process strategy**, which specifies the pattern of decisions made in managing processes so that the processes will achieve their competitive priorities. Process strategy guides a variety of process decisions, and in turn is guided by operations strategy and the organization's ability to obtain the resources necessary to support them. We begin by defining four basic process decisions: (1) process structure (including layout), (2) customer involvement, (3) resource flexibility, and (4) capital intensity. We discuss these decisions for both service and manufacturing processes. We pay particular attention to ways in which these decisions fit together, depending on factors such as competitive priorities, customer contact, and volume. We conclude with two basic change strategies for analyzing and modifying processes: (1) process reengineering and (2) process improvement.

Three principles concerning process strategy are particularly important:

1. The key to successful process decisions is to make choices that fit the situation and that make sense together. They should not work at cross-purposes, with one process optimized at the expense of other processes. A more effective process is one that matches key process characteristics and has a close *strategic fit.*

2. Although this section of the text focuses on individual processes, they are the building blocks that eventually create the firm's whole supply chain. The cumulative effect on customer satisfaction and competitive advantage is huge.

3. Whether processes in the supply chain are performed internally or by outside suppliers and customers, management must pay particular attention to the interfaces between processes. Dealing with these interfaces underscores the need for cross-functional coordination.

Process Strategy across the Organization

As we explained in Chapter 1, processes are everywhere and are the basic unit of work. Consider the following two major points: (1) supply chains have processes and (2) processes are found throughout the whole organization, and not just in operations.

Supply Chains Have Processes

Parts 2 and 3 of this book about operations management are titled Managing Processes and Managing Supply Chains. If you infer that both Parts are essential aspects of operations management, you are correct. If you also infer that only Part 2 deals with processes, you would be wrong. The correct conclusion is that Part 3 is about managing <mark>supply chain processes,</mark> which are business processes that have external customers or suppliers. Table 3.1 illustrates some common supply chain processes.

supply chain processes
Business processes that have external customers or suppliers.

TABLE 3.1 | **SUPPLY CHAIN PROCESS EXAMPLES**

Process	Description	Process	Description
Outsourcing	Exploring available suppliers for the best options to perform processes in terms of price, quality, delivery time, environmental issues	**Customer Service**	Providing information to answer questions or resolve problems using automated information services as well as voice-to-voice contact with customers
Warehousing	Receiving shipments from suppliers, verifying quality, placing in inventory, and reporting receipt for inventory records	**Logistics**	Selecting transportation mode (train, ship, truck, airplane, or pipeline) scheduling both inbound and outbound shipments, and providing intermediate inventory storage
Sourcing	Selecting, certifying and evaluating suppliers and managing supplier contracts	**Cross-docking**	Packing of products of incoming shipments so they can be easily sorted more economically at intermediate warehouses for outgoing shipments to their final destination

These supply chain processes should be documented and analyzed for improvement, examined for quality improvement and control, and assessed in terms of capacity and bottlenecks. These topics are not repeated in Part 3, which instead examines the broader issues in managing supply chains, but still are essential to managing supply chain processes. Supply chain processes will be only as good as the processes within the organization that have only internal suppliers and customers. Each process in the chain, from suppliers to customers, must be designed to achieve its competitive priorities and add value to the work performed.

Processes Are Not Just in Operations

Processes are found in accounting, finance, human resources, management information systems, and marketing. Organizational structure throughout the many diverse industries varies, but for the most part, all organizations perform similar business processes. Table 3.2 lists a sample of them that are outside the operations area. All of these processes must be managed. What you learn in Part 2 applies to all of these business functions.

TABLE 3.2 | **ILLUSTRATIVE BUSINESS PROCESSES OUTSIDE OF OPERATIONS**

Activity based costing	Employee-development	Payroll
Asset management	Employee-recruiting	Records management
Billing budget	Employee-training	Research and development
Complaint handling	Engineering	Sales
Credit management	Environment	Help desks
Customer-satisfaction	External communications	Disaster recovery
Employee-benefits	Finance	Waste management
Employee-compensation	Security management	Warranty

Managers of these processes must make sure that they are adding as much customer value as possible. They must understand that many processes cut across organizational lines, regardless of whether the firm is organized along functional, product, regional, or process lines.

Process Strategy Decisions

A process involves the use of an organization's resources to provide something of value. No service can be provided and no product can be made without a process, and no process can exist without at least one service or product. One recurring question in managing processes is

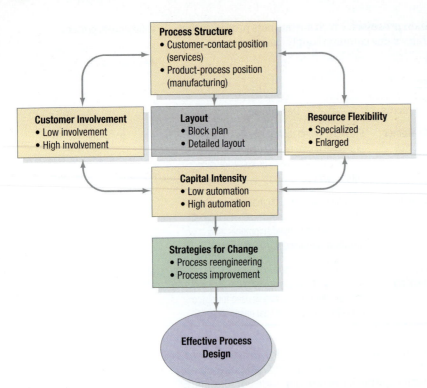

process structure

The process type relative to
the kinds of resources needed,
how resources are partitioned
between them, and their key
characteristics.

layout

The physical arrangement of
operations created by the various
processes.

customer involvement

The ways in which customers become part of the process and the
extent of their participation.

resource flexibility

The ease with which employees
and equipment can handle a wide
variety of products, output levels,
duties, and functions.

capital intensity

The mix of equipment and human
skills in a process.

deciding *how* to provide services or make products. Many different choices are available in
selecting human resources, equipment, outsourced services, materials, work flows, and
methods that transform inputs into outputs.
Another choice is which processes are to be
done in-house, and which processes are to be
outsourced—that is, done outside the firm and
purchased as materials and services. This decision helps to define the supply chain, and is
covered more fully in subsequent chapters. Supplement A, "Decision Making," also introduces
make-or-buy break-even.

Process decisions directly affect the process itself and indirectly the services and the
products that it provides. Whether dealing
with processes for offices, service providers, or
manufacturers, operations managers must consider four common process decisions. Figure 3.1
shows that they are all important steps toward
an effective process design. These four decisions
are best understood at the process or subprocess level, rather than at the firm level.

- **Process structure** determines the process type relative to the kinds of resources
needed, how resources are partitioned be-
tween them, and their key characteristics. A **layout**, which is the physical arrangement of operations created from the various processes, puts these decisions into tangible form.

- **Customer involvement** reflects the ways in which customers become part of the process and
the extent of their participation.

- **Resource flexibility** is the ease with which employees and equipment can handle a wide variety of products, output levels, duties, and functions.

- **Capital intensity** is the mix of equipment and human skills in a process. The greater the relative cost of equipment, the greater is the capital intensity.

The concepts that we develop around these four decisions establish a framework within
which we can address the appropriate process design in every situation. There is no "how to" element here in this chapter. That comes in subsequent chapters that show how to actually design
the processes or modify existing ones. Instead, we establish the patterns of choices that create
a good fit between the four decisions. For example, if you walk through a manufacturing facility
where materials flow smoothly from one work station to the next (which we will define later to be a
line process), you would be tempted to conclude that all processes should be line processes. They
seem so efficient and organized. However, if volumes are low and the products made are customized, converting to a line process would be a big mistake. When volumes are low and products are
customized, resources must be more flexible to handle a variety of products. The result is a more
disorganized appearance with jobs crisscrossing in many different directions depending on the
product being made. Despite appearances, this process is the best choice.

Process Structure in Services

One of the first decisions a manager makes in designing a well-functioning process is to choose a
process type that best achieves the competitive priorities for that process. Strategies for designing
processes can be quite different, depending on whether a service is being provided or a product is
being manufactured. We begin with service processes, given their huge implication for workforce
resources in industrialized countries.

Nature of Service Processes: Customer Contact

A process strategy that gets customers in and out of a fast-food restaurant quickly would not be
the right process strategy for a five-star restaurant, where customers seek a leisurely dining experience. To gain insights, we must start at the process level and recognize key contextual variables associated with the process. A good process strategy for a service process depends first and foremost

on the type and amount of customer contact. **Customer contact** is the extent to which the customer is present, is actively involved, and receives personal attention during the service process. Face-to-face interaction, sometimes called a *moment of truth* or *service encounter,* brings the customer and service providers together. At that time, customer attitudes about the quality of the service provided are shaped. Table 3.3 shows several dimensions of customer contact. Many levels are possible on each of the five dimensions. The nested-process concept applies to customer contact, because some parts of a process can have low contact and other parts of a process can have high contact.

TABLE 3.3 | DIMENSIONS OF CUSTOMER CONTACT IN SERVICE PROCESSES

Dimension	High Contact	Low Contact
Physical presence	Present	Absent
What is processed	People	Possessions or information
Contact intensity	Active, visible	Passive, out of sight
Personal attention	Personal	Impersonal
Method of delivery	Face-to-face	Regular mail or e-mail

Customer-Contact Matrix

The customer-contact matrix, shown in Figure 3.2, brings together three elements: (1) the degree of customer contact, (2) customization, and (3) process characteristics. The matrix is the starting point for evaluating and improving a process.

Customer Contact and Customization   The horizontal dimension of the matrix represents the service provided to the customer in terms of customer contact and competitive priorities. A key competitive priority is how much customization is needed. Positions on the left side of the matrix represent high customer contact and highly customized services. The customer is more likely to be present and active. The process is more likely to be visible to the customer, who receives more personal attention. The right side of the matrix represents low customer contact, passive involvement, less personalized attention, and a process out of the customer's sight.

Process Divergence and Flow   The vertical dimension of the customer-contact matrix deals with two characteristics of the process itself: (1) process divergence and (2) flow. Each process can be analyzed on these two dimensions.

Process divergence is the extent to which the process is highly customized with considerable latitude as to how its tasks are performed. If the process changes with each customer, virtually every performance of the service is unique. Examples of highly divergent service processes where many steps in them change with each customer are found in consulting, law, and architecture. A service with low divergence, on the other hand, is repetitive and standardized. The work is performed exactly the same with all customers, and tends to be less complex. Certain hotel services and telephone services are highly standardized to assure uniformity.

Closely related to divergence is how the customer, object, or information being processed flows through the service facility. Work progresses through the sequence of steps in a process, which could range from highly diverse to linear. When divergence is considerable, the work flow tends to be more flexible. A **flexible flow** means that the customers, materials, or information move in diverse ways,

customer contact

The extent to which the customer is present, is actively involved, and receives personal attention during the service process.

process divergence

The extent to which the process is highly customized with considerable latitude as to how its tasks are performed.

flexible flow

The customers, materials, or information move in diverse ways, with the path of one customer or job often crisscrossing the path that the next one takes.

◀ **FIGURE 3.2**

Customer-Contact Matrix for Service Processes

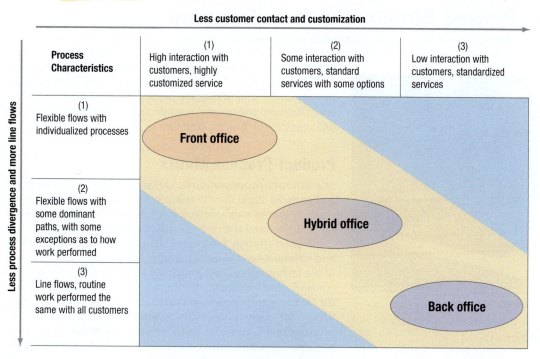

Myrleen Ferguson Cate/PhotoEdit

A financial consultant discusses options with a young couple at their home. This process scores high on customer contact, because the customers are present, take an active part in creating the service, receive personal attention, and have face-to-face contact.

line flow

The customers, materials, or information move linearly from one operation to the next, according to a fixed sequence.

front office

A process with high customer contact where the service provider interacts directly with the internal or external customer.

hybrid office

A process with moderate levels of customer contact and standard services with some options available.

back office

A process with low customer contact and little service customization.

with the path of one customer or job often criss-crossing the path that the next one takes. Each one can follow a carefully preplanned path, even though the first impression is one of disorganized, jumbled flows. Such an appearance goes naturally with high process divergence. A **line flow** means that the customers, materials, or information move linearly from one operation to the next, according to a fixed sequence. When diversity is low and the process standardized, line flows are a natural consequence.

Service Process Structuring

Figure 3.2 shows several desirable positions in the matrix that effectively connect the service product with the process. The manager has three process structures, which form a continuum, to choose from: (1) front office, (2) hybrid office, and (3) back office. It is unlikely that a process can be a top performer if a process lies too far from one of these diagonal positions, occupying instead one of the extreme positions represented by the light blue triangles in the matrix (refer to Figure 3.2). Such positions represent too much of a disconnect between the service provided and process characteristics.

Front Office   A **front-office** process has high customer contact where the service provider interacts directly with the internal or external customer. Because of the customization of the service and variety of service options, many of the steps in it have considerable divergence. Work flows are flexible, and they vary from one customer to the next. The high-contact service process tends to be adapted or tailored to each customer.

Hybrid Office A hybrid office tends to be in the middle of the five dimensions in Table 3.3, or perhaps high on some contact measures and low on others. A **hybrid-office** process has moderate levels of customer contact and standard services, with some options available from which the customer chooses. The work flow progresses from one workstation to the next, with some dominant paths apparent.

Back Office   A **back-office** process has low customer contact and little service customization. The work is standardized and routine, with line flows from one service provider to the next until the service is completed. Preparing the monthly client fund balance reports in the financial services industry is a good example. It has low customer contact, low divergence, and a line flow.

Process Structure in Manufacturing

Many processes at a manufacturing firm are actually services to internal or external customers, and so the previous discussion on services applies to them. Similarly, manufacturing processes can be found in service firms. Clarity comes when viewing work at the process level, rather than the organizational level. Here we focus instead on the manufacturing processes. Because of the differences between service and manufacturing processes, we need a different view on process structure.

Bill Varie/Alamy

An employee discusses work with her supervisor. Each employee in this series of work stations are in a back office, because they have low customer contact and little service customization.

Product-Process Matrix

The product-process matrix, shown in Figure 3.3, brings together three elements: (1) volume, (2) product customization, and (3) process characteristics. It synchronizes the product to be manufactured with the manufacturing process itself.

A good strategy for a manufacturing process depends first and foremost on volume. Customer contact, a primary feature of the customer-contact matrix for services, normally is not a consideration for manufacturing processes (although it *is* a factor for the many service processes throughout manufacturing firms). For many manufacturing processes, high product customization means lower volumes for many of the steps in the process. The vertical dimension of the product-process matrix

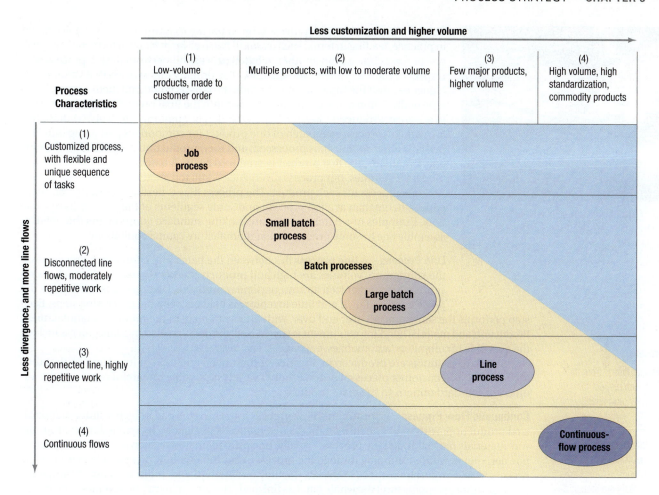

Less customization and higher volume

Product-Process Matrix for
Manufacturing Processes

deals with the same two characteristics in the customer-contact matrix: process divergence and flow. Each manufacturing process should be analyzed on these two dimensions, just as was done for a service process.

Manufacturing Process Structuring

Figure 3.3 shows several desirable positions (often called *process choices*) in the product-process matrix that effectively connect the manufactured product with the process. **Process choice** is the way of structuring the process by organizing resources around the process or organizing them around the products. Organizing around the process means, for example, that all milling machines are grouped together and process all products or parts needing that kind of transformation. Organizing around the product means bringing together all the different human resources and equipment needed for a specific product and dedicating them to producing just that product. The manager has four process choices, which form a continuum, to choose from: (1) job process, (2) batch process, (3) line process, and (4) continuous-flow process. As with the customer-contact matrix, it is unlikely that a manufacturing process can be a top performer if its position is too far from the diagonal. The fundamental message in Figure 3.3 is that the best choice for a manufacturing process depends on the volume and degree of customization required of the process. The process choice might apply to an entire manufacturing process or just one subprocess nested within it.

process choice

A way of structuring the process by organizing resources around the process or organizing them around the products.

Job Process   A **job process** creates the flexibility needed to produce a wide variety of products in significant quantities, with considerable divergence in the steps performed. Customization is high and volume for any one product is low. The workforce and equipment are flexible to handle considerable task divergence. Companies choosing job processes often bid for work. Typically, they make products to order and do not produce them ahead of time. Each new order is handled as a single unit—as a job. Examples are machining a metal casting for a customized order or producing customized cabinets.

With a job process, all equipment and workers capable of certain types of work are located together. Because customization is high and most jobs have a different sequence of steps, this process choice creates flexible flows through the operations rather than a line flow.

job process

A process with the flexibility needed to produce a wide variety of products in significant quantities, with considerable divergence in the steps performed.

A batch of apple fritters roll off one of the pastry lines at the King Soopers's Bakery in Denver to be packaged for transportation. The pastry line is a batch process, and a different kind of pastry will be made next.

Batch Process The batch process is by far the most common process choice found in practice, leading to terms such as *small batch* or *large batch* to further distinguish one process choice from another. A **batch process** differs from the job process with respect to volume, variety, and quantity. The primary difference is that volumes are higher because the same or similar products or parts going into them are produced repeatedly. Some of the components going into the final product may be processed in advance. Production lots are handled in larger quantities (or *batches*) than they are with job processes. A batch of one product (or component part going into it or perhaps other products) is processed, and then production is switched to the next one. Eventually, the first product is produced again. A batch process has average or moderate volumes, but process divergence is still too great to warrant dedicating a separate process for each product. The process flow is flexible, but more dominant paths emerge than at a job process, and some segments of the process have a line flow. Examples of a batch process are making standard components that feed an assembly line or some processes that manufacture capital equipment.

Line Process A **line process** lies between the batch and continuous processes on the continuum; volumes are high and products are standardized, which allows resources to be organized around particular products. Divergence is minimal in the process or line flows, and little inventory is held between the processing steps. Each step performs the same process over and over, with little variability in the products manufactured. Production and material handling equipment is specialized. Products created by a line process include the assembly of computers, automobiles, appliances, and toys.

Standard products are produced in advance of their need and held in inventory so that they are ready when a customer places an order. Product variety is possible by careful control of the addition of standard options to the main product.

Continuous Flow Process A **continuous flow process** is the extreme end of high-volume standardized production, with rigid line flows. Process divergence is negligible. Its name derives from the way materials move through the process. Usually, one primary material (such as a liquid, a gas, or a powder) moves without stopping through the process. A continuous-flow process differs from a line process in one important respect: Materials (be they undifferentiated or discrete) flow through the process without stopping until the whole batch is finished. The time span can be several shifts or even several months. Examples of a continuous flow process are petroleum refining; chemical processes; and processes making steel, soft drinks, and food (such as Borden's huge pasta-making plant).

Production and Inventory Strategies

Strategies for manufacturing processes differ from those in services, not only because of low customer contact and involvement, but also because of the ability to use inventories[1]. Make-to-order, assemble-to-order, and make-to-stock strategies are three approaches to inventory that should be coordinated with process choice.

Make-to-Order Strategy Manufacturers that make products to customer specifications in low volumes tend to use the **make-to-order strategy**, coupling it with job or small batch processes. It is a more complex process than assembling a final product from standard components. This strategy provides a high degree of customization and typically uses job or small batch processes. The processes have high divergence. Specialized medical equipment, castings, and expensive homes are suited to the make-to-order strategy.

Assemble-to-Order Strategy The **assemble-to-order strategy** is an approach to producing a wide variety of products from relatively few subassemblies and components after the customer orders are received. Typical competitive priorities are variety and fast delivery times. The assemble-to-order strategy often involves a line process for assembly and a batch process for fabrication. Because they are devoted to manufacturing standardized components and subassemblies in high volumes, the fabrication processes focus on creating appropriate amounts of component inventories for the assembly processes. Once the specific order from the customer is received, the assembly processes create the product from standardized components and subassemblies produced by the fabrication processes.

Stocking finished products would be economically prohibitive because the numerous possible options make forecasting relatively inaccurate. Thus, the principle of **postponement** is applied, whereby the final activities in the provision of a product are delayed until the orders are

batch process

A process that differs from the job process with respect to volume, variety, and quantity.

line process

A process that lies between the batch and continuous processes on the continuum; volumes are high and products are standardized, which allows resources to be organized around particular products.

continuous flow process

The extreme end of high-volume standardized production and rigid line flows, with production not starting and stopping for long time intervals.

make-to-order strategy

A strategy used by manufacturers that make products to customer specifications in low volumes.

assemble-to-order strategy

A strategy for producing a wide variety of products from relatively few subassemblies and components after the customer orders are received.

postponement

The strategy of delaying final activities in the provision of a product until the orders are received.

---

[1]Service firms also hold inventories, but only as purchased material. Manufacturing firms have the additional flexibility of holding inventories as subassemblies or finished products.

received. The assemble-to-order strategy is also linked to **mass customization**, where highly divergent processes generate a wide variety of customized products at reasonably low costs. Both postponement and mass customization are covered more fully in Chapter 10, "Supply Chain Design."

Make-to-Stock Strategy   Manufacturing firms that hold items in stock for immediate delivery, thereby minimizing customer delivery times, use a **make-to-stock strategy**. This strategy is feasible for standardized products with high volumes and reasonably accurate forecasts. It is the inventory strategy of choice for line or continuous-flow processes. Examples of products produced with a make-to-stock strategy include garden tools, electronic components, soft drinks, and chemicals.

Combining a line process with the make-to-stock strategy is sometimes called **mass production.** It is what the popular press commonly envisions as the classical manufacturing process, because the environment is stable and predictable, with workers repeating narrowly defined tasks with low divergence.

A Chinese manufacturing firm using the make-to-stock strategy has considerable inventory stacked on pallets with rows of shelving racks in the background.

James Hardy/Altopress/Newscom

Layout

Selecting process structures for the various processes housed in a facility is a strategic decision, but must be followed by a more tactical decision—creating a layout. A *layout* is the physical arrangement of operations (or departments) created from the various processes and puts them in tangible form. For organizational purposes, processes tend to be clustered together into operations or departments. An *operation* is a group of human and capital resources performing all or part of one or more processes. For example, an operation could be several customer service representatives in a customer reception area; a group of machines and workers producing cell phones; or a marketing department. Regardless of how processes are grouped together organizationally, many of them cut across departmental boundaries. The flows across departmental lines could be informational, services, or products. Process structures that create more flows across departmental lines, as with job or batch processes, are the most challenging layout problems.

Here we demonstrate an approach to layout design that positions those departments close together that have strong interactions between them. It involves three basic steps, whether the design is for a new layout or for revising an existing layout: (1) gather information, (2) develop a block plan, and (3) design a detailed layout. We illustrate these steps with the Office of Budget Management (OBM), which is a major division in a large state government.

Gather Information

OBM consists of 120 employees assigned to six different departments. Workloads have expanded to the extent that 30 new employees must be hired and somehow housed in the space allocated to OBM. The goal is to improve communication among people who must interact with each other effectively, creating a good work environment.

Three types of information are needed to begin designing the revised layout for OBM: (1) space requirements by center, (2) available space, and (3) closeness factors. OBM has grouped its processes into six different departments: (1) administration, (2) social services, (3) institutions, (4) accounting, (5) education, and (6) internal audit. The exact space requirements of each department, in square feet, are as follows:

Department	Area Needed (ft²)
1. Administration	3,500
2. Social services	2,600
3. Institutions	2,400
4. Accounting	1,600
5. Education	1,500
6. Internal audit	3,400
	Total 15,000

mass customization

The strategy that uses highly divergent processes to generate a wide variety of customized products at reasonably low costs.

make-to-stock strategy

A strategy that involves holding items in stock for immediate delivery, thereby minimizing customer delivery times.

mass production

A term sometimes used in the popular press for a line process that uses the make-to-stock strategy.

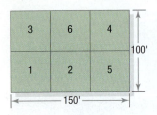

▲ FIGURE 3.4
Current Block Plan for the
Office of Budget Management

block plan

A plan that allocates space and
indicates placement of each
operation.

closeness matrix

A table that gives a measure of
the relative importance of each
pair of operations being located
close together.

Management must tie space requirements to capacity and staffing plans; calculate the specific equipment and space needs for each center; and allow circulation space, such as aisles and the like. At OBM, a way must be found to include all 150 employees in its assigned area. Consulting with the managers and employees involved can help avoid excessive resistance to change and make the transition smoother.

A **block plan** allocates space and indicates placement of each operation. To describe a new facility layout, the plan need only provide the facility's dimensions and space allocations. When an existing facility layout is being modified, the current block plan is also needed. OBM's available space is 150 feet by 100 feet, or 15,000 square feet. The designer could begin the design by dividing the total amount of space into six equal blocks (2,500 square feet each). The equal-space approximation shown in Figure 3.4 is sufficient until the detailed layout stage, when larger departments (such as administration) are assigned more space than smaller departments.

The layout designer must also know which operations need to be located close to one another. The table below shows OBM's **closeness matrix**, which gives a measure of the relative importance of each pair of operations being located close together. The metric used depends on the type of processes involved and the organizational setting. It can be a qualitative judgment on a scale from 0 to 10 that the manager uses to account for multiple performance criteria, as in the OBM's case. Only the right-hand portion of the matrix is used. The closeness factors are indicators of the need for proximity based on an analysis of information flows and the need for face-to-face meetings. They give clues as to which departments should be located close together. For example, the most important interaction is between the administration and internal audit departments for OBM, with a score of 10. This closeness factor is given in the first row and last column. Thus, the designer should locate departments 1 and 6 close together, which is not the arrangement in the current layout. Entries in both the columns and rows result in five factor scores for each department.

CLOSENESS FACTORS

Department	1	2	3	4	5	6
1. Administration	—	3	6	5	6	10
2. Social services		—	8	1	1	
3. Institutions			—	3	9	
4. Accounting				—	2	
5. Education					—	1
6. Internal audit						—

At a manufacturing plant, the closeness factor could be the number of trips (or some other measure of materials movement) between each pair of operations per day. This information can be gleaned by conducting a statistical sampling, polling supervisors and materials handlers, or using the routings and ordering frequencies for typical items made at the plant.

Finally, the information gathered for OBM includes performance criteria that depend not on the relative location of department pairs, but the *absolute* location of a single department. OBM has two such criteria.

1. Education (department 5) should remain where it is because it is next to the office library.

2. Administration (department 1) should remain where it is because that location has the largest conference room, which administration uses often. Relocating the conference room would be costly.

Develop a Block Plan

Having gathered the needed information, the next step is to develop a block plan that best satisfies performance criteria and area requirements. The most elementary way to do so is by trial and error. Because success depends on the designer's ability to spot patterns in the data, this approach does not guarantee the selection of the best or even a nearly best solution. When supplemented by the use of a computer to evaluate solutions, however, research shows that such an approach compares quite favorably with more sophisticated computerized techniques.

weighted-distance method

A mathematical model used to
evaluate layouts (of facility locations) based on closeness factors.

Applying the Weighted-Distance Method

When *relative* locations are a primary concern, such as for effective information flow, communication, material handling, and stockpicking, the weighted-distance method can be used to compare alternative block plans. The **weighted-distance method** is a mathematical model used to evaluate layouts based on closeness factors. A similar approach, sometimes called the *load-distance method*, can be used to evaluate facility locations. The objective is to select a layout (or facility location) that minimizes the total weighted distances. The distance between two points is expressed by assigning the points to grid coordinates on a block diagram or map. An alternative approach is to use time rather than distance.

For a rough calculation, which is all that is needed for the weighted-distance method, either a Euclidean or rectilinear distance measure may be used. **Euclidean distance** is the straight-line distance, or shortest possible path, between two points. To calculate this distance, we create a graph. The distance between two points, say, points A and B, is

$$d_{AB} = \sqrt{(x_A - x_B)^2 + (y_A - y_B)^2}$$

where

d_{AB} = distance between points A and B
x_A = x-coordinate of point A
y_A = y-coordinate of point A
x_B = x-coordinate of point B
y_B = y-coordinate of point B

Rectilinear distance measures the distance between two points with a series of 90-degree turns, as along city blocks. The distance traveled in the x-direction is the absolute value of the difference between the x-coordinates. Adding this result to the absolute value of the difference between the y-coordinates gives

$$d_{AB} = |x_A - x_B| + |y_A - y_B|$$

For assistance in calculating distances using either measure, see Tutor 3.1 in OM Explorer.

The layout designer seeks to minimize the weighted-distance (*wd*) score by locating centers that have high-closeness ratings close together. To calculate a layout's *wd* score, we use either of the distance measures and simply multiply the proximity scores by the distances between centers. The sum of those products becomes the layout's final *wd* score—the lower the better. The location of a center is defined by its x-coordinate and y-coordinate.

Euclidean distance

The straight-line distance, or shortest possible path, between two points.

rectilinear distance

The distance between two points with a series of 90-degree turns, as along city blocks.

MyOMLab

Tutor 3.1 in MyOMLab provides an example to calculate both Euclidean and rectilinear distance measures.

EXAMPLE 3.1	**Calculating the Weighted-Distance Score**

The block plan in Figure 3.5 was developed using trial and error. A good place to start was to fix Departments 1 and 5 in their current locations. Then, the department pairs that had the largest closeness factors were located. The rest of the layout fell into place rather easily.

How much better, in terms of the *wd* score, is the proposed block plan shown in Figure 3.5 than the current plan shown in Figure 3.4? Use the rectilinear distance measure.

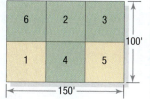

▲ **FIGURE 3.5**
Proposed Block Plan

SOLUTION
The accompanying table lists each pair of departments that has a nonzero closeness factor in the closeness matrix. For the third column, calculate the rectilinear distances between the departments in the current layout. For example, departments 3 and 5 in the current plan are in the upper-left corner and bottom-right corner of the building, respectively. The distance between the centers of these blocks is three units (two horizontally and one vertically). For the fourth column, we multiply the weights (closeness factors) by the distances, and then add the results for a total *wd* score of 112 for the current plan. Similar calculations for the proposed plan produce a *wd* score of only 82. For example, between departments 3 and 5 is just one unit of distance (one vertically and zero horizontally).

MyOMLab

Active Model 3.1 in MyOMLab allows evaluation of the impact of swapping OBM departmental positions.

Current Plan

3	6	4
1	2	5

Proposed Plan

6	2	3
1	4	5

Department Pair	Closeness Factor (*w*)	CURRENT PLAN Distance (*d*)	CURRENT PLAN Weighted-Distance Score (*wd*)	PROPOSED PLAN Distance (*d*)	PROPOSED PLAN Weighted-Distance Score (*wd*)
1, 2	3	1	3	2	6
1, 3	6	1	6	3	18
1, 4	5	3	15	1	5
1, 5	6	2	12	2	12
1, 6	10	2	20	1	10
2, 3	8	2	16	1	8
2, 4	1	2	2	1	1

Department Pair	Closeness Factor (w)	CURRENT PLAN		PROPOSED PLAN	
		Distance (d)	Weighted-Distance Score (wd)	Distance (d)	Weighted-Distance Score (wd)
2, 5	1	1	1	2	2
3, 4	3	2	6	2	6
3, 5	9	3	27	1	9
4, 5	2	1	2	1	2
5, 6	1	2	2	3	3
			Total 112		Total 82

To be exact, we could multiply the two wd total scores by 50 because each unit of distance represents 50 feet. However, the relative difference between the two totals remains unchanged.

DECISION POINT

The wd score for the proposed layout makes a sizeable drop from 112 to 82, but management is not sure the improvement outweighs the cost of relocating four of the six departments (i.e., all departments but 1 and 5).

⦿ Rectilinear Distances   ◯ Euclidean Distances

Department Pair	Closeness Factor	Distance	Score
1, 6	10	1	10
3, 5	9	1	9
2, 3	8	1	8
1, 3	6	1	6
1, 5	6	2	12
1, 4	5	3	15
1, 2	3	2	6
3, 4	3	2	6
4, 5	2	1	2
2, 4	1	1	1
2, 5	1	2	2
5, 6	1	3	3
	Total		80

6	2	4
1	3	5

▲ **FIGURE 3.6**
Second Proposed Block Plan
(Analyzed with *Layout* Solver)

Although the wd score for the proposed layout in Example 3.1 represents an almost 27 percent improvement, the designer may be able to improve on this solution. Furthermore, the manager must determine whether the revised layout is worth the cost of relocating four of the six departments. If relocation costs are too high, a less-expensive proposal must be found.

OM Explorer and POM for Windows can help identify some even more attractive proposals. For example, one option is to modify the proposed plan by switching the locations of departments 3 and 4. OM Explorer's output in Figure 3.6 shows that the wd score for this second revision not only drops to 80, but requires that only three departments be relocated compared with the original layout in Figure 3.4. Perhaps this second proposed plan is the best solution.

Design a Detailed Layout

After finding a satisfactory block plan, the final step translates it into a detailed representation, showing the exact size and shape of each center; the arrangement of elements (e.g., desks, machines, and storage areas); and the location of aisles, stairways, and other service space. These visual representations can be two-dimensional drawings, three-dimensional models, or computer-aided graphics. This step helps decision makers discuss the proposal and problems that might otherwise be overlooked. Such visual representations can be particularly important when evaluating high customer-contact processes.

Customer Involvement

Having covered process structure decisions and how they are translated into a layout, we now turn to a second major decision—customer involvement—shown in Figure 3.1. Customer involvement reflects the ways in which customers become part of the process and the extent of their participation. It is especially important for many service processes, particularly if customer contact is (or should be) high.

While eBay devised one way to accommodate the variability created by customer involvement, Starbucks faces a different kind of customer variability. The coffee shop chain allows customers to choose among many permutations of sizes, flavors, and preparation techniques in its beverages.

In order to fill orders accurately and efficiently, Starbucks trains its counter clerks to call out orders to beverage makers in a particular sequence. It is even better when customers themselves can do so. Starbucks attempts to teach customers its ordering protocol. First, it provides a "guide-to-ordering pamphlet" for customers to look over. Second, it trains clerks to repeat the order in the correct sequence for the beverage makers, which may not be how the customer presented it. This process not only makes it easier for the beverage makers, but also indirectly "trains" the customers in how to place their orders.

The detailed layout becomes a reality. Shown here is part of the office in one of the ABB facilities, a global leader in power and automation technologies. Here the workstations are small and semiprivate, but "outposts" are available. This common area has easy chairs fitted with arms that provide a surface for writing or laptops. People meet comfortably for face-to-face talks, rather than communicating by e-mail.

Possible Disadvantages

Customer involvement is not always a good idea. In some cases, giving the customer more active contact in a service process will just be disruptive, making the process less efficient. Managing the timing and volume of customer demands becomes more challenging if the customer is physically present and expects prompt delivery. Exposing the facilities and employees to the customer can have important quality implications (favorable or unfavorable). Such changes make interpersonal skills a prerequisite to the service provider's job, but higher skill levels come at a cost. Revising the facility layout might be a necessary investment, now that managing customer perceptions becomes an important part of the process. It also might mean having many smaller decentralized facilities closer to the various customer concentration areas if the customer comes to the service providers.

Possible Advantages

Despite these possible disadvantages, the advantages of a more customer-focused process might increase the net value to the customer. Some customers seek active participation in and control over the service process, particularly if they will enjoy savings in both price and time. The manager must assess whether advantages outweigh disadvantages, judging them in terms of the competitive priorities and customer satisfaction. More customer involvement can mean better quality, faster delivery, greater flexibility, and even lower cost. Self-service is the choice of many retailers, such as gasoline stations, supermarkets, and bank services. Manufacturers of products (such as toys, bicycles, and furniture) may also prefer to let the customer perform the final assembly because product, shipping, and inventory costs frequently are lower. Customer involvement can also help coordinate across the supply chain (see Chapter 12, "Supply Chain Integration"). Emerging technologies allow companies to engage in an active dialogue with customers and make them partners in creating value and forecasting future demand. Companies can also revise some of their traditional processes, such as pricing and billing systems, to account for their customers' new role. For example, in business-to-business relationships, the Internet changes the roles that companies play with other businesses. Suppliers to automobile companies can be close collaborators in the process of developing new vehicles and no longer are passive providers of materials and services. The same is true for distributors. Walmart does more than just distribute Procter & Gamble's products: It shares daily sales information and works with Procter & Gamble in managing inventories and warehousing operations.

A customer at Starbucks, a large coffee shop chain, places his order in the correct way. By structuring the ordering process for counter clerks and customers, Starbucks can deal efficiently with the variety in products offered, and with no hit on the service experience.

ABB, Inc.

Technicians in the parts repair department of this ABB facility must be flexible enough to repair many different parts for automation equipment installed at customer locations in the field. This operation has 30 different workstations configured to perform different types of processes. Workers are cross-trained to move from one station to another, depending on what needs to be done.

Resource Flexibility

Just as managers must account for customer contact when making customer involvement decisions, so must they account for process divergence and diverse process flows when making resource flexibility decisions in Figure 3.1. High task divergence and flexible process flows require more flexibility of the process's resources—its employees, facilities, and equipment. Employees need to perform a broad range of duties, and equipment must be general purpose. Otherwise, resource utilization will be too low for economical operations.

Workforce

Operations managers must decide whether to have a **flexible workforce**. Members of a flexible workforce are capable of doing many tasks, either at their own workstations or as they move from one workstation to another. However, such flexibility often comes at a cost, requiring greater skills and thus more training and education. Nevertheless, benefits can be large: Worker flexibility can be one of the best ways to achieve reliable customer service and alleviate capacity bottlenecks. Resource flexibility helps to absorb the feast-or-famine workloads in individual operations that are caused by low-volume production, divergent tasks, flexible flows, and fluid scheduling.

flexible workforce

A workforce whose members are capable of doing many tasks, either at their own workstations or as they move from one workstation to another.

The type of workforce required also depends on the need for volume flexibility. When conditions allow for a smooth, steady rate of output, the likely choice is a permanent workforce that expects regular full-time employment. If the process is subject to hourly, daily, or seasonal peaks and valleys in demand, the use of part-time or temporary employees to supplement a smaller core of full-time employees may be the best solution. However, this approach may not be practical if knowledge and skill requirements are too high for a temporary worker to grasp quickly.

Equipment

MyOMLab

Tutor 3.2 in MyOMLab demonstrates how to do break-even analysis for equipment selection.

Low volumes mean that process designers should select flexible, general-purpose equipment. Figure 3.7 illustrates this relationship by showing the total cost lines for two different types of equipment that can be chosen for a process. Each line represents the total annual cost of the process at different volume levels. It is the sum of fixed costs and variable costs (see Supplement A, "Decision Making"). When volumes are low (because customization is high), process 1 is the better choice. It calls for inexpensive general-purpose equipment, which keeps investment in equipment low and makes fixed costs (F_1) small. Its variable unit cost is high, which gives its total cost line a relatively steep slope. Process 1 does the job, but not at peak efficiency.

Conversely, process 2 is the better choice when volumes are high and customization is low. Its advantage is low variable unit cost, as reflected in the flatter total cost line. This efficiency is possible when customization is low because the equipment can be designed for a narrow range of products or tasks. Its disadvantage is high equipment investment and, thus, high fixed costs (F_2). When annual volume produced is high enough, spreading these fixed costs over more units produced, the advantage of low variable costs more than compensates for the high fixed costs.

▼ **FIGURE 3.7**
Relationship Between Process Costs and Product Volume

The break-even quantity in Figure 3.7 is the quantity at which the total costs for the two alternatives are equal. At quantities beyond this point, the cost of process 1 exceeds that of process 2. Unless the firm expects to sell more than the break-even amount, which is unlikely with high customization and low volume, the capital investment of process 2 is not warranted.

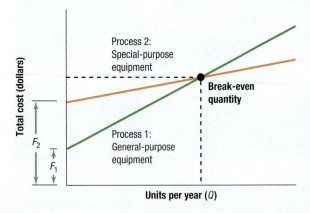

Capital Intensity

Capital intensity is the mix of equipment and human skills in the process; the greater the relative cost of equipment, the greater is the capital intensity. As the capabilities of technology increase and its costs decrease, managers face an ever-widening range of choices, from operations utilizing very little automation to those requiring task-specific equipment and little human intervention. **Automation** is a system, process, or piece of equipment that is self-acting and self-regulating. Although automation is often thought to be necessary to gain competitive advantage, it has both advantages and disadvantages. Thus, the automation decision requires careful examination.

Automating Manufacturing Processes

Substituting labor-saving capital equipment and technology for labor has been a classic way of improving productivity and quality consistency in manufacturing processes. If investment costs are large, automation works best when volume is high, because more customization typically means reduced volume. Gillette, for example, spent $750 million on the production lines and robotics that gave it a capacity to make 1.2 billion razor cartridges a year. The equipment is complicated and expensive. Only with such high volumes could this line process produce the product at a price low enough that consumers could afford to buy it.

One big disadvantage of capital intensity can be the prohibitive investment cost for low-volume operations (see Figure 3.7). Generally, capital-intensive operations must have high utilization to be justifiable. Also, automation does not always align with a company's competitive priorities. If a firm offers a unique product or high-quality service, competitive priorities may indicate the need for hand labor and individual attention rather than new technology. A case in point is the downstream processes in Gillette's supply chain that package and store the razor cartridges. It customizes the packaging for different regions of the world, so that volumes for any one type of package are much lower. As a result of the low volumes, Gillette does not use expensive automation for these processes. In fact, it outsources them. Producing razor cartridges to stock using highly automated processes, and then packaging them in customized fashion at remote locations on demand, is also a good example of the principle of postponement.

Volkswagen aspires to become a full-line manufacturer of cars ranging from the smallest compacts to the largest luxury models. In the United States, for example, it offers the Jetta, GTI, Golf, Passat, CC, Routan, Tiguan, Touareg and Eos. The new VW Phaeton is the latest example of the brand's higher end. It uses a line process dedicated strictly to Phaeton cars. Because of its focus on Phaetons, which are made nowhere else, it enjoys high volumes. This volume justifies the high automation invested in this $208 million plant. The building is located in the heart of Dresden, a city known for its arts and craftsmanship. Its walls are made almost exclusively of glass, and it floors are covered entirely in Canadian maple. There are no smokestacks, no noises, and no toxic byproducts. Parts arrive and luxury cars depart. Of course, such a plant is not always possible, as with a steel mill or foundry. Shown in the photo is the arrival of the body structure that is painted in a plant about 60 miles from Dresden. It is what it looks like when it arrives, before the Phaeton assembly begins. For a photo tour, see **http://forums.vwvortex.com/showthread.php?1837641**.

Fixed Automation Manufacturers use two types of automation: (1) fixed and (2) flexible (or programmable). Particularly appropriate for line and continuous-flow process choices, **fixed automation** produces one type of part or product in a fixed sequence of simple operations. Operations managers favor fixed automation when demand volumes are high, product designs are stable, and product life cycles are long. These conditions compensate for the process's two primary drawbacks: (1) large initial investment cost and (2) relative inflexibility. However, fixed automation maximizes efficiency and yields the lowest variable cost per unit if volumes are high.

Flexible Automation **Flexible (or programmable) automation** can be changed easily to handle various products. The ability to reprogram machines is useful for both low-customization and high-customization processes. In the case of high customization, a machine that makes a variety of products in small batches can be programmed to alternate between products. When a machine has been dedicated to a particular product or family of products, as in the case of low customization and a line flow, and the product is at the end of its life cycle, the machine can simply be reprogrammed with a new sequence of tasks for a new product. An **industrial robot**, which is a versatile, computer-controlled machine programmed to perform various tasks, is a classic example of flexible automation. These "steel-collar" workers operate independently of human control. A robot's arm has up to six standard movements. The robot's "hand" actually does the work. The hand can be changed to perform different tasks, such as materials handling, assembly, and testing. Managerial Practice 3.1 describes how R.R. Donnelley benefits from more flexible automation, allowing for quick change-overs from one customer order to the next.

Automating Service Processes

Using capital inputs as a labor-saving device is also possible for service processes. In educational services, for example, long-distance learning technology now can supplement or even replace the traditional classroom experience by using books, computers, Web sites, and videos as facilitating goods that go with the service. Justifying technology need not be limited to cost reduction. Sometimes, it can actually allow more task divergence by making available a wide menu of choices to the customer.

automation
A system, process, or piece of equipment that is self-acting and self-regulating.

fixed automation
A manufacturing process that produces one type of part or product in a fixed sequence of simple operations.

flexible (or programmable) automation
A manufacturing process that can be changed easily to handle various products.

industrial robot
Versatile, computer-controlled machine programmed to perform various tasks.

Flexible Automation at Just Born, Inc., a candy company in Pennsylvania (**www.justborn.com**).

On the left, a robot picks up the PEEPS® brand marshmallow yellow bunnies with great speed in groups of four and places them into preformed trays, which then move to automatic shrink-wrapping machines. The trays are then boxed into cases for shipment. Upstream in this line process, the bunnies were extruded in shape on a belt with about 1/2 inch of sugar on it and then cooled as the conveyor moves along. The robot is regularly reprogrammed based on the marshmallow configuration being produced.

On the right, the robot using vacuum cups on the "hand" of its arm picks up boxes of five-pound MIKE AND IKE® bags that are packaged with six bags in a box. The robot reads the bar code, knows the pallet configuration, picks up and places the pallet and a thin cardboard liner on the conveyor, and stacks the boxes until complete. The pallet then comes toward the camera to be picked off the line by a fork truck, and then shrink-wrapped (automatically) and loaded into a 54-foot trailer. PEEPS® and MIKEANDIKE® are registered trademarks of Just Born, Inc. Used with permission.

Technology in the future will surely make possible even a greater degree of customization and variety in services that currently only human providers can now deliver. Beyond cost and variety considerations, management must understand the customer and how much close contact is valued. If the customers seek a visible presence and personal attention, technologies reduced to sorting through a variety of options on the Internet or over the telephone might be a poor choice.

The need for volume to justify expensive automation is just as valid for service processes as for manufacturing processes. Increasing the volume lowers the cost per dollar of sales. Volume is essential for many capital-intensive processes in the transportation, communications, and utilities industries.

Economies of Scope

economies of scope

Economies that reflect the ability to produce multiple products more cheaply in combination than separately.

If capital intensity is high, resource flexibility usually is low. In certain types of manufacturing operations, such as machining and assembly, programmable automation breaks this inverse relationship between resource flexibility and capital intensity. It makes possible both high capital intensity and high resource flexibility, creating economies of scope. **Economies of scope** reflect the ability to produce multiple products more cheaply in combination than separately. In such situations, two conflicting competitive priorities—customization and low price—become more compatible. However, taking advantage of economies of scope requires that a family of parts or products have enough collective volume to utilize equipment fully.

Economies of scope also apply to service processes. Consider, for example, Disney's approach to the Internet. When the company's managers entered the volatile Internet world, their businesses were only weakly tied together. Disney's Infoseek business, in fact, was not even fully owned. However, once its Internet markets became more crystallized, managers at Disney moved to reap the benefits of economies of scope. They aggressively linked their Internet processes with one another and with other parts of Disney. A flexible technology that handles many services together can be less expensive than handling each one separately, particularly when the markets are not too volatile.

Strategic Fit

The manager should understand how the four major process decisions tie together, so as to spot ways of improving poorly designed processes. The choices should fit the situation and each other. When the fit is more *strategic*, the process will be more effective. We examine services and manufacturing processes, looking for ways to test for strategic fit.

MANAGERIAL PRACTICE 3.1   Flexible Automation at R.R. Donnelley

R.R. Donnelley & Sons Company is the largest commercial printer in the United States and the number one printer of books. The industry makes huge capital investments in its printing presses to help drive down the variable unit cost of a book (see Figure 3.7). Its uses a make-to-order strategy, with customers such as book publishers placing new orders as their inventories became too low. However, the "make-ready" time to prepare for the new order and change over the presses for the next customer order was time-consuming. Keeping such expensive equipment idle for change-overs is costly. These high costs force customers, such as book publishers, to make large, infrequent orders for their books.

Flexible automation at its Roanoke, Virginia, plant allows R.R. Donnelley to take a different course, and it is reaping big rewards. The new process begins when the contents of a book arrive via the Internet as a PDF (portable document format) file and go to the plant's prepress department. The intricate manual operations required to prepare text and pictures for printing traditionally caused the biggest bottlenecks. Roanoke now makes its plates digitally instead of from photographic film. With the elimination of steps, such as duplicating and cleaning the file, a job that once took hours can now be completed in 12 minutes. The all-digital workflow also makes possible the creation of electronic instructions, known as ink presets, which improve productivity and quality. Cleaner and sharper plates are created for the presses because, unlike film, electronic type does not have to be repeatedly handled.

With more flexible automation, the Roanoke plant produces 75 percent of its titles in 2 weeks or less, compared with 4 to 6 weeks for a 4-color book using traditional technology. Management created a culture of continuous improvement at the plant, home of some 300 workers. Overall, Roanoke increased throughput by 20 percent without having to buy an additional press

R.R. Donnelly has been able to achieve flexible automation by receiving books digitally and preparing them to go on press electronically. This allows the company to put books on press more quickly and print smaller, more manageable quantities in a single print run.

and binding line, a savings of $15 million. Its presses run around the clock producing 3.5 million books a month; productivity rose 20 percent, and service improved. Book publishers now enjoy a just-in-time product when they want it.

Source: Gene Bylinsky, "Two of America's Best Have Found New Life Using Digital Tech," *Fortune*, vol. 148, no. 4 (2003), pp. 54–55. © 2004 Time Inc. All rights reserved; **www.rrdonnelley.com**, March 21, 2011

Decision Patterns for Service Processes

After analyzing a process and determining its position on the customer-contact matrix in Figure 3.2, it may be apparent that it is improperly positioned, either too far to the left or right, or too far to the top or bottom. Opportunities for improvement become apparent. Perhaps, more customization and customer contact is needed than the process currently provides. Perhaps, instead, the process is too divergent, with unnecessarily flexible flows. Reducing divergence might reduce costs and improve productivity.

The process should reflect its desired competitive priorities. Front offices generally emphasize top quality and customization, whereas back offices are more likely to emphasize low-cost operation, consistent quality, and on-time delivery. The process structure selected then points the way to appropriate choices on customer involvement, resource flexibility, and capital intensity. Figure 3.8 shows how these key process decisions are tied to customer contact. High customer contact at a front-office service process means:

1. *Process Structure.* The customer (internal or external) is present, actively involved, and receives personal attention. These conditions create processes with high divergence and flexible process flows.

2. *Customer Involvement.* When customer contact is high, customers are more likely to become part of the process. The service created for each customer is unique.

3. *Resource Flexibility.* High process divergence and flexible process flows fit with more flexibility from the process's resources—its workforce, facilities, and equipment.

4. *Capital Intensity.* When volume is higher, automation and capital intensity are more likely. Even though higher volume is usually assumed to be found in the back office, it is just as likely to be in the front office for financial services. Information technology is a major type of automation at many service processes, which brings together both resource flexibility and automation.

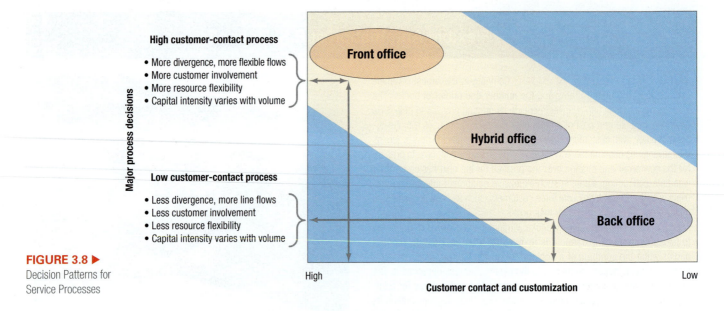

FIGURE 3.8 ▶
Decision Patterns for
Service Processes

Of course, this list provides general tendencies rather than rigid prescriptions. Exceptions can be found, but these relationships provide a way of understanding how service process decisions can be linked coherently.

Decision Patterns for Manufacturing Processes

Just as a service process can be repositioned in the customer-contact matrix, a manufacturing process can also be moved in the product-process matrix. Changes can be made either in the horizontal direction of Figure 3.3 by changing the degree of customization and volume, or they can be moved in the vertical direction by changing process divergence. The production and inventory strategy can also be changed. Competitive priorities must be considered when translating strategy into specific manufacturing processes. Figure 3.9 shows some usual tendencies found in practice. Job and small batch processes are usual choices if top quality, on-time delivery, and flexibility (customization, variety, and volume flexibility) are given primary emphasis. Large batch, line, and continuous-flow processes match up with an emphasis on low-cost operations, consistent quality, and delivery speed.

For production and inventory strategies, the make-to-order strategy matches up with flexibility (particularly customization) and top quality. Because delivery speed is more difficult, meeting due dates and on-time delivery get the emphasis on the time dimension. The assemble-to-order strategy allows delivery speed and flexibility (particularly variety) to be achieved, whereas the make-to-stock strategy is the usual choice if delivery speed and low-cost operations are emphasized. Keeping an item in stock assures quick delivery because it is generally available when needed, without delays in producing it. High volumes open up opportunities to reduce costs.

The process structure selected once again points the way to appropriate choices on customer involvement, resource flexibility, and capital intensity. Figure 3.10 summarizes the relationships between volume and the four key process decisions. High volumes per part type at a manufacturing process typically mean:

1. *Process Structure.* High volumes, combined with a standard product, make a line flow possible. It is just the opposite where a job process produces to specific customer orders.

2. *Customer Involvement.* Customer involvement is not a factor in most manufacturing processes, except for choices made on product variety and customization. Less discretion is allowed with line or continuous-flow processes in order to avoid the unpredictable demands required by customized orders.

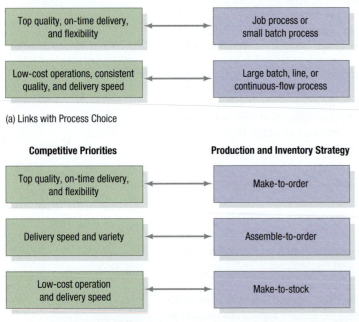

(a) Links with Process Choice

(b) Links with Production and Inventory Strategy

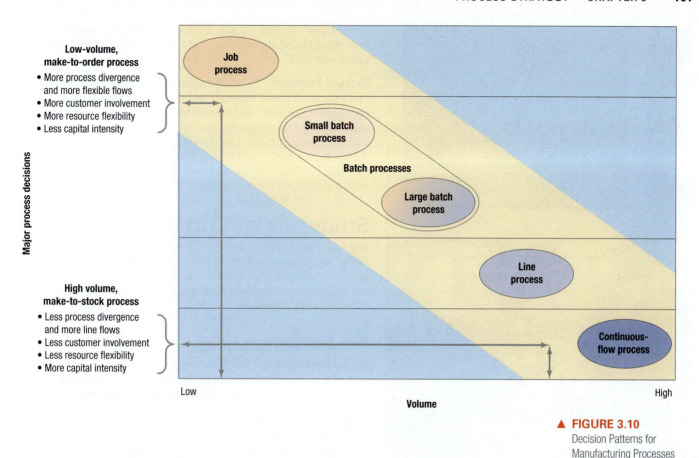

Low-volume,
make-to-order process
- More process divergence and more flexible flows
- More customer involvement
- More resource flexibility
- Less capital intensity

High volume,
make-to-stock process
- Less process divergence and more line flows
- Less customer involvement
- Less resource flexibility
- More capital intensity

Major process decisions

Job process

Small batch process

Batch processes

Large batch process

Line process

Continuous-flow process

Low   High
Volume

▲ **FIGURE 3.10**
Decision Patterns for
Manufacturing Processes

3. *Resource Flexibility.* When volumes are high and process divergence is low, flexibility is not needed to utilize resources effectively, and specialization can lead to more efficient processes.

4. *Capital Intensity.* High volumes justify the large fixed costs of an efficient operation. The King Soopers's bread line (see *The Big Picture* and the video in MyOMLab) is capital-intensive. It is automated from dough mixing to placement of the product on shipping racks. Expanding this process would be expensive. By way of contrast, the King Soopers's custom cake process is labor-intensive and requires little investment to equip the workers.

MyOMLab

Gaining Focus

In the past, new services or products often were added to a facility in the name of better utilizing fixed costs and keeping everything under the same roof. The result was a jumble of competitive priorities, process structures, and technologies. In the effort to do everything, nothing was done well.

Focus by Process Segments   A facility's operations often can neither be characterized nor actually designed for one set of competitive priorities and one process choice. King Soopers (see *The Big Picture* and video in MyOMLab) had three processes under one roof, but management segmented them into three separate operations that were relatively autonomous. At a services facility, some parts of the process might seem like a front office and other parts like a back office. Such arrangements can be effective, provided that sufficient focus is given to each process.

MyOMLab

Plants within plants (PWPs) are different operations within a facility with individualized competitive priorities, processes, and workforces under the same roof. Boundaries for PWPs may be established by physically separating subunits or simply by revising organizational relationships. At each PWP, customization, capital intensity volume, and other relationships are crucial and must be complementary. The advantages of PWPs are fewer layers of management, greater ability to rely on team problem solving, and shorter lines of communication between departments.

plants within plants (PWPs)
Different operations within a facility with individualized competitive priorities, processes, and workforces under the same roof.

Focused Service Operations   Service industries also implement the concepts of focus and PWPs. Specialty retailers opened stores with smaller, more accessible spaces. These focused facilities generally chipped away at the business of large department stores. Using the same philosophy,

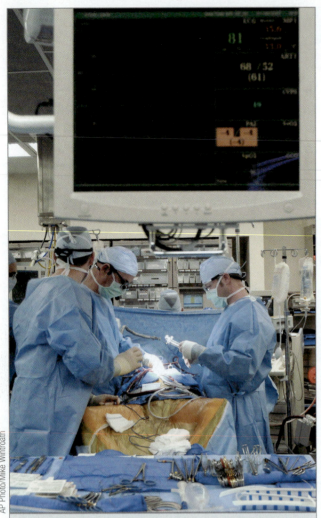

AP Photo/Mike Wintroath

Focused factories are not just found in manufacturing. This single-speciality facility focuses just on heart surgery and has all of the advanced resources needed that cannot be provided by a general hospital. Another example is the Toronto-based Shouldice Clinic, which focuses just on hernias.

some department stores now focus on specific customers or products. Remodeled stores create the effect of many small boutiques under one roof.

Focused Factories Hewlett-Packard, Rolls-Royce, Japan's Ricoh and Mitsubishi, and Britain's Imperial Chemical Industries PLC are some of the firms that created **focused factories**, splitting large plants that produced all the company's products into several specialized smaller plants. The theory is that narrowing the range of demands on a facility will lead to better performance because management can concentrate on fewer tasks and lead a workforce toward a single goal.

Strategies for Change

The four major process decisions represent broad, strategic issues. Decisions that are made must be translated into actual process designs or redesigns. We conclude with two different but complementary philosophies for process design: (1) process reengineering and (2) process improvement.

Process Reengineering

Reengineering is the fundamental rethinking and radical redesign of processes to improve performance dramatically in terms of cost, quality, service, and speed. Process reengineering is about reinvention, rather than incremental improvement. It is strong medicine and not always needed or successful. Pain, in the form of layoffs and large cash outflows for investments in information technology, almost always accompanies massive change. However, reengineering processes can have big payoffs. Table 3.4 lists the key elements of the overall approach.

Reengineering has led to many successes and will continue to do so. However, it is not simple or easily done, nor is it appropriate for all processes or all organizations. The best understanding of a process, and how to improve it, often lies with the people who perform the work each day, not with cross-functional teams or top management.

focused factories

The result of a firm's splitting large plants that produced all the company's products into several specialized smaller plants.

reengineering

The fundamental rethinking and radical redesign of processes to improve performance dramatically in terms of cost, quality, service, and speed.

TABLE 3.4 | **KEY ELEMENTS OF REENGINEERING**

Element	Description
Critical processes	The emphasis of reengineering should be on core business processes. Normal process-improvement activities can be continued with the other processes.
Strong leadership	Senior executives must provide strong leadership for reengineering to be successful. Otherwise, cynicism, resistance ("we tried that before"), and boundaries between departments can block radical changes.
Cross-functional teams	A team, consisting of members from each functional area affected by the process change, is charged with carrying out a reengineering project. Self-managing teams and employee empowerment are the rule rather than the exception.
Information technology	Information technology is a primary enabler of process engineering. Most reengineering projects design processes around information flows, such as customer order fulfillment.
Clean-slate philosophy	Reengineering requires a "clean-slate" philosophy—that is, starting with the way the customer wants to deal with the company. To ensure a customer orientation, teams begin with internal and external customer objectives for the process.
Process analysis	Despite the clean-slate philosophy, a reengineering team must understand things about the current process: what it does, how well it performs, and what factors affect it. The team must look at every procedure involved in the process throughout the organization.

Process Improvement

Process improvement is the systematic study of the activities and flows of each process to improve it. Its purpose is to "learn the numbers," understand the process, and dig out the details. Once a process is really understood, it can be improved. The relentless pressure to provide better quality at a lower price means that companies must continually review all aspects of their operations. Process improvement goes on, whether or not a process is reengineered. There is always a better way.

process improvement
The systematic study of the activities and flows of each process to improve it.

An individual or a whole team examines the process, using the tools described in the next chapter. One must look for ways to streamline tasks, eliminate whole processes entirely, cut expensive materials or services, improve the environment, or make jobs safer. One must find the ways to trim costs and delays and to improve customer satisfaction.

LEARNING GOALS IN REVIEW

1 **Explain why processes exist everywhere in all organizations.** The "Process Strategy Across the Organization" section, pp. 90–91, demonstrates that supply chain process exist throughout the supply chain and actually exist in all places throughout the organization. Pay particular attention to Tables 3.1 and 3.2.

2 **Discuss the four major process decisions.** The "Process Strategy Decisions" section, pp. 91–92, identifies the four key decisions around which we develop a vocabulary for understanding operations. Figure 3.1 shows how they interact in creating an effective process design.

3 **Position a process on the customer-contact matrix or product-process matrix.** The Customer-Contact Matrix for service processes in Figure 3.2 shows how the degree of customer contact and customization are linked with process divergence and line flows. Three natural positions emerge: the front office, hybrid office, and back office. In manufacturing, the key drivers are customization and volume, which are linked with line flows and the extent of repetitive work. Figure 3.3 shows these relationships in the form of the Product-Process Matrix, with natural positions ranging from the job process to the continuous flow process.

4 **Configure operations into layouts.** The "Layout" section, pp. 99–102, puts the process structure into a physical form by showing here each operation is located within the facility.

Example 3.1 shows how to develop a block plan, and evaluate it with the help of the *Layout* Solver of OM Explorer. See Solved Problem 3.1 for another example.

5 **Define customer involvement, resource flexibility, capital intensity, and economies of scope.** These topics are covered in the "Customer Involvement," "Resource Flexibility," "Capital Intensity," and "Economies of Scope" sections on pp. 100–104. Note that customer involvement has advantages and disadvantages, resource flexibility applies to both workforce and equipment, and economies of scope in certain situations can break the inverse relationship between resource flexibility and capital intensity.

6 **Discuss how process decisions should fit together.** The "Strategic Fit" section, pp. 104–108, describes how the four major process decisions should tie together. Figures 3.8 and 3.9 show the decision patterns in pictorial form. The section concludes with a way to achieve these patterns by gaining focus, either with focused factories or gaining focus by process segments.

7 **Define process reengineering and process improvement.** The "Strategies for Change," pp. 108–109, describe both approaches to finding better process designs. Table 3.4 gives the key elements of reengineering. Process improvement is more of an incremental approach which uses the tools described in the next chapter.

MyOMLab helps you develop analytic skills and assesses your progress with multiple problems on the break-even analysis in choosing between two different processes, the weighted-distance method, and layout.

MyOMLab Resources	Titles	Link to the Book
Videos	*King Soopers Bakery: Process Choice*	Manufacturing Process Structuring
	Process Choice: Pearson Education Information Technology	Capital Intensity
OM Explorer Solver	Layout	Layout
OM Explorer Tutors	3.1 Distance Measures	Applying the Weighted-Distance Method Example 3.1 (pp. 99–100)
	3.2 Breakeven for Equipment Selection	Resource Flexibility Figure 3.7 (p. 102)
POM for Windows: Layout	Layout	Example 3.1 (pp. 99–100)

MyOMLab Resources	Titles	Link to the Book
Tutor Exercises	3.1—Mt. Mudge	Applying the Weighted-Distance Method Example 3.1 (pp. 99–100)
	3.2—Break-Even for Equipment Selection	Resource Flexibility Figure 3.7 (p. 102)
SmartDraw	Often used to prepare detailed layouts and floor plans	Detailed Layout
Virtual Tours	1. Leannie Company Doll Factory	Process Choice; Production and Inventory Strategy
	2. LA Aluminum Casting Company	Process Choice; Production and Inventory Strategy
Internet Exercises	3.1—United Parcel Service	Service Strategy; Customer Involvement
	3.2—Carnival and Twilight	
	3.3—Timbuk2	
Advanced Problems	3.1 CCI Electronics	Layout
	3.2 Getwell Hospital	Layout
Additional Cases	Car Lube Operations	Layout
	Hightech, Inc.	Layout
	The Pizza Connection	Design a Detailed layout
	Bill's Hardware	Capital Intensity
	The Big Picture: Process Choice at King Soopers Bakery	Manufacturing Process Structuring
Key Equations		
Image Library		

Key Equations

1. Euclidean distance: $d_{AB} = \sqrt{(x_A - x_B)^2 + (y_A - y_B)^2}$
2. Rectilinear distance: $d_{AB} = |x_A - x_B| + |y_A - y_B|$

Key Terms

assemble-to-order strategy 96
automation 103
back office 94
batch process 96
block plan 98
capital intensity 92
closeness matrix 98
continuous flow process 96
customer contact 93
customer involvement 92
economies of scope 104
Euclidean distance 99
fixed automation 103
flexible (or programmable) automation 103
flexible flow 93

flexible workforce 102
focused factories 108
front office 94
hybrid office 94
industrial robot 103
job process 95
layout 92
line flow 94
line process 96
make-to-order strategy 96
make-to-stock strategy 97
mass customization 97
mass production 97

plants within plants (PWPs) 107
postponement 96
process choice 95
process divergence 93
process improvement 109
process strategy 90
process structure 92
rectilinear distance 99
reengineering 108
resource flexibility 92
supply chain processes 91
weighted-distance method 98

Solved Problem 1

A defense contractor is evaluating its machine shop's current layout. Figure 3.11 shows the current layout, and the table shows the closeness matrix for the facility measured as the number of trips per day between department pairs. Safety and health regulations require departments E and F to remain at their current locations.

	TRIPS BETWEEN DEPARTMENTS					
Department	A	B	C	D	E	F
A	—	8	3		9	5
B		—		3		
C			—		8	9
D				—		3
E					—	3
F						—

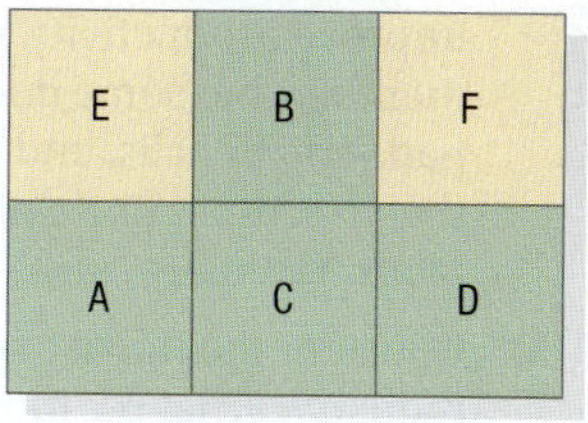

▲ FIGURE 3.11
Current Layout

a. Use trial and error to find a better layout.

b. How much better is your layout than the current layout in terms of the *wd* score? Use rectilinear distance.

SOLUTION

a. In addition to keeping departments E and F at their current locations, a good plan would locate the following department pairs close to each other: A and E, C and F, A and B, and C and E. Figure 3.12 was worked out by trial and error and satisfies all these requirements. Start by placing E and F at their current locations. Then, because C must be as close as possible to both E and F, put C between them. Place A below E, and B next to A. All of the heavy traffic concerns have now been accommodated. Department D, located in the remaining space, does not need to be relocated.

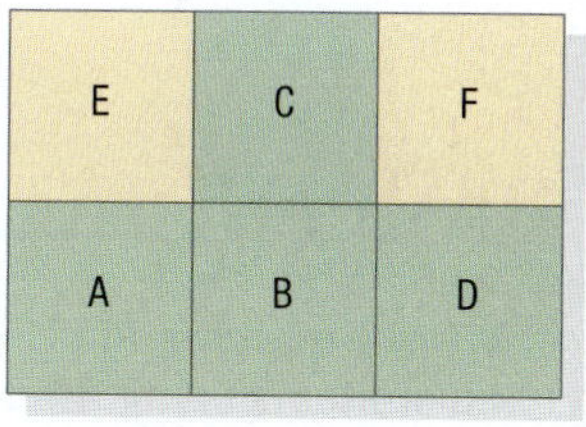

▲ FIGURE 3.12
Proposed Layout

Department Pair	Number of Trips (1)	CURRENT PLAN		PROPOSED PLAN	
		Distance (2)	*wd* Score (1) × (2)	Distance (3)	*wd* Score (1) × (3)
A, B	8	2	16	1	8
A, C	3	1	3	2	6
A, E	9	1	9	1	9
A, F	5	3	15	3	15
B, D	3	2	6	1	3
C, E	8	2	16	1	8
C, F	9	2	18	1	9
D, F	3	1	3	1	3
E, F	3	2	6	2	6
			wd = 92		*wd* = 67

b. The table reveals that the *wd* score drops from 92 for the current plan to 67 for the revised plan, a 27 percent reduction.

Discussion Questions

1. What processes at manufacturing firms are really service processes that involve considerable customer contact? Can customer contact be high, even if the process only has internal customers?

2. Consider this sign seen in a local restaurant: "To-go orders do NOT include complimentary chips and salsa. If you have any questions, see our management, NOT our employees." What impact does this message have on its employees, their service processes, and customer satisfaction? Contrast this approach with the one taken by a five-star restaurant. Are the differences primarily due to different competitive priorities?

3. How do the process strategies of eBay and McDonald's differ, and how do their choices relate to customer-introduced variability?

4. Medical technology can outfit a patient with an artificial heart, or cure vision defects with the touch of a laser. However, hospitals still struggle with their back-office processes, such as getting X-ray files from radiology on the fourth floor to the first-floor view boxes in the emergency room without having to send a runner. More than 90 percent of the estimated 30 billion health transactions each year are conducted by telephone, fax, or mail. To what extent, and how, can information technology

improve productivity and quality for such processes? Remember that some doctors are not ready to give up their pads and pencils, and many hospitals have strong lines drawn around its departments, such as pharmacy, cardiology, radiology, and pediatrics.

5. Consider the range of processes in the financial services industry. What position on the customer-contact matrix would the process of selling financial services to municipalities occupy? The process of preparing monthly fund balance reports? Explain why they would differ.

6. Performance criteria important in creating a layout can go well beyond communication and materials handling. Identify the types of layout performance criteria that might be most important in the following settings.

   a. Airport

   b. Bank

   c. Classroom

   d. Product designers' office

   e. Law firm

7. Rate operators at a call center, who respond to queries from customers who call in about the company's product, on each of the five dimensions of customer contact in Table 3.3. Use a seven-point scale, where 1 = very low and 7 = very high. For example, the operators newer are physically present with the customer, and so they would get a score of 1 for physical presence. Explain your ratings, and then calculate a combined score for the overall customer contact. Did you use equal weights in calculating the combined score? Why or why not? Where is your process positioned on the customer-contact matrix? Is it properly aligned? Why or why not?

8. Select one of the three processes shown in the MyOMLab's video for King Soopers (bread, pastry, or custom cakes). What kind of transformation process, process choice, and inventory strategy are involved? Is the process properly aligned? Explain.

Problems

The OM Explorer and POM for Windows software is available to all students using the 10th edition of this textbook. Go to **www.pearsonhighered.com/krajewski** to download these computer packages. If you purchased MyOMLab, you also have access to Active Models software and significant help in doing the following problems. Check with your instructor on how best to use these resources. In many cases, the instructor wants you to understand how to do the calculations by hand. At the least, the software provides a check on your calculations. When calculations are particularly complex and the goal is interpreting the results in making decision, the software entirely replaces the manual calculations.

Problems 1 and 2 apply break-even analysis (discussed in Supplement A, "Decision Making") to process decisions.

1. Dr. Gulakowicz is an orthodontist. She estimates that adding two new chairs will increase fixed costs by $150,000, including the annual equivalent cost of the capital investment and the salary of one more technician. Each new patient is expected to bring in $3,000 per year in additional revenue, with variable costs estimated at $1,000 per patient. The two new chairs will allow Dr. Gulakowicz to expand her practice by as many as 200 patients annually. How many patients would have to be added for the new process to break even?

2. Two different manufacturing processes are being considered for making a new product. The first process is less capital-intensive, with fixed costs of only $50,000 per year and variable costs of $700 per unit. The second process has fixed costs of $400,000, but has variable costs of only $200 per unit.

   a. What is the break-even quantity, beyond which the second process becomes more attractive than the first?

   b. If the expected annual sales for the product is 800 units, which process would you choose?

3. Baker Machine Company is a job shop that specializes in precision parts for firms in the aerospace industry. Figure 3.13 shows the current block plan for the key manufacturing centers of the 75,000-square-foot facility. Refer to the following closeness matrix and use rectilinear

distance (the current distance from inspection to shipping and receiving is three units) to calculate the change in the weighted distance, wd, score if Baker exchanges the locations of the tool crib and inspection.

CLOSENESS MATRIX

Department	Trips Between Departments					
	1	2	3	4	5	6
1. Burr and grind	—	8	3		9	5
2. Numerically controlled (NC) equipment		—		3		
3. Shipping and receiving			—		8	9
4. Lathes and drills				—		3
5. Tool crib					—	3
6. Inspection						—

◀ **FIGURE 3.13**
Current Layout

4. Baker Machine (see Problem 3) is considering two alternative layouts. Compare the wd scores using rectilinear distance of the following two block plans to determine which alternative layout is better.

◀ **FIGURE 3.13(a)**
Alternative Layout 1

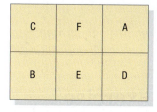

◀ **FIGURE 3.13(b)**
Alternative Layout 2

5. The head of the information systems group at Conway Consulting must assign six new analysts to offices. The following closeness matrix shows the expected frequency of contact between analysts. The block plan in Figure 3.14 shows the available office locations (1–6) for the six analysts (A–F). Assume equal-sized offices and rectilinear distance.

CLOSENESS MATRIX

| Analyst | Contacts Between Analysts | | | | | |
	A	B	C	D	E	F
Analyst A	—		6			
Analyst B		—		12		
Analyst C			—	2	7	
Analyst D				—		4
Analyst E					—	
Analyst F						—

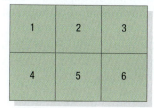

◀ **FIGURE 3.14**
Conway Consulting's Block Plan

Evaluate the *wd* scores of the following three alternative layouts, again assuming rectilinear distance, and determine which is best.

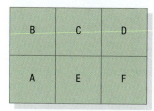

◀ **FIGURE 3.14(a)**
Alternative Layout 1

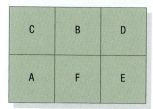

◀ **FIGURE 3.14(b)**
Alternative Layout 2

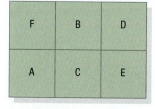

◀ **FIGURE 3.14(c)**
Alternative Layout 3

6. Richard Garber is the head designer for Matthews and Novak Design Company. Garber has been called in to design the layout for a newly constructed office building. From statistical samplings over the past three months, Garber developed the following closeness matrix for daily trips between the department's offices.

CLOSENESS MATRIX

| Department | Trips Between Departments | | | | | |
	A	B	C	D	E	F
A	—	25	90			185
B		—			105	
C			—		125	125
D				—	25	
E					—	105
F						—

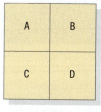

◀ **FIGURE 3.15**
Alternative Block Plan

a. If other factors are equal, which two offices should be located closest together?

b. Figure 3.15 shows an alternative layout for the department. What is the total weighted-distance score for this plan based on rectilinear distance and assuming that offices A and B are three units of distance apart?

c. Use the explicit enumeration method of the POM for Windows software to find the block plan that minimize the total weighted-distance score.

7. A firm with four departments has the following closeness matrix and the current block plan shown in Figure 3.16.

a. What is the weighted-distance score for the current layout (assuming rectilinear distance)?

CLOSENESS MATRIX

| Department | Trips Between Departments | | | |
	A	B	C	D
A	—	12	10	8
B		—	20	6
C			—	0
D				—

◀ **FIGURE 3.16**
Current Block Plan

b. Develop a better layout. What is its total weighted-distance score?

8. The department of engineering at a university in New Jersey must assign six faculty members to their new offices. The following closeness matrix indicates the expected number of contacts per day between professors. The available office spaces (1–6) for the six faculty members are shown in Figure 3.17. Assume equal-sized offices. The distance between offices 1 and 2 (and between offices 1 and 3) is 1 unit, whereas the distance between offices 1 and 4 is 2 units.

a. Because of their academic positions, Professor A must be assigned to office 1, Professor C must be assigned to office 2, and Professor D must be assigned to office 6. Which faculty members should be assigned to offices 3, 4, and 5, respectively, to minimize the total weighted-distance score (assuming rectilinear distance)?

b. What is the weighted-distance score of your solution?

CLOSENESS MATRIX

Professor	A	B	C	D	E	F
		Contacts Between Professors				
A	—		4			
B		—		12		10
C			—		2	7
D				—		4
E					—	
F						—

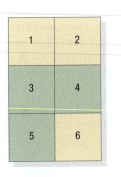

◀ **FIGURE 3.17**
Available Space

Active Model Exercise

This Active Model for Example 3.1 appears in MyOMLab. It allows you to see the effects of performing paired swaps of departments.

QUESTIONS

1. What is the current total weighted-distance score?

2. Use the swap button one swap at a time. If the swap helps, move to the next pair. If the swap does not help, hit the swap button once again to put the departments back. What is the minimum weighted-distance score after all swaps have been tried?

3. Look at the two data tables, and use the yellow-shaded column to put departments in spaces. What space assignments lead to the minimum cost? What is this cost?

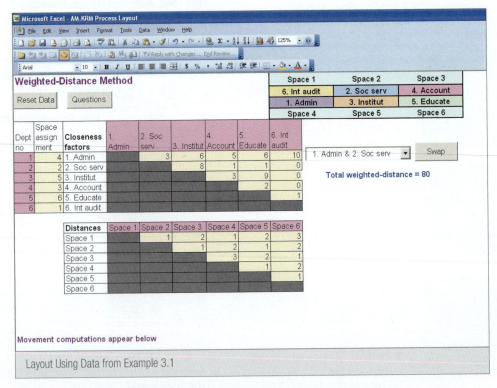

Layout Using Data from Example 3.1

CASE   Custom Molds, Inc.

Custom Molds, Inc., manufactures custom-designed molds for plastic parts and produces custom-made plastic connectors for the electronics industry. Located in Tucson, Arizona, Custom Molds was founded by the father-and-son-team of Tom and Mason Miller in 1987. Tom Miller, a mechanical engineer, had more than 20 years of experience in the connector industry with AMP, Inc., a large multinational producer of electronic connectors. Mason Miller graduated from the Arizona State University in 1986 with joint degrees in chemistry and chemical engineering.

The company was originally formed to provide manufacturers of electronic connectors with a source of high-quality, custom-designed molds for producing plastic parts. The market consisted mainly of the product design and development divisions of those manufacturers. Custom Molds worked closely with each customer to design and develop molds to be used in the customer's product development processes. Thus, virtually every mold had to meet exacting standards and was somewhat unique. Orders for multiple molds would arrive when customers moved from the design and pilot-run stage of development to large-scale production of newly designed parts.

As the years went by, Custom Molds's reputation grew as a designer and fabricator of precision molds. Building on this reputation, the Millers decided to expand into the limited manufacture of plastic parts. Ingredient-mixing facilities and injection-molding equipment were added, and by the mid-1990s, Custom Molds developed its reputation to include being a supplier of high-quality plastic parts. Because of limited capacity, the company concentrated its sales efforts on supplying parts that were used in limited quantities for research and development efforts and in preproduction pilot runs.

Production Processes

By 2000, operations at Custom Molds involved two distinct processes: one for fabricating molds and one for producing plastic parts. Although different, in many instances these two processes were linked, as when a customer would have Custom Molds both fabricate a mold and produce the necessary parts to support the customer's research and design efforts. All fabrication and production operations were housed in a single facility. The layout was characteristic of a typical job shop, with like processes and similar equipment grouped in various places in the plant. Figure 3.18 shows a layout of the plant floor. Multiple pieces of various types of high-precision machinery,

Dock		Dock	
Receiving raw materials inventory	Lunch room	Packing and shipping finished goods inventory	
Dry mix	Cut and trim	Testing and inspection	
Wet mix		Injection machines	
Assembly			
Offices		Mold fabrication	

▲ **FIGURE 3.18**
Plant Layout

including milling, turning, cutting, and drilling equipment, were located in the mold-fabrication area.

Fabricating molds is a skill-oriented, craftsman-driven process. When an order is received, a design team, comprising a design engineer and one of 13 master machinists, reviews the design specifications. Working closely with the customer, the team establishes the final specifications for the mold and gives them to the master machinist for fabrication. It is always the same machinist who was assigned to the design team. At the same time, the purchasing department is given a copy of the design specifications, from which it orders the appropriate raw materials and special tooling. The time needed to receive the ordered materials is usually three to four weeks. When the materials are received for a particular mold, the plant master scheduler reviews the workload of the assigned master machinist and schedules the mold for fabrication.

Fabricating a mold takes from two to four weeks, depending on the amount of work the machinist already has scheduled. The fabrication process itself takes only three to five days. Upon completion, the mold is sent to the testing and inspection area, where it is used to produce a small number of parts on one of the injection molding machines. If the parts meet the design specifications established by the design team, the mold is passed on to be cleaned and polished. It is then packed and shipped to the customer. One day is spent inspecting and testing the mold and a second day cleaning, polishing, packing, and shipping it to the customer. If the parts made by the mold do not meet design specifications, the mold is returned to the master machinist for retooling and the process starts over. Currently, Custom Molds has a published lead time of nine weeks for delivery of custom-fabricated molds.

The manufacturing process for plastic parts is somewhat different from that for mold fabrication. An order for parts may be received in conjunction with an order for a mold to be fabricated. In instances where Custom Molds has previously fabricated the mold and maintains it in inventory, an order may be just for parts. If the mold is already available, the order is reviewed by a design engineer, who verifies the part and raw material specifications. If the design engineer has any questions concerning the specifications, the customer is contacted and any revisions to specifications are mutually worked out and agreed upon.

Upon acceptance of the part and raw material specifications, raw material orders are placed and production is scheduled for the order. Chemicals and compounds that support plastic-parts manufacturing are typically ordered and received within one week. Upon receipt, the compounds are first dry-mixed and blended to achieve the correct composition. Then, the mixture is wet-mixed to the desired consistency (called *slurry*) for injection into molding machines. When ready, the slurry is transferred to the injection molding area by an overhead pipeline and deposited in holding tanks adjacent to the injection machines. The entire mixing process takes only one day.

When the slurry is staged and ready, the proper molds are secured—from inventory or from the clean and polish operation if new molds were fabricated for the order—and the parts are manufactured. Although different parts require different temperature and pressure settings, the time to produce a part is relatively constant. Custom Molds has the capacity to produce 5,000 parts per day in the injection-molding department; historically, however, the lead time for handling orders in this department has averaged one week. Upon completion of molding, the parts are taken to the cut and trim operation, where they are disconnected and leftover flashing is removed. After being inspected, the parts may be taken to assembly or transferred to the packing and shipping area for shipment to the customer. If assembly of the final parts is not required, the parts can be on their way to the customer two days after being molded.

Sometimes, the final product requires some assembly. Typically, this entails attaching metal leads to plastic connectors. If assembly is necessary, an additional three days are needed before the order can be shipped. Custom Molds is currently quoting a three-week lead time for parts not requiring fabricated molds.

The Changing Environment

In early 2009, Tom and Mason Miller began to realize that the electronics industry they supplied, along with their own business, was changing. Electronics manufacturers had traditionally manufactured their own component parts to reduce costs and ensure a timely supply of parts. By the 1990s, this trend had changed. Manufacturers were developing strategic partnerships with parts suppliers to ensure the timely delivery of high-quality, cost-effective parts. This approach allowed funds to be diverted to other uses that could provide a larger return on investment.

The impact on Custom Molds could be seen in sales figures over the past three years. The sales mix was changing. Although the number of orders per year for mold fabrication remained virtually constant, orders for multiple molds were declining, as shown in the following table:

NUMBER OF ORDERS			
Order Size	Molds 2006	Molds 2007	Molds 2008
1	80	74	72
2	60	70	75
3	40	51	55
4	5	6	5
5	3	5	4
6	4	8	5
7	2	0	1
8	10	6	4
9	11	8	5
10	15	10	5
Total orders	230	238	231

The reverse was true for plastic parts, for which the number of orders per year had declined, but for which the order sizes were becoming larger, as illustrated in the following table:

NUMBER OF ORDERS			
Order Size	Parts 2006	Parts 2007	Parts 2008
50	100	93	70
100	70	72	65
150	40	30	35
200	36	34	38
250	25	27	25
500	10	12	14
750	1	3	5
1,000	2	2	8
3,000	1	4	9
5,000	1	3	8
Total orders	286	280	277

During this same period, Custom Molds began having delivery problems. Customers were complaining that parts orders were taking four to five weeks instead of the stated three weeks and that the delays were disrupting production schedules. When asked about the situation, the master scheduler said that determining when a particular order could be promised for delivery was difficult. Bottlenecks were occurring during the production process, but where or when they would occur could not be predicted. The bottlenecks always seemed to be moving from one operation to another.

Tom Miller thought that he had excess labor capacity in the mold-fabrication area. So, to help push through those orders that were behind schedule, he assigned one of the master machinists the job of identifying and expediting those late orders. However, that tactic did not seem to help much. Complaints about late deliveries were still being received. To add to the problems, two orders had been returned recently because of the number of defective parts. The Millers knew that something had to be done. The question was "What?"

QUESTIONS

1. What are the major issues facing Tom and Mason Miller?
2. What are the competitive priorities for Custom Molds's processes and the changing nature of the industry?
3. What alternatives might the Millers pursue? What key factors should they consider as they evaluate these alternatives?

Source: This case was prepared by Dr. Brooke Saladin, Wake Forest University, as a basis for classroom discussion. Copyright © Brooke Saladin. Used with permission.

Selected References

Brink, Harold, Senthiah, and Rajan Naik. "A Better Way to Automate Service Operations." *McKinsey on Business Technology*, no. 20 (Summer, 2010), pp. 1–10.

Baghai, Ramin, Edward H. Levine, and Saumya S. Sutaria. "Service-Line Strategies for US Hospitals." *The McKinsey Quarterly* (July 2008), pp. 1–9.

Booth, Alan. "The Management of Technical Change: Automation in the UK and USA since 1950." *The Economic History Review*, vol. 62, no. 2 (May 2009), pp. 493–494.

Chase, Richard B. and Uday M. Apte. "A History of Research in Service Operations: What's the Big Idea?" *Journal of Operations Management*, vol. 25 (2007), pp. 375–386.

Fisher, Marshall L. "Bob Hayes: Forty Years of Leading Operations Management Into Uncharted Waters." *Production and Operations Management*, vol. 16, no. 2, (March–April 2007), pp. 159–168.

Grover, Varun, and Manoj K. Malhotra. "Business Process Reengineering: A Tutorial on the Concept, Evolution, Method, Technology, and Application." *Journal of Operations Management*, vol. 15, no. 3 (1997), pp. 194–213.

Hayes, Robert. "Operations, Strategy, and Technology: Pursuing the Competitive Edge." *Strategic Direction*, vol. 22, no. 7, (2006).

Johansson, Pontus and Jan Olhger. "Linking Product-Process Matrices for Manufacturing and Industrial Service Operations." *International Journal of Production Economics*, vol. 104 (2006), pp. 615–624.

Hammer, Michael. "Deep Change: How Operational Innovation Can Transform Your Company." *Harvard Business Review*, vol. 82, no. 4 (April 2004), pp. 85–93.

Hill, Terry. *Manufacturing Strategy: Text and Cases*, 3rd ed. Homewood, IL: Irwin/McGraw-Hill, 2000.

Jack, Eric, and John Collis. "Strengthen and Tone: A Flexible Approach to Operations Can Build Some Serious Muscle." *APICS Magazine* (June 2006), pp. 35–38.

Kung, Peter and Claus Hagen. "The Fruits of Business Process Management: An Experience Report from a Swiss Bank." *Business Process Management Journal*, vol. 13, no. 4 (2007), pp. 477–487.

Malhotra, Manoj K., and Larry P. Ritzman. "Resource Flexibility Issues in Multistage Manufacturing." *Decision Sciences*, vol. 21, no. 4 (1990), pp. 673–690.

Metters, Richard, Kathryn King-Metters, and Madeleine Pullman. *Successful Service Operations Management*. Mason, OH: South-Western, 2003.

Prajogo, Daniel. "The Implementation of Operations Management Techniques in Service Organisations." *International Journal of Operations & Production Management*, vol. 26, No. 12 (2006), pp. 1374–1390.

Rayport, Jeffrey F., and Bernard J. Jaworski. "Best Face Forward." *Harvard Business Review*, vol. 82, no. 12 (2003), pp. 47–58.

Safizadeh, M. Hossein, Joy M. Field, and Larry P. Ritzman. "An Empirical Analysis of Financial Services Processes with a Front-Office or Back-Office Orientation." *Journal of Operations Management*, vol. 21, no. 5 (2003), pp. 557–576.

Safizadeh, M. Hossein, Larry P. Ritzman, and Debasish Mallick. "Revisiting Alternative Theoretical Paradigms in Manufacturing." *Production and Operations Management*, vol. 9, no. 2 (2000), pp. 111–127.

Sehgal, Sanjay, B.S. Sahay, and S.K. Goyal. "Reengineering the Supply Chain in a Paint Company." *International Journal of Productivity and Performance Management*, vol. 55, no. 8 (2006), pp. 655–670.

Skinner, Wickham. "Operations Technology: Blind Spot in Strategic Management." *Interfaces*, vol. 14 (January–February 1984), pp. 116–125.

Swink, Morgan and Anand Nair. "Capturing the Competitive Advantages of AMT: Design-Manufacturing Integration as a Complementary Asset." *Journal of Operations Management*, vol. 25 (2007), pp. 736–754.

Zomerdijk, Leonieke G. and Jan de Vries. "Structuring Front Office and Back Office Work in Service Delivery Systems." *International Journal of Operations & Production Management*, vol. 27, no. 1 (2007), pp. 108–131.

paul prescott/Shutterstock.com

4

PROCESS ANALYSIS

McDonald's continually seeks ways to improve its processes so as to provide better quality at a lower cost, with more sustainable resources. This effort combined with innovative menu options pays off. In September, 2011 it delivered its 100th consecutive month of positive global comparable sales. Sales were up by 3.9% in the US and 2.7% in Europe.

McDonald's Corporation

System revenues (company-operated and franchised restaurants) at McDonald's reached a record-high $24 billion in 2010. It has more than 32,000 restaurants around the world and 62 million customers visit them each day. It employs 1.7 million people across the globe. Its stock price in October 2011 was $89.94. Things were not so good in 2002, when customer complaints were growing more frequent and bitter. Its stock price was only $16.08 at the end of 2002. McDonald's is now listening to the customers again, and changing its processes to reflect it. The board brought on a new CEO who had spent 20 years on the operational side of the business. With a zeal for measuring customer satisfaction and sharing the data freely with operators, he pulled off a turnaround that stunned everyone in the business with its speed and scope.

Initiatives were launched to collect performance measures and revamp McDonald's processes to meet customer expectations. McDonald's sends mystery shoppers to restaurants to conduct anonymous reviews using a hard-number scoring system. Mystery diners from outside survey firms jot down on a paper checklist their grades for speed of service; food temperature; presentation and taste; cleanliness of the counter, tables and condiment islands; even whether the counter crewperson smiles at diners. Trailing six-month and year-to-date results are posted on an internal McDonald's Web site so owners can compare their scores with regional averages. Operators could now pinpoint lingering problems, and performance measures focus operators' attention on needed

119

process changes. Customers now are encouraged to report their experience at a particular U.S. restaurant by e-mail, regular mail, or toll-free telephone call.

Another initiative was to send 900 operations missionaries into the field, each visiting stores multiple times to fine-tune processes while also conducting day-long seminars where store managers could share tips from corporate kitchen gurus—such as where to place staff—that would shave previous seconds off average service times. The process was changed back to toasting buns rather than microwaving them, giving them an even sweeter caramelized flavor. Other initiatives were taken on McDonald's fast lane. Every six seconds shaved off the wait time adds a percentage point to sales growth. Outdoor menu boards now have more pictures and fewer words. An LED display confirms what customers say, reducing confusion later on. Premium sandwiches are put in boxes rather than paper wrappers, saving a few seconds, and boxes are color coded by sandwich to improve speed and accuracy.

Processes are also being changed to be environment friendly, reaching back from the counters of its restaurants into its supply chain. The U.S. menu involves 330 unique consumer package designs, with 83 percent now made from paper or some other wood-fiber material. Its bulk cooking oil delivery system uses reusable containers, eliminating more than 1,500 pounds of packaging waste per restaurant per year. Its commitment to using sustainable resources has it working with its suppliers to improve coatings on its food packaging. It has shifted more than 18,000 metric tons of fish away from unsustainable sources over the past 5 years. It emphasizes reuse and recycling, managing electrical energy, and effective water management. It also seeks certified sustainable sources for its food. For example, it is piloting a three-year beef farm study to investigate the carbon emissions on 350 beef farms.

All in all, performance measurement and process analysis are increasing customer value and paying off on the bottom line.

Source: Daniel Kruger, "You Want Data with That?" *Forbes*, vol. 173, no. 6 (March 2004), pp. 58–60; **http://www.mcdonalds.com**, April 5, 2011.

LEARNING GOALS   *After reading this chapter, you should be able to:*

1. Explain a systematic way to analyze processes.
2. Define flowcharts, swim lane flowcharts, service blueprints, and process charts.
3. Describe the various work measurement techniques.
4. Identify metrics for process evaluation.
5. Describe Pareto charts, cause-and-effect diagrams, and process simulation.
6. Create better processes using benchmarking.
7. Identify keys for effective process management.

Processes are perhaps the least understood and managed aspect of a business. No matter how talented and motivated people are, a firm cannot gain competitive advantage with faulty processes. Just as Mark Twain said of the Mississippi River, a process just keeps rolling on—with one big difference. Most processes can be improved if someone thinks of a way and implements it effectively. Indeed, companies will either adapt processes to the changing needs of customers or cease to exist. Long-term success comes from managers and employees who really understand their businesses. But all too often, highly publicized efforts that seem to offer quick-fix solutions fail to live up to expectations over the long haul, be they programs for conceptualizing a business vision, conducting culture transformation campaigns, or providing leadership training.

Within the field of operations management, many important innovations over the past several decades include work-simplification or better-methods programs, statistical process control, optimization techniques, statistical forecasting techniques, material requirements planning, flexible automation, lean manufacturing, total quality management, reengineering, Six Sigma programs, enterprise resource planning, and e-commerce. We cover these important approaches in the following chapters because they can add significant customer value to a process. However, they are best viewed as just part of a total system for the effective management of work processes, rather than cure-alls.

Of course, process analysis is needed for both reengineering and process improvement, but it is also part of monitoring performance over time. In this chapter, we begin with a systematic approach for analyzing a process that identifies opportunities for improvement, documents the current process, evaluates the process to spot performance gaps, redesigns the process to eliminate the gaps, and implements the desired changes. The goal is continual improvement.

Four supporting techniques—(1) flowcharts, (2) service blueprints, (3) work measurement techniques, and (4) process charts—can give good insights into the current process and the proposed changes. Data analysis tools, such as checklists, bar charts, Pareto charts, and cause-and-effect diagrams, allow the analyst to go from problem symptoms to root causes. Simulation is a more advanced technique to evaluate process performance. We conclude with some of the keys to managing processes effectively, to ensuring that changes are implemented and an infrastructure is set up for making continuous improvements. Process analysis, however, extends beyond the analysis of individual processes. It is also a tool for improving the operation of supply chains.

Process Analysis across the Organization

All parts of an organization need to be concerned about process analysis simply because they are doing work, and process analysis focuses on how work is actually done. Are they providing the most value to their customers (internal or external), or can they be improved? Operations and sales departments are often the first areas that come to mind because they are so closely connected with the core processes. However, support processes in accounting, finance, and human resources are crucial to an organization's success as well. Top management also gets involved, as do other departments. During these handoffs of the "baton," disconnects are often the worst and opportunities for improvement the greatest.

A Systematic Approach

Figure 4.1 shows a six-step blueprint for process analysis. **Process analysis is** the documentation and detailed understanding of how work is performed and how it can be redesigned. Process analysis begins with identifying a new opportunity for improvement and ends with implementing a revised process. The last step goes back to the first step, thus creating a cycle of continual improvement. We introduce a closely related model in Chapter 5, "Quality and Performance" as part of the Six Sigma Improvement (DMAIC) Model. Other approaches to process improvement are reengineering in Chapter 3, "Process Strategy" and value stream mapping and other techniques in Chapter 8, "Lean Systems." We avoid overlap by covering each technique just once, while bringing out the essence of the approach covered in each chapter. The chapters do have a shared goal: better processes.

Step 1: Identify Opportunities

In order to identify opportunities, managers must pay particular attention to the four core processes: (1) supplier relationship, (2) new service/product development, (3) order fulfillment, and (4) the customer relationship. Each of these processes, and the subprocesses nested within them, are involved in delivering value to external customers. Are customers currently satisfied with the services or products they receive, or is there room for improvement? How about internal customers? Customer satisfaction must

---

Creating Value through Operations Management

Using Operations to Compete
Project Management

Managing Processes

Process Strategy
Process Analysis
Quality and Performance
Capacity Planning
Constraint Management
Lean Systems

Managing Supply Chains

Supply Chain Inventory Management
Supply Chain Design
Supply Chain Location Decisions
Supply Chain Integration
Supply Chain Sustainability and Humanitarian Logistics
Forecasting
Operations Planning and Scheduling
Resource Planning

process analysis
The documentation and detailed understanding of how work is performed and how it can be redesigned.

▼ **FIGURE 4.1**
Blueprint for Process Analysis

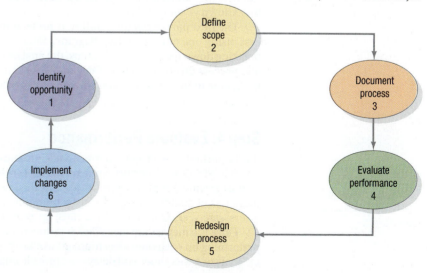

be monitored periodically, either with a formal measurement system or with informal checks or studies. Managers sometimes develop an inventory of their core and support processes that provide a guide for what processes need scrutiny.

Another way to identify opportunities is by looking at the strategic issues. Do gaps exist between a process's competitive priorities and its current competitive capabilities, as was found for the assessment of operations strategy at a credit card division in Chapter 1, "Using Operations to Compete"? Do multiple measures of cost, top quality, quality consistency, delivery speed, and on-time delivery meet or exceed expectations? Is there a good *strategic fit* in the process? If the process provides a service, does its position on the customer-contact matrix (see Figure 3.2) seem appropriate? How does the degree of customer contact match up with process structure, customer involvement, resource flexibility, and capital intensity (see Figure 3.8)? Similar questions should be asked about manufacturing processes regarding the strategic fit between process choice, volume, and product customization (see Figure 3.10).

Employees who actually perform the process or internal suppliers or customers should be encouraged to bring their ideas to managers and staff specialists (such as industrial engineers), or perhaps pass on their ideas through a formal suggestion system. A **suggestion system** is a voluntary system by which employees submit their ideas on process improvements. Usually, a specialist evaluates the proposals, makes sure worthy suggestions are implemented, and provides feedback to those who make the suggestions. Sometimes, the person or team making a good suggestion is rewarded with money or special recognition.

suggestion system

A voluntary system by which employees submit their ideas on process improvements.

Step 2: Define the Scope

Step 2 establishes the boundaries of the process to be analyzed. Is it a broad process that stretches across the whole organization, involving many steps and many employees, or a more narrowly bracketed nested subprocess that is just part of one person's job? A process's scope can be too narrow or too broad. For example, a broadly defined process that outstrips the resources available, sometimes called "trying to boil the ocean," is doomed because it will increase employee frustration without producing any results.

The resources that management assigns to improving or reengineering a process should match the scope of the process. For a small nested process involving only one employee, perhaps he or she is asked to redesign the process. For a project that deals with a major core process, managers typically establish one or more teams. A **design team** consists of knowledgeable, team-oriented individuals who work at one or more steps in the process, conduct the process analysis, and make the necessary changes. Other resources may be full-time specialists called internal or external *facilitators*. Facilitators know process analysis methodology, and they can guide and train the design team. If the process cuts across several departmental lines, it may benefit from a *steering team* of several managers from various departments, headed by a project manager who oversees the process analysis.

design team

A group of knowledgeable, team-oriented individuals who work at one or more steps in the process, conduct the process analysis, and make the necessary changes.

Step 3: Document the Process

Once scope is established, the analyst should document the process. Documentation includes making a list of the process's inputs, suppliers (internal or external), outputs, and customers (internal or external). This information then can be shown as a diagram, with a more detailed breakdown given in a table.

The next part of documentation is understanding the different steps performed in the process, using one or more of the diagrams, tables, and charts described later in this chapter. When breaking down the process into steps, the analyst notes the degrees and types of customer contact and process divergence along the various steps in the process. The analyst also notes what steps are visible to the customer and where in the process work is handed off from one department to the next.

Step 4: Evaluate Performance

It is important to have good performance measures to evaluate a process for clues on how to improve it. **Metrics** are performance measures for the process and the steps within it. A good place to start is with competitive priorities, but they need to be specific. The analyst creates multiple measures of quality, customer satisfaction, time to perform each step or the whole process, cost, errors, safety, environmental measures, on-time delivery, flexibility, and the like.

Once the metrics are identified, it is time to collect information on how the process is currently performing on each one. Measurement can be rough-cut estimates or quite extensive. Techniques for analyzing wait times and delays can provide important information (see Supplement B, "Waiting

metrics

Performance measures that are established for a process and the steps within it.

Lines" and MyOMLab Supplement E, "Simulation"). Work measurement techniques are also more extensive and are previewed in a later section of this chapter.

Step 5: Redesign the Process

A careful analysis of the process and its performance on the selected metrics should uncover *disconnects*, or gaps, between actual and desired performance. Performance gaps can be caused by illogical, missing, or extraneous steps. They can be caused by metrics that reinforce the silo mentality of individual departments when the process spans across several departments. The analyst or design team should dig deep to find the root causes of performance gaps.

Using analytical and creative thinking, the design team generates a long list of ideas for improvements. These ideas are then sifted and analyzed. Ideas that are justifiable, where benefits outweigh costs, are reflected in a new process design. The new design should be documented "as proposed." Combining the new process design with the documentation of the current process gives the analysts clear before and after pictures. The new documentation should make clear how the revised process will work and the performance expected for the various metrics used.

McDonald's uses mystery shoppers to evaluate its stores. It also sends operations "emissaries" to its stores to help managers fine-tune their processes, while revising processes and its supply chain to be more environmentally friendly.

Step 6: Implement Changes

Implementation is more than developing a plan and carrying it out. Many processes have been redesigned effectively, but never get implemented. People resist change: "We have always done it that way" or "we tried that before." Widespread participation in process analysis is essential, not only because of the work involved but also because it builds commitment. It is much easier to implement something that is partly your own idea. In addition, special expertise may be needed, such as for developing software. New jobs and skills may be needed, involving training and investments in new technology. Implementation brings to life the steps needed to bring the redesigned process online. Management or the steering committee must make sure that the implementation project goes according to schedule.

In the remainder of this chapter, we examine steps in process analysis in detail.

Documenting the Process

Five techniques are effective for documenting and evaluating processes: (1) flowcharts, (2) swim lane flowcharts, (3) service blueprints, (4) work measurement techniques, and (5) process charts. They allow you to "lift the lid and peer inside" to see how an organization does its work. You can see how a process operates, at any level of detail, and how well it is performing. Trying to create one of these charts might even reveal a lack of any established process. It may not be a pretty picture, but it is how work actually gets done. Techniques for documenting the process lend themselves to finding performance gaps, generating ideas for process improvements, and documenting the look of a redesigned process.

flowchart

A diagram that traces the flow of information, customers, equipment, or materials through the various steps of a process.

Flowcharts

A **flowchart** traces the flow of information, customers, equipment, or materials through the various steps of a process. Flowcharts are also known as flow diagrams, process maps, relationship maps, or blueprints. Flowcharts have no precise format and typically are drawn with boxes (with a brief description of the step inside), and with lines and arrows to show sequencing. The rectangle (□) shape is the usual choice for a box, although other shapes (O, ◯, ◇, ▽, or ▱) can differentiate between different types of steps (e.g., operation, delay, storage, inspection, and so on). Colors and shading can also call attention to different types of steps, such as those particularly high on process divergence. Divergence is also communicated when an outgoing arrow from a step splits into two or more arrows that lead to different boxes. Although many representations are

A consultant discusses the proposal for a new organizational development program with clients during a follow-up meeting. The use of flowcharts can help in understanding this step as just one part of the overall sales process for a consulting company.

acceptable, there must be agreement on the conventions used. They can be given as a key somewhere in the flowchart, and/or described in accompanying text. It is also important to communicate *what* (e.g., information, customer order, customer, materials, and so on) is being tracked.

You can create flowcharts with several programs. Microsoft PowerPoint offers many different formatting choices for flowcharts (see the Flowchart submenu under AutoShapes). The tutorials "Flowcharting in Excel" and "Flowcharting in PowerPoint" in MyOMLab offer other options, and its live demonstrations of flowcharting in Figures 4.2 and 4.3 are instructive. Other powerful software packages for flowcharting and drawing diagrams (such as organization charts and decision trees) are SmartDraw (**www.smartdraw.com**), Microsoft Visio (**www.microsoft.com/office/visio**), and Micrografx (**www.micrografx.com**). Often, free downloads are available at such sites on a trial basis.

Flowcharts can be created for several levels in the organization. For example, at the strategic level, they could show the core processes and their linkages, as in Figure 1.4. At this level, the flowcharts would not have much detail; however, they would give a bird's eye view of the overall business. Just identifying a core process is often helpful. Let us now turn to the process level, where we get into the details of the process being analyzed. Figure 4.2 shows such a process, which consists of many steps that have subprocesses nested within them. Rather than representing everything in

MyOMLab

FIGURE 4.2 ▶
Flowchart of the Sales Process for a Consulting Company

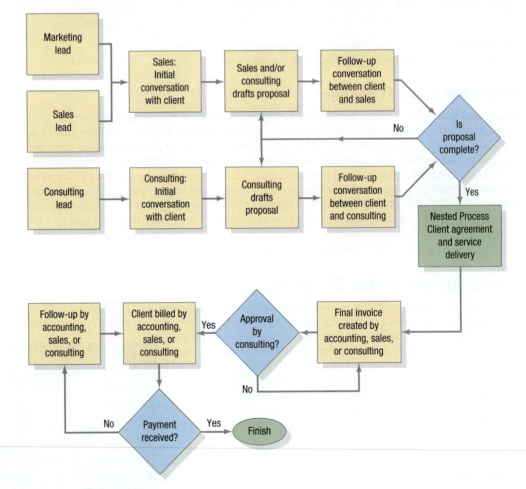

FIGURE 4.3 ▶
Flowchart of the Nested Subprocess of Client Agreement and Service Delivery

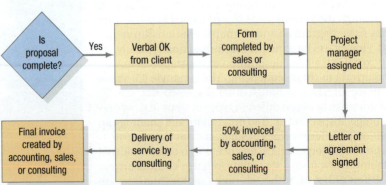

one flowchart, Figure 4.2 presents an overview of the whole process. It describes the sales process for a consulting firm that specializes in organizational development and corporate education programs. Four different departments (accounting, consulting, marketing, and sales) interact with the external customer (client). The process goes through three main phases: (1) generating business leads, (2) client agreement and service delivery, and (3) billing and collection.

Figure 4.2 illustrates one other feature. The diamond shape (◇) represents a yes/no decision or outcome, such as the results of an inspection or a recognition of different kinds of customer requirements. In Figure 4.2, the diamond represents three yes/no decision points: (1) whether the proposal is complete, (2) whether consulting approves the invoice, and (3) whether payment is received. These yes/no decision points are more likely to appear when a process is high in divergence.

Sometimes, it is impossible to get the whole flowchart on one page. Figures 4.2 and 4.3 show how to create nested processes for steps that can be more aggregated. For example, Figure 4.3 flowcharts a nested process within the client agreement and service delivery step in Figure 4.2. Figure 4.3 brings out more details, such as invoicing the customer for 50 percent of the total estimated cost of the service before the service is delivered, and then putting together a final invoice after the service is finished. This nesting approach often becomes a practical necessity because only so much detail can be shown in any single flowchart.

Swim Lane Flowcharts

The **swim lane flowchart** is a visual representation that groups functional areas responsible for different sub-processes into lanes. It is most appropriate when the business process spans several department boundaries, and where each department or a functional area is separated by parallel lines similar to lanes in a swimming pool. Swim lanes are labeled according to the functional groups they represent, and can be arranged either horizontally or vertically.

The swim lane flowchart in Figure 4.4 illustrates the order placement and acceptance process at a manufacturing company. The process starts when an order is generated by a customer and

swim lane flowchart

A visual representation that groups functional areas responsible for different sub-processes into lanes. It is most appropriate when the business process spans several department boundaries.

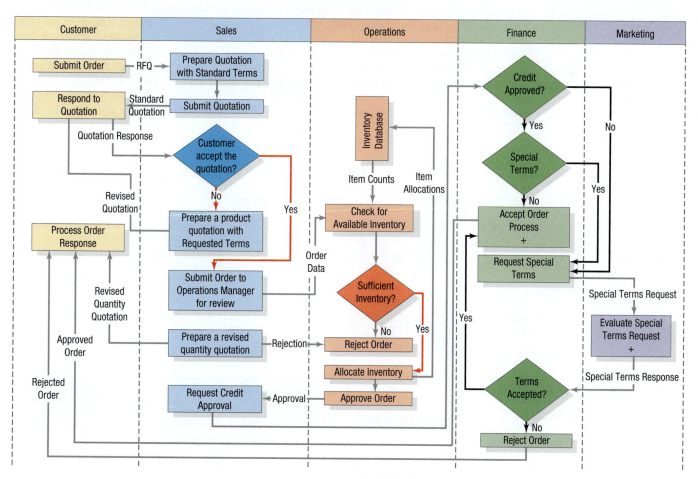

▲ **FIGURE 4.4**
Swim Lane Flowchart of the Order-Filling Process Showing Handoffs Between Departments
Source: D. Kroenke, *Using MIS*, 4th ed., 2012, p. 336. Reprinted by permission of Pearson, Upper Saddle River, NJ.

ends when the order is actually rejected, modified, or approved by the company in consultation with the customer. All functions contributing to this process are included in the flowchart. The columns represent different departments or functional areas, and the steps appear in the department column where they are performed. The customer is also shown as one of the column headings. This approach shows the *handoffs* from one department to another when the outgoing arrow from a step goes to another column. Special dotted-line arrows are one way to show handoffs. Handoffs are points where cross-functional coordination is at particular risk due to the silo mentality. Misunderstandings, backlogs, and errors are more likely at these points.

Flowcharts allow the process analyst and managers to look at the horizontal organization, rather than the vertical organization and departmental boundaries implied by a typical organizational chart. Flowcharts show how organizations produce their outputs through cross-functional work processes, and allow the design team to see all the critical interfaces between functions and departments.

Service Blueprints

service blueprint

A special flowchart of a service process that shows which steps have high customer contact.

A good design for service processes depends first and foremost on the type and amount of customer contact. A **service blueprint** is a special flowchart of a service process that shows which steps have high customer contact. It uses a line of visibility to identify which steps are visible to the customer (and thus. more of a front-office process) and those that are not (back office process).

Another approach to creating a service blueprint is to create three levels. The levels clarify how much control the customer has over each step. For example, consider a customer driving into a Fast Lube shop to have their car serviced. Level 1 would be when the customer is in control, such as driving in for service or paying the bill at the end. Level 2 could be when the customer interacts with the service provider, such as making the initial service request, or being notified on what needs to be done. Level 3 could be when the service is removed from the customer's control, such as when the work is performed and the invoice is prepared.

Figure 4.5 illustrates a fairly complex service blueprint. It not only shows steps with its customers, but also with its consumer's customers. It shows the steps taken by a consulting company that specializes in inventory appraisals and inventory liquidations. Its external customers are large banks that make asset-based loans. The bank's customers, in turn, are customers seeking a loan based on the value of their assets (including inventories). Figure 4.5 describes the consulting company's current evaluation and appraisal process. This service blueprint not only shows the steps in its current inventory evaluation and appraisal process, but also which steps are visible to its external customers (the banks) and its customers' customers (the company seeking a loan). The steps visible to the banks (salmon boxes) are partitioned with the vertical lines of visibility. The steps

▼ **FIGURE 4.5**
Service Blueprint of
Consulting Company's
Inventory Appraisal Process

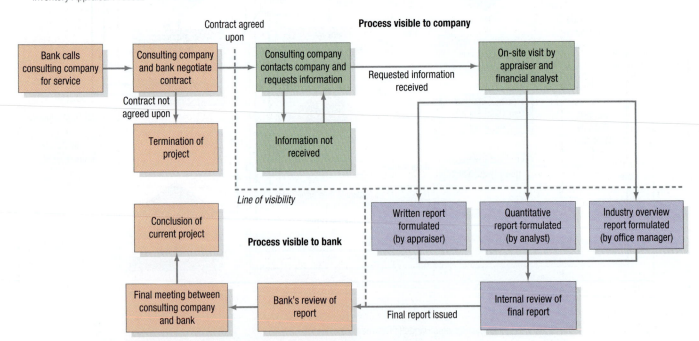

visible to the company seeking a loan (green boxes) are partitioned off by the top left vertical line of visibility and the horizontal line of visibility. The steps in purple are performed by the consulting company and not visible to external customers.

The process begins with a call from a bank seeking the service by the consulting company. There are three main steps in the overall process.

1. The bank contacts the consulting company and they agree on the contract.

2. The consulting company performs the inventory evaluation on the site of the company seeking a loan from the bank.

3. The consulting company prepares the final report and presents it to the bank.

Of course, visibility is just one aspect of customer contact, and it may not adequately capture how actively the customer is involved or how much personal attention is required. A service blueprint can use colors, shading, or box shapes, instead of the lines of visibility, to show the extent and type of customer contact. Another approach to service blueprinting is to tag each step with a number, and then have an accompanying table that describes in detail the customer contact for each numbered step. There is no one "right way" to create a flow chart or service blueprint.

Work Measurement Techniques

Process documentation would not be complete without estimates of the average time each step in the process would take. Time estimates are needed not just for process improvement efforts, but for capacity planning, constraint management, performance appraisal, and scheduling. Estimating task times can be as simple as making a reasoned guess, asking a knowledgeable person, or taking notes while observing the process. More extensive studies involve collecting data for several weeks, consulting cost accounting data, or checking data recorded in information systems.

Formal techniques are also available that rely on the judgment of skilled observers: (1) the time study method, (2) the elemental standard data approach, (3) the predetermined data approach, and (4) work sampling. A fifth method, (5) learning curve analysis, is particularly appropriate when a new product or process is introduced and the time per unit produced has not yet stabilized. The method chosen depends on the purpose of the data, process type (job or line), and degree of product customization. A more comprehensive treatment of these techniques is provided in MyOMLab Supplement H, "Measuring Output Rates" and MyOMLab Supplement I, "Learning Curve Analysis."

Time Study Method   Time study uses a trained analyst to perform four basic steps in setting a time standard for a job or process: (1) selecting the work elements (steps in a flowchart or process chart) within the process to be studied, (2) timing the elements, (3) determining the sample size, and (4) setting the final standard. It is essentially the average time observed, adjusted for normal effort and making an allowance for breaks, unavoidable delays, and the like. The analyst records time spent on each element of the process being studied using a stopwatch, and records the time spent on each element for several repetitions. The analyst assigns a performance rating for each element to adjust for normal effort. Some elements may be performed faster or slower than normal, in the analyst's judgment. The allowance is expressed as a proportion or percent of the total *normal* time.

Elemental Standard Data Approach   Another approach is needed when products or services are highly customized, job processes prevail, and process divergence is great. Elemental standard data is a database of standards compiled by a firm's analysts for basic elements that they can draw on later to estimate the time required for a particular job. This approach works well when work elements within certain jobs are similar to those in other jobs. Sometimes, the time required for a work element depends on variable characteristics of the jobs, such as the amount of metal to be deposited for a welding process. In such cases, an equation that relates these characteristics to the time required is also stored in the database. Another method, such as time study or past records, still must be used to compile the normal times (before the allowance is added) stored in the database.

MyOMLab

time study

A work measurement method using a trained analyst to perform four basic steps in setting a time standard for a job or process: selecting the work elements (or nested processes) within the process to be studied, timing the elements, determining the sample size, and setting the final standard.

elemental standard data

A database of standards compiled by a firm's analysts for basic elements that they can draw on later to estimate the time required for a particular job, which is most appropriate when products or services are highly customized, job processes prevail, and process divergence is great.

EXAMPLE 4.1	Time Study of Watch Assembly Process

A process at a watch assembly plant has been changed. The process is divided into three work elements. A time study has been performed with the following results. The time standard for the process previously was 14.5 minutes. Based on the new time study, should the time standard be revised?

SOLUTION

The new time study had an initial sample of four observations, with the results shown in the following table. The performance rating factor (RF) is shown for each element (to adjust for normal effort), and the allowance for the whole process is 18 percent of the total *normal* time.

	Obs 1	Obs 2	Obs 3	Obs 4	Average (min)	RF	Normal Time
Element 1	2.60	2.34	3.12	2.86	2.730	1.0	2.730
Element 2	4.94	4.78	5.10	4.68	4.875	1.1	5.363
Element 3	2.18	1.98	2.13	2.25	2.135	0.9	1.922
					Total Normal Time = **10.015 minutes**		

The normal time for an element in the table is its average time, multiplied by the RF. The total normal time for the whole process is the sum of the normal times for the three elements, or 10.015 minutes. To get the standard time (ST) for the process, just add in the allowance, or

$$ST = 10.015(1 + 0.18) = \textbf{11.82} \text{ minutes/watch}$$

DECISION POINT

The time to assemble a watch appears to have decreased considerably. However, based on the precision that management wants, the analyst decided to increase the sample size before setting a new standard. MyOMLab Supplement H, "Measuring Output Rates," gives more information on determining the number of additional observations needed.

MyOMLab

---

predetermined data approach

A database approach that divides each work element into a series of micromotions that make up the element. The analyst then consults a published database that contains the normal times for the full array of possible micromotions.

work sampling

A process that estimates the proportion of time spent by people or machines on different activities, based on observations randomized over time.

FIGURE 4.6 ▼
Work Sampling Study of Admission Clerk at Health Clinic Using OM Explorer's *Time Study* Solver.

Predetermined Data Approach The **predetermined data approach** divides each work element even more, into a series of micromotions that make up the element. The analyst then consults a published database that contains the normal times for the full array of possible micromotions. A process's normal time can then be calculated as the sum of the times given in the database for the elements performed in the process. This approach makes most sense for highly repetitive processes with little process divergence and line flows. The micromotions (such as reach, move, or apply pressure) are very detailed.

Work Sampling Method **Work sampling** estimates the proportion of time spent by people or machines on different activities, based on observations randomized over time. Examples of these activities include working on a service or product, doing paperwork, waiting for instructions, waiting for maintenance, or being idle. Such data can then be used to assess a process's productivity, estimate the allowances needed to set standards for other work measurement methods, and spot areas for process improvement. It is best used when the processes are highly divergent with flexible flows. Figure 4.6 shows the input data and numerical results for one week of observations. Figure 4.6 shows an idle time of 23.81 percent for the week. It also reports that 237 more observations are needed to achieve the confidence and precision levels required with the input data. How these conclusions are reached is explained in MyOMLab Supplement H, "Measuring Output Rates".

(a) Input Data and Numerical Results

Increase Observations		Remove An Observation	

Confidence z	1.96	Precision p	0.05

Observation Period	Times Busy	Times Idle	Observations
Monday	6	1	7
Tuesday	5	2	7
Wednesday	7	0	7
Thursday	9	2	11
Friday	5	5	10
Total	32	10	42

(b) Idle Time and Observations Required

Portion of idle times	0.2381
Total observations required	279
Additional observations required	237

Learning Curve Analysis   The time estimation techniques just covered assume that the process is stable. If the process is revised, then just repeat the method for the revised process after it stabilizes. Learning curve analysis, on the other hand, takes into account that learning takes place on an ongoing basis, such as when new products or services are introduced frequently. With instruction and repetition, workers learn to perform jobs more efficiently, process improvements are identified, and better administration methods are created. These learning effects can be anticipated with a **learning curve**, a line that displays the relationship between processing time and the cumulative quantity of a product or service produced. The time required to produce a unit or create a service decreases as more units or customers are processed. The learning curve for a process depends on the rate of learning and the actual or estimated time for the first unit processed. Figure 4.7 demonstrates the learning curve assuming an 80 percent learning rate, with the first unit taking 120,000 hours and the cumulative average time for the first 10 units produced. The learning rate deals with each *doubling* of the output total. The time for the second unit is 80 percent of the first (or 120,000 × .80 = 96,000 hours), the time for the fourth unit is 80 percent of the second unit (or 96,000 × .80 = 76,800 hours), and so on. Finding the time estimate for a unit that is not an exact doubling (such as the fifth unit), and also the cumulative average time for the first 10 units, is explained in MyOMLab Supplement I, "Learning Curve Analysis".

learning curve

A line that displays the relationship between processing time and the cumulative quantity of a product or service produced.

MyOMLab

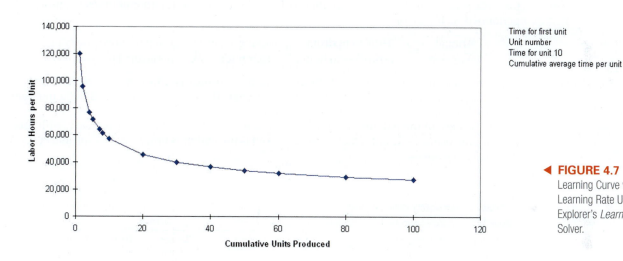

Time for first unit	120,000
Unit number	10
Time for unit 10	57,172
Cumulative average time per unit	75,784

◄ **FIGURE 4.7**
Learning Curve with 80% Learning Rate Using OM Explorer's *Learning Curves* Solver.

Process Charts

A **process chart** is an organized way of documenting all the activities performed by a person or group of people at a workstation, with a customer, or working with certain materials. It analyzes a process using a table, and provides information about each step in the process. In contrast to flowcharts, swim lane flowcharts, and service blueprints, it requires the time estimates (see work measurement techniques covered in the last section). Often it is used to drill down to the job level for an individual person, a team, or a focused nested process. It can have many formats. Here, we group the type of activities for a typical process into five categories:

process chart

An organized way of documenting all the activities performed by a person or group of people, at a workstation, with a customer, or on materials.

- *Operation.* Changes, creates, or adds something. Drilling a hole or serving a customer are examples of operations.

- *Transportation.* Moves the study's subject from one place to another (sometimes called *materials handling*). The subject can be a person, a material, a tool, or a piece of equipment. A customer walking from one end of a counter to the other, a crane hoisting a steel beam to a location, and a conveyor carrying a partially completed product from one workstation to the next are examples of transportation. It could also be the shipment of a finished product to the customer or a warehouse.

- *Inspection.* Checks or verifies something but does not change it. Getting customer feedback, checking for blemishes on a surface, weighing a product, and taking a temperature reading are examples of inspections.

- *Delay.* Occurs when the subject is held up awaiting further action. Time spent waiting for a server; time spent waiting for materials or equipment; cleanup time; and time that workers, machines, or workstations are idle because they have no work to complete are examples of delays.

- *Storage.* Occurs when something is put away until a later time. Supplies unloaded and placed in a storeroom as inventory, equipment put away after use, and papers put in a file cabinet are examples of storage.

Depending on the situation, other categories can be used. For example, subcontracting for outside services might be a category, temporary storage and permanent storage, or environmental waste might be three separate categories. Choosing the right category for each activity requires taking the perspective of the subject charted. A delay for the equipment could be inspection or transportation for the operator.

To complete a chart for a new process, the analyst must identify each step performed. If the process is an existing one, the analyst can actually observe the steps and categorize each step according to the subject being studied. The analyst then records the distance traveled and the time taken to perform each step. After recording all the activities and steps, the analyst summarizes the steps, times, and distances data. Figure 4.8 shows a process chart prepared using OM Explorer's *Process Chart* Solver. It is for a patient with a twisted ankle being treated at a hospital. The process begins at the entrance and ends with the patient exiting after picking up the prescription.

After a process is charted, the analyst sometimes estimates the annual cost of the entire process. It becomes a benchmark against which other methods for performing the process can be evaluated. Annual labor cost can be estimated by finding the product of (1) time in hours to perform the process each time, (2) variable costs per hour, and (3) number of times the process is performed each year, or

$$\frac{\text{Annual}}{\text{labor cost}} = \left(\begin{array}{c}\text{Time to perfrom}\\\text{the process in hours}\end{array}\right)\left(\begin{array}{c}\text{Variable costs}\\\text{per hour}\end{array}\right)\left(\begin{array}{c}\text{Number of times process}\\\text{performed per year}\end{array}\right)$$

MyOMLab

Tutor 4.1 in MyOMLab provides a new example to practice creating process charts.

For example, if the average time to serve a customer is 4 hours, the variable cost is $25 per hour, and 40 customers are served per year, then the labor cost is $4,000 per year (or 4 hrs/customer × $25/hr × 40 customers/yr).

In the case of the patient in Figure 4.8, this conversion would not be necessary, with total patient time being sufficient. What is being tracked is the patient's time, not the time and costs of the service providers.

FIGURE 4.8 ▶

Process Chart for Emergency Room Admission

Process:	Emergency room admission
Subject:	Ankle injury patient
Beginning:	Enter emergency room
Ending:	Leave hospital

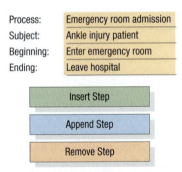

Summary

Activity	Number of Steps	Time (min)	Distance (ft)
Operation ●	5	23.00	
Transport ➡	9	11.00	815
Inspect ■	2	8.00	
Delay ◗	3	8.00	
Store ▼	—	—	

Step No.	Time (min)	Distance (ft)	●	➡	■	◗	▼	Step Description
1	0.50	15.0		X				Enter emergency room, approach patient window
2	10.00		X					Sit down and fill out patient history
3	0.75	40.0		X				Nurse escorts patient to ER triage room
4	3.00				X			Nurse inspects injury
5	0.75	40.0		X				Return to waiting room
6	1.00					X		Wait for available bed
7	1.00	60.0		X				Go to ER bed
8	4.00					X		Wait for doctor
9	5.00				X			Doctor inspects injury and questions patient
10	2.00	200.0		X				Nurse takes patient to radiology
11	3.00		X					Technician x-rays patient
12	2.00	200.0		X				Return to bed in ER
13	3.00					X		Wait for doctor to return
14	2.00		X					Doctor provides diagnosis and advice
15	1.00	60.0		X				Return to emergency entrance area
16	4.00		X					Check out
17	2.00	180.0		X				Walk to pharmacy
18	4.00		X					Pick up prescription
19	1.00	20.0		X				Leave the building

You can design your own process chart spreadsheets to bring out issues that are particularly important for the process you are analyzing, such as categories for customer contact, process divergence, and the like. You can also track performance measures other than time and distance traveled, such as error rates. In addition, you can also create a different version of the process chart spreadsheet that examines processes much as done with flowcharts, except now in the form of a table. The columns that categorize the activity type could be replaced by one or more columns reporting different metrics of interest, rather than trying to fit them into a flowchart. Although it might not look as elegant, it could be just as informative—and easier to create.

Evaluating Performance

Metrics and performance information complete the documentation of a process (see step 3 in Figure 4.1). Metrics can be displayed in various ways. Sometimes, they can be added directly on the flowchart or process chart. When the number of metrics gets unwieldy, another approach is to create a supporting table for the chart. Its rows are the steps in the flowchart, swim lane flowchart, service blueprint, or process chart. The columns are the current performance, goals, and performance gaps for various metrics.

The leader of a design team presents several charts that document a process in their office that they are analyzing. He is identifying several areas of substandard performance across a range of different metrics. The next step will be to redesign the process. The flipchart on the right will be quite useful in generating rapid fire ideas from the team on how the process might be improved.

The specific metrics analysts choose depends on the process being analyzed and on the competitive priorities. Good starting points are the per-unit processing time and cost at each step, and the time elapsed from beginning to end of the process. Capacity utilization, environmental issues, and customer (or job) waiting times reveal where in the process delays are most likely to occur. Customer satisfaction measures, error rates, and scrap rates identify possible quality problems. We introduce many such metrics in subsequent chapters. Figure 4.9 shows the chapter and/or supplement that relates to some basic metrics. Only when these subsequent chapters are understood do we really complete our discussion of process analysis.

Chapter 5, Quality and Performance

- Customer satisfaction measures
- Error rate
- Rework or scrap rate
- Internal failure cost

Chapter 6, Capacity Planning; Supplement B, Waiting Lines; MyOMLab Supplement H, Measuring Output Rates; MyOMLab Supplement I, Learning Curve Analysis

- Processing time
- Setup time
- Capacity utilization
- Average waiting time
- Average number of customers or jobs waiting in line

Chapter 7, Constraint Management

- Total time from start to finish (throughput time)
- Setup time
- Cycle time
- Throughput time
- Idle time
- Operating expenses

Chapter 8, Lean Systems

- Setup time
- Takt time
- Average waiting time
- Waste

◀ **FIGURE 4.9**

Metrics for Flowcharts, Process Charts, and Accompanying Tables

Data Analysis Tools

Metrics may reveal a performance gap. Various tools are available to help you understand the causes of the problem[1]. Here we present six tools: (1) checklists, (2) histograms and bar charts, (3) Pareto charts, (4) scatter diagrams, (5) cause-and-effect diagrams, and (6) graphs. Many of them were developed initially to analyze quality issues, but they apply equally well to the full range of performance measures.

Checklists   Data collection through the use of a checklist is often the first step in the analysis of a metric. A **checklist** is a form used to record the frequency of occurrence of certain process failures. A **process failure** is any performance shortfall, such as error, delay, environmental waste, rework, and the like. The characteristics may be measurable on a continuous scale (e.g., weight, customer satisfaction on a 1-to-7 scale, unit cost, scrap loss percentage, time, or length) or on a yes-or-no basis (e.g., customer complaint, posting error, paint discoloration, or inattentive servers).

Histograms and Bar Charts   Data from a checklist often can be presented succinctly and clearly with histograms or bar charts. A **histogram** summarizes data measured on a continuous scale, showing the frequency distribution of some process failure (in statistical terms, the central tendency and dispersion of the data). Often the mean of the data is indicated on the histogram. A **bar chart** (see Figure 4.10) is a series of bars representing the frequency of occurrence of data characteristics measured on a yes-or-no basis. The bar height indicates the number of times a particular process failure was observed.

Pareto Charts   When managers discover several process problems that need to be addressed, they have to decide which should be attacked first. Vilfredo Pareto, a nineteenth-century Italian scientist whose statistical work focused on inequalities in data, proposed that most of an "activity" is caused by relatively few of its factors. In a restaurant quality problem, the activity could be customer complaints and the factor could be "discourteous server." For a manufacturer, the activity could be product defects and the factor could be "missing part." Pareto's concept, called the 80–20 rule, is that 80 percent of the activity is caused by 20 percent of the factors. By concentrating on the 20 percent of the factors (the "vital few"), managers can attack 80 percent of the process failure problems. Of course, the exact percentages vary with each situation, but inevitably relatively few factors cause most of the performance shortfalls.

The few vital factors can be identified with a **Pareto chart,** a bar chart on which the factors are plotted along the horizontal axis in decreasing order of frequency (see Figure 4.11). The chart has two vertical axes, the one on the left showing frequency (as in a histogram) and the one on the right showing the cumulative percentage of frequency. The cumulative frequency curve identifies the few vital factors that warrant immediate managerial attention.

Sidebar Glossary

checklist

A form used to record the frequency of occurrence of certain process failures.

process failure

Any performance shortfall, such as error, delay, environmental waste, rework, and the like.

histogram

A summarization of data measured on a continuous scale, showing the frequency distribution of some process failure (in statistical terms, the central tendency and dispersion of the data).

bar chart

A series of bars representing the frequency of occurrence of data characteristics measured on a yes-or-no basis.

Pareto chart

A bar chart on which factors are plotted along the horizontal axis in decreasing order of frequency.

EXAMPLE 4.2	Pareto Chart for a Restaurant

MyOMLab

Active Model 4.1 in MyOMLab provides additional insights on this Pareto chart example and its extensions.

MyOMLab

Tutor 4.2 in MyOMLab provides a new example on creating Pareto charts.

The manager of a neighborhood restaurant is concerned about the lower numbers of customers patronizing his eatery. Complaints have been rising, and he would like to find out what issues to address and present the findings in a way his employees can understand.

SOLUTION

The manager surveyed his customers over several weeks and collected the following data:

Complaint	Frequency
Discourteous server	12
Slow service	42
Cold dinner	5
Cramped tables	20
Atmosphere	10

---

[1]Several of these tools, particularly Pareto charts and cause-and-effect diagrams, are closely affiliated with Chapter 5, "Quality and Performance." We introduce them here because they apply to process failures in general, and not just to quality rejects.

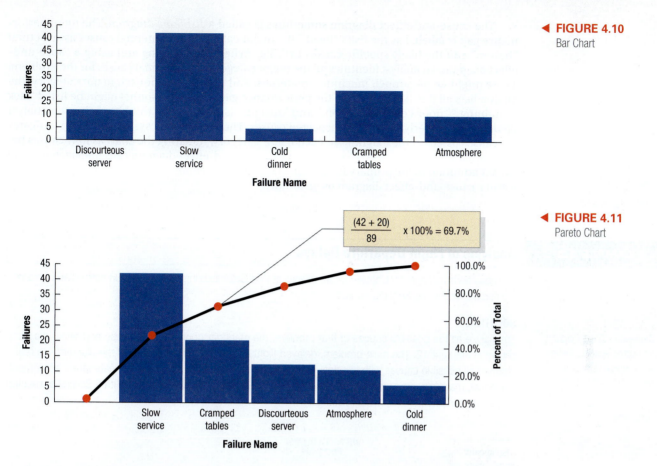

◀ **FIGURE 4.10**
Bar Chart

◀ **FIGURE 4.11**
Pareto Chart

$$\frac{(42 + 20)}{89} \times 100\% = 69.7\%$$

Figure 4.10 is a bar chart and Figure 4.11 is a Pareto chart, both created with OM Explorer's *Bar, Pareto, and Line Charts* Solver. They present the data in a way that shows which complaints are more prevalent (the vital few). You can reformat these charts for any "yes-or-no" metrics by unprotecting the spreadsheet and then making your revisions. For example, if you are using Microsoft Excel 2010, just click on Home/Format/Protection/Unprotect Sheet. Another approach is to create your own spreadsheets from scratch. More advanced software with point-and-click interfaces include Minitab (**www.minitab.com/index.htm**), SAS (**www.sas.com/rnd/app/qc.html**), and Microsoft Visio (**www.microsoft.com/office/visio**).

DECISION POINT

It was clear to the manager (and all employees) which complaints, if rectified, would cover most of the process failure problems in the restaurant. First, slow service will be addressed by training the existing staff, adding another server, and improving the food preparation process. Removing some decorative furniture from the dining area and spacing the tables better will solve the problem with cramped tables. The Pareto chart shows that these two problems, if rectified, will account for almost 70 percent of the complaints.

Scatter Diagrams   Sometimes managers suspect that a certain factor is causing a particular process failure. A **scatter diagram**, which is a plot of two variables showing whether they are related, can be used to verify or negate the suspicion. Each point on the scatter diagram represents one data observation. For example, the manager of a castings shop may suspect that casting defects are a function of the diameter of the casting. A scatter diagram could be constructed by plotting the number of defective castings found for each diameter of casting produced. After the diagram is completed, any relationship between diameter and number of process failures will be clear.

scatter diagram

A plot of two variables showing whether they are related.

Cause-and-Effect Diagrams   An important aspect of process analysis is linking each metric to the inputs, methods, and process steps that build a particular attribute into the service or product. One way to identify a design problem is to develop a **cause-and-effect diagram** that relates a key performance problem to its potential causes. First developed by Kaoru Ishikawa, the diagram helps management trace disconnects directly to the operations involved. Processes that have no bearing on a particular problem are not shown on the diagram.

cause-and-effect diagram

A diagram that relates a key performance problem to its potential causes.

The cause-and-effect diagram sometimes is called a *fishbone diagram.* The main performance gap is labeled as the fish's "head," the major categories of potential causes as structural "bones," and the likely specific causes as "ribs." When constructing and using a cause-and-effect diagram, an analyst identifies all the major categories of potential causes for the problem. These might be personnel, machines, materials, and processes. For each major category, the analyst lists all the likely causes of the performance gap. Under personnel might be listed "lack of training," "poor communication," and "absenteeism." Creative thinking helps the analyst identify and properly classify all suspected causes. The analyst then systematically investigates the causes listed on the diagram for each major category, updating the chart as new causes become apparent. The process of constructing a cause-and-effect diagram calls management and worker attention to the primary factors affecting process failures. Example 4.3 demonstrates the use of a cause-and-effect diagram by an airline.

| EXAMPLE 4.3 | **Analysis of Flight Departure Delays** |

The operations manager for Checker Board Airlines at Port Columbus International Airport noticed an increase in the number of delayed flight departures.

▼ **FIGURE 4.12**

Cause-and-Effect Diagram for Flight Departure Delays

Source: Adopted from D. Daryl Wyckoff, "New Tools for Achieving Service Quality," *The Cornell H.R.A. Quarterly.* Used by permission. All rights reserved.

SOLUTION

To analyze all the possible causes of that problem, the manager constructed a cause-and-effect diagram, shown in Figure 4.12. The main problem, delayed flight departures, is the "head" of the diagram. He brainstormed all possible causes with his staff, and together they identified several major categories: equipment, personnel, materials, procedures, and "other factors" that are beyond managerial control. Several suspected causes were identified for each major category.

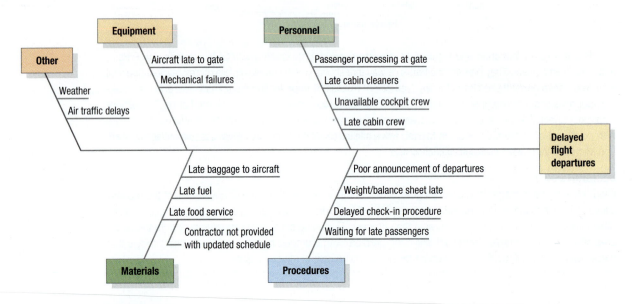

DECISION POINT

The operations manager, having a good understanding of the process, suspected that most of the flight delays were caused by problems with materials. Consequently, he had food service, fueling, and baggage-handling operations examined. He learned that the number of tow trucks for the baggage-transfer operations was insufficient and that planes were delayed waiting for baggage from connecting flights.

graphs

Representations of data in a variety of pictorial forms, such as line charts and pie charts.

Graphs   Graphs represent data in a variety of pictorial formats, such as line charts and pie charts. *Line charts* represent data sequentially with data points connected by line segments to highlight trends in the data. Line charts are used in control charts (see Chapter 5, "Quality and Performance") and forecasting (see Chapter 14, "Forecasting"). Pie charts represent process factors as slices of a pie; the size of each slice is in proportion to the number of occurrences of the factor.

Pie charts are useful for showing data from *a group of factors* that can be represented as percentages totaling 100 percent.

Each of the tools for improving quality may be used independently, but their power is greatest when they are used together. In solving a process-related problem, managers often must act as detectives, sifting data to clarify the issues involved and deducing the causes. We call this process *data snooping*. Example 4.4 demonstrates how the tools for improving quality can be used for data snooping.

A simulation model goes one step further than data analysis tools, because it can show how the process dynamically changes over time. **Process simulation** is the act of reproducing the behavior of a process, using a model that describes each step. Once the process is modeled, the analyst can make changes in the model to measure the impact on certain metrics, such as response time, waiting lines, resource utilization, and the like. To learn more about how simulation works, see MyOMLab Supplement E, "Simulation". A more advanced capability is possible using SimQuick, found in MyOMLab (**www.nd.edu/~dhartvig/simquick/top.html**). Other software packages include Extend (**http://www.extendsim.com//**), SIMPROCESS (**www.caciasl.com**), ProModel (**www.promodel.com**), and Witness (**www.lanner.com**).

process simulation

The act of reproducing the behavior of a process, using a model that describes each step.

MyOMLab

Redesigning the Process

A doctor pinpoints an illness after a thorough examination of the patient, and then the doctor recommends treatments based on the diagnosis; so it is with processes. After a process is documented, metrics data collected, and disconnects identified, the process analyst or design team puts together a set of changes that will make the process better. At this step, people directly involved in the process are brought in to get their ideas and inputs.

Generating Ideas: Questioning and Brainstorming

Sometimes, ideas for reengineering or improving a process become apparent after documenting the process and carefully examining the areas of substandard performance, handoffs between departments, and steps where customer contact is high. Example 4.4 illustrates how such documentation pointed to a better way of handling the fiber boards through better training. In other cases, the better solution is less evident. Ideas can be uncovered (because there is always a better way) by asking six questions about each step in the process, and about the process as a whole:

1. *What* is being done?
2. *When* is it being done?
3. *Who* is doing it?
4. *Where* is it being done?
5. *How* is it being done?
6. *How* **well** does it do on the various metrics of importance?

Answers to these questions are challenged by asking still another series of questions. *Why* is the process even being done? *Why* is it being done where it is being done? *Why* is it being done when it is being done?

Creativity can also be stimulated by **brainstorming,** letting a group of people knowledgeable about the process propose ideas for change by saying whatever comes to mind. A facilitator records the ideas on a flipchart, so that all can see. Participants are discouraged from evaluating any of the ideas generated during the session. The purpose is to encourage creativity and to get as many ideas as possible, no matter how far-fetched the ideas may seem. The participants of a brainstorming session need not be limited to the design team as long as they have seen or heard the process documentation. A growing number of big companies, such as Sun Life Financial and Georgia-Pacific, are taking advantage of the Internet and specially designed software to run brainstorming sessions that allow people at far-flung locations to "meet" online and hash out solutions to particular problems. The technology lets employees see, and build on, one another's ideas, so that one person's seed of a notion can grow into a practical plan.

brainstorming

Letting a group of people, knowledgeable about the process, propose ideas for change by saying whatever comes to mind.

EXAMPLE 4.4	**Identifying Causes of Poor Headliner Process Failures**

The Wellington Fiber Board Company produces headliners, the fiberglass components that form the inner roof of passenger cars. Management wanted to identify which process failures were most prevalent and to find the cause.

Step 1. Checklist

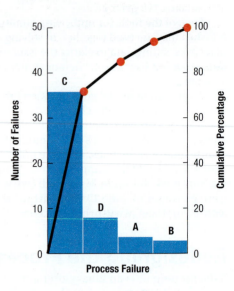

Step 2. Pareto Chart

Headliner failures

Process failure	Tally	Total
A. Tears in fabric	IIII	4
B. Discolored fabric	III	3
C. Broken fiber board	IIII IIII IIII IIII IIII IIII IIII I	36
D. Ragged edges	IIII II	7
		Total 50

Step 3. Cause-and-Effect Diagram

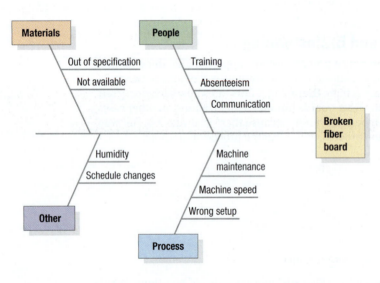

Step 4. Bar Chart

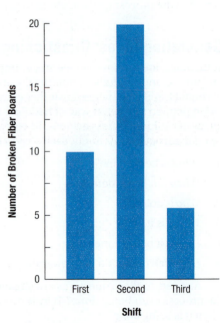

▲ **FIGURE 4.13**

Application of the Tools
for Improving Quality

SOLUTION

Figure 4.13 shows the sequential application of several tools for improving quality.

Step 1: A checklist of different types of process failures was constructed from last month's production records.

Step 2: A Pareto chart prepared from the checklist data indicated that broken fiber board accounted for 72 percent of the process failures.

Step 3: A cause-and-effect diagram for broken fiber board identified several potential causes for the problem. The one strongly suspected by the manager was employee training.

Step 4: The manager reorganized the production reports into a bar chart according to shift because the personnel on the three shifts had varied amounts of experience.

DECISION POINT

The bar chart indicated that the second shift, with the least experienced workforce, had most of the process failures. Further investigation revealed that workers were not using proper procedures for stacking the fiber boards after the press operation, which caused cracking and chipping. The manager set up additional training sessions focused on board handling. Although the second shift was not responsible for all the process failures, finding the source of many of the failures enabled the manager to improve the performance of her operations.

After the brainstorming session is over, the design team moves into the "get real" phase: They evaluate the different ideas. The team identifies the changes that give the best payoffs for process redesign. The redesign could involve issues of capacity, layout, technology, or even location, all of which are discussed in more detail in the following chapters.

The redesigned process is documented once again, this time as the "after" view of the process. Expected payoffs are carefully estimated, along with risks. For changes involving investments, the time value of money must be considered (see MyOMLab Supplement F, "Financial Analysis,"). The impact on people (skills, degree of change, training requirements, and resistance to change) must also be factored into the evaluation of the new design.

MyOMLab

Managerial Practice 4.1 describes how Baptist Memorial Hospital analyzed its processes to solve its capacity problem and improve patient satisfaction at the same time without any addition of new resources.

Benchmarking

Benchmarking can be another valuable source for process redesign. **Benchmarking** is a systematic procedure that measures a firm's processes, services, and products against those of industry leaders. Companies use benchmarking to better understand how outstanding companies do things so that they can improve their own processes.

benchmarking

A systematic procedure that measures a firm's processes, services, and products against those of industry leaders.

MANAGERIAL PRACTICE 4.1   Baptist Memorial Hospital

Baptist Memorial Hospital–Memphis is a 760-bed tertiary care hospital. It had a capacity problem, or so it seemed, with occupancy routinely exceeding 90 percent. However, it solved the problem with process improvement, rather than adding staff or bed capacity. Administration, nurses, and physicians centralized bed assignments and added a new bed-tracking system to provide bed information in real time. They then focused on improving processes at the emergency department (ED). An express admission unit (EAU), a 21-bed dedicated area that processes direct and emergency department admissions, was opened to remove responsibility for a particularly time-intensive activity from busy unit nurses. The new processes were less divergent and had more of a line flow. They then began testing process improvement ideas for change on a small scale, altering processes to improve them, and spreading the processes to other areas when they are successful. They began to fax reports from the ED to the receiving unit, shifted more nurses to work during peak periods, began lab and X-ray diagnostic procedures at triage when the EU was at capacity, took patients directly to a room when one became available with bedside registration, and segmented the urgent care population within the ED.

Redesigned processes reduced patient delays. Turnaround time for the overall ED was reduced by 9 percent, even while the ED volume was increasing. Length of stay was reduced by 2 days, the equivalent of building 12 Intensive care unit (ICU) beds. The mortality rate decreased, volume increased by 20 percent, and patient satisfaction improved significantly. What first appeared to be a capacity problem was resolved without adding staff or the number of beds—it was solved with redesigned processes.

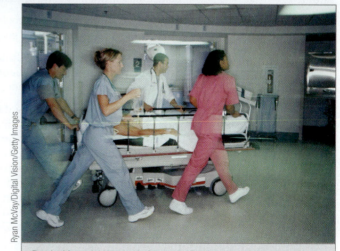

Ryan McVay/Digital Vision/Getty Images

Baptist Memorial Hospital in Memphis, Tennessee, holds "huddle meetings" at least three times a day seeking out process improvements. The meetings bring together the hospital's house supervisor, housekeeping supervisor, and key nurses. Improvements have been dramatic. In 2011, the hospital was ranked in the top 5 percent nationally for emergency medicine.

Source: Suzanne S. Horton, "Increasing Capacity While Improving the Bottom Line," *Frontiers of Health Services Management,* vol. 20, no. 4 (Summer 2004), pp. 17–23; Richard S. Zimmerman, "Hospital Capacity, Productivity, and Patient Safety—It All Flows Together," *Frontiers of Health Services Management,* vol. 20, no. 4 (Summer 2004), pp. 33–38, *baptistonline.org,* April, 2011.

Gerard Vesey/Alamy

Omgeo is a behind-the-scenes company that settles trades between financial services firms. The process used to involve dozens of scribbled faxes, telexes, and telephone calls made for the typical trade costs from $10 to $12. Now, the process costs only 20 cents to $1 per trade—and investment managers essentially got the service free. A key change was using the Internet and new information technology solutions. Information goes into a central database that the broker, investment manager, and custodian banks all have access to in real time. The details of the trades are automatically compared to eliminate errors.

Benchmarking focuses on setting quantitative goals for improvement. *Competitive* benchmarking is based on comparisons with a direct industry competitor. *Functional* benchmarking compares areas such as administration, customer service, and sales operations with those of outstanding firms in any industry. For instance, Xerox benchmarked its distribution function against L.L. Bean's because L.L. Bean is renowned as a leading retailer in distribution efficiency and customer service.

Internal benchmarking involves using an organizational unit with superior performance as the benchmark for other units. This form of benchmarking can be advantageous for firms that have several business units or divisions. All forms of benchmarking are best applied in situations where you are looking for a long-term program of continuous improvement.

Typical measures used in benchmarking include cost per unit, service upsets (breakdowns) per customer, processing time per unit, customer retention rates, revenue per unit, return on investment, and customer satisfaction levels.

Benchmarking consists of four basic steps:

Step 1. Planning. Identify the process, service, or product to be benchmarked and the firm(s) to be used for comparison; determine the performance metrics for analysis; collect the data.

Step 2. Analysis. Determine the gap between the firm's current performance and that of the benchmark firm(s); identify the causes of significant performance gaps.

Step 3. Integration. Establish goals and obtain the support of managers who must provide the resources for accomplishing the goals.

Step 4. Action. Develop cross-functional teams of those most affected by the changes; develop action plans and team assignments; implement the plans; monitor progress; recalibrate benchmarks as improvements are made.

Collecting benchmarking data can sometimes be a challenge. Internal benchmarking data is surely the most accessible. One way of benchmarking is always available—tracking the performance of a process over time. Functional benchmarking data are often collected by professional associations or consulting firms. Several corporations and government organizations have agreed to share and standardize performance benchmarks. The American Productivity and Quality Center, a nonprofit organization, created thousands of measures, as Figure 4.14 illustrates. A full range of metrics can be explored at **www.apqc.org**. Another source is the Supply-Chain Council, which has defined key metrics in its Supply-Chain Operations Reference (SCOR) model (see Chapter 12, "Supply Chain Integration").

Managing and Implementing Processes

Failure to manage processes is failure to manage the business. Implementing a beautifully redesigned process is only the beginning to continually monitoring and improving processes. Metrics goals must be continually evaluated and reset to fit changing requirements. Avoid the following seven mistakes when managing processes:[2]

1. *Not Connecting with Strategic Issues.* Is particular attention being paid to core processes, competitive priorities, impact of customer contact and volume, and strategic fit during process analysis?

---

[2]Geary A. Rummler and Alan P. Brache, *Improving Performance,* 2nd ed. (San Francisco: Jossey-Bass, 1995), pp. 126–133.

Customer Relationship Process

- Total cost of "enter, process, and track orders" per $1,000 revenue
- System costs of process per $100,000 revenue
- Value of sales order line item not fulfilled due to stockouts, as percentage of revenue
- Percentage of finished goods sales value that is returned
- Average time from sales order receipt until manufacturing or logistics is notified
- Average time in direct contact with customer per sales order line item
- Energy consumed in transporting product
- Total distance travelled for products
- Green house gas emissions

◀ **FIGURE 4.14**
Illustrative Benchmarking
Metrics by Type of Process

Order Fulfillment Process

- Value of plant shipments per employee
- Finished goods inventory turnover
- Reject rate as percentage of total orders processed
- Percentage of orders returned by customers due to quality problems
- Standard customer lead time from order entry to shipment
- Percentage of orders shipped on time
- Use of non-renewable energy sources
- Use of toxic ingredients
- Safe and healthy work environment

New Service/Product Development Process

- Percentage of sales due to services/products launched last year
- Cost of "generate new services/products" process per $1,000 revenue
- Ratio of projects entering the process to projects completing the process
- Time to market for existing service/product improvement project
- Time to market for new service/product project
- Time to profitability for existing service/product improvement project

Supplier Relationship Process

- Cost of "select suppliers and develop/maintain contracts" process per $1,000 revenue
- Number of employees per $1,000 of purchases
- Percentage of purchase orders approved electronically
- Average time to place a purchase order
- Total number of active vendors per $1,000 of purchases
- Percentage of value of purchased material that is supplier certified
- Amount of toxic chemicals used in supplies production process
- Energy consumed in transporting raw materials and parts
- Total distance travelled for raw materials and parts
- Green house gas emissions
- Supplier's use of toxic chemicals in production process
- Percentage of child labor used by supplier

Support Process

- Systems cost of finance function per $1,000 revenue
- Percentage of finance staff devoted to internal audit
- Total cost of payroll processes per $1,000 revenue
- Number of accepted jobs as percentage of job offers
- Total cost of "source, recruit, and select" process per $1,000 revenue
- Average employee turnover rate

2. *Not Involving the Right People in the Right Way.* Does process analysis closely involve the people performing the process, or those closely connected to it as internal customers and suppliers?

3. *Not Giving the Design Teams and Process Analysts a Clear Charter, and then Holding Them Accountable.* Does management set expectations for change and maintain pressure for results? Does it allow paralysis in process improvement efforts by requiring excessive analysis?

4. *Not Being Satisfied Unless Fundamental "Reengineering" Changes are Made.* Is the radical change from process reengineering the expectation? If so, the cumulative effect of many small improvements that could be made incrementally could be lost. Process management efforts should not be limited to downsizing or to reorganization only, even though jobs may be eliminated or the structure changed. It should not be limited to big technological innovation projects, even though technological change occurs often.

5. *Not Considering the Impact on People.* Are the changes aligned with the attitudes and skills of the people who must implement the redesigned process? It is crucial to understand and deal with the *people side* of process changes.

6. *Not Giving Attention to Implementation.* Are processes redesigned, but never implemented? A great job of flowcharting and benchmarking is of only academic interest if the proposed changes are not implemented. Sound project management practices are required.

7. *Not Creating an Infrastructure for Continuous Process Improvement.* Is a measurement system in place to monitor key metrics over time? Is anyone checking to see whether anticipated benefits of a redesigned process are actually being realized?

Managers must make sure that their organization spots new performance gaps in the continual search for process improvements. Process redesign efforts need to be part of periodic reviews and even annual plans. Measurement is the particular focus of the next chapter. It covers how a performance tracking system is the basis for feedback and improvement efforts. The essence of a learning organization is the intelligent use of such feedback.

LEARNING GOALS IN REVIEW

① Explain a systematic way to analyze processes. The section "A Systematic Approach" on pp. 121–123, gives six steps to analysis. Focus on Figure 4.1 for the sequence of these steps.

② Define flowcharts, swim lane flowcharts, and service blueprints. The "Documenting the Process" section, pp. 123–127, demonstrates these three techniques for documenting and evaluating processes. More than one flowchart can be used to handle nested processes. Service blueprints show the line of visibility where there is customer contact.

③ Describe the various work measurement techniques. The time study method, elemental standard data approach, predetermined data approach, work sampling method, and learning curve analysis are briefly described in the "Work Measurement Techniques" section, pp. 127–129. For a more complete description, see MyOMLab Supplement H, "Measuring Output Rates"

④ Identify metrics for process evaluation. The "Evaluating Performance" section, p. 131, identifies a variety of performance measures. Figure 4.9 breaks them out as they relate to the remaining chapters of Part 2.

⑤ Describe Pareto charts, cause-and-effect diagrams, and process simulation. These techniques, described in the "Data Analysis Tools" section on pp. 132–135, help you to understand the causes of performance gaps. Process simulation is a more advanced tool and described in more depth in MyOMLab Supplement E.

⑥ Create better processes using benchmarking. "Benchmarking," pp. 137–138, whether it is functional, internal, or competitive, is a systematic procedure that measures a firm's processes or products against those in other areas. Figure 4.14 provides an array of metrics that can be used, depending on the process being evaluated.

⑦ Identify keys for effective process management. The "Managing and Implementing Processes" section, pp. 138–139, gives seven mistakes that can be made. There must be a continual search for process improvements.

MyOMLab helps you develop analytical skills and assesses your progress with multiple problems on process charts, standard times, learning curves, bar charts, scatter diagrams, Pareto charts, and histograms.

MyOMLab Resources	Titles	Link to the Book
Videos	*Process Analysis at Starwood*	Entire chapter
Active Model Exercise	4.1 Pareto Chart	Active Model Exercise: 4.1 Pareto Chart; Evaluating Performance; Example 4.2 (pp. 132–133); Example 4.4 (p. 135–137)
OM Explorer Solvers	Process Charts Bar, Pareto, and Line Charts	Documenting the Process; Figure 4.8 (p. 130) Evaluating the Process; Example 4.2 (pp. 132–133); Example 4.4 (pp. 135–137)
OM Explorer Tutors	4.1 Process Charts 4.2 Pareto Charts	Process Charts; Figure 4.8 (p. 130); Solved Problem 2 (pp. 142–143) Evaluating Performance; Example 4.2 (pp. 132–133); Example 4.4 (pp. 135–137)
OM Explorer Tutor Exercises	4.1 Process Chart of your choosing 4.2 Pareto Chart	Documenting the Process; Figure 4.8 (p. 130) Evaluating Performance; Example 4.2 (pp. 132–133); Example 4.4 (pp. 135–137)

MyOMLab Resources	Titles	Link to the Book
Flowchart Tutorials	4.1 Flowcharting in Excel 4.2 Flowcharting in PowerPoint 4.3 Live Flowcharting for Figures 4.2–4.3	Documenting the Process Documenting the Process Documenting the Process
SmartDraw	Often used in practice to create flowcharts	Get free trial version online
Virtual Tours	1. Anrosia and Hershey Foods Corporation	Strategic Fit
MyOMLab Supplements	F. Financial Analysis H. Measuring Output Rates I. Learning Curve Analysis	Redesigning the Process Work Measurement Techniques Learning Curve Analysis
Internet Exercises	1. BIC Stationary, BIC Lighters, BIC Shavers 2. Fender Guild Guitars	Documenting the Process Benchmarking
Additional Case	The Facilities Maintenance Problem at Midwest University	Entire chapter
Key Equations		
Image Library		

Key Terms

bar chart 132
benchmarking 137
brainstorming 135
cause-and-effect diagram 133
checklist 132
design team 122
elemental standard data 127
flowchart 123

graphs 134
histogram 132
learning curve 129
metrics 122
Pareto chart 132
predetermined data approach 128
process analysis 121
process chart 129

process failure 132
process simulation 135
scatter diagram 133
service blueprint 126
suggestion system 122
swim lane flowchart 125
time study 127
work sampling 128

Solved Problem 1

Create a flowchart for the following telephone-ordering process at a retail chain that specializes in selling books and music CDs. It provides an ordering system via the telephone to its time-sensitive customers besides its regular store sales.

First, the automated system greets customers and identifies whether they have a tone or pulse phone. Customers choose 1 if they have a tone phone; otherwise, they wait for the first available service representative to process their request. If customers have a tone phone, they complete their request by choosing options on the phone. First, the system checks to see whether customers have an existing account. Customers choose 1 if they have an existing account or choose 2 if they want to open a new account. Customers wait for the service representative to open a new account if they choose 2.

Next, customers choose between the options of making an order, canceling an order, or talking to a customer representative for questions and/or complaints. If customers choose to make an order, then they specify the order type as a book or a music CD, and a specialized customer representative for books or music CDs picks up the phone to get the order details. If customers choose to cancel an order, then they wait for the automated response. By entering the order code via phone, customers can cancel the order. The automated system says the name of the ordered item and asks for the confirmation of the customer. If the customer validates the cancellation of the order, then the system cancels the order; otherwise, the system asks the customer to input the order code again. After responding to the request, the system asks whether the customer has additional requests; if not, the process terminates.

SOLUTION

Figure 4.15 shows the flowchart.

FIGURE 4.15 ▶

Flowchart of Telephone Ordering Process

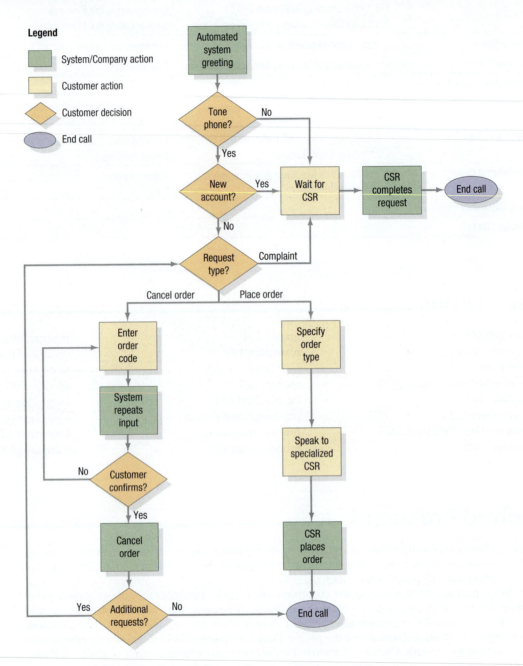

Solved Problem 2

An automobile service is having difficulty providing oil changes in the 29 minutes or less mentioned in its advertising. You are to analyze the process of changing automobile engine oil. The subject of the study is the service mechanic. The process begins when the mechanic directs the customer's arrival and ends when the customer pays for the services.

SOLUTION

Figure 4.16 shows the completed process chart. The process is broken into 21 steps. A summary of the times and distances traveled is shown in the upper right-hand corner of the process chart.

The times add up to 28 minutes, which does not allow much room for error if the 29-minute guarantee is to be met and the mechanic travels a total of 420 feet.

Process: Changing engine oil
Subject: Mechanic
Beginning: Direct customer arrival
Ending: Total charges, receive payment

Insert Step

Append Step

Remove Step

Summary

Activity		Number of Steps	Time (min)	Distance (ft)
Operation	●	7	16.50	
Transport	➡	8	5.50	420
Inspect	■	4	5.00	
Delay	❱	1	0.70	
Store	▼	1	0.30	

Step No.	Time (min)	Distance (ft)	●	➡	■	❱	▼	Step Description
1	0.80	50.0		X				Direct customer into service bay
2	1.80		X					Record name and desired service
3	2.30				X			Open hood, verify engine type, inspect hoses, check fluids
4	0.80	30.0		X				Walk to customer in waiting area
5	0.60		X					Recommend additional services
6	0.70					X		Wait for customer decision
7	0.90	70.0		X				Walk to storeroom
8	1.90		X					Look up filter number(s), find filter(s)
9	0.40				X			Check filter number(s)
10	0.60	50.0		X				Carry filter(s) to service pit
11	4.20		X					Perform under-car services
12	0.70	40.0	X	X				Climb from pit, walk to automobile
13	2.70		X					Fill engine with oil, start engine
14	1.30				X			Inspect for leaks
15	0.50	40.0		X				Walk to pit
16	1.00				X			Inspect for leaks
17	3.00		X					Clean and organize work area
18	0.70	80.0		X				Return to auto, drive from bay
19	0.30					X		Park the car
20	0.50	60.0		X				Walk to customer waiting area
21	2.30		X					Total charges, receive payment

Solved Problem 3

What improvement can you make in the process shown in Figure 4.16?

SOLUTION

Your analysis should verify the following three ideas for improvement. You may also be able to come up with others.

a. **Move Step 17 to Step 21.** Customers should not have to wait while the mechanic cleans the work area.

b. **Store Small Inventories of Frequently Used Filters in the Pit.** Steps 7 and 10 involve travel to and from the storeroom. If the filters are moved to the pit, a copy of the reference material must also be placed in the pit. The pit will have to be organized and well lighted.

c. **Use Two Mechanics.** Steps 10, 12, 15, and 17 involve running up and down the steps to the pit. Much of this travel could be eliminated. The service time could be shortened by having one mechanic in the pit working simultaneously with another working under the hood.

Solved Problem 4

Vera Johnson and Merris Williams manufacture vanishing cream. Their packaging process has four steps: (1) mix, (2) fill, (3) cap, and (4) label. They have had the reported process failures analyzed, which shows the following:

Process failure		Frequency
Lumps of unmixed product		7
Over- or underfilled jars		18
Jar lids did not seal		6
Labels rumpled or missing		29
	Total	60

Draw a Pareto chart to identify the vital failures.

SOLUTION

Defective labels account for 48.33 percent of the total number of failures:

$$\frac{29}{60} \times 100\% = 48.33\%$$

Improperly filled jars account for 30 percent of the total number of failures:

$$\frac{18}{60} \times 100\% = 30.00\%$$

The cumulative percent for the two most frequent failures is

$$48.33\% + 30.00\% = 78.33\%$$

Lumps represent $\frac{7}{60} \times 100\% = 11.67\%$ of failures; the cumulative percentage is

$$78.33\% + 11.67\% = 90.00\%$$

Defective seals represent $\frac{6}{60} \times 100\% = 10\%$ of failures; the cumulative percentage is

$$10\% + 90\% = 100.00\%$$

The Pareto chart is shown in Figure 4.17.

FIGURE 4.17 ▶
Pareto Chart

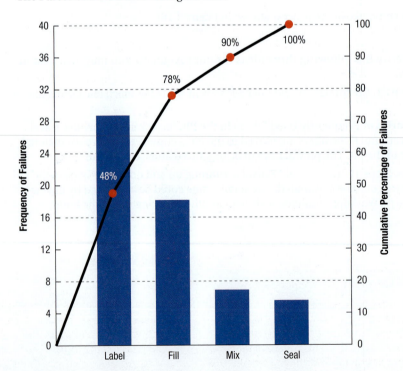

Discussion Questions

1. Continuous improvement recognizes that many small improvements add up to sizable benefits. Will continuous improvement take a company at the bottom of an industry to the top? Explain.

2. The Hydro-Electric Company (HEC) has three sources of power. A small amount of hydroelectric power is generated by damming wild and scenic rivers; a second source of power comes from burning coal, with emissions that create acid rain and contribute to global warming; the third source of power comes from nuclear fission. HEC's coal-fired plants use obsolete pollution-control technology, and an investment of several hundred million dollars would be required to update it. Environmentalists urge HEC to promote conservation and purchase power from suppliers that use the cleanest fuels and technology.

   However, HEC is already suffering from declining sales, which have resulted in billions of dollars invested in idle equipment. Its large customers are taking advantage of laws that permit them to buy power from low-cost suppliers. HEC must cover the fixed costs of idle capacity by raising rates charged to its remaining customers or face defaulting on bonds (bankruptcy). The increased rates motivate even more customers to seek low-cost suppliers, the start of a death spiral for HEC. To prevent additional rate increases, HEC implements a cost-cutting program and puts its plans to update pollution controls on hold.

   Form sides and discuss the ethical, environmental, and political issues and trade-offs associated with HEC's strategy.

3. Paul O'Neill, former U.S. Treasury Secretary, estimates that arguably half of the $2 trillion a year that Americans spend on health care is needlessly wasted. Brainstorm up to 10 blue-sky ideas to solve the following problems:

   a. A typical retail pharmacy spends 20 percent of its time playing telephone tag with doctors trying to find out what the intent was for a given prescription.

   b. After the person responsible for filling the prescription determines what they think they are supposed to do, errors can be made even in filling the prescription. For example, administering an adult dose (rather than the dose for a premature baby) of Heparin in a preemie ICU is fatal.

   c. Drugs get distributed at a hospital on a batch basis. For example, carts can be filled on Monday, Wednesday, and Friday. A huge volume of drugs can come back on Monday because they are not consumed on the wards between Friday and Monday, patient conditions changed, or the doctor decided on a different intervention. A technician spends the rest of the day restocking the shelves with the returns and 40 percent of the intravenous materials prepared on Friday morning are poured down the drain.

   d. Sometimes the administration of the drug was not done on the agreed schedule, because the nurses were busy doing something else.

   e. For every bed in an acute care hospital system, someone falls during the year. Most falls occur after 11 P.M. and before 6 A.M. Sometimes a bone is fractured, leading to immobilization and then pneumonia.

   f. One in every 14 people who goes to a U.S. hospital gets an infection they did not bring with them.

Problems

The OM Explorer and POM for Windows software is available to all students using the 10th edition of this textbook. Go to **www.pearsonhighered.com/krajewski** to download these computer packages. If you purchased MyOMLab, you also have access to Active Models software and significant help in doing the following problems. Check with your instructor on how best to use these resources. In many cases, the instructor wants you to understand how to do the calculations by hand. At the least, the software provides a check on your calculations. When calculations are particularly complex and the goal is interpreting the results in making decisions, the software replaces entirely the manual calculations.

1. Consider the Custom Molds, Inc. case at the end of Chapter 3, "Process Strategy." Prepare a flowchart of the mold fabrication process and the parts manufacturing process, showing how they are linked. For a good tutorial on how to create flowcharts, see **http://www.hci.com.au/hcisite5/library/materials/Flowcharting.htm.** Also check out the Flowcharting Tutor in Excel in MyOMLab.

2. Do Problem 1 using a process chart spreadsheet of your own design, one that differs from the *Process Chart* Solver in OM Explorer. It should have one or more columns to record information or metrics that you think are relevant, be it external customer contact, time delays, completion times, percent rework, costs, capacity, and/or demand rates. Your entries should show what information you would collect, even though only part of it is available in the case.

3. Founded in 1970, ABC is one of the world's largest insurance companies with locations in 28 countries. Given the following description, flowchart the new policy setup process as it existed in 1970:

   Individual customers who wanted to set up a new policy would visit one of ABC's 70 branch offices or make contact with an agent. They would then fill out an application and sometimes attach a check. The branch office then sent the application package through company mail to the XYZ division in London. In addition, a customer might also fill out the application at home and send it directly to any number of ABC locations, which would then transfer it to the London operation. Once received, XYZ separated the various parts of the application, then scanned and digitized it. The electronic image was then retrieved from a server and delivered to an associate's desktop client computer. The associate was responsible for entering the information on the form into the appropriate database. If the information supplied on the application was complete, a confirmation notice was automatically printed and sent to the customer. If the information was incomplete, then another associate, trained to deal with customers on the telephone, would call the customer to obtain the additional information. If the customer noticed something wrong on the confirmation notice received, she or he would either call a toll-free number or send in a letter describing the problem. The Customer Problem Resolution division dealt with problems arising at this point. An updated confirmation notice was sent to the customer. If the information was correct, the application transaction was complete.

4. Do Problem 3 using a process chart spreadsheet of your own design, one that differs from the *Process Chart* Solver in OM Explorer. It should have one or more columns to record information or metrics that you think should be collected to analyze the process (see Problem 2).

5. Prepare a flowchart of the field service division process at DEF, as described here. Start from the point where a call is received and end when a technician finishes the job.

DEF was a multibillion dollar company that manufactured and distributed a wide variety of electronic, photographic, and reprographic equipment used in many engineering and medical system applications. The Field Service Division employed 475 field service technicians, who performed maintenance and warranty repairs on the equipment sold by DEF. Customers would call DEF's National Service Center (NSC), which received about 3,000 calls per day. The NSC staffed its call center with about 40 call-takers. A typical incoming service call was received at the NSC and routed to one of the call-takers, who entered information about the machine, the caller's name, and the type of problem into DEF's mainframe computer. In some cases, the call-taker attempted to help the customer fix the problem. However, call-takers were currently only able to avoid about 10 percent of the incoming emergency maintenance service calls. If the service call could not be avoided, the call-taker usually stated the following script: "Depending upon the availability of our technicians, you should expect to see a technician sometime between now and (now 2 X)." ("X" was the target response time based on the model number and the zone.) This information was given to the customer because many customers wanted to know when a tech would arrive on site.

Call-takers entered service call information on DEF's computer system, which then sent the information electronically to the regional dispatch center assigned to that customer location. (DEF had four regional dispatch centers with a total of about 20 dispatchers.) Service call information was printed on a small card at the dispatch center. About every hour, cards were ripped off the printer and given to the dispatcher assigned to that customer location. The dispatcher placed each card on a magnetic board under the name of a tech that the dispatcher believed would be the most likely candidate for the service call, given the location of the machine, the current location of the tech, and the tech's training profile. After completing a service call, techs called the dispatcher in the regional dispatch center, cleared the call, and received a new call assigned by the dispatcher. After getting the service call from a dispatcher, a tech called the customer to give an expected time of arrival, drove to the customer site, diagnosed the problem, repaired the machine if parts were available in the van, and then telephoned the dispatcher for the next call. If the tech did not have the right parts for a repair, the tech informed the NSC and the part was express mailed to the customer; the repair was done the next morning.

6. Big Bob's Burger Barn would like to graphically depict the interaction among its lunch-ordering customers and its three employees. Customers come into the restaurant and eat there, rather than drive through and eat in the car. Using the brief process descriptions below, develop a service blueprint.

Fry Employee: receive customer order from counter employee, retrieve uncooked food, drop food into fry vat, wrap cooked food into special packaging, place wrapped items on service counter.

Grill Employee: receive customer order from counter employee, retrieve uncooked food, place food onto grill, build sandwich with requested condiments, deliver sandwich to Counter Employee.

Counter Employee: take order from customer, transmit appropriate orders to Fry and Grill Employee, transact payment, retrieve drinks, wrap sandwich, package order, and deliver order to customer.

7. After viewing the *Process Choice at the King Soopers Bakery* video in MyOMLab, prepare a flowchart for the three processes at King Soopers. For additional information on the processes, see the *Big Picture* for Chapter 3 that is also in MyOMLab.

8. Your class has volunteered to work for Referendum 13 on the November ballot, which calls for free tuition and books for all college courses except Operations Management. Support for the referendum includes assembling 10,000 yard signs (preprinted water-resistant paper signs to be glued and stapled to a wooden stake) on a fall Saturday. Construct a flowchart and a process chart for yard sign assembly. What inputs in terms of materials, human effort, and equipment are involved? Estimate the amount of volunteers, staples, glue, equipment, lawn and garage space, and pizza required.

9. Suppose you are in charge of a large mailing to the alumni of your college, inviting them to contribute to a scholarship fund. The letters and envelopes have been individually addressed (mailing labels were not used). The letters are to be processed (matched with correct envelope, time estimated to be 0.2 minutes each), folded (0.12 minutes each), and stuffed into the correct envelope (0.10 minutes each). The envelopes are to be sealed (0.05 minutes each), and a large commemorative stamp is to be placed in the upper right-hand corner of each envelope (0.10 minutes each).

   a. Make a process chart for this activity, assuming that it is a one-person operation.

   b. Estimate how long it will take to stuff, seal, and stamp 2,000 envelopes. Assume that the person doing this work is paid $8 per hour. How much will it cost to process 2,000 letters?

   c. Consider each of the following process changes. Which changes would reduce the time and cost of the current process?

   ■ Each letter has the same greeting "Dear Alumnus or Alumna," instead of the person's name.

   ■ Mailing labels are used and have to be put on the envelopes (0.10 minutes each).

- Prestamped envelopes are used.
- Envelopes are stamped by a postage meter which can stamp 200 letters per minute.
- Window envelopes are used.
- A preaddressed envelope is included with each letter for contributions (adds 0.05 minutes to stuffing step).

d. Would any of these changes be likely to reduce the effectiveness of the mailing? If so, which ones? Why?

e. Would the changes that increase time and cost be likely to increase the effectiveness of the mailing? Why or why not?

10. Diagrams of two self-service gasoline stations, both located on corners, are shown in Figure 4.18 (a) and (b). Both have two rows of four pumps and a booth at which an attendant receives payment for the gasoline. At neither station is it necessary for the customer to pay in advance. The exits and entrances are marked on the diagrams. Analyze the flows of cars and people through each station.

a. Which station has the more efficient flows from the standpoint of the customer?

b. Which station is likely to lose more potential customers who cannot gain access to the pumps because another car is headed in the other direction?

c. At which station can a customer pay without getting out of the car?

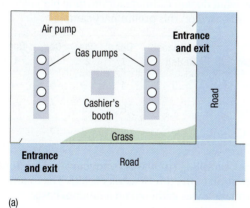

(a)

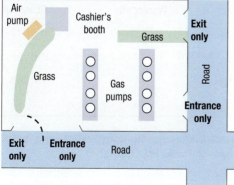

(b)

▲ **FIGURE 4.18**
Two Self-Service Gasoline Stations

11. The management of the Just Like Home Restaurant has asked you to analyze some of its processes. One of these processes is making a single-scoop ice cream cone. Cones can be ordered by a server (for table service) or by a customer (for takeout).

Figure 4.19 illustrates the process chart for this operation.

- The ice cream counter server earns $10 per hour (including variable fringe benefits).
- The process is performed 10 times per hour (on average).
- The restaurant is open 363 days a year, 10 hours a day.

a. Complete the Summary (top-right) portion of the chart.

b. What is the total labor cost associated with the process?

c. How can this operation be made more efficient? Make a process chart using OM Explorer's *Process Charts* Solver of the improved process. What are the annual labor savings if this new process is implemented?

12. As a graduate assistant, your duties include grading and keeping records for Operations Management course homework assignments. Five sections for 40 students each are offered each semester. A few graduate students attend sections 3 and 4. Graduate students must complete some extra work to higher standards for each assignment. Every student delivers (or is supposed to deliver) directly to (under) the door of your office one homework assignment every Tuesday. Your job is to correct the homework, record grades, sort the papers by class section, sort by student last name in alphabetical order, and return the homework papers to the appropriate instructors (not necessarily in that order). There are some complications. A fair majority of the students sign their names legibly, others identify work with their correct ID number, and a few do neither. Rarely do students identify their section number or graduate status. Prepare a list of process chart steps and place them in an efficient sequence.

13. At the Department of Motor Vehicles (DMV), the process of getting license plates for your car begins when you enter the facility and take a number. You walk 50 feet to the waiting area. During your wait, you count about 30 customers waiting for service. You notice that many customers become discouraged and leave. When a number is called, if a customer stands, the ticket is checked by a uniformed person, and the customer is directed to the available clerk. If no one stands, several minutes are lost while the same number is called repeatedly. Eventually, the next number is called, and more often than not, that customer has left too. The DMV clerk has now been idle for several minutes but does not seem to mind.

After 4 hours, your number is called and checked by the uniformed person. You walk 60 feet to the clerk, and the process of paying city sales taxes is completed in four minutes. The clerk then directs you to the waiting area for paying state personal property tax, 80 feet away. You take a different number and sit down with some different customers who are just renewing licenses. A 1-hour, 40-minute wait this time, and after a walk of 25 feet you pay property taxes in a process that takes two minutes. Now that you have paid taxes, you are eligible to pay registration and license fees. That department is 50 feet away, beyond the employees' cafeteria.

FIGURE 4.19 ▶
Process Chart for Making Ice
Cream Cones

Process:	Making one ice cream cone
Subject:	Server at counter
Beginning:	Walk to cone storage area
Ending:	Give it to server or customer

| | Summary | | | |
Activity	Number of Steps	Time (min)	Distance (ft)
Operation ●			
Transport ➡			
Inspect ■			
Delay ▶			
Store ▼			

Insert Step

Append Step

Remove Step

Step No.	Time (min)	Distance (ft)	●	➡	■	▶	▼	Step Description
1	0.20	5.0		X				Walk to cone storage area
2	0.05		X					Remove empty cone
3	0.10	5.0		X				Walk to counter
4	0.05		X					Place cone in holder
5	0.20	8.0		X				Walk to sink area
6	0.50						X	Ask dishwasher to wash scoop
7	0.15	8.0		X				Walk to counter with clean scoop
8	0.05		X					Pick up empty cone
9	0.10	2.5		X				Walk to flavor ordered
10	0.75		X					Scoop ice cream from container
11	0.75		X					Place ice cream in cone
12	0.25					X		Check for stability
13	0.05	2.5		X				Walk to order placement area
14	0.05		X					Give server or customer the cone

The registration and license customers are called in the same order in which personal property taxes were paid. There is only a ten-minute wait and a three-minute process. You receive your license plates, take a minute to abuse the license clerk, and leave exactly six hours after arriving.

Make a process chart using OM Explorer's *Process Charts* Solver to depict this process, and suggest improvements.

14. Refer to the process chart for the automobile oil change in Solved Problem 2. Calculate the annual labor cost if:

■ The mechanic earns $40 per hour (including variable fringe benefits).

■ The process is performed twice per hour (on average).

■ The shop is open 300 days a year, 10 hours a day.

a. What is the total labor cost associated with the process?

b. If steps 7, 10, 12, and 15 were eliminated, estimate the annual labor savings associated with implementing this new process.

15. A time study of an employee assembling peanut valves resulted in the following set of observations. What is the standard time, given a performance rating of 95 percent and an allowance of 20 percent of the total normal time?

Average Time (seconds)	Observations
15	14
20	12
25	15

16. An initial time study was done on a process with the following results (in minutes). Based on the data obtained so far, assuming an allowance of 20 percent of the normal

time, what do you estimate for the time per customer served, based on this preliminary sample?

Element	Performance Rating	Obs 1	Obs 2	Obs 3	Obs 4	Obs 5
Element 1	70	4	3	5	4	3
Element 2	110	8	10	9	11	10
Element 3	90	6	8	7	7	6

17. A work sampling study was conducted to determine the proportion of the time a worker is idle. The following information was gathered on a random basis:

Day	Number of Times Worker Idle	Total Number of Observations
Monday	17	44
Tuesday	18	56
Wednesday	14	48
Thursday	16	60

a. Based on these preliminary results, what percent of the time is the worker working?

b. If idle time is judged to be excessive, what additional categories might you add to a follow-up work sampling study to identify the root causes?

18. A contractor is preparing a bid to install swimming pools at a new housing addition. The estimated time to build the first pool is 35 hours. The contractor estimates an 85 percent learning rate. Without using the computer:

a. How long do you estimate the time required to install the second pool?

b. How long do you estimate the time required to install the fourth pool?

19. Return to Problem 18. Using OM Explorer's *Learning Curves* Solver, how long do you estimate the time required to install the fifth pool? What is your estimate of the total time for all five pools?

20. The manager of Perrotti's Pizza collects data concerning customer complaints about pizza delivery. Either the pizza arrives late, or the wrong pizza is delivered.

Problem	Frequency
Topping is stuck to box lid	17
Pizza arrives late	35
Wrong topping or combination	9
Wrong style of crust	6
Wrong size	4
Pizza is partially eaten	3
Pizza never arrives	6

a. Use a Pareto chart to identify the "vital few" delivery problems. Comment on potential root causes of these problems and identify any especially egregious quality failures.

b. The manager of Perrotti's Pizza is attempting to understand the root causes of late pizza delivery and has asked each driver to keep a log of specific difficulties that create late deliveries. After one week, the logs included the following entries:

delivery vehicle broke down, couldn't make it across town to deliver second pizza in time, couldn't deliver four pizzas to four different customers in time, kitchen was late in producing order, got lost, order ticket was lost in production, couldn't read address on ticket and went to wrong house.

Organize these causes into a cause-and-effect diagram.

21. Smith, Schroeder, and Torn (SST) is a short-haul household furniture moving company. SST's labor force, selected from the local community college football team, is temporary and part-time. SST is concerned with recent complaints, as tabulated on the following tally sheet:

Complaint	Tally
Broken glass	ҢＨ ҢＨ III
Delivered to wrong address	ҢＨ IIII
Furniture rubbed together while on truck	ҢＨ ҢＨ ҢＨ ҢＨ
Late delivery	ҢＨ
Late arrival for pickup	ҢＨ ҢＨ ҢＨ III
Missing items	ҢＨ ҢＨ ҢＨ ҢＨ ҢＨ I
Nicks and scratches from rough handling	ҢＨ ҢＨ
Soiled upholstery	ҢＨ III

a. Draw a bar chart and a Pareto chart using OM Explorer to identify the most serious moving problems.

b. The manager of Smith, Schroeder, and Torn is attempting to understand the root causes of complaints. He has compiled the following list of issues that occurred during problem deliveries.

truck broke down, ran out of packing boxes, multiple deliveries in one day caused truck to be late, no furniture pads, employee dropped several items, drive got lost on route to address, ramp into truck was bent, no packing tape, new employee doesn't know how to pack, moving dolly has broken wheel, employee late to work

Organize these causes into a cause-and-effect diagram.

22. Rick DeNeefe, manager of the Golden Valley Bank credit authorization department, recently noticed that a major competitor was advertising that applications for equity loans could be approved within two working days. Because fast credit approval was a competitive priority, DeNeefe wanted to see how well his department was doing relative to the competitor's. Golden Valley stamps each application with the date and time it is received and again when a decision is made. A total of 104 applications were received in March. The time required for each decision, rounded to the nearest hour, is shown in the following table. Golden Valley's employees work 8 hours per day.

Decision Process Time (hours)	Frequency
8	8
11	19
14	28
17	10
20	25
23	4
26	10
Total	104

a. Draw a bar chart for these data.

b. Analyze the data. How is Golden Valley Bank doing with regard to this competitive priority?

23. Last year, the manager of the service department at East Woods Ford instituted a customer opinion program to find out how to improve service. One week after service on a vehicle was performed, an assistant would call the customer to find out whether the work had been done satisfactorily and how service could be improved. After one year of gathering data, the assistant discovered that the complaints could be grouped into the following five categories:

Complaint	Frequency
Unfriendly atmosphere	5
Long wait for service	17
Price too high	20
Incorrect bill	8
Needed to return to correct problem	50
Total	100

a. Use OM Explorer to draw a bar chart and a Pareto chart to identify the significant service problems.

b. Categorize the following causes of complaints into a cause-and-effect diagram: tools, scheduling, defective parts, training, billing system, performance measures, diagnostic equipment, and communications.

24. Oregon Fiber Board makes roof liners for the automotive industry. The manufacturing manager is concerned about product quality. She suspects that one particular failure, tears in the fabric, is related to production-run size. An assistant gathers the following data from production records:

Run	Size	Failures (%)	Run	Size	Failures (%)
1	1,000	3.5	11	6,500	1.5
2	4,100	3.8	12	1,000	5.5
3	2,000	5.5	13	7,000	1.0
4	6,000	1.9	14	3,000	4.5
5	6,800	2.0	15	2,200	4.2
6	3,000	3.2	16	1,800	6.0
7	2,000	3.8	17	5,400	2.0
8	1,200	4.2	18	5,800	2.0
9	5,000	3.8	19	1,000	6.2
10	3,800	3.0	20	1,500	7.0

a. Draw a scatter diagram for these data.

b. Does there appear to be a relationship between run size and percent failures? What implications does this data have for Oregon Fiber Board's business?

25. Grindwell, Inc., a manufacturer of grinding tools, is concerned about the durability of its products, which depends on the permeability of the sinter mixtures used in production. Suspecting that the carbon content might be the source of the problem, the plant manager collected the following data:

Carbon Content (%)	Permeability Index
5.5	16
3.0	31
4.5	21
4.8	19
4.2	16
4.7	23
5.1	20
4.4	11
3.6	20

a. Draw a scatter diagram for these data.

b. Is there a relationship between permeability and carbon content?

c. If low permeability is desirable, what does the scatter diagram suggest with regard to the carbon content?

26. The operations manager for Superfast Airlines at Chicago's O'Hare Airport noticed an increase in the number of delayed flight departures. She brainstormed possible causes with her staff:

■ Aircraft late to gate

■ Acceptance of late passengers

■ Passengers arriving late at gate

■ Passenger processing delays at gate

■ Late baggage to aircraft

■ Other late personnel or unavailable items

■ Mechanical failures

Draw a cause-and-effect diagram to organize the possible causes of delayed flight departures into the following major categories: equipment, personnel, material, procedures, and "other factors" beyond managerial control. Provide a detailed set of causes for each major cause identified by the operations manager, and incorporate them in your cause-and-effect diagram.

27. Plastomer, Inc. specializes in the manufacture of high-grade plastic film used to wrap food products. Film is rejected and scrapped for a variety of reasons (e.g., opacity, high carbon content, incorrect thickness or gauge, scratches, and so on). During the past month, management collected data on the types of rejects and the amount of scrap generated by each type. The following table presents the results:

Type of Failure	Amount of Scrap (lbs.)
Air bubbles	500
Bubble breaks	19,650
Carbon content	150
Unevenness	3,810
Thickness or gauge	27,600
Opacity	450
Scratches	3,840
Trim	500
Wrinkles	10,650

Draw a Pareto chart to identify which type of failure management should attempt to eliminate first.

28. Management of a shampoo bottling company introduced a new 13.5-ounce pack and used an existing machine, with some modifications, to fill it. To measure filling consistency by the modified machine (set to fill 13.85 ounces), an analyst collected the following data (volume in ounces) for a random sample of 100 bottles:

a. Draw a histogram for these data.

b. Bottles with less than 12.85 ounces or more than 14.85 ounces are considered to be out of specification. Based on the sample data, what percentage of the bottles filled by the machine will be out of specification?

Bottle Volume (ounces)									
13.0	13.3	13.6	13.2	14.0	12.9	14.2	12.9	14.5	13.5
14.1	14.0	13.7	13.4	14.4	14.3	14.8	13.9	13.5	14.3
14.2	14.1	14.0	13.9	13.9	14.0	14.5	13.6	13.3	12.9
12.8	13.1	13.6	14.5	14.6	12.9	13.1	14.4	14.0	14.4
13.1	14.1	14.2	12.9	13.3	14.0	14.1	13.1	13.6	13.7
14.0	13.6	13.2	13.4	13.9	14.5	14.0	14.4	13.9	14.6
12.9	14.3	14.0	12.9	14.2	14.8	14.5	13.1	12.7	13.9
13.6	14.4	13.1	14.5	13.5	13.3	14.0	13.6	13.5	14.3
13.2	13.8	13.7	12.8	13.4	13.8	13.3	13.7	14.1	13.7
13.7	13.8	13.4	13.7	14.1	12.8	13.7	13.8	14.1	14.3

Advanced Problems

29. This problem should be solved as a team exercise:

Shaving is a process that most men perform each morning. Assume that the process begins at the bathroom sink with the shaver walking (say, 5 feet) to the cabinet (where his shaving supplies are stored) to pick up bowl, soap, brush, and razor. He walks back to the sink, runs the water until it gets warm, lathers his face, shaves, and inspects the results. Then, he rinses the razor, dries his face, walks over to the cabinet to return the bowl, soap, brush, and razor, and comes back to the sink to clean it up and complete the process.

a. Develop a process chart for shaving. (Assume suitable values for the time required for the various activities involved in the process.)

b. Brainstorm to generate ideas for improving the shaving process. Having fewer than 20 ideas is unacceptable. (Do not try to evaluate the ideas until the group has compiled as complete a list as possible. Otherwise, judgment will block creativity.)

30. At Conner Company, a custom manufacturer of printed circuit boards, the finished boards are subjected to a final inspection prior to shipment to its customers. As Conner's quality assurance manager, you are responsible for making a presentation to management on quality problems at the beginning of each month. Your assistant has analyzed the reject memos for all the circuit boards that were rejected during the past month. He has given you a summary statement listing the reference number of the circuit board and the reason for rejection from one of the following categories:

A = Poor electrolyte coverage

B = Improper lamination

C = Low copper plating

D = Plating separation

E = Improper etching

For 50 circuit boards that had been rejected last month, the summary statement showed the following:

C B C C D E C C B A D A C C C B C A C D C A C C B

A C A C B C C A C A A C C D A C C C E C C A B A C

a. Prepare a tally sheet (or checklist) of the different reasons for rejection.

b. Develop a Pareto chart to identify the more significant types of rejection.

c. Examine the causes of the most significant type of defect, using a cause-and-effect diagram.

Active Model Exercise

This Active Model appears in MyOMLab. Continuing on with Example 4.2, it allows you to evaluate the structure of a Pareto chart.

QUESTIONS

1. What percentage of overall complaints does discourteous service account for?

2. What percentage of overall complaints do the three most common complaints account for?

3. How does it affect the chart if we eliminate discourteous service?

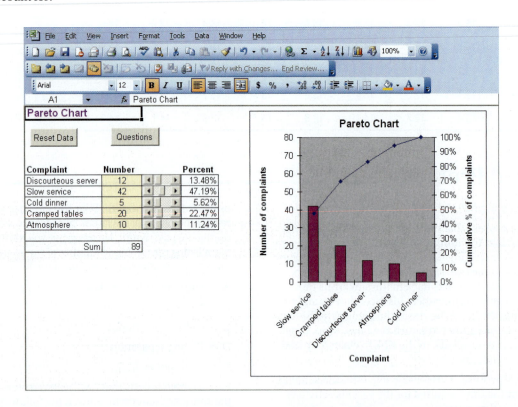

The features and layout of The Phoenician property of Starwood Hotels and Resorts at Scottsdale, Arizona, are shown in the following figure. Starwood Hotels and Resorts is no stranger to process improvement. In fact, the president's letter in a recent annual report stated that through "… benchmarking, Six Sigma, and recognition of excellence, [Starwood is] driving results in a virtual cycle of self-improvement at all levels of the Company." Recognizing that improved processes in one department of a single hotel, if rolled out across the organization, could lead to significant improvements, the company recently created a program called the "Power of Innovation," or POI.

The Power of Innovation program in Starwood seeks to capture best practices that exist throughout hotels across all brands in North America. An internal team with expertise in kitchen preparation and production, laundry, stewarding, front office, and housekeeping works with individual properties to build upon and maximize the existing knowledge of local property management teams. The team usually spends about a week on property entrenched in operations to really see day-to-day activity over an extended period. Of particular interest is scheduling the workforce to meet the demand of each hotel's individual operations while streamlining operations processes.

At the Westin Galleria-Oaks in Houston, Texas, for example, the POI team helped management achieve a 6 percent productivity improvement in the kitchen preparation and production job, with a reduction of 2,404 hours used and $23,320 in annual payroll savings alone. At the same time, other POI projects at the hotel generated an additional $14,400 in annual payroll savings.

The Phoenician in Scottsdale also had a visit from the POI team. One area the team focused on was stewarding. The typical stewarding process includes the following duties: dishwashing, kitchen trash removal, polishing silver, and assisting with banquet meal food prep lines. Stewards support eight kitchens and two bakeries, and work with housekeeping in keeping public areas, such as restrooms and pool cabanas, clean.

A flowchart that diagrams the existing stewarding process that the team documented is shown in the figure. In any given day, a particular steward may provide support to more than one kitchen, and be called upon to do a variety of tasks.

Before the POI team arrived, stewards were dedicated to a particular kitchen or area during their shift. Each kitchen required stewarding coverage as outlined by the executive chef, so more than one steward may be assigned to an area. A certain amount of stewarding work could be forecast

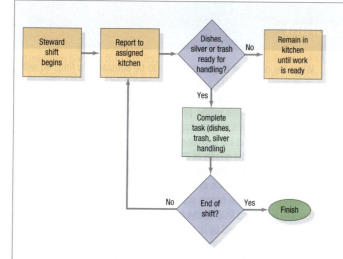

Selecte

Andersen, Bjør
Milwaukee, V

Ahire, Sanjay L.
Process Imp
Journal of Gi
no. 2 (Januar

Bhuiyan, Nadjia
Improvemer
tional Journa
no. 8 (2006),

Carey, Susan. "C
(August 9, 20

Davenport, Tho
Harvard Bus

Edmondson, A
Harvard Bus

Fisher, Anne. "G
no. 11 (Nove

Fleming, John
Human Sigr
pp. 101–108.

Greasley, A. "Us
to Support a
Organisation

Grosskopf, Alex
Business Pro
Kiffer Press,

took to perform each stewarding task. Some restaurants and kitchens did not require full-time coverage by a steward, so the steward would be assigned multiple kitchens to fill a work shift. In the case of coverage between the 19th Hole restaurant on one side of the resort, and the Canyon Building on the other side, that steward would walk one-half mile, one way, to take care of duties in both locations because they lacked enough work for a dedicated steward in each location.

Often, stewards had downtime as they waited for banquet dishes to be cleared, or kitchen pots and utensils to be brought in for cleaning. Some restaurants had china with special cleaning requirements, meaning those dishes had to be handwashed instead of being placed in an automated sanitizing dishwasher. This situation required a dedicated steward to perform that task.

Time studies revealed how long it took stewards to move from one kitchen to the next. The studies also helped the POI team understand how long it took to wash dishes in the five-star restaurant versus the casual poolside dining area's kitchen. Additionally, the studies uncovered building design and landscaping limitations that prevented staff from moving between kitchens quickly. In some cases, a maze of corridors added miles to the distances covered each day, and thick privacy hedges barred entry to sidewalk shortcuts.

by the food and beverage manager, based on scheduled banquets, afternoon teas, conference buffets, and restaurant reservations. Considerable uncertainty also arose from traffic generated by leisure travelers and local clientele, meaning that stewards assigned to designated areas periodically did not have a steady flow of work.

On a weekly basis, activity levels for the dedicated stewarding staff were determined, based on executive chef input. Other factors considered in the weekly planning included prior year activity, special events and holidays, and number of children. With this information, the executive steward created a summary of all meals, called covers, by location, date, and time of day. Then, an Excel spreadsheet template was used to create the schedule for deployment of stewarding staff throughout the resort's kitchens and restaurants.

In performing its analysis, the POI team examined staff availability, banquet events, restaurants, occupied room counts, and other drivers of business to areas supported by stewards. Time studies were done to determine how far stewards were traveling throughout the property, and how long it

QUESTIONS

1. How can the management specifically improve the stewarding process at The Phoenician? Using the information provided, create a flowchart illustrating the new process.

2. What are the benefits that the POI program can bring to Starwood? Can these benefits be extended to other processes and properties within the Starwood system?

3. Of the seven mistakes organizations can make when managing processes (see last section of this chapter), which ones might Starwood be most at risk of making? Why?

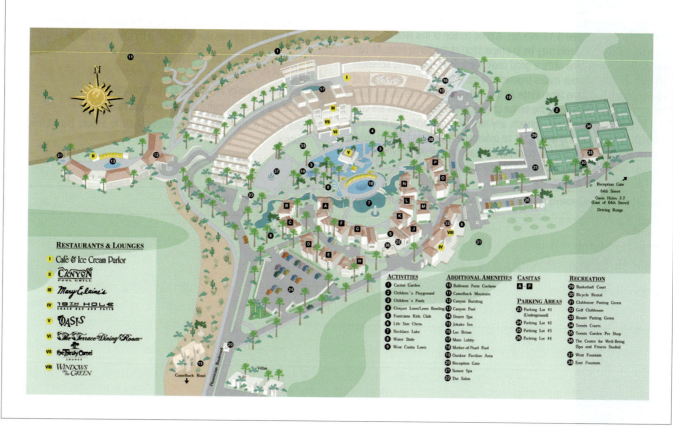

RESTAURANTS & LOUNGES

I Café & Ice Cream Parlor
II CANYON POOL GRILL
III Mary Elaine's
IV 19TH HOLE GRACE BAR AND PATIO
V OASIS
VI The Terrace Dining Room
VII the Thirsty Camel LOUNGE
VIII WINDOWS ON THE GREEN

ACTIVITIES
1 Cactus Garden
2 Children's Playground
3 Children's Pools
4 Croquet Lawn/Lawn Bowling
5 Funicians Kids Club
6 Life Size Chess
7 Necklace Lake
8 Water Slide
9 West Casita Lawns

ADDITIONAL AMENITIES
10 Ballroom Porte Cochere
11 Camelback Mountain
12 Canyon Building
13 Canyon Pool
14 Desert Spa
15 Jokake Inn
16 Las Brisas
17 Main Lobby
18 Mother-of-Pearl Pool
19 Outdoor Pavilion Area
20 Reception Gate
21 Sunset Spa
22 The Salon

CASITAS
A–P

PARKING AREAS
23 Parking Lot #1 (Underground)
24 Parking Lot #2
25 Parking Lot #3
26 Parking Lot #4

RECREATION
29 Basketball Court
30 Bicycle Rental
31 Clubhouse Putting Green
32 Golf Clubhouse
33 Resort Putting Green
34 Tennis Courts
35 Tennis Garden Pro Shop
36 The Centre for Well-Being (Spa and Fitness Studio)
37 West Fountain
38 East Fountain

CASE

"Two bean tacos
please." Ivan Ka
the beverage or
customers mear
income was gre

José's is
range of Mexica
It is located in
large metropoli
limited free off-
the Mexican the
with serapes, t
and mariachi al

Patrons e
rectly into the
patrons are gre
expected wait.
Saturday nights
Because space
until their party

After sea
with water. If st
the patrons with
can restaurant
announces the
the beverages,

The men
stocks (chicken
lettuce) and a
and spices). E
stocks so that
the requested
once it has bee
cooking, so se
be imagined,
production of
approximately
been complete
rect and pleas
touches. Wher
ers them to the
to detect wher

Source: This cas

A Baseline Engineer for Verizon readies his computer that will control a bank of cell phones making and receiving calls on different networks. He travels through northern Virginia, Washington, DC, and Maryland with a truck outfitted to test the services of Verizon and its competitors to see where faults lie in Verizon's system.

5

QUALITY AND PERFORMANCE

Verizon Wireless

Anyone who owns a cell phone knows the agony of a dropped call. Did you know that the reason for the dropped call may be the phone itself, and not the strength of the signal? Verizon Wireless serves more than 62 million customers in the United States and, along with the other major carriers, it knows that if the phone does not work, the company, and not the manufacturer, will likely take the blame from the customer. Verizon touts the reliability of its services and can ill afford the failure of cell phones due to the quality of manufacture. Verizon expects manufacturers such as Motorola, Samsung, and LG Electronics to provide defect-free phones; however, experience has indicated that extensive testing by Verizon employees is also needed.

In addition to a tear-down analysis that looks for weaknesses in a phone's hardware and components, the device is tested for its ability to withstand temperature extremes, vibration, and stress. Beyond these physical tests, Verizon uses two approaches to assess a phone's capability to receive cellular signals and clearly communicate to the caller. First, Verizon hires 98 test personnel who drive $300,000 specially equipped vans more than 1 million miles a year to measure network performance using prospective new cell phones. They make more than 3 million voice call attempts and 16 million data tests annually. The tests check the coverage of the network as well as the capability of the cell phones to pick up the signals and clearly communicate to the caller. Second, Verizon uses Mr. Head, a robotic mannequin, who has a recorded voice and is electronically equipped with a rubber ear that evaluates how well the phone's mouthpiece transmits certain phonetics. Mr. Head utters what sounds like gibberish; however, it actually covers the range of sounds in normal speech patterns. Other systems monitor the tests and summarize results.

Some phones spend so much time in the test phase that ultimately they never make it to the market. Clearly, in those cases, the cost of poor quality to the manufacturer is very high.

Source: Amol Sharma, "Testing, Testing," *Wall Street Journal* (October 23, 2007); Janet Hefler, "Verizon Tester Checks Vineyard Networks," *The Martha's Vineyard Times* (August 30, 2007); Jon Gales, "Ride Along With a Verizon Wireless Test Man," *Mobile Tracker* (April 4, 2005) **http:// investor.verizon.com** (2007).

LEARNING GOALS *After reading this chapter, you should be able to:*

1 Define the four major costs of quality.

2 Describe the role of ethics in the quality of services and products.

3 Explain the basic principles of TQM programs.

4 Explain the basic principles of Six Sigma programs.

5 Describe how to construct control charts and use them to determine whether a process is out of statistical control.

6 Describe how to determine whether a process is capable of producing a service or product to specifications.

Creating Value through Operation Management

Using Operations to Compete
Project Management

Managing Processes

Process Strategy
Process Analysis
Quality and Performance
Capacity Planning
Constraint Management
Lean Systems

Managing Supply Chains

Supply Chain Inventory Management
Supply Chain Design
Supply Chain Location Decisions
Supply Chain Integration
Supply Chain Sustainability and Humanitarian Logistics
Forecasting
Operations Planning and Scheduling
Resource Planning

The challenge for businesses today is to satisfy their customers through the exceptional performance of their processes. Verizon Wireless is one example of a company that met the challenge by designing and managing processes that provide customers with total satisfaction. Evaluating process performance is important if this is to happen.

Evaluating process performance is also necessary for managing supply chains. For example, at Verizon Wireless, the process of delivering cell phone communications to the customer might be measured on the consistency of service and the sound quality of the voice transmissions. The procurement process, which involves selecting the suppliers for the cell phones and evaluating how they deliver their products, might be measured in terms of the quality of the cell phones delivered to Verizon, the on-time delivery performance of the suppliers, and the cost of the cell phones. Ultimately, the evaluation of the supply chain consisting of these two processes and many others will depend on how well it satisfies the customers of Verizon, who consider the value of the service to be how well it meets or exceeds expectations. The performance of these individual processes must be consistent with the performance measures for the supply chain.

Quality and Performance across the Organization

Quality and performance should be everybody's concern. Take for example QVC, a $7.4 billion televised shopping service. QVC airs 24 hours a day, all year round. QVC sells some 60,000 items ranging from jewelry, tools, cookware, clothing, and gourmet food to computers and annually ships more than 166 million packages worldwide.

QVC's processes, which span all the functional areas, spring into action with a customer order: Order taking and delivery date promising, billing, and order delivery all ensue once an order is placed. QVC operates four call centers that handle 179 million calls annually from customers who want to order something, complain about a problem, or just get product information. The call center representative's demeanor and skill are critical to achieving a successful customer encounter. QVC management keeps track of productivity, quality, and customer satisfaction measures for all processes. When the measures slip, problems are addressed aggressively. Knowing how to assess whether the process is performing well and when to take action are key skills QVC managers must have. In this chapter, we first address the costs of quality and then focus on Total Quality Management and Six Sigma, two philosophies and supporting tools that many companies embrace to evaluate and improve quality and performance.

Costs of Quality

When a process fails to satisfy a customer, the failure is considered a **defect**. For example, according to the California Academy of Family Physicians, defects for the processes in a doctor's practice are defined as "anything that happened in my office that should not have happened, and that I absolutely do not want to happen again." Obviously, this definition covers process failures that the patient sees, such as poor communication and errors in prescription dosages. It also includes failures the patient does not see, such as incorrect charting.

Many companies spend significant time, effort, and expense on systems, training, and organizational changes to improve the quality and performance of their processes. They believe that it is important to be able to gauge current levels of performance so that any process gaps can be determined. Gaps reflect potential dissatisfied customers and additional costs for the firm. Most experts estimate that the costs of quality range from 20 to 30 percent of gross sales. These costs can be broken down into four major categories: (1) prevention, (2) appraisal, (3) internal failure, and (4) external failure.

defect
Any instance when a process fails to satisfy its customer.

Prevention Costs

Prevention costs are associated with preventing defects before they happen. They include the costs of redesigning the process to remove the causes of poor performance, redesigning the service or product to make it simpler to produce, training employees in the methods of continuous improvement, and working with suppliers to increase the quality of purchased items or contracted services. In order to prevent problems from happening, firms must invest additional time, effort, and money.

prevention costs
Costs associated with preventing defects before they happen.

Appraisal Costs

Appraisal costs are incurred when the firm assesses the level of performance of its processes. As the costs of prevention increase and performance improves, appraisal costs decrease because fewer resources are needed for quality inspections and the subsequent search for causes of any problems that are detected.

appraisal costs
Costs incurred when the firm assess the performance level of its processes.

Internal Failure Costs

Internal failure costs result from defects that are discovered during the production of a service or product. Defects fall into two main categories: (1) *rework,* which is incurred if some aspect of a service must be performed again or if a defective item must be rerouted to some previous operation(s) to correct the defect; and (2) *scrap,* which is incurred if a defective item is unfit for further processing. For example, an analysis of the viability of acquiring a company might be sent back to the mergers and acquisitions department if an assessment of the company's history of environmental compliance is missing. The proposal for the purchase of the company may be delayed, which may result in the loss of the purchase opportunity.

internal failure costs
Costs resulting from defects that are discovered during the production of a service or product.

External Failure Costs

External failure costs arise when a defect is discovered after the customer receives the service or product. Dissatisfied customers talk about bad service or products to their friends, who in turn tell others. If the problem is bad enough, consumer protection groups may even alert the media. The potential impact on future profits is difficult to assess, but without doubt external failure costs erode market share and profits. Encountering defects and correcting them after the product is in the customer's hands is costly.

external failure costs
Costs that arise when a defect is discovered after the customer receives the service or product.

External failure costs also include warranty service and litigation costs. A **warranty** is a written guarantee that the producer will replace or repair defective parts or perform the service to the customer's satisfaction. Usually, a warranty is given for some specified period. For example, television repairs are usually guaranteed for 90 days and new automobiles for 5 years or 50,000 miles, whichever comes first. Warranty costs must be considered in the design of new services or products.

warranty
A written guarantee that the producer will replace or repair defective parts or perform the service to the customer's satisfaction.

Ethics and Quality

The costs of quality go beyond the out-of-pocket costs associated with training, appraisal, scrap, rework, warranties, litigation, or the lost sales from dissatisfied customers. There is a greater societal effect that must be factored into decision making involving the production of services or products, which often requires balancing the traditional measures of quality performance and the overall benefits to society. For example, in the health care industry, aiming for zero complications in cardiac surgery might sound good; however, if it comes at the cost of turning down high-risk patients, is society being served in the best way? Or, how much time, energy, and money should go into delivering vaccines or preventing complications? These are questions that often do not have clear answers.

Deceptive business practices are another source of concern for service or product quality. Deceptive business practice involves three elements: (1) the conduct of the provider is intentional and motivated by a desire to exploit the customer; (2) the provider conceals the truth based upon what is actually known to the provider; and (3) the transaction is intended to generate a disproportionate economic benefit to the provider at the expense of the customer. This behavior is unethical, diminishes the quality of the customers' experience, and may impose a substantial cost on society. Quality is all about increasing the satisfaction of customers. When a firm engages in unethical behavior and the customer finds out about it, the customer is unlikely to favorably assess the quality of his or her experience with that firm or to return as a customer.

Firms that produce better quality services or products can expect to earn a premium for that higher quality. They can also expect to grow and prosper over time because of their ability to create true value for customers. Firms that engage in deception, however, undermine the ability and competence of their employees and demean their relationship with external customers. The unfortunate message these firms send to their employees, who are also their internal customers, is that management views them as being less capable of producing quality services or products than their counterparts in ethical firms. Under these conditions employees are also less likely to be motivated to put forth their best effort. The message unethical firms send to their external customers is that their product or service cannot effectively compete with that of others and so they must engage in deception in order to be profitable. Employees of firms that attempt to profit by deceiving customers are less likely to create true value for customers through product or service improvements that can enhance the customers' experience. That erodes a firm's ability to compete now and in the future.

Ethical behavior falls on the shoulders of all employees of an organization. It is not ethical to knowingly deceive customers and pass defective services or products to internal or external customers. The well-being of all stakeholders, such as stockholders, customers, employees, partners, and creditors, should be considered.

The quality costs of prevention, assessment, internal failure, and external failure must be balanced with ethical considerations to arrive at the appropriate processes and approaches to manage them. Nonetheless, developing the cultural environment for ethical behavior is not cost-free. Employees must be educated in how ethics interfaces with their jobs. The firm may organize an ethics task force or an ethics public relations group to provide an interface between the firm and society. Documentation may be required. We now turn to a discussion of Total Quality Management and Six Sigma, two philosophies companies use to evaluate and improve quality and process performance along technical, service, and ethical dimensions.

total quality management (TQM)

A philosophy that stresses three principles for achieving high levels of process performance and quality: (1) customer satisfaction, (2) employee involvement, and (3) continuous improvement in performance.

quality

A term used by customers to describe their general satisfaction with a service or product.

Total Quality Management

Total quality management (TQM) is a philosophy that stresses three principles for achieving high levels of process performance and quality. These principles are related to (1) customer satisfaction, (2) employee involvement, and (3) continuous improvement in performance. As Figure 5.1 indicates, TQM also involves a number of other important elements. We have covered tools and process analysis techniques useful for process problem solving, redesign, and improvement in Chapter 4. Service/product design and purchasing are covered later in this text. Here, we just focus on the three main principles of TQM.

Customer Satisfaction

Customers, internal or external, are satisfied when their expectations regarding a service or product have been met or exceeded. Often, customers use the general term **quality** to describe their level of satisfaction with a service or product. Quality has multiple dimensions in the mind of the customer, which cut across the nine competitive priorities we introduced in Chapter 1, "Using Operations to Compete." One or more of the following five definitions apply at any one time.

▲ **FIGURE 5.1**
TQM Wheel

Conformance to Specifications   Although customers evaluate the service or product they receive, it is the processes that produced the service or product that are really being judged. In this case, a process failure would be the process's inability to meet certain advertised or implied performance standards. Conformance to specifications may relate to consistent quality, on-time delivery, or delivery speed.

Value   Another way customers define quality is through value, or how well the service or product serves its intended purpose at a price customers are willing to pay. The service/product development process plays a role here, as do the firm's competitive priorities relating to top quality versus low-cost operations. The two factors must be balanced to produce value for the customer. How much value a service or product has in the mind of the customer depends on the customer's expectations before purchasing it.

Fitness for Use   When assessing how well a service or product performs its intended purpose, the customer may consider the convenience of a service, the mechanical features of a product, or other aspects such as appearance, style, durability, reliability, craftsmanship, and serviceability. For example, you may define the quality of the entertainment center you purchased on the basis of how easy it was to assemble and its appearance and styling.

Support   Often the service or product support provided by the company is as important to customers as the quality of the service or product itself. Customers get upset with a company if its financial statements are incorrect, responses to its warranty claims are delayed, its advertising is misleading, or its employees are not helpful when problems are incurred. Good support once the sale has been made can reduce the consequences of quality failures.

Psychological Impressions   People often evaluate the quality of a service or product on the basis of psychological impressions: atmosphere, image, or aesthetics. In the provision of services where the customer is in close contact with the provider, the appearance and actions of the provider are especially important. Nicely dressed, courteous, friendly, and sympathetic employees can affect the customer's perception of service quality.

Call centers provide support for a firm's products or services as well as contribute to the psychological impression of the customer regarding the experience. Calls to the center are often monitored to ensure that the customer is satisfied.

Bernhard Classen/Alamy

   Attaining quality in all areas of a business is a difficult task. To make things even more difficult, consumers change their perceptions of quality. In general, a business's success depends on the accuracy of its perceptions of consumer expectations and its ability to bridge the gap between those expectations and operating capabilities. Good quality pays off in higher profits. High-quality services and products can be priced higher and yield a greater return. Poor quality erodes the firm's ability to compete in the marketplace and increases the costs of producing its service or product. Managerial Practice 5.1 shows how Steinway & Sons balanced consumer expectations for high-end pianos with its capability to meet those expectations.

Employee Involvement

One of the important elements of TQM is employee involvement, as shown in Figure 5.1. A program in employee involvement includes changing organizational culture and encouraging teamwork.

Cultural Change   One of the main challenges in developing the proper culture for TQM is to define *customer* for each employee. In general, customers are internal or external. *External customers* are the people or firms who buy the service or product. Some employees, especially those having little contact with external customers, may have difficulty seeing how their jobs contribute to the whole effort.

   It is helpful to point out to employees that each employee also has one or more *internal customers*—employees in the firm who rely on the output of other employees. All employees must do a good job of serving their internal customers if external customers ultimately are to be satisfied. They will be satisfied only if each internal customer demands value be added that the external customer will recognize and pay for. The notion of internal customers applies to all parts of a firm and enhances cross-functional coordination. For example, accounting must prepare accurate and timely reports for management, and purchasing must provide high-quality materials on time for operations.

MANAGERIAL PRACTICE 5.1   Quality and Performance at Steinway & Sons

A specialist adjusts the levers and dampers of a grand concert piano at the Steinway & Sons factory in Hamburg.

Christian Charisius/Reuters/Corbis

The first contestant in the Van Cliburn International Piano Competition is about to play Tchaikovsky Piano Concerto No. 1 before a packed audience in Fort Worth, Texas. The tension mounts as his fingers approach the keyboard of the Steinway & Sons grand concert piano; both the contestant and piano perform admirably much to the relief of the contestant and the operations manager of the concert. Why was the Steinway piano chosen for such a visible event? It is one of the highest-quality grand pianos you can buy. In addition, Steinway has a market share of over 95 percent in concert halls and it is the piano of choice for professional musicians from Van Cliburn to Billy Joel.

Steinway began operations in the 1880s. Today, the company blends the art of hand crafting, which uses methods essentially the same as when the company started, with twenty-first-century manufacturing technology to produce about 3,100 grand pianos a year. Some 12,000 parts are fashioned, mostly in-house, and assembled for each piano; it takes 9 months to a year compared to 20 days for a mass-produced piano. Eight different species of wood go into every grand piano, each selected for its physical properties and aesthetic characteristics. The craft-oriented production process is painstaking to ensure quality at each step. For example, each board for a piano is hand

selected for a given part. In a time-consuming process, craftsmen bend 17 laminations of the piano's hard maple rim into place with clamps. The Alaska Sitka spruce soundboard is hand-planed so it is arched, thicker at its center than its tapered edges, to withstand the 1,000 pounds of pressure from the more than 200 strings. The piano's "action," which contains keys (88 of them), whippens, shanks, and hammers, uses 100 parts, manufactured on numerical control machines, to sound each note and is pieced together at 30 different desks. Quality is checked at each operation to avoid passing defective parts downstream.

There are six characteristics of quality in Steinway pianos:

- **Sound** Tone and pitch contribute to the fullness and roundness of the sound from the piano. In a process called "voicing," minute adjustments are made to the felt pad of each hammer in the piano's action to either mellow the tone or increase its brilliance. Then a tone regulator listens to the piano's pitch and turns the tuning pins to adjust string tension. Steinways are world renowned for their sound; however, because of the natural characteristics of the wood, each piano will have its own personality.

- **Finish** Wood veneers are selected for their beauty. Boards not meeting standards are discarded, creating a large amount of scrap.

- **Feel** Each of the 88 keys must require the same amount of pressure to activate. In a process called "action weigh-off," lead is added to each key so that there is a consistent feel. Action parts are held to tolerances within +/−0.0005 inch.

- **Durability** The piano must have a long life and perform up to expectations throughout.

- **Image** There is a certain mystique associated with the Steinway brand. Some people attribute a cult-like experience to owning a Steinway.

- **Service** Steinway will go out of its way to service a piano that is inoperative, even to the extent of providing a loaner for a major concert.

The six characteristics link to four of our definitions of quality: (1) conformance to specifications (*feel*), (2) fitness for use (*sound, finish, durability*); (3) support (*service*); and (4) psychological impressions (*image*). As for value, our fifth definition of quality, Steinway grand pianos cost anywhere from $47,000 to $165,000 unless you want a nine-foot recreation of the famous Alma-Tadema piano built in 1887, in which case it will cost $675,000. Want to buy one?

Sources: Andy Serwer, "Happy Birthday, Steinway," *Fortune*, vol. 147, no. 5 (March 17, 2003), pp. 94–97; Leo O'Connor, "Engineering on a Grand scale," *Mechanical Engineering*, vol. 116, no. 10 (October, 1994), pp. 52–58; Steinway Musical Instruments, Inc. Annual Report 2006, **www.steinwaymusical.com**; **www.steinway.com/factory/tour.shtml**, 2007.

quality at the source

A philosophy whereby defects are caught and corrected where they were created.

teams

Small groups of people who have a common purpose, set their own performance goals and approaches, and hold themselves accountable for success.

In TQM, everyone in the organization must share the view that quality control is an end in itself. Errors or defects should be caught and corrected at the source, not passed along to an internal or external customer. For example, a consulting team should make sure its billable hours are correct before submitting them to the accounting department. This philosophy is called **quality at the source**. In addition, firms should avoid trying to "inspect quality into the product" by using inspectors to weed out unsatisfactory services or defective products after all operations have been performed. By contrast, in some manufacturing firms, workers have the authority to stop a production line if they spot quality problems.

Teams   Employee involvement is a key tactic for improving processes and quality. One way to achieve employee involvement is by the use of **teams**, which are small groups of people who have a common purpose, set their own performance goals and approaches, and hold themselves accountable for success.

The three approaches to teamwork most often used are (1) problem-solving teams, (2) special-purpose teams, and (3) self-managed teams. All three use some amount of **employee empowerment**, which moves responsibility for decisions further down the organizational chart—to the level of the employee actually doing the job.

First introduced in the 1920s, *problem-solving teams*, also called **quality circles**, became popular in the late 1970s after the Japanese used them successfully. Problem-solving teams are small groups of supervisors and employees who meet to identify, analyze, and solve process and quality problems. Employees take more pride and interest in their work if they are allowed to help shape it. Although problem-solving teams can successfully reduce costs and improve quality, they die if management fails to implement many of the suggestions they generate.

An outgrowth of the problem-solving teams, **special-purpose teams** address issues of paramount concern to management, labor, or both. For example, management may form a special-purpose team to design and introduce new work policies or new technologies or to address customer service problems. Essentially, this approach gives workers a voice in high-level decisions. Special-purpose teams first appeared in the United States in the early 1980s.

The **self-managed team** approach takes worker participation to its highest level: A small group of employees work together to produce a major portion, or sometimes all, of a service or product. Members learn all the tasks involved in the operation, rotate from job to job, and take over managerial duties such as work and vacation scheduling, ordering supplies, and hiring. In some cases, team members design the process and have a high degree of latitude as to how it takes shape. Self-managed teams essentially change the way work is organized because employees have control over their jobs. Some self-managed teams have increased productivity by 30 percent or more in their firms.

Process measurement is the key to quality improvement. Here a quality inspector measures the diameter of holes in a machined part.

Continuous Improvement

Continuous improvement, based on a Japanese concept called *kaizen*, is the philosophy of continually seeking ways to improve processes. Continuous improvement involves identifying benchmarks of excellent practice and instilling a sense of employee ownership in the process. The focus of continuous improvement projects is to reduce waste, such as reducing the length of time required to process requests for loans at a bank, the amount of scrap generated at a milling machine, or the number of employee injuries at a construction site. The basis of the continuous improvement philosophy are the beliefs that virtually any aspect of a process can be improved and that the people most closely associated with a process are in the best position to identify the changes that should be made. The idea is not to wait until a massive problem occurs before acting.

Employees should be given problem-solving tools, such as the statistical process control (SPC) methods we discuss later in this chapter, and a sense of ownership of the process to be improved. A sense of operator ownership emerges when employees feel a responsibility for the processes and methods they use and take pride in the quality of the service or product they produce. It comes from participation on work teams and in problem-solving activities, which instill in employees a feeling that they have some control over their workplace and tasks.

Most firms actively engaged in continuous improvement train their work teams to use the **plan-do-study-act cycle** for problem solving. Another name for this approach is the Deming Wheel, named after the renowned statistician W. Edwards Deming who taught quality improvement techniques to the Japanese after World War II. Figure 5.2 shows this cycle, which lies at the heart of the continuous improvement philosophy. The cycle comprises the following steps:

1. *Plan.* The team selects a process (an activity, method, machine, or policy) that needs improvement. The team then documents the selected process, usually by analyzing related data; sets qualitative goals for improvement; and discusses various ways to achieve the goals. After assessing the benefits and costs of the alternatives, the team develops a plan with quantifiable measures for improvement.

2. *Do.* The team implements the plan and monitors progress. Data are collected continuously to measure the improvements in the process. Any changes in the process are documented, and further revisions are made as needed.

employee empowerment

An approach to teamwork that moves responsibility for decisions further down the organizational chart—to the level of the employee actually doing the job.

quality circles

Another name for problem-solving teams; small groups of supervisors and employees who meet to identify, analyze, and solve process and quality problems.

special-purpose teams

Groups that address issues of paramount concern to management, labor, or both.

self-managed team

A small group of employees who work together to produce a major portion, or sometimes all, of a service or product.

continuous improvement

The philosophy of continually seeking ways to improve processes based on a Japanese concept called *kaizen*.

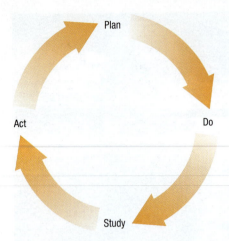

3. *Study.* The team analyzes the data collected during the *do* step to find out how closely the results correspond to the goals set in the *plan* step. If major shortcomings exist, the team reevaluates the plan or stops the project.

4. *Act.* If the results are successful, the team documents the revised process so that it becomes the standard procedure for all who may use it. The team may then instruct other employees in the use of the revised process.

Problem-solving projects often focus on those aspects of processes that do not add value to the service or product. Value is added in processes such as machining a part or serving a customer through a Web page. No value is added in activities such as inspecting parts for defects or routing requests for loan approvals to several different departments. The idea of continuous improvement is to reduce or eliminate activities that do not add value and, thus, are wasteful.

▲ **FIGURE 5.2**
Plan-Do-Study-Act Cycle

plan-do-study-act cycle

A cycle, also called the Deming Wheel, used by firms actively engaged in continuous improvement to train their work teams in problem solving.

Six Sigma

Six Sigma, which relies heavily on the principles of TQM, is a comprehensive and flexible system for achieving, sustaining, and maximizing business success by minimizing defects and variability in processes. Six Sigma has a different focus than TQM: It is driven by a close understanding of customer needs; the disciplined use of facts, data, and statistical analysis; and diligent attention to managing, improving, and reinventing business processes. Figure 5.3 shows how Six Sigma focuses on reducing variation in processes as well as centering processes on their target measures of performance. Either flaw—too much variation or an off-target process—degrades performance of the process. For example, a mortgage loan department of a bank might advertise loan approval decisions in 2 days. If the actual performance ranges from 1 day to 5 days, with an average of 2 days, those customers who had to wait longer than 2 days would be upset. Process variability causes customer dissatisfaction. Similarly, if actual performance consistently produced loan decisions in 3 days, all customers would be dissatisfied. In this case, the process is consistent, but off the target. Six Sigma is a rigorous approach to align processes with their target performance measures with low variability.

Six Sigma

A comprehensive and flexible system for achieving, sustaining, and maximizing business success by minimizing defects and variability in processes.

The name Six Sigma, originally developed by Motorola for its manufacturing operations, relates to the goal of achieving low rates of defective output by developing processes whose mean output for a performance measure is +/− six standard deviations (sigma) from the limits of the design specifications for the service or product. We will discuss variability and its implications on the capability of a process to perform at acceptable levels when we present the tools of statistical process control.

Although Six Sigma was rooted in an effort to improve manufacturing processes, credit General Electric with popularizing the application of the approach to non-manufacturing processes such as sales, human resources, customer service, and financial services. The concept of eliminating defects is the same, although the definition of "defect" depends on the process involved. For example, a human resource department's failure to meet a hiring target counts as a defect. Six Sigma has been successfully applied to a host of service processes, including financial services, human resource processes, marketing processes, and health care administrative processes.

▼ **FIGURE 5.3**
Six Sigma Approach Focuses on Reducing Spread and Centering the Process

Six Sigma Improvement Model

Figure 5.4 shows the Six Sigma Improvement Model, a five-step procedure that leads to improvements in process performance. The model bears a lot of similarity to Figure 4.1, the Blueprint for Process Analysis, for good reason: Both models strive for process improvement. Either model can be applied to projects involving incremental improvements to processes or to projects requiring major changes, including a redesign of an existing process or the development of a new process. The Six Sigma Improvement Model, however, is heavily reliant on statistical process control. The following steps comprise the model:

- *Define.* Determine the characteristics of the process's output that are critical to customer satisfaction and identify any gaps between these characteristics and the

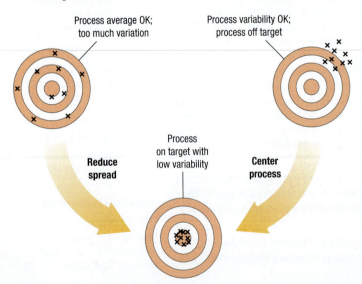

Process average OK;
too much variation

Process variability OK;
process off target

Reduce spread

Process on target with low variability

Center process

process's capabilities. Get a picture of the current process by documenting it using *flowcharts* and *process charts.*

- *Measure.* Quantify the work the process does that affects the gap. Select what to measure, identify data sources, and prepare a data collection plan.

- *Analyze.* Use the data on measures to perform process analysis, applying tools such as Pareto charts, scatter diagrams, and cause-and-effect diagrams and the statistical process control (SPC) tools in this chapter to determine where improvements are necessary. Whether or not major redesign is necessary, establish procedures to make the desired outcome routine.

- *Improve.* Modify or redesign existing methods to meet the new performance objectives. Implement the changes.

- *Control.* Monitor the process to make sure that high performance levels are maintained. Once again, data analysis tools such as Pareto charts, bar charts, scatter diagrams, as well as the statistical process control tools can be used to control the process.

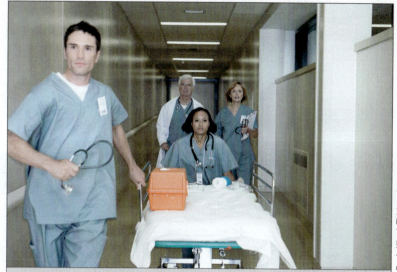

Hospital personnel rush to help a patient in an emergency. Six Sigma can be used to improve service processes in a hospital.

Successful users of Six Sigma have found that it is essential to rigorously follow the steps in the Six Sigma Improvement Model, which is sometimes referred to as the *DMAIC process* (whose name comes from using the first letter of each step in the model). To accomplish the goals of Six Sigma, employees must be trained in the "whys" and the "how-tos" of quality and what it means to customers, both internal and external. Successful firms using Six Sigma develop a cadre of internal teachers who then are responsible for teaching and assisting teams involved in a process improvement project. These teachers have different titles depending on their experience and level of achievement. **Green Belts** devote part of their time to teaching and helping teams with their projects and the rest of their time to their normally assigned duties. **Black Belts** are full-time teachers and leaders of teams involved in Six Sigma projects. Finally, **Master Black Belts** are full-time teachers who review and mentor Black Belts.

Acceptance Sampling

Before any internal process can be evaluated for performance, the inputs to that process must be of good quality. **Acceptance sampling,** which is the application of statistical techniques to determine if a quantity of material from a supplier should be accepted or rejected based on the inspection or test of one or more samples, limits the buyer's risk of rejecting good-quality materials (and unnecessarily delaying the production of goods or services) or accepting bad-quality materials (and incurring downtime due to defective materials or passing bad products to customers). Relative to the specifications for the material the buyer is purchasing, the buyer specifies an **acceptable quality level (AQL),** which is a statement of the proportion of defective items (outside of specifications) that the buyer will accept in a shipment. These days, that proportion is getting very small, often measured in parts per ten-thousand. The idea of acceptance sampling is to take a sample, rather than testing the entire quantity of material, because that is often less expensive. Therein lies the risk—the sample may not be representative of the entire lot of goods from the supplier. The basic procedure is straightforward.

1. A random sample is taken from a large quantity of items and tested or measured relative to the specifications or quality measures of interest.

2. If the sample passes the test (low number of defects), the entire quantity of items is accepted.

3. If the sample fails the test, either (a) the entire quantity of items is subjected to 100 percent inspection and all defective items repaired or replaced or (b) the entire quantity is returned to the supplier.

In a supply chain, any company can be both a producer of goods purchased by another company and a consumer of goods or raw materials supplied by another company. Figure 5.5 shows

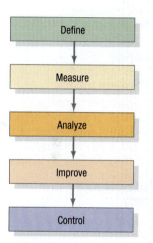

▲ **FIGURE 5.4**
Six Sigma Improvement Model

Green Belt

An employee who achieved the first level of training in a Six Sigma program and spends part of his or her time teaching and helping teams with their projects.

Black Belt

An employee who reached the highest level of training in a Six Sigma program and spends all of his or her time teaching and leading teams involved in Six Sigma projects.

Master Black Belt

Full-time teachers and mentors to several Black Belts.

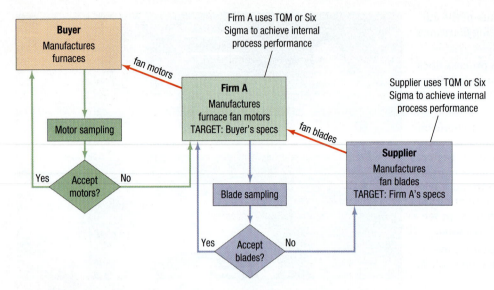

▲ FIGURE 5.5
Interface of Acceptance Sampling and Process Performance Approaches in a Supply Chain

acceptable quality level (AQL)

The quality level desired by the consumer.

acceptance sampling

The application of statistical techniques to determine whether a quantity of material should be accepted or rejected based on the inspection or test of a sample.

statistical process control (SPC)

The application of statistical techniques to determine whether a process is delivering what the customer wants.

a flowchart of how acceptance sampling and internal process performance (TQM or Six Sigma) interface in a supply chain. From the perspective of the supply chain, the buyer's specifications for various dimensions of quality become the targets the supplier shoots for in a supply contract. The supplier's internal processes must be up to the task; TQM or Six Sigma can help achieve the desired performance. The buyer's sampling plan will provide a high probability of accepting AQL (or better). MyOMLab Supplement G, "Acceptance Sampling Plans," shows how to design an acceptance sampling plan that meets the level of risk desired.

Statistical Process Control

Regardless of whether a firm is producing a service or a product, it is important to ensure that the firm's processes are providing the quality that customers want. A key element of TQM or Six Sigma is building the capability to monitor the performance of processes so that corrective action can be initiated in a timely fashion. Evaluating the performance of processes requires a variety of data gathering approaches. We already discussed checklists, histograms and bar charts, Pareto charts, scatter diagrams, cause-and-effect diagrams, and graphs (see Chapter 4, "Process Analysis"). All of these tools can be used with TQM or Six Sigma. Here. we focus on the powerful statistical tools that can be used to monitor and manage repetitive processes.

Statistical process control (SPC) is the application of statistical techniques to determine whether a process is delivering what customers want. In SPC, tools called control charts are used primarily to detect defective services or products or to indicate that the process has changed and that services or products will deviate from their design specifications, unless something is done to correct the situation. SPC can also be used to inform management of improved process changes. Examples of process changes that can be detected by SPC include the following:

- A decrease in the average number of complaints per day at a hotel
- A sudden increase in the proportion of defective gear boxes
- An increase in the time to process a mortgage application
- A decline in the number of scrapped units at a milling machine
- An increase in the number of claimants receiving late payment from an insurance company

Let us consider the last situation. Suppose that the manager of the accounts payable department of an insurance company notices that the proportion of claimants receiving late payments rose from an average of 0.01 to 0.03. The first question is whether the rise is a cause for alarm or just a random occurrence. Statistical process control can help the manager decide whether further action should be taken. If the rise in the proportion is not just a random occurrence, the manager should seek explanations of the poor performance. Perhaps the number of claims significantly increased, causing an overload on the employees in the department. The decision might be to hire more personnel. Or perhaps the procedures being used are ineffective or the training of employees is inadequate. SPC is an integral part of TQM and Six Sigma.

Variation of Outputs

No two services or products are exactly alike because the processes used to produce them contain many sources of variation, even if the processes are working as intended. Nonetheless, it is important to minimize the variation in outputs because frequently variation is what the customer sees and feels. Suppose a physicians' clinic submits claims on behalf of its patients to a particular insurance company. In this situation, the physicians' clinic is the customer of the insurance company's bill payment process. In some cases, the clinic receives payment in 4 weeks, and in other cases 20 weeks. The time to process a request for payment varies because of the load on the insurance company's processes, the medical history of the patient, and the skills and attitudes of the employees. Meanwhile, the clinic must cover its expenses while it waits for payment. Regardless of whether the process is producing services or products, nothing can be done to eliminate variation in output completely; however, management should investigate the *causes* of the variation in order to minimize it.

Performance Measurements   Performance can be evaluated in two ways. One way is to measure **variables**—that is, service or product characteristics, such as weight, length, volume, or time, that can be *measured*. The advantage of using performance variables is that if a service or product misses its performance specifications, the inspector knows by how much. The disadvantage is that such measurements typically involve special equipment, employee skills, exacting procedures, and time and effort.

Another way to evaluate performance is to measure **attributes**; service or product characteristics that can be quickly *counted* for acceptable performance. This method allows inspectors to make a simple "yes/no" decision about whether a service or product meets the specifications. Attributes often are used when performance specifications are complex and measurement of variables is difficult or costly. Some examples of attributes that can be counted are the number of insurance forms containing errors that cause underpayments or overpayments, the proportion of airline flights arriving within 15 minutes of scheduled times, and the number of stove-top assemblies with spotted paint.

The advantage of counting attributes is that less effort and fewer resources are needed than for measuring variables. The disadvantage is that, even though attribute counts can reveal that process performance has changed, they do not indicate by how much. For example, a count may determine that the proportion of airline flights arriving within 15 minutes of their scheduled times declined, but the result does not show how much beyond the 15-minute allowance the flights are arriving. For that, the actual deviation from the scheduled arrival, a variable, would have to be measured.

Sampling   The most thorough approach to inspection is to inspect each service or product at each stage of the process for quality. This method, called *complete inspection*, is used when the costs of passing defects to an internal or external customer outweigh the inspection costs. Firms often use automated inspection equipment that can record, summarize, and display data. Many companies find that automated inspection equipment can pay for itself in a reasonably short time.

A well-conceived **sampling plan** can approach the same degree of protection as complete inspection. A sampling plan specifies a **sample size**, which is a quantity of randomly selected observations of process outputs, the time between successive samples, and decision rules that determine when action should be taken. Sampling is appropriate when inspection costs are high because of the special knowledge, skills, procedures, and expensive equipment that are required to perform the inspections, or because the tests are destructive.

Sampling Distributions   Relative to a performance measure, a process will produce output that can be described by a *process distribution*, with a mean and variance that will be known only with a complete inspection with 100 percent accuracy. The purpose of sampling, however, is to estimate a variable or attribute measure for the output of the process without doing a complete inspection. That measure is then used to assess the performance of the process itself. For example, the time required to process specimens at an intensive care unit lab in a hospital (a variable measure) will vary. If you measured the time to complete an analysis of a large number of patients and plotted the results, the data would tend to form a pattern that can be described as a process distribution. With sampling, we try to estimate the parameters of the process distribution using statistics such as the sample mean and the sample range or standard deviation.

1. The *sample mean* is the sum of the observations divided by the total number of observations:

$$\bar{x} = \frac{\sum_{i=1}^{n} x_i}{n}$$

variables

Service or product characteristics, such as weight, length, volume, or time, that can be measured.

attributes

Service or product characteristics that can be quickly counted for acceptable performance.

sampling plan

A plan that specifies a sample size, the time between successive samples, and decision rules that determine when action should be taken.

sample size

A quantity of randomly selected observations of process outputs.

Wine production is an example of a situation where complete inspection is not an option. Here a quality inspector draws a sample of white wine from a stainless steel maturation tank.

where

$$x_i = \text{observation of a quality characteristic (such as time)}$$

$$n = \text{total number of observations}$$

$$\bar{x} = \text{mean}$$

2. The *range* is the difference between the largest observation in a sample and the smallest. The *standard deviation* is the square root of the variance of a distribution. An estimate of the process standard deviation based on a sample is given by

$$\sigma = \sqrt{\frac{\sum_{i=1}^{n}(x_i - \bar{x})^2}{n-1}} \quad \text{or} \quad \sigma = \sqrt{\frac{\sum_{i=1}^{n}x^2 - \frac{\left(\sum_{i=1}^{n}x_i\right)^2}{n}}{n-1}}$$

where

$$\sigma = \text{standard deviation of a sample}$$

$$n = \text{total number of observations in the sample}$$

$$\bar{x} = \text{mean}$$

$$x_i = \text{observation of a quality characteristic}$$

Relatively small values for the range or the standard deviation imply that the observations are clustered near the mean.

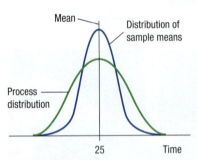

▲ FIGURE 5.6

Relationship Between the Distribution of Sample Means and the Process Distribution

These sample statistics have their own distribution, which we call a *sampling distribution*. For example, in the lab analysis process, an important performance variable is the time it takes to get results to the critical care unit. Suppose that management wants results available in an average of 25 minutes. That is, it wants the process distribution to have a mean of 25 minutes. An inspector periodically taking a sample of five analyses and calculating the sample mean could use it to determine how well the process is doing. Suppose that the process is actually producing the analyses with a mean of 25 minutes. Plotting a large number of these sample means would show that they have their own sampling distribution with a mean centered on 25 minutes, as does the process distribution mean, but with much less variability. The reason is that the sample means offset the highs and lows of the individual times in each sample. Figure 5.6 shows the relationship between the sampling distribution of sample means and the process distribution for the analysis times.

Some sampling distributions (e.g., for means with sample sizes of four or more and proportions with sample sizes of 20 or more) can be approximated by the normal distribution, allowing the use of the normal tables (see Appendix 1, "Normal Distribution"). For example, suppose you wanted to determine the probability that a sample mean will be more than 2.0 standard deviations higher than the process mean. Go to Appendix 1 and note that the entry in the table for $z = 2.0$ standard deviations is 0.9772. Consequently, the probability is $1.0000 - 0.9772 = 0.0228$, or 2.28 percent. The probability that the sample mean will be more than 2.0 standard deviations lower than the process mean is also 2.28 percent because the normal distribution is symmetric to the mean. The ability to assign probabilities to sample results is important for the construction and use of control charts.

common causes of variation

The purely random, unidentifiable sources of variation that are unavoidable with the current process.

Common Causes   The two basic categories of variation in output include common causes and assignable causes. **Common causes of variation** are the purely random, unidentifiable sources of variation that are unavoidable with the current process. A process distribution can be characterized by its *location*, *spread*, and *shape*. Location is measured by the *mean* of the distribution, while spread is measured by the *range* or *standard deviation*. The shape of process distributions can be characterized as either symmetric or skewed. A *symmetric* distribution has the same number of observations above and below the mean. A *skewed* distribution has a greater number of observations either above or below the mean. If process variability results solely from common causes of variation, a typical assumption is that the distribution is symmetric, with most observations near the center.

assignable causes of variation

Any variation-causing factors that can be identified and eliminated.

Assignable Causes   The second category of variation, **assignable causes of variation**, also known as *special causes,* includes any variation-causing factors that can be identified and eliminated. Assignable causes of variation include an employee needing training or a machine needing repair. Let us return to the example of the lab analysis process. Figure 5.7 shows how assignable causes can change the distribution of output for the analysis process. The **green** curve

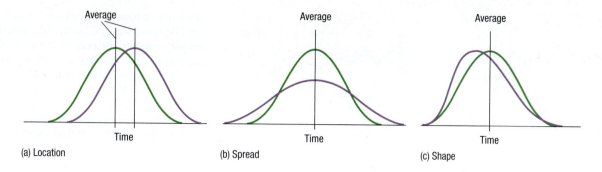

(a) Location   (b) Spread   (c) Shape

is the process distribution when only common causes of variation are present. The **purple** curves depict a change in the distribution because of assignable causes. In Figure 5.7(a), the **purple** curve indicates that the process took more time than planned in many of the cases, thereby increasing the average time of each analysis. In Figure 5.7(b), an increase in the variability of the time for each case affected the spread of the distribution. Finally, in Figure 5.7(c), the **purple** curve indicates that the process produced a preponderance of the tests in less than average time. Such a distribution is skewed, or no longer symmetric to the average value. A process is said to be in statistical control when the location, spread, or shape of its distribution does not change over time. After the process is in statistical control, managers use SPC procedures to detect the onset of assignable causes so that they can be addressed.

▲ **FIGURE 5.7**
Effects of Assignable Causes on the Process Distribution for the Lab Analysis Process

control chart

A time-ordered diagram that is used to determine whether observed variations are abnormal.

Control Charts

To determine whether observed variations are abnormal, we can measure and plot the performance measure taken from the sample on a time-ordered diagram called a **control chart**. A control chart has a nominal value, or central line, which can be the process's historic average or a target that managers would like the process to achieve, and two control limits based on the sampling distribution of the quality measure. The control limits are used to judge whether action is required. The larger value represents the *upper control limit* (UCL), and the smaller value represents the *lower control limit* (LCL). Figure 5.8 shows how the control limits relate to the sampling distribution. A sample statistic that falls between the UCL and the LCL indicates that the process is exhibiting common causes of variation. A statistic that falls outside the control limits indicates that the process is exhibiting assignable causes of variation.

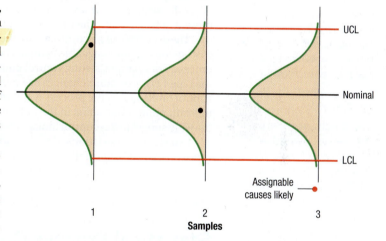

▲ **FIGURE 5.8**
How Control Limits Relate to the Sampling Distribution: Observations from Three Samples

Observations falling outside the control limits do not always mean poor quality. For example, in Figure 5.8 the assignable cause may be a new billing process introduced to reduce the number of incorrect bills sent to customers. If the proportion of incorrect bills, that is, the performance measure from a sample of bills, falls *below* the LCL of the control chart, the new procedure likely changed the billing process for the better, and a new control chart should be constructed.

Managers or employees responsible for evaluating a process can use control charts in the following way:

1. Take a random sample from the process and calculate a variable or attribute performance measure.

2. If the statistic falls outside the chart's control limits or exhibits unusual behavior, look for an assignable cause.

3. Eliminate the cause if it degrades performance; incorporate the cause if it improves performance. Reconstruct the control chart with new data.

4. Repeat the procedure periodically.

Sometimes, problems with a process can be detected even though the control limits have not been exceeded. Figure 5.9 contains four examples of control charts. Chart (a) shows a process that is in statistical control. No action is needed. However, chart (b) shows a pattern called a *run* or a sequence of observations with a certain characteristic. A typical rule is to take remedial action

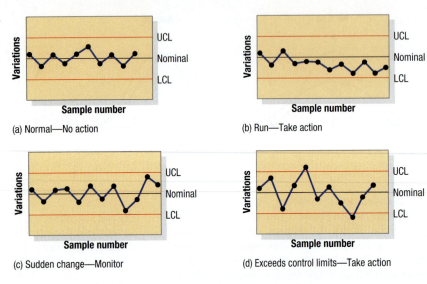

▲ FIGURE 5.9
Control Chart Examples

(a) Normal—No action

(b) Run—Take action

(c) Sudden change—Monitor

(d) Exceeds control limits—Take action

when five or more observations show a downward or upward trend, even if the points have not yet exceeded the control limits. Here, nine sequential observations are below the mean and show a downward trend. The probability is low that such a result could take place by chance.

Chart (c) shows that the process takes a sudden change from its normal pattern. The last four observations are unusual: The first drops close to the LCL, the next two rise toward the UCL, and the fourth remains above the nominal value. Managers or employees should monitor processes with such sudden changes even though the control limits have not been exceeded. Finally, chart (d) indicates that the process went out of control twice because two sample results fell outside the control limits. The probability that the process distribution has changed is high. We discuss more implications of being out of statistical control when we discuss process capability later in this chapter.

Control charts are not perfect tools for detecting shifts in the process distribution because they are based on sampling distributions. Two types of error are possible with the use of control charts. A **type I error** occurs when the conclusion is made that the process is out of control based on a sample result that falls outside the control limits, when in fact it was due to pure randomness. A **type II error** occurs when the conclusion is that the process is in control and only randomness is present, when actually the process is out of statistical control.

These errors can be controlled by the choice of control limits. The choice would depend on the costs of looking for assignable causes when none exist versus the cost of not detecting a shift in the process. For example, setting control limits at +/− three standard deviations from the mean reduces the type I error because chances are only 0.26 percent that a sample result will fall outside of the control limits unless the process is out of statistical control. However, the type II error may be significant; more subtle shifts in the nature of the process distribution will go undetected because of the wide spread in the control limits. Alternatively, the spread in the control limits can be reduced to +/− two standard deviations, thereby increasing the likelihood of sample results from a non-faulty process falling outside of the control limits to 4.56 percent. Now, the type II error is smaller, but the type I error is larger because employees are likely to search for assignable causes when the sample result occurred solely by chance. As a general rule, use wider limits when the cost for searching for assignable causes is large relative to the cost of not detecting a shift in the process distribution.

type I error

An error that occurs when the employee concludes that the process is out of control based on a sample result that falls outside the control limits, when in fact it was due to pure randomness.

type II error

An error that occurs when the employee concludes that the process is in control and only randomness is present, when actually the process is out of statistical control.

Statistical Process Control Methods

Statistical process control (SPC) methods are useful for both measuring the current process performance and detecting whether the process has changed in a way that will affect future performance. In this section, we first discuss mean and range charts for variable measures of performance and then consider control charts for attributes measures.

Control Charts for Variables

Control charts for variables are used to monitor the mean and the variability of the process distribution.

R-chart

A chart used to monitor process variability.

***R*-Chart** A range chart, or ***R*-chart**, is used to monitor process variability. To calculate the range of a set of sample data, the analyst subtracts the smallest from the largest measurement in each sample. If any of the ranges fall outside the control limits, the process variability is not in control.

The control limits for the *R*-chart are

$$UCL_R = D_4\overline{R} \text{ and } LCL_R = D_3\overline{R}$$

where

$\overline{R}$ = average of several past R values and the central line of the control chart

D_3, D_4 = constants that provide three standard deviation (three-sigma) limits for a given sample size

Notice that the values for D_3 and D_4 shown in Table 5.1 change as a function of the sample size. Notice, too, that the spread between the control limits narrows as the sample size increases. This change is a consequence of having more information on which to base an estimate for the process range.

TABLE 5.1 | **FACTORS FOR CALCULATING THREE-SIGMA LIMITS FOR THE**
$\bar{x}$-CHART AND R-CHART

Size of Sample (n)	Factor for UCL and LCL for $\bar{x}$-Chart (A_2)	Factor for LCL for R-Chart (D_3)	Factor for UCL for R-Chart (D_4)
2	1.880	0	3.267
3	1.023	0	2.575
4	0.729	0	2.282
5	0.577	0	2.115
6	0.483	0	2.004
7	0.419	0.076	1.924
8	0.373	0.136	1.864
9	0.337	0.184	1.816
10	0.308	0.223	1.777

Source: Reprinted with permission from *ASTM Manual on Quality Control of Materials,* copyright © ASTM International, 100 Barr Harbor Drive, West Conshohocken, PA 19428.

$\bar{x}$-Chart   An $\bar{x}$-**Chart** (read "x-bar chart") is used to see whether the process is generating output, on average, consistent with a target value set by management for the process or whether its current performance, with respect to the average of the performance measure, is consistent with its past performance. A target value is useful when a process is completely redesigned and past performance is no longer relevant. When the assignable causes of process variability have been identified and the process variability is in statistical control, the analyst can then construct an $\bar{x}$-chart. The control limits for the $\bar{x}$-chart are

$$\text{UCL}_{\bar{x}} = \bar{\bar{x}} + A_2\bar{R} \quad \text{and} \quad \text{LCL}_{\bar{x}} = \bar{\bar{x}} - A_2\bar{R}$$

where

$\bar{\bar{x}} =$ central line of the chart, which can be either the average of past sample means or a target value set for the process

$A_2 =$ constant to provide three-sigma limits for the sample mean

The values for A_2 are contained in Table 5.1. Note that the control limits use the value of $\bar{R}$; therefore, the $\bar{x}$-chart must be constructed *after* the process variability is in control.

To develop and use $\bar{x}$- and R-charts, do the following:

Step 1.   Collect data on the variable quality measurement (such as time, weight, or diameter) and organize the data by sample number. Preferably, at least 20 samples of size n should be taken for use in constructing a control chart.

Step 2.   Compute the range for each sample and the average range, $\bar{R}$, for the set of samples.

Step 3.   Use Table 5.1 to determine the upper and lower control limits of the R-chart.

Step 4.   Plot the sample ranges. If all are in control, proceed to step 5. Otherwise, find the assignable causes, correct them, and return to step 1.

Step 5.   Calculate $\bar{x}$ for each sample and determine the central line of the chart, $\bar{\bar{x}}$.

Step 6.   Use Table 5.1 to determine the parameters for $\text{UCL}_{\bar{x}}$ and $\text{LCL}_{\bar{x}}$ and construct the $\bar{x}$-chart.

Step 7.   Plot the sample means. If all are in control, the process is in statistical control in terms of the process average and process variability. Continue

$\bar{x}$-chart

A chart used to see whether the process is generating output, on average, consistent with a target value set by management for the process or whether its current performance, with respect to the average of the performance measure, is consistent with past performance.

An analyst measures the diameter of a part with a micrometer. After he measures the sample, he plots the range on the control chart.

to take samples and monitor the process. If any are out of control, find the assignable causes, address them, and return to step 1. If no assignable causes are found after a diligent search, assume that the out-of-control points represent common causes of variation and continue to monitor the process.

EXAMPLE 5.1	## Using $\bar{x}$- and R-Charts to Monitor a Process

MyOMLab

Active Model 5.1 in MyOMLab provides additional insight on the x-bar and R-charts and their uses for the metal screw problem.

MyOMLab

Tutor 5.1 in MyOMLab provides a new example to practice the use of x-bar and R-charts.

The management of West Allis Industries is concerned about the production of a special metal screw used by several of the company's largest customers. The diameter of the screw is critical to the customers. Data from five samples appear in the accompanying table. The sample size is 4. Is the process in statistical control?

SOLUTION

Step 1: For simplicity, we use only 5 samples. In practice, more than 20 samples would be desirable. The data are shown in the following table.

DATA FOR THE $\bar{x}$- AND R-CHARTS: OBSERVATIONS OF SCREW DIAMETER (IN.)

Sample Number	Observations				R	$\bar{x}$
	1	2	3	4		
1	0.5014	0.5022	0.5009	0.5027	0.0018	0.5018
2	0.5021	0.5041	0.5024	0.5020	0.0021	0.5027
3	0.5018	0.5026	0.5035	0.5023	0.0017	0.5026
4	0.5008	0.5034	0.5024	0.5015	0.0026	0.5020
5	0.5041	0.5056	0.5034	0.5047	0.0022	0.5045
				Average	0.0021	0.5027

Step 2: Compute the range for each sample by subtracting the lowest value from the highest value. For example, in sample 1 the range is $0.5027 - 0.5009 = 0.0018$ in. Similarly, the ranges for samples 2, 3, 4, and 5 are 0.0021, 0.0017, 0.0026, and 0.0022 in., respectively. As shown in the table, $\bar{R} = 0.0021$.

Step 3: To construct the R-chart, select the appropriate constants from Table 5.1 for a sample size of 4. The control limits are

$$\text{UCL}_R = D_4\bar{R} = 2.282(0.0021) = 0.00479 \text{ in.}$$

$$\text{LCL}_R = D_3\bar{R} = 0(0.0021) = 0 \text{ in.}$$

Step 4: Plot the ranges on the R-chart, as shown in Figure 5.10. None of the sample ranges falls outside the control limits. Consequently, the process variability is in statistical control. If any of the sample ranges fall outside of the limits, or an unusual pattern appears (see Figure 5.9), we would search for the causes of the excessive variability, address them, and repeat step 1.

FIGURE 5.10 ▶

Range Chart from the *OM Explorer $\bar{x}$- and R-Chart* Solver, Showing that the Process Variability Is In Control

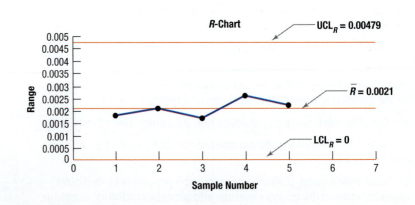

Step 5: Compute the mean for each sample. For example, the mean for sample 1 is

$$\frac{0.5014 + 0.5022 + 0.5009 + 0.5027}{4} = 0.5018 \text{ in.}$$

Similarly, the means of samples 2, 3, 4, and 5 are 0.5027, 0.5026, 0.5020, and 0.5045 in., respectively. As shown in the table, $\bar{\bar{x}} = 0.5027$.

Step 6: Now, construct the $\bar{x}$-chart for the process average. The average screw diameter is 0.5027 in., and the average range is 0.0021 in., so use $\bar{\bar{x}} = 0.5027$, $\bar{R} = 0.0021$, and A_2 from Table 5.1 for a sample size of 4 to construct the control limits:

$$UCL_{\bar{x}} = \bar{\bar{x}} + A_2\bar{R} = 0.5027 + 0.729(0.0021) = 0.5042 \text{ in.}$$

$$LCL_{\bar{x}} = \bar{\bar{x}} - A_2\bar{R} = 0.5027 - 0.729(0.0021) = 0.5012 \text{ in.}$$

Step 7: Plot the sample means on the control chart, as shown in Figure 5.11.

The mean of sample 5 falls above the UCL, indicating that the process average is out of statistical control and that assignable causes must be explored, perhaps using a cause-and-effect diagram.

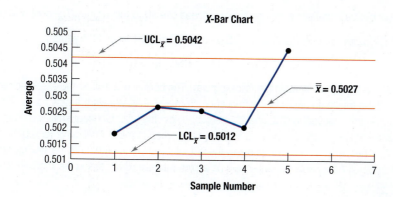

◀ **FIGURE 5.11**

The x-bar Chart from the *OM Explore* $\bar{x}$- and R-Chart Solver for the Metal Screw, Showing that Sample 5 Is Out of Control

DECISION POINT

A new employee operated the lathe machine that makes the screw on the day sample 5 was taken. To solve the problem, management initiated a training session for the employee. Subsequent samples showed that the process was back in statistical control.

If the standard deviation of the process distribution is known, another form of the $\bar{x}$-chart may be used:

$$UCL_{\bar{x}} = \bar{\bar{x}} + z\sigma_{\bar{x}} \quad \text{and} \quad LCL_{\bar{x}} = \bar{\bar{x}} - z\sigma_{\bar{x}}$$

where

$\sigma_{\bar{x}} = \sigma/\sqrt{n} = $ standard deviation of sample means

$\sigma = $ standard deviation of the process distribution

$n = $ sample size

$\bar{\bar{x}} = $ central line of the chart, which can be either the average of past sample means or a target value set for the process

$z = $ normal deviate (number of standard deviations from the average)

The analyst can use an R-chart to be sure that the process variability is in control before constructing the $\bar{x}$-chart. The advantage of using this form of the $\bar{x}$-chart is that the analyst can adjust the spread of the control limits by changing the value of z. This approach can be useful for balancing the effects of type I and type II errors.

EXAMPLE 5.2	**Designing an $\bar{x}$-Chart Using the Process Standard Deviation**

The Sunny Dale Bank monitors the time required to serve customers at the drive-by window because it is an important quality factor in competing with other banks in the city. After analyzing the data gathered in an extensive study of the window operation, bank management determined that the mean time to process a customer at the peak demand period is 5 minutes, with a standard deviation of 1.5 minutes. Management wants to monitor the mean time to process a customer by periodically using a sample size of six customers. Assume that the process variability is in statistical control. Design an $\bar{x}$-chart that has a type I error of 5 percent. That is, set the control limits so that there is a 2.5 percent chance a sample result will fall below the LCL and a 2.5 percent chance that a sample result will fall above the UCL. After several weeks of sampling, two successive samples came in at 3.70 and 3.68 minutes, respectively. Is the customer service process in statistical control?

SOLUTION

$$\bar{\bar{x}} = 5.0 \text{ minutes}$$

$$\sigma = 1.5 \text{ minutes}$$

$$n = 6 \text{ customers}$$

$$z = 1.96$$

The process variability is in statistical control, so we proceed directly to the $\bar{x}$-chart. The control limits are

$$UCL_{\bar{x}} = \bar{\bar{x}} + z\sigma/\sqrt{n} = 5.0 + 1.96(1.5)/\sqrt{6} = 6.20 \text{ minutes}$$

$$LCL_{\bar{x}} = \bar{\bar{x}} - z\sigma/\sqrt{n} = 5.0 - 1.96(1.5)/\sqrt{6} = 3.80 \text{ minutes}$$

The value for z can be obtained in the following way. The normal distribution table (see Appendix 1) gives the proportion of the total area under the normal curve from $-\infty$ to z. We want a type I error of 5 percent, or 2.5 percent of the curve above the UCL and 2.5 percent below the LCL. Consequently, we need to find the z value in the table that leaves only 2.5 percent in the upper portion of the normal curve (or 0.9750 in the table). The value is 1.96. The two new samples are below the LCL of the chart, implying that the average time to serve a customer has dropped. Assignable causes should be explored to see what caused the improvement.

DECISION POINT

Management studied the time period over which the samples were taken and found that the supervisor of the process was experimenting with some new procedures. Management decided to make the new procedures a permanent part of the customer service process. After all employees were trained in the new procedures, new samples were taken and the control chart reconstructed.

Control Charts for Attributes

Two charts commonly used for performance measures based on attributes measures are the *p*- and *c*-chart. The *p*-chart is used for controlling the proportion of defects generated by the process. The *c*-chart is used for controlling the number of defects when more than one defect can be present in a service or product.

p-Charts   The **p-chart** is a commonly used control chart for attributes. The performance characteristic is counted rather than measured, and the entire service or item can be declared good or defective. For example, in the banking industry, the attributes counted might be the number of nonendorsed deposits or the number of incorrect financial statements sent to customers. The method involves selecting a random sample, inspecting each item in it, and calculating the sample proportion defective, *p*, which is the number of defective units divided by the sample size.

Sampling for a *p*-chart involves a "yes/no" decision: The process output either is or is not defective. The underlying statistical distribution is based on the binomial distribution. However, for large sample sizes, the normal distribution provides a good approximation to it. The standard deviation of the distribution of proportion defectives, σ_p, is

$$\sigma_p = \sqrt{\bar{p}(1 - \bar{p})/n}$$

p-chart

A chart used for controlling the proportion of defective services or products generated by the process.

where

n = sample size
$\bar{p}$ = central line on the chart, which can be either the historical average population proportion defective or a target value

We can use σ_p to arrive at the upper and lower control limits for a p-chart:

$$\text{UCL}_p = \bar{p} + z\sigma_p \ \text{ and } \ \text{LCL}_p = \bar{p} - z\sigma_p$$

where

z = normal deviate (number of standard deviations from the average)

The chart is used in the following way. Periodically, a random sample of size n is taken, and the number of defective services or products is counted. The number of defectives is divided by the sample size to get a sample proportion defective, p, which is plotted on the chart. When a sample proportion defective falls outside the control limits, the analyst assumes that the proportion defective generated by the process has changed and searches for the assignable cause. Observations falling below the LCL_p indicate that the process may actually have improved. The analyst may find no assignable cause because it is always possible that an out-of-control proportion occurred randomly. However, if the analyst discovers assignable causes, those sample data should not be used to calculate the control limits for the chart.

EXAMPLE 5.3   Using a p-Chart to Monitor a Process

The operations manager of the booking services department of Hometown Bank is concerned about the number of wrong customer account numbers recorded by Hometown personnel. Each week a random sample of 2,500 deposits is taken, and the number of incorrect account numbers is recorded. The results for the past 12 weeks are shown in the following table. Is the booking process out of statistical control? Use three-sigma control limits, which will provide a Type I error of 0.26 percent.

MyOMLab
Active Model 5.2 in MyOMLab provides additional insight on the p-chart and its uses for the booking services department.

MyOMLab
Tutor 5.2 in MyOMLab provides a new example to practice the use of the p-chart.

Sample Number	Wrong Account Numbers	Sample Number	Wrong Account Numbers
1	15	7	24
2	12	8	7
3	19	9	10
4	2	10	17
5	19	11	15
6	4	12	3
			Total 147

SOLUTION

Step 1: Using this sample data to calculate $\bar{p}$

$$\bar{p} = \frac{\text{Total defectives}}{\text{Total number of observations}} = \frac{147}{12(2,500)} = 0.0049$$

$$\sigma_p = \sqrt{\bar{p}(1-\bar{p})/n} = \sqrt{0.0049(1-0.0049)/2,500} = 0.0014$$

$$\text{UCL}_p = \bar{p} + z\sigma_p = 0.0049 + 3(0.0014) = 0.0091$$

$$\text{LCL}_p = \bar{p} - z\sigma_p = 0.0049 - 3(0.0014) = 0.0007$$

Step 2: Calculate each sample proportion defective. For sample 1, the proportion of defectives is 15/2,500 = 0.0060.

Step 3: Plot each sample proportion defective on the chart, as shown in Figure 5.12.

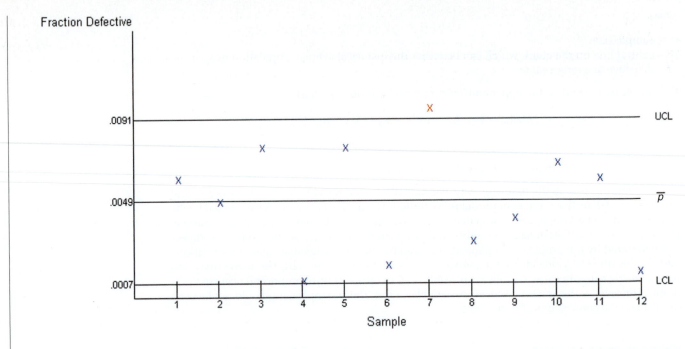

▲ **FIGURE 5.12**
The *p*-Chart from POM
for Windows for Wrong
Account Numbers,
Showing that Sample 7
Is Out of Control

Sample 7 exceeds the UCL; thus, the process is out of control and the reasons for the poor performance that week should be determined.

DECISION POINT

Management explored the circumstances when sample 7 was taken. The encoding machine used to print the account numbers on the checks was defective that week. The following week the machine was repaired; however, the recommended preventive maintenance on the machine was not performed for months prior to the failure. Management reviewed the performance of the maintenance department and instituted changes to the maintenance procedures for the encoding machine. After the problem was corrected, an analyst recalculated the control limits using the data without sample 7. Subsequent weeks were sampled, and the booking process was determined to be in statistical control. Consequently, the *p*-chart provides a tool to indicate when a process needs adjustment.

c-Charts   Sometimes services or products have more than one defect. For example, a roll of carpeting may have several defects, such as tufted or discolored fibers or stains from the production process. Other situations in which more than one defect may occur include accidents at a particular intersection, bubbles in a television picture face panel, and complaints from a patron at a hotel. When management is interested in reducing the number of defects per unit or service encounter, another type of control chart, the **c-chart**, is useful.

c-chart

A chart used for controlling the number of defects when more than one defect can be present in a service or product.

The underlying sampling distribution for a *c*-chart is the Poisson distribution. The Poisson distribution is based on the assumption that defects occur over a continuous region on the surface of a product or a continuous time interval during the provision of a service. It further assumes that the probability of two or more defects at any one location on the surface or at any instant of time is negligible. The mean of the distribution is $\bar{c}$ and the standard deviation is $\sqrt{\bar{c}}$. A useful tactic is to use the normal approximation to the Poisson so that the central line of the chart is $\bar{c}$ and the control limits are

$$\mathrm{UCL}_c = \bar{c} + z\sqrt{\bar{c}} \ \text{ and } \ \mathrm{LCL}_c = \bar{c} - z\sqrt{\bar{c}}$$

EXAMPLE 5.4 Using a *c*-Chart to Monitor Defects per Unit

The Woodland Paper Company produces paper for the newspaper industry. As a final step in the process, the paper passes through a machine that measures various product quality characteristics. When the paper production process is in control, it averages 20 defects per roll.

a. Set up a control chart for the number of defects per roll. For this example, use two-sigma control limits.
b. Five rolls had the following number of defects: 16, 21, 17, 22, and 24, respectively. The sixth roll, using pulp from a different supplier, had 5 defects. Is the paper production process in control?

MyOMLab

Tutor 5.3 in MyOMLab provides a new example to practice the use of the *c*-chart.

SOLUTION

a. The average number of defects per roll is 20. Therefore

$$UCL_c = \bar{c} + z\sqrt{\bar{c}} = 20 + 2(\sqrt{20}) = 28.94$$

$$LCL_c = \bar{c} - z\sqrt{\bar{c}} = 20 - 2(\sqrt{20}) = 11.06$$

The control chart is shown in Figure 5.13.

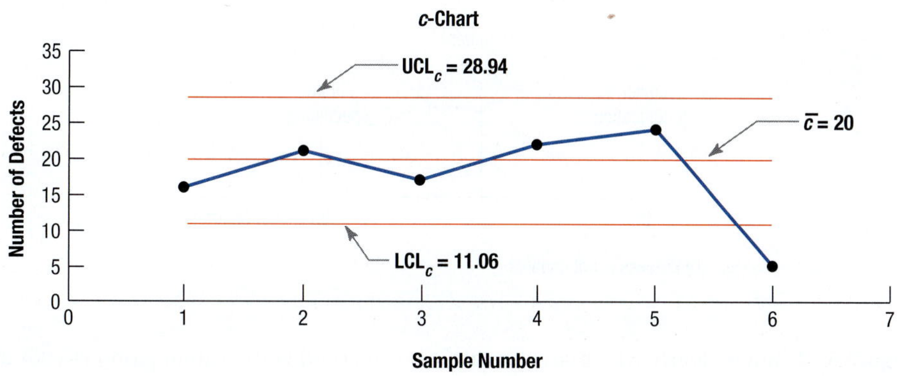

◀ **FIGURE 5.13** The *c*-Chart from the *OM Explorer c-Chart* Solver for Defects per Roll of Paper

b. Because the first five rolls had defects that fell within the control limits, the process is still in control. The sixth roll's five defects, however, is below than the LCL, and therefore, the process is technically "out of control." The control chart indicates that something good has happened.

DECISION POINT

The supplier for the first five samples has been used by Woodland Paper for many years. The supplier for the sixth sample is new to the company. Management decided to continue using the new supplier for a while, monitoring the number of defects to see whether it stays low. If the number remains below the LCL for 20 consecutive samples, management will make the switch permanent and recalculate the control chart parameters.

Process Capability

Statistical process control techniques help managers achieve and maintain a process distribution that does not change in terms of its mean and variance. The control limits on the control charts signal when the mean or variability of the process changes. However, a process that is in statistical control may not be producing services or products according to their design specifications because the control limits are based on the mean and variability of the *sampling distribution*, not the design specifications. **Process capability** refers to the ability of the process to meet the design specifications for a service or product. Design specifications often are expressed as a **nominal value**, or target, and a **tolerance**, or allowance above or below the nominal value.

nominal value

A target for design specifications.

process capability

The ability of the process to meet the design specifications for a service or product.

tolerance

An allowance above or below the nominal value.

For example, the administrator of an intensive care unit lab might have a nominal value for the turnaround time of results to the attending physicians of 25 minutes and a tolerance of ±5 minutes because of the need for speed under life-threatening conditions. The tolerance gives an *upper specification* of 30 minutes and a *lower specification* of 20 minutes. The lab process must be capable of providing the results of analyses within these specifications; otherwise, it will produce a certain proportion of "defects." The administrator is also interested in detecting occurrences of turnaround times of less than 20 minutes because something might be learned that can be built into the lab process in the future. For the present, the physicians are pleased with results that arrive within 20 to 30 minutes.

Defining Process Capability

Figure 5.14 shows the relationship between a process distribution and the upper and lower specifications for the lab process turnaround time under two conditions. In Figure 5.14(a), the process is capable because the extremes of the process distribution fall within the upper and lower specifications. In Figure 5.14(b), the process is not capable because the lab process produces too many reports with long turnaround times.

FIGURE 5.14 ▶

The Relationship Between a
Process Distribution and Upper
and Lower Specifications

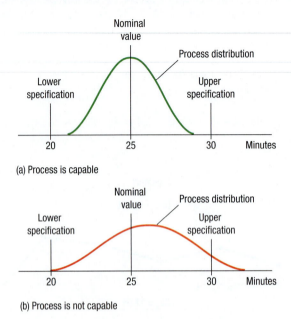

(a) Process is capable

(b) Process is not capable

Figure 5.14 shows clearly why managers are so concerned with reducing process variability. The less variability—represented by lower standard deviations—the less frequently bad output is produced. Figure 5.15 shows what reducing variability implies for a process distribution that is a normal probability distribution. The firm with two-sigma performance (the specification limits equal the process distribution mean ± 2 standard deviations) produces 4.56 percent defects, or 45,600 defects per million. The firm with four-sigma performance produces only 0.0063 percent defects, or 63 defects per million. Finally, the firm with six-sigma performance produces only 0.0000002 percent defects, or 0.002 defects per million.[1]

FIGURE 5.15 ▼

Effects of Reducing Variability on
Process Capability

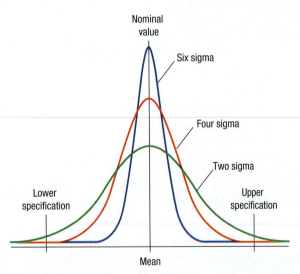

How can a manager determine quantitatively whether a process is capable? Two measures commonly are used in practice to assess the capability of a process: the process capability index and the process capability ratio.

Process Capability Index   The **process capability index**, C_{pk}, is defined as

$$C_{pk} = \text{Minimum of} \left[\frac{\overline{\overline{x}} - \text{Lower specification}}{3\sigma}, \frac{\text{Upper specification} - \overline{\overline{x}}}{3\sigma} \right]$$

where

$\sigma = $ standard deviation of the process distribution

The process capability index measures how well the process is centered as well as whether the variability is acceptable. As a general rule, most values of any process distribution fall within ± 3 standard deviations of the mean. Consequently, ± 3 standard deviations is used as the benchmark. Because the process capability index is concerned with how well the process distribution is centered relative to the specifications, it checks to see if the process average is at least three standard deviations

---

[1]Our discussion assumes that the process distribution has no assignable causes. Six Sigma programs, however, define defect performance with the assumption that the process average has moved 1.5 standard deviations. In such a case, there would be 3.4 defects per million. See **www.isixsigma.com** for the rationale behind that assumption.

from the upper and lower specifications. We take the minimum of the two ratios because it gives the *worst-case* situation.

The process capability index must be compared to a critical value to judge whether a process is capable. Firms striving to achieve three-sigma performance use a critical value for the ratio of 1.0. A firm targeting four-sigma performance will use 1.33 (or 4/3), a firm targeting five-sigma performance will use 1.67 (or 5/3), and a firm striving for six-sigma performance will use 2.00 (or 6/3). Processes producing services or products with less than three-sigma performance will have C_{pk} values less than 1.0.

If a process passes the process capability index test, we can declare the process is capable. Suppose a firm desires its processes to produce at the level of four-sigma performance. If C_{pk} is greater than or equal to the critical value of 1.33, we can say the process is capable. If C_{pk} is less than the critical value, either the process average is too close to one of the tolerance limits and is generating defective output, or the process variability is too large. To find out whether the variability is the culprit, we need another test.

Process Capability Ratio   If a process fails the process capability *index* test, we need a quick test to see if the process variability is causing the problem. If a process is *capable*, it has a process distribution whose extreme values fall within the upper and lower specifications for a service or product. For example, if the process distribution is normal, 99.74 percent of the values fall within ±3 standard deviations. In other words, the range of values of the quality measure generated by a process is approximately 6 standard deviations of the process distribution. Hence, if a process is capable at the three-sigma level, the difference between the upper and lower specification, called the *tolerance width*, must be greater than 6 standard deviations. The **process capability ratio**, C_p, is defined as

$$C_p = \frac{\text{Upper specification} - \text{Lower specification}}{6\sigma}$$

Suppose management wants four-sigma capability in their processes, and a process just failed the process capability index test at that level. A C_p value of 1.33, say, implies that the variability of the process is at the level of four-sigma quality and that the process is capable of consistently producing outputs within specifications, assuming that the process is centered. Because C_p passed the test, but C_{pk} did not, we can assume that the problem is that the process is not centered adequately.

Using Continuous Improvement to Determine the Capability of a Process

To determine the capability of a process to produce outputs within the tolerances, use the following steps.

Step 1.   Collect data on the process output, and calculate the mean and the standard deviation of the process output distribution.

Step 2.   Use the data from the process distribution to compute process control charts, such as an $\bar{x}$- and an R-chart.

Step 3.   Take a series of at least 20 consecutive random samples of size n from the process and plot the results on the control charts. If the sample statistics are within the control limits of the charts, the process is in statistical control. If the process is not in statistical control, look for assignable causes and eliminate them. Recalculate the mean and standard deviation of the process distribution and the control limits for the charts. Continue until the process is in statistical control.

Step 4.   Calculate the process capability *index*. If the results are acceptable, the process is capable and document any changes made to the process; continue to monitor the output by using the control charts. If the results are unacceptable, calculate the process capability *ratio*. If the results are acceptable, the process variability is fine and management should focus on centering the process. If the results of the process capability ratio are unacceptable, management should focus on reducing the variability in the process until it passes the test. As changes are made, recalculate the mean and standard deviation of the process distribution and the control limits for the charts and return to step 3.

Quality Engineering

Successful quality performance is often more than process improvement; it also involves service/product design. Originated by Genichi Taguchi, **quality engineering** is an approach that involves

EXAMPLE 5.5   Assessing the Process Capability of the Intensive Care Unit Lab

A doctor examines a specimen through his microscope in a lab at St. Vincent's Hospital.

The intensive care unit lab process has an average turnaround time of 26.2 minutes and a standard deviation of 1.35 minutes. The nominal value for this service is 25 minutes with an upper specification limit of 30 minutes and a lower specification limit of 20 minutes. The administrator of the lab wants to have four-sigma performance for her lab. Is the lab process capable of this level of performance?

SOLUTION

The administrator began by taking a quick check to see if the process is capable by applying the process capability index:

$$\text{Lower specification calculation} = \frac{26.2 - 20.0}{3(1.35)} = 1.53$$

$$\text{Upper specification calculation} = \frac{30.0 - 26.2}{3(1.35)} = 0.94$$

$$C_{pk} = \text{Minimum of } [1.53, 0.94] = 0.94$$

Since the target value for four-sigma performance is 1.33, the process capability index told her that the process was not capable. However, she did not know whether the problem was the variability of the process, the centering of the process, or both. The options available to improve the process depended on what is wrong.

She next checked the process variability with the process capability ratio:

$$C_p = \frac{30.0 - 20.0}{6(1.35)} = 1.23$$

The process variability did not meet the four-sigma target of 1.33. Consequently, she initiated a study to see where variability was introduced into the process. Two activities, report preparation and specimen slide preparation, were identified as having inconsistent procedures. These procedures were modified to provide consistent performance. New data were collected and the average turnaround was now 26.1 minutes with a standard deviation of 1.20 minutes. She now had the process variability at the four-sigma level of performance, as indicated by the process capability ratio:

$$C_p = \frac{30.0 - 20.0}{6(1.20)} = 1.39$$

However, the process capability index indicated additional problems to resolve:

$$C_{pk} = \text{Minimum of } \left[\frac{(26.1 - 20.0)}{3(1.20)}, \frac{(30.0 - 26.1)}{3(1.20)}\right] = 1.08$$

DECISION POINT

The lab process was still not at the level of four-sigma performance on turnaround time. The lab administrator searched for the causes of the off-center turnaround time distribution. She discovered periodic backlogs at a key piece of testing equipment. Acquiring a second machine provided the capacity to reduce the turnaround times to four-sigma capability.

MyOMLab

Active Model 5.3 in MyOMLab provides additional insight on the process capability problem at the intensive care unit lab.

MyOMLab

Tutor 5.4 in MyOMLab provides a new example to practice the process capability measures.

quality loss function

The rationale that a service or product that barely conforms to the specifications is more like a defective service or product than a perfect one.

combining engineering and statistical methods to reduce costs and improve quality by optimizing product design and manufacturing processes. Taguchi believes that unwelcome costs are associated with *any* deviation from a quality characteristic's target value. Taguchi's view is that the **quality loss function** is zero when the quality characteristic of the service or product is exactly on the target value, and that the quality loss function value rises exponentially as the quality characteristic gets closer to the specification limits. The rationale is that a service or product that barely conforms to the specifications is more like a defective service or product than a perfect one. Figure 5.16 shows Taguchi's quality loss function schematically. Taguchi concluded that managers should continually search for ways to reduce *all* variability from the target value in the production process and not be content with merely adhering to specification limits. See **http://elsmar.com/Taguchi.html** for a detailed discussion and animation of the Taguchi Loss Function.

International Quality Documentation Standards

Once a company has gone through the effort of making its processes capable, it must document its level of quality so as to better market its services or products. This documentation of quality is especially important in international trade. However, if each country had its own set of standards, companies selling in international markets would have difficulty complying with quality documentation standards in each country where they did business. To overcome this problem, the International Organization for Standardization devised a family of standards called ISO 9000 for companies doing business in the European Union. Subsequently, ISO 14000 was devised for environmental management systems and ISO 26000 for guidance on social responsibility.

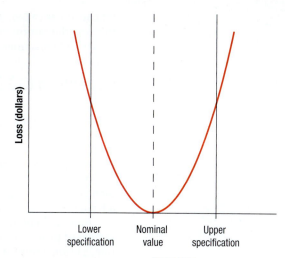

▲ **FIGURE 5.16**
Taguchi's Quality Loss Function

The ISO 9001:2008 Documentation Standards

ISO 9001:2008 is the latest update of the ISO 9000 standards governing documentation of a quality program. According to the International Organization for Standardization, the ISO 9001:2008 standards address *quality management* by specifying what the firm does to fulfill the customer's quality requirements and applicable regulatory requirements, while aiming to enhance customer satisfaction and achieve continual improvement of its performance in pursuit of these objectives. Companies become certified by proving to a qualified external examiner that they comply with all the requirements. Once certified, companies are listed in a directory so that potential customers can see which companies are certified and to what level. Compliance with ISO 9001:2008 standards says *nothing* about the actual quality of a product. Rather, it indicates to customers that companies can provide documentation to support whatever claims they make about quality. As of 2009, more than 1 million organizations worldwide have been certified in the ISO 9000 family of documentation standards.

> **ISO 9001:2008**
> A set of standards governing documentation of a quality program.

ISO 14000:2004 Environmental Management System

The **ISO 14000:2004** standards require documentation of a firm's environmental program. According to the International Organization for Standardization, the ISO 14000:2004 family addresses *environmental management* by specifying what the firm does to minimize harmful effects on the environment caused by its activities, and to achieve continual improvement of its environmental performance. The documentation standards require participating companies to keep track of their raw materials use and their generation, treatment, and disposal of hazardous wastes. Although not specifying what each company is allowed to emit, the standards require companies to prepare a plan for ongoing improvement in their environmental performance. ISO 14000:2004 covers a number of areas, including the following:

> **ISO 14000:2004**
> Documentation standards that require participating companies to keep track of their raw materials use and their generation, treatment, and disposal of hazardous wastes.

- *Environmental Management System.* Requires a plan to improve performance in resource use and pollutant output.
- *Environmental Performance Evaluation.* Specifies guidelines for the certification of companies.
- *Environmental Labeling.* Defines terms such as *recyclable*, *energy efficient*, and *safe for the ozone layer*.
- *Life-Cycle Assessment.* Evaluates the lifetime environmental impact from the manufacture, use, and disposal of a product.

To maintain their certification, companies must be inspected by outside, private auditors on a regular basis. As of 2010, more than 200,000 organizations in 155 countries have been certified for ISO 14000.

ISO 26000:2010 Social Responsibility Guidelines

The **ISO 26000:2010** guidelines, according to the International Organization for Standards, provide harmonized, globally relevant guidance on social responsibility for private and public sector organizations based on international consensus among experts. A firm does not get certified in ISO 26000; the guidelines are voluntary and are intended to promote best practice in ethical behavior in business. The seven core subjects of social responsibility covered in the guidelines are (1) human rights, (2) labor practices, (3) the environment, (4) fair operating practices,

> **ISO 26000:2010**
> International guidelines for organizational social responsibility.

(5) consumer issues, (6) community involvement and development, and (7) the organization. In this way the international community is encouraging ethical business behavior between businesses and consumers.

Benefits of ISO Certification

Completing the certification process can take as long as 18 months and involve many hours of management and employee time. The cost of certification can exceed $1 million for large companies. Despite the expense and commitment involved in ISO certification, it bestows significant external and internal benefits. The external benefits come from the potential sales advantage that companies in compliance have. Companies looking for a supplier will more likely select a company that has demonstrated compliance with ISO documentation standards, all other factors being equal. Consequently, more and more firms are seeking certification to gain a competitive advantage.

Internal benefits can be substantial. Registered companies report an average of 48 percent increased profitability and 76 percent improvement in marketing. The British Standards Institute, a leading third-party auditor, estimates that most ISO 9000-registered companies experience a 10 percent reduction in the cost of producing a product because of the quality improvements they make while striving to meet the documentation requirements. Certification in ISO 9001:2008 requires a company to analyze and document its procedures, which is necessary in any event for implementing continuous improvement, employee involvement, and similar programs. The guidelines and requirements of the ISO documentation standards provide companies with a jump-start in pursuing TQM programs.

Baldrige Performance Excellence Program

Regardless of where a company does business, it is clear that all organizations have to produce high-quality products and services if they are to be competitive. To emphasize that point, in August 1987 the U.S. Congress signed into law the Malcolm Baldrige National Quality Improvement Act, creating the Malcolm Baldrige National Quality Award, which is now entitled the **Baldrige Performance Excellence Program** (**www.quality.nist.gov**). Named for the late secretary of commerce, who was a strong proponent of enhancing quality as a means of reducing the trade deficit, the award promotes, recognizes, and publicizes quality strategies and achievements.

The application and review process for the Baldrige award is rigorous. However, the act of preparing the application itself is often a major benefit to organizations because it helps firms define what *quality* means for them. According to the U.S. Commerce Department's National Institute of Standards and Technology (NIST), investing in quality principles and performance excellence pays off in increased productivity, satisfied employees and customers, and improved profitability, both for customers and investors. The seven major criteria for the award are the following:

1. *Leadership.* Describes how senior leaders' actions guide and sustain the organization and how they communicate with the workforce and encourage high performance.

2. *Strategic Planning.* Describes how the organization establishes its strategy to address its strategic challenges, leverage its strategic advantages, and summarizes the organization's key strategic objectives and their related goals.

3. *Customer Focus.* Describes how the organization determines its service or product offerings and the mechanisms to support the customers' use of them.

4. *Measurement, Analysis, and Knowledge Management.* Describes how the organization measures, analyzes, reviews, and improves its performance through the use of data and information at all levels of the organization.

5. *Workforce Focus.* Describes how the organization engages, compensates, and rewards its workers and how they are developed to achieve high performance.

6. *Operations Focus.* Describes how the organization designs its work systems and determines its key processes to deliver customer value, prepare for potential emergencies, and achieve organizational success and sustainability.

7. *Results.* Describe the organization's performance and improvement in five categories: products and processes, customer focus, workforce focus, leadership and governance, and financial and market.

Customer satisfaction underpins these seven criteria. Criterion 7, Results, is given the most weight in selecting winners.

Baldrige Performance Excellence Program

A program named for the late secretary of commerce, Malcolm Baldrige, who was a strong proponent of enhancing quality as a means of reducing the trade deficit; organizations vie for an award that promotes, recognizes, and publicizes quality strategies and achievements.

LEARNING GOALS IN REVIEW

1 **Define the four major costs of quality.** See the section "Costs of Quality," pp. 159.

2 **Describe the role of ethics in the quality of services and products.** We explain how deceptive business practices can affect a customer's experiences and why the costs of quality should be balanced with ethical considerations in the section "Ethics and Quality," pp. 159–160.

3 **Explain the basic principles of TQM programs.** See the section "Total Quality Management," pp. 160–164. Focus on the five customer definitions of quality, Managerial Practice 5.1, which shows how one company matched processes to the five definitions, the importance of employee involvement, and how continuous improvement works. The key figures are Figures 5.1 and 5.2.

4 **Explain the basic principles of Six Sigma Programs.** We have summarized the essence of these important programs in the section "Six Sigma," pp. 164–165. Be sure to understand

Figure 5.3, which shows the goals of Six Sigma, and Figure 5.4, which provides the improvement model. Figure 5.5 shows how TQM or Six Sigma works in a supply chain through the tactic of acceptance sampling.

5 **Describe how to construct control charts and use them to determine whether a process is out of statistical control.** See the section "Statistical Process Control," pp. 166–170. Understanding Figures 5.6 and 5.7 is key to understanding the methods to follow. The section "Statistical Process Control Methods," pp. 170–177, shows you how to determine if a process is in statistical control. Study Examples 5.1 to 5.5 as well as Solved Problems 1 to 3.

6 **Describe how to determine whether a process is capable of producing a service or product to specifications.** The major take-away in the chapter is found in the section "Process Capability," pp. 177–180. Be sure you understand Figures 5.4 and 5.5; study Example 5.5 and Solved Problem 4.

MyOMLab helps and assesses students with 16 problems on *x*-bar and *R*-bar Charts, *p*-charts, *c*-Charts, and Process Capability.

MyOMLab Resources	Titles	Link to the Book
Video	*Starwood: Process Performance and Quality*	Costs of Quality; Total Quality Management; Six Sigma
	Christchurch Parkroyal TQM	Costs of Quality; Total Quality Management
Active Model Exercises	5.1 *x*-bar and *R*-Charts	Control Charts for Variables; Example 5.1 (pp. 172–173)
	5.2 *p*-Chart	Control Charts for Attributes; Example 5.3 (pp. 175–176)
	5.3 Process Capability	Process Capability; Example 5.5 (p. 180)
OM Explorer Solvers	*c*-Charts	Control Charts for Attributes; Example 5.4 (p. 177); Figure 5.13 (p. 177); Solved Problem 3 (p. 187)
	p-Charts	Control Charts for Attributes; Example 5.3 (pp. 175–176); Solved Problem 2 (pp. 186–187)
	Process Capability	Process Capability; Example 5.5 (p. 180); Solved Problem 4 (p. 188)
	R- and *x*-bar Charts	Control Charts for Variables; Example 5.1 (p. 172); Figure 5.10 and Figure 5.11 (pp. 172–173); Solved Problem 1 (pp. 185–186)
OM Explorer Tutors	5.1 *R*- and *x*-bar Charts	Control Charts for Variables; Example 5.1 (pp. 172–173)
	5.2 *p*-Charts	Control Charts for Attributes; Example 5.3 (pp. 175–176)
	5.3 *c*-Charts	Control Charts for Attributes; Example 5.4 (p. 177)
	5.4 Process Capability	Process Capability; Example 5.5 (p. 180)
POM for Windows	*p*-Charts	Control Charts for Attributes; Example 5.3 (pp. 175–176); Figure 5.12 (p. 176); Solved Problem 2 (pp. 186–187)
	x-bar Charts	Control Charts for Variables; Example 5.1 (pp. 172–173); Example 5.2 (p. 174); Solved Problem 1 (pp. 185–186)
	c-Charts	Control Charts for Attributes; Example 5.4 (p. 177); Solved Problem 3 (p. 187)
	Process Capability	Process Capability; Example 5.5 (p. 180); Solved Problem 4 (p. 188)
	Acceptance Sampling	Acceptance Sampling
SimQuick Simulation Exercises	Circuit Board Process	Six-Sigma Improvement Model

MyOMLab Resources	Titles	Link to the Book
Tutor Exercises	5.1 x-bar and R-Chart with Target Weight of 7.04 oz.	Control Charts for Variables
	5.2 p-Chart When Changes in Sample Values	Control Charts for Attributes
	5.3 Process Capability with a Change in Process Average and Variability	Process Capability
Virtual Tours	1. Steinway Factory, Verne O. Powell Flutes	Total Quality Management; Six Sigma
	2. Beach Beat Surfboards	Total Quality Management
MyOMLab Supplements	G. Acceptance Sampling Plans	Acceptance Sampling
Internet Exercise	1. National Institute of Standards and Technology, and International Organization for Standardization	Baldrige Performance Excellence Award International Quality Documentation Standards
	2. SAS Scandinavian Airline	Total Quality Management
	3. Jack in the Box	Customer Satisfaction
	4. Maybach	Customer Satisfaction
	5. Bureau of Transportation Statistics	Statistical Process Control
Key Equations		
Image Library		

Key Equations

1. Sample mean: $\bar{x} = \dfrac{\sum_{i=1}^{n} x_i}{n}$

2. Standard deviation of a sample:

$$\sigma = \sqrt{\frac{\sum_{i=1}^{n}(x_i - \bar{x})^2}{n-1}} \quad \text{or} \quad \sigma = \sqrt{\frac{\sum_{i=1}^{n} x_i^2 - \dfrac{(\sum x_i)^2}{n}}{n-1}}$$

3. Control limits for variable process control charts

    a. R-chart, range of sample:

    $$\text{Upper control limit} = \text{UCL}_R = D_4\bar{R}$$
    $$\text{Lower control limit} = \text{LCL}_R = D_3\bar{R}$$

    b. $\bar{x}$-chart, sample mean:

    $$\text{Upper control limit} = \text{UCL}_{\bar{x}} = \bar{\bar{x}} + A_2\bar{R}$$
    $$\text{Lower control limit} = \text{LCL}_{\bar{x}} = \bar{\bar{x}} - A_2\bar{R}$$

    c. When the standard deviation of the process distribution, σ, is known:

    $$\text{Upper control limit} = \text{UCL}_{\bar{x}} = \bar{\bar{x}} + z\sigma_{\bar{x}}$$
    $$\text{Lower control limit} = \text{LCL}_{\bar{x}} = \bar{\bar{x}} - z\sigma_{\bar{x}}$$

    where

    $$\sigma_{\bar{x}} = \frac{\sigma}{\sqrt{n}}$$

4. Control limits for attribute process control charts

   a. *p*-chart, proportion defective:

$$\text{Upper control limit} = \text{UCL}_p = \bar{p} + z\sigma_p$$
$$\text{Lower control limit} = \text{LCL}_p = \bar{p} - z\sigma_p$$

   where

$$\sigma_p = \sqrt{\bar{p}(1 - \bar{p})/n}$$

   b. *c*-chart, number of defects:

$$\text{Upper control limit} = \text{UCL}_c = \bar{c} + z\sqrt{\bar{c}}$$
$$\text{Lower control limit} = \text{LCL}_c = \bar{c} - z\sqrt{\bar{c}}$$

5. Process capability index:

$$C_{pk} = \text{Minimum of} \left[\frac{\bar{\bar{x}} - \text{Lower specification}}{3\sigma}, \frac{\text{Upper specification} - \bar{\bar{x}}}{3\sigma} \right]$$

6. Process capability ratio:

$$C_p = \frac{\text{Upper specification} - \text{Lower specification}}{6\sigma}$$

Key Terms

acceptable quality level (AQL) 165	internal failure costs 159	quality loss function 180
acceptance sampling 165	ISO 9001:2008 181	*R*-chart 170
appraisal costs 159	ISO 14000:2004 181	sample size 167
assignable causes of variation 168	ISO 26000:2010 181	sampling plan 167
attributes 167	Master Black Belt 165	self-managed team 163
Baldrige Performance Excellence	nominal value 177	Six Sigma 164
Program 182	*p*-chart 174	special-purpose teams 163
Black Belt 165	plan-do-study-act cycle 163	statistical process control (SPC) 166
c-chart 176	prevention costs 159	teams 162
common causes of variation 168	process capability 177	tolerance 177
continuous improvement 163	process capability index, C_{pk} 178	total quality management (TQM) 160
control chart 169	process capability ratio, C_p 179	type I error 170
defect 159	quality 160	type II error 170
employee empowerment 163	quality at the source 162	variables 167
external failure costs 159	quality circles 163	warranty 159
Green Belt 165	quality engineering 179	$\bar{x}$-chart 171

Solved Problem 1

The Watson Electric Company produces incandescent lightbulbs. The following data on the number of lumens for 40-watt lightbulbs were collected when the process was in control.

	OBSERVATION			
Sample	**1**	**2**	**3**	**4**
1	604	612	588	600
2	597	601	607	603
3	581	570	585	592
4	620	605	595	588
5	590	614	608	604

   a. Calculate control limits for an *R*-chart and an $\bar{x}$-chart.

   b. Since these data were collected, some new employees were hired. A new sample obtained the following readings: 570, 603, 623, and 583. Is the process still in control?

SOLUTION

a. To calculate $\bar{x}$, compute the mean for each sample. To calculate R, subtract the lowest value in the sample from the highest value in the sample. For example, for sample 1,

$$\bar{x} = \frac{604 + 612 + 588 + 600}{4} = 601$$

$$R = 612 - 588 = 24$$

Sample	$\bar{x}$	R
1	601	24
2	602	10
3	582	22
4	602	32
5	604	24
Total	2,991	112
Average	$\bar{\bar{x}} = 598.2$	$\bar{R} = 22.4$

The R-chart control limits are

$$\text{UCL}_R = D_4\bar{R} = 2.282(22.4) = 51.12$$

$$\text{LCL}_R = D_3\bar{R} = 0(22.4) = 0$$

The $\bar{x}$-chart control limits are

$$\text{UCL}_{\bar{x}} = \bar{\bar{x}} + A_2\bar{R} = 598.2 + 0.729(22.4) = 614.53$$

$$\text{LCL}_{\bar{x}} = \bar{\bar{x}} - A_2\bar{R} = 598.2 - 0.729(22.4) = 581.87$$

b. First check to see whether the variability is still in control based on the new data. The range is 53 (or 623 − 570), which is outside the UCL for the R-chart. Since the process variability is out of control, it is meaningless to test for the process average using the current estimate for $\bar{R}$. A search for assignable causes inducing excessive variability must be conducted.

Solved Problem 2

The data processing department of the Arizona Bank has five data entry clerks. Each working day their supervisor verifies the accuracy of a random sample of 250 records. A record containing one or more errors is considered defective and must be redone. The results of the last 30 samples are shown in the table. All were checked to make sure that none was out of control.

Sample	Number of Defective Records	Sample	Number of Defective Records	Sample	Number of Defective Records	Sample	Number of Defective Records
1	7	9	6	17	12	24	7
2	5	10	13	18	4	25	13
3	19	11	18	19	6	26	10
4	10	12	5	20	11	27	14
5	11	13	16	21	17	28	6
6	8	14	4	22	12	29	11
7	12	15	11	23	6	30	9
8	9	16	8			Total	300

a. Based on these historical data, set up a *p*-chart using $z = 3$.

b. Samples for the next 4 days showed the following:

Sample	Number of Defective Records
Tues	17
Wed	15
Thurs	22
Fri	21

What is the supervisor's assessment of the data-entry process likely to be?

SOLUTION

a. From the table, the supervisor knows that the total number of defective records is 300 out of a total sample of 7,500 [or 30(250)]. Therefore, the central line of the chart is

$$\bar{p} = \frac{300}{7,500} = 0.04$$

The control limits are

$$UCL_p = \bar{p} + z\sqrt{\frac{\bar{p}(1 - \bar{p})}{n}} = 0.04 + 3\sqrt{\frac{0.04(0.96)}{250}} = 0.077$$

$$LCL_p = \bar{p} - z\sqrt{\frac{\bar{p}(1 - \bar{p})}{n}} = 0.04 - 3\sqrt{\frac{0.04(0.96)}{250}} = 0.003$$

b. Samples for the next 4 days showed the following:

Sample	Number of Defective Records	Proportion
Tues	17	0.068
Wed	15	0.060
Thurs	22	0.088
Fri	21	0.084

Samples for Thursday and Friday are out of control. The supervisor should look for the problem and, upon identifying it, take corrective action.

Solved Problem 3

The Minnow County Highway Safety Department monitors accidents at the intersection of Routes 123 and 14. Accidents at the intersection have averaged three per month.

a. Which type of control chart should be used? Construct a control chart with three-sigma control limits.

b. Last month, seven accidents occurred at the intersection. Is this sufficient evidence to justify a claim that something has changed at the intersection?

SOLUTION

a. The safety department cannot determine the number of accidents that did *not* occur, so it has no way to compute a proportion defective at the intersection. Therefore, the administrators must use a *c*-chart for which

$$UCL_c = \bar{c} + z\sqrt{\bar{c}} = 3 + 3\sqrt{3} = 8.20$$

$$LCL_c = \bar{c} - z\sqrt{\bar{c}} = 3 - 3\sqrt{3} = -2.196, \text{ adjusted to } 0$$

There cannot be a negative number of accidents, so the LCL in this case is adjusted to zero.

b. The number of accidents last month falls within the UCL and LCL of the chart. We conclude that no assignable causes are present and that the increase in accidents was due to chance.

Solved Problem 4

Pioneer Chicken advertises "lite" chicken with 30 percent fewer calories. (The pieces are 33 percent smaller.) The process average distribution for "lite" chicken breasts is 420 calories, with a standard deviation of the population of 25 calories. Pioneer randomly takes samples of six chicken breasts to measure calorie content.

a. Design an $\bar{x}$-chart using the process standard deviation. Use three-sigma limits.

b. The product design calls for the average chicken breast to contain 400 ± 100 calories. Calculate the process capability index (target = 1.33) and the process capability ratio. Interpret the results.

SOLUTION

a. For the process standard deviation of 25 calories, the standard deviation of the sample mean is

$$\sigma_{\bar{x}} = \frac{\sigma}{\sqrt{n}} = \frac{25}{\sqrt{6}} = 10.2 \text{ calories}$$

$$\text{UCL}_{\bar{x}} = \bar{\bar{x}} + z\sigma_{\bar{x}} = 420 + 3(10.2) = 450.6 \text{ calories}$$

$$\text{LCL}_{\bar{x}} = \bar{\bar{x}} - z\sigma_{\bar{x}} = 420 - 3(10.2) = 389.4 \text{ calories}$$

b. The process capability index is

$$C_{pk} = \text{Minimum of} \left[\frac{\bar{\bar{x}} - \text{Lower specification}}{3\sigma}, \frac{\text{Upper specification} - \bar{\bar{x}}}{3\sigma} \right]$$

$$= \text{Minimum of} \left[\frac{420 - 300}{3(25)} = 1.60, \frac{500 - 420}{3(25)} = 1.07 \right] = 1.07$$

The process capability ratio is

$$C_p = \frac{\text{Upper specification} - \text{Lower specification}}{6\sigma} = \frac{500 \text{ calories} - 300 \text{ calories}}{6(25)} = 1.33$$

Because the process capability ratio is 1.33, the process should be able to produce the product reliably within specifications. However, the process capability index is 1.07, so the current process is not centered properly for four-sigma performance. The mean of the process distribution is too close to the upper specification.

Discussion Questions

1. Consider Managerial Practice 5.1 and the discussion of Steinway's approach to achieving top quality. To get a better idea of the craft-oriented production process, visit **www.steinway.com/factory/tour.shtml**. However, Steinway also uses automation to produce the action mechanisms, a critical assembly in the grand pianos. Given the overall image of a Steinway piano, a very pricey hand-crafted object of beauty, what do you think of the use of automated equipment? Do you think it is a mistake to use automation in this way?

2. Recently, the Polish General Corporation, well-known for manufacturing appliances and automobile parts, initiated a $13 billion project to produce automobiles. A great deal of learning on the part of management and employees was required. Even though pressure was mounting to get a new product to market in early 2012, the production manager of the newly formed automobile division insisted on almost a year of trial runs before sales started because workers have to do their jobs 60 to 100 times before they can memorize the right sequence. The launch date was set for early 2013. What are the consequences of using this approach to enter the market with a new product?

3. Explain how unethical business practices degrade the quality of the experience a customer has with a service or product. How is the International Organization for Standardization trying to encourage ethical business behavior?

Problems

The OM Explorer and POM for Windows software is available to all students using the 10th edition of this textbook. Go to **www.pearsonhighered.com/krajewski** to download these computer packages. If you purchased MyOMLab, you also have access to Active Models software and significant help in doing the following problems. Check with your instructor on how best to use these resources. In many cases, the instructor wants you to understand how to do the calculations by hand. At the least, the software provides a check on your calculations. When calculations are particularly complex and the goal is interpreting the results in making decisions, the software replaces entirely the manual calculations. The software also can be a valuable resource well after your course is completed.

1. At Quickie Car Wash, the wash process is advertised to take less than 7 minutes. Consequently, management has set a target average of 390 seconds for the wash process. Suppose the average range for a sample of 9 cars is 10 seconds. Use Table 5.1 to establish control limits for sample means and ranges for the car wash process.

2. At Isogen Pharmaceuticals, the filling process for its asthma inhaler is set to dispense 150 milliliters (ml) of steroid solution per container. The average range for a sample of 4 containers is 3 ml. Use Table 5.1 to establish control limits for sample means and ranges for the filling process.

3. Garcia's Garage desires to create some colorful charts and graphs to illustrate how reliably its mechanics "get under the hood and fix the problem." The historic average for the proportion of customers that return for the same repair within the 30-day warranty period is 0.10. Each month, Garcia tracks 100 customers to see whether they return for warranty repairs. The results are plotted as a proportion to report progress toward the goal. If the control limits are to be set at two standard deviations on either side of the goal, determine the control limits for this chart. In March, 8 of the 100 customers in the sample group returned for warranty repairs. Is the repair process in control?

4. The Canine Gourmet Company produces delicious dog treats for canines with discriminating tastes. Management wants the box-filling line to be set so that the process average weight per packet is 45 grams. To make sure that the process is in control, an inspector at the end of the filling line periodically selects a random box of 10 packets and weighs each packet. When the process is in control, the range in the weight of each sample has averaged 6 grams.

    a. Design an R- and an $\bar{x}$-chart for this process.

    b. The results from the last 5 samples of 10 packets are

Sample	$\bar{x}$	R
1	44	9
2	40	2
3	46	5
4	39	8
5	48	3

Is the process in control? Explain.

5. Aspen Plastics produces plastic bottles to customer order. The quality inspector randomly selects four bottles from the bottle machine and measures the outside diameter of the bottle neck, a critical quality dimension that determines whether the bottle cap will fit properly. The dimensions (in.) from the last six samples are

	BOTTLE			
Sample	1	2	3	4
1	0.594	0.622	0.598	0.590
2	0.587	0.611	0.597	0.613
3	0.571	0.580	0.595	0.602
4	0.610	0.615	0.585	0.578
5	0.580	0.624	0.618	0.614
6	0.585	0.593	0.607	0.569

    a. Assume that only these six samples are sufficient, and use the data to determine control limits for an R- and an $\bar{x}$-chart.

    b. Suppose that the specification for the bottle neck diameter is 0.600 ± 0.050 and the population standard deviation is 0.013 in. What is the Process Capability Index? The Process Capability Ratio?

    c. If the firm is seeking four-sigma performance, is the process capable of producing the bottle?

6. In an attempt to judge and monitor the quality of instruction, the administration of Mega-Byte Academy devised an examination to test students on the basic concepts that all should have learned. Each year, a random sample of 10 graduating students is selected for the test. The average score is used to track the quality of the educational process. Test results for the past 10 years are shown in Table 5.2.

    Use these data to estimate the center and standard deviation for this distribution. Then, calculate the two-sigma control limits for the process average. What comments would you make to the administration of the Mega-Byte Academy?

7. As a hospital administrator of a large hospital, you are concerned with the absenteeism among nurses' aides. The issue has been raised by registered nurses, who feel they often have to perform work normally done by their aides. To get the facts, absenteeism data were gathered for the last 3 weeks, which is considered a representative period for future conditions. After taking random samples of 64 personnel files each day, the following data were produced:

TABLE 5.2 | **TEST SCORES ON EXIT EXAM**

					STUDENT						
Year	1	2	3	4	5	6	7	8	9	10	Average
1	63	57	92	87	70	61	75	58	63	71	69.7
2	90	77	59	88	48	83	63	94	72	70	74.4
3	67	81	93	55	71	71	86	98	60	90	77.2
4	62	67	78	61	89	93	71	59	93	84	75.7
5	85	88	77	69	58	90	97	72	64	60	76.0
6	60	57	79	83	64	94	86	64	92	74	75.3
7	94	85	56	77	89	72	71	61	92	97	79.4
8	97	86	83	88	65	87	76	84	81	71	81.8
9	94	90	76	88	65	93	86	87	94	63	83.6
10	88	91	71	89	97	79	93	87	69	85	84.9

Day	Aides Absent	Day	Aides Absent
1	4	9	7
2	3	10	2
3	2	11	3
4	4	12	2
5	2	13	1
6	5	14	3
7	3	15	4
8	4		

Because your assessment of absenteeism is likely to come under careful scrutiny, you would like a type I error of only 1 percent. You want to be sure to identify any instances of unusual absences. If some are present, you will have to explore them on behalf of the registered nurses.

a. Design a *p*-chart.

b. Based on your *p*-chart and the data from the last 3 weeks, what can you conclude about the absenteeism of nurses' aides?

8. A textile manufacturer wants to set up a control chart for irregularities (e.g., oil stains, shop soil, loose threads, and tears) per 100 square yards of carpet. The following data were collected from a sample of twenty 100-square-yard pieces of carpet:

Sample	1	2	3	4	5	6	7	8	9	10
Irregularities	11	8	9	12	4	16	5	8	17	10
Sample	11	12	13	14	15	16	17	18	19	20
Irregularities	11	5	7	12	13	8	19	11	9	10

a. Using these data, set up a *c*-chart with $z = 3$.

b. Suppose that the next five samples had 15, 18, 12, 22, and 21 irregularities. What do you conclude?

9. The IRS is concerned with improving the accuracy of tax information given by its representatives over the telephone. Previous studies involved asking a set of 25 questions of a large number of IRS telephone representatives to determine the proportion of correct responses. Historically, the average proportion of correct responses has been 72 percent. Recently, IRS representatives have been receiving more training. On April 26, the set of 25 tax questions were again asked of 20 randomly selected IRS telephone representatives. The numbers of correct answers were 18, 16, 19, 21, 20, 16, 21, 16, 17, 10, 25, 18, 25, 16, 20, 15, 23, 19, 21, and 19.

a. What are the upper and lower control limits for the appropriate *p*-chart for the IRS? Use $z = 3$.

b. Is the tax information process in statistical control?

10. A travel agency is concerned with the accuracy and appearance of itineraries prepared for its clients. Defects can include errors in times, airlines, flight numbers, prices, car rental information, lodging, charge card numbers, and reservation numbers, as well as typographical errors. As the possible number of errors is nearly infinite, the agency measures the number of errors that do occur. The current process results in an average of three errors per itinerary.

a. What are the two-sigma control limits for these defects?

b. A client scheduled a trip to Dallas. Her itinerary contained six errors. Interpret this information.

11. Jim's Outfitters, Inc., makes custom fancy shirts for cowboys. The shirts could be flawed in various ways, including flaws in the weave or color of the fabric, loose buttons or decorations, wrong dimensions, and uneven stitches. Jim randomly examined 10 shirts, with the following results:

Shirt	Defects
1	8
2	0
3	7
4	12
5	5
6	10
7	2
8	4
9	6
10	6

a. Assuming that 10 observations are adequate for these purposes, determine the three-sigma control limits for defects per shirt.

b. Suppose that the next shirt has 13 flaws. What can you say about the process now?

12. The Big Black Bird Company produces fiberglass camper tops. The process for producing the tops must be controlled so as to keep the number of dimples low. When the process was in control, the following defects were found in 10 randomly selected camper tops over an extended period of time:

Top	Dimples
1	7
2	9
3	14
4	11
5	3
6	12
7	8
8	4
9	7
10	6

a. Assuming 10 observations are adequate for this purpose, determine the three-sigma control limits for dimples per camper top.

b. Suppose that the next camper top has 15 dimples. What can you say about the process now?

13. The production manager at Sunny Soda, Inc., is interested in tracking the quality of the company's 12-ounce bottle filling line. The bottles must be filled within the tolerances set for this product because the dietary information on the label shows 12 ounces as the serving size.

The design standard for the product calls for a fill level of 12.00 ± 0.10 ounces. The manager collected the following sample data (in fluid ounces per bottle) on the production process:

Sample	OBSERVATION 1	2	3	4
1	12.00	11.97	12.10	12.08
2	11.91	11.94	12.10	11.96
3	11.89	12.02	11.97	11.99
4	12.10	12.09	12.05	11.95
5	12.08	11.92	12.12	12.05
6	11.94	11.98	12.06	12.08
7	12.09	12.00	12.00	12.03
8	12.01	12.04	11.99	11.95
9	12.00	11.96	11.97	12.03
10	11.92	11.94	12.09	12.00
11	11.91	11.99	12.05	12.10
12	12.01	12.00	12.06	11.97
13	11.98	11.99	12.06	12.03
14	12.02	12.00	12.05	11.95
15	12.00	12.05	12.01	11.97

a. Are the process average and range in statistical control?

b. Is the process capable of meeting the design standard at four-sigma quality? Explain.

14. The Money Pit Mortgage Company is interested in monitoring the performance of the mortgage process. Fifteen samples of 5 completed mortgage transactions each were taken during a period when the process was believed to be in control. The times to complete the transactions were measured. The means and ranges of the mortgage process transaction times, measured in days, are as follows:

Sample	1	2	3	4	5	6	7	8	9	10	11	12	13	14	15
Mean	17	14	8	17	12	13	15	16	13	14	16	9	11	9	12
Range	6	11	4	8	9	14	12	15	10	10	11	6	9	11	13

Subsequently, samples of size 5 were taken from the process every week for the next 10 weeks. The times were measured and the following results obtained:

Sample	16	17	18	19	20	21	22	23	24	25
Mean	11	14	9	15	17	19	13	22	20	18
Range	7	11	6	4	12	14	11	10	8	6

a. Construct the control charts for the mean and the range, using the original 15 samples.

b. On the control charts developed in part (a), plot the values from samples 16 through 25 and comment on whether the process is in control.

c. In part (b), if you concluded that the process was out of control, would you attribute it to a drift in the mean, or an increase in the variability, or both? Explain your answer.

15. The Money Pit Mortgage Company of Problem 14 made some changes to the process and undertook a process capability study. The following data were obtained for 15 samples of size 5. Based on the individual observations, management estimated the process standard deviation to be 4.21 (days) for use in the process capability analysis. The lower and upper specification limits (in days) for the mortgage process times were 5 and 25.

Sample	1	2	3	4	5	6	7	8	9	10	11	12	13	14	15
Mean	11	12	8	16	13	12	17	16	13	14	17	9	15	14	9
Range	9	13	4	11	10	9	8	15	14	11	6	6	12	10	11

a. Calculate the process capability index and the process capability ratio values.

b. Suppose management would be happy with three-sigma performance. What conclusions is management likely to draw from the capability analysis? Can valid conclusions about the process be drawn from the analysis?

c. What remedial actions, if any, do you suggest that management take?

16. Webster Chemical Company produces mastics and caulking for the construction industry. The product is blended in large mixers and then pumped into tubes and capped. Management is concerned about whether the filling process for tubes of caulking is in statistical control. The process should be centered on 8 ounces per tube. Several samples of eight tubes were taken, each tube was weighed, and the weights in Table 5.3 were obtained.

a. Assume that only six samples are sufficient and develop the control charts for the mean and the range.

b. Plot the observations on the control chart and comment on your findings.

17. Management at Webster, in Problem 16, is now concerned as to whether caulking tubes are being properly capped. If a significant proportion of the tubes are not being sealed, Webster is placing its customers in a messy situation. Tubes are packaged in large boxes of 144. Several boxes are inspected, and the following numbers of leaking tubes are found:

Sample	Tubes	Sample	Tubes	Sample	Tubes
1	3	8	6	15	5
2	5	9	4	16	0
3	3	10	9	17	2
4	4	11	2	18	6
5	2	12	6	19	2
6	4	13	5	20	1
7	2	14	1	Total	72

Calculate p-chart three-sigma control limits to assess whether the capping process is in statistical control.

18. At Webster Chemical Company, lumps in the caulking compound could cause difficulties in dispensing a smooth bead from the tube. Even when the process is in control, an average of four lumps per tube of caulk will remain. Testing for the presence of lumps destroys the product, so an analyst takes random samples. The following results are obtained:

Tube No.	Lumps	Tube No.	Lumps	Tube No.	Lumps
1	6	5	6	9	5
2	5	6	4	10	0
3	0	7	1	11	9
4	4	8	6	12	2

Determine the c-chart two-sigma upper and lower control limits for this process. Is the process in statistical control?

TABLE 5.3 | **OUNCES OF CAULKING PER TUBE**

	TUBE NUMBER							
Sample	1	2	3	4	5	6	7	8
1	7.98	8.34	8.02	7.94	8.44	7.68	7.81	8.11
2	8.33	8.22	8.08	8.51	8.41	8.28	8.09	8.16
3	7.89	7.77	7.91	8.04	8.00	7.89	7.93	8.09
4	8.24	8.18	7.83	8.05	7.90	8.16	7.97	8.07
5	7.87	8.13	7.92	7.99	8.10	7.81	8.14	7.88
6	8.13	8.14	8.11	8.13	8.14	8.12	8.13	8.14

19. Janice Sanders, CEO of Pine Crest Medical Clinic, is concerned over the number of times patients must wait more than 30 minutes beyond their scheduled appointments. She asked her assistant to take random samples of 64 patients to see how many in each sample had to wait more than 30 minutes. Each instance is considered a defect in the clinic process. The table below contains the data for 15 samples.

Sample	Number of Defects
1	5
2	2
3	1
4	3
5	1
6	5
7	2
8	3
9	6
10	3
11	9
12	9
13	5
14	2
15	3

a. Assuming Janice Sanders is willing to use three-sigma control limits, construct a p-chart.

b. Based on your p-chart and the data in the table, what can you conclude about the waiting time of the patients?

20. Representatives of the Patriot Insurance Company take medical information over the telephone from prospective policy applicants prior to a visit to the applicant's place of residence by a registered nurse who takes vital sign measurements. When the telephone interview has incorrect or incomplete information, the entire process of approving the application is unnecessarily delayed and has the potential of causing loss of business. The following data were collected to see how many applications contain errors. Each sample has 200 randomly selected applications.

Sample	Defects	Sample	Defects
1	20	16	15
2	18	17	40
3	29	18	35
4	12	19	21
5	14	20	24
6	11	21	9
7	30	22	20
8	25	23	17
9	27	24	28
10	16	25	10
11	25	26	17
12	18	27	22
13	25	28	14
14	16	29	19
15	20	30	20

a. What are the upper and lower control limits of a p-chart for the number of defective applications? Use $z = 3$.

b. Is the process in statistical control?

21. The Digital Guardian Company issues policies that protect clients from downtime costs due to computer system failures. It is very important to process the policies quickly because long cycle times not only put the client at risk, they could also lose business for Digital Guardian. Management is concerned that customer service is degrading because of long cycle times, measured in days. The following table contains the data from five samples, each sample consisting of eight random observations.

Sample	OBSERVATION (DAYS)							
	1	2	3	4	5	6	7	8
1	13	9	4	8	8	15	8	6
2	7	15	8	10	10	14	10	15
3	8	11	4	11	8	12	9	15
4	12	7	12	9	11	8	12	8
5	8	12	6	12	11	5	12	8

a. What is your estimate of the process average?

b. What is your estimate of the average range?

c. Construct an R- and an $\bar{x}$-chart for this process. Are assignable causes present?

22. The Farley Manufacturing Company prides itself on the quality of its products. The company is engaged in competition for a very important project. A key element is a part that ultimately goes into precision testing equipment. The specifications are 8.000 ± 3.000 millimeters.

Management is concerned about the capability of the process to produce that part. The following data (shown below) were randomly collected during test runs of the process:

Sample	\multicolumn{8}{c}{OBSERVATION (MILLIMETERS)}
	1	2	3	4	5	6	7	8
1	9.100	8.900	8.800	9.200	8.100	6.900	9.300	9.100
2	7.600	8.000	9.000	10.100	7.900	9.000	8.000	8.800
3	8.200	9.100	8.200	8.700	9.000	7.000	8.800	10.800
4	8.200	8.300	7.900	7.500	8.900	7.800	10.100	7.700
5	10.000	8.100	8.900	9.000	9.300	9.000	8.700	10.000

Assume that the process is in statistical control. Is the process capable of producing the part at the three-sigma level? Explain.

23. A critical dimension of the service quality of a call center is the wait time of a caller to get to a sales representative. Periodically, random samples of three customer calls are measured for time. The results of the last four samples are in the following table:

Sample	\multicolumn{3}{c}{Time (Sec)}
1	495	501	498
2	512	508	504
3	505	497	501
4	496	503	492

a. Assuming that management is willing to use three-sigma control limits, and using only the historical information contained in the four samples, show that the call center access time is in statistical control.

b. Suppose that the standard deviation of the process distribution is 5.77. If the specifications for the access time are 500 ± 18 sec., is the process capable? Why or why not? Assume three-sigma performance is desired.

24. An automatic lathe produces rollers for roller bearings, and the process is monitored by statistical process control charts. The central line of the chart for the sample means is set at 8.50 and for the range at 0.31 mm. The process is in control, as established by samples of size 5. The upper and lower specifications for the diameter of the rollers are $(8.50 + 0.25)$ and $(8.50 - 0.25)$ mm, respectively.

a. Calculate the control limits for the mean and range charts.

b. If the standard deviation of the process distribution is estimated to be 0.13 mm, is the process capable of meeting specifications? Assume four-sigma performance is desired.

c. If the process is not capable, what percent of the output will fall outside the specification limits? (*Hint*: Use the normal distribution.)

Advanced Problems

25. Canine Gourmet Super Breath dog treats are sold in boxes labeled with a net weight of 12 ounces (340 grams) per box. Each box contains 8 individual 1.5-ounce packets. To reduce the chances of shorting the customer, product design specifications call for the packet-filling process average to be set at 43.5 grams so that the average net weight per box of 8 packets will be 348 grams. Tolerances are set for the box to weigh 348 ± 12 grams. The standard deviation for the *packet-filling* process is 1.01 grams. The target process capability ratio is 1.33. One day, the packet-filling process average weight drifts down to 43.0 grams. Is the packaging process capable? Is an adjustment needed?

26. The Precision Machining Company makes hand-held tools on an assembly line that produces one product every minute. On one of the products, the critical quality dimension is the diameter (measured in thousandths of an inch) of a hole bored in one of the assemblies. Management wants to detect any shift in the process average diameter from 0.015 in. Management considers the variance in the process to be in control. Historically, the average range has been 0.002 in., regardless of the process average. Design an $\bar{x}$-chart to control this process, with a center line at 0.015 in. and the control limits set at three sigmas from the center line.

Management provided the results of 80 minutes of output from the production line, as shown in Table 5.4. During these 80 minutes, the process average changed once. All measurements are in thousandths of an inch.

a. Set up an $\bar{x}$-chart with $n = 4$. The frequency should be sample four and then skip four. Thus, your first sample would be for minutes $1 - 4$, the second would be for minutes $9 - 12$, and so on. When would you stop the process to check for a change in the process average?

b. Set up an $\bar{x}$-chart with $n = 8$. The frequency should be sample eight and then skip four. When would you stop the process now? What can you say about the desirability of large samples on a frequent sampling interval?

27. Using the data from Problem 26, continue your analysis of sample size and frequency by trying the following plans.

a. Using the $\bar{x}$-chart for $n = 4$, try the frequency sample four, then skip eight. When would you stop the process in this case?

b. Using the $\bar{x}$-chart for $n = 8$, try the frequency sample eight, then skip eight. When would you consider the process to be out of control?

TABLE 5.4 | **SAMPLE DATA FOR PRECISION MACHINING COMPANY**

Minutes	Diameter (thousandths of an inch)											
1–12	15	16	18	14	16	17	15	14	14	13	16	17
13–24	15	16	17	16	14	14	13	14	15	16	15	17
25–36	14	13	15	17	18	15	16	15	14	15	16	17
37–48	18	16	15	16	16	14	17	18	19	15	16	15
49–60	12	17	16	14	15	17	14	16	15	17	18	14
61–72	15	16	17	18	13	15	14	14	16	15	17	18
73–80	16	16	17	18	16	15	14	17				

TABLE 5.5 | **SAMPLE DATA FOR DATA TECH CREDIT CARD SERVICE**

Samples	Number of Errors in Sample of 250									
1–10	3	8	5	11	7	1	12	9	0	8
11–20	3	5	7	9	11	3	2	9	13	4
21–30	12	10	6	2	1	7	10	5	8	4

c. Using your results from parts (a) and (b), determine what trade-offs you would consider in choosing between them.

28. The manager of the customer service department of Data Tech Credit Card Service Company is concerned about the number of defects produced by the billing process. Every day a random sample of 250 statements was inspected for errors regarding incorrect entries involving account numbers, transactions on the customer's account, interest charges, and penalty charges. Any statement with one or more of these errors was considered a defect. The study lasted 30 days and yielded the data in Table 5.5.

a. Construct a p-chart for the billing process.

b. Is there any nonrandom behavior in the billing process that would require management attention?

29. Red Baron Airlines serves hundreds of cities each day, but competition is increasing from smaller companies affiliated with major carriers. One of the key competitive priorities is on-time arrivals and departures. Red Baron defines *on time* as any arrival or departure that takes place within 15 minutes of the scheduled time. To stay on top of the market, management set the high standard of 98 percent on-time performance. The operations department was put in charge of monitoring the performance of the airline. Each week, a random sample of 300 flight arrivals and departures was checked for schedule performance. Table 5.6 contains the numbers of arrivals and departures over the last 30 weeks that did not meet Red Baron's definition of on-time service. What can you tell management about the quality of service? Can you identify any nonrandom behavior in the process? If so, what might cause the behavior?

30. Beaver Brothers, Inc., is conducting a study to assess the capability of its 150-gram bar soap production line. A critical quality measure is the weight of the soap bars after stamping. The lower and upper specification limits

are 162 and 170 grams, respectively. As a part of an initial capability study, 25 samples of size 5 were collected by the quality assurance group and the observations in Table 5.7 were recorded.

After analyzing the data by using statistical control charts, the quality assurance group calculated the process capability ratio, C_p, and the process capability index, C_{pk}. It then decided to improve the stamping process, especially the feeder mechanism. After making all the changes that were deemed necessary, 18 additional samples were collected. The summary data for these samples are

$$\bar{\bar{x}} = 163 \text{ grams}$$

$$\bar{R} = 2.326 \text{ grams}$$

$$\sigma = 1 \text{ gram}$$

All sample observations were within the control chart limits. With the new data, the quality assurance group recalculated the process capability measures. It was pleased with the improved C_p but felt that the process should be centered at 166 grams to ensure that everything was in order. Its decision concluded the study.

a. Draw the control charts for the data obtained in the initial study and verify that the process was in statistical control.

b. What were the values obtained by the group for C_p and C_{pk} for the initial capability study? Comment on your findings and explain why further improvements were necessary.

c. What are the C_p and C_{pk} after the improvements? Comment on your findings, indicating why the group decided to change the centering of the process.

d. What are the C_p and C_{pk} if the process were centered at 166? Comment on your findings.

TABLE 5.6 | SAMPLE DATA FOR RED BARON AIRLINES

Samples	Number of Late Planes in Sample of 300 Arrivals and Departures									
1–10	3	8	5	11	7	2	12	9	1	8
11–20	3	5	7	9	12	5	4	9	13	4
21–30	12	10	6	2	1	8	4	5	8	2

TABLE 5.7 | SAMPLE DATA FOR BEAVER BROTHERS, INC.

Sample	OBS.1	OBS.2	OBS.3	OBS.4	OBS.5
1	167.0	159.6	161.6	164.0	165.3
2	156.2	159.5	161.7	164.0	165.3
3	167.0	162.9	162.9	164.0	165.4
4	167.0	159.6	163.7	164.1	165.4
5	156.3	160.0	162.9	164.1	165.5
6	164.0	164.2	163.0	164.2	163.9
7	161.3	163.0	164.2	157.0	160.6
8	163.1	164.2	156.9	160.1	163.1
9	164.3	157.0	161.2	163.2	164.4
10	156.9	161.0	163.2	164.3	157.3
11	161.0	163.3	164.4	157.6	160.6
12	163.3	164.5	158.4	160.1	163.3
13	158.2	161.3	163.5	164.6	158.7
14	161.5	163.5	164.7	158.6	162.5
15	163.6	164.8	158.0	162.4	163.6
16	164.5	158.5	160.3	163.4	164.6
17	164.9	157.9	162.3	163.7	165.1
18	155.0	162.2	163.7	164.8	159.6
19	162.1	163.9	165.1	159.3	162.0
20	165.2	159.1	161.6	163.9	165.2
21	164.9	165.1	159.9	162.0	163.7
22	167.6	165.6	165.6	156.7	165.7
23	167.7	165.8	165.9	156.9	165.9
24	166.0	166.0	165.6	165.6	165.5
25	163.7	163.7	165.6	165.6	166.2

Active Model Exercise

This Active Model appears in MyOMLab. It allows you to see the effects of sample size and z-values on control charts.

QUESTIONS

1. Has the booking process been in statistical control?

2. Suppose we use a 95 percent p-chart. How do the upper and lower control limits change? What are your conclusions about the booking process?

3. Suppose that the sample size is reduced to 2,000 instead of 2,500. How does this affect the chart?

4. What happens to the chart as we reduce the z-value?

5. What happens to the chart as we reduce the confidence level?

p-Chart

Reset Data   Questions

Number of samples	12
Sample size	2500
z value	3.0000
Confidence	99.73%

Total sample size	30000	Upper Control Limit	0.0091
Total defects	147	Center Line	0.0049
Percentage defects	0.0049	Lower Control Limit	0.0007
Std dev of p-bar	0.0014		

	# Defects	Fraction Defective
Sample 1	15	0.0060
Sample 2	12	0.0048
Sample 3	19	0.0076
Sample 4	2	0.0008
Sample 5	19	0.0076
Sample 6	4	0.0016
Sample 7	24	0.0096 Above UCL
Sample 8	7	0.0028
Sample 9	10	0.0040
Sample 10	17	0.0068
Sample 11	15	0.0060
Sample 12	3	0.0012

p-Chart Using Data from Example 5.3

VIDEO CASE   Process Performance and Quality at Starwood Hotels & Resorts

Starwood Hotels & Resorts is no stranger to quality measurement. In the most recent year, Starwood properties around the globe held 51 of approximately 700 spots on Condé Nast's Gold List of the world's best places to stay. Its spa and golf programs have consistently been ranked among the best in the world.

At Starwood, processes and programs are driven by the work of its team of Six Sigma experts, called Black Belts. Developed by Motorola more than 20 years ago, Six Sigma is a comprehensive and flexible system for achieving, sustaining, and maximizing business success by driving out defects and variability in a process. Starwood uses the five-step DMAIC process: (1) define, (2) measure, (3) analyze, (4) improve, and (5) control.

Clearly, understanding customer needs is paramount. To this end, Starwood collects data from customers on its Guest Satisfaction Index survey, called the "Voice of the Customer." The survey covers every department guests may have encountered during their stay, from the front desk and hotel room, to restaurants and concierge. Past surveys indicated that how well

problems were resolved during the guest stay was a key driver in high guest satisfaction scores. To increase its scores for problem resolution, the Sheraton brand of Starwood launched the Sheraton Service Promise program in the United States and Canada. The program was designed to give guests a single point of contact for reporting any problems. It was intended to focus associate (employee) attention on taking care of service issues during the guest's stay within 15 minutes of first receiving notice.

However, although scores did increase, they did not increase by enough. Consequently, Sheraton brought in its Six Sigma team to see what it could do. The team employed the basic Six Sigma model of define-measure-analyze-improve-control to guide its work. To define the problem, the Six Sigma team worked with data collected and analyzed by an independent survey organization, National Family Opinion. The study indicated that three key factors are needed in problem resolution: (1) speed, (2) empathy, and (3) efficiency. All three must be met in order for the guests to be satisfied and the Sheraton Service Promise fulfilled. Then, the team looked at the

Sharp/Getty Images News/Getty Images

6

CAPACITY PLANNING

Lines of LCD TVs at the 150 billion yen plant of Sharp electronics company in Kameyama, Mie Prefecture, Central Japan. These top notch LCD TV sets, equipped with 6.22 mega-dot LCD panel on its very thin 8.65 cm in thickness body, are designed to meet the surging demand for bigger but slimmer TVs.

① what is Sharp corporation?

Sharp Corporation

创造商
经销商

Sharp Corporation, founded in 1912 in Japan, is a global manufacturer and distributor of consumer and information products such as LCD TVs and projectors, DVD recorders, mobile communication handsets, point-of-sale systems, home appliances such as refrigerators and microwave ovens, and electronic components such as flash memories, LCD panels, and optical sensors, among others. It has manufacturing, sales, and R&D presence in 25 different countries and regions of the world. For production of its large size LCDs needed in television sets, Sharp increased the capacity of its Kameyama No. 2 plant in Japan from 15,000 LCD sheets per month in August 2006 to 30,000 LCD sheets per month during the second phase in January 2007. Capacity enhancements in the third phase (July 2007) were 60,000 LCD sheets per month. The world's largest LCD TV screen at that time, measuring 108 inch, was made at the Kameyama plant No. 2 from eight generation glass substrates.

Is such a rapid expansion of long-term capacity a good idea? Opinions vary. Even though Sharp had thrived in recent years due to its LCD business, the decision to expand was made because production capacity had reached its limit and Sharp was unable to meet its market demand. As a result, Sharp had been overtaken in global LCD television sales, and its share in the second quarter of 2006 was only 10.8 percent, fourth behind Sony, Samsung, and Philips Electronics. However, the competition did not stay idle even as Sharp quickly expanded its capacity for LCD sheets. A joint company of Sony and Samsung planned for opening a new plant in autumn of 2007 that would have the same capacity as

Sharp's Kameyama plant. Critics fear that if all three major players continue to add capacity faster than the rate at which the market can grow, it will lead to a mature industry with too much capacity and too little profitability.

Meanwhile, Sharp is betting that costs for flat panel LCDs will decline rapidly due to mass production, a strategy that it is also using to increase the capacity of thin-film type solar cells at the Katsuragi plant that can support electricity generation output from 15 MW per year to 160 MW per year. Sharp is also hoping that large capacity investments in LCD sheets will help enhance its leadership position in the LCD industry well into the future, even though it may come at the expense of lowered profitability for the entire industry.

Source: William Trent, "LCD Producers Continue to Add Too Much Capacity As They Battle for Market Share," http://seekingalpha.com/article/19038-lcd-producers-continue-to-add-too-much-capacity-as-they-battle-for-market-share, October 2006; http://www.sharp-world.com/corporate/info/ci/g_organization/index.html, and http://www.sharp-world.com/corporate/ir/event/policy_meeting/pdf/shar080108e_1.pdf, May 31, 2011.

LEARNING GOALS *After reading this chapter, you should be able to:*

1. Define capacity and utilization.
2. Describe economies and diseconomies of scale.
3. Identify different capacity timing and sizing strategies.
4. Identify a systematic approach to capacity planning.
5. Describe how waiting-line models, simulation, and decision trees assist capacity decisions.

Creating Value through Operations Management

Using Operations to Compete
Project Management

Managing Processes

Process Strategy
Process Analysis
Quality and Performance
Capacity Planning
Constraint Management
Lean Systems

Managing Supply Chains

Supply Chain Inventory Management
Supply Chain Design
Supply Chain Location Decisions
Supply Chain Integration
Supply Chain Sustainability and Humanitarian Logistics
Forecasting
Operations Planning and Scheduling
Resource Planning

Capacity is the maximum rate of output of a process or a system. Managers are responsible for ensuring that the firm has the capacity to meet current and future demand. Otherwise, the organization will miss out on opportunities for growth and profits. Making adjustments to decrease capacity, or to overcome capacity shortfalls as Sharp Corporation did when trailing its nearest competitors for LCD TVs, is therefore an important part of the job. Acquisition of new capacity requires extensive planning, and often involves significant expenditure of resources and time. Bringing new capacity online can take several years, for instance in the semiconductor industry or in the construction of new nuclear power plants.

Capacity decisions related to a process need to be made in light of the role the process plays within the organization and the supply chain as a whole, because changing the capacity of a process will have an impact on other processes in the chain. Increasing or decreasing capacity by itself is not as important as ensuring that the entire supply chain, from order entry to delivery, is designed for effectiveness. Capacity decisions must be made in light of several long-term issues such as the firm's economies and diseconomies of scale, capacity cushions, timing and sizing strategies, and trade-offs between customer service and capacity utilization.

This chapter focuses on how managers can best revise capacity levels and best determine when to add or reduce capacity for the long term. The type of capacity decisions differ for different time horizons. Both long-term as well as short-term issues associated with planning capacity and managing constraints are important, and must be understood in conjunction with one another. While we deal here with the long-term decisions shown in the capacity management framework, short-term decisions centered on making the most of existing capacity by managing constraints are more fully explored in Chapter 7, "Constraint Management."

Capacity management

Capacity planning (long-term)
- Economies and diseconomies of scale
- Capacity timing and sizing strategies
- Systematic approach to capacity decisions

Constraint management (short-term)
- Theory of constraints
- Identification and management of bottlenecks
- Product mix decisions using bottlenecks
- Managing constraints in a line process

Planning Capacity across the Organization

Capacity decisions have implications for different functional areas throughout the organization. Accounting needs to provide the cost information needed to evaluate capacity expansion decisions. Finance performs the financial analysis of proposed capacity expansion investments and raises funds to support them. Marketing provides demand forecasts needed to identify capacity gaps. Management information systems designs the electronic infrastructure that is needed to make data such as cost information, financial performance measures, demand forecasts, and work standards available to those needing it to analyze capacity options. Operations is involved in the selection of capacity strategies that can be implemented to effectively meet future demand. Purchasing facilitates acquisition of outside capacity from suppliers. Finally, human resources focuses on hiring and training employees needed to support internal capacity plans. So all departments in a firm get involved with and are affected by long-term capacity planning decisions.

capacity

The maximum rate of output of a process or a system.

Planning Long-Term Capacity

Long-term capacity plans deal with investments in new facilities and equipment at the organizational level, and require top management participation and approval because they are not easily reversed. These plans cover at least two years into the future, but construction lead times can sometimes be longer and result in longer planning time horizons.

As already seen in our opening vignette, long-term capacity planning is central to the success of an organization. Too much capacity can be as agonizing as too little. Often entire industries can fluctuate over time between too much and too little capacity, as evidenced in the airline and cruise ship industry over the past 20 years. When choosing a capacity strategy, managers must consider questions such as the following: How much of a cushion is needed to handle variable, or uncertain, demand? Should we expand capacity ahead of demand, or wait until demand is more certain? Even before these questions can be answered, a manager needs to be able to measure a process's capacity. So a systematic approach is needed to answer these and similar questions and to develop a capacity strategy appropriate for each situation.

Measures of Capacity and Utilization

No single capacity measure is best for all situations. A retailer measures capacity as annual sales dollars generated per square foot, whereas an airline measures capacity as available seat-miles (ASMs) per month. A theater measures capacity as number of seats, while a job shop measures capacity as number of machine hours. In general, capacity can be expressed in one of two ways: in terms of output measures or input measures.

Output Measures of Capacity   Output measures of capacity are best utilized when applied to individual processes within the firm, or when the firm provides a relatively small number of standardized services and products. High-volume processes, such as those in a car manufacturing plant, are a good example. In this case, capacity would be measured in terms of the number of cars produced per day. However, many processes produce more than one service or product. As the amount of customization and variety in the product mix increases, output-based capacity measures become less useful. Then input measures of capacity become the usual choice for measuring capacity.

Sonia A. Tellez

For the first time since 1992, Intel built a new 2.5 billion chip fabrication plant in the high-tech Jinzhou New District north of the port city of Dalian. With more than 1700 employees, it covers an area of 163,000 square meters, which is roughly the same as 23 soccer fields. Started in September 2007, the plant started production in October 2010 of 300-millimeter integrated wafers and chip sets for laptop computers, high-performance desktop PCs and powerful servers.

Input Measures of Capacity   Input measures are generally used for low-volume, flexible processes, such as those associated with a custom furniture maker. In this case, the furniture maker might measure capacity in terms of inputs such as number of workstations or number of workers. The problem with input measures is that demand is invariably expressed as an output rate. If the furniture maker wants to keep up with demand, he or she must convert the business's annual demand for furniture into labor hours and number of employees required to fulfill those hours. We will explain precisely how this input–output conversion is done later in the chapter.

utilization

The degree to which equipment, space, or the workforce is currently being used, and is measured as the ratio of average output rate to maximum capacity (expressed as a percent).

Utilization   Utilization is the degree to which a resource such as equipment, space, or the workforce is currently being used, and is measured as the ratio of average output rate to maximum capacity (expressed as a percent). The average output rate and the capacity must be measured in the same terms—that is, time, customers, units, or dollars. The utilization rate indicates the need for adding extra capacity or eliminating unneeded capacity.

$$\text{Utilization} = \frac{\text{Average output rate}}{\text{Maximum capacity}} \times 100\%$$

Here, we refer to maximum capacity as the greatest level of output that a process can reasonably sustain for a longer period, using realistic employee work schedules and the equipment currently in place. In some processes, this capacity level implies a one-shift operation; in others, it implies a three-shift operation. A process can be operated above its capacity level using marginal methods of production, such as overtime, extra shifts, temporarily reduced maintenance activities, overstaffing, and subcontracting. Although they help with temporary peaks, these options cannot be sustained for long. For instance, being able to handle 40 customers for a one-week peak is quite different from sustaining it for six months. Employees do not want to work excessive overtime for extended periods, so quality drops. In addition, the costs associated with overtime drive up the firm's costs. So operating processes close to (or even temporarily above) their maximum capacity can result in low customer satisfaction, minimal profits, and even losing money despite high sales levels. Such was the case with U.S. aircraft manufacturers in the late 1980s, which culminated in Boeing acquiring McDonnell Douglas in 1997 in order to shore up skyrocketing costs and plummeting profits.

Economies of Scale

economies of scale

A concept that states that the average unit cost of a service or good can be reduced by increasing its output rate.

Deciding on the best level of capacity involves consideration for the efficiency of the operations. A concept known as economies of scale states that the average unit cost of a service or good can be reduced by increasing its output rate. Four principal reasons explain why economies of scale can drive costs down when output increases: (1) Fixed costs are spread over more units; (2) construction costs are reduced; (3) costs of purchased materials are cut; and (4) process advantages are found.

Spreading Fixed Costs   In the short term, certain costs do not vary with changes in the output rate. These fixed costs include heating costs, debt service, and managers' salaries. The depreciation of plant and equipment already owned is also a fixed cost in the accounting sense. When the average output rate—and, therefore, the facility's utilization rate—increases, the average unit cost drops because fixed costs are spread over more units.

Reducing Construction Costs   Certain activities and expenses are required to build small and large facilities alike: building permits, architects' fees, and rental of building equipment. Doubling the size of the facility usually does not double construction costs.

Cutting Costs of Purchased Materials   Higher volumes can reduce the costs of purchased materials and services. They give the purchaser a better bargaining position and the opportunity to take advantage of quantity discounts. Retailers such as Walmart reap significant economies of scale because their national and international stores buy and sell huge volumes of each item.

Finding Process Advantages   High-volume production provides many opportunities for cost reduction. At a higher output rate, the process shifts toward a line process, with resources dedicated to individual products. Firms may be able to justify the expense of more efficient technology or more specialized equipment. The benefits from dedicating resources to individual services or products may include speeding up the learning effect, lowering inventory, improving process and job designs, and reducing the number of changeovers.

Diseconomies of Scale

diseconomies of scale

Occurs when the average cost per unit increases as the facility's size increases.

Bigger is not always better, however. At some point, a facility can become so large that diseconomies of scale set in; that is, the average cost per unit increases as the facility's size increases. The reason is that excessive size can bring complexity, loss of focus, and inefficiencies that raise the average unit cost of a service or product. Too many layers of employees and bureaucracy can cause management to lose touch with employees and customers. A less agile organization loses the flexibility needed to respond to changing demand. Many large companies become so involved in analysis and planning that they innovate less and avoid risks. The result is that small companies outperform corporate giants in numerous industries.

Figure 6.1 illustrates the transition from economies of scale to diseconomies of scale. The 500-bed hospital shows economies of scale because the average unit cost at its *best operating level*, represented by the blue dot, is less than that of the 250-bed hospital. However, assuming that sufficient demand exists, further expansion to a 750-bed hospital leads to higher average unit costs and diseconomies of scale. One reason the 500-bed hospital enjoys greater economies of scale than the 250-bed hospital is that the cost of building and equipping it is less than twice the cost for the smaller hospital. The 750-bed facility would enjoy similar savings. Its higher average unit costs can be explained only by diseconomies of scale, which outweigh the savings realized in construction costs.

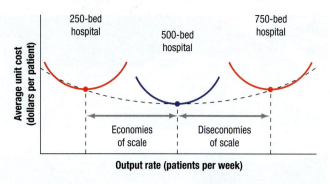

▲ **FIGURE 6.1**
Economies and Diseconomies of Scale

Figure 6.1 does not mean that the optimal size for all hospitals is 500 beds. Optimal size depends on the number of patients per week to be served. On the one hand, a hospital serving a small community could have lower costs by choosing a 250-bed capacity rather than the 500-bed capacity. On the other hand, a large community might be served more efficiently by two 500-bed hospitals than by one 1,000-bed facility if diseconomies of scale exist at the bigger size.

Capacity Timing and Sizing Strategies

Operations managers must examine three dimensions of capacity strategy before making capacity decisions: (1) sizing capacity cushions, (2) timing and sizing expansion, and (3) linking process capacity and other operating decisions.

Sizing Capacity Cushions

Average utilization rates for any resource should not get too close to 100 percent over the long term, though it may occur for some processes from time to time in the short run. If the demand keeps increasing over time, then long-term capacity must be increased as well to provide some buffer against uncertainties. When average utilization rates approach 100 percent, it is usually a signal to increase capacity or decrease order acceptance to avoid declining productivity. The **capacity cushion** is the amount of reserve capacity a process uses to handle sudden increases in demand or temporary losses of production capacity; it measures the amount by which the average utilization (in terms of total capacity) falls below 100 percent. Specifically,

$$\text{Capacity cushion} = 100\ (\%) - \text{Average Utilization rate}\ (\%)$$

The appropriate size of the cushion varies by industry. In the capital-intensive paper industry, where machines can cost hundreds of millions of dollars each, cushions well under 10 percent are preferred. The less capital-intensive hotel industry breaks even with a 60 to 70 percent utilization (40 to 30 percent cushion), and begins to suffer customer-service problems when the cushion drops to 20 percent. The more capital-intensive cruise ship industry prefers cushions as small as 5 percent. Large cushions are particularly vital for front-office processes where customers expect fast service times.

Businesses find large cushions appropriate when demand varies. In certain service industries (the grocery industry, for example), demand on some days of the week is predictably higher than on other days, and even hour-to-hour changes are typical. Long customer waiting times are not acceptable because customers grow impatient if they have to wait in a supermarket checkout line for more than a few minutes. Prompt customer service requires supermarkets to maintain a capacity cushion large enough to handle peak demand. Large cushions also are necessary when future demand is uncertain, particularly if resource flexibility is low. Simulation and waiting-line analysis (see Supplement B, "Waiting Lines") can help managers better anticipate the relationship between capacity cushion and customer service.

Another type of demand uncertainty occurs with a changing product mix. Though total demand measured in monetary terms might remain stable, the load can shift unpredictably from one workstation to another as the product mix changes. Supply uncertainty tied to delivery of purchased materials also makes large capacity cushions helpful. Capacity often comes in large increments because a complete machine has to be purchased even if only a fraction of its available capacity is needed, which in turn creates a large cushion. Firms also need to build in excess capacity to allow for employee absenteeism, vacations, holidays, and any other delays. If a firm is experiencing high overtime costs and frequently needs to rely on subcontractors, it perhaps needs to increase its capacity cushions.

capacity cushion

The amount of reserve capacity a process uses to handle sudden increases in demand or temporary losses of production capacity; it measures the amount by which the average utilization (in terms of total capacity) falls below 100 percent.

Jeff Kowalsky/Bloomberg/Getty Images

Dawn Bacon works on a General Motors Co. 2012 Opel Ampera, GM's European version of the Volt, at Detroit-Hamtramck Assembly Plant in Detroit, Michigan, U.S. GM will invest $69 million and add 2,500 jobs to start making two new models at the Detroit plant that builds the Chevrolet Volt plug-in hybrid as the automaker boosts U.S. production.

The argument in favor of small cushions is simple: Unused capacity costs money. For capital-intensive firms, minimizing the capacity cushion is vital. Studies indicate that businesses with high capital intensity achieve a low return on investment when the capacity cushion is high. This strong correlation does not exist for labor-intensive firms, however. Their return on investment is about the same because the lower investment in equipment makes high utilization less critical. Small cushions have other advantages. By implementing a small cushion, a company can sometimes uncover inefficiencies that were difficult to detect when cushions were larger. These inefficiencies might include employee absenteeism or unreliable suppliers. Once managers and workers identify such problems, they often can find ways to correct them.

Timing and Sizing Expansion

The second issue of capacity strategy concerns when to adjust capacity levels and by how much. At times, capacity expansion can be done in response to changing market trends. General Motors decided to increase production capacity of the four-seat series hybrid car Chevrolet Volt from 30,000 units to 45,000 units in 2012 because of strong public interest. While we deal with this issue from the perspective of capacity expansion in greater detail here, it must be noted that firms may not always be looking to expand capacity, but at times may be forced to retrench as evidenced by the situation in the airlines industry, where all major airlines have consolidated routes and reduced the total number of flights in the face of increasing oil costs. Some of this consolidation has been achieved through mergers like United Airlines and Continental to create the world's largest airline company, as well as between Delta and Northwest Airlines.

Figure 6.2 illustrates two extreme strategies for expanding capacity: the *expansionist strategy*, which involves large, infrequent jumps in capacity, and the *wait-and-see strategy*, which involves smaller, more frequent jumps.

FIGURE 6.2 ▶
Two Capacity Strategies

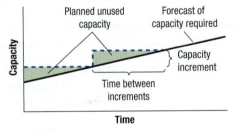

(a) Expansionist strategy

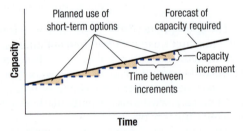

(b) Wait-and-see strategy

The timing and sizing of expansion are related; that is, if demand is increasing and the time between increments increases, the size of the increments must also increase. The expansionist strategy, which stays ahead of demand, minimizes the chance of sales lost to insufficient capacity. The wait-and-see strategy lags behind demand. To meet any shortfalls, it relies on short-term options, such as use of overtime, temporary workers, subcontractors, stockouts, and the postponement of preventive maintenance on equipment.

Several factors favor the expansionist strategy. Expansion can result in economies of scale and a faster rate of learning, thus helping a firm reduce its costs and compete on price. This strategy might increase the firm's market share or act as a form of preemptive marketing. By making a large capacity expansion or announcing that one is imminent, the firm can preempt the expansion of other firms. These other firms must sacrifice some of their market share or risk burdening the industry with overcapacity. To be successful, however, the preempting firm must have the credibility to convince the competition that it will carry out its plans—and must signal its plans before the competition can act. Managerial Practice 6.1 illustrates the use of expansionist strategy in the ethanol industry.

MANAGERIAL PRACTICE 6.1 — Expansionist Capacity Strategy in the Ethanol Industry

A search for renewable energy and alternative sources of fuel has made ethanol production a big growth industry in the United States. The state of Nebraska alone has capital investments of more than $5 billion in 24 ethanol production plants, which together produce more than 2 billion gallons of ethanol each year. While Nebraska serves the ethanol markets in western United States, midwestern states like Iowa have the largest invested capacity. Many of these plants have gone into operation since 2007.

But troubling signs may be on the horizon. Companies and farm cooperatives have built so many distilleries so quickly that the ethanol market is suddenly plagued by a glut and oversupply is putting downward pressure on price. At the same time, scarcity and price of raw materials needed to feed ethanol plants is becoming an issue. One bushel of corn produces approximately 2.6 gallons of ethanol, and the ethanol plants in the state of Nebraska alone require over 769 million bushels of grain in the process every year. As a result, the price of corn has risen steadily from under $2 per bushel in 2006 to $7.76 per bushel in April 2011. The price per bushel doubled in one year alone between 2010 and 2011, though it is expected to come down in the future as more corn is planted and the corn reserves rise due to lesser exports. These price increases have further eroded profitability. Despite the steep increase in prices of crude oil, demand for ethanol has not increased proportionately because it is less efficient and constitutes 10 percent or less in most gasoline blends.

Capacity was also expanded without taking into consideration the transportation needs of ethanol. Unlike gasoline, traditional pipelines cannot be used due to the fact that ethanol is corrosive and contains traces of water. So it must be transported by specialized ethanol rail cars, trucks, and barges. Capacity of such a transportation network has not kept up with the surge in surplus ethanol production. Consequently, demand for ethanol along coastal regions of the country cannot be met quickly and efficiently from the heartland locations of Corn Belt ethanol plants.

While the government subsidies may keep the output of ethanol growing, experts believe that the expansion of the industry has been poorly managed. Some companies are modifying future plans and canceling new plant construction in fear of consolidation and going out of business.

Andy Kropa/Redux

The Lincolnway Energy plant, a midsize distillery in Nevada, Iowa, converts corn into ethanol in a largely automated process. Lincolnway Energy was once virtually alone in the region. Today, though, competing distilleries are operating and pouring even more ethanol onto the market.

Source: Clifford Krauss, "Ethanol's Boom Stalling as Glut Depresses Price," *New York Times* (September 30, 2007); Christopher Leonard, "Corn Reserves Expected to Rise, Easing Food Prices," *The Associated Press,* May 12, 2011; Nebraska Ethanol Board, "Ethanol Plants in Nebraska," **http://www.ne-ethanol.org/industry/ethplants.htm** (May 31, 2011).

The conservative wait-and-see strategy is to expand in smaller increments, such as by renovating existing facilities rather than building new ones. Because the wait-and-see strategy follows demand, it reduces the risks of overexpansion based on overly optimistic demand forecasts, obsolete technology, or inaccurate assumptions regarding the competition.

However, this strategy has its own risks, such as being preempted by a competitor or being unable to respond if demand is unexpectedly high. Critics claim the wait-and-see strategy is a short-term strategy typical of some U.S. management styles. Managers on the fast track to corporate advancement tend to take fewer risks. They earn promotions by avoiding the big mistakes and maximizing short-term profits and return on investment. The wait-and-see strategy fits this short-term outlook but can erode market share over the long run.

Management may choose one of these two strategies or one of the many between these extremes. With strategies in the more moderate middle, firms can expand more frequently (on a smaller scale) than they can with the expansionist strategy without lagging behind demand as with the wait-and-see strategy. An intermediate strategy could be to *follow the leader,* expanding when others do. If others are right, so are you, and nobody gains a competitive advantage. If others make a mistake and over expand, so do you, but everyone shares in the agony of overcapacity. Such a situation was noted for the airlines industry, and may yet occur in the LCD industry due to large capacity expansions by Sharp Corporation, Sony, and Samsung as described in the opening vignette.

Linking Capacity and Other Decisions

Capacity decisions should be closely linked to processes and supply chains throughout the organization. When managers make decisions about designing processes, determining degree of resource flexibility and inventory, and locating facilities, they must consider its impact on capacity cushions. Capacity cushions in the long run buffer the organization against uncertainty, as do resource flexibility, inventory, and longer customer lead times. If a change is made in any one decision area, the capacity cushion may also need to be changed to compensate. For example, capacity cushions for a process can be lowered if less emphasis is placed on fast deliveries (*competitive priorities*), yield losses (*quality*) drop, or if investment in capital-intensive equipment increases or worker flexibility increases (*process design*). Capacity cushions can also be lowered if the company is willing to smooth the output rate by raising prices when inventory is low and decreasing prices when it is high.

A Systematic Approach to Long-Term Capacity Decisions

Long-term decisions for capacity would typically include whether to add a new plant or warehouse or to reduce the number of existing ones, how many workstations a given department should have, or how many workers are needed to staff a given process. Some of these decisions can take years to become operational. Hence, a systematic approach is needed to plan for long-term capacity decisions.

Although each situation is somewhat different, a four-step procedure generally can help managers make sound capacity decisions. (In describing this procedure, we assume that management already performed the preliminary steps of determining the process's existing capacity and assessing whether its current capacity cushion is appropriate.)

1. Estimate future capacity requirements.
2. Identify gaps by comparing requirements with available capacity.
3. Develop alternative plans for reducing the gaps.
4. Evaluate each alternative, both qualitatively and quantitatively, and make a final choice.

Step 1: Estimate Capacity Requirements

capacity requirement

What a process's capacity should be for some future time period to meet the demand of customers (external or internal), given the firm's desired capacity cushion.

planning horizon

The set of consecutive time periods considered for planning purposes.

A process's **capacity requirement** is what its capacity should be for some future time period to meet the demand of the firm's customers (external or internal), given the firm's desired capacity cushion. Larger requirements are practical for processes or workstations that could potentially be bottlenecks in the future, and management may even plan for larger cushions than normal.

Capacity requirements can be expressed in one of two ways: with an output measure or with an input measure. Either way, the foundation for the estimate is forecasts of demand, productivity, competition, and technological change. These forecasts normally need to be made for several time periods in a **planning horizon**, which is the set of consecutive time periods considered for planning purposes. Long-term capacity plans need to consider more of the future (perhaps, a whole decade) than do short-term plans. Unfortunately, the further ahead you look, the more chance you have of making an inaccurate forecast.

Using Output Measures   The simplest way to express capacity requirements is as an output rate. As discussed earlier, output measures are appropriate for high-volume processes with little product variety or process divergence. Here, demand forecasts for future years are used as a basis for extrapolating capacity requirements into the future. If demand is expected to double in the next five years, then the capacity requirements also double. For example, if a process's current demand is 50 customers per day, then the demand in five years would be 100 customers per day. If the desired capacity cushion is 20 percent, management should plan for enough capacity to serve $[100/(1 - 0.2)] = 125$ customers in five years.

Using Input Measures   Output measures may be insufficient in the following situations:

- Product variety and process divergence is high.
- The product or service mix is changing.
- Productivity rates are expected to change.
- Significant learning effects are expected.

In such cases, it is more appropriate to calculate capacity requirements using an input measure, such as the number of employees, machines, computers, or trucks. Using an input measure for the capacity requirement brings together demand forecasts, process time estimates, and the desired capacity cushion. When just one service or product is processed at an operation and the time period is a particular year, the capacity requirement, M, is

$$\text{Capacity requirement} = \frac{\text{Processing hours required for year's demand}}{\substack{\text{Hours available from a single capacity unit (such as an employee} \\ \text{or machine) per year, after deducting desired cushion}}}$$

$$M = \frac{Dp}{N[1 - (C/100)]}$$

where

D = demand forecast for the year (number of customers served or units produced)

p = processing time (in hours per customer served or unit produced)

N = total number of hours per year during which the process operates

C = desired capacity cushion (expressed as a percent)

M is the number of input units required and should be calculated for each year in the time horizon. The processing time, p, depends on the process and methods selected to do the work. The denominator is the total number of hours, N, available for the year from one unit of capacity (an employee or machine), multiplied by a proportion that accounts for the desired capacity cushion, C. The proportion is simply $1.0 - C/100$, where C is converted from a percent to a proportion by dividing by 100. For example, a 20 percent capacity cushion means that $1.0 - C/100 = 0.80$.

Setups may be involved if multiple products are being manufactured. **Setup time is** the time required to change a process or an operation from making one service or product to making another. The total setup time is found by dividing the number of units forecast per year, D, by the number of units made in each lot, Q, (number of units processed between setups), which gives the number of setups per year, and then multiplying by the time per setup, s. For example, if the annual demand is 1,200 units and the average lot size is 100, there are $1,200/100 = 12$ setups per year. Accounting for both processing and setup times for multiple services (products), we get

setup time

The time required to change a process or an operation from making one service or product to making another.

$$\text{Capacity requirement} = \frac{\substack{\text{Processing and setup hours required for} \\ \text{year's demand, summed over all services or products}}}{\substack{\text{Hours available from a single capacity unit per year,} \\ \text{after deducting desired cushion}}}$$

$$M = \frac{[Dp + (D/Q)s]_{\text{product 1}} + [Dp + (D/Q)s]_{\text{product 2}} + \cdots + [Dp + (D/Q)s]_{\text{product } n}}{N[1-(C/100)]}$$

where

Q = number of units in each lot

s = setup time (in hours) per lot

What to do when M is not an integer depends on the situation. For example, it is impossible to buy a fractional machine. In this case, round up the fractional part, unless it is cost efficient to use short-term options, such as overtime or stockouts, to cover any shortfalls. If, instead, the capacity unit is the number of employees at a process, a value of 23.6 may be achieved using just 23 employees and a modest use of overtime (equivalent to having 60 percent of another full-time person). Here, the fractional value should be retained as useful information.

EXAMPLE 6.1	**Estimating Capacity Requirements When Using Input Measures**

A copy center in an office building prepares bound reports for two clients. The center makes multiple copies (the lot size) of each report. The processing time to run, collate, and bind each copy depends on, among other factors, the number of pages. The center operates 250 days per year, with one 8-hour shift. Management believes that a capacity cushion of 15 percent (beyond the allowance built into time standards) is best. It currently has three copy machines. Based on the following table of information, determine how many machines are needed at the copy center.

Item	Client X	Client Y
Annual demand forecast (copies)	2,000	6,000
Standard processing time (hour/copy)	0.5	0.7
Average lot size (copies per report)	20	30
Standard setup time (hours)	0.25	0.40

SOLUTION

$$M = \frac{[Dp + (D/Q)s]_{\text{product 1}} + [Dp + (D/Q)s]_{\text{product 2}} + \cdots + [Dp + (D/Q)s]_{\text{product } n}}{N\,[1-(C/100)]}$$

$$= \frac{[2,000(0.5) + (2,000/20)(0.25)]_{\text{client X}} + [6,000(0.7) + (6,000/30)(0.40)]_{\text{client Y}}}{[(250 \text{ day/year})(1 \text{ shift/day})(8 \text{ hours/day})][1.0 - (15/100)]}$$

$$= \frac{5,305}{1,700} = 3.12$$

Rounding up to the next integer gives a requirement of four machines.

DECISION POINT

The copy center's capacity is being stretched and no longer has the desired 15 percent capacity cushion with the existing three machines. Not wanting customer service to suffer, management decided to use overtime as a short-term solution to handle past-due orders. If demand continues at the current level or grows, it will acquire a fourth machine.

Step 2: Identify Gaps

capacity gap

Positive or negative difference between projected demand and current capacity.

A **capacity gap** is any difference (positive or negative) between projected capacity requirements (M) and current capacity. Complications arise when multiple operations and several resource inputs are involved. Expanding the capacity of some operations may increase overall capacity. However, as we will learn later in Chapter 7, "Constraint Management," if one operation is more constrained than others, total process capacity can be expanded only if the capacity of the constrained operation is expanded.

Step 3: Develop Alternatives

base case

The act of doing nothing and losing orders from any demand that exceeds current capacity, or incur costs because capacity is too large.

The next step is to develop alternative plans to cope with projected gaps. One alternative, called the **base case**, is to do nothing and simply lose orders from any demand that exceeds current capacity or incur costs because capacity is too large. Other alternatives if expected demand exceeds current capacity are various timing and sizing options for adding new capacity, including the expansionist and wait-and-see strategies illustrated in Figure 6.2. Additional possibilities include expanding at a different location and using short-term options, such as overtime, temporary workers, and subcontracting. Alternatives for reducing capacity include the closing of plants or warehouses, laying off employees, or reducing the days or hours of operation.

Step 4: Evaluate the Alternatives

In this final step, the manager evaluates each alternative, both qualitatively and quantitatively.

Qualitative Concerns   Qualitatively, the manager looks at how each alternative fits the overall capacity strategy and other aspects of the business not covered by the financial analysis. Of particular concern might be uncertainties about demand, competitive reaction, technological change, and cost estimates. Some of these factors cannot be quantified and must be assessed on the basis of judgment and experience. Others can be quantified, and the manager can analyze each alternative by using different assumptions about the future. One set of assumptions could represent a worst case, in which demand is less, competition is greater, and construction costs are higher than expected. Another set of assumptions could represent the most optimistic view of the future. This type of "what-if" analysis allows the manager to get an idea of each alternative's implications before making a final choice.

Qualitative factors would tend to dominate when a business is trying to enter new markets or change the focus of its business strategy. For instance, Dell is planning a $1 billion investment in cloud computing (on-demand provision of computational resources for data and software through computer networks rather than local computers) and virtualization (creating a virtual rather than an actual version of an operating system or a storage device) in the 2011 fiscal year. Little hard data may be available to guide the exact size and timing of the significant expansion of data center capacity that must be undertaken to support this diversification strategy, which will also include a deeper focus on sales training and expertise.

Quantitative Concerns   Quantitatively, the manager estimates the change in cash flows for each alternative over the forecast time horizon compared to the base case. **Cash flow** is the difference between the flows of funds into and out of an organization over a period of time, including revenues, costs, and changes in assets and liabilities. The manager is concerned here only with calculating the cash flows attributable to the project.

Dell's Green Energy-saving and Environment-protecting Data Center will monitor the energy consumption index of 54 government offices in Beijing, China

cash flow

The difference between the flows of funds into and out of an organization over a period of time, including revenues, costs, and changes in assets and liabilities.

EXAMPLE 6.2	**Evaluating the Alternatives**

Grandmother's Chicken Restaurant is experiencing a boom in business. The owner expects to serve 80,000 meals this year. Although the kitchen is operating at 100 percent capacity, the dining room can handle 105,000 diners per year. Forecasted demand for the next 5 years is 90,000 meals for next year, followed by a 10,000-meal increase in each of the succeeding years. One alternative is to expand both the kitchen and the dining room now, bringing their capacities up to 130,000 meals per year. The initial investment would be $200,000, made at the end of this year (year 0). The average meal is priced at $10, and the before-tax profit margin is 20 percent. The 20 percent figure was arrived at by determining that, for each $10 meal, $8 covers variable costs and the remaining $2 goes to pretax profit.

What are the pretax cash flows from this project for the next 5 years compared to those of the base case of doing nothing?

MyOMLab

Tutor 6.2 in MyOMLab provides a new example to practice projecting cash flows for capacity decisions.

SOLUTION

Recall that the base case of doing nothing results in losing all potential sales beyond 80,000 meals. With the new capacity, the cash flow would equal the extra meals served by having a 130,000-meal capacity, multiplied by a profit of $2 per meal. In year 0, the only cash flow is −$200,000 for the initial investment. In year 1, the 90,000-meal demand will be completely satisfied by the expanded capacity, so the incremental cash flow is $(90,000 − 80,000)($2) = $20,000$. For subsequent years, the figures are as follows:

Year 2: Demand = 100,000; Cash flow = (100,000 − 80,000)$2 = $40,000

Year 3: Demand = 110,000; Cash flow = (110,000 − 80,000)$2 = $60,000

Year 4: Demand = 120,000; Cash flow = (120,000 − 80,000)$2 = $80,000

Year 5: Demand = 130,000; Cash flow = (130,000 − 80,000)$2 = $100,000

If the new capacity were smaller than the expected demand in any year, we would subtract the base case capacity from the new capacity (rather than the demand). The owner should account for the time value of money, applying such techniques as the net present value or internal rate of return methods (see MyOMLab Supplement F, "Financial Analysis"). For instance, the net present value (NPV) of this project at a discount rate of 10 percent is calculated here, and equals $13,051.76.

MyOMLab

$NPV = -200{,}000 + [(20{,}000/1.1)] + [40{,}000/(1.1)^2] + [60{,}000/(1.1)^3] + [80{,}000/(1.1)^4] + [100{,}000/(1.1)^5]$

$= -\$200{,}000 + \$18{,}181.82 + \$33{,}057.85 + \$45{,}078.89 + \$54{,}641.07 + \$62{,}092.13$

$= \$13{,}051.76$

DECISION POINT

Before deciding on this capacity alternative, the owner should also examine the qualitative concerns, such as future location of competitors. In addition, the homey atmosphere of the restaurant may be lost with expansion. Furthermore, other alternatives should be considered (see Solved Problem 2).

Tools for Capacity Planning

Capacity planning requires demand forecasts for an extended period of time. Unfortunately, forecast accuracy declines as the forecasting horizon lengthens. In addition, anticipating what competitors will do increases the uncertainty of demand forecasts. Demand during any period of time may not be evenly distributed; peaks and valleys of demand may (and often do) occur within the time period. These realities necessitate the use of capacity cushions. In this section, we introduce three tools that deal more formally with demand uncertainty and variability: (1) waiting-line models, (2) simulation, and (3) decision trees. Waiting-line models and simulation account for the random, independent behavior of many customers, in terms of both their time of arrival and their processing needs. Decision trees allow anticipation of events, such as competitors' actions, which requires a sequence of decisions regarding capacities.

Waiting-Line Models

Waiting-line models often are useful in capacity planning, such as selecting an appropriate capacity cushion for a high customer-contact process. Waiting lines tend to develop in front of a work center, such as an airport ticket counter, a machine center, or a central computer. The reason is that the arrival time between jobs or customers varies, and the processing time may vary from one customer to the next. Waiting-line models use probability distributions to provide estimates of average customer wait time, average length of waiting lines, and utilization of the work center. Managers can use this information to choose the most cost-effective capacity, balancing customer service and the cost of adding capacity.

Supplement B, "Waiting Lines," follows this chapter and provides a fuller treatment of these models. It introduces formulas for estimating important characteristics of a waiting line, such as average customer waiting time and average facility utilization for different facility designs. For example, a facility might be designed to have one or multiple lines at each operation and to route customers through one or multiple operations. Given the estimating capability of these formulas and cost estimates for waiting and idle time, managers can select cost-effective designs and capacity levels that also provide the desired level of customer service.

Figure 6.3 shows output from POM for Windows for waiting lines. A professor meeting students during office hours has an arrival rate of three students per hour and a service rate of six students per hour. The output shows that the capacity cushion is 50 percent (1 − average server utilization of 0.50). This result is expected because the processing rate is double the arrival rate. What might not be expected is that a typical student spends 20 minutes either in line or talking with the professor, and the probability of having two or more students at the office is 0.25. These numbers might be surprisingly high, given such a large capacity cushion.

FIGURE 6.3 ▼
POM for Windows Output for
Waiting Lines during Office Hours

Waiting Lines Results

Parameter	Value	Parameter	Value	Minutes	Seconds
Single-Server Model		Average server utilization	.5		
Arrival rate(lambda)	3	Average number in the line(Lq)	.5		
Service rate(mu)	6	Average number in the system(L)	1		
Number of servers	1	Average time in the line(Wq)	.17	10	600
		Average time in the system(W)	.33	20	1200

Table of Probabilities

k	Prob (num in sys = k)	Prob (num in sys <= k)	Prob (num in sys >k)
0	.5	.5	.5
1	.25	.75	.25
2	.13	.88	.13
3	.06	.94	.06
4	.03	.97	.03
5	.02	.98	.02
6	.01	1	.01
7	.0	1	.0

Simulation

More complex waiting-line problems must be analyzed with simulation. It can identify the process's bottlenecks and appropriate capacity cushions, even for complex processes with random demand patterns and predictable surges in demand during a typical day. The SimQuick simulation package, provided in MyOMLab, allows you to build dynamic models and systems. Other simulation packages can be found with Extend, Simprocess, ProModel, and Witness.

MyOMLab

Decision Trees

A decision tree can be particularly valuable for evaluating different capacity expansion alternatives when demand is uncertain and sequential decisions are involved (see Supplement A, "Decision Making"). For example, the owner of Grandmother's Chicken Restaurant (see Example 6.2) may expand the restaurant now, only to discover in year 4 that demand growth is much higher than forecasted. In that case, she needs to decide whether to expand further. In terms of construction costs and downtime, expanding twice is likely to be much more expensive than building a larger facility from the outset. However, making a large expansion now, when demand growth is low, means poor facility utilization. Much depends on the demand.

Figure 6.4 shows a decision tree for this view of the problem, with new information provided. Demand growth can be either low or high, with probabilities of 0.40 and 0.60, respectively. The initial expansion in year 1 (square node 1) can either be small or large. The second decision node (square node 2), whether to expand at a later date, is reached only if the initial expansion is small and demand turns out to be high. If demand is high and if the initial expansion was small, a decision must be made about a second expansion in year 4. Payoffs for each branch of the tree are estimated. For example, if the initial expansion is large, the financial benefit is either $40,000 or $220,000, depending on whether demand is low or high. Weighting these payoffs by the probabilities yields an expected value of $148,000. This expected payoff is higher than the $109,000 payoff for the small initial expansion, so the better choice is to make a large expansion in year 1.

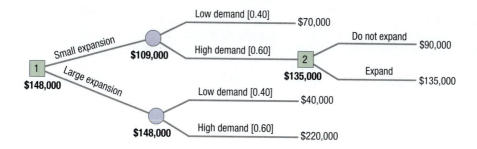

◀ **FIGURE 6.4**
A Decision Tree for Capacity Expansion

LEARNING GOALS IN REVIEW

1. **Define capacity and utilization.** Review the section "Measures of Capacity and Utilization," pp. 203–204, and understand why and how capacity is measured in high volume processes is different from its measurement in low volume flexible processes.

2. **Describe economies and diseconomies of scale.** See the section on "Planning Long-Term Capacity," pp. 203–205. Figure 6.1 illustrates the relationship between average unit cost and output rate, and shows different output ranges over which economies and diseconomies of scale can occur.

3. **Identify different capacity timing and sizing strategies.** The section "Capacity Timing and Sizing Strategies," pp. 205–208, and Figure 6.2 differentiates between expansionist and

wait-and-see strategies. Understand the notion of capacity cushions, and how they link to other decisions in the firm.

4. **Identify a systematic approach to capacity planning.** The section "A Systematic Approach to Long-Term Capacity Decisions," pp. 208–211, shows you how capacity requirements can be estimated for both input based as well as output based measures. Focus on how different alternatives can be developed to fill the capacity gaps between requirements and current capacity.

5. **Describe how waiting-line models, simulation, and decision trees assist capacity decisions.** The section "Tools for Capacity Planning," pp. 212–213, illustrate how these different methods can be used to arrive at capacity decisions.

MyOMLab helps you develop analytical skills and assesses your progress with multiple problems on capacity utilization, capacity requirements, future pretax cash flows, with and without expansion, decisions trees, NPV, and payback periods.

MyOMLab Resources	Titles	Link to the Book
Video	*Gate Turnaround at Southwest Airlines*	Planning Long-Term Capacity; Capacity Timing and Sizing Strategies
	1st Bank Villa Italia: Waiting Lines	Supplement B; Waiting-Line Models
OM Explorer Solvers	Capacity Requirements	A Systematic Approach to Long-Term Capacity Decisions
OM Explorer Tutors	6.1 Capacity Requirements	A Systematic Approach to Long-Term Capacity Decisions; Example 6.1 (pp. 209–210)
	6.2 Projecting Cash Flows	Evaluate the Alternatives; Example 6.2 (p. 211)
Tutor Exercises	6.1 Plowton Industries	Planning Long-Term Capacity; Capacity Timing and Sizing Strategies; Solved Problem 1 (pp. 215–216)
Virtual Tours	FaucetCraft Manufacturing	Planning Long-Term Capacity; Capacity Timing and Sizing Strategies; Example 6.1 (pp. 209–210)
	Portland Bolt & Manufacturing	Planning Long-Term Capacity; Capacity Timing and Sizing Strategies; Example 6.1 (pp. 209–210)
MyOMLab Supplements	F. Financial Analysis	Evaluating the Alternatives
	H. Measuring Output Rates	MyOMLab Supplement H
	I. Learning Curve Analysis	MyOMLab Supplement I
Internet Exercise	Vintak Intex Co., Ltd.	Planning Long-Term Capacity; Capacity Timing and Sizing Strategies; Figure 6.2 (p. 206)
	Design Homes, Inc.	Capacity Timing and Sizing Strategies; Example 6.1 (pp. 209–210); Example 6.2 (p. 211)
Additional Case	Fitness Plus (B)	Example 6.1 (pp. 209–210); Example 6.2 (p. 211); Solved Problem 2 (pp. 216–217)
Key Equations		
Image Library		

Key Equations

1. Utilization, expressed as a percent:

$$\text{Utilization} = \frac{\text{Average output rate}}{\text{Maximum capacity}} \times 100\%$$

2. Capacity cushion, C, expressed as a percent:

$$C = 100\% - \text{Average Utilization rate}(\%)$$

a. Capacity requirement for one service or product:

$$M = \frac{Dp}{N[1 - (C/100)]}$$

b. Capacity requirement for multiple services or products:

$$M = \frac{[Dp + (D/Q)s]_{\text{product 1}} + [Dp + (D/Q)s]_{\text{product 2}} + \cdots + [Dp + (D/Q)s]_{\text{product } n}}{N[1 - (C/100)]}$$

Key Terms

base case 210
capacity 202
capacity cushion 205
capacity gap 210

capacity requirement 208
cash flow 211
diseconomies of scale 204
economies of scale 204

planning horizon 208
setup time 209
utilization 204

Solved Problem 1

You have been asked to put together a capacity plan for a critical operation at the Surefoot Sandal Company. Your capacity measure is number of machines. Three products (men's, women's, and children's sandals) are manufactured. The time standards (processing and setup), lot sizes, and demand forecasts are given in the following table. The firm operates two 8-hour shifts, 5 days per week, 50 weeks per year. Experience shows that a capacity cushion of 5 percent is sufficient.

| Product | TIME STANDARDS | | Lot Size (pairs/lot) | Demand Forecast (pairs/yr) |
	Processing (hr/pair)	Setup (hr/pair)		
Men's sandals	0.05	0.5	240	80,000
Women's sandals	0.10	2.2	180	60,000
Children's sandals	0.02	3.8	360	120,000

a. How many machines are needed?

b. If the operation currently has two machines, what is the capacity gap?

SOLUTION

a. The number of hours of operation per year, N, is $N = (2 \text{ shifts/day})(8 \text{ hours/shifts})(250 \text{ days/machine-year}) = 4{,}000 \text{ hours/machine-year}$

The number of machines required, M, is the sum of machine-hour requirements for all three products divided by the number of productive hours available for one machine:

$$M = \frac{[Dp + (D/Q)s]_{\text{men}} + [Dp + (D/Q)s]_{\text{women}} + [Dp + (D/Q)s]_{\text{children}}}{N[1-(C/100)]}$$

$$= \frac{\begin{array}{c}[80{,}000(0.05) + (80{,}000/240)0.5] + [60{,}000(0.10) + (60{,}000/180)2.2] \\ + [120{,}000(0.02) + (120{,}000/360)3.8]\end{array}}{4{,}000[1 - (5/100)]}$$

$$= \frac{14{,}567 \text{ hours/year}}{3{,}800 \text{ hours/machine} - \text{year}} = 3.83 \text{ or } 4 \text{ machines}$$

b. The capacity gap is 1.83 machines ($3.83 - 2$). Two more machines should be purchased, unless management decides to use short-term options to fill the gap.

The *Capacity Requirements* Solver in OM Explorer confirms these calculations, as Figure 6.5 shows, using only the "Expected" scenario for the demand forecasts.

FIGURE 6.5 ▶

Using the *Capacity Requirements* Solver for Solved Problem 1

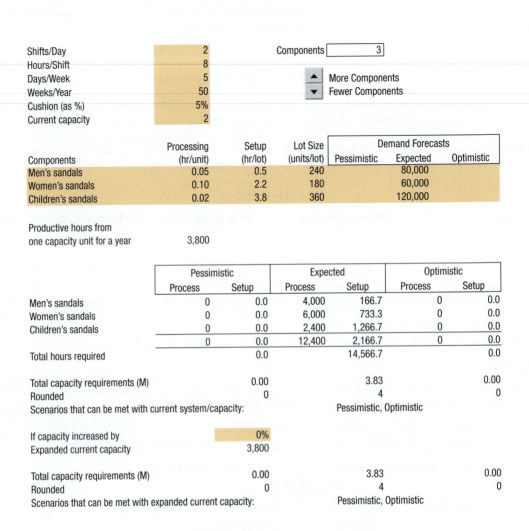

		Processing (hr/unit)	Setup (hr/lot)	Lot Size (units/lot)	Demand Forecasts Pessimistic	Demand Forecasts Expected	Demand Forecasts Optimistic
Shifts/Day	2						
Hours/Shift	8						
Days/Week	5						
Weeks/Year	50						
Cushion (as %)	5%						
Current capacity	2						

Components = 3

More Components / Fewer Components

Components	Processing (hr/unit)	Setup (hr/lot)	Lot Size (units/lot)	Pessimistic	Expected	Optimistic
Men's sandals	0.05	0.5	240		80,000	
Women's sandals	0.10	2.2	180		60,000	
Children's sandals	0.02	3.8	360		120,000	

Productive hours from one capacity unit for a year   3,800

	Pessimistic Process	Pessimistic Setup	Expected Process	Expected Setup	Optimistic Process	Optimistic Setup
Men's sandals	0	0.0	4,000	166.7	0	0.0
Women's sandals	0	0.0	6,000	733.3	0	0.0
Children's sandals	0	0.0	2,400	1,266.7	0	0.0
	0	0.0	12,400	2,166.7	0	0.0
Total hours required		0.0		14,566.7		0.0

	Pessimistic	Expected	Optimistic
Total capacity requirements (M)	0.00	3.83	0.00
Rounded	0	4	0
Scenarios that can be met with current system/capacity:		Pessimistic, Optimistic	

If capacity increased by	0%		
Expanded current capacity	3,800		

	Pessimistic	Expected	Optimistic
Total capacity requirements (M)	0.00	3.83	0.00
Rounded	0	4	0
Scenarios that can be met with expanded current capacity:		Pessimistic, Optimistic	

Solved Problem 2

The base case for Grandmother's Chicken Restaurant (see Example 6.2) is to do nothing. The capacity of the kitchen in the base case is 80,000 meals per year. A capacity alternative for Grandmother's Chicken Restaurant is a two-stage expansion. This alternative expands the kitchen at the end of year 0, raising its capacity from 80,000 meals per year to that of the dining area (105,000 meals per year). If sales in year 1 and 2 live up to expectations, the capacities of both the kitchen and the dining room will be expanded at the *end* of year 3 to 130,000 meals per year. This upgraded capacity level should suffice up through year 5. The initial investment would be $80,000 at the end of year 0 and an additional investment of $170,000 at the end of year 3. The pretax profit is $2 per meal. What are the pretax cash flows for this alternative through year 5, compared with the base case?

SOLUTION

Table 6.1 shows the cash inflows and outflows. The year 3 cash flow is unusual in two respects. First, the cash inflow from sales is $50,000 rather than $60,000. The increase in sales over the base is 25,000 meals ($105,000 - 10,000$) instead of 30,000 meals ($110,000 - 80,000$) because the restaurant's capacity falls somewhat short of demand. Second, a cash outflow of $170,000 occurs at the end of year 3, when the second-stage expansion occurs. The net cash flow for year 3 is $50,000 - $170,000 = -$120,000$.

For comparison purposes, the NPV of this project at a discount rate of 10 percent is calculated as follows, and equals negative $2,184.90.

$$\text{NPV} = -80{,}000 + (20{,}000/1.1) + [40{,}000/(1.1)^2] - [120{,}000/(1.1)^3] + [80{,}000/(1.1)^4] + [100{,}000/(1.1)^5]$$

$$= -\$80{,}000 + \$18{,}181.82 + \$33{,}057.85 - \$90{,}157.77 + \$54{,}641.07 + \$62{,}092.13$$

$$= -\$2{,}184.90$$

On a purely monetary basis, a single-stage expansion seems to be a better alternative than this two-stage expansion. However, other qualitative factors as mentioned earlier must be considered as well.

TABLE 6.1 | **CASH FLOWS FOR TWO-STAGE EXPANSION AT GRANDMOTHER'S CHICKEN RESTAURANT**

Year	Projected Demand (meals/yr)	Projected Capacity (meals/yr)	Calculation of Incremental Cash Flow Compared to Base Case (80,000 meals/yr)	Cash Inflow (outflow)
0	80,000	80,000	Increase kitchen capacity to 105,000 meals =	($80,000)
1	90,000	105,000	90,000 − 80,000 = (10,000 meals)($2/meal) =	$20,000
2	100,000	105,000	100,000 − 80,000 = (20,000 meals)($;2/meal) =	$40,000
3	110,000	105,000	105,000 − 80,000 = (25,000 meals)($2/meal) =	$50,000
			Increase total capacity to 130,000 meals =	($170,000)
				($120,000)
4	120,000	130,000	120,000 − 80,000 = (40,000 meals)($2/meal) =	$80,000
5	130,000	130,000	130,000 − 80,000 = (50,000 meals)($2/meal) =	$100,000

Discussion Questions

1. What are the economies of scale in college class size? As class size increases, what symptoms of diseconomies of scale appear? How are these symptoms related to customer contact?

2. A young boy sets up a lemonade stand on the corner of College Street and Air Park Boulevard. Temperatures in the area climb to 100°F. during the summer. The intersection is near a major university and a large construction site. Explain to this young entrepreneur how his business might benefit from economies of scale. Explain also some conditions that might lead to diseconomies of scale.

Problems

The OM Explorer and POM for Windows software is available to all students using the 10th edition of this textbook. Go to **www.pearsonhighered.com/krajewski** to download these computer packages. If you purchased MyOMLab, you also have access to Active Models software and significant help in doing the following problems. Check with your instructor on how best to use these resources. In many cases, the instructor wants you to understand how to do the calculations by hand. At the least, the software provides a check on your calculations. When calculations are particularly complex and the goal is interpreting the results in making decisions, the software replaces entirely the manual calculations.

Problems 12, 14, 17, 18, 19 and 22 require reading of Supplement A, "Decision Making." Problems 11, 15, 16, 19 and 22 require reading of MyOMLab Supplement F, "Financial Analysis."

1. The Dahlia Medical Center has 30 labor rooms, 15 combination labor and delivery rooms, 3 delivery rooms, and 1 special delivery room reserved for complicated births. All of these facilities operate around the clock. Time spent in labor rooms varies from hours to days, with an average of about a day. The average uncomplicated delivery requires about 1 hour in a delivery room.

During an exceptionally busy 3-day period, 109 healthy babies were born at Dahlia Medical Center. Sixty babies were born in separate labor and delivery rooms, 45 were born in combined labor and delivery rooms, and only 4 babies required a labor room and the complicated-delivery room. Which of the facilities (labor rooms, combination labor and delivery rooms, or delivery rooms) had the greatest utilization rate?

2. A process currently services an average of 50 customers per day. Observations in recent weeks show that its utilization is about 90 percent, allowing for just a 10 percent capacity cushion. If demand is expected to be 75 percent of the current level in 5 years and management wants to have a capacity cushion of just 5 percent, what capacity requirement should be planned?

3. An airline company must plan its fleet capacity and its long-term schedule of aircraft usage. For one flight

segment, the average number of customers per day is 70, which represents a 65 percent utilization rate of the equipment assigned to the flight segment. If demand is expected to increase to 84 customers for this flight segment in 3 years, what capacity requirement should be planned? Assume that management deems that a capacity cushion of 25 percent is appropriate.

4. Macon Controls produces three different types of control units used to protect industrial equipment from overheating.

Each of these units must be processed by a machine that Macon considers to be their process bottleneck. The plant operates on two 8-hour shifts, 5 days per week, 52 weeks per year. Table 6.2 provides the time standards at the bottleneck, lot sizes, and demand forecasts for the three units. Because of demand uncertainties, the operations manager obtained three demand forecasts (pessimistic, expected, and optimistic). The manager believes that a 20 percent capacity cushion is best.

TABLE 6.2 | **CAPACITY INFORMATION FOR MACON CONTROLS**

| Component | TIME STANDARD | | Lot Size (units/lot) | DEMAND FORECAST | | |
	Processing (hr/unit)	Setup (hr/lot)		Pessimistic	Expected	Optimistic
A	0.05	1.0	60	15,000	18,000	25,000
B	0.20	4.5	80	10,000	13,000	17,000
C	0.05	8.2	120	17,000	25,000	40,000

a. How many machines are required to meet minimum (Pessimistic) demand, expected demand, maximum (Optimistic) demand?

b. How many machines are required if the operations manager decides to double lot sizes?

c. If the operations manager has three machines and believes that the plant can reduce setup time by 20 percent through process improvement initiatives, does that plant have adequate capacity to meet all demand scenarios without increasing lot sizes?

5. Up, Up, and Away is a producer of kites and wind socks. Relevant data on a bottleneck operation in the shop for the upcoming fiscal year are given in the following table:

Item	Kites	Wind Socks
Demand forecast	30,000 units/year	12,000 units/year
Lot size	20 units	70 units
Standard processing time	0.3 hour/unit	1.0 hour/unit
Standard setup time	3.0 hours/lot	4.0 hours/lot

The shop works two shifts per day, 8 hours per shift, 200 days per year. Currently, the company operates four machines, and desires a 25 percent capacity cushion. How many machines should be purchased to meet the upcoming year's demand without resorting to any short-term capacity solutions?

6. Tuff-Rider, Inc., manufactures touring bikes and mountain bikes in a variety of frame sizes, colors, and component combinations. Identical bicycles are produced in lots of 100. The projected demand, lot size, and time standards are shown in the following table:

Item	Touring	Mountain
Demand forecast	5,000 units/year	10,000 units/year
Lot size	100 units	100 units
Standard processing time	.25 hour/unit	.50 hour/unit
Standard setup time	2 hours/lot	3 hours/lot

The shop currently works 8 hours a day, 5 days a week, 50 weeks a year. It operates five workstations, each producing one bicycle in the time shown in the table. The shop maintains a 15 percent capacity cushion. How many workstations will be required next year to meet expected demand without using overtime and without decreasing the firm's current capacity cushion?

7. Arabelle is considering expanding the floor area of her high-fashion import clothing store, The French Prints of Arabelle, by increasing her leased space in the upscale Cherry Creek Mall from 2,000 square feet to 3,000 square feet. The Cherry Creek Mall boasts one of the country's highest ratios of sales value per square foot. Rents (including utilities, security, and similar costs) are $110 per square foot per year. Salary increases related to French Prints' expansion are shown in the following table, along with projections of sales per square foot. The purchase cost of goods sold averages 70 percent of the sales price. Sales are seasonal, with an important peak during the year-end holiday season.

Year	Quarter	Sales (per sq ft)	Incremental Salaries
1	1	$ 90	$12,000
	2	60	8,000
	3	110	12,000
	4	240	24,000
2	1	99	12,000
	2	66	8,000
	3	121	12,000
	4	264	24,000

a. If Arabelle expands French Prints at the end of year 0, what will her quarterly pretax cash flows be through year 2?

b. Project the quarterly pretax cash flows assuming that the sales pattern (10 percent annually compounded increase) continues through year 3.

8. The Astro World amusement park has the opportunity to expand its size now (the end of year 0) by purchasing adjacent property for $250,000 and adding attractions at a cost of $550,000. This expansion is expected to increase attendance by 30 percent over projected attendance without expansion. The price of admission is $30, with a $5 increase planned for the beginning of year 3. Additional operating costs are expected to be $100,000 per year. Estimated attendance for the next 5 years, *without expansion*, is as follows:

Year	1	2	3	4	5
Attendance	30,000	34,000	36,250	38,500	41,000

   a. What are the pretax combined cash flows for years 0 through 5 that are attributable to the park's expansion?

   b. Ignoring tax, depreciation, and the time value of money, determine how long it will take to recover (pay back) the investment.

9. Kim Epson operates a full-service car wash, which operates from 8 A.M. to 8 P.M. 7 days a week. The car wash has two stations: an automatic washing and drying station and a manual interior cleaning station. The automatic washing and drying station can handle 30 cars per hour. The interior cleaning station can handle 200 cars per day. Based on a recent year-end review of operations, Kim estimates that future demand for the interior cleaning station for the 7 days of the week, expressed in average number of cars per day, would be as follows:

Day	Mon.	Tues.	Wed.	Thurs.	Fri.	Sat.	Sun.
Cars	160	180	150	140	280	300	250

   By installing additional equipment (at a cost of $50,000), Kim can increase the capacity of the interior cleaning station to 300 cars per day. Each car wash generates a pretax contribution of $4.00. Should Kim install the additional equipment if she expects a pretax payback period of 3 years or less?

10. Roche Brothers is considering a capacity expansion of its supermarket. The landowner will build the addition to suit in return for $200,000 upon completion and a 5-year lease. The increase in rent for the addition is $10,000 per month. The annual sales projected through year 5 follow. The current effective capacity is equivalent to 500,000 customers per year. Assume a 2 percent pretax profit on sales.

Year	1	2	3	4	5
Customers	560,000	600,000	685,000	700,000	715,000
Average Sales per Customer	$50.00	$53.00	$56.00	$60.00	$64.00

   a. If Roche expands its capacity to serve 700,000 customers per year now (end of year 0), what are the projected annual incremental pretax cash flows attributable to this expansion?

   b. If Roche expands its capacity to serve 700,000 customers per year at the end of year 2, the landowner will build the same addition for $240,000 and a 3-year lease

at $12,000 per month. What are the projected annual incremental pretax cash flows attributable to this expansion alternative?

11. MKM International is seeking to purchase a new CNC machine in order to reduce costs. Two alternative machines are in consideration. Machine 1 costs $500,000, but yields a 15 percent savings over the current machine used. Machine 2 costs $900,000, but yields a 25 percent savings over the current machine used. In order to meet demand, the following forecasted cost information for the current machine is also provided.

   a. Based on the NPV of the cash flows for these 5 years, which machine should MKM International Purchase? Assume a discount rate of 12 percent.

   b. If MKM International lowered its required discount rate to 8 percent, what machine would it purchase?

Year	Projected Cost
1	1,000,000
2	1,350,000
3	1,400,000
4	1,450,000
5	2,550,000

12. Dawson Electronics is a manufacturer of high-tech control modules for lawn sprinkler systems. Denise, the CEO, is trying to decide if the company should develop one of the two potential new products, the Water Saver 1000 or the Greener Grass 5000. With each product, Dawson can capture a bigger market share if it chooses to expand capacity by buying additional machines. Given different demand scenarios, their probabilities of occurrence, and capacity expansion versus no change in capacity, the potential sales of each product are summarized in Table 6.3.

TABLE 6.3 | DEMAND AND SALES INFORMATION FOR DAWSON ELECTRONICS

	Water Saver 1000 Dollar Sales ($1000)	Greener Grass 5000 Dollar Sales ($1000)	Probability of Occurrence
With Capacity Expansion			
Low Demand	1000	2500	0.25
Medium Demand	2000	3000	0.50
High Demand	3000	5000	0.25
Without Capacity Expansion			
Low Demand	700	1000	0.25
Medium Demand	1000	2000	0.50
High Demand	2000	3000	0.25

   a. What is the expected payoff for Water Saver 1000 and the Greener Grass 5000, with and without capacity expansion?

   b. Which product should Denise choose to produce, and with which capacity expansion option?

Advanced Problems

13. Knott's Industries manufactures standard and super premium backyard swing sets. Currently it has four identical swing-set-making machines, which are operated 250 days per year and 8 hours each day. A capacity cushion of 20 percent is desired. The following information is also known:

	Standard Model	Super Premium Model
Annual Demand	20,000	10,000
Standard Processing Time	7 min	20 min
Average Lot Size	50	30
Standard Setup Time per Lot	30 min	45 min

   a. Does Knott's have sufficient capacity to meet annual demand?

   b. If Knott's was able to reduce the setup time for the Super Premium Model from 45 minutes to 30 minutes, would there be enough current capacity to produce 20,000 units of each type of swing set?

14. A manager is trying to decide whether to buy one machine or two. If only one machine is purchased and demand proves to be excessive, the second machine can be purchased later. Some sales would be lost, however, because the lead time for delivery of this type of machine is 6 months. In addition, the cost per machine will be lower if both machines are purchased at the same time. The probability of low demand is estimated to be 0.30 and that of high demand to be 0.70. The after-tax NPV of the benefits from purchasing two machines together is $90,000 if demand is low and $170,000 if demand is high.

   If one machine is purchased and demand is low, the NPV is $120,000. If demand is high, the manager has three options: (1) doing nothing, which has an NPV of $120,000; (2) subcontracting, with an NPV of $140,000; and (3) buying the second machine, with an NPV of $130,000.

   a. Draw a decision tree for this problem.

   b. What is the best decision and what is its expected payoff?

15. Several years ago, River City built a water purification plant to remove toxins and filter the city's drinking water. Because of population growth, the demand for water next year will be more than the plant's capacity of 120 million gallons per year. Therefore, the city must expand the facility. The estimated demand over the next 20 years is given in Table 6.4.

   The city planning commission is considering three alternatives.

   - *Alternative 1:* Expand enough at the end of year 0 to last 20 years, which means an 80 million gallon increase (200 –120).
   - *Alternative 2:* Expand at the end of year 0 and at the end of year 10.
   - *Alternative 3:* Expand at the end of years 0, 5, 10, and 15.

   Each alternative would provide the needed 200 million gallons per year at the end of 20 years, when the value of the plant would be the same regardless of the alternative chosen. Significant economies of scale can be achieved in

   construction costs: A 20 million gallon expansion would cost $18 million; a 40 million gallon expansion, $30 million; and an 80 million gallon expansion, only $50 million. The level of future interest rates is uncertain, leading to uncertainty about the hurdle rate. The city believes that it could be as low as 12 percent and as high as 16 percent (see MyOMLab Supplement F, "Financial Analysis").

   a. Compute the cash flows for each alternative, compared to a base case of doing nothing. (*Note*: As a municipal utility, the operation pays no taxes.)

   b. Which alternative minimizes the present value of construction costs over the next 20 years if the discount rate is 12 percent? Sixteen percent?

   c. Because the decision involves public policy and compromise, what political considerations does the planning commission face?

TABLE 6.4 | **WATER DEMAND**

Year	Demand	Year	Demand	Year	Demand
0	120	7	148	14	176
1	124	8	152	15	180
2	128	9	156	16	184
3	132	10	160	17	188
4	136	11	164	18	192
5	140	12	168	19	196
6	144	13	172	20	200

16. Mars Incorporated is interested in going to market with a new fuel savings device that attaches to electrically powered industrial vehicles. The device, code named "Python," promises to save up to 15 percent of the electrical power required to operate the average electric forklift. Mars expects that modest demand expected during the introductory year will be followed by a steady increase in demand in subsequent years. The extent of this increase in demand will be based on customer's expectations regarding the future cost of electricity, and which is shown in Table 6.5. Mars expects to sell the device for $500 each, and does not expect to be able to raise its price over the foreseeable future.

TABLE 6.5 | **DEMAND FOR PYTHON POWER SAVING DEVICE**

	EXPECTED DEMAND OF THE DEVICE IN UNITS/YEAR	
Year	Small Increases in the Cost of Electrical Power	Large Increases in the Cost of Electrical Power
1	1,000	10,000
2	5,000	8,000
3	1,000	15,000
4	15,000	20,000
5	18,000	30,000

Mars is faced with two alternatives:

- *Alternative 1:* Make the device themselves, which requires an initial outlay of $250,000 in plant and equipment and a variable cost of $75 per unit.
- *Alternative 2:* Outsource the production, which requires no initial investment, but incurs a per unit cost of $300.

a. Assuming small increases in the cost of electrical power, compute the cash flows for each alternative. Over the next 5 years, which alternative maximizes the NPV of this project if the discount rate is 10 percent?

b. Assuming large increases in the cost of electrical power, compute the cash flows for each alternative. Over the next 5 years, which alternative maximizes the NPV of this project if the discount rate is 10 percent?

17. Acme Steel Fabricators experienced booming business for the past 5 years. The company fabricates a wide range of steel products, such as railings, ladders, and light structural steel framing. The current manual method of materials handling is causing excessive inventories and congestion. Acme is considering the purchase of an overhead rail-mounted hoist system or a forklift truck to increase capacity and improve manufacturing efficiency.

The annual pretax payoff from the system depends on future demand. If demand stays at the current level, the probability of which is 0.50, annual savings from the overhead hoist will be $10,000. If demand rises, the hoist will save $25,000 annually because of operating efficiencies in addition to new sales. Finally, if demand falls, the hoist will result in an estimated annual loss of $65,000. The probability is estimated to be 0.30 for higher demand and 0.20 for lower demand.

If the forklift is purchased, annual payoffs will be $5,000 if demand is unchanged, $10,000 if demand rises, and −$25,000 if demand falls.

a. Draw a decision tree for this problem and compute the expected value of the payoff for each alternative.

b. Which is the best alternative, based on the expected values?

18. Referring to problem 4, the operations manager at Macon Controls believes that pessimistic demand has a probability of 20 percent, expected demand has a probability of 50 percent, and optimistic demand has a probability of 30 percent. Currently, new machines must be purchased at a cost of $500,000 a piece, the price charged for each control unit is $110, and the variable cost of production is $50 per unit. (Hint: since the price and variable cost for each control unit are the same, the profit maximizing product mix will be the same as the mix that maximizes the total number of units produced).

a. Draw a decision tree for this problem.

b. How many machines should the company purchase, and what is the expected payoff?

19. Darren Mack owns the "Gas n' Go" convenience store and gas station. After hearing a marketing lecture, he realizes that it might be possible to draw more customers to his high-margin convenience store by selling his gasoline at a lower price. However, the "Gas n' Go" is unable to qualify for volume discounts on its gasoline purchases, and

therefore cannot sell gasoline for profit if the price is lowered. Each new pump will cost $95,000 to install, but will increase customer traffic in the store by 1,000 customers per year. Also, because the "Gas n' Go" would be selling its gasoline at no profit, Darren plans on increasing the profit margin on convenience store items incrementally over the next 5 years. Assume a discount rate of 8 percent. The projected convenience store sales per customer and the projected profit margin for the next 5 years are as follows:

Year	Projected Convenience Store Sales Per Customer	Projected Profit Margin
1	$ 5.00	20%
2	$ 6.50	25%
3	$ 8.00	30%
4	$10.00	35%
5	$11.00	40%

a. What is the NPV of the next 5 years of cash flows if Darren had four new pumps installed?

b. If Darren required a payback period of 4 years, should he go ahead with the installation of the new pumps?

20. Mackelprang, Inc., is in the initial stages of building the premier planned community in the greater Phoenix, Arizona, metropolitan area. The main selling point will be the community's lush golf courses. Homes with golf course views will generate premiums far larger than homes with no golf course views, but building golf courses is expensive and takes up valuable space that non-view homes could be built upon. Mackelprang, Inc., has limited land capacity. In order to maximize its profits, it is faced with a decision as to how many golf courses it should build, which, in turn, will impact how many homes with and without golf course views it will be able to construct. Mackelprang, Inc., realizes that this decision is directly related to the premium buyers will be willing to spend to buy homes with golf course views. Mackelprang, Inc., is required to build at least one golf course, but has enough space to build up to three golf courses. The following table indicates the costs and potential revenues for each course:

a. Which golf course or courses should Mackelprang, Inc., build?

b. What is the expected payoff for this project?

	Indian River	The Cactus	Wildwood
Cost	$2.6M	$1.25M	$2.5M
Highest Possible Revenue	$4M	$2M	$2M
Probability of High Revenue	0.3	0.2	0.3
Likely Revenue	$2.5M	$1.5M	$4M
Probability of Likely Revenue	0.4	0.5	0.5
Lowest Possible Revenue	$1M	$1M	$1M
Probability of Low Revenue	0.3	0.3	0.2

21. Two new alternatives have come up for expanding Grandmother's Chicken Restaurant (see Solved Problem 2). They involve more automation in the kitchen and feature a special cooking process that retains the original-recipe taste of the chicken. Although the process is more capital-intensive, it would drive down labor costs, so the pretax profit for *all* sales (not just the sales from the capacity added) would go up from 20 to 22 percent. This gain would increase the pretax profit by 2 percent of each sales dollar through $800,000 (80,000 meals × $10) and by 22 percent of each sales dollar between $800,000 and the new capacity limit. Otherwise, the new alternatives are much the same as those in Example 6.2 and Solved Problem 2.

   - *Alternative 1:* Expand both the kitchen and the dining area now (at the end of year 0), raising the capacity to 130,000 meals per year. The cost of construction, including the new automation, would be $336,000 (rather than the earlier $200,000).

   - *Alternative 2:* Expand only the kitchen now, raising its capacity to 105,000 meals per year. At the end of year 3, expand both the kitchen and the dining area to the 130,000 meals-per-year volume. Construction and equipment costs would be $424,000, with $220,000 at the end of year 0 and the remainder at the end of year 3. As with alternative 1, the contribution margin would go up to 22 percent.

   With both new alternatives, the salvage value would be negligible. Compare the cash flows of all alternatives. Should Grandmother's Chicken Restaurant expand with the new or the old technology? Should it expand now or later?

22. The vice president of operations at Dintell Corporation, a major supplier of passenger-side automotive air bags, is considering a $50 million expansion at the firm's Fort Worth, Texas, production complex. The most recent economic projections indicate a 0.60 probability that the overall market will be $400 million per year over the next 5 years and a 0.40 probability that the market will be only $200 million per year during the same period. The marketing department estimates that Dintell has a 0.50 probability of capturing 40 percent of the market and an equal probability of obtaining only 30 percent of the market. The cost of goods sold is estimated to be 70 percent of sales. For planning purposes, the company currently uses a 12 percent discount rate, a 40 percent tax rate, and the MACRS depreciation schedule. The criteria for investment decisions at Dintell are (1) the net expected present value must be greater than zero; (2) there must be at least a 70 percent chance that the net present value will be positive; and (3) there must be no more than a 10 percent chance that the firm will lose more than 20 percent of the initial value.

   a. Based on the stated criteria, determine whether Dintell should fund the project.

   b. What effect will a probability of 0.70 of capturing 40 percent of the market have on the decision?

   c. What effect will an increase in the discount rate to 15 percent have on the decision? A decrease to 10 percent?

   d. What effect will the need for another $10 million in the third year have on the decision?

VIDEO CASE | Gate Turnaround at Southwest Airlines

Rollin King and Herb Kelleher started Southwest Airlines in 1971 with this idea: if they could take airline passengers where they want to go, on time, at the lowest possible price, and have a good time while doing it, people would love to fly their airline. The result? No other airline in the industry's history has enjoyed the customer loyalty and extended profitability for which Southwest is now famous. The company now flies more than 3,400 times each day to over 64 destinations across the United States.

There's more to the story, however, than making promises and hoping to fulfill them. A large part of Southwest Airlines' success lies in its ability to plan long-term capacity to better match demand, as also improving the utilization of its fleet by turning around an aircraft at the gate faster than its competitors. Capacity at Southwest is measured in seat-miles, and even a single minute reduction in aircraft turnaround time system wide means additional seat-miles being added to the available capacity of Southwest Airlines.

As soon as an aircraft calls "in range" at one of Southwest's airport locations, called a station, the local operations manager notifies the ground operations team so that they can start mobilizing all the parties involved in servicing the aircraft in preparation for its next departure. The grounds operations team consists of a baggage transfer driver who has responsibility for getting connecting flight bags to their proper planes, a local baggage driver who moves bags to baggage claim for passenger pick-up, a lavatory truck driver who handles restroom receptacle drainage, a lead gate agent to handle baggage carts and track incoming and outgoing bag counts, and a bin agent to manage baggage and cargo inside the plane. The ground operations team

Baggage transfer starts less than 40 seconds after engine shutdown at Southwest Airlines.

knows it must turn the plane around in 25 minutes or less. The clock starts when the pilot sets the wheel brakes.

Inbound and outbound flights are coordinated by the supervisors between Southwest's 64 airport stations through the company's Operations Terminal Information System (OTIS). Each local supervisor is able to keep track of their flights and manage any delays or problems that may have crept into the system by keeping in touch with headquarters in Dallas for system-wide issues that may impact a local station, along with using the OTIS information coming from stations sending flights their way.

Just what, exactly, does it take to turn around an aircraft? In-bound flight 3155 from Phoenix to Dallas' Love Field is a good example. In Phoenix, the operations coordinators and ground operations team push back the plane as scheduled at 9:50 A.M. The flight is scheduled to arrive at 3:35 P.M. in Dallas. The Phoenix team enters into OTIS the information the ground operations team will need in Dallas, such as wheelchairs, gate-checked baggage, cargo bin locator data, and other data needed to close out the flight on their end. This action lets the Dallas station know what to expect when the plane lands.

In Dallas, the local ground operations coordinators have been monitoring all 110 inbound flights and now see Phoenix flight 3155 in the system, scheduled for an on-time arrival. When the pilot calls "in-range" as it nears Dallas, the ground crew prepares for action.

As the plane is guided to its "stop mark" at the gate, the lead agent waits for the captain's signal that the engines have been turned off and brakes set. Within just 10 seconds, the provisioning truck pulls up to open the back door for restocking supplies such as drinks and snacks. The waiting fuel truck extends its hose to the underwing connection, and in less than 2 minutes, picks up refueling instructions and starts to load fuel. As soon as the aircraft is in position, the operations team steers the jetway into position and locks it against the aircraft. The door is opened, the in-flight crew is greeted, and passengers start to deplane.

Outside, less than 40 seconds after engine shutdown, baggage is rolling off the plane and gets placed onto the first cart. Any transfer bags get sent to their next destination, and gate-checked bags are delivered to the top of the jetway stairs for passenger pick-up.

While passengers make their way out of the plane, the in-flight crew helps clean up and prepare the cabin for the next flight. If all goes well, the last passenger will leave the plane after only 8 minutes. By this time, passengers waiting to board have already lined up in their designated positions for boarding. The gate agent confirms that the plane is ready for passenger boarding, and calls for the first group to turn in their boarding passes and file down the jetway.

At the completion of boarding, the operations agent checks the fuel invoice, cargo bin loading schedule with actual bag counts in their bins from the baggage agents, and a lavatory service record confirming that cleaning has taken place. Final paperwork is given to the captain. The door to the aircraft is closed, and the jetway is retracted. Thirty seconds later, the plane is pushed back and the operations agent gives a traditional salute to the captain to send the flight on its way. Total elapsed time: less than 25 minutes.

Managing Southwest's capacity has been somewhat simplified by strategic decisions made early on in the company's life. First, the company's fleet of aircraft is all Boeing 737's. This single decision impacts all areas of operations — from crew training to aircraft maintenance. The single-plane configuration also provides Southwest with crew scheduling flexibility. Since pilots and flight crews can be deployed across the entire fleet, there are no constraints with regard to training and certification pegged to specific aircraft types.

The way Southwest has streamlined its operations for tight turnarounds means it must maintain a high capacity cushion to accommodate variability in its daily operations. Anything from weather delays to unexpected maintenance issues at the gate can slow down the flow of operations to a crawl. To handle these unplanned but anticipated challenges, Southwest builds into its schedules enough cushion to manage these delays yet not so much that employees and planes are idle. Additionally, the company encourages discussion to keep on top of what's working and where improvements can be made. If a problem is noted at a downstream station, say bags were not properly loaded, this information quickly travels back up to the originating station for correction so that it does not happen again.

Even with the tightly managed operations Southwest Airlines enjoys, company executives know that continued improvement is necessary if the company is to remain profitable into the future. Company executives know when they have achieved their goals when internal and external metrics are reached. For example, the Department of Transportation (DOT) tracks on-time departures, customer complaints, and mishandled baggage for all airlines. The company sets targets for achievement on these dimensions and lets employees know on a monthly basis how the company is doing against those metrics and the rest of the industry. Regular communication with all employees is delivered via meetings, posters, and newsletters. Rewards such as prizes and profit sharing are given for successful achievement.

As for the future, Bob Jordan, Southwest's Executive Vice President for Strategy and Planning, puts it this way: "We make money when our planes are in the air, not on the ground. If we can save one minute off every turn system-wide, that's like putting five additional planes in the air. If a single plane generates annual revenue of $25 million, there's $125 million in profit potential from those time savings."

QUESTIONS

1. How can capacity and utilization be measured at an airline such as Southwest Airlines?
2. Which factors can adversely impact turn-around times at Southwest Airlines?
3. How does Southwest Airlines know they are achieving their goals?
4. What are the important long-term issues relevant for managing capacity, revenue, and customer satisfaction for Southwest Airlines?

CASE | Fitness Plus, Part A

Fitness Plus, Part B, explores alternatives to expanding a new downtown facility and is included in the Instructor's Resource Manual. If you are interested in this topic, ask your instructor for a preview.

Fitness Plus is a full-service health and sports club in Greensboro, North Carolina. The club provides a range of facilities and services to support three primary activities: fitness, recreation, and relaxation. Fitness activities generally take place in four areas of the club: the (1) aerobics room, which can accommodate 35 people per class; a (2) room equipped with free weights; a (3) workout room with 24 pieces of Nautilus equipment; and a (4) large workout room containing 29 pieces of cardiovascular equipment. This equipment includes nine stairsteppers, six treadmills, six life-cycle bikes, three airdyne bikes, two cross-aerobics machines, two rowing machines, and one climber. Recreational facilities comprise eight racquetball courts, six tennis courts, and a large outdoor pool. Fitness Plus also sponsors softball, volleyball, and swim teams in city recreation leagues. Relaxation is accomplished through yoga classes held twice a week in the aerobics room, whirlpool tubs located in each locker room, and a trained massage therapist.

Situated in a large suburban office park, Fitness Plus opened its doors in 1995. During the first 2 years, membership was small and use of the facilities was light. By 1997, membership had grown as fitness began to play a large role in more and more people's lives. Along with this growth came increased use of club facilities. Records indicate that in 2000, an average of 15 members per hour checked into the club during a typical day. Of course, the actual number of members per hour varied by both day and time. On some days during a slow period, only six to eight members would check in per hour. At a peak time, such as Mondays from 4:00 P.M. to 7:00 P.M., the number would be as high as 40 per hour.

The club was open from 6:30 A.M. to 11:00 P.M. Monday through Thursday. On Friday and Saturday, the club closed at 8:00 P.M., and on Sunday the hours were 12:00 P.M. to 8:00 P.M.

As the popularity of health and fitness continued to grow, so did Fitness Plus. By May 2005, the average number of members arriving per hour during a typical day had increased to 25. The lowest period had a rate of 10 members per hour; during peak periods, 80 members per hour checked in to use the facilities. This growth brought complaints from members about overcrowding and unavailability of equipment. Most of these complaints centered on the Nautilus, cardiovascular, and aerobics fitness areas. The owners began to wonder whether the club was indeed too small for its membership. Past research indicated that individuals work out an average of 60 minutes per visit. Data collected from member surveys showed the following facilities usage pattern: 30 percent of the members do aerobics, 40 percent use the cardiovascular equipment, 25 percent use the Nautilus machines, 20 percent use the free weights, 15 percent use the racquetball courts, and 10 percent use the tennis courts. The owners wondered whether they could use this information to estimate how well existing capacity was being utilized.

If capacity levels were being stretched, now was the time to decide what to do. It was already May, and any expansion of the existing facility would take at least 4 months. The owners knew that January was always a peak membership enrollment month and that any new capacity needed to be ready by then. However, other factors had to be considered. The area was growing both in terms of population and geographically. The downtown area just received a major facelift, and many new offices and businesses were moving back to it, causing a resurgence in activity.

With this growth came increased competition. A new YMCA was offering a full range of services at a low cost. Two new health and fitness facilities had opened within the past year in locations 10 to 15 minutes from Fitness Plus. The first, called the Oasis, catered to the young adult crowd and restricted the access of children under 16 years old. The other facility, Gold's Gym, provided excellent weight and cardiovascular training only.

As the owners thought about the situation, they had many questions: Were the capacities of the existing facilities constrained, and if so, where? If capacity expansion was necessary, should the existing facility be expanded? Because of the limited amount of land at the current site, expansion of some services might require reducing the capacity of others. Finally, owing to increased competition and growth downtown, was now the time to open a facility to serve that market? A new facility would take 6 months to renovate, and the financial resources were not available to do both.

QUESTIONS

1. What method would you use to measure the capacity of Fitness Plus? Has Fitness Plus reached its capacity?

2. Which capacity strategy would be appropriate for Fitness Plus? Justify your answer.

3. How would you link the capacity decision being made by Fitness Plus to other types of operating decisions?

Selected References

Bakke, Nils Arne, and Ronald Hellberg. "The Challenges of Capacity Planning." *International Journal of Production Economics*, vols. 31–30 (1993), pp. 243–264.

Bower, J.L., and C.G. Gilbert, "How Managers' Everyday Decisions Create or Destroy Your Company's Strategy." *Harvard Business Review*, vol. 85, no. 2 (2007), pp. 72–79.

Hartvigsen, David. *SimQuick: Process Simulation with Excel*, 2d ed. Upper Saddle River, NJ: Prentice Hall, 2004.

"Intel's $10 Billion Gamble." *Fortune* (November 11, 2002), pp. 90–102.

Tenhiala, A. "Contingency Theory of Capacity Planning: The Link between Process Types and Planning Methods." *Journal of Operations Management*, vol. 29 (2011), pp. 65–77.

Klassen, Robert D., and Larry J. Menor. "The Process Management Triangle: An Empirical Investigation of Process Trade-offs." *Journal of Operations Management*, vol. 25 (2007), pp. 1015–1034.

Leonard, Christopher. "Corn Reserves Expected to Rise, Easing Food Prices," *The Associated Press*, May 12, 2011.

Ritzman, Larry P., and M. Hossein Safizadeh. "Linking Process Choice with Plant-Level Decisions About Capital and Human Resources." *Production and Operations Management*, vol. 8, no. 4 (1999), pp. 374–392.

Anyone who has ever waited at a stoplight, at McDonald's, or at the registrar's office has experienced the dynamics of waiting lines. Perhaps one of the best examples of effective management of waiting lines is that of Walt Disney World. One day the park may have only 25,000 customers, but on another day the numbers may top 90,000. Careful analysis of process flows, technology for people-mover (materials handling) equipment, capacity, and layout keeps the waiting times for attractions to acceptable levels.

The analysis of waiting lines is of concern to managers because it affects process design, capacity planning, process performance, and ultimately, supply chain performance. In this supplement we discuss why waiting lines form, the uses of waiting-line models in operations management, and the structure of waiting-line models. We also discuss the decisions managers address with the models. Waiting lines can also be analyzed using computer simulation. Software such as SimQuick, a simulation package included in MyOMLab, or Excel spreadsheets can be used to analyze the problems in this supplement.

MyOMLab

Why Waiting Lines Form

A **waiting line** is one or more "customers" waiting for service. The customers can be people or inanimate objects, such as machines requiring maintenance, sales orders waiting for shipping, or inventory items waiting to be used. A waiting line forms because of a temporary imbalance between the demand for service and the capacity of the system to provide the service. In most real-life waiting-line problems, the demand rate varies; that is, customers arrive at unpredictable intervals. Most often, the rate of producing the service also varies, depending on customer needs. Suppose that bank customers arrive at an average rate of 15 per hour throughout the day and that

waiting line

One or more "customers" waiting for service.

LEARNING GOALS *After reading this supplement, you should be able to:*

① Identify the elements of a waiting-line problem in a real situation.

② Describe the single-server, multiple-server, and finite-source models.

③ Explain how to use waiting-line models to estimate the operating characteristics of a process.

④ Describe the situations where simulation should be used for waiting line analysis and the nature of the information that can be obtained.

⑤ Explain how waiting-line models can be used to make managerial decisions.

the bank can process an average of 20 customers per hour. Why would a waiting line ever develop? The answers are that the customer arrival rate varies throughout the day and the time required to process a customer can vary. During the noon hour, 30 customers may arrive at the bank. Some of them may have complicated transactions requiring above-average process times. The waiting line may grow to 15 customers for a period of time before it eventually disappears. Even though the bank manager provided for more than enough capacity on average, waiting lines can still develop.

Waiting lines can develop even if the time to process a customer is constant. For example, a subway train is computer controlled to arrive at stations along its route. Each train is programmed to arrive at a station, say, every 15 minutes. Even with the constant service time, waiting lines develop while riders wait for the next train or cannot get on a train because of the size of the crowd at a busy time of the day. Consequently, variability in the rate of demand determines the sizes of the waiting lines in this case. In general, if no variability in the demand or service rate occurs and enough capacity is provided, no waiting lines form.

Uses of Waiting-Line Theory

Waiting-line theory applies to service as well as manufacturing firms, relating customer arrival and service-system processing characteristics to service-system output characteristics. In our discussion, we use the term *service* broadly—the act of doing work for a customer. The service system might be hair cutting at a hair salon, satisfying customer complaints, or processing a production order of parts on a certain machine. Other examples of customers and services include lines of theatergoers waiting to purchase tickets, trucks waiting to be unloaded at a warehouse, machines waiting to be repaired by a maintenance crew, and patients waiting to be examined by a physician. Regardless of the situation, waiting-line problems have several common elements.

Structure of Waiting-Line Problems

Analyzing waiting-line problems begins with a description of the situation's basic elements. Each specific situation will have different characteristics, but four elements are common to all situations:

1. An input, or **customer population**, that generates potential customers
2. A waiting line of customers
3. The **service facility**, consisting of a person (or crew), a machine (or group of machines), or both necessary to perform the service for the customer
4. A **priority rule**, which selects the next customer to be served by the service facility

Figure B.1 shows these basic elements. The triangles, circles, and squares are intended to show a diversity of customers with different needs. The **service system** describes the number of lines and the arrangement of the facilities. After the service has been performed, the served customers leave the system.

Customer Population

A customer population is the source of input to the service system. If the potential number of new customers for the service system is appreciably affected by the number of customers already in the system, the input source is said to be *finite*. For example, suppose that a maintenance crew is assigned responsibility for the repair of 10 machines. The customer population for the

customer population

An input that generates potential customers.

service facility

A person (or crew), a machine (or group of machines), or both necessary to perform the service for the customer.

priority rule

A rule that selects the next customer to be served by the service facility.

service system

The number of lines and the arrangement of the facilities.

FIGURE B.1 ▶
Basic Elements of Waiting-Line Models

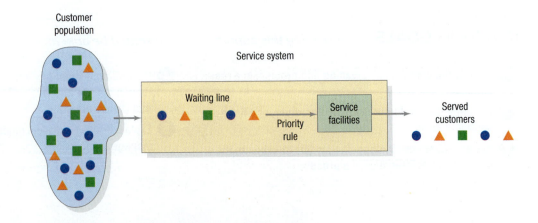

maintenance crew is 10 machines in working order. The population generates customers for the maintenance crew as a function of the failure rates for the machines. As more machines fail and enter the service system, either waiting for service or being repaired, the customer population becomes smaller and the rate at which it can generate another customer falls. Consequently, the customer population is said to be finite.

Alternatively, an *infinite* customer population is one in which the number of customers in the system does not affect the rate at which the population generates new customers. For example, consider a mail-order operation for which the customer population consists of shoppers who have received a catalog of products sold by the company. Because the customer population is so large and only a small fraction of the shoppers place orders at any one time, the number of new orders it generates is not appreciably affected by the number of orders waiting for service or being processed by the service system. In this case, the customer population is said to be infinite.

Customers in waiting lines may be *patient* or *impatient*, which has nothing to do with the colorful language a customer may use while waiting in line for a long time on a hot day. In the context of waiting-line problems, a patient customer is one who enters the system and remains there until being served; an impatient customer is one who either decides not to enter the system (balks) or leaves the system before being served (reneges). For the methods used in this supplement, we make the simplifying assumption that all customers are patient.

The Service System

The service system may be described by the number of lines and the arrangement of facilities.

Number of Lines   Waiting lines may be designed to be a *single line* or *multiple lines*. Figure B.2 shows an example of each arrangement. Generally, single lines are utilized at airline counters, inside banks, and at some fast-food restaurants; whereas multiple

Sometimes customers are not organized neatly into lines. Here ships wait to use the port facilities in Victoria Harbor, West Kowloon, Hong Kong.

lines are utilized in grocery stores, at drive-in bank operations, and in discount stores. When multiple servers are available and each one can handle general transactions, the single-line arrangement keeps servers uniformly busy and gives customers a sense of fairness. Customers believe that they are being served on the basis of when they arrived, and not on how well they guessed their waiting time when selecting a particular line. The multiple-line design is best when some of the servers provide a limited set of services. In this arrangement, customers select the services they need and wait in the line where that service is provided, such as at a grocery store that provides special lines for customers paying with cash or having fewer than 10 items.

Sometimes customers are not organized neatly into "lines." Machines that need repair on the production floor of a factory may be left in place, and the maintenance crew comes to them. Nonetheless, we can think of such machines as forming a single line or multiple lines, depending on the number of repair crews and their specialties. Likewise, passengers who telephone for a taxi also form a line even though they may wait at different locations.

Arrangement of Service Facilities   Service facilities consist of the personnel and equipment necessary to perform the service for the customer. Service facility arrangement is described by the number of channels and phases. A **channel** is one or more facilities required to perform a given service. A **phase** is a single step in providing the service. Some services require a single phase, while others require a sequence of phases. Consequently, a service facility uses some combination

channel
One or more facilities required to perform a given service.

phase
A single step in providing a service.

Service facilities

(a) Single line

Service facilities

(b) Multiple lines

FIGURE B.2 ▲
Waiting-Line Arrangements

of channels and phases. Managers should choose an arrangement based on customer volume and the nature of services provided. Figure B.3 shows examples of the five basic types of service facility arrangements.

In the *single-channel, single-phase* system, all services demanded by a customer can be performed by a single-server facility. Customers form a single line and go through the service facility one at a time. Examples are a drive-through car wash and a machine that must process several batches of parts.

The *single-channel, multiple-phase* arrangement is used when the services are best performed in sequence by more than one facility, yet customer volume or other constraints limit the design to one channel. Customers form a single line and proceed sequentially from one service facility to the next. An example of this arrangement is a McDonald's drive-through, where the first facility takes the order, the second takes the money, and the third provides the food.

The *multiple-channel, single-phase* arrangement is used when demand is large enough to warrant providing the same service at more than one facility or when the services offered by the facilities are different. Customers form one or more lines, depending on the design. In the single-line design, customers are served by the first available server, as in the lobby of a bank. If each channel has its own waiting line, customers wait until the server for their line can serve them, as at a bank's drive-through facilities.

The *multiple-channel, multiple-phase* arrangement occurs when customers can be served by one of the first-phase facilities but then require service from a second-phase facility, and so on. In some cases, customers cannot switch channels after service has begun; in others they can. An example of this arrangement is a laundromat. Washing machines are the first-phase facilities, and dryers are the second-phase facilities. Some of the washing machines and dryers may be designed for extra-large loads, thereby providing the customer a choice of channels.

The most complex waiting-line problem involves customers who have unique sequences of required services; consequently, service cannot be described neatly in phases. A *mixed* arrangement is used in such a case. In the mixed arrangement, waiting lines can develop in front of each facility, as in a medical center, where a patient goes to an exam room for a nurse to take his or her blood pressure and weight, goes back to the waiting room until the doctor can see him or her, and

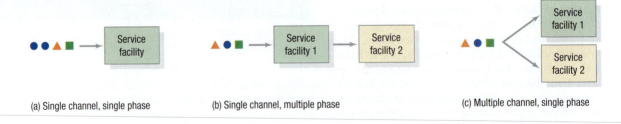

(a) Single channel, single phase   (b) Single channel, multiple phase   (c) Multiple channel, single phase

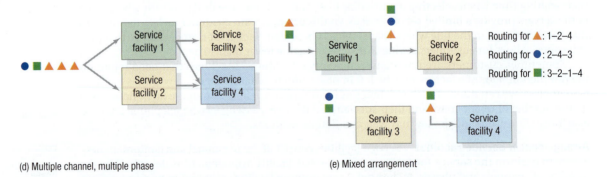

(d) Multiple channel, multiple phase   (e) Mixed arrangement

Routing for ▲: 1–2–4
Routing for ●: 2–4–3
Routing for ■: 3–2–1–4

FIGURE B.3 ▲
Examples of Service Facility Arrangements

after consultation proceeds to the laboratory to give a blood sample, radiology to have an X–ray taken, or the pharmacy for prescribed drugs, depending on specific needs.

Priority Rule

The priority rule determines which customer to serve next. Most service systems that you encounter use the first-come, first-served (FCFS) rule. The customer at the head of the waiting line has the highest priority, and the customer who arrived last has the lowest priority. Other priority disciplines might take the customer with the earliest promised due date (EDD) or the customer with the shortest expected processing time (SPT).[1]

A **preemptive discipline** is a rule that allows a customer of higher priority to interrupt the service of another customer. For example, in a hospital emergency room, patients with the most life-threatening injuries receive treatment first, regardless of their order of arrival. Modeling of systems having complex priority disciplines is usually done using computer simulation.

preemptive discipline
A rule that allows a customer of higher priority to interrupt the service of another customer.

Probability Distributions

The sources of variation in waiting-line problems come from the random arrivals of customers and the variations in service times. Each of these sources can be described with a probability distribution.

Arrival Distribution

Customers arrive at service facilities randomly. The variability of customer arrivals often can be described by a Poisson distribution, which specifies the probability that n customers will arrive in T time periods:

$$P_n = \frac{(\lambda T)^n}{n!} e^{-\lambda T} \text{ for } n = 0,1,2,\dots$$

where

P_n = probability of n arrivals in T time periods

λ = average number of customer arrivals per period

e = 2.7183

The mean of the Poisson distribution is λT, and the variance also is λT. The Poisson distribution is a discrete distribution; that is, the probabilities are for a specific number of arrivals per unit of time.

EXAMPLE B.1	**Calculating the Probability of Customer Arrivals**

Management is redesigning the customer service process in a large department store. Accommodating four customers is important. Customers arrive at the desk at the rate of two customers per hour. What is the probability that four customers will arrive during any hour?

SOLUTION

In this case $\lambda = 2$ customers per hour, $T = 1$ hour, and $n = 4$ customers. The probability that four customers will arrive in any hour is

$$P_4 = \frac{[2(1)]^4}{4!} e^{-2(1)} = \frac{16}{24} e^{-2} = 0.090$$

DECISION POINT

The manager of the customer service desk can use this information to determine the space requirements for the desk and waiting area. There is a relatively small probability that four customers will arrive in any hour. Consequently, seating capacity for two or three customers should be more than adequate unless the time to service each customer is lengthy. Further analysis on service times is warranted.

---

[1]We focus on FCFS in this supplement. See Chapter 15, "Operations Planning and Scheduling," for additional discussion of FCFS and EDD. See also Supplement J, "Operations Scheduling," for SPT and additional rules.

interarrival times

The time between customer arrivals.

Another way to specify the arrival distribution is to do it in terms of customer **interarrival times**—that is, the time between customer arrivals. If the customer population generates customers according to a Poisson distribution, the *exponential distribution* describes the probability that the next customer will arrive in the next T time periods. As the exponential distribution also describes service times, we discuss the details of this distribution in the next section.

Service Time Distribution

The exponential distribution describes the probability that the service time of the customer at a particular facility will be no more than T time periods. The probability can be calculated by using the formula

$$P(t \le T) = 1 - e^{-\mu T}$$

where

μ = average number of customers completing service per period

t = service time of the customer

T = target service time

The mean of the service time distribution is $1/\mu$, and the variance is $(1/\mu)^2$. As T increases, the probability that the customer's service time will be less than T approaches 1.0.

For simplicity, let us look at a single-channel, single-phase arrangement.

EXAMPLE B.2	**Calculating the Service Time Probability**

The management of the large department store in Example B.1 must determine whether more training is needed for the customer service clerk. The clerk at the customer service desk can serve an average of three customers per hour. What is the probability that a customer will require less than 10 minutes of service?

SOLUTION
We must have all the data in the same time units. Because μ = 3 customers per hour, we convert minutes of time to hours, or T = 10 minutes = 10/60 hour = 0.167 hour. Then

$$P(t \le T) = 1 - e^{-\mu T}$$

$$P(t \le 0.167 \text{ hour}) = 1 - e^{-3(0.167)} = 1 - 0.61 = 0.39$$

DECISION POINT
The probability that the customer will require only 10 minutes or fewer is not high, which leaves the possibility that customers may experience lengthy delays. Management should consider additional training for the clerk so as to reduce the time it takes to process a customer request.

Some characteristics of the exponential distribution do not always conform to an actual situation. The exponential distribution model is based on the assumption that each service time is independent of those that preceded it. In real life, however, productivity may improve as human servers learn about the work. Another assumption underlying the model is that very small, as well as very large, service times are possible. However, real-life situations often require a fixed-length start-up time, some cutoff on total service time, or nearly constant service time.

Using Waiting-Line Models to Analyze Operations

Operations managers can use waiting-line models to balance the gains that might be made by increasing the efficiency of the service system against the costs of doing so. In addition, managers should consider the costs of *not* making improvements to the system: Long waiting lines or long waiting times may cause customers to balk or renege. Managers should therefore be concerned about the following operating characteristics of the system.

1. *Line Length.* The number of customers in the waiting line reflects one of two conditions. Short lines could mean either good customer service or too much capacity. Similarly, long lines could indicate either low server efficiency or the need to increase capacity.

2. *Number of Customers in System.* The number of customers in line and being served also relates to service efficiency and capacity. A large number of customers in the system causes congestion and may result in customer dissatisfaction, unless more capacity is added.

3. *Waiting Time in Line.* Long lines do not always mean long waiting times. If the service rate is fast, a long line can be served efficiently. However, when waiting time seems long, customers perceive the quality of service to be poor. Managers may try to change the arrival rate of customers or design the system to make long wait times seem shorter than they really are. For example, at Walt Disney World, customers in line for an attraction are entertained by videos and also are informed about expected waiting times, which seems to help them endure the wait.

4. *Total Time in System.* The total elapsed time from entry into the system until exit from the system may indicate problems with customers, server efficiency, or capacity. If some customers are spending too much time in the service system, it may be necessary to change the priority discipline, increase productivity, or adjust capacity in some way.

5. *Service Facility Utilization.* The collective utilization of service facilities reflects the percentage of time that they are busy. Management's goal is to maintain high utilization and profitability without adversely affecting the other operating characteristics.

The best method for analyzing a waiting-line problem is to relate the five operating characteristics and their alternatives to dollars. However, placing a dollar figure on certain characteristics (such as the waiting time of a shopper in a grocery store) is difficult. In such cases, an analyst must weigh the cost of implementing the alternative under consideration against a subjective assessment of the cost of *not* making the change.

We now present three models and some examples showing how waiting-line models can help operations managers make decisions. We analyze problems requiring the single-server, multiple-server, and finite-source models, all of which are single phase. References to more advanced models are cited at the end of this supplement.

Single-Server Model

The simplest waiting-line model involves a single server and a single line of customers. To further specify the model, we make the following assumptions:

1. The customer population is infinite and all customers are patient.

2. The customers arrive according to a Poisson distribution, with a mean arrival rate of λ.

3. The service distribution is exponential, with a mean service rate of μ.

4. The mean service rate exceeds the mean arrival rate.

5. Customers are served on a first-come, first-served basis.

6. The length of the waiting line is unlimited.

With these assumptions, we can apply various formulas to describe the operating characteristics of the system:

Visitors to Disney MGM Studios, Disney World, Orlando, Florida patiently wait in line for the Aerosmith Rock N Roller Coaster ride, which is an example of a single-channel, single-phase system.

Melvyn Longhurst/Alamy

$$\rho = \text{Average utilization of the system}$$
$$= \frac{\lambda}{\mu}$$

$$P_n = \text{Probability that } n \text{ customers are in the system}$$
$$= (1 - \rho)\rho^n$$

$$L = \text{Average number of customers in the service system}$$
$$= \frac{\lambda}{\mu - \lambda}$$

$$L_q = \text{Average number of customers in the waiting line}$$
$$= \rho L$$

$$W = \text{Average time spent in the system, including service}$$
$$= \frac{1}{\mu - \lambda}$$

$$W_q = \text{Average waiting time in line}$$
$$= \rho W$$

Calculating the Operating Characteristics of a Single-Channel, Single-Phase System

MyOMLab

Active Model B.1 in MyOMLab provides additional insight on the single-server model and its uses for this problem.

The manager of a grocery store in the retirement community of Sunnyville is interested in providing good service to the senior citizens who shop in her store. Currently, the store has a separate checkout counter for senior citizens. On average, 30 senior citizens per hour arrive at the counter, according to a Poisson distribution, and are served at an average rate of 35 customers per hour, with exponential service times. Find the following operating characteristics:

a. Probability of zero customers in the system

b. Average utilization of the checkout clerk

c. Average number of customers in the system

d. Average number of customers in line

e. Average time spent in the system

f. Average waiting time in line

SOLUTION

The checkout counter can be modeled as a single-channel, single-phase system. Figure B.4 shows the results from the *Waiting-Lines* Solver from OM Explorer. Manual calculations of the equations for the *single-server model* are demonstrated in the Solved Problem at the end of the supplement.

FIGURE B.4 ▶

Waiting-Lines Solver for Single-Channel, Single-Phase System

Servers	(Number of servers s assumed to be 1 in single-serve model)
Arrival Rate (λ)	30
Service Rate (μ)	35

Probability of zero customers in the system (P_0)	0.1429
Probability of [exactly ▼] 0 customers in the system	0.1429
Average utilization of the server (p)	0.8571
Average number of customers in the system (L)	6.0000
Average number of customers in line (L_q)	5.1429
Average waiting/service time in the system (W)	0.2000
Average waiting time in line (W_q)	0.1714

Both the average waiting time in the system (*W*) and the average time spent waiting in line (*W_q*) are expressed in hours. To convert the results to minutes, simply multiply by 60 minutes/hour. For example, $W = 0.20(60) = 12.00$ minutes, and $W_q = 0.1714(60) = 10.28$ minutes.

Analyzing Service Rates with the Single-Server Model

MyOMLab

Tutor B.1 in MyOMLab provides a new example to practice the single-server model.

The manager of the Sunnyville grocery in Example B.3 wants answers to the following questions:

a. What service rate would be required so that customers averaged only 8 minutes in the system?

b. For that service rate, what is the probability of having more than four customers in the system?

c. What service rate would be required to have only a 10 percent chance of exceeding four customers in the system?

SOLUTION

The *Waiting-Lines* Solver from OM Explorer could be used iteratively to answer the questions. Here we show how to solve the problem manually.

a. We use the equation for the average time in the system and solve for μ.

$$W = \frac{1}{\mu - \lambda}$$

$$8 \text{ minutes} = 0.133 \text{ hour} = \frac{1}{\mu - 30}$$

$$0.133\mu - 0.133(30) = 1$$

$$\mu = 37.52 \text{ customers/hour}$$

b. The probability of more than four customers in the system equals 1 minus the probability of four or fewer customers in the system.

$$P = 1 - \sum_{n=0}^{4} P_n$$

$$= 1 - \sum_{n=0}^{4}(1 - \rho)\rho^n$$

and

$$\rho = \frac{30}{37.52} = 0.80$$

Then,

$$P = 1 - 0.2(1 + 0.8 + 0.8^2 + 0.8^3 + 0.8^4)$$

$$= 1 - 0.672 = 0.328$$

Therefore, there is a nearly 33 percent chance that more than four customers will be in the system.

c. We use the same logic as in part (b), except that μ is now a decision variable. The easiest way to proceed is to find the correct average utilization first, and then solve for the service rate.

$$P = 1 - (1 - \rho)(1 + \rho + \rho^2 + \rho^3 + \rho^4)$$

$$= 1 - (1 + \rho + \rho^2 + \rho^3 + \rho^4) + \rho(1 + \rho + \rho^2 + \rho^3 + \rho^4)$$

$$= 1 - 1 - \rho - \rho^2 - \rho^3 - \rho^4 + \rho + \rho^2 + \rho^3 + \rho^4 + \rho^5$$

$$= \rho^5$$

or

$$\rho = P^{1/5}$$

If $P = 0.10$,

$$\rho = (0.10)^{1/5} = 0.63$$

Therefore, for a utilization rate of 63 percent, the probability of more than four customers in the system is 10 percent. For $\lambda = 30$, the mean service rate must be

$$\frac{30}{\mu} = 0.63$$

$$\mu = 47.62 \text{ customers/hour}$$

DECISION POINT
The service rate would only have to increase modestly to achieve the 8-minute target. However, the probability of having more than four customers in the system is too high. The manager must now find a way to increase the service rate from 35 per hour to approximately 48 per hour. She can increase the service rate in several different ways, ranging from employing a high school student to help bag the groceries to installing self checkout stations.

Multiple-Server Model

With the multiple-server model, customers form a single line and choose one of s servers when one is available. The service system has only one phase. We make the following assumptions in addition to those for the single-server model: There are s identical servers, and the service distribution for each server is exponential, with a mean service time of $1/\mu$. It should always be the case that $s\mu$ exceeds λ.

EXAMPLE B.5	**Estimating Idle Time and Hourly Operating Costs with the Multiple-Server Model**

The management of the American Parcel Service terminal in Verona, Wisconsin, is concerned about the amount of time the company's trucks are idle (not delivering on the road), which the company defines as waiting to be unloaded and being unloaded at the terminal. The terminal operates with four unloading bays. Each bay requires a crew of two employees, and each crew costs $30 per hour. The estimated cost of an idle truck is $50 per hour. Trucks arrive at an average rate of three per hour, according to a Poisson distribution. On average, a crew can unload a semitrailer rig in one hour, with exponential service times. What is the total hourly cost of operating the system?

MyOMLab

Tutor B.2 in MyOMLab provides a new example to practice the multiple-server model.

MyOMLab

Active Model B.2 in
MyOMLab provides
additional insight on the
multiple-server model and
its uses for this problem.

SOLUTION

The *multiple-server model* for $s = 4$, $\mu = 1$, and $\lambda = 3$ is appropriate. To find the total cost of labor and idle trucks, we must calculate the average number of trucks in the system at all times.

Figure B.5 shows the results for the American Parcel Service problem using the *Waiting-Lines* Solver from OM Explorer. The results show that the four-bay design will be utilized 75 percent of the time and that the average number of trucks either being serviced or waiting in line is 4.53 trucks. That is, on average at any point in time, we have 4.53 idle trucks. We can now calculate the hourly costs of labor and idle trucks:

Labor cost:	$30(s) = $30(4) = 120.00
Idle truck cost:	$50(L) = $50(4.53) = 226.50
	Total hourly cost = $346.50

FIGURE B.5 ▶

Waiting-Lines Solver for
Multiple-Server Model

Servers	4
Arrival Rate (λ)	3
Service Rate (μ)	1

Probability of zero customers in the system (P_0)	0.0377
Probability of exactly ▼ 0 customers in the system	0.0377
Average utilization of the servers (ρ)	0.7500
Average number of customers in the system (L)	4.5283
Average number of customers in line (L_q)	1.5283
Average waiting/service time in the system (W)	1.5094
Average waiting time in line (W_q)	0.5094

DECISION POINT

Management must now assess whether $346.50 per day for this operation is acceptable. Attempting to reduce costs by eliminating crews will only increase the waiting time of the trucks, which is more expensive per hour than the crews. However, the service rate can be increased through better work methods; for example, L can be reduced and daily operating costs will be less.

Little's Law

Little's law

A fundamental law that relates
the number of customers in a
waiting-line system to the arrival
rate and waiting time of
customers.

One of the most practical and fundamental laws in waiting-line theory is **Little's law**, which relates the number of customers in a waiting-line system to the arrival rate and the waiting time of customers. Using the same notation we used for the single-server model, Little's law can be expressed as $L = \lambda W$ or $L_q = \lambda W_q$. However, this relationship holds for a wide variety of arrival processes, service-time distributions, and numbers of servers. The practical advantage of Little's law is that you only need to know two of the parameters to estimate the third. For example, consider the manager of a motor vehicle licensing facility who receives many complaints about the time people must spend either having their licenses renewed or getting new license plates. It would be difficult to obtain data on the times individual customers spend at the facility. However, the manager can have an assistant monitor the number of people who arrive at the facility each hour and compute the average (λ). The manager also could periodically count the number of people in the sitting area and at the stations being served and compute the average (L). Using Little's law, the manager can then estimate W, the average time each customer spent in the facility. For example, if 40 customers arrive per hour and the average number of customers being served or waiting is 30, the average time each customer spends in the facility can be computed as

$$\text{Average time in the facility} = W = \frac{L \text{ customers}}{\lambda \text{ customers/ hour}} = \frac{30}{40} = 0.75 \text{ hours, or 45 minutes}$$

If the time a customer spends at the facility is unreasonable, the manager can focus on either adding capacity or improving the work methods to reduce the time spent serving the customers.

Likewise, Little's law can be used for manufacturing processes. Suppose that a production manager knows the average time a unit of product spends at a manufacturing process (W) and the average number of units per hour that arrive at the process (λ). The production manager can then estimate the average work-in-process (L) using Little's law. *Work-in-process* (WIP) consists of items, such as components or assemblies, needed to produce a final product in manufacturing.

Cars line up at the Triborough Bridge toll, New York City. This is an example of a multiple-channel, single-phase system, where some channels are devoted to special services.

For example, if the average time a gear case used for an outboard marine motor spends at a machine center is 3 hours, and an average of five gear cases arrive at the machine center per hour, the average number of gear cases waiting and being processed (or work-in-process) at the machine center can be calculated as

$$\text{Work-in-process} = L = \lambda W = 5 \text{ gear cases/hour (3 hours)} = 15 \text{ gear cases}$$

Knowing the relationship between the arrival rate, the lead time, and the work-in-process, the manager has a basis for measuring the effects of process improvements on the work-in-process at the facility. For example, adding some capacity to a bottleneck in the process can reduce the average lead time of the product at the process, thereby reducing the work-in-process inventory.

Even though Little's law is applicable in many situations in both service and manufacturing environments, it is not applicable in situations where the customer population is finite, which we address next.

Finite-Source Model

We now consider a situation in which all but one of the assumptions of the single-server model are appropriate. In this case, the customer population is finite, having only N potential customers. If N is greater than 30 customers, the single-server model with the assumption of an infinite customer population is adequate. Otherwise, the finite-source model is the one to use.

EXAMPLE B.6	Analyzing Maintenance Costs with the Finite-Source Model

The Worthington Gear Company installed a bank of 10 robots about 3 years ago. The robots greatly increased the firm's labor productivity, but recently attention has focused on maintenance. The firm does no preventive maintenance on the robots because of the variability in the breakdown distribution. Each machine has an exponential breakdown (or interarrival) distribution with an average time between failures of 200 hours. Each machine hour lost to downtime costs $30, which means that the firm has to react quickly to machine failure. The firm employs one maintenance person, who needs 10 hours on average to fix a robot. Actual maintenance times are exponentially distributed. The wage rate is $10 per hour for the maintenance person, who can be put to work productively elsewhere when not fixing robots. Determine the daily cost of labor and robot downtime.

MyOMLab

Tutor B.3 in MyOMLab provides a new example to practice the finite-source model.

SOLUTION

The *finite-source model* is appropriate for this analysis because the customer population consists of only 10 machines and the other assumptions are satisfied. Here, $\lambda = 1/200$, or 0.005 break-down per hour, and $\mu = 1/10 = 0.10$ robot per hour. To calculate the cost of labor and robot downtime, we need to estimate the average utilization of the maintenance person and L, the average number of robots in the maintenance system at any time. Either OM Explorer or POM for Windows can be used to help with the calculations. Figure B.6

MyOMLab

Active Model B.3 in MyOMLab provides additional insight on the finite-source model and its uses for this problem.

shows the results for the Worthington Gear Problem using the *Waiting-Lines* Solver from OM Explorer. The results show that the maintenance person is utilized only 46.2 percent of the time, and the average number of robots waiting in line or being repaired is 0.76 robot. However, a failed robot will spend an average of 16.43 hours in the repair system, of which 6.43 hours of that time is spent waiting for service. While an individual robot may spend more than 2 days with the maintenance person, the maintenance person has a lot of idle time with a utilization rate of only 42.6 percent. That is why there is only an average of 0.76 robot being maintained at any point of time.

FIGURE B.6 ▶
Waiting-Lines Solver for
Finite-Source Model

Customers	10
Arrival Rate (λ)	0.005
Service Rate (μ)	0.1

Probability of zero customers in the system (P_0)	0.5380
Probability of [fewer than ▼] 0 customers in the system	#N/A
Average utilization of the server (p)	0.4620
Average number of customers in the system (L)	0.7593
Average number of customers in line (L_q)	0.2972
Average waiting/service time in the system (W)	16.4330
Average waiting time in line (W_q)	6.4330

The daily cost of labor and robot downtime is

Labor cost:	($10/hour)(8 hours/day)(0.462 utilization)	= $ 36.96
Idle robot cost:	(0.76 robot)($30/robot hour)(8 hours/day)	= 182.40
	Total daily cost	= $219.36

DECISION POINT
The labor cost for robot repair is only 20 percent of the idle cost of the robots. Management might consider having a second repair person on call in the event two or more robots are waiting for repair at the same time.

Waiting Lines and Simulation

For each of the problems we analyzed with the waiting-line models, the arrivals had a Poisson distribution (or exponential interarrival times), the service times had an exponential distribution, the service facilities had a simple arrangement, the waiting line was unlimited, and the priority discipline was first-come, first-served. Waiting-line theory has been used to develop other models in which these criteria are not met, but these models are complex. For example, POM for Windows includes a finite system-size model in which limits can be placed on the size of the system (waiting line and server capacity). It also has several models that relax assumptions on the service time distribution. Nonetheless, many times the nature of the customer population, the constraints on the line, the priority rule, the service-time distribution, and the arrangement of the facilities are such that waiting-line theory is no longer useful. In these cases, simulation often is used. MyOMLab Supplement E, "Simulation," discusses simulation programming languages and powerful PC-based packages. Here we illustrate process simulation with the SimQuick software (also provided in MyOMLab).

SimQuick   SimQuick is an easy-to-use package that is simply an Excel spreadsheet with some macros. Models can be created for a variety of simple processes, such as waiting lines, inventory control, and projects. Here, we consider the passenger security process at one terminal of a medium-sized airport between the hours of 8 A.M. and 10 A.M. The process works as follows. Passengers arriving at the security area immediately enter a single line. After waiting in line, each passenger goes through one of two inspection stations, which involves walking through a metal detector and running any carry-on baggage through a scanner. After completing this inspection, 10 percent of the passengers are randomly selected for an additional inspection, which involves a pat-down and a more thorough search of the person's

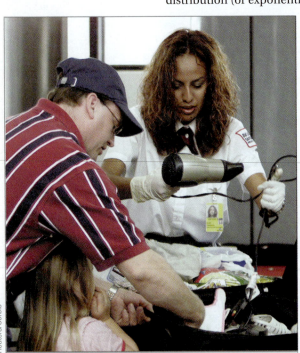

A passenger, randomly selected for additional screening, helps LA International Airport security personnel examine his luggage. The airport security process is a multiple-channel, multiple-phase system.

carry-on baggage. Two stations handle this additional inspection, and selected passengers go through only one of them. Management is interested in examining the effect of increasing the percentage of passengers who undergo the second inspection. In particular, they want to compare the waiting times for the second inspection when 10 percent, then 15 percent, and then 20 percent of the passengers are randomly selected for this inspection. Management also wants to know how opening a third station for the second inspection would affect these waiting times.

A first step in simulating this process with SimQuick is to draw a flowchart of the process using SimQuick's building blocks. SimQuick has five building blocks that can be combined in a wide variety of ways. Four of these types are used to model this process. An *entrance* is used to model the arrival of passengers at the security process. A *buffer* is used to model each of the two waiting lines, one before each type of inspection, as well as the passengers that have finished the process. Each of the four inspection stations is modeled with a *workstation*. Finally, the random selection of passengers for the second inspection is modeled with a *decision point*. Figure B.7 shows the flowchart.

Information describing each building block is entered into SimQuick tables. In this model, three key types of information are entered: (1) when people arrive at the entrance, (2) how long inspections take at the four stations, and (3) what percentage of passengers are randomly selected for the additional inspection. All of this information must be entered into SimQuick in the form of statistical distributions. The first two types of information are determined by observing the real process from 8 A.M. and 10 A.M. The third type of information is a policy decision (10 percent, 15 percent, or 20 percent).

The original model is run 30 times, simulating the arrival of passengers during the hours from 8 A.M. to 10 A.M. Statistics are collected by SimQuick and summarized. Figure B.8 provides some key results for the model of the present process as output by SimQuick (many other statistics are collected, but not displayed here).

The numbers shown are averages across the 30 simulations. The number 237.23 is the average number of passengers that enter line 1 during the simulated two hours. The two mean inventory statistics tell us, on average, 5.97 simulated passengers were standing in line 1 and 0.10 standing in line 2. The two statistics on *cycle time*, interpreted here as the time a passenger spends in one or more SimQuick building blocks, tell us that the simulated passengers in line 1 waited an average of 3.12 minutes, while those in line 2 waited 0.53 minutes. The final inventory statistic tells us that, on average, 224.57 simulated passengers passed through the security process in the simulated two hours. The next step is to change the percentage of simulated passengers selected for the second inspection to 15 percent, and then to 20 percent, and rerun the model. Of course, these process changes will increase the average waiting time for the second inspection, but by how much? The final step is to rerun these simulations with one more workstation and see its effect on the waiting time for the second inspection. All the details for this model (as well as many others) appear in the book *SimQuick: Process Simulation with Excel*, which is included, along with the SimQuick software, in MyOMLab.

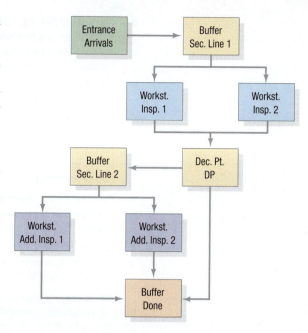

FIGURE B.7 ▲
Flowchart of Passenger Security Process

Element Types	Element Names	Statistics	Overall Means
Entrance(s)	Door	Objects entering process	237.23
Buffer(s)	Line 1	Mean inventory	5.97
		Mean cycle time	3.12
	Line 2	Mean inventory	0.10
		Mean cycle time	0.53
	Done	Final inventory	224.57

FIGURE B.8 ▲
Simulation Results of Passenger Security Process

MyOMLab

Decision Areas for Management

After analyzing a waiting-line problem, management can improve the service system by making changes in one or more of the following areas.

1. *Arrival Rates.* Management often can affect the rate of customer arrivals, λ, through advertising, special promotions, or differential pricing. For example, hotels in the Caribbean will reduce their room rates during the hot, rainy season to attract more customers and increase their utilization.

2. *Number of Service Facilities.* By increasing the number of service facilities, such as tool cribs, toll booths, or bank tellers, or by dedicating some facilities in a phase to a unique set of services, management can increase system capacity.

3. *Number of Phases.* Managers can decide to allocate service tasks to sequential phases if they determine that two sequential service facilities may be more efficient than one. For instance, in assembly lines a decision concerns the number of phases or workers needed along the assembly line. Determining the number of workers needed on the line also involves assigning a certain set of work elements to each one. Changing the facility arrangement can increase the service rate, μ, of each facility and the capacity of the system.

4. *Number of Servers per Facility.* Managers can influence the service rate by assigning more than one person to a service facility.

5. *Server Efficiency.* By adjusting the capital-to-labor ratio, devising improved work methods, or instituting incentive programs, management can increase the efficiency of servers assigned to a service facility. Such changes are reflected in μ.

6. *Priority Rule.* Managers set the priority rule to be used, decide whether to have a different priority rule for each service facility, and decide whether to allow preemption (and, if so, under what conditions). Such decisions affect the waiting times of the customers and the utilization of the servers.

7. *Line Arrangement.* Managers can influence customer waiting times and server utilization by deciding whether to have a single line or a line for each facility in a given phase of service.

Obviously, these factors are interrelated. An adjustment in the customer arrival rate might have to be accompanied by an increase in the service rate, λ, in some way. Decisions about the number of facilities, the number of phases, and waiting-line arrangements also are related.

LEARNING GOALS IN REVIEW

① **Identify the elements of a waiting-line problem in a real situation.** The section "Structure of Waiting-Line Problems," pp. 226–229, defines the four elements of every waiting-line problem. Figures B.1, B.2, and B.3 depict these elements and various service facility arrangements.

② **Describe the single-server, multiple-server, and finite-source models.** See the section "Using Waiting-Line Models to Analyze Operations," pp. 230–236, for a description and demonstration of these three models. Examples B.3, B.4 and the Solved Problem at the end of the supplement apply the single-server model. Example B.5 shows the multiple-server model and Example B.6 applies the finite-source model.

③ **Explain how to use waiting-line models to estimate the operating characteristics of a process.** Examples B.3 through

B.6 show how to obtain estimates for the important operating characteristics of processes using waiting-line models.

④ **Describe the situations where simulation should be used for waiting line analysis and the nature of the information that can be obtained.** The section "Waiting Lines and Simulation," pp. 236–237, explains when simulation must be used and discusses an example that demonstrates the nature of the managerial information that can be obtained from that analysis.

⑤ **Explain how waiting-line models can be used to make managerial decisions.** The section "Decision Areas for Management," pp. 237–238, describes seven decision areas that can be analyzed with waiting-line models.

MyOMLab helps you develop analytical skills and assesses your progress with multiple problems on utilization rate, probability of more than *n* customers in the system, number of customers waiting, probability of no customers in the system, average number of customers in the system, and service rate to keep average number of customers to a certain level.

MyOMLab Resources	Titles	Link to the Book
Video	*1st Bank Villa Italia: Waiting Lines*	Entire supplement
Active Model Exercise	B.1 Single-Server Model B.2 Multiple-Server Model with Costs B.3 Finite-Source Model with Costs	Single-Server Model; Example B.3 (p. 232) Multiple-Server Model; Example B.5 (p. 233) Finite-Source Model; Example B.6 (p. 235)
OM Explorer Solvers	Single-Server Model Multiple-Server Model Finite-Source Model	Single-Server Model; Example B.3 (p. 232); Figure B.4 (p. 232) Multiple-Server Model; Example B.5 (p. 233); Figure B.5 (p. 234) Finite-Source Model; Example B.6 (p. 235); Figure B.6 (p. 236)

MyOMLab Resources	Titles	Link to the Book
OM Explorer Tutors	B.1 Single-Server Waiting-Line Model B.2 Multi-Server Model B.3 Finite Source	Single-Server Model; Example B.3 (p. 232) Multiple-Server Model; Example B.5 (p. 233) Finite-Source Model; Example B.6 (p. 235)
POM for Windows	B.1 Single-Server Model B.2 Multiple-Server Model with Costs B.3 Finite-Source Model with Costs B.4 Finite System-Size Model	Single-Server Model; Example B.3 (p. 232) Multiple-Server Model; Example B.5 (p. 233) Finite-Source Model; Example B.6 (p. 235) Using Waiting-Line Models to Analyze Operations
Virtual Tours	New York City Fire Department	Entire supplement
Internet Exercise	Surfing the net on Google	Structure of Waiting Lines
Online Text	SimQuick: Process Simulation with Excel, 2e	Waiting Lines and Simulation; Figure B.7 (p. 237); Figure B.8 (p. 237)
Key Equations		
Image Library		

Key Equations

1. Customer arrival Poisson distribution:

$$P_n = \frac{(\lambda T)^n}{n!} e^{-\lambda T}$$

2. Service time exponential distribution:

$$P(t \le T) = 1 - e^{-\mu T}$$

3. Average utilization of the system:

$$\rho = \frac{\lambda}{\mu}$$

4. Probability that n customers are in the system:

$$P_n = (1 - \rho)\rho^n$$

5. Probability that zero customers are in the system:

$$P_0 = 1 - \rho$$

6. Average number of customers in the service system:

$$L = \frac{\lambda}{\mu - \lambda}$$

7. Average number of customers in the waiting line:

$$L_q = \rho L$$

8. Average time spent in the system, including service:

$$W = \frac{1}{\mu - \lambda}$$

9. Average waiting time in line:

$$W_q = \rho W$$

10. Little's Law

$$L = \lambda W$$

Key Terms

channel 227
customer population 226
interarrival times 230
Little's law 234

phase 227
preemptive discipline 229
priority rule 226
service facility 226

service system 226
waiting line 225

Solved Problem

A photographer takes passport pictures at an average rate of 20 pictures per hour. The photographer must wait until the customer smiles, so the time to take a picture is exponentially distributed. Customers arrive at a Poisson-distributed average rate of 19 customers per hour.

a. What is the utilization of the photographer?

b. How much time will the average customer spend with the photographer?

SOLUTION

a. The assumptions in the problem statement are consistent with a single-server model. Utilization is

$$\rho = \frac{\lambda}{\mu} = \frac{19}{20} = 0.95$$

b. The average customer time spent with the photographer is

$$W = \frac{1}{\mu - \lambda} = \frac{1}{20 - 19} = 1 \text{ hour}$$

Problems

The OM Explorer and POM for Windows software is available to all students using the 10th edition of this textbook. Go to **www.pearsonhighered.com/krajewski** to download these computer packages. If you purchased MyOMLab, you also have access to Active Models software and significant help in doing the following problems. Check with your instructor on how best to use these resources. In many cases, the instructor wants you to understand how to do the calculations by hand. At the least, the software provides a check on your calculations. When calculations are particularly complex and the goal is interpreting the results in making decisions, the software entirely replaces the manual calculations.

1. The Solomon, Smith, and Samson law firm produces many legal documents that must be word processed for clients and the firm. Requests average eight pages of documents per hour, and they arrive according to a Poisson distribution. The secretary can word process 10 pages per hour on average according to an exponential distribution.

   a. What is the average utilization rate of the secretary?

   b. What is the probability that more than four pages are waiting or being word processed?

   c. What is the average number of pages waiting to be word processed?

2. Benny's Arcade has six video game machines. The average time between machine failures is 50 hours. Jimmy, the maintenance engineer, can repair a machine in 15 hours on average. The machines have an exponential failure distribution, and Jimmy has an exponential service-time distribution.

   a. What is Jimmy's utilization?

   b. What is the average number of machines out of service, that is, waiting to be repaired or being repaired?

   c. What is the average time a machine is out of service?

3. Moore, Aiken, and Payne is a critical care dental clinic serving the emergency needs of the general public on a first-come, first-served basis. The clinic has five dental chairs, three of which are currently staffed by a dentist. Patients in distress arrive at the rate of five per hour, according to a Poisson distribution, and do not balk or renege. The average time required for an emergency treatment is 30 minutes, according to an exponential distribution. Use POM for Windows or OM Explorer to answer the following questions:

   a. If the clinic manager would like to ensure that patients do not spend more than 15 minutes on average waiting to see the dentist, are three dentists on staff adequate? If not, how many more dentists are required?

   b. From the current state of three dentists on staff, what is the change in each of the following operating characteristics when a fourth dentist is placed on staff:

   - Average utilization

   - Average number of customers in line

   - Average number of customers in the system

   c. From the current state of three dentists on staff, what is the change in each of the following operating characteristics when a fifth dentist is placed on staff:

- Average utilization
- Average number of customers in line
- Average number of customers in the system

4. Fantastic Styling Salon is run by three stylists, Jenny Perez, Jill Sloan, and Jerry Tiller, each capable of serving four customers per hour, on average. Use POM for Windows or OM Explorer to answer the following questions:

   During busy periods of the day, when nine customers on average arrive per hour, all three stylists are on staff.

   a. If all customers wait in a common line for the next available stylist, how long would a customer wait in line, on average, before being served?

   b. Suppose that each customer wants to be served by a specific stylist, 1/3 want Perez, 1/3 want Sloan, 1/3 want Tiller. How long would a customer wait in line, on average, before being served?

   During less busy periods of the day, when six customers on average arrive per hour, only Perez and Sloan are on staff.

   c. If all customers wait in a common line for the next available stylist, how long would a customer wait in line, on average, before being served?

   d. Suppose that each customer wants to be served by a specific stylist, 60 percent want Perez and 40 percent want Sloan. How long would a customer wait in line, on average, before being served by Perez? By Sloan? Overall?

5. You are the manager of a local bank where three tellers provide services to customers. On average, each teller takes 3 minutes to serve a customer. Customers arrive, on average, at a rate of 50 per hour. Having recently received complaints from some customers that they waited a long time before being served, your boss asks you to evaluate the service system. Specifically, you must provide answers to the following questions:

   a. What is the average utilization of the three-teller service system?

   b. What is the probability that no customers are being served by a teller or are waiting in line?

   c. What is the average number of customers waiting in line?

   d. On average, how long does a customer wait in line before being served?

   e. On average, how many customers would be at a teller's station and in line?

6. Pasquist Water Company (PWC) operates a 24-hour facility designed to efficiently fill water-hauling tanker trucks. Trucks arrive randomly to the facility and wait in line to access a wellhead pump. Since trucks vary in size and the filling operation is manually performed by the truck driver, the time to fill a truck is also random.

   a. If the manager of PWC uses the "multiple-server model" to calculate the operating characteristics of the facility's waiting line, list three assumptions she must make regarding the behavior of waiting trucks and the truck arrival process.

   b. Suppose an average of 336 trucks arrive each day, there are four wellhead pumps, and each pump can serve an average of four trucks per hour.

- What is the probability that exactly 10 trucks will arrive between 1:00 P.M. and 2:00 P.M. on any given day?

- How likely is it that once a truck is in position at a wellhead, the filling time will be less than 15 minutes?

   c. Contrast and comment on the performance differences between:

- One waiting line feeding all four stations.

- One waiting line feeding two wellhead pumps and a second waiting line feeding two other wellhead pumps. Assume that drivers cannot see each line and must choose randomly between them. Further, assume that once a choice is made, the driver cannot back out of the line.

7. The supervisor at the Precision Machine Shop wants to determine the staffing policy that minimizes total operating costs. The average arrival rate at the tool crib, where tools are dispensed to the workers, is eight machinists per hour. Each machinist's pay is $20 per hour. The supervisor can staff the crib either with a junior attendant who is paid $5 per hour and can process 10 arrivals per hour or with a senior attendant who is paid $12 per hour and can process 16 arrivals per hour. Which attendant should be selected, and what would be the total estimated hourly cost?

8. The daughter of the owner of a local hamburger restaurant is preparing to open a new fast-food restaurant called Hasty Burgers. Based on the arrival rates at her father's outlets, she expects customers to arrive at the drive-up window according to a Poisson distribution, with a mean of 20 customers per hour. The service rate is flexible; however, the service times are expected to follow an exponential distribution. The drive-in window is a single-server operation.

   a. What service rate is needed to keep the average number of customers in the service system (waiting line and being served) to four?

   b. For the service rate in part (a), what is the probability that more than four customers are in line and being served?

   c. For the service rate in part (a), what is the average waiting time in line for each customer? Does this average seem satisfactory for a fast-food business?

9. The manager of a branch office of Banco Mexicali observed that during peak hours an average of 20 customers arrives per hour and that there is an average of four customers in the branch office at any time. How long does the average customer spend waiting in line and being serviced?

10. Paula Caplin is manager of a major electronics repair facility owned by Fisher Electronics. Recently, top management expressed concern over the growth in the number of repair jobs in process at the facility. The average arrival rate is 120 jobs per day. The average job spends 4 days at the facility.

   a. What is the current work-in-process level at the facility?

   b. Suppose that top management has put a limit of one-half the current level of work-in-process. What goal must Paula establish and how might she accomplish it?

Advanced Problems

11. Failsafe Textiles employs three highly skilled maintenance workers who are responsible for repairing the numerous industrial robots used in its manufacturing process. A worker can fix one robot every 8 hours on average, with an exponential distribution. An average of one robot fails every 3 hours, according to a Poisson distribution. Each down robot costs the company $100.00 per hour in lost production. A new maintenance worker costs the company $80.00 per hour in salary, benefits, and equipment. Should the manager hire any new personnel? If so, how many people? What would you recommend to the manager, based on your analysis?

12. The College of Business and Public Administration at Benton University has a copy machine on each floor for faculty use. Heavy use of the five copy machines causes frequent failures. Maintenance records show that a machine fails every 2.5 days (or $\lambda = 0.40$ failure/day). The college has a maintenance contract with the authorized dealer of the copy machines. Because the copy machines fail so frequently, the dealer has assigned one person to the college to repair them. The person can repair an average of 2.5 machines per day. Using the finite-source model, answer the following questions:

   a. What is the average utilization of the maintenance person?

   b. On average, how many copy machines are being repaired or waiting to be repaired?

   c. What is the average time spent by a copy machine in the repair system (waiting and being repaired)?

13. You are in charge of a quarry that supplies sand and stone aggregates to your company's construction sites. Empty trucks from construction sites arrive at the quarry's huge piles of sand and stone aggregates and wait in line to enter the station, which can load either sand or aggregate. At the station, they are filled with material, weighed, checked out, and proceed to a construction site. Currently, nine empty trucks arrive per hour, on average. Once a truck has entered a loading station, it takes 6 minutes for it to be filled, weighed, and checked out. Concerned that trucks are spending too much time waiting and being filled, you are evaluating two alternatives to reduce the average time the trucks spend in the system. The first alternative is to add side boards to the trucks (so that more material could be loaded) and to add a helper at the loading station (so that filling time could be reduced) at a total cost of $50,000. The arrival rate of trucks would change to six per hour, and the filling time would be reduced to 4 minutes. The second alternative is to add another loading station identical to the current one at a cost of $80,000. The trucks would wait in a common line and the truck at the front of the line would move to the next available station.

   Which alternative would you recommend if you want to reduce the current average time the trucks spend in the system, including service?

Selected References

Cooper, Robert B. *Introduction to Queuing Theory*, 2nd ed. New York: Elsevier-North Holland, 1980.

Hartvigsen, David. *SimQuick: Process Simulation with Excel*, 2nd ed. Upper Saddle River, NJ: Prentice Hall, 2004.

Hillier, F.S., and G.S. Lieberman. *Introduction to Operations Research*, 2nd ed. San Francisco: Holden-Day, 1975.

Little, J.D.C. "A Proof for the Queuing Formula: $L = \lambda M$." *Operations Research*, vol. 9, (1961), pp. 383–387.

Moore, P.M. *Queues, Inventories and Maintenance.* New York: John Wiley & Sons, 1958.

Saaty, T.L. *Elements of Queuing Theory with Applications.* New York: McGraw-Hill, 1961.

VARLEY/SIPA/Newscom

Oil containment hard boom collecting foaming sea water at Queen Bess Island near Grand
Isle, Louisiana

British Petroleum Oil Spill in Gulf of Mexico

British Petroleum (BP) is one of the world's leading international oil, gas, and petrochemical products company, with operations in 29 countries, 79,000 employees, and 2010 sales of nearly $30 billion. It operates over 22,000 retail sites. On April 20, 2010, there was an explosion and fire on Transocean Ltd's Deepwater Horizon drilling rig that had been licensed to BP. It sank two days later in 5,000 feet of water, and released as many as 4.9 billion barrels of oil into the Gulf of Mexico before the damaged well was finally capped in mid-July 2010. The resulting oil spill closed down fisheries and threatened the delicate coastline and its fragile ecosystems. Pinnacle Strategies was one of the firms hired by BP to help in boosting the output of spill-fighting equipment like boats, ships, and rigs, as well as supplies of critical resources like containment booms, skimmers, and decontamination suits.

A boom is an inflatable floating device that can be used to trap oil downwind on a body of water. This oil can then be pumped into containers by skimming equipment. Limited production capacities of booms, however, represented a daunting challenge. Prestige Products in Walker, Michigan, could only make 500 feet of boom a day, whereas a single order of the size requested by BP would exceed the combined capacity of every boom manufacturer in the United States. Despite increasing the staff from 5 to 75 and raising production to 12,800 feet daily, the Prestige plant felt that it had reached its limit. That is where Ed Kincer from Pinnacle Strategies stepped in. He noticed that the boom was assembled in a flurry, with little to do in-between for several minutes. Cutters sliced boom by cutting one side, then walking 100 feet to cut the other side. Workers also sat idle while waiting for a welding machine. Waste occurred in the form of excessive walks, waiting for machines, and changing production rhythms. Kincer identified the constraints in the process, found ways to manage them, and more than tripled capacity. Prestige eventually ended up making more than a million feet of boom for BP.

Theory of constraints is the scientific approach that was used by Pinnacle to boost throughput for BP's other key suppliers as well. Kvichak Marine in Seattle quadrupled output of oil skimmers, while Illinois-based Elastec increased production from 4 skimmers a week to 26. Abasco, a Houston-based boom manufacturer, increased production by 20 percent due to rebalancing staff such that the welding operation kept going even during the breaks. At Supply Pro, a Texan manufacturer of absorbent boom, capacity increased several fold by using cellulose instead of scarce polypropylene. In six months, Pinnacle more than doubled the supply of skimmers, booms, and other critical resources by identifying bottlenecks at dozens of factories and working around them. These capacity enhancements throughout BP's supply chain ensured that lack of materials did not end up constraining the clean-up operations in the fight against the oil spill.

Source: Brown, A. "Theory of Constraints Tapped to Accelerate BP's Gulf of Mexico Cleanup." *Industry Week* (March 18, 2011); **http://www.newsweek.com/photo/2010/05/22/oil-spill-timeline.html**; **http://www.bp.com/**, May 5, 2011.

LEARNING GOALS *After reading this chapter, you should be able to:*

1 Explain the theory of constraints.

2 Understand linkage of capacity constraints to financial performance measures.

3 Identify bottlenecks.

4 Apply theory of constraints to product mix decisions.

5 Describe how to manage constraints in an assembly line.

constraint

Any factor that limits the performance of a system and restricts its output. In linear programming, a limitation that restricts the permissible choices for the decision variables.

bottleneck

A capacity constraint resource (CCR) whose available capacity limits the organization's ability to meet the product volume, product mix, or demand fluctuation required by the marketplace.

Suppose one of a firm's processes was recently reengineered, and yet results were disappointing. Costs were still high or customer satisfaction still low. What could be wrong? The answer might be constraints that remain in one or more steps in the firm's processes. A **constraint** is any factor that limits the performance of a system and restricts its output, while *capacity* is the maximum rate of output of a process or a system. When constraints exist at any step, as they did at suppliers of BP, capacity can become imbalanced—too high in some departments and too low in others. As a result, the overall performance of the system suffers.

Constraints can occur up or down the supply chain, with either the firm's suppliers or customers, or within one of the firm's processes like service/product development or order fulfillment. Three kinds of constraints can generally be identified: physical (usually machine, labor, or workstation capacity or material shortages, but could be space or quality), market (demand is less than capacity), or managerial (policy, metrics, or mind-sets that create constraints that impede work flow). A **bottleneck**[1] is a special type of a constraint that relates to the capacity shortage of a process, and is defined as any resource whose available capacity limits the organization's ability to meet the service or product volume, product mix, or fluctuating requirements demanded by the marketplace. A business system or a process would have at least one constraint or a bottleneck; otherwise, its output would be limited only by market demand. The experience of BP and other firms in the health care, banking, and manufacturing industries demonstrates how important managing constraints can be to an organization's future.

---

[1] Under certain conditions, a bottleneck is also called a *capacity constrained resource* (CCR). The process with the least capacity is called a bottleneck if its output is less than the market demand, or called a CCR if it is the least capable resource in the system but still has higher capacity than the market demand.

Managing Constraints across the Organization

Firms must manage their constraints and make appropriate capacity choices at the individual-process level, as well as at the organization level. Hence, this process involves inter-functional cooperation. Detailed decisions and choices made within each of these levels affect where resource constraints or bottlenecks show up, both within and across departmental lines. Relieving a bottleneck in one part of an organization might not have the desired effect unless a bottleneck in another part of the organization is also addressed. A bottleneck could be the sales department not getting enough sales or the loan department not processing loans fast enough. The constraint could be a lack of capital or equipment, or it could be planning and scheduling.

Managers throughout the organization must understand how to identify and manage bottlenecks in all types of processes, how to relate the capacity and performance measures of one process to another, and how to use that information to determine the firm's best service or product mix. This chapter explains how managers can best make these decisions.

Traffic bottleneck in Beijing, China.

Lou-Foto/Alamy

The Theory of Constraints

The **theory of constraints (TOC)** is a systematic management approach that focuses on actively managing those constraints that impede a firm's progress toward its goal of maximizing profits and effectively using its resources. The theory was developed nearly three decades ago by Eli Goldratt, a well-known business systems analyst. It outlines a deliberate process for identifying and overcoming constraints. The process focuses not just on the efficiency of individual processes, but also on the bottlenecks that constrain the system as a whole. Pinnacle Strategies in the opening vignette followed this theory to improve BP's operations.

TOC methods increase the firm's profits more effectively by focusing on making materials flow rapidly through the entire system. They help firms look at the big picture—how processes can be improved to increase overall work flows, and how inventory and workforce levels can be reduced while still effectively utilizing critical resources. To do this, it is important to understand the relevant performance and capacity measures at the operational level, as well as their relationship to the more broadly understood financial measures at the firm level. These measures and relationships, so critical in successfully applying the principles of the TOC, are defined in Table 7.1.

theory of constraints (TOC)

A systematic management approach that focuses on actively managing those constraints that impede a firm's progress toward its goal.

> **Creating Value through Operations Management**

Using Operations to Compete
Project Management

> **Managing Processes**

Process Strategy
Process Analysis
Quality and Performance
Capacity Planning
Constraint Management
Lean Systems

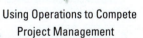
> **Managing Supply Chains**

Supply Chain Inventory Management
Supply Chain Design
Supply Chain Location Decisions
Supply Chain Integration
Supply Chain Sustainability and Humanitarian Logistics
Forecasting
Operations Planning and Scheduling
Resource Planning

TABLE 7.1 | HOW THE FIRM'S OPERATIONAL MEASURES RELATE TO ITS FINANCIAL MEASURES

Operational Measures	TOC View	Relationship to Financial Measures
Inventory (I)	All the money invested in a system in purchasing things that it intends to sell	A decrease in I leads to an increase in net profit, ROI, and cash flow.
Throughput (T)	Rate at which a system generates money through sales	An increase in T leads to an increase in net profit, ROI, and cash flows.
Operating Expense (OE)	All the money a system spends to turn inventory into throughput	A decrease in OE leads to an increase in net profit, ROI, and cash flows.
Utilization (U)	The degree to which equipment, space, or workforce is currently being used, and is measured as the ratio of average output rate to maximum capacity, expressed as a percentage	An increase in U at the bottleneck leads to an increase in net profit, ROI, and cash flows.

According to the TOC view, every capital investment in the system, including machines and work-in-process materials, represents inventory because they could all potentially be sold to make money. Producing a product or a service that does not lead to a sale will not increase a firm's throughput, but will increase its inventory and operating expenses. It is always best to manage the system so that utilization at the bottleneck resource is maximized in order to maximize throughput.

Key Principles of the TOC

The chief concept behind the TOC is that the bottlenecks should be scheduled to maximize their throughput of services or products while adhering to promised completion dates. The underlying assumption is that demand is greater or equal to the capacity of the process that produces the service or product, otherwise instead of internal changes, marketing must work towards promoting increasing its demand. For example, manufacturing a garden rake involves attaching a bow to the rake's head. Rake heads must be processed on the blanking press, welded to the bow, cleaned, and attached to the handle to make the rake, which is packaged and finally shipped to Sears, Home Depot, or Walmart, according to a specific delivery schedule. Suppose that the delivery commitments for all styles of rakes for the next month indicate that the welding station is loaded at 105 percent of its capacity, but that the other processes will be used at only 75 percent of their capacities. According to the TOC, the welding station is the bottleneck resource, whereas the blanking, cleaning, handle attaching, packaging, and shipping processes are nonbottleneck resources. Any idle time at the welding station must be eliminated to maximize throughput. Managers should therefore focus on the welding schedule.

Seven key principles of the TOC that revolve around the efficient use and scheduling of bottlenecks and improving flow and throughput are summarized in Table 7.2.

TABLE 7.2 | SEVEN KEY PRINCIPLES OF THE THEORY OF CONSTRAINTS

1. The focus should be on balancing flow, not on balancing capacity.

2. Maximizing the output and efficiency of every resource may not maximize the throughput of the entire system.

3. An hour lost at a bottleneck or a constrained resource is an hour lost for the whole system. In contrast, an hour saved at a nonbottleneck resource is a mirage because it does not make the whole system more productive.

4. Inventory is needed only in front of the bottlenecks in order to prevent them from sitting idle, and in front of assembly and shipping points in order to protect customer schedules. Building inventories elsewhere should be avoided.

5. Work, which can be materials, information to be processed, documents, or customers, should be released into the system only as frequently as the bottlenecks need it. Bottleneck flows should be equal to the market demand. Pacing everything to the slowest resource minimizes inventory and operating expenses.

6. Activating a nonbottleneck resource (using it for improved efficiency that does not increase throughput) is not the same as utilizing a bottleneck resource (that does lead to increased throughput). Activation of nonbottleneck resources cannot increase throughput, nor promote better performance on financial measures outlined in Table 7.1.

7. Every capital investment must be viewed from the perspective of its global impact on overall throughput (T), inventory (I), and operating expense (OE).

Bal Seal Engineering is a designer and manufacturer of custom seals and canted-coil™ springs for aerospace, automotive, transportation, medical and other industries. By applying many modern management principles including the theory of constraints (TOC), the company has been able to grow and improve customer satisfaction.

Practical application of the TOC involves the implementation of the following steps.

1. *Identify the System Bottleneck(s).* For the rake example, the bottleneck is the welding station because it is restricting the firm's ability to meet the shipping schedule and, hence, total value-added funds. Other ways of identifying the bottleneck will be looked at in more detail a little later in this chapter.

2. *Exploit the Bottleneck(s).* Create schedules that maximize the throughput of the bottleneck(s). For the rake example, schedule the welding station to maximize its utilization while meeting the shipping commitments to the extent possible. Also make sure that only good quality parts are passed on to the bottleneck.

3. *Subordinate All Other Decisions to Step 2.* Nonbottleneck resources should be scheduled to support the schedule of the bottleneck and not produce more than the bottleneck can handle. That is, the blanking press should not produce more than the welding station can handle, and the activities of the cleaning and subsequent operations should be based on the output rate of the welding station.

4. *Elevate the Bottleneck(s).* After the scheduling improvements in steps 1–3 have been exhausted and the bottleneck is still a constraint to throughput, management should consider increasing the capacity of the bottleneck. For example, if the welding station is still a constraint after exhausting schedule improvements, consider increasing its capacity by adding another shift or another welding machine. Other mechanisms are also available for increasing bottleneck capacity, and we address them a little later.

5. *Do Not Let Inertia Set In.* Actions taken in steps 3 and 4 will improve the welder throughput and may alter the loads on other processes. Consequently, the system constraint(s) may shift. Then, the practical application of steps 1–4 must be repeated in order to identify and manage the new set of constraints.

Because of its potential for improving performance dramatically, many manufacturers have applied the principles of the theory of constraints. All manufacturers implementing TOC principles can also dramatically change the mind-set of employees and managers. Instead of focusing solely on their own functions, they can see the "big picture" and where other improvements in the system might lie.

Identification and Management of Bottlenecks

Bottlenecks can both be internal or external to the firm, and typically represent a process, a step, or a workstation with the lowest capacity. **Throughput time** is the total elapsed time from the start to the finish of a job or a customer being processed at one or more workcenters. Where a bottleneck lies in a given service or manufacturing process can be identified in two ways. A workstation in a process is a bottleneck if (1) it has the highest total time per unit processed, or (2) it has the highest average utilization and total workload

throughput time

Total elapsed time from the start to the finish of a job or a customer being processed at one or more workcenters.

Managing Bottlenecks in Service Processes

Example 7.1 illustrates how a bottleneck step or activity can be identified for a loan approval process at a bank.

| EXAMPLE 7.1 | Identifying the Bottleneck in a Service Process |

Managers at the First Community Bank are attempting to shorten the time it takes customers with approved loan applications to get their paperwork processed. The flowchart for this process, consisting of several different activities, each performed by a different bank employee, is shown in Figure 7.1. Approved loan applications first arrive at activity or step 1, where they are checked for completeness and put in order. At step 2, the loans are categorized into different classes according to the loan amount and whether they are being requested for personal or commercial reasons. While credit checking commences at step 3, loan application data are entered in parallel into the information system for record-keeping purposes at step 4. Finally, all paperwork for setting up the new loan is finished at step 5. The time taken in minutes is given in parentheses.

Which single step is the bottleneck, assuming that market demand for loan applications exceeds the capacity of the process? The management is also interested in knowing the maximum number of approved loans this system can process in a 5-hour work day.

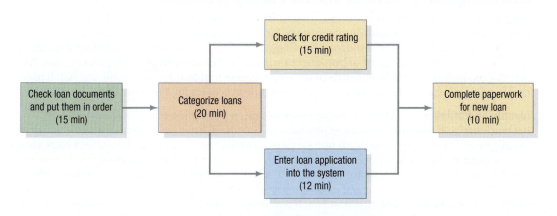

▲ **FIGURE 7.1**

Processing Credit Loan Applications at First Community Bank

SOLUTION

We define the bottleneck as step 2, which has the highest time per loan processed. The throughput time to complete an approved loan application is 15 + 20 + max (15, 12) + 10 = **60** minutes. Although we assume no waiting time in front of any step, in practice such a smooth process flow is not always the case. So the actual time taken for completing an approved loan will be longer than 60 minutes due to nonuniform arrival of applications, variations in actual processing times, and the related factors.

The capacity for loan completions is derived by translating the "minutes per customer" at the bottleneck step to "customer per hour." At First Community Bank, it is 3 customers per hour because the bottleneck step 2 can process only 1 customer every 20 minutes (60/3).

DECISION POINT

Step 2 is the bottleneck constraint. The bank will be able to complete a maximum of only 3 loan accounts per hour, or 15 new loan accounts in a 5-hour day. Management can increase the flow of loan applications by increasing the capacity of step 2 up to the point where another step becomes the bottleneck.

Due to constrained resources like doctors, nurses, and equipment, patients wait for medical care in a crowded waiting room at South Central Family Health Center in Los Angeles, California

A front-office process with high customer contact and divergence does not enjoy the simple line flows shown in Example 7.1. Its operations may serve many different customer types, and the demands on any one operation could vary considerably from one day to the next. However, bottlenecks can still be identified by computing the average utilization of each operation. However, the variability in workload also creates *floating bottlenecks*. One week the mix of work may make operation 1 a bottleneck, and the next week it may make operation 3 the bottleneck. This type of variability increases the complexity of day-to-day scheduling. In this situation, management prefers lower utilization rates, which allow greater slack to absorb unexpected surges in demand.

TOC principles outlined here are fairly broad-based and widely applicable. They can be useful for evaluating individual processes as well as large systems for both manufacturers as well as service providers. Service organizations, such as Delta Airlines, United Airlines, and major hospitals across the United States, including the U.S. Air Force health care system, use the TOC to their advantage.

Managing Bottlenecks in Manufacturing Processes

Bottlenecks can exist in all types of manufacturing processes, including the job process, batch process, line process, and continuous process. Since these processes differ in their design, strategic intent, and allocation of resources (see Chapter 3, "Process Strategy," for additional details), identification and management of bottlenecks will also differ accordingly with process type. We first discuss in this section issues surrounding management of bottlenecks in job and batch processes, while relegating constraint management in line processes for a later section.

Identifying Bottlenecks   Manufacturing processes often pose some complexities when identifying bottlenecks. If multiple services or products are involved, extra setup time at a workstation is usually needed to change over from one service or product to the next, which in turn increases the overload at the workstation being changed over. *Setup times* and their associated costs affect the size of the lots traveling through the job or batch processes. Management tries to reduce setup times because they represent unproductive time for workers or machines and thereby allow for smaller, more economic, batches. Nonetheless, whether setup times are significant or not, one way to identify a bottleneck operation is by its utilization. Example 7.2 illustrates how a bottleneck can be identified in a manufacturing setting where setups are negligible.

| EXAMPLE 7.2 | Identifying the Bottleneck in a Batch Process |

Diablo Electronics manufactures four unique products (A, B, C, and D) that are fabricated and assembled in five different workstations (V, W, X, Y, and Z) using a small batch process. Each workstation is staffed by a worker who is dedicated to work a single shift per day at an assigned workstation. Batch setup times have been reduced to such an extent that they can be considered negligible. A flowchart denotes the path each product follows through the manufacturing process as shown in Figure 7.2, where each product's price, demand per week, and processing times per unit are indicated as well. Inverted triangles represent purchased parts and raw materials consumed per unit at different workstations. Diablo can make and sell up to the limit of its demand per week, and no penalties are incurred for not being able to meet all the demand.

Which of the five workstations (V, W, X, Y, or Z) has the highest utilization, and thus serves as the bottle-neck for Diablo Electronics?

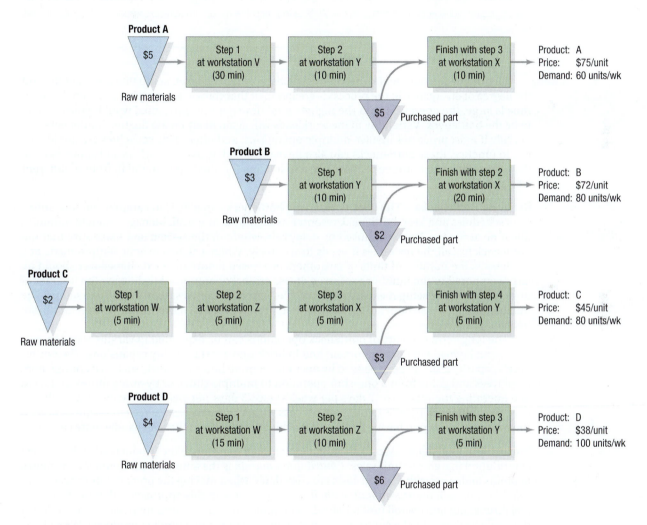

▲ **FIGURE 7.2**
Flowchart for Products A, B, C, and D

SOLUTION

Because the denominator in the utilization ratio is the same for every workstation, with one worker per machine at each step in the process, we can simply identify the bottleneck by computing aggregate workloads at each workstation.

The firm wants to satisfy as much of the product demand in a week as it can. Each week consists of 2,400 minutes of available production time. Multiplying the processing time at each station for a given product with the number of units demanded per week yields the workload represented by that product. These loads are summed across all products going through a workstation to arrive at the total load for the workstation, which is then compared with the others and the existing capacity of 2,400 minutes.

Workstation	Load from Product A	Load from Product B	Load from Product C	Load from Product D	Total Load (min)
V	60 × 30 = 1800	0	0	0	1,800
W	0	0	80 × 5 = 400	100 × 15 = 1,500	1,900
X	60 × 10 = 600	80 × 20 = 1,600	80 × 5 = 400	0	2,600
Y	60 × 10 = 600	80 × 10 = 800	80 × 5 = 400	100 × 5 = 500	2,300
Z	0	0	80 × 5 = 400	100 × 10 = 1,000	1,400

DECISION POINT

Workstation X is the bottleneck for Diablo Electronics because the aggregate workload at X is larger than the aggregate workloads of workstations V, W, Y, and Z and the maximum available capacity of 2,400 minutes per week.

Identifying the bottlenecks becomes considerably harder when setup times are lengthy and the degree of divergence in the process is greater than that shown in Example 7.2. When the setup time is large, the operation with the highest total time per unit processed would typically tend to be the bottleneck. Variability in the workloads will again likely create floating bottlenecks, especially if most processes involve multiple operations, and often their capacities are not identical. In practice, these bottlenecks can also be determined by asking workers and supervisors in the plant where the bottlenecks might lie and looking for piled up material in front of different workstations.

Relieving Bottlenecks   The key to preserving bottleneck capacity is to carefully monitor short-term schedules and keep bottleneck resource as busy as is practical. Managers should minimize idle time at the bottlenecks, caused by delays elsewhere in the system and make sure that the bottleneck has all the resources it needs to stay busy. When a changeover or setup is made at a bottleneck, the number of units or customers processed before the next changeover should be large compared to the number processed at less critical operations. Maximizing the number of units processed per setup means fewer setups per year and, thus, less total time lost to setups. The number of setups also depends on the required product variety; more variety necessitates more frequent changeovers.

The long-term capacity of bottleneck operations can be expanded in various ways. Investments can be made in new equipment and in brick-and-mortar facility expansions. The bottleneck's capacity also can be expanded by operating it more hours per week, such as by hiring more employees and going from a one-shift operation to multiple shifts, or by hiring more employees and operating the plant 6 or 7 days per week versus 5 days per week. Managers also might relieve the bottleneck by redesigning the process, either through *process reengineering* or *process improvement*, or by purchasing additional machines or machines that can handle more capacity.

Product Mix Decisions   Managers might be tempted to produce the products with the highest contribution margins or unit sales. *Contribution margin* is the amount each product contributes to profits and overhead; no fixed costs are considered when making the product mix decision. We call this approach the *traditional method*. The problem with this approach is that the firm's actual throughput and overall profitability depend more upon the contribution margin generated at the bottleneck than by the contribution margin of each individual product produced. We call this latter approach the *bottleneck method*. Example 7.3 illustrates both of these methods.

EXAMPLE 7.3	Determining the Product Mix Using Contribution Margin

The senior management at Diablo Electronics (see Exercise 7.2) wants to improve profitability by accepting the right set of orders, and so collected some additional financial data. Variable overhead costs are $8,500 per week. Each worker is paid $18 per hour and is paid for an entire week, regardless of how much the worker is used. Consequently, labor costs are fixed expenses. The plant operates one 8-hour shift per day, or 40 hours each week. Currently, decisions are made using the traditional method, which is to accept as much of the highest contribution margin product as possible (up to the limit of its demand), followed by the next highest contribution margin product, and so on until no more capacity is available. Pedro Rodriguez, the newly hired production supervisor, is knowledgeable about the theory of constraints and bottleneck-based scheduling.

He believes that profitability can indeed be improved if bottleneck resources were exploited to determine the product mix. What is the change in profits if, instead of the traditional method used by Diablo Electronics, the bottleneck method advocated by Pedro is used to select the product mix?

SOLUTION

Decision Rule 1: Traditional Method

Select the best product mix according to the highest overall contribution margin of each product.

Step 1: Calculate the contribution margin per unit of each product as shown here.

	A	B	C	D
Price	$75.00	$72.00	$45.00	$38.00
Raw material and purchased parts	−10.00	−5.00	−5.00	−10.00
= Contribution margin	$65.00	$67.00	$40.00	$28.00

When ordered from highest to lowest, the contribution margin per unit sequence of these products is B, A, C, D.

Step 2: Allocate resources V, W, X, Y, and Z to the products in the order decided in step 1. Satisfy each demand until the bottleneck resource (workstation X) is encountered. Subtract minutes away from 2,400 minutes available for each week at each stage.

Work Center	Minutes at the Start	Minutes Left After Making 80 B	Minutes Left After Making 60 A	Can Only Make 40 C	Can Still Make 100 D
V	2,400	2,400	600	600	600
W	2,400	2,400	2,400	2,200	700
X	2,400	800	200	0	0
Y	2,400	1,600	1,000	800	300
Z	2,400	2,400	2,400	2,200	1,200

The best product mix according to this traditional approach is then 60 A, 80 B, 40 C, and 100 D.

Step 3: Compute profitability for the selected product mix.

Profits		
Revenue	(60 × $75) + (80 × $72) + (40 × $45) + (100 × $38)	= $15,860
Materials	(60 × $10) + (80 × $5) + (40 × $5) + (100 × $10)	= −$2,200
Labor	(5 workers) × (8 hours/day) × (5 days/week) × ($18/hour)	= −$3,600
Overhead		= −$8,500
Profit		= $1,560

Manufacturing the product mix of 60 A, 80 B, 40 C, and 100 D will yield a profit of $1,560 per week.

Decision Rule 2: Bottleneck Method

Select the best product mix according to the dollar contribution margin per minute of processing time at the bottleneck workstation X. This method would take advantage of the principles outlined in the theory of constraints and get the most dollar benefit from the bottleneck.

Step 1: Calculate the contribution margin/minute of processing time at bottleneck workstation X:

	Product A	Product B	Product C	Product D
Contribution margin	$65.00	$67.00	$40.00	$28.00
Time at bottleneck	10 minutes	20 minutes	5 minutes	0 minutes
Contribution margin per minute	$6.50	$3.35	$8.00	Not defined

When ordered from highest to lowest contribution margin/minute at the bottleneck, the manufacturing sequence of these products is D, C, A, B, which is reverse of the earlier order. Product D is scheduled first because it does not consume any resources at the bottleneck.

Step 2:   Allocate resources V, W, X, Y, and Z to the products in the order decided in step 1. Satisfy each demand until the bottleneck resource (workstation X) is encountered. Subtract minutes away from 2,400 minutes available for each week at each stage.

Work Center	Minutes at the Start	Minutes Left After Making 100 D	Minutes Left After Making 80 C	Minutes Left After Making 60 A	Can Only Make 70 B
V	2,400	2,400	2,400	600	600
W	2,400	900	500	500	500
X	2,400	2,400	2,000	1,400	0
Y	2,400	1,900	1,500	900	200
Z	2,400	1,400	1,000	1,000	1,000

The best product mix according to this bottleneck-based approach is then 60 A, 70 B, 80 C, and 100 D.

Step 3:   Compute profitability for the selected product mix.

Profits		
Revenue	$(60 \times \$75) + (70 \times \$72) + (80 \times \$45) + (100 \times \$38)$	= $16,940
Materials	$(60 \times \$10) + (70 \times \$5) + (80 \times \$5) + (100 \times \$10)$	= −$2,350
Labor	$(5 \text{ workers}) \times (8 \text{ hours/day}) \times (5 \text{ days/week}) \times (\$18/\text{hour})$	= −$3,600
Overhead		= −$8,500
Profit		= $2,490

Manufacturing the product mix of 60 A, 70 B, 80 C, and 100 D will yield a profit of $2,490 per week.

DECISION POINT

By focusing on the bottleneck resources in accepting customer orders and determining the product mix, the sequence in which products are selected for production is reversed from **B, A, C, D** to **D, C, A, B**. Consequently, the product mix is changed from 60 A, 80 B, 40 C, and 100 D to 60 A, 70 B, 80 C, and 100 D. The increase in profits by using the bottleneck method is $930, ($2,490 − $1,560), or almost 60 percent over the traditional approach.

Linear programming (see Supplement D) could also be used to find the best product mix in Example 7.3. It must be noted, however, that the problem in Example 7.3 did not involve significant setup times. Otherwise, they must be taken into consideration for not only identifying the bottleneck, but also in determining the product mix. The experiential learning exercise of Min-Yo Garment Company at the end of this chapter provides an interesting illustration of how the product mix can be determined when setup times are significant. In this way, the principles behind the theory of constraints can be exploited for making better decisions about a firm's most profitable product mix.

drum-buffer-rope (DBR)

A planning and control system that regulates the flow of work-in-process materials at the bottleneck or the capacity constrained resource (CCR) in a productive system.

Drum-Buffer-Rope Systems   **Drum-Buffer-Rope (DBR)** is a planning and control system based on the theory of constraints that is often used in manufacturing firms to plan and schedule production. It works by regulating the flow of work-in-process materials at the bottleneck or the capacity constrained resource (CCR). The bottleneck schedule is the *drum* because it sets the beat or the production rate for the entire plant and is linked to the market demand. The *buffer* is a time buffer that plans early flows to the bottleneck and thus protects it from disruption. It also ensures that the bottleneck is never starved for work. A finished-goods inventory buffer can also be placed in front of the shipping point in order to protect customer shipping schedules. Finally, the *rope* represents the tying of material release to the drum beat, which is the rate at which the bottleneck controls the throughput of the entire plant. It is thus a communication device to ensure that raw material is not introduced into the system at a rate faster than what the bottleneck can handle. Completing the loop, *buffer management* constantly monitors the execution of incoming bottleneck work. Working together, the drum, the buffer, and the rope can help managers create a production schedule that reduces lead times and inventories while simultaneously increasing throughput and on-time delivery.

To better understand the drum-buffer-rope system, consider the schematic layout shown in Figure 7.3. Process B, with a capacity of only 500 units per week, is the bottleneck because the upstream Process A and downstream Process C have capacities of 800 units per week and 700 units per week, respectively, and the market demand is 650 units per week, on average. In this case, because the capacity at process B is less than the market demand, it is the bottleneck. A constraint time buffer, which can be in the form of materials arriving earlier than needed, is placed right in front of the bottleneck (Process B). A shipping buffer, in the form of finished goods inventory, can also be placed prior to the shipping schedule in order to protect customer orders that are firm. Finally, a rope ties the material release schedule to match the schedule, or drum beat, at the bottleneck. The material flow is pulled forward by the drum beat prior to the bottleneck, while it is pushed downstream toward the customer subsequent to the bottleneck.

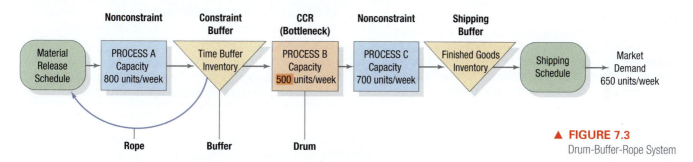

▲ FIGURE 7.3
Drum-Buffer-Rope System

DBR specifically strives to improve throughput by better utilizing the bottleneck resource and protecting it from disruption through the time buffer and protective buffer capacity elsewhere. So while the process batch in the DBR is any size that minimizes setups and improves utilization at the bottleneck, at nonconstrained resources the process batches are equal to what is needed for production at that time. The material can consequently be released in small batches known as transfer batches at the release point, which then combine at the constraint buffer to make a full process batch at the bottleneck. Transfer batches can be as small as one unit each, to allow a downstream workstation to start work on a batch before it is completely finished at the prior process. Using transfer batches typically facilitates a reduction in overall lead time.

DBR can be an effective system to use when the product the firm produces is relatively simple and the production process has more line flows. Planning is greatly simplified in this case and primarily revolves around scheduling the constrained resource and triggering other points to meet that bottleneck's schedule. Effectively implementing a DBR system requires an understanding of the TOC principles. However, such a system can be utilized in many different kinds of manufacturing and service organizations, either by itself or in conjunction with other planning and control systems. Managerial Practice 7.2 illustrates how the use of a DBR system improved the performance of the Marine Corps Maintenance Center in Albany, Georgia.

MANAGERIAL PRACTICE 7.1 — The Drum-Buffer-Rope System at a U.S. Marine Corps Maintenance Center

The U.S. Marine Corps Maintenance Center in Albany, Georgia, overhauls and repairs vehicles used by the corps, such as fuel tankers, trucks, earthmoving equipment, amphibious vehicles, and light armored vehicles. The overhaul process starts with the disassembly of each vehicle to determine the amount and nature of work that needs to be performed. The type and duration of the repair work can vary tremendously for even the same type of vehicle. Faced with such uncertainty, the Center was struggling until four years ago to complete its equipment repairs on time, and it had an increasing backlog to boot. For instance, the Center was able to repair only about 5 MK-48s (heavy-duty haulers) per month, when twice as many MK-48s—10 per month—typically

needed repair. Different units of the corps were threatening to divert their orders to the private-sector repair companies.

TOC principles were used to identify the bottlenecks on the shop floor. After the Center's operations were studied in depth, however, contrary to everyone's expectations, it was discovered that more than enough capacity was available to repair and overhaul 10 MK-48s per month. The problem was not capacity; it was the Center's scheduling system. Products were being pushed onto the shop floor without regard for the status of the resources on the floor. Thus, what the Center had was a policy constraint related to the scheduling process, not an actual physical resource constraint.

In order to improve the Center's performance, its managers implemented a simplified form of a drum-buffer-rope system as shown in Figure 7.3. Since the Marine Corps Maintenance Center was not constrained by any internal resource, the drum in such a simplified system was based on firm orders. As orders came in, a quick check was done to measure the total load the Center's least-capable resource was handling. If the resource was not too heavily loaded, the order was accepted and released onto the shop floor for processing. The rope tied the shipping schedule directly to the material release schedule instead of the bottleneck schedule, and the only buffer maintained was the shipping buffer. Such a simplified DBR system did not require any specialized software. It focused simply on the market demand for repairs.

The Center's results following the change were impressive. Repair cycle times were reduced from an average of 167 days to 58 days, work-in-process levels were reduced from 550 percent of demand to 140 percent, and the cost to repair products went down by 25 to 30 percent due to an increased throughput. The Center's ability to repair MK-48s became much more flexible, too. In fact, it can now repair as many as 23 MK-48s per month. The Center is on schedule for 99 percent of the production lines where the TOC principles have been implemented, and the repair costs have decreased by 25 percent. Carrying out these simple improvements made the Albany Maintenance Center a world-class overhaul and repair operation.

Repairs to assault vehicles can vary tremendously at the U.S. Marine Corps Maintenance Center in Albany, Georgia. The center struggled to keep up with its repairs until managers implemented the simplified form of a drum-buffer-rope system. The result? Repair times fell from 167 days to just 58 days, on average.

Source: Mandyam Srinivasan, Darren Jones, and Alex Miller, "Applying Theory of Constraints Principles and Lean Thinking at the Marine Corps Maintenance Center," *Defense Acquisition Review Journal,* August–November 2004; M. Srinivasan, Darren Jones, and Alex Miller, "Corps Capabilities," APICS Magazine (March 2005), pp. 46–50.

Managing Constraints in a Line Process

As noted in Chapter 3, "Process Strategy," products created by a line process include the assembly of computers, automobiles, appliances, and toys. Such assembly lines can exist in providing services as well. For instance, putting together a standardized hamburger with a fixed sequence of steps is akin to operating an assembly line. While the product mix or demand volumes do not change as rapidly for line processes as for job or batch processes, the load can shift between work centers in a line as the end product being assembled is changed or the total output rate of the line is altered. Constraints arising out of such actions can be managed by balancing the workload between different stations in a line, which we explain next in greater detail.

Line Balancing

line balancing

The assignment of work to stations in a line process so as to achieve the desired output rate with the smallest number of workstations.

Line balancing is the assignment of work to stations in a line process so as to achieve the desired output rate with the smallest number of workstations. Normally, one worker is assigned to a station. Thus, the line that produces at the desired pace with the fewest workers is the most efficient one. Achieving this goal is much like the theory of constraints, because both approaches are concerned about bottlenecks. Line balancing differs in how it addresses bottlenecks. Rather than (1) taking on new customer orders to best use bottleneck capacity or (2) scheduling so that bottleneck resources are conserved, line balancing takes a third approach. It (3) creates workstations with workloads as evenly balanced as possible. It seeks to create workstations so that the capacity utilization for the bottleneck is not much higher than for the other workstations in the line. Another difference is that line balancing applies only to line processes that do assembly work, or to work that can be bundled in many ways to create the jobs for each workstation in the line. The latter situation can be found both in manufacturing and service settings.

Line balancing must be performed when a line is set up initially, when a line is rebalanced to change its hourly output rate, or when a product or process changes. The goal is to obtain workstations with well-balanced workloads (e.g., every station takes roughly 3 minutes per customer in a cafeteria line with different food stations).

work elements

The smallest units of work that can be performed independently.

The analyst begins by separating the work into **work elements**, which are the smallest units of work that can be performed independently. The analyst then obtains the time standard for each element and identifies the work elements, called **immediate predecessors**, which must be done before the next element can begin.

immediate predecessors

Work elements that must be done before the next element can begin.

Precedence Diagram   Most lines must satisfy some technological precedence requirements; that is, certain work elements must be done before the next can begin. However, most lines also allow for some latitude and more than one sequence of operations. To help you better visualize immediate predecessors, let us run through the construction of a **precedence diagram**.[2] We denote the work elements by circles, with the time required to perform the work shown below each circle. Arrows lead from immediate predecessors to the next work element. Example 7.4 illustrates a manufacturing process, but a back office line-flow process in a service setting can be approached similarly.

precedence diagram

A diagram that allows one to visualize immediate predecessors better; work elements are denoted by circles, with the time required to perform the work shown below each circle.

EXAMPLE 7.4	Constructing a Precedence Diagram

Green Grass, Inc., a manufacturer of lawn and garden equipment, is designing an assembly line to produce a new fertilizer spreader, the Big Broadcaster. Using the following information on the production process, construct a precedence diagram for the Big Broadcaster.

Work Element	Description	Time (sec)	Immediate Predecessor(s)
A	Bolt leg frame to hopper	40	None
B	Insert impeller shaft	30	A
C	Attach axle	50	A
D	Attach agitator	40	B
E	Attach drive wheel	6	B
F	Attach free wheel	25	C
G	Mount lower post	15	C
H	Attach controls	20	D, E
I	Mount nameplate	18	F, G
		Total 244	

SOLUTION

Figure 7.4 shows the complete diagram. We begin with work element A, which has no immediate predecessors. Next, we add elements B and C, for which element A is the only immediate predecessor. After entering time standards and arrows showing precedence, we add elements D and E, and so on. The diagram simplifies interpretation. Work element F, for example, can be done anywhere on the line after element C is completed. However, element I must await completion of elements F and G.

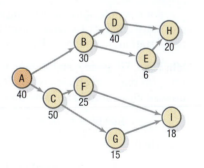

◀ **FIGURE 7.4**
Precedence Diagram for Assembling the Big Broadcaster

DECISION POINT

Management now has enough information to develop a line-flow layout that clusters work elements to form workstations, with a goal being to balance the workloads and, in the process, minimize the number of workstations required.

Desired Output Rate   The goal of line balancing is to match the output rate to the staffing or production plan. For example, if the plan calls for 4,800 units or customers per week and the line operates 80 hours per week, the desired output rate ideally would be 60 units or customers (4,800/80) per hour. Matching output to the plan ensures on-time delivery and prevents buildup of unwanted inventory or customer delays. However, managers should avoid rebalancing a line too frequently because each time a line is rebalanced many workers' jobs on the line must be redesigned, temporarily hurting productivity and sometimes even requiring a new detailed layout for some stations.

---

[2]Precedence relationships and precedence diagrams are also important in the entirely different context of project management covered in Chapter 2.

cycle time

The maximum time allowed for work on a unit at each station.

Cycle Time   After determining the desired output rate for a line, the analyst can calculate the line's cycle time. A line's **cycle time** is the maximum time allowed for work on a unit at each station.[3] If the time required for work elements at a station exceeds the line's cycle time, the station will be a bottleneck, preventing the line from reaching its desired output rate. The target cycle time is the reciprocal of the desired hourly output rate:

$$c = \frac{1}{r}$$

where

$$c = \text{cycle time in hours per unit}$$

$$r = \text{desired output rate in units per hour}$$

For example, if the line's desired output rate is 60 units per hour, the cycle time is $c = 1/60$ hour per unit, or 1 minute.

Theoretical Minimum   To achieve the desired output rate, managers use line balancing to assign every work element to a station, making sure to satisfy all precedence requirements and to minimize the number of stations, n, formed. If each station is operated by a different worker, minimizing n also maximizes worker productivity. Perfect balance is achieved when the sum of the work-element times at each station equals the cycle time, c, and no station has any idle time. For example, if the sum of each station's work-element times is 1 minute, which is also the cycle time, the line achieves perfect balance. Although perfect balance usually is unachievable in practice, owing to the unevenness of work-element times and the inflexibility of precedence requirements, it sets a benchmark, or goal, for the smallest number of stations possible. The **theoretical minimum (TM)** for the number of stations is

theoretical minimum (TM)

A benchmark or goal for the smallest number of stations possible, where the total time required to assemble each unit (the sum of all work-element standard times) is divided by the cycle time.

$$\text{TM} = \frac{\Sigma t}{c}$$

where

$$\Sigma t = \text{total time required to assemble each unit (the sum of all work-element standard times)}$$

$$c = \text{cycle time}$$

For example, if the sum of the work-element times is 15 minutes and the cycle time is 1 minute, TM = 15/1, or 15 stations. Any fractional values obtained for TM are rounded up because fractional stations are impossible.

Idle Time, Efficiency, and Balance Delay   Minimizing n automatically ensures (1) minimal idle time, (2) maximal efficiency, and (3) minimal balance delay. Idle time is the total unproductive time for all stations in the assembly of each unit:

$$\text{Idle time} = nc - \Sigma t$$

where

$$n = \text{number of stations}$$

$$c = \text{cycle time}$$

$$\Sigma t = \text{total standard time required to assemble each unit}$$

Efficiency is the ratio of productive time to total time, expressed as a percent:

$$\text{Efficiency (\%)} = \frac{\Sigma t}{nc}(100)$$

balance delay

The amount by which efficiency falls short of 100 percent.

Balance delay is the amount by which efficiency falls short of 100 percent:

$$\text{Balance delay (\%)} = 100 - \text{Efficiency}$$

As long as c is fixed, we can optimize all three goals by minimizing n.

---

[3]Except in the context of line balancing, *cycle time* has a different meaning. It is the elapsed time between starting and completing a job. Some researchers and practitioners prefer the term *lead time*.

Skoda Automotive assembly line in Mlada Boleslav. Since 1991, Skoda has been a part of Volkswagen for over 20 years.

Nataliya Hora/Shutterstock.com

EXAMPLE 7.5 | **Calculating the Cycle Time, Theoretical Minimum, and Efficiency**

Green Grass's plant manager just received marketing's latest forecasts of Big Broadcaster sales for the next year. She wants its production line to be designed to make 2,400 spreaders per week for at least the next 3 months. The plant will operate 40 hours per week.

a. What should be the line's cycle time?

b. What is the smallest number of workstations that she could hope for in designing the line for this cycle time?

c. Suppose that she finds a solution that requires only five stations. What would be the line's efficiency?

MyOMLab
Tutor 7.1 in MyOMLab provides another example to calculate these line-balancing measures.

SOLUTION

a. First, convert the desired output rate (2,400 units per week) to an hourly rate by dividing the weekly output rate by 40 hours per week to get $r = 60$ units per hour. Then, the cycle time is

$$c = 1/r = 1/60 \text{ (hour/unit)} = 1 \text{ minute/unit} = 60 \text{ seconds/unit}$$

b. Now, calculate the theoretical minimum for the number of stations by dividing the total time, Σt, by the cycle time, $c = 60$ seconds. Assuming perfect balance, we have

$$\text{TM} = \frac{\Sigma t}{c} = \frac{244 \text{ seconds}}{60 \text{ seconds}} = 4.067 \text{ or } 5 \text{ stations}$$

c. Now, calculate the efficiency of a five-station solution, assuming for now that one can be found:

$$\text{Efficiency (\%)} = \frac{\Sigma t}{nc}(100) = \frac{244}{5(60)}(100) = 81.3\%$$

DECISION POINT

If the manager finds a solution with five stations that satisfies all precedence constraints, then that is the optimal solution; it has the minimum number of stations possible. However, the efficiency (sometimes called the *theoretical maximum efficiency*) will be only 81.3 percent. Perhaps the line should be operated less than 40 hours per week (thereby adjusting the cycle time) and the employees transferred to other kinds of work when the line does not operate.

Finding a Solution Often, many assembly-line solutions are possible, even for such simple problems as Green Grass's. The goal is to cluster the work elements into workstations so that (1) the number of workstations required is minimized, and (2) the precedence and cycle-time requirements are not violated. The idea is to assign work elements to workstations subject to the precedence requirements so that the work content for the station is equal (or nearly so, but less than) the cycle time for the line. In this way, the number of workstations will be minimized.

Here we use the trial-and-error method to find a solution, although commercial software packages are also available. Most of these packages use different decision rules in picking which work element to assign next to a workstation being created. The ones used by POM for Windows are described in Table 7.3. The solutions can be examined for improvement, because there is no guarantee that they are optimal or even feasible. Some work elements cannot be assigned to the same station, some changes can be made to reduce the number of stations, or some shifts can provide better balance between stations.

TABLE 7.3 | **HEURISTIC DECISION RULES IN ASSIGNING THE NEXT WORK ELEMENT TO A WORKSTATION BEING CREATED**

Create one station at a time. For the station now being created, identify the unassigned work elements that qualify for assignment: They are candidates if

1. All of their predecessors have been assigned to this station or stations already created.
2. Adding them to the workstation being created will not create a workload that exceeds the cycle time.

Decision Rule	Logic
Longest work element	Picking the candidate with the longest time to complete is an effort to fit in the most difficult elements first, leaving the ones with short times to "fill out" the station.
Shortest work element	This rule is the opposite of the longest work element rule because it gives preference in workstation assignments to those work elements that are quicker. It can be tried because no single rule guarantees the best solution. It might provide another solution for the planner to consider.
Most followers	When picking the next work element to assign to a station being created, choose the element that has the most *followers* (due to precedence requirements). In Figure 7.4, item C has three followers (F, G, and I) whereas item D has only one follower (H). This rule seeks to maintain flexibility so that good choices remain for creating the last few workstations at the end of the line.
Fewest followers	Picking the candidate with the fewest followers is the opposite of the most followers rule.

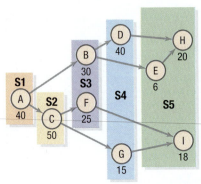

▲ **FIGURE 7.5**
Big Broadcaster Precedence
Diagram Solution

pacing

The movement of product from one station to the next as soon as the cycle time has elapsed.

Figure 7.5 shows a solution that creates just five workstations. We know that five is the minimum possible, because five is the theoretical minimum found in Example 5. All of the precedence and cycle-time requirements are also satisfied. Consequently, the solution is optimal for this problem. Each worker at each station must perform the work elements in the proper sequence. For example, workstation S5 consists of one worker who will perform work elements E, H, and I on each unit that comes along the assembly line. The processing time per unit is 44 seconds (6 + 20 + 18) which does not exceed the cycle time of 60 seconds (see Example 5). Furthermore, the immediate predecessors of these three work elements are assigned to this workstation or upstream workstations, so their precedence requirements are satisfied. The worker at workstation S5 can do element I at any time but will not start element H until element E is finished.

Managerial Considerations

In addition to balancing a line for a given cycle time, managers have four other considerations: (1) pacing, (2) behavioral factors, (3) number of models produced, and (4) different cycle times.

Pacing The movement of product from one station to the next as soon as the cycle time has elapsed is called **pacing**. Pacing manufacturing processes allows materials handling to be automated and requires less inventory storage area. However, it is less flexible in handling unexpected delays that require either slowing down the entire line or pulling the unfinished work off the line to be completed later.

Behavioral Factors The most controversial aspect of line-flow layouts is behavioral response. Studies show that installing production lines increases absenteeism, turnover, and grievances.

Paced production and high specialization (say, cycle times of less than 2 minutes) lower job satisfaction. Workers generally favor inventory buffers as a means of avoiding mechanical pacing. One study even showed that productivity increased on unpaced lines.

Number of Models Produced   A line that produces several items belonging to the same family is called a **mixed-model line**. In contrast, a single-model line produces one model with no variations. Mixed-model production enables a plant to achieve both high-volume production *and* product variety. However, it complicates scheduling and increases the need for good communication about the specific parts to be produced at each station.

mixed-model line

A production line that produces several items belonging to the same family.

Cycle Times   A line's cycle time depends on the desired output rate (or sometimes on the maximum number of workstations allowed). In turn, the maximum line efficiency varies considerably with the cycle time selected. Thus, exploring a range of cycle times makes sense. A manager might go with a particularly efficient solution even if it does not match the desired output rate. The manager can compensate for the mismatch by varying the number of hours the line operates through overtime, extending shifts, or adding shifts. Multiple lines might even be the answer.

LEARNING GOALS IN REVIEW

1 **Explain the theory of constraints.** Constraints or bottlenecks can exist in the form of internal resources or market demand in both manufacturing and service organizations, and in turn play an important role in determining system performance. See the section on "The Theory of Constraints (TOC)," pp. 245–247. Review opening vignette on BP Oil Spill clean-up for an application of TOC.

2 **Understand linkage of capacity constraints to financial performance measures.** Review and understand Table 7.1 on page 245.

3 **Identify bottlenecks.** The TOC provides guidelines on how to identify and manage constraints. The section "Identification and Management of Bottlenecks," pp. 247–254, shows you how to identify bottlenecks in both service as well as manufacturing firms.

4 **Apply theory of constraints to product mix decisions.** Review Example 7.3 on pages 250–251 to understand how using a bottleneck based method for allocating resources and determining the product mix leads to greater profits.

5 **Describe how to manage constraints in an assembly line.** Assembly line balancing, as a special form of a constraint in managing a line process within both manufacturing and services, can also be an effective mechanism for matching output to a plan and running such processes more efficiently. The section "Managing Constraints in a Line Process," page 254–259, shows you how to balance assembly lines and create work stations. Review Solved Problem 2 on page 261 for an application of line balancing principles.

MyOMLab helps you develop analytical skills and assesses your progress with multiple problems on processing time for average customer, bottleneck activity, maximum customers served at bottleneck per hour, average capacity of system, theoretical maximum of stations in an assembly line, longest work element decision rule, line's efficiency, and cycle time.

MyOMLab Resources	Titles	Link to the Book
Video	*Constraint Management at Southwest Airlines*	Managing Constraints Across the Organization; The Theory of Constraints
	1st Bank Villa Italia: Waiting Lines	Identification and Management of Bottlenecks
OM Explorer Solvers	Min-Yo Garment Company Spreadsheet	Estimate Capacity Requirements; Example 6.1 (pp. 209–210); Managing Bottlenecks in Manufacturing Processes; Example 7.2 (pp. 249–250)
OM Explorer Tutors	7.1 Line Balancing	Line Balancing; Example 7.5 (p. 257); Solved Problem 2 (p. 261)
POM for Windows	Line Balancing	Line Balancing; Example 7.5 (p. 257); Solved Problem 2 (p. 261)
SimQuick Simulation Exercises	Simulating process of making jewelry boxes and making choice on investing in new machine	The Theory of Constraints; Example 7.1 (pp. 247–248)
Internet Exercise	Granite Rock and Chevron	The Theory of Constraints; Identification and Management of Bottlenecks
Key Equations		
Image Library		

Key Equations

1. Cycle time: $c = \dfrac{1}{r}$

2. Theoretical minimum number of workstations: $\text{TM} = \dfrac{\Sigma t}{c}$

3. Idle time: $nc - \Sigma t$

4. Efficiency(%): $\dfrac{\Sigma t}{nc}(100)$

5. Balance delay (%): $100 - \text{Efficiency}$

Key Terms

balance delay 256
bottleneck 244
constraint 244
cycle time 256
drum-buffer-rope (DBR) 252

immediate predecessors 254
line balancing 254
mixed-model line 259
pacing 258
precedence diagram 255

theoretical minimum (TM) 256
theory of constraints (TOC) 245
throughput time 247
work elements 254

Solved Problem 1

Bill's Car Wash offers two types of washes: Standard and Deluxe. The process flow for both types of customers is shown in the following chart. Both wash types are first processed through steps A1 and A2. The Standard wash then goes through steps A3 and A4 while the Deluxe is processed through steps A5, A6, and A7. Both offerings finish at the drying station (A8). The numbers in parentheses indicate the minutes it takes for that activity to process a customer.

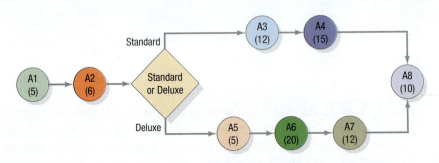

a. Which step is the bottleneck for the Standard car wash process? For the Deluxe car wash process?

b. What is the capacity (measured as customers served per hour) of Bill's Car Wash to process Standard and Deluxe customers? Assume that no customers are waiting at step A1, A2, or A8.

c. If 60 percent of the customers are Standard and 40 percent are Deluxe, what is the average capacity of the car wash in customers per hour?

d. Where would you expect Standard wash customers to experience waiting lines, assuming that new customers are always entering the shop and that no Deluxe customers are in the shop? Where would the Deluxe customers have to wait, assuming no Standard customers?

SOLUTION

a. Step A4 is the bottleneck for the Standard car wash process, and Step A6 is the bottleneck for the Deluxe car wash process, because these steps take the longest time in the flow.

b. The capacity for Standard washes is 4 customers per hour because the bottleneck step A4 can process 1 customer every 15 minutes (60/15). The capacity for Deluxe car washes is 3 customers per hour (60/20). These capacities are derived by translating the "minutes per customer" of each bottleneck activity to "customers per hour."

c. The average capacity of the car wash is $(0.60 \times 4) + (0.40 \times 3) = 3.6$ customers per hour.

d. Standard wash customers would wait before steps A1, A2, A3, and A4 because the activities that immediately precede them have a higher rate of output (i.e., smaller processing times). Deluxe wash customers would experience a wait in front of steps A1, A2, and A6 for the same reasons. A1 is included for both types of washes because the arrival rate of customers could always exceed the capacity of A1.

Solved Problem 2

A company is setting up an assembly line to produce 192 units per 8-hour shift. The following table identifies the work elements, times, and immediate predecessors:

Work Element	Time (Sec)	Immediate Predecessor(s)
A	40	None
B	80	A
C	30	D, E, F
D	25	B
E	20	B
F	15	B
G	120	A
H	145	G
I	130	H
J	115	C, I
	Total 720	

a. What is the desired cycle time (in seconds)?

b. What is the theoretical minimum number of stations?

c. Use trial and error to work out a solution, and show your solution on a precedence diagram.

d. What are the efficiency and balance delay of the solution found?

SOLUTION

a. Substituting in the cycle-time formula, we get

$$c = \frac{1}{r} = \frac{8 \text{ hours}}{192 \text{ units}}(3{,}600 \text{ seconds/hour}) = 150 \text{ seconds/unit}$$

b. The sum of the work-element times is 720 seconds, so

$$\text{TM} = \frac{\Sigma t}{c} = \frac{720 \text{ seconds/unit}}{150 \text{ seconds/unit-station}} = 4.8 \text{ or } 5 \text{ stations}$$

which may not be achievable.

c. The precedence diagram is shown in Figure 7.6. Each row in the following table shows work elements assigned to each of the five workstations in the proposed solution.

d. Calculating the efficiency, we get

$$\text{Efficiency} = \frac{\Sigma t}{nc}(100) = \frac{720 \text{ seconds/unit}}{5[150 \text{ seconds/unit}]} = 96\%$$

Thus, the balance delay is only 4 percent (100–96).

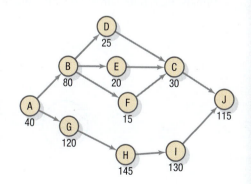

▲ **FIGURE 7.6**
Precedence Diagram

Station	Candidate(s)	Choice	Work-Element Time (Sec)	Cumulative Time (Sec)	Idle Time (c = 150 Sec)
S1	A	A	40	40	110
	B	B	80	120	30
	D, E, F	D	25	145	5
S2	E, F, G	G	120	120	30
	E, F	E	20	140	10
S3	F, H	H	145	145	5
S4	F, I	I	130	130	20
	F	F	15	145	5
S5	C	C	30	30	120
	J	J	115	145	5

Discussion Questions

1. Take a process that you encounter on a daily basis, such as the lunch cafeteria or the journey from your home to school/work, and identify the bottlenecks that limit the throughput of this process.

2. Using the same process as in question 1, identify conditions that would lead to the bottlenecks changing or shifting away from the existing bottleneck.

3. How could the efficiency of the redesigned process be improved further?

Problems

The OM Explorer and POM for Windows software is available to all students using the 10th edition of this textbook. Go to **www.pearsonhighered.com/krajewski** to download these computer packages. If you purchased MyOMLab, you also have access to Active Models software and significant help in doing the following problems. Check with your instructor on how best to use these resources. In many cases, the instructor wants you to understand how to do the calculations by hand. At the least, the software provides a check on your calculations. When calculations are particularly complex and the goal is interpreting the results in making decision, the software entirely replaces the manual calculations.

1. Bill's Barbershop has two barbers available to cut customers' hair. Both barbers provide roughly the same

experience and skill, but one is just a little bit slower than the other. The process flow in Figure 7.7 shows that all customers go through steps B1 and B2 and then can be served at either of the two barbers at step B3. The process ends for all customers at step B4. The numbers in parentheses indicate the minutes it takes that activity to process a customer.

a. How long does it take the average customer to complete this process?

b. What single activity is the bottleneck for the entire process?

c. How many customers can this process serve in an hour?

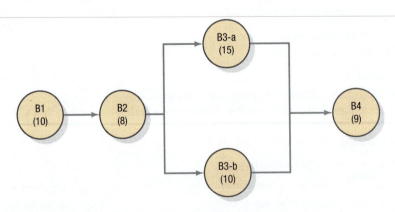

▲ FIGURE 7.7
Process Flow for Bill's Barbershop

2. Figure 7.8 details the process flow for two types of customers who enter Barbara's Boutique shop for customized dress alterations. After step T1, Type A customers proceed to step T2 and then to any of the three workstations at T3, followed by steps T4 and T7. After step T1, Type B customers proceed to step T5 and then steps T6 and T7. The numbers in parentheses are the minutes it takes to process a customer.

   a. What is the capacity of Barbara's shop in terms of the numbers of Type A customers who can be served

   in an hour? Assume no customers are waiting at steps T1 or T7.

   b. If 30 percent of the customers are Type A customers and 70 percent are Type B customers, what is the average capacity of Barbara's shop in customers per hour?

   c. Assuming that the arrival rate is greater than five customers per hour, when would you expect Type A customers to experience waiting lines, assuming no Type B customers in the shop? Where would the Type B customers have to wait, assuming no Type A customers?

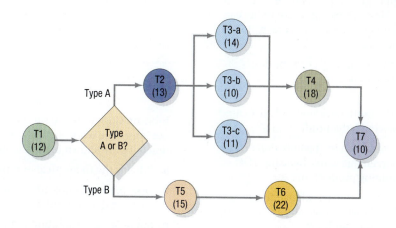

▲ **FIGURE 7.8**
Process Flow for Barbara's Boutique Customers

3. Canine Kernels Company (CKC) manufactures two different types of dog chew toys (A and B, sold in 1,000-count boxes) that are manufactured and assembled on three different workstations (W, X, and Y) using a small-batch process (see Figure 7.9). Batch setup times are negligible. The flowchart denotes the path each product follows through the manufacturing process, and each product's price, demand per week, and processing times per unit are indicated as well. Purchased parts and raw materials consumed during production are represented by inverted triangles. CKC can make and sell up to the limit of its demand per week; no penalties are incurred for not being able to meet all the demand. Each workstation is staffed by a worker who is dedicated to work on that workstation alone and is paid $6 per hour. Total labor costs per week are fixed. Variable overhead costs are $3,500/week. The plant operates one 8-hour shift per day, or 40 hours/week. Which of the three workstations, W, X, or Y, has the highest aggregate workload, and thus serves as the bottleneck for CKC?

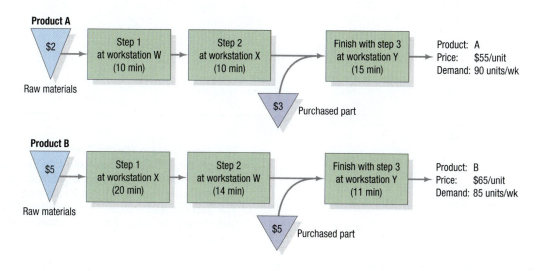

▲ **FIGURE 7.9**
Flowchart for Canine Kernels Company (CKC)

4. The senior management at Canine Kernels Company (CKC) is concerned with the existing capacity limitation, so they want to accept the mix of orders that maximizes the company's profits. Traditionally, CKC has utilized a method whereby decisions are made to produce as much of the product with the highest contribution margin as possible (up to the limit of its demand), followed by the next highest contribution margin product, and so on until no more capacity is available. Because capacity is limited, choosing the proper product mix is crucial. Troy Hendrix, the newly hired production supervisor, is an avid follower of the theory of constraints philosophy and the bottleneck method for scheduling. He believes that profitability can indeed be approved if bottleneck resources are exploited to determine the product mix.

   a. What is the profit if the traditional contribution margin method is used for determining CKC's product mix?

   b. What is the profit if the bottleneck method advocated by Troy is used for selecting the product mix?

   c. Calculate the profit gain, both in absolute dollars as well as in terms of percentage gains, by using TOC principles for determining product mix.

5. Use the longest work element rule to balance the assembly line described in the following table and Figure 7.10 so that it will produce 40 units per hour.

   a. What is the cycle time?

   b. What is the theoretical minimum number of workstations?

   c. Which work elements are assigned to each workstation?

   d. What are the resulting efficiency and balance delay percentages?

   e. Use the shortest work element rule to balance the assembly line. Do you note any changes in solution?

Work Element	Time (Sec)	Immediate Predecessor(s)
A	40	None
B	80	A
C	30	A
D	25	B
E	20	C
F	15	B
G	60	B
H	45	D
I	10	E, G
J	75	F
K	15	H, I, J
	Total 415	

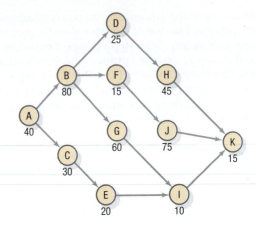

▲ **FIGURE 7.10**
Precedence Diagram

6. Johnson Cogs wants to set up a line to serve 60 customers per hour. The work elements and their precedence relationships are shown in the following table.

   a. What is the theoretical minimum number of stations?

   b. How many stations are required using the longest work element decision rule?

   c. Suppose that a solution requiring five stations is obtained. What is its efficiency?

Work Element	Time (Sec)	Immediate Predecessor(s)
A	40	None
B	30	A
C	50	A
D	40	B
E	6	B
F	25	C
G	15	C
H	20	D, E
I	18	F, G
J	30	H, I
	Total 274	

7. The *trim line* at PW is a small subassembly line that, along with other such lines, feeds into the final chassis line. The entire assembly line, which consists of more than 900 workstations, is to make PW's new E cars. The trim line itself involves only 13 work elements and must handle 20 cars per hour. Work-element data are as follows:

Work Element	Time (Sec)	Immediate Predecessor(s)
A	1.8	None
B	0.4	None
C	1.6	None
D	1.5	A
E	0.7	A
F	0.5	E
G	0.8	B
H	1.4	C
I	1.4	D
J	1.4	F, G
K	0.5	H
L	1.0	J
M	0.8	I, K, L

a. Draw a precedence diagram.

b. What cycle time (in minutes) results in the desired output rate?

c. What is the theoretical minimum number of stations?

d. Use the longest work element decision rule to balance the line and calculate the efficiency of your solution.

e. Use the most followers work element decision rule to balance the line and calculate the efficiency of your solution.

8. In order to meet holiday demand, Penny's Pie Shop requires a production line that is capable of producing 50 pecan pies per week, while operating only 40 hours per week. There are only 4 steps required to produce a single pecan pie with respective processing times of 5 min, 5 min, 45 min, and 15 min.

a. What should be the line's cycle time?

b. What is the smallest number of workstations Penny could hope for in designing the line considering this cycle time?

c. Suppose that Penny finds a solution that requires only four stations. What would be the efficiency of this line?

Advanced Problems

9. Melissa's Photo Studio offers both individual and group portrait options. The process flow diagram in Figure 7.11 shows that all customers must first register and then pay at one of two cashiers. Then, depending on whether they want a single or group portrait they go to different rooms. Finally, everyone picks up their own finished portrait.

a. How long does it take to complete the entire process for a group portrait?

b. What single activity is the bottleneck for the entire process, assuming the process receives equal amounts of both groups and individuals?

c. What is the capacity of the bottleneck for both groups and individuals?

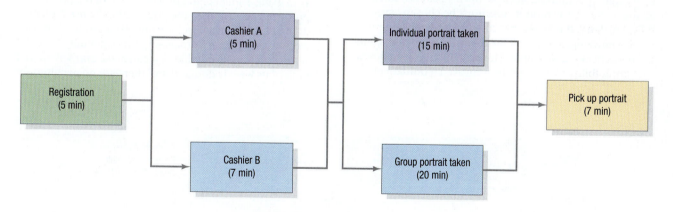

▲ **FIGURE 7.11**
Melissa's Photo Studio

10. Yost-Perry Industries (YPI) manufactures a mix of affordable guitars (A, B, C) that are fabricated and assembled at four different processing stations (W, X, Y, Z). The operation is a batch process with small setup times that can be considered negligible. The product information (price, weekly demand, and processing times) and process sequences are shown in Figure 7.12. Raw materials and purchased parts (shown as a per-unit consumption rate) are represented by inverted triangles. YPI is able to make and sell up to the limit of its demand per week with no penalties incurred for not meeting the full demand. Each workstation is staffed by one highly skilled worker who is dedicated to work on that workstation alone and is paid $15 per hour. The plant operates one 8-hour shift per day and operates on a 5-day work week (i.e., 40 hours of production per person per week). Overhead costs are $9,000/week. Which of the four workstations, W, X, Y, or Z, has the highest aggregate workload, and thus serves as the bottleneck for YPI?

▶ **FIGURE 7.12**

Flowchart for Yost-Perry
Industries (YPI)

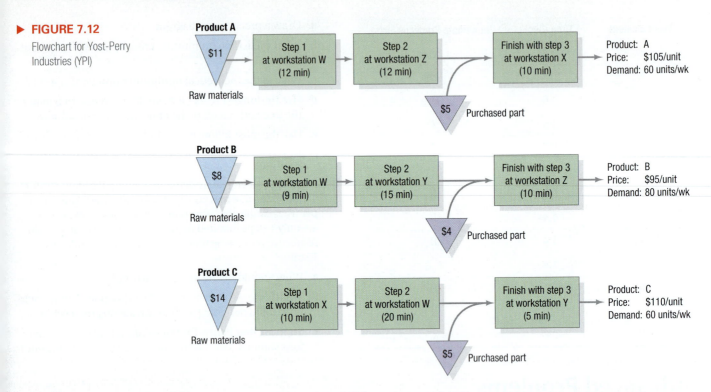

11. Yost-Perry Industries' (YPI) senior management team wants to improve the profitability of the firm by accepting the right set of orders. Currently, decisions are made using the traditional method, which is to accept as much of the highest contribution margin product as possible (up to the limit of its demand), followed by the next highest contribution margin product, and so on until all available capacity is utilized. Because the firm cannot satisfy all the demand, the product mix must be chosen carefully. Jay Perry, the newly promoted production supervisor, is knowledgeable about the theory of constraints and the bottleneck-based method for scheduling. He believes that profitability can indeed be improved if bottleneck resources are exploited to determine the product mix. What is the change in profits if, instead of the traditional method that YPI has used thus far, the bottleneck method advocated by Jay is used for selecting the product mix?

12. A.J.'s Wildlife Emporium manufactures two unique birdfeeders (Deluxe and Super Duper) that are manufactured and assembled in up to three different workstations (X, Y, Z) using a small batch process. Each of the products is produced according to the flowchart in Figure 7.13. Additionally, the flowchart indicates each product's price, weekly demand, and processing times per unit. Batch setup times are negligible. A.J. can make and sell up to the limit of its weekly demand and there are no penalties for

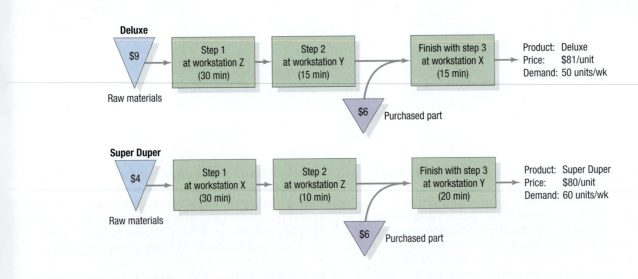

▲ **FIGURE 7.13**

A.J.'s Wildlife Emporium Flowchart

not being able to meet all of the demand. Each workstation is staffed by a worker who is dedicated to work on that workstation alone and is paid $16 per hour. The plant operates 40 hours per week, with no overtime. Overhead costs are $2,000 per week. Based on the information provided, as well as the information contained in the flowchart, answer the following questions.

a. Using the traditional method, which bases decisions solely on a product's contribution to profits and overhead, what is the optimal product mix and what is the overall profitability?

b. Using the bottleneck-based method, what is the optimal product mix and what is the overall profitability?

13. Cooper River Glass Works (CRGW) produces four different models of desk lamps as shown in Figure 7.14. The operations manager knows that total monthly demand exceeds the capacity available for production. Thus, she is interested in determining the product mix which will maximize profits. Each model's price, routing, processing times, and material cost is provided in Figure 7.14. Demand next month is estimated to be 200 units of model Alpha, 250 units of model Bravo, 150 units of model Charlie, and 225 units of model Delta. CRGW operates only one 8-hour shift per day and is scheduled to work 20 days next month (no overtime). Further, each station requires a 10 percent capacity cushion.

a. Which station is the bottleneck?

b. Using the traditional method, which bases decisions solely on a product's contribution to profits and overhead, what is the optimal product mix and what is the overall profitability?

c. Using the bottleneck-based method, what is the optimal product mix and what is the overall profitability?

14. The senior management at Davis Watercraft would like to determine if it is possible to improve firm profitability by changing their existing product mix. Currently, the product mix is determined by giving resource priority to the highest contribution margin watercraft. Davis Watercraft always has a contingent of 10 workers on hand; each worker is paid $25 per hour. Overhead costs are $35,000 per week. The plant operates 18 hours per day and 6 days per week. Labor is considered a fixed expense because workers are paid for their time regardless of their utilization. The production manager has determined that workstation 1 is the bottleneck. Detailed production information is provided below.

	Model		
	A	B	C
Price	$450	$400	$500
Material Cost	$50	$40	$110
Weekly Demand	100	75	40
Processing Time Station 1	60 min	0 min	30 min
Processing Time Station 2	0 min	0 min	60 min
Processing Time Station 3	10 min	60 min	0 min
Processing Time Station 4	20 min	30 min	40 min

a. Using the traditional method, which bases decisions solely on a product's contribution to profits and overhead, what is the product mix that yields the highest total profit? What is the resulting profit?

b. Using the bottleneck-based method, what is the product mix that yields the highest total profit? What is the resulting profit?

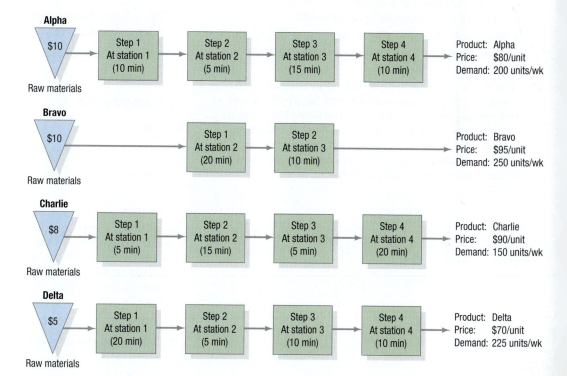

▶ FIGURE 7.14

Cooper River Glass Works Flowchart

Alpha

$10 Raw materials

Step 1 At station 1 (10 min) → Step 2 At station 2 (5 min) → Step 3 At station 3 (15 min) → Step 4 At station 4 (10 min)

Product: Alpha
Price: $80/unit
Demand: 200 units/wk

Bravo

$10 Raw materials

Step 1 At station 2 (20 min) → Step 2 At station 3 (10 min)

Product: Bravo
Price: $95/unit
Demand: 250 units/wk

Charlie

$8 Raw materials

Step 1 At station 1 (5 min) → Step 2 At station 2 (15 min) → Step 3 At station 3 (5 min) → Step 4 At station 4 (20 min)

Product: Charlie
Price: $90/unit
Demand: 150 units/wk

Delta

$5 Raw materials

Step 1 At station 1 (20 min) → Step 2 At station 2 (5 min) → Step 3 At station 3 (10 min) → Step 4 At station 4 (10 min)

Product: Delta
Price: $70/unit
Demand: 225 units/wk

15. A paced assembly line has been devised to manufacture calculators, as the following data show:

Station	Work Element Assigned	Work Element Time (min)
S1	A	2.7
S2	D, E	0.6, 0.9
S3	C	3.0
S4	B, F, G	0.7, 0.7, 0.9
S5	H, I, J	0.7, 0.3, 1.2
S6	K	2.4

a. What is the maximum hourly output rate from this line? (*Hint:* The line can go only as fast as its slowest workstation.)

b. What cycle time corresponds to this maximum output rate?

c. If a worker is at each station and the line operates at this maximum output rate, how much idle time is lost during each 10-hour shift?

d. What is the line's efficiency?

16. Jane produces custom greeting cards using six distinct work elements. She would like to produce 10 cards in each 8-hour card-making session. Figure 7.15 details each work element and its associated durations in minutes as well as their precedence relationships.

a. What cycle time is required to satisfy the required output rate?

b. What is the theoretical minimum number of workstations required?

c. If Jane identifies a five-station solution, what is the associated efficiency and balance delay?

d. If the cycle time increased by 100 percent, would the theoretical minimum number of workstations also increase by 100 percent?

17. Greg Davis, a business major at the University of South Carolina (USC), has opened Six Points Saco (SPS), a specialty subs–taco restaurant, at the rim of the USC campus. SPS has grown in popularity over the one year that it has been in operation, and Greg is trying to perfect the business model before making it into a franchise. He wants to maximize the productivity of his staff, as well as serve customers well in a timely fashion. One area of concern is the drive-thru operation during the 11:30 A.M. to 12:30 P.M. lunch hour.

The process of fulfilling an order involves fulfilling the tasks listed below.

Greg is interested in getting a better understanding of the staffing patterns that will be needed in order to operate his restaurant. After taking a course in operations management at the university, he knows that fulfilling a customer order at SPS is very similar to operating an assembly line. He has also used the POM for Windows software before, and wants to apply it for examining different demand scenarios for serving his customers.

a. If all the seven tasks are handled by one employee, how many customers could be served per hour?

b. If Greg wants to process 45 customers per hour, how many employees will he need during the peak period?

c. With the number of employees determined in part b, what is the maximum number of customers who could be served every hour (i.e., what is the maximum output capacity)?

d. Assuming that no task is assigned to more than one employee, what is the "maximum output capacity" from this assembly line? How many employees will be needed to actually accomplish this maximum output capacity?

e. Beyond the output accomplished in part d, if Greg decides to add one additional worker to help out with a bottleneck task, where should he add that worker? With that addition, would he be able to process more customers per hour? If so, what is the new maximum output capacity for the drive-thru?

	Task	Time (Seconds)	Immediate Predecessors
A.	Take an order at the booth. Most orders are for a taco and a sub.	25	
B.	Collect money at the window.	20	A
C.	Gather drinks.	35	B
D.	Assemble taco order.	32	B
E.	Assemble sub order.	30	B
F.	Put drinks, taco, and sub in a bag.	25	C, D, E
G.	Give the bag to the customer.	10	F

18. Refer back to problem 7. Suppose that in addition to the usual precedence constraints, there are two zoning constraints within the trim line. First, work elements K and L should be assigned to the same station; both use a common component, and assigning them to the same station conserves storage space. Second, work elements H and J cannot be performed at the same station.

a. Using trial and error, balance the line as best you can.

b. What is the efficiency of your solution?

FIGURE 7.15 ▶

Precedence Diagram for Custom Greeting Cards

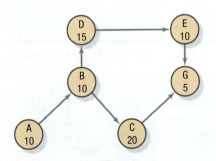

EXPERIENTIAL LEARNING   Min-Yo Garment Company

The Min-Yo Garment Company is a small firm in Taiwan that produces sportswear for sale in the wholesale and retail markets. Min-Yo's garments are unique because they offer fine embroidery and fabrics with a variety of striped and solid patterns. Over the 20 years of its existence, the Min-Yo Garment Company has become known as a quality producer of sports shirts with dependable deliveries. However, during that same period, the nature of the apparel industry has undergone change. In the past, firms could be successful producing standardized shirts in high volumes with few pattern or color choices and long production lead times. Currently, with the advent of regionalized merchandising and intense competition at the retail level, buyers of the shirts are looking for shorter lead times and much more variety in patterns and colors. Consequently, many more business opportunities are available today than ever before to a respected company such as Min-Yo.

Even though the opportunity for business success seemed bright, the management meeting last week was gloomy. Min-Yo Lee, president and owner of Min-Yo Garment, expressed concerns over the performance of the company: "We are facing strong competition for our products. Large apparel firms are driving prices down on high-volume licensed brands. Each day more firms enter the customized shirt business. Our profits are lower than expected, and delivery performance is deteriorating. We must reexamine our capabilities and decide what we can do best."

Products

Min-Yo has divided its product line into three categories: licensed brands, subcontracted brands, and special garments.

Licensed Brands

Licensed brands are brands that are owned by one company but, through a licensing agreement, are produced by another firm that also markets the brand in a specific geographic region. The licenser may have licensees all over the world. The licensee pays the licenser a fee for the privilege of marketing the brand in its region, and the licenser agrees to provide some advertising for the product, typically through media outlets that have international exposure. A key aspect of the licensing agreement is that the licensee must agree to provide sufficient quantities of product at the retail level. Running out of stock hurts the image of the brand name.

Currently, only one licensed brand is manufactured by Min-Yo. The brand, called the Muscle Shirt, is owned by a large "virtual corporation" in Italy that has no manufacturing facilities of its own. Min-Yo has been licensed to manufacture Muscle Shirts and sell them to large retail chains in Taiwan. The retail chains require prompt shipments at the end of each week. Because of competitive pressures from other licensed brands, low prices are important. Min-Yo sells each Muscle Shirt to retail chains for $6.

The demand for Muscle Shirts averages 900 shirts per week. The following demand for Muscle Shirts has been forecasted for the next 12 weeks.

Min-Yo's forecasts of Muscle Shirts are typically accurate to within ±200 shirts per week. If demand exceeds supply in any week, the excess demand is lost. No backorders are taken, and Min-Yo incurs no cost penalty for lost sales.

Subcontracted Brands

Manufacturers in the apparel industry often face uncertain demand. To maintain level production at their plants, many manufacturers seek subcontractors to produce their brands. Min-Yo is often considered a subcontractor because of its reputation in the industry. Although price is a consideration, the owners of subcontracted brands emphasize dependable delivery and the ability of the subcontractor to adjust order quantities on short notice.

Week	Demand	Week	Demand
1*	700	7	1,100
2	800	8	1,100
3	900	9	900
4	900	10	900
5	1,000	11	800
6	1,100	12	700

*In other words, the company expects to sell 700 Muscle Shirts at the end of week 1.

Currently, Min-Yo manufactures only one subcontracted brand, called the Thunder Shirt because of its bright colors. Thunder Shirts are manufactured to order for a company in Singapore. Min-Yo's price to this company is $7 per shirt. When orders are placed, usually twice a month, the customer specifies the delivery of certain quantities in each of the next 2 weeks. The last order the customer placed is overdue, forcing Min-Yo to pay a penalty charge. To avoid another penalty, 200 shirts must be shipped in week 1. The Singapore company is expected to specify the quantities it requires for weeks 2 and 3 at the beginning of week 1. The delivery schedule containing the orders for weeks 4 and 5 is expected to arrive at the beginning of week 3, and so on. The customer has estimated its average weekly needs for the year to be 200 shirts per week, although its estimates are frequently inaccurate.

Because of the importance of this large customer to Min-Yo and the lengthy negotiations of the sales department to get the business, management always tries to satisfy its needs. Management believes that if Min-Yo Garment ever refuses to accept an order from this customer, Min-Yo will lose the Thunder Shirt business. Under the terms of the sales contract, Min-Yo agreed to pay this customer $1 for every shirt not shipped on time for each week the shipment of the shirt is delinquent. Delinquent shipments must be made up.

Special Garments

Special garments are made only to customer order because of their low volume and specialized nature. Customers come to Min-Yo Garment to manufacture shirts for special promotions or special company occasions. Min-Yo's special garments are known as Dragon Shirts because of the elaborate embroidery and oriental flair of the designs. Because each shirt is made to a particular customer's specifications and requires a separate setup, special garments cannot be produced in advance of a firm customer order.

Although price is not a major concern for the customers of special garments, Min-Yo sells Dragon Shirts for $8 a shirt to ward off other companies seeking to enter the custom shirt market. Its customers come to Min-Yo because the company can produce almost any design with high quality and deliver an entire order on time. When placing an order for a Dragon Shirt, a customer specifies the design of the shirt (or chooses from Min-Yo's catalog), supplies specific designs for logos, and specifies the quantity of the order and the delivery date. In the past, management checked to see whether such an order would fit into the schedule, and then either accepted or rejected it on that basis. If Min-Yo accepts an order for delivery at the *end* of a certain week and fails to meet this commitment, it pays a penalty of $2 per shirt for each week delivery is delayed. This penalty is incurred weekly until the delinquent order is delivered. The company tried to forecast demand for specific designs of Dragon Shirts but has given up. Last week, Min-Yo had four Dragon Shirt opportunities of 50, 75, 200, and 60 units but chose not to accept any of the orders. Dragon Shirt orders in the past ranged from 50 units to 300 units with varying lead times.

Figure 7.16, Min-Yo's current open-order file, shows that in some prior week Min-Yo accepted an order of 400 Thunder Shirts for delivery last week. The open-order file is important because it contains the commitment management made to customers. Commitments are for a certain quantity and a date of delivery. As customer orders are accepted, management enters the quantity in the green cell representing the week that they are due. Because Dragon Shirts are unique unto themselves, they each have their own order number for future use. No Dragon Shirt orders appear in the open-order file because Min-Yo has not committed to any in the past several weeks.

Manufacturing

Process

The Min-Yo Garment Company has the latest process technology in the industry—a machine, called a garment maker, that is run by one operator on each of three shifts. This single machine process can make every garment Min-Yo produces; however, the changeover times consume a substantial amount of capacity. Company policy is to run the machine three shifts a day, five days a week. If business is insufficient to keep the machine busy, the workers are idle because Min-Yo is committed to never fire or lay off a worker. By the same token, the firm has a policy of never working on weekends. Thus, the capacity of the process is 5 days × 24 hours = 120 hours per week. The hourly wage is $10 per hour, so the firm is committed to a fixed labor cost of $10 × 120 = $1,200 per week. Once the machine has been set up to make a particular type of garment, it can produce

that garment at the rate of 10 garments per hour, regardless of type. The cost of the material in each garment, regardless of type, is $4. Raw materials are never a problem and can be obtained overnight.

Scheduling the Garment Maker

Scheduling at Min-Yo is done once each week, after production for the week has been completed and shipped, after new orders from customers have arrived, and before production for the next week has started. Scheduling results in two documents.

The first is a production schedule, shown in Figure 7.17. The schedule shows what management wants the garment maker process to produce in a given week. Two spreadsheet entries are required for each product that is to be produced in a given week. They are in the green shaded cells. The first is the production quantity. In Figure 7.17, the schedule shows that Min-Yo produced quantities of 800 units for Muscle and 200 units for Thunder last week. The second input is a "1" if the machine is to be set up for a given product or a "blank" if no changeover is required. Figure 7.17 shows that last week changeovers were required for the Muscle and Thunder production runs. The changeover information is important because, at the end of a week, the garment maker process will be set up for the last product produced. If the same product is to be produced first the following week, no new changeover will be required. Management must keep track of the sequence of production each week to take advantage of this savings. The only exception to this rule is Dragon Shirts, which are unique orders that always require a changeover. In week 0, Min-Yo did not produce any Dragon Shirts; however, it did produce 800 Muscle Shirts, followed by 200 Thunder Shirts. Finally, the spreadsheet calculates the hours required for the proposed

MIN-YO GARMENT COMPANY

Open Order File (Record of commitments)

Product	Week Order is Due									
	1	2	3	4	5	6	7	8	9	10
Thunder Orders	400									
Dragon Order 1										
Dragon Order 2										
Dragon Order 3										
Dragon Order 4										
Dragon Order 5										
Dragon Order 6										
Dragon Order 7										
Dragon Order 8										
Dragon Order 9										
Dragon Order 10										
Dragon Order 11										
Dragon Order 12										
Dragon Order 13										
Dragon Order 14										
Dragon Order 15										

| ▶ ▶| | Intro | Open Order File | Week 1 | Week 2 | Week 3 | Week 4 | Week 5 | Week 6 | Week 7 | Week 8 | Week 9 | Week 10 | Su ◀ |

▲ **FIGURE 7.16**

Min-Yo's Open Order File

Note: All orders are to be delivered at the end of the week indicated, after production for the week has been completed and before next week's production is started.

MIN-YO GARMENT COMPANY

PRODUCTION SCHEDULE

PRODUCT		
	Changeover	Quantity
Muscle	1	800
Hours		88
Thunder	1	200
Hours		30

The two inputs to the Production Schedule table are:
1. The quantity you decide to produce this time period
2. Whether there is a setup/changover required (1 or 0)

				Changeover	Quantity		Changeover	Quantity
Dragon Order 1			Dragon Order 11			Dragon Order 21		
Dragon Order 2			Dragon Order 12			Dragon Order 22		
Dragon Order 3			Dragon Order 13			Dragon Order 23		
Dragon Order 4			Dragon Order 14			Dragon Order 24		
Dragon Order 5			Dragon Order 15			Dragon Order 25		
Dragon Order 6			Dragon Order 16			Dragon Order 26		
Dragon Order 7			Dragon Order 17			Dragon Order 27		
Dragon Order 8			Dragon Order 18			Dragon Order 28		
Dragon Order 9			Dragon Order 19			Dragon Order 29		
Dragon Order 10			Dragon Order 20			Dragon Order 30		
Total Dragon Hours	0							
Total Dragon Production	0							
Total Hours scheduled	118							

Is production within capacity?   Yes

▲ **FIGURE 7.17**
Min-Yo's Production Schedule

schedule. Changeover times for Muscle, Thunder, and Dragon Shirts are 8, 10, and 25 hours respectively. Because the garment maker process produces 10 garments per hour regardless of type, the production hours required for Muscle Shirts is 8 + 800/10 = 88 hours, and the production hours for Thunder Shirts is 10 + 200/10 = 30 hours, as shown in Figure 7.17. The total time spent on the garment maker process on all products in a week cannot exceed 120 hours. The spreadsheet will not allow you to proceed if this constraint is violated.

The second document is a weekly profit and loss (P&L) statement that factors in sales and production costs, including penalty charges and inventory carrying costs, as shown in Figure 7.18. The inventory carrying cost for *any type of product* is $0.10 per shirt per week left in inventory after shipments for the week have been made. The spreadsheet automatically calculates the P&L statement, which links to the open-order file and the production schedule, after the demand for Muscle Shirts is known. Figure 7.18 shows that the actual demand for Muscle Shirts last week was 750 shirts.

P&L STATEMENT

Product	Price	Beg Inv	Production	Available	Demand	Sales	End Inv	Inv/Past due costs
Muscle	$6	550	800	1350	750	4500	600	60
Thunder	$7		200	200	400	1400	-200	200
Dragon Orders	$8		0	0	0	0	0	0
Totals			1000			5900		260

		Current	Cumulative
Sales Total		$5,900	$5,900
Labor	$1,200		
Materials	$4,000		
Inv/Past due	$260		
Total Cost		$5,460	
Profit Contribution		$440	$440

▲ **FIGURE 7.18**
Min-Yo's P&L Schedule

Notes

- The past due quantity of shirts are those shirts not shipped as promised, and appear as a negative number in the "End Inv" column.

- Available = Beginning inventory + Production

- Sales = Demand × Price when demand < available; Available × Price, otherwise

- Inventory cost = $0.10 times number of shirts in inventory. Past due cost equals past due quantity times the penalty ($1 for Thunder Shirts; $2 for Dragon Shirts). These costs are combined in the "Inv/Past Due Costs" column.

The Simulation

At Min-Yo Garment Company, the executive committee meets weekly to discuss the new order possibilities and the load on the garment maker process. The executive committee consists of top management representatives from finance, marketing, and operations. You will be asked to participate on a team and play the role of a member of the executive committee in class. During this exercise, you must decide how far into the future to plan. Some decisions, such as the markets you want to exploit, are long-term in nature. Before class, you may want to think about the markets and their implications for manufacturing. Other decisions are short-term and have an impact on the firm's ability to meet its commitments. In class, the simulation will proceed as follows.

1. Use the Min-Yo Tables spreadsheet in OM Explorer in MyOMLab. It is found in the Solver menu, under Constraint Management. You will start by specifying the production schedule for week 1, based on the forecasts for week 1 in the case narrative for Muscle Shirts and additional information on new and existing orders for the customized shirts from your instructor. *You may assume that your managerial predecessors left the garment machine set up for Thunder Shirts.* The production schedule decision is to be made in collaboration with your executive committee colleagues in class.

2. When all the teams have finalized their production plans for week 1, the instructor will supply the actual demands for Muscle Shirts in week 1. Enter that quantity in the P&L statement in the spreadsheet for week 1.

3. After the P&L statement for week 1 is completed, the instructor will announce the new order requests for Thunder Shirts and Dragon Shirts to be shipped in week 2 and the weeks beyond.

4. You should look at your order requests, accept those that you want, and reject the rest. Add those that you accept for delivery in future periods to your open-order file. Enter the quantity in the cell representing the week the order is due. You are then irrevocably committed to them and their consequences.

5. You should then make out a new production schedule, specifying what you want your garment-maker process to do in the next week (it will be for week 2 at that time).

6. The instructor will impose a time limit for each period of the simulation. When the time limit for one period has been reached, the simulation will proceed to the next week. Each week the spreadsheet will automatically update your production and financial information in the Summary Sheet.

VIDEO CASE   Constraint Management at Southwest Airlines

What if you could take a commercial airline flight any time, and anywhere you wanted to go? Just show up at the airport without the need to consider time schedules or layovers. Aside from the potentially cost-prohibitive nature of such travel, there are also constraints in the airline system that preclude this kind of operation. From the lobby check-in process through to boarding at the gate and processing plane turnaround, the process of operating the airline is filled with constraints that must be managed in order for them to be successful and profitable. Flight schedules are tightly orchestrated and controlled, departure and arrival gates at airports are limited, and individual aircraft have seating capacities in each section of the plane, to name a few.

Southwest Airlines is one company that has figured out how to manage its constraints and generate positive customer experiences in the process. No other airline can claim the same level of profitability and customer satisfaction Southwest regularly achieves. What is its secret?

Talk to any loyal Southwest customer and you will hear rave reviews about its low fares, great customer service, and lack of assigned seating that gives customers a chance to choose who they sit next to onboard. From an operations perspective, it is much more than what the customer sees. Behind the scenes, operations managers carefully manage and execute—3,400 times a day in over 60 cities in the United States—a process designed to manage all potential bottleneck areas.

Southwest's famous rapid gate-turnaround of 25 minutes or less demonstrates how attention to the activities that ground operations must complete to clean, fuel, and prepare a plane for flight can become bottlenecks if not properly scheduled. In the terminal at the gate, passenger boarding also can be a bottleneck if the boarding process itself is not carefully managed. Since the individual mix of passengers present a different set of issues with each flight that often are not evident until the passengers actually arrive

Pearson

Passengers boarding a Southwest Airlines flight.

at the gate, ranging from families with kids and strollers to large quantities of carry-on bags and passengers needing wheelchair assistance, operations managers must be ready for any and all situations to avoid a boarding bottleneck while also assuring a pleasant and stress-free gate experience for all passengers.

In 2007, as part of the company's continuous improvement activities, Southwest focused its attention on the passenger boarding process to determine whether there was a better way to board. Its existing process consisted of three groups, A, B, C, with no assigned seating. Depending on passenger check-in and arrival time, passengers were given a spot in a group.

Those first to check-in received choice places in the A group. Last to check-in ended up in the C group, and usually had a choice of only middle seats in the back of the plane upon boarding. As passengers arrived at the gate, they queued up in their respective boarding group areas to await the boarding call.

Seven different alternate boarding scenarios were designed and tested. They included

- New family pre-boarding behind the "A" group of first-to-board passengers
- Family pre-boarding before anyone else, but seating choices limited on-board to behind the wing
- Six boarding groups (within A-B-C groups) instead of the original three A-B-C groups
- Assigned boarding gate line positions based on both boarding group and gate arrival time
- Single boarding chute at the gate, but up to nine groups all in one queue

- Boarding with a countdown clock to give customers an incentive to get in line and board quickly; incentives given out if everyone was on time
- Educational boarding video to make the boarding process fun, inform passengers how to board efficiently, and provide the company another way to promote its brand.

QUESTIONS

1. Analyze Southwest's passenger boarding process using the Theory of Constraints.
2. Which boarding scenario among the different ones proposed would you recommend for implementation? Why?
3. How should Southwest evaluate the gate boarding and plane turnaround process?
4. How will Southwest know that the bottleneck had indeed been eliminated after the change in the boarding process?

Selected References

Brown, A. "Theory of Constraints Tapped to Accelerate BP's Gulf of Mexico Cleanup." *Industry Week* (March 18, 2011).

Corominas, Albert, Rafael Pastor, and Joan Plans. "Balancing Assembly Line with Skilled and Unskilled Workers." *Omega*, vol. 36, no. 6 (2008), pp. 1126–1132.

Goldratt, E.M., and J. Cox. *The Goal*, 3rd rev. ed. New York: North River Press, 2004.

McClain, John O., and L. Joseph Thomas. "Overcoming the Dark Side of Worker Flexibility." *Journal of Operations Management*, vol. 21, (2003), pp. 81–92.

Srikanth, Mokshagundam L., and Michael Umble. *Synchronous Management: Profit-Based Manufacturing for the 21st Century*, vol. 1. Guilford, CT: Spectrum Publishing Company, 1997.

Srinivasan, Mandyam, Darren Jones, and Alex Miller. 2004. "Applying Theory of Constraints Principles and Lean Thinking at the Marine Corps Maintenance Center." *Defense Acquisition Rev. Quart.* (August–November 2004), pp. 134–145.

Srinivasan, Mandyam, Darren Jones, and Alex Miller. "Corps Capabilities." *APICS Magazine* (March 2005), pp. 46–50.

Steele, Daniel C., Patrick R. Philipoom, Manoj K. Malhotra, and Timothy D. Fry. "Comparisons Between Drum-Buffer-Rope and Material Requirements Planning: A Case Study." *International Journal of Production Research*, vol. 43, no. 15 (2005), pp. 3181–3208.

Umble, M., E. Umble, and S. Murakami. "Implementing Theory of Constraints in a Traditional Japanese Manufacturing Environment: The Case of Hitachi Tool Engineering." *International Journal of Production Research*, vol. 44, no. 15 (2006), pp. 1863–1880.

EVERETT KENNEDY BROWN/epa/Corbis

8

LEAN SYSTEMS

Panasonic company employees work on the disassembly line of CRT television sets at the company's Panasonic Eco Technology Center in Kato city, Hyogo prefecture, Japan. In conjunction with Panasonic's efforts to develop easy-to-recycle products, the company is also developing new recycling technologies and focusing on building lean operations. To prevent waste, an annual total of 30,700 tons of home appliances are recycled at the Panasonic facility, including 29,400 television units, 135,700 air conditioners, 173,300 washing machines and 149,100 refrigerators.

Panasonic Corporation

Panasonic Corporation, which was originally founded as Matsushita Corporation in 1918 to produce lamps, has grown to become one of the largest electronic manufacturing firms in the world. With over 384,000 employees and 680 consolidated companies, it is the largest producer in Japan of over 15,000 electronic products under the brand names of Panasonic, National, Quasar, and Technics, among others. Renowned for its global focus on efficiency and lean operations, its pursuit of excellence is exemplified nowhere better than at the Matsushita Electric Company's factory in Saga on Japan's southern island of Kyushu, where cordless phones, fax machines, and security cameras are made in record time by machines in a spotless facility.

Even though the plant's efficiency had doubled over a four-year span, managers saw opportunities for "trimming the fat" and improving further. The plant's conveyor belts were replaced by a cluster of robots that could seamlessly hand off work to one another, flexibly substitute for a broken robot, and synchronize production using software. As a result, throughput time declined from 2.5 days to 40 minutes, allowing the Saga plant to make twice as many phones per week, which in turn allowed a reduction in inventory because components such as chips and circuit boards spend much less time in the factory. Being able to make things faster means that the plant can quickly change the product mix even as customer demands shift and new products are introduced, thus allowing Panasonic to keep ahead of low cost rivals in Korea, China, and other Asian countries.

Panasonic has used the lessons learned from the Saga mother plant to change layouts and setups at six other plants in China, Malaysia, Mexico, and Great Britain.

These plants have been able to similarly cut their inventories and improve productivity, even as ideas from local staff at each plant were incorporated into the change effort. The next sets of improvements are focused on breaking assembly lines into cells and better utilizing the idle robots. In addition, standardized circuit board designs that are common to a large variety of end products are being used in order to minimize the retooling of robots for every type of board. By relentlessly focusing on minimizing waste and continuously improving efficiency, Panasonic has posted record profit growths and has become a model for other electronic firms.

Source: Kenji Hall, "No One Does Lean like the Japanese," *Business Week* (July 10, 2006), pp. 40–41; http://panasonic .net/corporate/, April 25, 2011.

LEARNING GOALS *After reading this chapter, you should be able to:*

1 Describe how lean systems can facilitate the continuous improvement of processes.

2 Identify the characteristics and strategic advantages of lean systems.

3 Understand value stream mapping and its role in waste reduction.

4 Understand *kanban* systems for creating a production schedule in a lean system.

5 Explain the implementation issues associated with the application of lean systems.

Creating Value through Operations Management

Using Operations to Compete
Project Management

Managing Processes

Process Strategy
Process Analysis
Quality and Performance
Capacity Planning
Constraint Management
Lean Systems

Managing Supply Chains

Supply Chain Inventory Management
Supply Chain Design
Supply Chain Location Decisions
Supply Chain Integration
Supply Chain Sustainability and Humanitarian Logistics
Forecasting
Operations Planning and Scheduling
Resource Planning

Panasonic Corporation is a learning organization and an excellent example of an approach for designing supply chains known as **lean systems**, which allow firms like Panasonic to continuously improve its operations and spread the lessons learned across the entire corporation. Lean systems are operations systems that maximize the value added by each of a company's activities by removing waste and delays from them. They encompass the company's operations strategy, process design, quality management, constraint management, layout design, supply chain design, and technology and inventory management, and can be used by both service and manufacturing firms. Like a manufacturer, each service business takes an order from a customer, delivers the service, and then collects revenue. Each service business purchases services or items, receives and pays for them, and hires and pays employees. Each of these activities bears considerable similarity to those in manufacturing firms. They also typically contain huge amounts of waste. In the first two parts of the text, we have discussed many ways to improve processes, regardless of whether they are manufacturing or nonmanufacturing processes. These same principles can be applied to make service processes lean, whether they are front-office, hybrid-office, or back-office designs. We conclude the second part of the text by showing how process improvement techniques can be used to make a firm lean.

We begin by discussing the continuous improvement aspect of lean systems, followed by a discussion of the characteristics of lean systems and the design of layouts needed to achieve these characteristics. We also address different types of lean systems used in practice, and some of the implementation issues that companies face.

Lean Systems across the Organization

Lean systems affect a firm's internal linkages between its core and supporting processes and its external linkages with its customers and suppliers. The design of supply chains using the lean systems approach is important to various departments and functional areas across the organization. Marketing relies on lean systems to deliver high-quality services or products on time and at reasonable prices. Human resources must put in place the right incentive systems that reward teamwork, and also recruit, train, and evaluate the employees needed to create a flexible workforce that can successfully operate a lean system. Engineering must design products that use more common parts, so that fewer setups are required and focused factories can be used. Operations is responsible for maintaining close ties with suppliers, designing the lean system, and using it in the production of services or goods. Accounting must adjust its billing and cost accounting practices to provide the support needed to manage lean systems. Finally, top management must embrace the lean philosophy and make it a part of organizational culture and learning, as was done by Panasonic in the opening vignette.

Continuous Improvement Using a Lean Systems Approach

One of the most popular systems that incorporate the generic elements of lean systems is the just-in-time (JIT) system. According to Taiichi Ohno, one of the earlier pioneers at Toyota Corporation, the **just-in-time (JIT) philosophy is** simple but powerful—*eliminate waste* or *muda* by cutting excess capacity or inventory and removing non-value-added activities. Table 8.1 shows the eight types of waste that often occur in firms in an interrelated fashion, and which must be eliminated in implementing lean systems.

The goals of a lean system are thus to eliminate these eight types of waste, produce services and products only as needed, and to continuously improve the value-added benefits of operations. A **JIT system** organizes the resources, information flows, and decision rules that enable a firm to realize the benefits of JIT principles.

lean systems
Operations systems that maximize the value added by each of a company's activities by removing waste and delays from them.

just-in-time (JIT) philosophy
The belief that waste can be eliminated by cutting unnecessary capacity or inventory and removing non-value-added activities in operations.

TABLE 8.1 | THE EIGHT TYPES OF WASTE OR MUDA[1]

Waste	Definition
1. Overproduction	Manufacturing an item before it is needed, making it difficult to detect defects and creating excessive lead times and inventory.
2. Inappropriate Processing	Using expensive high precision equipment when simpler machines would suffice. It leads to overutilization of expensive capital assets. Investment in smaller flexible equipment, immaculately maintained older machines, and combining process steps where appropriate reduce the waste associated with inappropriate processing.
3. Waiting	Wasteful time incurred when product is not being moved or processed. Long production runs, poor material flows, and processes that are not tightly linked to one another can cause over 90 percent of a product's lead time to be spent waiting.
4. Transportation	Excessive movement and material handling of product between processes, which can cause damage and deterioration of product quality without adding any significant customer value.
5. Motion	Unnecessary effort related to the ergonomics of bending, stretching, reaching, lifting, and walking. Jobs with excessive motion should be redesigned.
6. Inventory	Excess inventory hides problems on the shop floor, consumes space, increases lead times, and inhibits communication. Work-in-process inventory is a direct result of overproduction and waiting.
7. Defects	Quality defects result in rework and scrap, and add wasteful costs to the system in the form of lost capacity, rescheduling effort, increased inspection, and loss of customer good will.
8. Underutilization of Employees	Failure of the firm to learn from and capitalize on its employees' knowledge and creativity impedes long-term efforts to eliminate waste.

By spotlighting areas that need improvement, lean systems lead to continuous improvement in quality and productivity. The Japanese term for this approach to process improvement is *kaizen.* The key to *kaizen* is the understanding that excess capacity or inventory hides underlying problems with the processes that produce a service or product. Lean systems provide the mechanism for management to reveal the problems by systematically lowering capacities or inventories until the problems are exposed. For example, Figure 8.1 characterizes the philosophy behind continuous improvement with lean systems. In services, the water surface represents service system capacity, such as staff levels. In manufacturing, the water surface represents product and component inventory levels. The rocks represent problems encountered in the fulfillment of services or products. When the water surface is high enough, the boat passes over the rocks because the high level of capacity or inventory covers up problems. As capacity or inventory shrinks, rocks are exposed. Ultimately, the boat will hit a rock if the water surface falls far enough. Through lean systems, workers, supervisors, engineers, and analysts apply methods for continuous improvement to demolish the exposed rock. The coordination required to achieve smooth material flows in lean systems identifies problems in time for corrective action to be taken.

Maintaining low inventories, periodically stressing the system to identify problems, and focusing on the elements of the lean system lie at the heart of continuous improvement. For example, a Kawasaki plant in Lincoln, Nebraska, periodically cuts its safety stocks almost to zero. The problems at the plant are exposed, recorded, and later assigned to employees as improvement projects. After improvements are made, inventories are permanently cut to the new level.

JIT system
A system that organizes the resources, information flows, and decision rules that enable a firm to realize the benefits of JIT principles.

---

[1] David McBride, "The Seven Manufacturing Wastes," August 29, 2003, **http://www.emsstrategies.com** by permission of EMS Consulting Group, Inc. © 2003.

▲ FIGURE 8.1
Continuous Improvement with
Lean Systems

Many firms use this trial-and-error process to develop more efficient manufacturing operations. In addition, workers using special presses often fabricate parts on the assembly line in exactly the quantities needed. Service processes, such as scheduling, billing, order taking, accounting, and financial planning, can be improved with lean systems, too. In service operations, a common approach used by managers is to place stress on the system by reducing the number of employees doing a particular activity or series of activities until the process begins to slow or come to a halt. The problems can be identified, and ways for overcoming them explored. Other *kaizen* tactics can be used as well. Eliminating the problem of too much scrap might require improving the firm's work processes, providing employees with additional training, or finding higher-quality suppliers. Eliminating capacity imbalances might involve revising the firm's master production schedule and improving the flexibility of its workforce. Irrespective of which problem is solved, there are always new ones that can be addressed to enhance system performance.

Oftentimes, continuous improvement occurs with the ongoing involvement and input of new ideas from employees, who play an important role in implementing the JIT philosophy. In 2007 alone, about 740,000 corporate-wide improvement suggestions were received at Toyota. A large majority of them got implemented, and employees making those suggestions received rewards ranging from 500 yen (about $5) to upwards of 50,000 yen (about $500) depending upon their bottom line impact.

Supply Chain Considerations in Lean Systems

In this section, we discuss the two salient characteristics of lean systems that are related to creating and managing material flows in a supply chain: close supplier ties and small lot sizes.

Close Supplier Ties

Because lean systems operate with low levels of capacity slack or inventory, firms that use them need to have a close relationship with their suppliers. Supplies must be shipped frequently, have short lead times, arrive on schedule, and be of high quality. A contract might even require a supplier to deliver goods to a facility as often as several times per day.

The lean system philosophy is to look for ways to improve efficiency and reduce inventories throughout the supply chain. Close cooperation between companies and their suppliers can be a win–win situation for everyone. Better communication of component requirements, for example, enables more efficient inventory planning and delivery scheduling by suppliers, thereby improving supplier profit margins. Customers can then negotiate lower component prices. Close supplier relations cannot be established and maintained if companies view their suppliers as adversaries whenever contracts are negotiated. Rather, they should consider suppliers to be partners in a venture, wherein both parties have an interest in maintaining a long-term, profitable relationship. Consequently, one of the first actions undertaken when a lean system is implemented is to pare down the number of suppliers, and make sure they are located in close geographic proximity in order to promote strong partnerships and better synchronize product flows.

A particularly close form of supplier partnerships through lean systems is the JIT II system, which was conceived and implemented by Bose Corporation, a producer of high-quality professional sound and speaker systems. In a JIT II system, the supplier is brought into the plant to be an active member of the purchasing office of the customer. The *in-plant representative* is on site full-time at the supplier's expense and is empowered to plan and schedule the replenishment of materials from the supplier. Thus, JIT II fosters extremely close interaction with suppliers. The qualifications for a supplier to be included in the program are stringent.

In general, JIT II can offer benefits to both buyers and suppliers because it provides the organizational structure needed to improve supplier coordination by integrating the logistics, production, and purchasing processes together. We have more to say about supplier relationships in Chapter 12, "Supply Chain Integration."

Small Lot Sizes

Lean systems use lot sizes that are as small as possible. A **lot is a** quantity of items that are processed together. Small lots have the advantage of reducing the average level of inventory relative to large lots. Small lots pass through the system faster than large lots since they do not keep materials waiting. In addition, if any defective items are discovered, large lots cause longer delays because the entire lot must be examined to find all the items that need rework. Finally, small lots help achieve a uniform workload on the system and prevent overproduction. Large lots consume large chunks of capacity at workstations and, therefore, complicate scheduling. Small lots can be juggled more effectively, enabling schedulers to efficiently utilize capacities.

Although small lots are beneficial to operations, they have the disadvantage of increased setup frequency. A *setup* is the group of activities needed to change or readjust a process between successive lots of items, sometimes referred to as a *changeover*. This changeover in itself is a process that can be made more efficient. Setups involve trial runs, and the material waste can be substantial as the machines are fine tuned for the new parts. Typically, a setup takes the same time regardless of the size of the lot. Consequently, many small lots, in lieu of several large lots, may result in waste in the form of idle employees, equipment, and materials. Setup times must be brief to realize the benefits of small-lot production.

Achieving brief setup times often requires close cooperation among engineering, management, and labor. For example, changing dies on large presses to form automobile parts from sheet metal can take 3 to 4 hours. At Honda's Marysville, Ohio, plant—where four stamping lines stamp all the exterior and major interior body panels for Accord production—teams worked on ways to reduce the changeover time for the massive dies. As a result, a complete change of dies for a giant 2,400-ton press now takes less than 8 minutes. The goal of **single-digit setup** means having setup times of less than 10 minutes. Some techniques used to reduce setup times at the Marysville plant include using conveyors for die storage, moving large dies with cranes, simplifying dies, enacting machine controls, using microcomputers to automatically feed and position work, and preparing for changeovers while a job currently in production is still being processed.

lot
A quantity of items that are processed together.

single-digit setup
The goal of having a setup time of less than 10 minutes.

Process Considerations in Lean Systems

In this section, we discuss the following characteristics of lean systems: pull method of work flow, quality at the source, uniform workstation loads, standardized components and work methods, flexible workforce, automation, Five S (5S) practices, and total preventive maintenance (TPM).

Pull Method of Work Flow

Managers have a choice as to the nature of the material flows in a process or supply chain. Most firms using lean operations use the **pull method,** in which customer demand activates the production of a good or service. In contrast, a method often used in conventional systems that do not emphasize lean systems is the **push method,** which involves using forecasts of demand and producing the item before the customer orders it. To differentiate between these two methods, let us use a service example that involves a favorite pastime, eating.

For an illustration of the pull method, consider a five-star restaurant in which you are seated at a table and offered a menu of exquisite dishes, appetizers, soups, salads, and desserts. You can choose from filet mignon, porterhouse steak, yellow fin tuna, grouper, and lamb chops. Your choice of several salads is prepared at your table. Although some appetizers, soups, and desserts can be prepared in advance and brought to temperature just before serving, the main course and salads cannot. Your order for the salad and the main course signals the chef to begin preparing your specific requests. For these items, the restaurant is using the *pull method*. Firms using the pull method must be able to fulfill the customer's demands within an acceptable amount of time.

For an understanding of the push method, consider a cafeteria on a busy downtown corner. During the busy periods around 12 P.M. and 5 P.M. lines develop, with hungry patrons eager to eat and then move on to other activities. The cafeteria offers choices of chicken (roasted or deep fried), roast beef, pork chops, hamburgers, hot dogs, salad, soup (chicken, pea, and clam chowder), bread (three types), beverages, and desserts (pies, ice cream, and cookies). Close coordination is required between the cafeteria's "front office," where its employees interface with customers; and its "back office," the kitchen, where the food is prepared and then placed along the cafeteria's buffet line.

pull method
A method in which customer demand activates production of the service or item.

push method
A method in which production of the item begins in advance of customer needs.

Diners fill their plates at a restaurant buffet. Because the food items must be prepared in advance, the restaurant uses a push method of work flow.

Because it takes substantial time to cook some of the food items, the cafeteria uses a *push method*. The cafeteria would have a difficult time using the pull method because it could not wait until a customer asked for an item before asking the kitchen to begin processing it. After all, shortages in food could cause riotous conditions (recall that customers are hungry), whereas preparing an excess amount of food will be wasteful because it will go uneaten. To make sure that neither of these conditions occurs, the cafeteria must accurately forecast the number of customers it expects to serve.

The choice between the push and pull methods is often situational. Firms using an assemble-to-order strategy sometimes use both methods: the push method to produce the standardized components and the pull method to fulfill the customer's request for a particular combination of the components.

Quality at the Source

Consistently meeting the customer's expectations is an important characteristic of lean systems. One way to achieve this goal is by adhering to a practice called *quality at the source*, which is a philosophy whereby defects are caught and corrected where they are created. The goal for workers is to act as their own quality inspectors and never pass on defective units to the next process. Automatically stopping the process when something is wrong and then fixing the problems on the line itself as they occur is also known as *jidoka*. *Jidoka* tends to separate worker and machine activities by freeing workers from tending to machines all the time, thus allowing them to staff multiple operations simultaneously. *Jidoka* represents a visual management system whereby status of the system in terms of safety, quality, delivery, and cost performance relative to the goals for a given fabrication cell or workstation in an assembly line is clearly visible to workers on the floor at all times.

An alternative to *jidoka* or quality at the source is the traditional practice of pushing problems down the line to be resolved later. This approach is often ineffective. For example, a soldering operation at the Texas Instruments antenna department had a defect rate that varied from 0 to 50 percent on a daily basis, averaging about 20 percent. To compensate, production planners increased the lot sizes, which only increased inventory levels and did nothing to reduce the number of defective items. The company's engineers then discovered through experimentation that gas temperature was a critical variable in producing defect-free items. They subsequently devised statistical control charts for the firm's equipment operators to use to monitor the temperature and adjust it themselves. Process yields immediately improved and stabilized at 95 percent, and Texas Instruments was eventually able to implement a lean system.

One successful approach for implementing quality at the source is to use *poka-yoke*, or mistake-proofing methods aimed at designing fail-safe systems that attack and minimize human error. *Poka-yoke* systems work well in practice. Consider, for instance, a company that makes modular products. The company could use the poka-yoke method by making different parts of the modular product in such a way that allows them to be assembled in only one way—the correct way. Similarly, a company's shipping boxes could be designed to be packed only in a certain way to minimize damage and eliminate all chances of mistakes. At Toyota plants, every vehicle being assembled is accompanied by an RFID chip containing information on how many nuts and bolts need to be tightened on that vehicle for an operation at a given workstation. A green light comes on when the right number of nuts have been tightened. Only then does the vehicle move forward on the assembly line.

Another tool for implementing quality at the source is *andon*, which is a system that gives machines and machine operators the ability to signal the occurrence of any abnormal condition such as tool malfunction, shortage of parts, or the product being made outside the desired specifications. It can take the form of audio alarms, blinking lights, LCD text displays, or chords that can be pulled by workers to ask for help or stop the production line if needed. Stopping a production line can, however, cost a company thousands of dollars each minute production is halted. Needless to say, management must realize the enormous responsibility this method puts on employees and must prepare them properly.

jidoka

Automatically stopping the process when something is wrong and then fixing the problems on the line itself as they occur.

poka-yoke

Mistake-proofing methods aimed at designing fail-safe systems that minimize human error.

Owen Franken/CORBIS

Uniform Workstation Loads

A lean system works best if the daily load on individual workstations is relatively uniform. Service processes can achieve uniform workstation loads by using reservation systems. For example, hospitals schedule surgeries in advance of the actual service so that the facilities and facilitating goods can be ready when the time comes. The load on the surgery rooms and surgeons can be evened out to make the best use of these resources. Another approach is to use differential pricing of the service to manage the demand for it. Uniform loads are the rationale behind airlines promoting weekend travel or red-eye flights that begin late in the day and end in the early morning. Efficiencies can be realized when the load on the firm's resources can be managed.

For manufacturing processes, uniform loads can be achieved by assembling the same type and number of units each day, thus creating a uniform daily demand at all workstations. Capacity planning, which recognizes capacity constraints at critical workstations, and line balancing are used to develop the master production schedule. For example, at Toyota the production plan may call for 4,500 vehicles per week for the next month. That requires two full shifts, 5 days per week, producing 900 vehicles each day, or 450 per shift. Three models are produced: Camry (C), Avalon (A), and Solara (S). Suppose that Toyota needs 200 Camrys, 150 Avalons, and 100 Solaras per shift to satisfy market demand. To produce 450 units in one shift of 480 minutes, the line must roll out a vehicle every $480/450 = 1.067$ minutes. The 1.067 minutes, or 64 seconds, represents the **takt time** of the process, defined as the cycle time needed to match the rate of production to the rate of sales or consumption.

With traditional big-lot production, all daily requirements of a model are produced in one batch before another model is started. The sequence of 200 Cs, 150 As, and 100 Ss would be repeated once per shift. Not only would these big lots increase the average inventory level, but they also would cause lumpy requirements on all the workstations feeding the assembly line.

But there are other two options for devising a production schedule for the vehicles. These options are based on the Japanese concept of *heijunka*, which is the leveling of production load by both volume and product mix. It does not build products according to the actual flow of customer orders, but levels out the total volume of orders in a period so that the same amount and mix are being made each day.[2]

Let us explore two possible *heijunka* options. The first option uses leveled **mixed-model assembly**, producing a mix of models in smaller lots. Note that the production requirements at Toyota are in the ratio of 4 Cs to 3 As to 2 Ss, found by dividing the model's production requirements by the greatest common divisor, or 50. Thus, the Toyota planner could develop a production cycle consisting of 9 units: 4 Cs, 3 As, and 2 Ss. The cycle would repeat in $9(1.067) = 9.60$ minutes, for a total of 50 times per shift (480 min/9.60 min = 50).

The second *heijunka* option uses a lot size of one, such as the production sequence of C–S–C–A–C–A–C–S–A repeated 50 times per shift. The sequence would achieve the same total output as the other options; however, it is feasible only if the setup

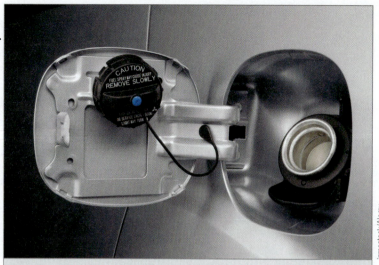

An example of poka-yoke is the design of new fuel doors in automobiles. They are mistake proof since the filling pipe insert keeps larger, leaded-fuel nozzle from being inserted. In addition, a gas cap tether does not allow the motorist to drive off without the cap, and is also fitted with a ratchet to signal proper tightness and prevent over-tightening.

izmostock/Alamy

takt time

Cycle time needed to match the rate of production to the rate of sales or consumption.

heijunka

The leveling of production load by both volume and product mix.

mixed-model assembly

A type of assembly that produces a mix of models in smaller lots.

Toyota Motor Manufacturing, Indiana (TMMI) is building the Sienna minivan and the Sequoia sport utility vehicle on a new mixed-model assembly line in the expanded plant.

John Hillery/REUTERS

---

[2]David McBride, "Heijunka, Leveling the Load," September 1, 2004, **http://www.emsstrategies.com**

times are brief. The sequence generates a steady rate of component requirements for the various models and allows the use of small lot sizes at the feeder workstations. Consequently, the capacity requirements at those stations are greatly smoothed. These requirements can be compared to actual capacities during the planning phase, and modifications to the production cycle, production requirements, or capacities can be made as necessary.

Standardized Components and Work Methods

In highly repetitive service operations, great efficiencies can be gained by analyzing work methods and documenting the improvements for all employees to use. For example, UPS consistently monitors its work methods, from sorting packages to delivering them, and revises them as necessary to improve service. In manufacturing, the standardization of components increases the total quantity that must be produced for that component. For example, a firm producing 10 products from 1,000 different components could redesign its products so that they consist of only 100 different components with larger daily requirements. Panasonic did similar standardization with its printed circuit boards in the opening vignette. Because the requirements per component increase, each worker performs a standardized task or work method more often each day. Productivity tends to increase because workers learn to do their tasks more efficiently with increased repetition. Standardizing components and work methods help a firm achieve the high-productivity, low-inventory objectives of a lean system.

Flexible Workforce

The role of workers is elevated in lean systems. Workers in flexible workforces can be trained to perform more than one job. A benefit of flexibility is the ability to shift workers among workstations to help relieve bottlenecks as they arise without the need for inventory buffers—an important aspect of the uniform flow of lean systems. Also, workers can step in and do the job for those who are on vacation or who are out sick. Although assigning workers to tasks they do not usually perform can temporarily reduce their efficiency, some job rotation tends to relieve boredom and refreshes workers. At some firms that have implemented lean systems, cross-trained workers may switch jobs every 2 hours.

The more customized the service or product is, the greater the firm's need for a multiskilled workforce. For example, stereo repair shops require broadly trained personnel who can identify a wide variety of component problems when the customer brings the defective unit into the shop and who then can repair the unit. Alternatively, back-office designs, such as the mail-processing operations at a large post office, have employees with more narrowly defined jobs because of the repetitive nature of the tasks they must perform. These employees do not have to acquire as many alternative skills. In some situations, shifting workers to other jobs may require them to undergo extensive, costly training.

Automation

Automation plays a big role in lean systems and is a key to low-cost operations. Money freed up because of inventory reductions or other efficiencies can be invested in automation to reduce costs. The benefits, of course, are greater profits, greater market share (because prices can be cut), or both. Automation can play a big role when it comes to providing lean services. For example, banks offer ATMs that provide various bank services on demand 24 hours a day. Automation should be planned carefully, however. Many managers believe that if some automation is good, more is better, which is not always the case. At times, humans can do some jobs better than robots and automated assembly systems.

Molly McMillin/MCT/Newscom

Spirit AeroSystems' plant in Prestwick, Scotland, has seen some big changes. For one, it has invested in automation on the A320 production line.

Five S Practices

Five S (5S) is a methodology for organizing, cleaning, developing, and sustaining a productive work environment. It represents five related terms, each beginning with an *S*, that describe workplace practices conducive to visual controls and lean production. As shown in Figure 8.2, these five practices of sort, straighten, shine, standardize, and sustain build upon one another and are done systematically to achieve lean systems. These practices are interconnected, and are not something that can be done as a stand-alone program. As such, they serve as an enabler and an essential foundation of lean systems. Table 8.2 shows the terms[3] that represent the 5S and what they imply.

It is commonly accepted that 5S forms an important cornerstone of waste reduction and removal of unneeded tasks, activities, and materials. 5S practices can enable workers to visually see everything differently, prioritize tasks, and achieve a greater degree of focus. They can also be applied to a diverse range of manufacturing and service settings including organizing work spaces, offices, tool rooms, shop floors, and the like. Implementation of 5S practices have been shown to lead to lowered costs, improved on-time delivery and productivity, higher product quality, better use of floor space, and a safe working environment. It also builds the discipline needed to make the lean systems work well.

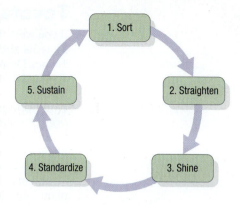

▲ **FIGURE 8.2**
5 S Practices

five S (5S)

A methodology consisting of five workplace practices—sorting, straightening, shining, standardizing, and sustaining—that are conducive to visual controls and lean production.

TABLE 8.2 | **5S DEFINED**

5S Term	Definition
1. Sort	Separate needed items from unneeded items (including tools, parts, materials, and paperwork), and discard the unneeded.
2. Straighten	Neatly arrange what is left, with a place for everything and everything in its place. Organize the work area so that it is easy to find what is needed.
3. Shine	Clean and wash the work area and make it shine.
4. Standardize	Establish schedules and methods of performing the cleaning and sorting. Formalize the cleanliness that results from regularly doing the first three S practices so that perpetual cleanliness and a state of readiness are maintained.
5. Sustain	Create discipline to perform the first four S practices, whereby everyone understands, obeys, and practices the rules when in the plant. Implement mechanisms to sustain the gains by involving people and recognizing them through a performance measurement system.

Total Preventive Maintenance (TPM)

Because lean systems emphasize finely tuned flows of work and little capacity slack or buffer inventory between workstations, unplanned machine downtime can be disruptive. Total Preventive Maintenance (TPM), which is also sometimes referred to as total productive maintenance, can reduce the frequency and duration of machine downtime. After performing their routine maintenance activities, technicians can test other machine parts that might need to be replaced. Replacing parts during regularly scheduled maintenance periods is easier and quicker than dealing with machine failures during production. Maintenance is done on a schedule that balances the cost of the preventive maintenance program against the risks and costs of machine failure. Routine preventive maintenance is important for service businesses that rely heavily on machinery, such as the rides at Walt Disney World or Universal Studios.

Another tactic is to make workers responsible for routinely maintaining their own equipment, which will develop employee pride in keeping the machines in top condition. This tactic, however, typically is limited to general housekeeping chores, minor lubrication, and adjustments. Maintaining high-tech machines requires trained specialists. Nonetheless, performing even simple maintenance tasks goes a long way toward improving the performance of machines.

For long-term improvements, data can be collected for establishing trends in failure pattern of machines, which can subsequently be analyzed to establish better standards and procedures for preventive maintenance. The data can also provide failure history and costs incurred to maintain the systems.

---

[3]The Japanese words for these 5S terms are *seiri, seiton, seiso, seiketsu,* and *shitsuke,* respectively.

Toyota Production System

If you were to select one company that regularly invokes the above mentioned features of Lean Systems and also exemplifies excellence in automobile manufacturing, it would probably be Toyota. Despite its recent problems with quality and product recalls, as well as component shortages and delayed new model launches caused by the Japanese earthquake in March 2011, Toyota has become one of the largest car manufacturers in the world and also one of its most admired. Worldwide in its presence, Toyota has 12 manufacturing plants in North America alone producing over 1.5 million vehicles per year. Much of this success is attributed to the famed Toyota Production System (TPS), which is one of the most admired lean manufacturing systems in existence. Replicating the system, however, is fraught with difficulties. What makes the system tick, and why has Toyota been able to use it so successfully in many different plants?

Most outsiders see the TPS as a set of tools and procedures that are readily visible during a plant tour. Even though they are important for the success of the TPS, they are not the key. What most people overlook is that through the process of continuous improvement, Toyota built a learning organization over the course of 50 years. Lean systems require constant improvements to increase efficiency and reduce waste. Toyota's system stimulates employees to experiment to find better ways to do their jobs. In fact, Toyota sets up all of its operations as "experiments" and teaches employees at all levels how to use the scientific method of problem solving.

Four principles form the basis of the TPS. First, all work must be completely specified as to content, sequence, timing, and outcome. Detail is important; otherwise, a foundation for improvements is missing. Second, every customer–supplier connection must be direct, unambiguously specifying the people involved, the form and quantity of the services or goods to be provided, the way the requests are made by each customer, and the expected time in which the requests will be met. Customer–supplier connections can be internal (employee to employee) or external (company to company). Third, the pathway for every service and product must be simple and direct. That is, services and goods do not flow to the next available person or machine, but to a specific person or machine. With this principle, employees can determine, for example, whether a capacity problem exists at a particular workstation and then analyze ways to solve it.

The first three principles define the system in detail by specifying how employees do work, interact with each other, and how the work flows are designed. However, these specifications actually are "hypotheses" about the way the system should work. For example, if something goes wrong at a workstation enough times, the hypothesis about the methods the employee uses to do work is rejected. The fourth principle, then, is that any improvement to the system must be made in accordance with the scientific method, under the guidance of a teacher, at the lowest possible organizational level. The scientific method involves clearly stating a verifiable hypothesis of the form, "If we make the following specific changes, we expect to achieve this specific outcome." The hypothesis must then be tested under a variety of conditions. Working with a teacher, who is often the employees' supervisor, is a key to becoming a learning organization. Employees learn the scientific method and eventually become teachers of others. Finally, making improvements at the lowest level of the organization means that the employees who are actually doing the work are actively involved in making the improvements. Managers are advised only to coach employees—not to fix their problems for them.

These four principles are deceptively simple. However, they are difficult but not impossible to replicate. Those organizations that successfully implement them enjoy the benefits of a lean system that adapts to change. Toyota's lean system made it an innovative leader in the auto industry and served as an important cornerstone of its success.

An employee helps assemble a vehicle in Toyota City, located in central Japan. Toyota's production system is among the most-admired lean manufacturing systems in the world.

Chang-Ran Kim/Reuters/Corbis

House of Toyota

Taiichi Ohno and Eiji Toyoda created a graphic representation shown in Figure 8.3 to define the Toyota Production System (TPS) to its employees and suppliers, and which is now known as the House of Toyota. It captures the four principles of TPS described above, and represents all the essential elements of lean systems that make the TPS work well. The house conveys stability. The roof, representing the primary goals of high quality, low cost, waste elimination, and short lead-times, is supported by the twin pillars of JIT and *jidoka*. Within JIT, TPS uses a pull system that focuses on one-piece work flow methods that can change and match the takt time of the process to the actual market demand because setup reductions and small changeover times are facilitated by cross-trained workers in cellular layouts. Implementing various tools of *jidoka* ensures that quality is built into the product rather than merely inspected at the end. Finally, within an environment of continuous improvement, operational stability to the House of Toyota is provided at the base by leveraging other lean concepts such as *heijunka*, standard work methods, 5S practices, total preventive maintenance, and elimination of waste throughout the supply chain within which the Toyota products flow to reach their eventual customers.

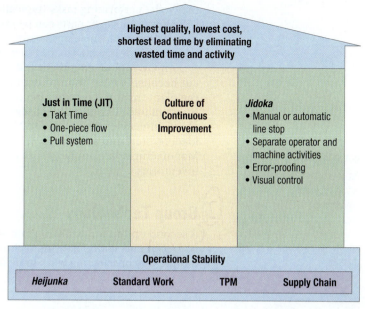

▲ FIGURE 8.3
House of Toyota[4]

4 Designing Lean System Layouts

Line flows are recommended in designing lean system layouts because they eliminate waste by reducing the frequency of setups. If volumes of specific products are large enough, groups of machines and workers can be organized into a line-flow layout to eliminate setups entirely. In a service setting, managers of back-office service processes can similarly organize their employees and equipment to provide uniform work flows through the process and, thereby, eliminate wasted employee time. Banks use this strategy in their check-processing operations, as does UPS in its parcel-sorting process.

When volumes are not high enough to justify dedicating a single line of multiple workers to a single customer type or product, managers still may be able to derive the benefits of line-flow layout—simpler materials handling, low setups, and reduced labor costs—by creating line-flow layouts in some portions of the facility. Two techniques for creating such layouts are one-worker, multiple-machines (OWMM) cells, and group technology (GT) cells.

one-worker, multiple-machines (OWMM) cell

A one-person cell in which a worker operates several different machines simultaneously to achieve a line flow.

▼ FIGURE 8.4
One-Worker, Multiple-Machines (OWMM) Cell

One Worker, Multiple Machines

If volumes are not sufficient to keep several workers busy on one production line, the manager might set up a line small enough to keep one worker busy. The **one-worker, multiple-machines (OWMM) cell** is a workstation in which a worker operates several different machines simultaneously to achieve a line flow. Having one worker operate several identical machines is not unusual. However, with an OWMM cell, several different machines are in the line.

Figure 8.4 illustrates a five-machine OWMM cell that is being used to produce a flanged metal part, with the machines encircling one operator in the center. (A U-shape also is common.) The operator moves around

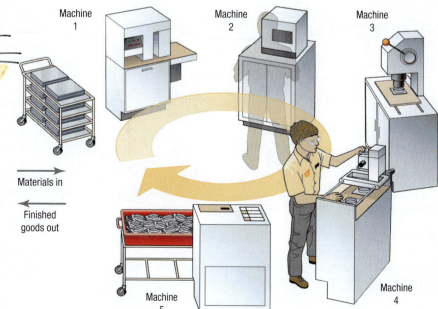

Materials in →
← Finished goods out

4 TBM Consulting Group; **http://www.tbmcg.com/about/ourroots/house_toyota.php**

the circle, performing tasks (typically loading and unloading) that have not been automated. Different products or parts can be produced in an OWMM cell by changing the machine setups. If the setup on one machine is especially time-consuming for a particular part, management can add a duplicate machine to the cell for use whenever that part is being produced.

An OWMM arrangement reduces both inventory and labor requirements. Inventory is cut because, rather than piling up in queues waiting for transportation to another part of the plant, materials move directly into the next operation. Labor is cut because more work is automated. The addition of several low-cost automated devices can maximize the number of machines included in an OWMM arrangement: automatic tool changers, loaders and unloaders, start and stop devices, and fail-safe devices that detect defective parts or products. Manufacturers are applying the OWMM concept widely because of their desire to achieve low inventories.

group technology (GT)

An option for achieving line-flow layouts with low volume processes; this technique creates cells not limited to just one worker and has a unique way of selecting work to be done by the cell.

② Group Technology

A second option for achieving line-flow layouts with low volume processes is **group technology (GT)**. This manufacturing technique creates cells not limited to just one worker and has a unique way of selecting work to be done by the cell. The GT method groups parts or products with similar characteristics into *families* and sets aside groups of machines for their production. Families may be based on size, shape, manufacturing or routing requirements, or demand. The goal is to identify a set of products with similar processing requirements and minimize machine changeover or setup. For example, all bolts might be assigned to the same family because they all require the same basic processing steps regardless of size or shape.

Once parts have been grouped into families, the next step is to organize the machine tools needed to perform the basic processes on these parts into separate cells. The machines in each cell require only minor adjustments to accommodate product changeovers from one part to the next in the same family. By simplifying product routings, GT cells reduce the time a job is in the shop. Queues of materials waiting to be worked on are shortened or eliminated. Frequently, materials handling is automated so that, after loading raw materials into the cell, a worker does not handle machined parts until the job has been completed.

Figure 8.5 compares process flows before and after creation of GT cells. Figure 8.5(a) shows a shop floor where machines are grouped according to function: lathing, milling, drilling, grinding, and assembly. After lathing, a part is moved to one of the milling machines, where it waits in line until it has a higher priority than any other job competing for the machine's capacity. When the milling operation on the part has been finished, the part is moved to a drilling machine, and so on. The queues can be long, creating significant time delays. Flows of materials are jumbled because the parts being processed in any one area of the shop have so many different routings.

By contrast, the manager of the shop shown in Figure 8.5(b) identified three product families that account for a majority of the firm's production. One family always requires two lathing operations followed by one operation at the milling machines. The second family always requires a milling operation followed by a grinding operation. The third family

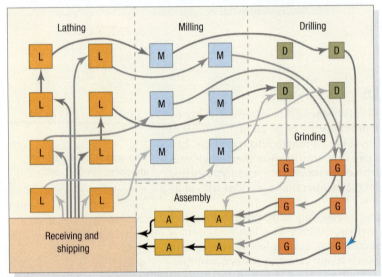

(a) Jumbled flows in a job shop without GT cells

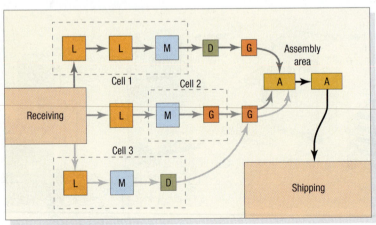

(b) Line flows in a job shop with three GT cells

▲ **FIGURE 8.5**

Process Flows Before and After the Use of GT Cells

Source: Mikell P. Groover. *Automation, Production Systems, and Computer-Aided Manufacturing.* 1st Edition, © 1980. Reprinted by permission of Pearson Education, Inc., Upper Saddle River, NJ.

requires the use of a lathe, a milling machine, and a drill press. For simplicity, only the flows of parts assigned to these three families are shown. The remaining parts are produced at machines outside the cells and still have jumbled routings. Some equipment might have to be duplicated, as when a machine is required for one or more cells and for operations outside the cells. However, by creating three GT cells, the manager has definitely created more line flows and simplified routings.

Value Stream Mapping

Value stream mapping (VSM) is a widely used qualitative lean tool aimed at eliminating waste or *muda*. Waste in many processes can be as high as 60 percent. Value stream mapping is helpful because it creates a visual "map" of every process involved in the flow of materials and information in a product's value chain. These maps consist of *a current state drawing*, a *future state drawing*, and an implementation plan. Value stream mapping spans the supply chain from the firm's receipt of raw materials or components to the delivery of the finished good to the customer. Thus, it tends to be broader in scope, displaying far more information than a typical process map or a flowchart used with Six Sigma process improvement efforts. Creating such a big picture representation helps managers identify the source of wasteful non-value-added activities.

Value stream mapping follows the steps shown in Figure 8.6. The first step is to focus on one product family for which mapping can be done. It is then followed by drawing a current state map of the existing production situation: Analysts start from the customer end and work upstream to draw the map by hand and record actual process times rather than rely on information not obtained by firsthand observation. Information for drawing the material and information flows can be gathered from the shop floor, including the data related to each process: cycle time (C/T), setup or changeover time (C/O), uptime (on-demand available machine time expressed as a percentage), production batch sizes, number of people required to operate the process, number of product variations, pack size (for moving the product to the next stage), working time (minus breaks), and scrap rate. Value stream mapping uses a standard set of icons for material flow, information flow, and general information (to denote operators, safety stock buffers, and so on). Even though the complete glossary is extensive, a representative set of these icons is shown in Figure 8.7. These icons provide a common language for describing in detail how a facility should operate to create a better flow.

value stream mapping (VSM)

A qualitative lean tool for eliminating waste or *muda* that involves a current state drawing, a future state drawing, and an implementation plan.

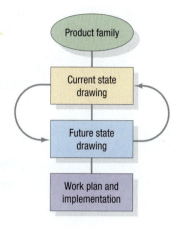

▲ **FIGURE 8.6**

Value Stream Mapping Steps

Source: Mike Rother and John Shook, *Learning to See* (Brookline, MA: The Lean Enterprise Institute, 2003), p. 9. Copyright © Lean Enterprise Institute, Inc. All right reserved, 2003. **www.lean.org**.

Material Flow Icons

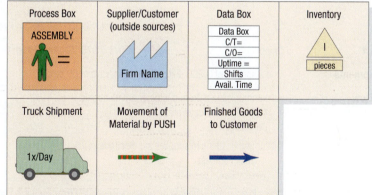

Information Flow Icons

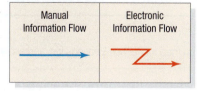

General Icons

▲ **FIGURE 8.7**

Selected Set of Value Stream Mapping Icons

EXAMPLE 8.1	Determining the Value Stream Map, Takt Time, and Total Capacity

Jensen Bearings Incorporated, a ball bearing manufacturing company located in Lexington, South Carolina, receives raw material sheets from Kline Steel Company once a week every Monday for a product family of retainers (casings in which ball bearings are held), and then ships its finished product on a daily basis to a second-tier automotive manufacturing customer named GNK Enterprises. The product family of the bearing manufacturing company under consideration consists of two types of retainers—large (L) and small (S)—that are packaged for shipping in returnable trays with 60 retainers in each tray. The manufacturing process consists of a cell containing pressing operation, a piercing and forming cell, and a finish grind operation, after which the two types of retainers are staged for shipping. The information collected by the operations manager at Jensen Bearings Inc. is shown in Table 8.3.

TABLE 8.3 | OPERATIONS DATA FOR A FAMILY OF RETAINERS AT JENSEN BEARINGS, INC.

Overall Process Attributes	Average demand: 3,200/day (1,000 "L"; 2,200 "S") Batch size: 1000 Number of shifts per day: 1 Availability: 8 hours per shift with two 30-minute lunch breaks	
Process Step 1	Press	Cycle time = 3 seconds Setup time = 2 hours Up time = 90% Operators = 1 Every Part Every = 1 week WIP = 5 days of sheets (Before Press)
Process Step 2	Pierce & Form	Cycle time = 22 seconds Setup time = 30 minutes Up time = 100% Operators = 1 WIP = 1,000 "L," 1,250 "S" (Before Pierce & Form)
Process Step 3	Finish Grind	Cycle time = 35 seconds Setup time = 45 minutes Up time = 100% Operators = 1 WIP = 1,050 "L," 2,300 "S" (Before Finish Grind)
Process Step 4	Shipping	WIP = 500 "L," 975 "S" (After Finish Grind)
Customer Shipments	One shipment of 3,200 units each day in trays of 60 pieces	

Information Flow	All communications from customer are electronic:
	   180/90/60/30/day Forecasts
	   Daily Order	

	All communications to supplier are electronic
	   4-week Forecast
	   Weekly Fax	

	There is a weekly schedule manually delivered to Press, Pierce & Form, and Finish Grind and a Daily Ship Schedule manually delivered to Shipping	
	All material is pushed	

a. Using data shown in Table 8.3; create a value stream map for Jensen Bearings Inc. and show how the data box values are calculated.

b. What is the takt time for this manufacturing cell?

c. What is the production lead time at each process in the manufacturing cell?

d. What is the total processing time of this manufacturing cell?

e. What is the capacity of this manufacturing cell?

SOLUTION

a.   We use the VSM icons to illustrate in Figure 8.8 what a current state map would look like for Jensen Bearings Inc. The process characteristics and inventory buffers in front of each process are shown in the current state map of Figure 8.8. One worker occupies each station. The process flows shown at the bottom of Figure 8.8 are similar to the flowcharts discussed in Chapter 4, "Process Analysis," except that more detailed information is presented here for each process. However, what really sets the value stream maps apart from flowcharts is the inclusion of information flows at the top of Figure 8.8, which plan and coordinate all the process activities. The value stream maps are more comprehensive than process flowcharts, and meld together planning and control systems (discussed in detail in Chapter 16, "Resource Planning") with detailed flowcharts (discussed in Chapter 4) to create a comprehensive supply chain view that includes both information and material flows between the firm and its suppliers and customers.

b.   The cell's takt time is the rate at which the cell must produce units in order to match demand.

Daily Demand = [(1,000 + 2,200) pieces per week]/5 working days per week = 640 pieces per day

Daily Availability = (7 hours per day) × (3,600 seconds per hour) = 25,200 seconds per day

Takt Time = Daily Availability/Daily Demand = (25,200 seconds per day)/640 pieces per day = 39.375 seconds per piece

c.   The production lead time (in days) is calculated by summing the inventory held between each processing step divided by daily demand.

Raw Material lead time = 5.0 days

WIP lead time between Press and Pierce & Form = (2,250/640) = 3.5 days

WIP lead time between Pierce & Form and Finish Grind = (3,350/640) = 5.2 days

WIP lead time between Finish Grind and Shipping = (1,475/640) = 2.3 days

Total Production Lead time = (5 + 3.5 + 5.2 + 2.3) = 16 days

d.   The cycle time at each process is added to compute total processing time. The manufacturing cell's total processing time is (3 + 22 + 35) = 60 seconds.

e.   The cell's capacity may be calculated by locating the bottleneck and computing the number of units that it can process in the available time per day at that bottleneck with the given batch size of 1,000 units.

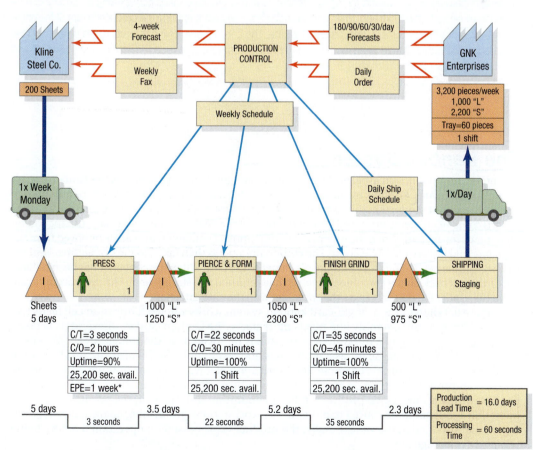

◀ **FIGURE 8.8**
Current State Map for a Family of Retainers at Jensen Bearings Incorporated

Capacity at Press	Capacity at Pierce & Form	Capacity at Finish Grind
Cycle time = 3 seconds	Cycle time = 22 seconds	Cycle time = 35 seconds
Setup Time = (2 hrs * 3,600 seconds per hour)/1000 units per batch = **7.2 seconds**	Setup Time = (30 minutes * 60 seconds per minute)/1000 units per batch = **1.8 seconds**	Setup Time = (45 minutes * 60 seconds per minute)/1,000 units per batch = **2.7 seconds**
Per Unit Processing Time = (3 + 7.2) = **10.2 seconds**	Per Unit Processing Time = (22 + 1.8) = **23.8 seconds**	Per Unit Processing Time = (35 + 2.7) = **37.7 seconds**

At a batch size of 1,000 units, Finish Grinding process is the bottleneck.

Availability at Grinding = 25,200 seconds per day

Time at bottleneck (with setup) = 37.7 seconds

Capacity (Availability/Time at bottleneck) = 25,200/37.7 = 668 units per day

DECISION POINT

Although the total processing time for each retainer is only 1 minute, it takes 16 days for the cumulative production lead time. Clearly *muda* or waste is present, and opportunities exist for reconfiguring the existing processes with the goal of eliminating inventories and reducing cumulative production lead time.

Once the current state map is done, the analysts can then use principles of lean systems to create a future state map with more streamlined product flows. The future state drawing highlights sources of waste and how to eliminate them. The developments of the current and future state maps are overlapping efforts. Finally, the last step is aimed at preparing and actively using an implementation plan to achieve the future state. It may take only a couple of days from the creation of a future state map to the point where implementation can begin for a single product family. At this stage, the future state map becomes a blueprint for implementing a lean system, and is fine-tuned as implementation progresses. As the future state becomes reality, a new future state map is drawn, thus denoting continuous improvement at the value stream level.

Unlike the theory of constraints (see Chapter 7, "Constraint Management"), which accepts the existing system bottlenecks and then strives to maximize the throughput given that set of constraint(s), value stream mapping endeavors to understand through current state and future state maps how existing processes can be altered to eliminate bottlenecks and other wasteful activities. The goal is to bring the production rate of the entire process closer to the customer's desired demand rate. The benefits of applying this tool to the waste-removal process include reduced lead times and work-in-process inventories, reduced rework and scrap rates, and lower indirect labor costs.

The *Kanban* System

kanban

A Japanese word meaning "card" or "visible record" that refers to cards used to control the flow of production through a factory.

One of the most publicized aspects of lean systems, and the TPS in particular, is the *kanban* system developed by Toyota. *Kanban*, meaning "card" or "visible record" in Japanese, refers to cards used to control the flow of production through a factory. In the most basic *kanban* system, a card is attached to each container of items produced. The container holds a given percent of the daily production requirements for an item. When the user of the parts empties a container, the card is removed from the container and put on a receiving post. The empty container is then taken to the storage area, and the card signals the need to produce another container of the part. When the container has been refilled, the card is put back on the container, which is then returned to a storage area. The cycle begins again when the user of the parts retrieves the container with the card attached.

Figure 8.9 shows how a single-card kanban system works when a fabrication cell feeds two assembly lines. As an assembly line needs more parts, the kanban card for those parts is taken to the receiving post, and a full container of parts is removed from the storage area. The receiving post accumulates cards for assembly lines and a scheduler sequences the production of replenishment parts. In this example, the fabrication cell will produce product 2 (red) before it produces product 1 (green). The cell consists of three different operations, but operation 2 has two workstations. Once production has been initiated in the cell, the product begins on operation 1, but could be routed to either of the workstations performing operation 2, depending on the workload at the time. Finally, the product is processed on operation 3 before being taken to the storage area.

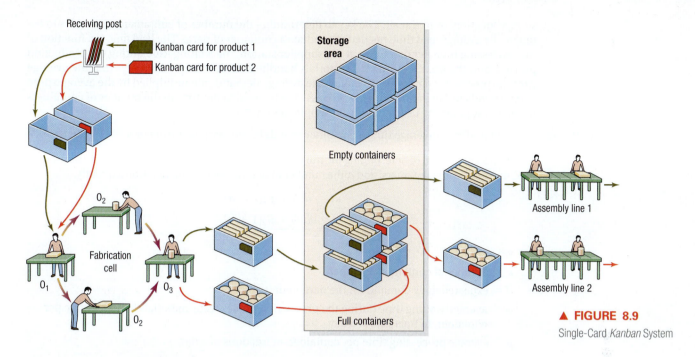

▲ **FIGURE 8.9**
Single-Card *Kanban* System

General Operating Rules

The operating rules for the single-card system are simple and are designed to facilitate the flow of materials while maintaining control of inventory levels.

1. Each container must have a card.

2. The assembly line always withdraws materials from the fabrication cell. The fabrication cell never pushes parts to the assembly line because, sooner or later, parts will be supplied that are not yet needed for production.

3. Containers of parts must never be removed from a storage area without a kanban first being posted on the receiving post.

4. The containers should always contain the same number of good parts. The use of nonstandard containers or irregularly filled containers disrupts the production flow of the assembly line.

5. Only nondefective parts should be passed along to the assembly line to make the best use of materials and worker's time. This rule reinforces the notion of building quality at the source, which is an important characteristic of lean systems.

6. Total production should not exceed the total amount authorized on the *kanbans* in the system.

Toyota uses a two-card system, based on a withdrawal card and a production-order card, to control inventory quantities more closely. The withdrawal card specifies the item and the quantity the user of the item should withdraw from the producer of the item, as well as the stocking locations for both the user and the producer. The production-order card specifies the item and the quantity to be produced, the materials required and where to find them, and where to store the finished item. Materials cannot be withdrawn without a withdrawal card, and production cannot begin without a production-order card. The cards are attached to containers when production commences. By manipulating the number of withdrawal and production cards in play at any time, management can control the flow of materials in the production system.

Determining the Number of Containers

The number of authorized containers in the TPS determines the amount of authorized inventory. Management must make two determinations: (1) the number of units to be held by each container, and (2) the number of containers flowing back and forth between the supplier station and the user station. The first decision amounts to determining the size of the production lot.

The number of containers flowing back and forth between two stations directly affects the quantities of work-in-process inventory, which includes any safety stock inventory to cover for unexpected requirements.[5] The containers spend some time in production, in a line waiting, in a

---

[5]We discuss safety stocks, and their use, in more detail in Chapter 9, "Supply Chain Inventory Management," and Chapter 10 "Supply Chain Design."

storage location, or in transit. The key to determining the number of containers required is to estimate the average lead time needed to produce a container of parts. The lead time is a function of the processing time per container at the supplier station, the waiting time during the production process, and the time required for materials handling. Little's Law, which says that the average work-in-process inventory (WIP) equals the average demand rate multiplied by the average time a unit spends in the manufacturing process, can be used to determine the number of containers needed to support the user station (see Supplement B, "Waiting Lines").

WIP = (average demand rate)(average time a container spends in the manufacturing process) + safety stock

In this application of determining the number of containers needed for a part, WIP is the product of κ, the number of containers, and c, the number of units in each container. Consequently,

$$\kappa c = \bar{d}(\bar{\omega} + \bar{\rho})(1 + \alpha)$$

$$\kappa = \frac{\bar{d}(\bar{\omega} + \bar{\rho})(1 + \alpha)}{c}$$

where

κ = number of containers for a part

$\bar{d}$ = expected daily demand for the part, in units

$\bar{\omega}$ = average waiting time during the production process plus materials handling time per container, in fractions of a day

$\bar{\rho}$ = average processing time per container, in fractions of a day

c = quantity in a standard container of the part

α = a policy variable that adds safety stock to cover for unexpected circumstances (Toyota uses a value of no more than 10 percent)

The number of containers must, of course, be an integer. Rounding κ up provides more inventory than desired, whereas rounding κ down provides less.

The container quantity, c, and the efficiency factor, α, are variables that management can use to control inventory. Adjusting c changes the size of the production lot, and adjusting α changes the amount of safety stock. The kanban system allows management to fine-tune the flow of

---

EXAMPLE 8.2	**Determining the Appropriate Number of Containers**

Solver-Number of Containers
Enter data in yellow-shaded area.

Daily Expected Demand	2000
Quantity in Standard Container	22
Container Waiting Time (days)	0.06
Processing Time (days)	0.02
Policy Variable	10%
Containers Required	8

▲ **FIGURE 8.10**
OM Explorer Solver for
Number of Containers

The Westerville Auto Parts Company produces rocker-arm assemblies for use in the steering and suspension systems of four-wheel-drive trucks. A typical container of parts spends 0.02 day in processing and 0.08 day in materials handling and waiting during its manufacturing cycle. The daily demand for the part is 2,000 units. Management believes that demand for the rocker-arm assembly is uncertain enough to warrant a *safety stock* equivalent of 10 percent of its authorized inventory.

a. If each container contains 22 parts, how many containers should be authorized?

b. Suppose that a proposal to revise the plant layout would cut materials handling and waiting time per container to 0.06 day. How many containers would be needed?

Solution

a. If $\bar{d}$ = 2,000 units/day, $\bar{\rho}$ = 0.02 day, α = 0.10, $\bar{\omega}$ = 0.08 day and c = 22 units,

$$\kappa = \frac{2,000(0.08 + 0.02)(1.0)}{22} = \frac{220}{22} = 10 \text{ containers}$$

b. Figure 8.10 from OM Explorer shows that the number of containers drops to 8.

DECISION POINT

The average lead time per container is $\bar{\omega} + \bar{\rho}$. With a lead time of 0.10 day, 10 containers are needed. However, if the improved facility layout reduces the materials handling time and waiting time to $\bar{\omega}$ 0.06 day, only 8 containers are needed. The maximum authorized inventory of the rocker-arm assembly is κc units. Thus, in part (a), the maximum authorized inventory is 220 units, but in part (b), it is only 176 units. Reducing $\bar{\omega} + \bar{\rho}$ by 20 percent reduces the inventory of the part by 20 percent. Management must balance the cost of the layout change (a one-time charge) against the long-term benefits of inventory reduction.

materials in the system in a straightforward way. For example, removing cards from the system reduces the number of authorized containers of the part, thus reducing the inventory of the part. Thus. a major benefit is the simplicity of the system, whereby product mix or volume changes can easily be accomplished by adjusting the number of *kanbans* in the system.

Other *Kanban* Signals

Cards are not the only way to signal the need for more production of a part. Other, less formal methods are possible, including container and containerless systems.

Container System Sometimes, the container itself can be used as a signal device: An empty container signals the need to fill it. Unisys took this approach for low-value items. The amount of inventory of the part is adjusted by adding or removing containers. This system works well when the container is specially designed for a particular part and no other parts could accidentally be put in the container. Such is the case when the container is actually a pallet or fixture used to position the part during precision processing.

Containerless System Systems requiring no containers have been devised. In assembly-line operations, operators use their own workbench areas to put completed units on painted squares, one unit per square. Each painted square represents a container, and the number of painted squares on each operator's bench is calculated to balance the line flow. When the subsequent user removes a unit from one of the producer's squares, the empty square signals the need to produce another unit. McDonald's uses a containerless system. Information entered by the order taker at the cash register is transmitted to the cooks and assemblers, who produce the sandwiches requested by the customer.

Managerial Practice 8.1 illustrates how the University of Pittsburgh Medical Center Shadyside used principles of *kanban* systems, 5S methodology, cellular layouts, and continuous flow processes to significantly improve performance in its pathology department.

Operational Benefits and Implementation Issues

To gain competitive advantage and to make dramatic improvements, a lean system can be the solution. Lean systems can be an integral part of a corporate strategy based on speed because they cut cycle times, improve inventory turnover, and increase labor productivity. Recent studies also show that practices representing different components of lean systems such as JIT, TQM, Six Sigma, total preventive maintenance (TPM), and human resource management (HRM), individually as well as cumulatively, improve the performance of manufacturing plants as well as service facilities. Lean systems also involve a considerable amount of employee participation through small-group interaction sessions, which have resulted in improvements in many aspects of operations, not the least of which is service or product quality.

Even though the benefits of lean systems can be outstanding, problems can still arise after a lean system has long been operational and which was witnessed recently in product recalls and a perceived shift away from tightly controlled quality that has always been the standard at Toyota. In addition, implementing a lean system can take a long time. We address below some of the issues managers should be aware of when implementing a lean system.

A worker moves a stack of partially assembled athletic shoes from her stitching station at the New Balance factory in Skowhegan, Maine. She and five other members of her team have worked out a plan so that each person is cross-trained in another's skills. Similar ideas for improving proficiency, which are discussed during biweekly meetings with workers and supervisors, have led to improved performance at New Balance.

Robert F. Bukaty/ASSOCIATED PRESS

Organizational Considerations

Implementing a lean system requires management to consider issues of worker stress, cooperation and trust among workers and management, and reward systems and labor classifications.

The Human Costs of Lean Systems Lean systems can be coupled with statistical process control (SPC) to reduce variations in output. However, this combination requires a high degree of regimentation and sometimes stresses the workforce. For example, in the Toyota Production System, workers must meet specified cycle times, and, with SPC, they must follow prescribed problem-solving methods. Such systems might make workers feel pushed and stressed, causing productivity losses or quality reductions. In addition, workers might feel a loss of some autonomy because of the close linkages in work flows between stations with little or no excess capacity or safety stocks. Managers can mitigate some of these effects by allowing for some slack in the system—either

MANAGERIAL PRACTICE 8.1   Lean Systems at the University of Pittsburgh Medical Center Shadyside

The University of Pittsburgh Medical Center (UPMC), comprising of 20 hospitals, serves more than 4 million people every year with 50,000 employees and 400 doctors' offices and outpatient sites. UPMC at Shadyside, a part of the UPMC system, is a 520-bed tertiary-care hospital with a medical staff of nearly 1,000 primary care physicians and specialists. Always seeking to improve, UPMC first applied principles of the Toyota Production System in 2001 in a 40-bed surgical unit and then systematized the concepts into a lean approach called the Clinical Design Initiative (CDI). This approach focuses on determining the root cause of a problem through direct observation, and then eliminating it by designing solutions that are visual, simple, and unambiguous. These solutions are then tested in a small area and improved until the desired clinical and cost outcomes, along with enhanced patient and staff satisfaction, are achieved. Once perfected, the improved process is rolled out to other areas of the hospital.

UPMC used the CDI methodology to speed up turnaround time in the pathology lab. The layout and work flows of the lab were based on a batch-and-queue push system that led to long lead times, complexity in tracking and moving large lots, delays in discovering quality problems, and high storage costs. Before making the transition to the lean system, UPMC ran a workshop on lean concepts for the staff members of the lab and followed it with a 5S exercise to better organize the department. Counter spaces were cleared so that the lab's equipment could be rearranged. Unneeded items were identified with red tags and removed. Visual controls were used to arrange the remaining items in a neat and easy-to-use manner.

The 5S exercise of cleaning house boosted staff morale. Kanban cards with reordering information were then attached to most items. Reordering supplies now takes only a few minutes a day. Stockouts and expensive rush orders have been eliminated and the overall inventory level of supplies has been reduced by 50 percent to 60 percent.

To move to a system based on line flows, equipment was moved around in the lab to create a cellular layout. The new arrangement allows tissue samples being processed to move through the lab cell from embedding,

wunkley/Alamy

After the pathology lab at the University of Pittsburgh Medical Center adopted a lean operations approach based on a line system versus a batch-and-queue system, the time it took to process samples dropped from days to just hours. Diagnoses were made more quickly as a result, and patients' stays at the hospital were shortened.

to cutting, to the oven, and slide staining. The samples move more quickly, and few or no samples end up waiting between steps. As a result, the overall time needed to prepare and analyze tissue samples fell from one or two days to less than a day. The reduction in turnaround time means doctors get pathology results quicker, which in turn speeds up diagnosis and leads to shorter stays for patients. Moreover, the lab does the same amount of work with 28 percent fewer people, and fewer errors because quality mistakes are discovered immediately.

Sources: "The Anatomy of Innovation," *Lean Enterprise Institute,* **www.lean.org; http://www.upmc.com**, April 25, 2011.

---

safety stock inventories or capacity slack—and by emphasizing work flows instead of worker pace. Managers also can promote the use of work teams and allow them to determine their task assignments within their domains of responsibility.

Cooperation and Trust In a lean system, workers and first-line supervisors must take on responsibilities formerly assigned to middle managers and support staff. Activities such as scheduling, expediting, and improving productivity become part of the duties of lower-level personnel. Consequently, the work relationships in the organization must be reoriented in a way that fosters cooperation and mutual trust between the workforce and management. However, this environment can be difficult to achieve, particularly in light of the historical adversarial relationship between the two groups.

Reward Systems and Labor Classifications In some instances, the reward system must be revamped when a lean system is implemented. At General Motors, for example, a plan to reduce stock at one plant ran into trouble because the production superintendent refused to cut back on the number of unneeded parts being made. Why? Because his or her salary was based on the plant's production volume.

The realignment of reward systems is not the only hurdle. Labor contracts traditionally crippled a company's ability to reassign workers to other tasks as the need arose. For example, a typical automobile plant in the United States has several unions and dozens of labor classifications.

Generally, the people in each classification are allowed to do only a limited range of tasks. In some cases, companies have managed to give these employees more flexibility by agreeing to other types of union concessions and benefits. In other cases, however, companies relocated their plants to take advantage of nonunion or foreign labor.

Process Considerations

Firms using lean systems typically have some dominant work flows. To take advantage of lean practices, firms might have to change their existing layouts. Certain workstations might have to be moved closer together, and cells of machines devoted to particular component families may have to be established. However, rearranging a plant to conform to lean practices can be costly. For example, many plants currently receive raw materials and purchased parts by rail, but to facilitate smaller and more frequent shipments, truck deliveries would be preferable. Loading docks might have to be reconstructed or expanded and certain operations relocated to accommodate the change in transportation mode and quantities of arriving materials.

Inventory and Scheduling

Manufacturing firms need to have stable master production schedules, short setups, and frequent, reliable supplies of materials and components to achieve the full potential of the lean systems concept.

Schedule Stability Daily production schedules in high-volume, make-to-stock environments must be stable for extended periods. At Toyota, the master production schedule is stated in fractions of days over a 3-month period and is revised only once a month. The first month of the schedule is frozen to avoid disruptive changes in the daily production schedule for each workstation; that is, the workstations execute the same work schedule each day of the month. At the beginning of each month, *kanbans* are reissued for the new daily production rate. Stable schedules are needed so that production lines can be balanced and new assignments found for employees who otherwise would be underutilized. Lean systems used in high-volume, make-to-stock environments cannot respond quickly to scheduling changes because little slack inventory or capacity is available to absorb these changes.

Setups If the inventory advantages of a lean system are to be realized, small lot sizes must be used. However, because small lots require a large number of setups, companies must significantly reduce setup times. Some companies have not been able to achieve short setup times and, therefore, have to use large-lot production, negating some of the advantages of lean practices. Also, lean systems are vulnerable to lengthy changeovers to new products because the low levels of finished goods inventory will be insufficient to cover demand while the system is down. If changeover times cannot be reduced, large finished goods inventories of the old product must be accumulated to compensate. In the automobile industry, every week that a plant is shut down for new-model changeover costs between $16 million and $20 million in pretax profits.

Purchasing and Logistics If frequent, small shipments of purchased items cannot be arranged with suppliers, large inventory savings for these items cannot be realized. For example, in the United States, such arrangements may prove difficult because of the geographic dispersion of suppliers.

The shipments of raw materials and components must be reliable because of the low inventory levels in lean systems. A plant can be shut down because of a lack of materials. Similarly, recovery becomes more prolonged and difficult in a lean system after supply chains are disrupted, which is what happened immediately after 9/11.

Process design and continuous improvement are key elements of a successful operations strategy. In concluding the second part of this text, we focused on lean systems as a directive for efficient process design and an approach to achieve continuous improvement. We showed how just-in-time systems (JIT), a popular lean systems approach, can be used for continuous improvement, and how a *kanban* system can be used to control the amount of work-in-process inventory. Transforming a current process design to one embodying a lean system's philosophy is a constant challenge for management, often fraught with implementation issues. However, the transformation can be facilitated by adopting appropriate management approaches as exemplified by firms like Panasonic and Toyota among others, and by using tools such as value stream mapping. The key point to remember is that the philosophy of lean systems, applicable at the process level, is also applicable at the supply chain level. In part 3 of the text, we focus on managing supply chains, which includes the design and integration of effective supply chains.

LEARNING GOALS IN REVIEW

1 **Describe how lean systems can facilitate the continuous improvement of processes.** See the section on "Continuous Improvement using a Lean Systems Approach," pp. 277–278. Review Figure 8.1 and the opening vignette on Panasonic Corporation.

2 **Identify the characteristics and strategic advantages of lean systems.** See the section on "Supply Chain Considerations in Lean Systems," pp. 278–279, and "Process Considerations in Lean Systems," pp. 279–283. The section on "Toyota Production System," pp. 284–285, illustrates how one firm implements Lean characteristics to gain strategic advantage over its competition.

3 **Understand value stream mapping and its role in waste reduction.** The section "Value Stream Mapping," pp. 287–290, shows you how to construct value stream maps and identify

waste in the processes. Review Example 8.1 for details on mapping and creating data boxes.

4 **Understand** *kanban* **systems for creating a production schedule in a lean system.** The section "The *Kanban* System," pp. 290–293, shows how firms like Toyota use simple visual systems to pull production and make exactly what the market demands. Example 8.2 shows how to calculate the number of *kanban* cards needed.

5 **Explain the implementation issues associated with the application of lean systems.** The section "Operational Benefits and Implementation Issues," pp. 293–295, reviews organizational and process considerations needed to successfully deploy lean systems and gain their benefits.

MyOMLab helps you develop analytical skills and assesses your progress with multiple problems on average cycle time, mixed model assembly, value stream mapping, number of containers, policy variable α, daily demand satisfied with given system, and authorized stock level.

MyOMLab Resources	Titles	Link to the Book
Video	*Lean Systems at Autoliv*	Lean Systems across the Organization; Process Considerations in Lean Systems; The Kanban System
	Versatile Buildings: Lean Systems	Continuous Improvement Using a Lean Systems Approach; Process Considerations in Lean Systems
OM Explorer Solvers	Number of Containers	The Kanban System; Example 8.2 (p. 292); Solved Problem 2 (p. 299)
OM Explorer Tutors	8.1 Number of Containers	The Kanban System; Example 8.2 (p. 292); Solved Problem 2 (p. 299)
Tutor Exercises	Number of Containers in Different Scenarios	The Kanban System; Example 8.2 (p. 292); Solved Problem 2 (p. 299)
Virtual Tours	Workhorse Custom Chassis	Designing Lean System Layouts
	Rieger Orgelbau Pipe Organ Factory	Inventory and Scheduling
Internet Exercise	NUMMI, Yellow and Roadway (YRC)	Continuous Improvement Using a Lean Systems Approach
Key Equations		
Image Library		

Key Equation

Number of containers:

$$\kappa = \frac{\overline{d}(\overline{\omega} + \overline{\rho})(1 + \alpha)}{c}$$

Key Terms

five S (5S) 283
group technology 286
heijunka 281
jidoka 280
JIT system 277
just-in-time (JIT) philosophy 277

kanban 290
lean systems 276
lot 279
mixed-model assembly 281
one worker, multiple machines (OWMM) cell 285

poka-yoke 280
pull method 279
push method 279
single-digit setup 279
takt time 281
value stream mapping (VSM) 287

Solved Problem 1

Metcalf, Inc. manufacturers engine assembly brackets for two major automotive customers. The manufacturing process for the brackets consists of a cell containing a forming operation, a drilling operation, a finish grinding operation, and packaging, after which the brackets are staged for shipping. The information collected by the operations manager at Metcalf, Inc. is shown in Table 8.4.

TABLE 8.4 | **OPERATIONS DATA FOR BRACKETS AT METCALF, INC.**

Overall Process Attributes	Average demand: 2700/day Batch size: 50 Number of shifts per day: 2 Availability: 8 hours per shift with a 30-minute lunch break	
Process Step 1	Forming	Cycle time = 11 seconds Setup time = 3 minutes Up time = 100% Operators = 1 WIP = 4000 units (Before Forming)
Process Step 2	Drilling	Cycle time = 10 seconds Setup time = 2 minutes Up time = 90% Operators = 1 WIP = 5,000 units (Before Drilling)
Process Step 3	Grinding	Cycle time = 17 seconds Setup time = 0 seconds Up time = 100% Operators = 1 WIP = 2,000 units (Before Grinding)
Process Step 4	Packaging	Cycle time = 15 seconds Setup time = 0 seconds Up time = 100% Operators = 1 WIP = 1,600 units (Before Packaging) WIP = 15,700 units (Before Shipping)
Customer Shipments	One shipment of 13,500 units each week	
Information Flow	All communications with customer are electronic There is a weekly order release to Forming All material is pushed	

a. Using data shown in Table 8.4; create a value stream map for Metcalf, Inc. and show how the data box values are calculated.

b. What is the takt time for this manufacturing cell?

c. What is the production lead time at each process in the manufacturing cell?

d. What is the total processing time of this manufacturing cell?

e. What is the capacity of this manufacturing cell?

SOLUTION

a. Figure 8.11, on the following page, shows the current value stream state map for Metcalf, Inc.

b. Daily Demand = 2,700 units per day

Daily Availability = (7.5 hours per day) × (3,600 seconds per hour) × (2 shifts per day) = 54,000 seconds per day

Takt Time = Daily Availability/Daily Demand = 54,000 seconds per day/2,700 pieces per day = 20 seconds per units

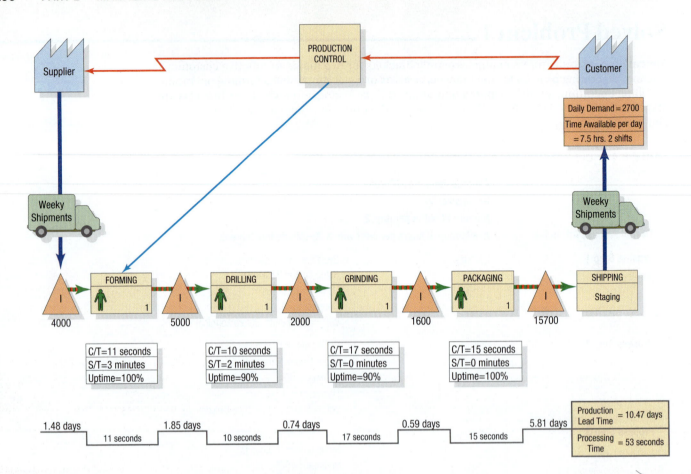

▲ **FIGURE 8.11**

Current State Value Stream Map for Metcalf, Inc.

c. The production lead time (in days) is calculated by summing the inventory held between each processing step divided by daily demand.

Raw Material lead time = [4,000/2,700] = 1.48 days

WIP lead time between Forming and Drilling = [5,000/2,700] = 1.85 days

WIP lead time between Drilling and Grinding = [2,000/2,700] = 0.74 day

WIP lead time between Grinding and Packaging = [1,600/2,700] = 0.59 day

Finished Goods lead time before Shipping = [15,700/2,700] = 5.81 days

The cell's total production lead time is: 1.48 + 1.85 + 0.74 + 0.59 + 5.81 = 10.47 days

d. The manufacturing cell's total processing time is (11 + 10 + 17 + 15) = 53 seconds

e. The cell's capacity may be calculated by locating the bottleneck and computing the number of units that it can process in the available time per day at that bottleneck.

Capacity at Forming	Capacity at Drilling	Capacity at Grinding	Capacity at Packaging
Cycle time = 11 seconds	Cycle time = 10 seconds	Cycle time = 17 seconds	Cycle time = 15 seconds
Setup Time = (3 minutes * 60 seconds per minute)/ 50 units per batch = **3.6 seconds**	Setup Time = (2 minutes * 60 seconds per minute)/ 50 units per batch = **2.4 seconds**	Setup Time = **zero seconds**	Setup Time = **zero seconds**
Per Unit Processing Time = (11 + 3.6) = **14.6 seconds**	Per Unit Processing Time = (10 + 2.4) = **12.4 seconds**	Per Unit Processing Time = (17 + 0) = **17.0 seconds**	Per Unit Processing Time = (15 + 0) = **15.0 seconds**

At a batch size of 50 units, Finish Grinding process is the bottleneck

Availability at Grinding = 54,000 seconds per day

Time at bottleneck (with setup) = 17.0 seconds

Capacity (Availability/Time at bottleneck) = 54,000/17 = 3,176 units per day

Solved Problem 2

A company using a *kanban* system has an inefficient machine group. For example, the daily demand for part L105A is 3,000 units. The average waiting time for a container of parts is 0.8 day. The processing time for a container of L105A is 0.2 day, and a container holds 270 units. Currently, 20 containers are used for this item.

a. What is the value of the policy variable, α?

b. What is the total planned inventory (work-in-process and finished goods) for item L105A?

c. Suppose that the policy variable, α, was 0. How many containers would be needed now? What is the effect of the policy variable in this example?

SOLUTION

a. We use the equation for the number of containers and then solve for α:

$$\kappa = \frac{\bar{d}(\bar{\omega} + \bar{\rho})(1 + \alpha)}{c}$$

$$20 = \frac{3,000(0.8 + 0.2)(1 + \alpha)}{270}$$

and

$$(1 + \alpha) = \frac{20(270)}{3,000(0.8 + 0.2)} = 1.8$$

$$\alpha = 1.8 - 1 = 0.8$$

b. With 20 containers in the system and each container holding 270 units, the total planned inventory is $20(270) = 5,400$ units.

c. If $\alpha = 0$

$$\kappa = \frac{3,000(0.8 + 0.2)(1 + 0)}{270} = 11.11, \text{ or } 12 \text{ containers}$$

The policy variable adjusts the number of containers. In this case, the difference is quite dramatic because $\bar{\omega} + \bar{\rho}$ is fairly large and the number of units per container is small relative to daily demand.

Discussion Questions

1. Compare and contrast the following two situations:

   a. A company's lean system stresses teamwork. Employees feel more involved and, therefore, productivity and quality increase at the company. The problem is that workers also experience a loss of individual autonomy.

   b. A humanities professor believes that all students want to learn. To encourage students to work together and learn from each other—thereby increasing the involvement, productivity, and the quality of the learning experience—the professor announces that all students in the class will receive the same grade and that it will be based on the performance of the group.

2. Which elements of lean systems would be most troublesome for manufacturers to implement? Why?

3. List the pressures that lean systems pose for supply chains, whether in the form of process failures due to inventory shortages or labor stoppages, etc. Reflect on how these pressures may apply to a firm which is actually implementing lean philosophy in their operations.

4. Identify a service or a manufacturing process that you are familiar with, and draw a current state value stream map to depict its existing information and material flows.

Problems

The OM Explorer and POM for Windows software is available to all students using the 10th edition of this textbook. Go to **www.pearsonhighered.com/krajewski** to download these computer packages. If you purchased MyOMLab, you also have access to Active Models software and significant help in doing the following problems. Check with your instructor on how best to use these resources. In many cases, the instructor wants you to understand how to do the calculations by hand. At the least, the software provides a check on your calculations. When calculations are particularly complex and the goal is interpreting the results in making decisions, the software replaces entirely the manual calculations.

1. The Harvey Motorcycle Company produces three models: the Tiger, a sure-footed dirt bike; the LX2000, a nimble cafe racer; and the Golden, a large interstate tourer. This month's master production schedule calls for the production of 54 Goldens, 42 LX2000s, and 30 Tigers per 7-hour shift.

   a. What average cycle time is required for the assembly line to achieve the production quota in 7 hours?

   b. If mixed-model scheduling is used, how many of each model will be produced before the production cycle is repeated?

   c. Determine a satisfactory production sequence for the ultimate in small-lot production: one unit.

   d. The design of a new model, the Cheetah, includes features from the Tiger, LX2000, and Golden models. The resulting blended design has an indecisive character and is expected to attract some sales from the other models. Determine a mixed-model schedule resulting in 52 Goldens, 39 LX2000s, 26 Tigers, and 13 Cheetahs per 7-hour shift. Although the total number of motorcycles produced per day will increase only slightly, what problem might be anticipated in implementing this change from the production schedule indicated in part (b)?

2. A fabrication cell at Spradley's Sprockets uses the pull method to supply gears to an assembly line. George Jitson is in charge of the assembly line, which requires 500 gears per day. Containers typically wait 0.20 day in the fabrication cell. Each container holds 20 gears, and one container requires 1.8 days in machine time. Setup times are negligible. If the policy variable for unforeseen contingencies is set at 5 percent, how many containers should Jitson authorize for the gear replenishment system?

3. You are asked to analyze the *kanban* system of LeWin, a French manufacturer of gaming devices. One of the workstations feeding the assembly line produces part M670N. The daily demand for M670N is 1,800 units. The average processing time per unit is 0.003 day. LeWin's records show that the average container spends 1.05 days waiting at the feeder workstation. The container for M670N can hold 300 units. Twelve containers are authorized for the part. Recall that $\bar{\rho}$ is the average processing time per container, not per individual part.

   a. Find the value of the policy variable, α, that expresses the amount of implied safety stock in this system.

   b. Use the implied value of α from part (a) to determine the required reduction in waiting time if one container was removed. Assume that all other parameters remain constant.

4. An assembly line requires two components: gadjits and widjits. Gadjits are produced by center 1 and widjits by center 2. Each unit of the end item, called a jit-together, requires 3 gadjits and 2 widjits, as shown in Figure 8.12. The daily production quota on the assembly line is 800 jit-togethers.

   The container for gadjits holds 80 units. The policy variable for center 1 is set at 0.09. The average waiting time for a container of gadjits is 0.09 day, and 0.06 day is needed to produce a container. The container for widjits holds 50 units, and the policy variable for

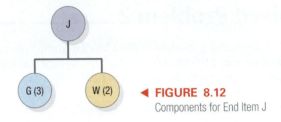

◄ **FIGURE 8.12**
Components for End Item J

center 2 is 0.08. The average waiting time per container of widgits is 0.14 day, and the time required to process a container is 0.20 day.

   a. How many containers are needed for gadjits?

   b. How many containers are needed for widjits?

5. Gestalt, Inc. uses a *kanban* system in its automobile production facility in Germany. This facility operates 8 hours per day to produce the Jitterbug, a replacement for the obsolete but immensely popular Jitney Beetle. Suppose that a certain part requires 150 seconds of processing at machine cell 33B and a container of parts average 1.6 hours of waiting time there. Management allows a 10 percent buffer for unexpected occurrences. Each container holds 30 parts, and 8 containers are authorized. How much daily demand can be satisfied with this system? (*Hint:* Recall that $\bar{\rho}$ is the average processing time per container, not per individual part.)

6. A U.S. Postal Service supervisor is looking for ways to reduce stress in the sorting department. With the existing arrangement, stamped letters are machine-canceled and loaded into tubs with 375 letters per tub. The tubs are then pushed to postal clerks, who read and key zip codes into an automated sorting machine at the rate of 1 tub per 375 seconds. To overcome the stress caused when the stamp canceling machine outpaces the sorting clerks, a pull system is proposed. When the clerks are ready to process another tub of mail, they will pull the tub from the canceling machine area. How many tubs should circulate between the sorting clerks and the canceling machine if 90,000 letters are to be sorted during an 8-hour shift, the safety stock policy variable, α, is 0.18, and the average waiting time plus materials handling time is 25 minutes per tub?

7. The production schedule at Mazda calls for 1,200 Mazdas to be produced during each of 22 production days in January and 900 Mazdas to be produced during each of 20 production days in February. Mazda uses a *kanban* system to communicate with Gesundheit, a nearby supplier of tires. Mazda purchases four tires per vehicle from Gesundheit. The safety stock policy variable, α, is 0.15. The container (a delivery truck) size is 200 tires. The average waiting time plus materials handling time is 0.16 day per container. Assembly lines are rebalanced at the beginning of each month. The average processing time per container in January is 0.10 day. February processing time will average 0.125 day per container. How many containers should be authorized for January? How many for February?

8. Jitsmart is a retailer of plastic action-figure toys. The action figures are purchased from Tacky Toys, Inc., and arrive in boxes of 48. Full boxes are stored on high shelves out of reach of customers. A small inventory is maintained

on child-level shelves. Depletion of the lower-shelf inventory signals the need to take down a box of action figures to replenish the inventory. A reorder card is then removed from the box and sent to Tacky Toys to authorize replenishment of a container of action figures. The average demand rate for a popular action figure, Agent 99, is 36 units per day. The total lead time (waiting plus processing) is 11 days. Jitsmart's safety stocky policy variable, α, is 0.25. What is the authorized stock level for Jitsmart?

9. Markland First National Bank of Rolla utilizes *kanban* techniques in its check processing facility. The following

information is known about the process. Each *kanban* container can hold 50 checks and spends 24 minutes a day in processing and 2 hours a day in materials handling and waiting. Finally, the facility operates 24 hours per day and utilizes a policy variable for unforeseen contingencies of 0.25.

a. If there are 20 *kanban* containers in use, what is the current daily demand of the check processing facility?

b. If the *muda* or the waste in the system were eliminated completely, how many containers would then be needed?

Advanced Problems

10. The Farm-4-Less tractor company produces a grain combine (GC) in addition to both a large (LT) and small size tractor (SM). Its production manager desires to produce to customer demand using a mixed model production line. The current sequence of production, which is repeated 30 times during a shift, is SM-GC-SM-LT-SM-GC-LT-SM. A new machine is produced every 2 minutes. The plant operates two 8-hour shifts. There is no downtime because the 4 hours between each shift are dedicated to maintenance and restocking raw material. Based on this information, answer the following questions.

a. How long does it take the production cycle to be completed?

b. How many of each type of machine does Farm-4-Less produce in a shift?

11. Figure 8.13 provides a new current state value stream map for the family of retainers at the Jensen Bearings, Inc. firm described in Example 8.1. This map depicts the value stream after Kline Steel agrees to accept daily orders for steel sheets and also agrees to deliver the finished goods on a daily basis.

Calculate each component of the new value stream's reduced lead time.

a. How many days of raw material does the Bearing's plant now hold?

b. How many days of work in process inventory is held between Press and Pierce & Form?

c. How many days of work in process inventory is held between Pierce & Form and Finish Grind?

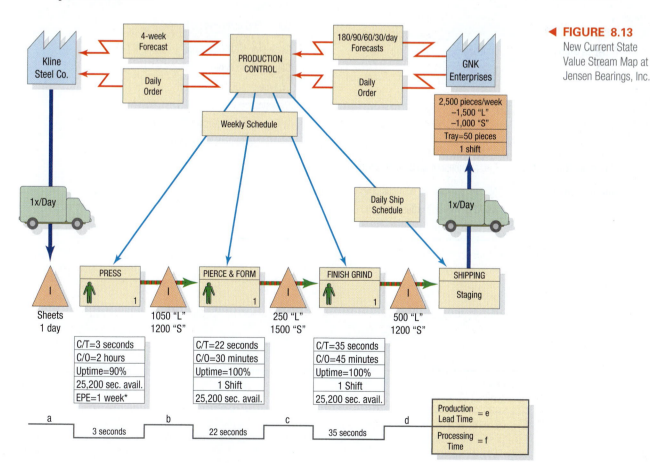

◀ **FIGURE 8.13**
New Current State Value Stream Map at Jensen Bearings, Inc.

d. How many days of work in process inventory is held between Finish Grind and Shipping?

e. What is the new value steam's production lead time?

f. What is the new value stream's processing time?

12. The manager at Ormonde, Inc. collected the value stream mapping data from the plant's most problematic manufacturing cell that fabricates parts for washing machines. This data is shown in Table 8.5. Using this data, calculate the current state performance of the cell and answer the following questions.

a. What is the cell's current inventory level?

b. What is the takt time for this manufacturing cell?

c. What is the production lead time at each process in the manufacturing cell?

d. What is the total processing time of this manufacturing cell?

e. What is the capacity of this manufacturing cell?

TABLE 8.5 | OPERATIONS DATA FOR ORMONDE, INC.

Overall Process Attributes	Average demand: 550/day Batch size: 20 Number of shifts per day: 3 Availability: 8 hours per shift with a 45-minute lunch break	
Process Step 1	Cutting	Cycle time = 120 seconds Setup time = 3 minutes Up time = 100% Operators = 1 WIP = 400 units (Before Cutting)
Process Step 2	Bending	Cycle time = 250 seconds Setup time = 5 minutes Up time = 99% Operators = 2 WIP = 500 units (Before Bending)
Process Step 3	Punching	Cycle time = 140 seconds Setup time = none Up time = 100% Operators = 1 WIP = 200 units (Before Punching) WIP = 1,000 units (After Punching)
Customer Shipments	One shipment of 2,750 units each week	
Information Flow	All communications with customer are electronic There is a weekly order release to Cutting All material is pushed	

VIDEO CASE   Lean Systems at Autoliv

Autoliv is a world-class example of lean manufacturing. This Fortune 500 company makes automotive safety components such as seat belts, airbags, and steering wheels, and has over 80 plants in more than 32 countries. Revenues in 2007 topped $6.7 billion. Autoliv's lean manufacturing environment is called the Autoliv Production System (APS), and is based on the principles of lean manufacturing pioneered by Toyota, one of the world's largest automobile manufacturers, and embodied in its Toyota Production System (TPS).

At the heart of Autoliv is a system that focuses on continuous improvement. Based on the "House of Toyota," Autoliv's Ogden, Utah, airbag module plant puts the concepts embodied in the house to work every day. The only difference between the Toyota house and the one at Autoliv is that the company has added a third pillar to its house to represent employee involvement in all processes because a culture of involvement, while the norm in Japan, is not always found in the United States.

Autoliv started its lean journey back in 1995. At that time, the Ogden plant was at manufacturing capacity with 22 work cells. Company managers acknowledge that, back then, Autoliv was "broken" and in need of significant and immediate change if it was to survive. This meant that everyone—from senior management to employees and suppliers—needed to be on-board with rebuilding the company. It was not that the company could not fulfill the needs of its automaker customers; however, with increasing demand for both reliable and cost-effective component supplies, pressure to change became obvious. Recognizing the value of Toyota's approach, senior management made the commitment to embark on its own journey to bring the transformative culture of lean manufacturing to Autoliv.

In 1998, *sensei* Takashi Harada arrived from Japan to spend three years teaching top company managers the principles, techniques, and culture of the lean system. This helped managers create an environment in which continuous improvement could be fostered and revered as an essential activity for long-term success. Because the environment was changing, it made it difficult at first for suppliers to meet Autoliv's constantly changing and unstable processes. It also made problems visible and forced the company to address and resolve the problems instead of finding ways to work around them as had been done in the past. Daily audits, monthly training, and more in-depth education programs were created to help focus attention on where changes needed to be made. Workers and management were organized into teams that were held accountable for common goals and tasked with working toward common success.

By 2004, the lean culture was integrated into the company, and it now hosts regular visits by other corporations who want to learn from Autoliv's journey and experiences. Compared to 1995, the space required for a typical work cell has been reduced by 88.5 percent, while the number of cells has grown over 400 percent. This has allowed Autoliv to dramatically increase its production capacity with minimal investment.

Lean concepts play out every day in the each plant. For example, everyone gathers at the start of the workday for pre-shift stretching and a brief meeting—this is part of the employee involvement pillar in the APS House. Then, workers head to one of 104 work cells on the plant floor. *Heijunka* Room team members deliver *heijunka* cards to each cell to communicate the work to be done in that cell. Lot sizes may vary with each card delivered to the cell. Everything the workers need to make the lot is in the cell and regularly replenished through the *kanban* card system. Every 24 minutes, another *heijunka* card comes to the cell to signal workers what they will build next. This is part of the JIT pillar in the house.

Since a culture of continuous improvement requires employees at every level to be responsible for quality, a worker may identify an "abnormal condition" during work execution that slows down the work of the cell, or

Autoliv employee folds an air bag in a Toyota-inspired production cell.

stops it altogether. This is embodied in the right pillar of the Toyota house—*jidoka*, which Autoliv interprets as "stop and fix." This is a rare occurrence, however, since both Autoliv and its suppliers are expected to deliver defect-free products. When a supplier is new, or has experienced quality issues, the supplier pays for inspection in Autoliv's receiving dock area until Autoliv is certain the supplier can meet quality expectations for all future deliveries. In this manner, workers in the cells know they can trust the integrity of the raw materials arriving through the *kanban* system into their cells for assembly. *Jidoka* may also come into play when a machine does not operate properly, or an employee notices a process that has deviated from the standard. When workers "stop and fix" a problem at the point of its creation, they save the company from added cost as well as lost confidence in the eyes of the customer.

To help focus worker efforts daily, Autoliv has a blue "communication wall" that everyone sees as they head to their work site. The wall contains the company's "policy deployment," which consists of company-wide goals for customer satisfaction, shareholder/financial performance, and safety and quality. The policy deployment begins with the company-wide goals, which then flow down to the plant level through the plant manager's goals, strategies, and actions for the facility. These linked activities assure that Autoliv achieves its goals. By communicating this information—and more—in a visual manner, the central pillar of the APS House is supported. Other visual communication and management methods are in place as well. For example, each cell has an overhead banner that states how that cell is doing each month in the areas of safety, quality, employee involvement, cost, and delivery. These all tie into the policy deployment shown on the communication wall.

Another visual communication method is to use a "rail" for the management of the *heijunka* cards in each cell. The rail has color-coded sections. As each card is delivered, it slides down a color-coded railing to the team. At the end nearest the cell, the rail is green, indicating that any cards that fall into this area can be completed within normal working hours. The middle of the rail is yellow, indicating overtime for the cell that day. The end is red, meaning weekend overtime is required to bring work processes back into harmony with customer demand. As a *heijunka* card slides down the rail, it stops when it hits the end or stacks up behind another card. If the cell is not performing at the required pace to meet customer demand, the cards will stack up on the rail and provide a very visual cue that the cell is not meeting expectations. This provides an opportunity for cell team members as well as management to implement immediate countermeasures to prevent required overtime if the situation is not remedied.

All aisles and walkways surrounding cells are to be clear of materials, debris, or other items. If anything appears in those areas, everyone can quickly see the abnormality. As team members work together to complete their day's work, the results of their efforts are displayed boldly on each cell's "communi-cube." This four-sided rotating display visually tells the story of the cell's productivity, quality, and 5S performance. The cube also contains a special section for the management of *kaizen* suggestions for the team itself. These *kaizens* enable the team to continuously improve the work environment as well as drive the achievement of team results.

Autoliv's lean journey embodied in the Autoliv Production System has led to numerous awards and achievement of its policy deployment goals. Product defects have been dramatically reduced, inventory levels are lower, and inventory turnover is approaching world-class levels of 50. Employee turnover is close to 5 percent, and remains well below that of other manufacturers in the industry. Yet the destination has not been reached. The company continues its emphasis on driving systemic improvement to avoid complacency and loss of competitive advantage. Best practices from sources beyond each immediate area of the organization are studied and integrated. And finding ways to engage and reward Autoliv's workforce in a maturing market is critical. Kaizen suggestions in the most recent year at the Ogden plant totaled 74,000, or nearly 60 per employee, indicating the culture of continuous improvement in Autoliv's APS House is alive and well.

QUESTIONS

1. Why is a visual management approach such an integral part of Autoliv's lean system?

2. Describe the JIT considerations presented in the chapter as they relate to Autoliv's manufacturing environment.

3. Which method of work flow is embodied in Autoliv's system? Why is this approach most suitable to its lean environment?

4. When Autoliv started its lean journey, a number of operational benefits and implementation issues had to be addressed. What were they, and how were they addressed?

CASE | Copper Kettle Catering

Copper Kettle Catering (CKC) is a full-service catering company that provides services ranging from box lunches for picnics or luncheon meetings to large wedding, dinner, or office parties. Established as a lunch delivery service for offices in 1972 by Wayne and Janet Williams, CKC has grown to be one of the largest catering businesses in Raleigh, North Carolina. The Williams's divide customer demand into two categories: *deliver only* and *deliver and serve*.

The deliver-only side of the business delivers boxed meals consisting of a sandwich, salad, dessert, and fruit. The menu for this service is limited to six sandwich selections, three salads or potato chips, and a brownie or fruit bar. Grapes and an orange slice are included with every meal, and iced tea can be ordered to accompany the meals. The overall level of demand for this service throughout the year is fairly constant, although the mix of menu items delivered varies. The planning horizon for this segment of the business is short: Customers usually call no more than a day ahead of time. CKC requires customers to call deliver-only orders in by 10:00 A.M. to guarantee delivery the same day.

The deliver-and-serve side of the business focuses on catering large parties, dinners, and weddings. The extensive range of menu items includes a full selection of hors d'oeuvres, entrées, beverages, and special-request items. The demand for these services is much more seasonal, with heavier demands occurring in the late spring–early summer for weddings and the late fall–early winter for holiday parties. However, this segment also has a longer planning horizon. Customers book dates and choose menu items weeks or months ahead of time.

CKC's food preparation facilities support both operations. The physical facilities layout resembles that of a job process. Five major work areas consist of a stove–oven area for hot food preparation, a cold area for salad preparation, an hors d'oeuvre preparation area, a sandwich preparation area, and an assembly area where deliver-only orders are boxed and deliver-and-serve orders are assembled and trayed. Three walk-in coolers store foods requiring refrigeration, and a large pantry houses nonperishable goods. Space limitations and the risk of spoilage limit the amount of raw materials and prepared food items that can be carried in inventory at any one time. CKC purchases desserts from outside vendors. Some deliver the desserts to CKC; others require CKC to send someone to pick up desserts at their facilities.

The scheduling of orders is a two-stage process. Each Monday, the Williamses develop the schedule of deliver-and-serve orders to be processed each day. CKC typically has multiple deliver-and-serve orders to fill each day of the week. This level of demand allows a certain efficiency in the preparation of multiple orders. The deliver-only orders are scheduled day to day, owing to the short-order lead times. CKC sometimes runs out of ingredients for deliver-only menu items because of the limited inventory space.

Wayne and Janet Williams have 10 full-time employees: 2 cooks and 8 food preparation workers, who also work as servers for the deliver-and-serve orders. In periods of high demand, the Williamses hire additional part-time servers. The position of cook is specialized and requires a high degree of training and skill. The rest of the employees are flexible and move between tasks as needed.

The business environment for catering is competitive. The competitive priorities are high-quality food, delivery reliability, flexibility, and cost—in that order. "The quality of the food and its preparation is paramount," states Wayne Williams. "Caterers with poor-quality food will not stay in business long." Quality is measured by both freshness and taste. Delivery reliability encompasses both on-time delivery and the time required to respond to customer orders (in effect, the order lead time). Flexibility focuses on both the range of catering requests that a company can satisfy and menu variety.

Recently, CKC began to notice that customers are demanding more menu flexibility and faster response times. Small specialty caterers who entered the market are targeting specific well-defined market segments. One example is a small caterer called Lunches-R-Us, which located a facility in the middle of a large office complex to serve the lunch trade and competes with CKC on cost.

Wayne and Janet Williams are impressed by the lean systems concept, especially the ideas related to increasing flexibility, reducing lead times, and lowering costs. They sound like what CKC needs to remain competitive. However, the Williamses wonder whether lean concepts and practices are transferable to a service business.

QUESTIONS

1. Are the operations of Copper Kettle Catering conducive to the application of lean concepts and practices? Explain.

2. What, if any, are the major barriers to implementing a lean system at Copper Kettle Catering?

3. What would you recommend that Wayne and Janet Williams do to take advantage of lean concepts in operating CKC?

Source: This case was prepared by Dr. Brooke Saladin, Wake Forest University, as a basis for classroom discussion. Copyright © Brooke Saladin. Used with permission.

Selected References

Ansberry, Clare. "Hurry-Up Inventory Method Hurts Where It Once Helped." *Wall Street Journal Online* (June 25, 2002).

Holweg, Matthias. "The Genealogy of Lean Production." *Journal of Operations Management*, vol. 25 (2007), pp. 420–437.

Klein, J. A. "The Human Costs of Manufacturing Reform." *Harvard Business Review* (March–April 1989), pp. 60–66.

Manufacturing Engineering Web site, www.mfgeng.com/5S.htm.

Mascitelli, Ron. "Lean Thinking: It's About Efficient Value Creation." *Target*, vol. 16, no. 2 (Second Quarter 2000), pp. 22–26.

McBride, David. "Toyota and Total Productive Maintenance." http://www.emsstrategies.com/dm050104article2.html; May 2004.

Millstein, Mitchell. "How to Make Your MRP System Flow." *APICS—The Performance Advantage* (July 2000), pp. 47–49.

Rother, Mike, and John Shook. *Learning to See.* Brookline, MA: The Lean Enterprise Institute, 2003.

Schaller, Jeff. "A 'Just Do It Now' Philosophy Rapidly Creates a Lean Culture, Produces Dramatic Results at Novametix Medical Systems." *Target*, vol. 18, no. 2 (Second Quarter 2002), pp. 48–54.

Schonberger, Richard J. "Japanese Production Management: An Evolution—With Mixed Success." *Journal of Operations Management*, vol. 25 (2007), pp. 403–419.

Shah, Rachna, and Peter T. Ward. "Defining and Developing Measures of Lean Production." *Journal of Operations Management*, vol. 25 (2007), pp. 785–805.

Spear, Steven, and H. Kent Bowen. "Decoding the DNA of the Toyota Production System." *Harvard Business Review* (September–October 1999), pp. 97–106.

Spear, Steven J. "Learning to Lead at Toyota." *Harvard Business Review* (May 2004), pp. 78–86.

Stewart, Douglas M., and John R. Grout. "The Human Side of Mistake Proofing." *Production and Operations Management*, vol. 10, no. 4 (Winter 2001), pp. 440–459.

Tonkin, Lea. "System Sensor's Lean Journey." *Target*, vol. 18, no. 2 (Second Quarter 2002), pp. 44–47.

Womack, James P., and Daniel T. Jones. "Lean Consumption" *Harvard Business Review* (March 2005) pp. 1–12.

Paul J. Milette/Palm Beach Post/ZUMA Press/Newscom

9

SUPPLY CHAIN INVENTORY MANAGEMENT

An employee stacks a shipment of pet supplies at a new Walmart distribution center in Fort Pierce, Florida. The 1.2 million square foot facility will serve 45 Walmart stores on the east coast of Florida.

Inventory Management at Walmart

In the market for shaver blade replacements? A printer? First-aid supplies? Dog food? Hair spray? If so, you expect that the store you shop at will have what you want. However, making sure that the shelves are stocked with tens of thousands of products is no simple matter for inventory managers at Walmart, which has 8,800 Walmart stores and Sam's Club locations in 15 markets, employs more than 2 million associates, serves 200 million customers per week worldwide, and uses 100,000 suppliers. You can imagine in an operation this large that some things can get lost. Linda Dillman, then CIO at Walmart, recounts the story of the missing hair spray at one of the stores. The shelf needed to be restocked with a specific hair spray; however, it took three days to find the case in the backroom. Most customers will not swap hair sprays, so Walmart lost three days of sales on that product.

Knowing what is in stock, in what quantity, and where it is being held is critical to effective inventory management. Without accurate inventory information, companies can make major mistakes by ordering too much, not enough, or shipping products to the wrong location. Companies can have large inventories and still have stockouts of product because they have too much inventory of some products and not enough of others. Walmart, a $405 billion company with inventories in excess of $33 billion, is certainly aware of the potential benefits from improved inventory management and is constantly experimenting with ways to reduce inventory investment. Knowing when to replenish inventory stocks and

inventory management

The planning and controlling of inventories in order to meet the competitive priorities of the organization.

lot size

The quantity of an inventory item management either buys from a supplier or manufactures using internal processes.

how much to order each time is critical when dealing with so much inventory investment. The application of technology is also important, such as using radio frequency identification (RFID) to track inventory shipments and stock levels at stores and warehouses throughout the supply chain. One handheld RFID reader could have found the missing case of hair spray in a few minutes.

Source: Laurie Sullivan, "Walmart's Way," Informationweek.com (September 27, 2004), pp. 36–50; Gus Whitcomb and Christi Gallagher, "Walmart Begins Roll-Out of Electronic Product Codes in Dallas/Fort Worth Area," **www.walmartstores.com** (February 2011), and Walmart 2010 Annual Report.

LEARNING GOALS   *After reading this chapter, you should be able to:*

1. Identify the advantages, disadvantages, and costs of holding inventory.

2. Define the different types of inventory and the roles they play in supply chains.

3. Explain the basic tactics for reducing inventories in supply chains.

4. Determine the items deserving most attention and tightest inventory control.

5. Calculate the economic order quantity and apply it to various situations.

6. Determine the order quantity and reorder point for a continuous review inventory control system.

7. Determine the review interval and target inventory level for a periodic review inventory control system.

8. Define the key factors that determine the appropriate choice of an inventory system.

Creating Value through Operations Management

Using Operations to Compete
Project Management

Managing Processes

Process Strategy
Process Analysis
Quality and Performance
Capacity Planning
Constraint Management
Lean Systems

Managing Supply Chains

Supply Chain Inventory Management

Supply Chain Design
Supply Chain Location Decisions
Supply Chain Integration
Supply Chain Sustainability and Humanitarian Logistics
Forecasting
Operations Planning and Scheduling
Resource Planning

Inventory management, the planning and controlling of inventories in order to meet the competitive priorities of the organization, is an important concern for managers in all types of businesses. Effective inventory management is essential for realizing the full potential of any supply chain. The challenge is not to pare inventories to the bone to reduce costs or to have plenty around to satisfy all demands, but to have the right amount to achieve the competitive priorities of the business most efficiently. This type of efficiency can only happen if the right amount of inventory is flowing through the supply chain—through suppliers, the firm, warehouses or distribution centers, and customers. These decisions were so important for Walmart that it decided to use RFID to improve the information flows in the supply chain. Much of inventory management involves *lot sizing*, which is the determination of how frequently and in what quantity to order inventory. We make ample reference to the term **lot size**, which is the quantity of an inventory item management either buys from a supplier or manufactures using internal processes. In this chapter, we focus on the decision-making aspects of inventory management. We begin with an overview of the importance of inventory management to the organization and how to choose the items most deserving of management attention. We then introduce the basics of inventory decision making by exploring the economic order quantity and how it can be used to balance inventory holding costs with ordering costs. A major segment of the chapter is devoted to retail and distribution inventory control systems and how to use them.

Inventory Management across the Organization

Inventories are important to all types of organizations, their employees, and their supply chains. Inventories profoundly affect everyday operations because they must be counted, paid for, used in operations, used to satisfy customers, and managed. Inventories require an investment of funds, as does the purchase of a new machine. Monies invested in inventory are not available for investment in other things; thus, they represent a drain on the cash flows of an organization. Nonetheless, companies realize that the availability of products is a key selling point in many markets and downright critical in many more.

So, is inventory a boon or a bane? Certainly, too much inventory on hand reduces profitability, and too little inventory on hand creates shortages in the supply chain and ultimately damages customer confidence. Inventory management, therefore, involves trade-offs. Let us discover how companies can effectively manage inventories across the organization.

Inventory and Supply Chains

The value of inventory management becomes apparent when the complexity of the supply chain is recognized. The performance of numerous suppliers determines the inward flow of materials and services to a firm. The performance of the firm determines the outward flow of services or products to the next stage of the supply chain. The flow of materials, however, determines inventory levels. **Inventory is a** stock of materials used to satisfy customer demand or to support the production of services or goods. Figure 9.1 shows how inventories are created at one node in a supply chain through the analogy of a water tank. The flow of water into the tank raises the water level. The inward flow of water represents input materials, such as steel, component parts, office supplies, or a finished product. The water level represents the amount of inventory held at a plant, service facility, warehouse, or retail outlet. The flow of water from the tank lowers the water level in the tank. The outward flow of water represents the demand for materials in inventory, such as customer orders for a Huffy bicycle or service requirements for supplies such as soap, food, or furnishings. The rate of the outward flow also reflects the ability of the firm to match the demand for services or products. Another possible outward flow is that of scrap, which also lowers the level of useable inventory. Together, the difference between input flow rate and the output flow rate determines the level of inventory. Inventories rise when more material flows into the tank than flows out; they fall when more material flows out than flows in. Figure 9.1 also shows clearly why firms utilize Six Sigma and total quality management (TQM) to reduce defective materials: The larger the scrap flows, the larger the input flow of materials required for a given level of output.

A fundamental question in supply chain management is how much inventory to have. The answer to this question involves a tradeoff between the advantages and disadvantages of holding inventory. Depending on the situation, the pressures for having small inventories may or may not exceed the pressures for having large inventories.

inventory

A stock of materials used to satisfy customer demand or to support the production of services or goods.

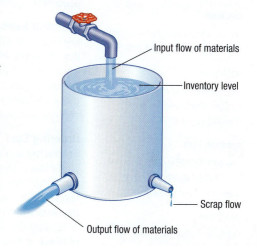

▲ **FIGURE 9.1**
Creation of Inventory

Pressures for Small Inventories

An inventory manager's job is to balance the advantages and disadvantages of both small and large inventories and find a happy medium between the two levels. The primary reason for keeping inventories small is that inventory represents a temporary monetary investment. As such, the firm incurs an opportunity cost, which we call the cost of capital, arising from the money tied up in inventory that could be used for other purposes. The **inventory holding cost** (or *carrying cost*) is the sum of the cost of capital plus the variable costs of keeping items on hand, such as storage and handling costs and taxes, insurance, and shrinkage costs. When these components change with inventory levels, so does the holding cost.

Companies usually state an item's holding cost per period of time as a percent of its value. The annual cost to maintain one unit in inventory typically ranges from 15 to 35 percent of its value. Suppose that a firm's holding cost is 20 percent. If the average value of total inventory is 20 percent of sales, the average annual cost to hold inventory is 4 percent [0.20(0.20)] of total sales. This cost is sizable in terms of gross profit margins, which often are less than 10 percent. Thus, the components of holding cost create pressures for small inventories.

inventory holding cost

The sum of the cost of capital and the variable costs of keeping items on hand, such as storage and handling, taxes, insurance, and shrinkage.

Cost of Capital   The cost of capital is the opportunity cost of investing in an asset relative to the expected return on assets of similar risk. Inventory is an asset; consequently, we should use a cost measure that adequately reflects the firm's approach to financing assets. Most firms use the *weighted average cost of capital (WACC)*, which is the average of the required return on a firm's stock equity and the interest rate on its debt, weighted by the proportion of equity and debt in its portfolio. The cost of capital usually is the largest component of holding cost, as high as 15 percent of inventory value, depending on the particular capitalization portfolio of the firm. Firms typically update the WACC on an annual basis because it is used to make many financial decisions.

Storage and Handling Costs   Inventory takes up space and must be moved into and out of storage. Storage and handling costs may be incurred when a firm rents space on either a long- or short-term basis. An inventory holding cost is incurred when a firm could use storage space productively in some other way.

Taxes, Insurance, and Shrinkage   More taxes are paid if end-of-year inventories are high, and the cost of insuring the inventories increases, too. Shrinkage takes three forms. The first, *pilferage*, or theft of inventory by customers or employees, is a significant percentage of sales for some businesses. The second form of shrinkage, called *obsolescence*, occurs when inventory cannot be used or sold at full value, owing to model changes, engineering modifications, or unexpectedly low demand. Obsolescence is a big expense in the retail clothing industry. Drastic discounts on seasonal clothing frequently must be offered on many of these products at the end of a season. Finally, *deterioration* through physical spoilage or damage due to rough or excessive material handling results in lost value. Food and beverages, for example, lose value and might even have to be discarded when their shelf life is reached. When the rate of deterioration is high, building large inventories may be unwise.

Pressures for Large Inventories

Given the costs of holding inventory, why not eliminate it altogether? Let us look briefly at the pressures related to maintaining large inventories.

stockout

An order that cannot be satisfied, resulting in a loss of the sale.

backorder

A customer order that cannot be filled when promised or demanded but is filled later.

ordering cost

The cost of preparing a purchase order for a supplier or a production order for manufacturing.

setup cost

The cost involved in changing over a machine or workspace to produce a different item.

quantity discount

A drop in the price per unit when an order is sufficiently large.

raw materials (RM)

The inventories needed for the production of services or goods.

Customer Service   Creating inventory can speed delivery and improve the firm's on-time delivery of goods. High inventory levels reduce the potential for stockouts and backorders, which are key concerns of wholesalers and retailers. A **stockout** is an order that cannot be satisfied, resulting in loss of the sale. A **backorder** is a customer order that cannot be filled when promised or demanded but is filled later. Customers do not like waiting for backorders to be filled. Many of them will take their business elsewhere. Sometimes, customers are given discounts for the inconvenience of waiting.

Ordering Cost   Each time a firm places a new order, it incurs an **ordering cost**, or the cost of preparing a purchase order for a supplier or a production order for manufacturing. For the same item, the ordering cost is the same, regardless of the order size. The purchasing agent must take the time to decide how much to order and, perhaps, select a supplier and negotiate terms. Time also is spent on paperwork, follow-up, and receiving the item(s). In the case of a production order for a manufactured item, a blueprint and routing instructions often must accompany the order. However, the Internet streamlines the order process and reduces the costs of placing orders.

Setup Cost   The cost involved in changing over a machine or workspace to produce a different item is the **setup cost**. It includes labor and time to make the changeover, cleaning, and sometimes new tools or equipment. Scrap or rework costs are also higher at the start of the production run. Setup cost also is independent of order size, which creates pressure to make or order a large supply of the items and hold them in inventory rather than order smaller batches.

Labor and Equipment Utilization   By creating more inventory, management can increase workforce productivity and facility utilization in three ways. First, placing larger, less frequent production orders reduces the number of unproductive setups, which add no value to a service or product. Second, holding inventory reduces the chance of the costly rescheduling of production orders because the components needed to make the product are not in inventory. Third, building inventories improves resource utilization by stabilizing the output rate when demand is cyclical or seasonal. The firm uses inventory built during slack periods to handle extra demand in peak seasons. This approach minimizes the need for extra shifts, hiring, layoffs, overtime, and additional equipment.

Transportation Cost   Sometimes, outbound transportation cost can be reduced by increasing inventory levels. Having inventory on hand allows more full-carload shipments to be made and minimizes the need to expedite shipments by more expensive modes of transportation. Inbound transportation costs can also be reduced by creating more inventory. Sometimes, several items are ordered from the same supplier. Placing these orders at the same time will increase inventories because some items will be ordered before they are actually needed; nonetheless, it may lead to rate discounts, thereby decreasing the costs of transportation and raw materials.

Metal saws, such as this one, require time to changeover from one product to the next. The depth and length of the cut must be adjusted and the blade itself may have to be changed.

Glowimages/Alamy

Payments to Suppliers   A firm often can reduce total payments to suppliers if it can tolerate higher inventory levels. Suppose that a firm learns that a key supplier is about to increase its prices. In this case, it might be cheaper for the firm to order a larger quantity than usual—in effect delaying the price increase—even though inventory will increase temporarily. A firm can also take advantage of quantity discounts this way. A **quantity discount,** whereby the price per unit drops when the order is sufficiently large, is an incentive to order larger quantities. Supplement C, "Special Inventory Models," shows how to determine order quantities in such a situation.

Types of Inventory

Inventory exists in three aggregate categories that are useful for accounting purposes. **Raw materials (RM)** are the inventories needed for the production of services or goods. They are considered to be inputs to the transformation processes of the firm. **Work-in-process (WIP)** consists of items, such as components or assemblies, needed to produce a final product in manufacturing. WIP is also present in some service operations, such as repair shops, restaurants, check-processing centers, and package delivery services. **Finished goods (FG)** in manufacturing plants, warehouses, and retail outlets are the items sold to the firm's customers. The finished goods of one firm may actually be the raw materials for another.

Figure 9.2 shows how inventory can be held in different forms and at various stocking points. In this example, raw materials—the finished goods of the supplier—are held both by the supplier and the manufacturer. Raw materials at the plant pass through one or more processes, which transform them into various levels of WIP inventory. Final processing of this inventory yields finished goods inventory. Finished goods can be held at the plant, the distribution center (which may be a warehouse owned by the manufacturer or the retailer), and retail locations.

Another perspective on inventory is to classify it by how it is created. In this context, inventory takes four forms: (1) cycle, (2) safety stock, (3) anticipation, and (4) pipeline. They cannot be identified physically; that is, an inventory manager cannot look at a pile of widgets and identify which ones are cycle inventory and which ones are safety stock inventory. However, conceptually, each of the four types comes into being in an entirely different way. Once you understand these differences, you can prescribe different ways to reduce inventory, which we discuss in the next section.

Raw materials, work-in-process, and finished goods inventories can all be stocked in the same facility. Modern warehouses allow for efficient inventory access.

Marcin Balcerzak/Shutterstock.com

work-in-process (WIP)

Items, such as components or assemblies, needed to produce a final product in manufacturing or service operations.

▼ **FIGURE 9.2**
Inventory of Successive Stocking Points

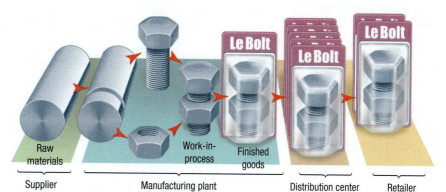

| Raw materials | Work-in-process | Finished goods |
| Supplier | Manufacturing plant | Distribution center | Retailer |

Cycle Inventory   The portion of total inventory that varies directly with lot size is called **cycle inventory.** Determining how frequently to order, and in what quantity, is called **lot sizing.** Two principles apply.

1. The lot size, Q, varies directly with the elapsed time (or cycle) between orders. If a lot is ordered every 5 weeks, the average lot size must equal 5 weeks' demand.

2. The longer the time between orders for a given item, the greater the cycle inventory must be.

At the beginning of the interval, the cycle inventory is at its maximum, or Q. At the end of the interval, just before a new lot arrives, cycle inventory drops to its minimum, or 0. The average cycle inventory is the average of these two extremes:

$$\text{Average cycle inventory} = \frac{Q + 0}{2} = \frac{Q}{2}$$

finished goods (FG)

The items in manufacturing plants, warehouses, and retail outlets that are sold to the firm's customers.

cycle inventory

The portion of total inventory that varies directly with lot size.

lot sizing

The determination of how frequently and in what quantity to order inventory.

Susan E. Benson/Stock Connection

Pipeline inventories result from moving items and materials from one location to another. Because trains offer an economical way to transport large quantities of goods, they are a favorite choice to reduce the costs of pipeline inventories.

This formula is exact only when the demand rate is constant and uniform. However, it does provide a reasonably good estimate even when demand rates are not constant. Factors other than the demand rate (e.g., scrap losses) also may cause estimating errors when this simple formula is used.

Safety Stock Inventory   To avoid customer service problems and the hidden costs of unavailable components, companies hold safety stock. **Safety stock inventory** is surplus inventory that protects against uncertainties in demand, lead time, and supply changes. Safety stocks are desirable when suppliers fail to deliver either the desired quantity on the specified date or items of acceptable quality, or when manufactured items require significant amounts of scrap or rework. Safety stock inventory ensures that operations are not disrupted when such problems occur, allowing subsequent operations to continue.

To create safety stock, a firm places an order for delivery earlier than when the item is typically needed.[1] The replenishment order therefore arrives ahead of time, giving a cushion against uncertainty. For example, suppose that the average lead time from a supplier is 3 weeks, but a firm orders 5 weeks in advance just to be safe. This policy creates a safety stock equal to a 2 weeks' supply (5 – 3).

Anticipation Inventory   Inventory used to absorb uneven rates of demand or supply, which businesses often face, is referred to as **anticipation inventory**. Predictable, seasonal demand patterns lend themselves to the use of anticipation inventory. Uneven demand can motivate a manufacturer to stockpile anticipation inventory during periods of low demand so that output levels do not have to be increased much when demand peaks. Anticipation inventory also can help when suppliers are threatened with a strike or have severe capacity limitations.

Pipeline Inventory   Inventory that is created when an order for an item is issued but not yet received is called **pipeline inventory**. This form of inventory exists because the firm must commit to enough inventory (on-hand plus in-transit) to cover the lead time for the order. Longer lead times or higher demands per week create more pipeline inventory. As such, the average pipeline inventory between two stocking points can be measured as the average demand during lead time, $\overline{D}_L$, which is the average demand for the item per period ($\overline{d}$) multiplied by the number of periods in the item's lead time (L) to move between the two points, or

$$\text{Pipeline inventory} = \overline{D}_L = \overline{d}L$$

The equation assumes that both $\overline{d}$ and L are constants and that L is not affected by the order or lot size, Q. Changing an item's lot size does not directly affect the average level of the pipeline inventory. Nonetheless, the lot size can *indirectly* affect pipeline inventory if it is related to the lead time. In such a case, pipeline inventory will change depending on the relationship of L to Q. Example 9.1 shows how this can happen.

safety stock inventory

Surplus inventory that a company holds to protect against uncertainties in demand, lead time, and supply changes.

anticipation inventory

Inventory used to absorb uneven rates of demand or supply.

pipeline inventory

Inventory that is created when an order for an item is issued but not yet received.

EXAMPLE 9.1	**Estimating Inventory Levels**

A plant makes monthly shipments of electric drills to a wholesaler in average lot sizes of 280 drills. The wholesaler's average demand is 70 drills a week, and the lead time from the plant is 3 weeks. The wholesaler must pay for the inventory from the moment the plant makes a shipment. If the wholesaler is willing to increase its purchase quantity to 350 units, the plant will give priority to the wholesaler and guarantee a lead time of only 2 weeks. What is the effect on the wholesaler's cycle and pipeline inventories?

---

[1]When orders are placed at fixed intervals, a second way to create safety stock is used. Each new order placed is larger than the quantity typically needed through the next delivery date.

SOLUTION

The wholesaler's current cycle and pipeline inventories are

$$\text{Cycle inventory} = \frac{Q}{2} = \frac{280}{2} = 140 \text{ drills}$$

$$\text{Pipeline inventory} = \overline{D}_L = \overline{d}L = (70 \text{ drills/week})(3 \text{ weeks}) = 210 \text{ drills}$$

Figure 9.3 shows the cycle and pipeline inventories if the wholesaler accepts the new proposal.

1. Enter the average lot size, average demand during a period, and the number of periods of lead time:

Average lot size	350
Average demand	70
Lead time	2

2. To compute cycle inventory, simply divide average lot size by 2. To compute pipeline inventory, multiply average demand by lead time:

Cycle inventory	175
Pipeline inventory	140

◀ **FIGURE 9.3**
Estimating Inventory Levels Using Tutor 9.1

DECISION POINT

The effect of the new proposal on cycle inventories is to increase them by 35 units, or 25 percent. The reduction in pipeline inventories, however, is 70 units, or 33 percent. The proposal would reduce the total investment in cycle and pipeline inventories. Also, it is advantageous to have shorter lead times because the wholesaler only has to commit to purchases 2 weeks in advance, rather than 3 weeks.

Inventory Reduction Tactics

Managers always are eager to find cost-effective ways to reduce inventory in supply chains. Later in this chapter we examine various ways for finding optimal lot sizes. Here, we discuss something more fundamental—the basic tactics (which we call *levers*) for reducing inventory in supply chains. A primary lever is one that must be activated if inventory is to be reduced. A secondary lever reduces the penalty cost of applying the primary lever and the need for having inventory in the first place.

Cycle Inventory   The primary lever to reduce cycle inventory is simply to reduce the lot sizes of items moving in the supply chain. However, making such reductions in Q without making any other changes can be devastating. For example, setup costs or ordering costs can skyrocket. If these changes occur, two secondary levers can be used:

1. Streamline the methods for placing orders and making setups in order to reduce ordering and setup costs and allow Q to be reduced. This may involve redesigning the infrastructure for information flows or improving manufacturing processes.

2. Increase repeatability to eliminate the need for changeovers. **Repeatability** is the degree to which the same work can be done again. Repeatability can be increased through high product demand; the use of specialization; the devotion of resources exclusively to a product; the use of the same part in many different products; the use of *flexible automation*; the use of the *one-worker, multiple-machines* concept; or through *group technology*. Increased repeatability may justify new setup methods, reduce transportation costs, and allow quantity discounts from suppliers.

repeatability

The degree to which the same work can be done again.

Safety Stock Inventory   The primary lever to reduce safety stock inventory is to place orders closer to the time when they must be received. However, this approach can lead to unacceptable customer service unless demand, supply, and delivery uncertainties can be minimized. Four secondary levers can be used in this case:

1. Improve demand forecasts so that fewer surprises come from customers. Design the mechanisms to increase collaboration with customers to get advanced warnings for changes in demand levels.

2. Cut the lead times of purchased or produced items to reduce demand uncertainty. For example, local suppliers with short lead times could be selected whenever possible.

3. Reduce supply uncertainties. Suppliers are likely to be more reliable if production plans are shared with them. Put in place the mechanisms to increase collaboration with suppliers.

Surprises from unexpected scrap or rework can be reduced by improving manufacturing processes. Preventive maintenance can minimize unexpected downtime caused by equipment failure.

4. Rely more on equipment and labor buffers, such as capacity cushions and cross-trained workers. These buffers are important to businesses in the service sector because they generally cannot inventory their services.

Anticipation Inventory   The primary lever to reduce anticipation inventory is simply to match demand rate with production rate. Secondary levers can be used to even out customer demand in one of the following ways:

1. Add new products with different demand cycles so that a peak in the demand for one product compensates for the seasonal low for another.

2. Provide off-season promotional campaigns.

3. Offer seasonal pricing plans.

Pipeline Inventory   An operations manager has direct control over lead times but not demand rates. Because pipeline inventory is a function of demand during the lead time, the primary lever is to reduce the lead time. Two secondary levers can help managers cut lead times:

1. Find more responsive suppliers and select new carriers for shipments between stocking locations or improve materials handling within the plant. Improving the information system could overcome information delays between a distribution center and retailer.

2. Change Q in those cases where the lead time depends on the lot size.

ABC Analysis

stock-keeping unit (SKU)

An individual item or product that has an identifying code and is held in inventory somewhere along the supply chain.

ABC analysis

The process of dividing SKUs into three classes, according to their dollar usage, so that managers can focus on items that have the highest dollar value.

Thousands of items, often referred to as stock-keeping units, are held in inventory by a typical organization, but only a small percentage of them deserve management's closest attention and tightest control. A **stock-keeping unit (SKU)** is an individual item or product that has an identifying code and is held in inventory somewhere along the supply chain. **ABC analysis** is the process of dividing SKUs into three classes according to their dollar usage so that managers can focus on items that have the highest dollar value. This method is the equivalent of creating a *Pareto chart* except that it is applied to inventory rather than to process errors. As Figure 9.4 shows, class A items typically represent only about 20 percent of the SKUs but account for 80 percent of the dollar usage. Class B items account for another 30 percent of the SKUs but only 15 percent of the dollar usage. Finally, 50 percent of the SKUs fall in class C, representing a mere 5 percent of the dollar usage. The goal of ABC analysis is to identify the class A SKUs so management can control their inventory levels.

The analysis begins by multiplying the annual demand rate for an SKU by the dollar value (cost) of one unit of that SKU to determine its dollar usage. After ranking the SKUs on the basis of dollar usage and creating the Pareto chart, the analyst looks for "natural" changes in slope. The dividing lines in Figure 9.4 between classes are inexact. Class A SKUs could be somewhat higher or lower than 20 percent of all SKUs but normally account for the bulk of the dollar usage.

Class A SKUs are reviewed frequently to reduce the average lot size and to ensure timely deliveries from suppliers. It is important to maintain high inventory turnover for these items. By contrast, class B SKUs require an intermediate level of control. Here, less frequent monitoring of suppliers coupled with adequate safety stocks can provide cost-effective coverage of demands. For class C SKUs, much looser control is appropriate. While a stockout of a class C SKU can be as crucial as for a class A SKU, the inventory holding cost of class C SKUs tends to be low. These features suggest that higher inventory levels can be tolerated and that more safety stock and larger lot sizes may suffice for class C SKUs. See Solved Problem 2 for a detailed example of ABC analysis.

Creating ABC inventory classifications is useless unless inventory records are accurate. Technology can help; many companies are tracking inventory wherever it exists in the supply chain. Chips imbedded in product packaging contain information on the product and send signals that can be accessed by sensitive receivers and transmitted to a central location for processing. There are other, less sophisticated approaches of achieving accuracy that can be used. One way is to assign responsibility to specific employees for issuing and receiving materials and accurately reporting each transaction. Another method is to secure inventory behind locked doors or gates to prevent unauthorized

▼ **FIGURE 9.4**
Typical Chart Using ABC Analysis

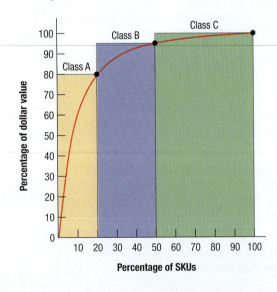

or unreported withdrawals. This method also guards against accidentally storing newly received inventory in the wrong locations, where it can be lost for months. **Cycle counting** can also be used, whereby storeroom personnel physically count a small percentage of the total number of SKUs each day, correcting errors that they find. Class A SKUs are counted most frequently. A final method, for computerized systems, is to make logic error checks on each transaction reported and fully investigate any discrepancies. The discrepancies can include (1) actual receipts when no receipts are scheduled, (2) disbursements that exceed the current on-hand inventory balance, and (3) receipts with an inaccurate (nonexistent) SKU number.

Now that we have identified the inventory items deserving of most attention, we devote the remainder of the chapter to the decisions of how much to order and when.

cycle counting

An inventory control method, whereby storeroom personnel physically count a small percentage of the total number of items each day, correcting errors that they find.

Economic Order Quantity

Supply chain managers face conflicting pressures to keep inventories low enough to avoid excess inventory holding costs but high enough to reduce ordering and setup costs. *Inventory holding cost* is the sum of the cost of capital and the variable costs of keeping items on hand, such as storage and handling, taxes, insurance, and shrinkage. *Ordering cost* is the cost of preparing a purchase order for a supplier or a production order for the shop, while *setup cost* is the cost of changing over a machine to produce a different item. In this section, we will address the *cycle inventory*, which is that portion of total inventory that varies directly with lot size. A good starting point for balancing these conflicting pressures and determining the best cycle-inventory level for an item is finding the **economic order quantity (EOQ),** which is the lot size that minimizes total annual cycle-inventory holding and ordering costs. The approach to determining the EOQ is based on the following assumptions:

economic order quantity (EOQ)

The lot size that minimizes total annual inventory holding and ordering costs.

1. The demand rate for the item is constant (for example, always 10 units per day) and known with certainty.

2. No constraints are placed (such as truck capacity or materials handling limitations) on the size of each lot.

3. The only two relevant costs are the inventory holding cost and the fixed cost per lot for ordering or setup.

4. Decisions for one item can be made independently of decisions for other items. In other words, no advantage is gained in combining several orders going to the same supplier.

5. The lead time is constant (e.g., always 14 days) and known with certainty. The amount received is exactly what was ordered and it arrives all at once rather than piecemeal.

The economic order quantity will be optimal when all five assumptions are satisfied. In reality, few situations are so simple. Nonetheless, the EOQ is often a reasonable approximation of the appropriate lot size, even when several of the assumptions do not quite apply. Here are some guidelines on when to use or modify the EOQ.

- **Do not use the EOQ**
  - If you use the "make-to-order" strategy and your customer specifies that the entire order be delivered in one shipment
  - If the order size is constrained by capacity limitations such as the size of the firm's ovens, amount of testing equipment, or number of delivery trucks
- **Modify the EOQ**
  - If significant quantity discounts are given for ordering larger lots
  - If replenishment of the inventory is not instantaneous, which can happen if the items must be used or sold as soon as they are finished without waiting until the entire lot has been completed (see Supplement C, "Special Inventory Models," for several useful modifications to the EOQ)
- **Use the EOQ**
  - If you follow a "make-to-stock" strategy and the item has relatively stable demand
  - If your carrying costs per unit and setup or ordering costs are known and relatively stable

The EOQ was never intended to be an optimizing tool. Nonetheless, if you need to determine a reasonable lot size, it can be helpful in many situations.

Calculating the EOQ

We begin by formulating the total cost for any lot size Q for a given SKU. Next, we derive the EOQ, which is the Q that minimizes total annual cycle-inventory cost. Finally, we describe how to convert the EOQ into a companion measure, the elapsed time between orders.

When the EOQ assumptions are satisfied, cycle inventory behaves as shown in Figure 9.5. A cycle begins with Q units held in inventory, which happens when a new order is received. During the cycle, on-hand inventory is used at a constant rate and, because demand is known with certainty and the lead time is a constant, a new lot can be ordered so that inventory falls to 0 precisely when the new lot is received. Because inventory varies uniformly between Q and 0, the average cycle inventory equals half the lot size, Q.

FIGURE 9.5 ▶
Cycle-Inventory Levels

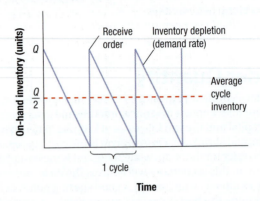

The annual holding cost for this amount of inventory, which increases linearly with Q, as Figure 9.6(a) shows, is

Annual holding cost = (Average cycle inventory)(Unit holding cost)

The annual ordering cost is

Annual ordering cost = (Number of orders/Year)(Ordering or setup cost)

The average number of orders per year equals annual demand divided by Q. For example, if 1,200 units must be ordered each year and the average lot size is 100 units, then 12 orders will be placed during the year. The annual ordering or setup cost decreases nonlinearly as Q increases, as shown in Figure 9.6(b), because fewer orders are placed.

FIGURE 9.6 ▶
Graphs of Annual Holding, Ordering, and Total Costs

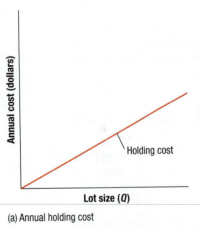

(a) Annual holding cost

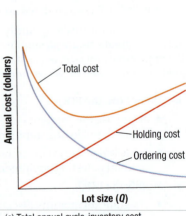

(b) Annual ordering cost

(c) Total annual cycle-inventory cost

The total annual cycle-inventory cost,[2] as graphed in Figure 9.6(c), is the sum of the two cost components:

Total cost = Annual holding cost + Annual ordering or setup cost[3]

$$C = \frac{Q}{2}(H) + \frac{D}{Q}(S)$$

---

[2] Expressing the total cost on an annual basis usually is convenient (although not necessary). Any time horizon can be selected as long as D and H cover the same time period. If the total cost is calculated on a monthly basis, D must be monthly demand and H must be the cost of holding a unit for 1 month.

[3] The number of orders actually placed in any year is always a whole number, although the formula allows the use of fractional values. However, rounding is not needed because what is being calculated is an average for multiple years. Such averages often are nonintegers.

where

C = total annual cycle-inventory cost

Q = lot size, in units

H = cost of holding one unit in inventory for a year, often expressed as a percentage of the item's value

D = annual demand, in units per year

S = cost of ordering or setting up one lot, in dollars per lot

---

EXAMPLE 9.2   **The Cost of a Lot-Sizing Policy**

A museum of natural history opened a gift shop two years ago. Managing inventories has become a problem. Low inventory turnover is squeezing profit margins and causing cash-flow problems.

One of the top-selling SKUs in the container group at the museum's gift shop is a bird feeder. Sales are 18 units per week, and the supplier charges $60 per unit. The cost of placing an order with the supplier is $45. Annual holding cost is 25 percent of a feeder's value, and the museum operates 52 weeks per year. Management chose a 390-unit lot size so that new orders could be placed less frequently. What is the annual cycle-inventory cost of the current policy of using a 390-unit lot size? Would a lot size of 468 be better?

SOLUTION

We begin by computing the annual demand and holding cost as

$$D = (18 \text{ units/week})(52 \text{ weeks/year}) = 936 \text{ units}$$

$$H = 0.25(\$60/\text{unit}) = \$15$$

The total annual cycle-inventory cost for the current policy is

$$C = \frac{Q}{2}(H) + \frac{D}{Q}(S)$$

$$= \frac{390}{2}(\$15) + \frac{936}{390}(\$45) = \$2,925 + \$108 = \$3,033$$

The total annual cycle-inventory cost for the alternative lot size is

$$C = \frac{468}{2}(\$15) + \frac{936}{468}(\$45) = \$3,510 + \$90 = \$3,600$$

DECISION POINT

The lot size of 468 units, which is a half-year supply, would be a more expensive option than the current policy. The savings in ordering costs are more than offset by the increase in holding costs. Management should use the total annual cycle-inventory cost function to explore other lot-size alternatives.

Figure 9.7 displays the impact of using several Q values for the bird feeder in Example 9.2. Eight different lot sizes were evaluated in addition to the current one. Both holding and ordering costs were plotted, but their sum—the total annual cycle-inventory cost curve—is the important feature. The graph shows that the best lot size, or EOQ, is the lowest point on the total annual cost curve, or between 50 and 100 units. Obviously, reducing the current lot-size policy ($Q = 390$) can result in significant savings.

A more efficient approach is to use the EOQ formula:

$$\text{EOQ} = \sqrt{\frac{2DS}{H}}$$

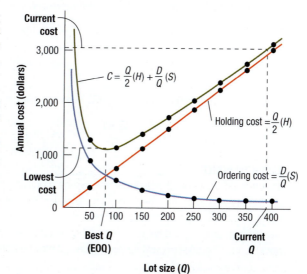

◀ **FIGURE 9.7**

Total Annual Cycle-Inventory Cost Function for the Bird Feeder

We use calculus to obtain the EOQ formula from the total annual cycle-inventory cost function. We take the first derivative of the total annual cycle-inventory cost function with respect to Q, set it equal to 0, and solve for Q. As Figure 9.7 indicates, the EOQ is the order quantity for which annual holding cost equals annual ordering cost. Using this insight, we can also obtain the EOQ formula by equating the formulas for annual ordering cost and annual holding cost and solving for Q. The graph in Figure 9.7 also reveals that when the annual holding cost for any Q exceeds the annual ordering cost, as with the 390-unit order, we can immediately conclude that Q is too high. A lower Q reduces holding cost and increases ordering cost, bringing them into balance. Similarly, if the annual ordering cost exceeds the annual holding cost, Q should be increased.

Sometimes, inventory policies are based on the time between replenishment orders, rather than on the number of units in the lot size. The **time between orders (TBO)** for a particular lot size is the average elapsed time between receiving (or placing) replenishment orders of Q units. Expressed as a fraction of a year, the TBO is simply Q divided by annual demand. When we use the EOQ and express time in terms of months, the TBO is

$$\text{TBO}_{\text{EOQ}} = \frac{\text{EOQ}}{D} \ (12 \ \text{months/year})$$

In Example 9.3, we show how to calculate TBO for years, months, weeks, and days.

time between orders (TBO)

The average elapsed time between receiving (or placing) replenishment orders of Q units for a particular lot size.

| **EXAMPLE 9.3** |

Finding the EOQ, Total Cost, and TBO

MyOMLab

Tutor 9.2 in MyOMLab provides a new example to practice the application of the EOQ model.

MyOMLab

Active Model 9.1 in MyOMLab provides additional insight on the EOQ model and its uses.

FIGURE 9.8 ▶
Total Annual Cycle-Inventory Costs Based on EOQ Using Tutor 9.3

For the bird feeder in Example 9.2, calculate the EOQ and its total annual cycle-inventory cost. How frequently will orders be placed if the EOQ is used?

SOLUTION

Using the formulas for EOQ and annual cost, we get

$$\text{EOQ} = \sqrt{\frac{2DS}{H}} = \sqrt{\frac{2(936)(45)}{15}} = 74.94, \ \text{or} \ 75 \ \text{units}$$

Figure 9.8 shows that the total annual cost is much less than the $3,033 cost of the current policy of placing 390-unit orders.

Parameters		Economic Order Quantity	
Current Lot Size (Q)	390		75
Demand (D)	936		
Order Cost (S)	$45		
Unit Holding Cost (H)	$15		

Annual Costs		Annual Costs based on EOQ	
Orders per Year	2.4	Orders per Year	12.48
Annual Ordering Cost	$108.00	Annual Ordering Cost	$561.60
Annual Holding Cost	$2,925.00	Annual Holding Cost	$562.50
Annual Inventory Cost	$3,033.00	Annual Inventory Cost	$1,124.10

When the EOQ is used, the TBO can be expressed in various ways for the same time period.

$$\text{TBO}_{\text{EOQ}} = \frac{\text{EOQ}}{D} = \frac{75}{936} = 0.080 \ \text{year}$$

$$\text{TBO}_{\text{EOQ}} = \frac{\text{EOQ}}{D}(12 \ \text{months/year}) = \frac{75}{936}(12) = 0.96 \ \text{month}$$

$$\text{TBO}_{\text{EOQ}} = \frac{\text{EOQ}}{D}(52 \ \text{weeks/year}) = \frac{75}{936}(52) = 4.17 \ \text{weeks}$$

$$\text{TBO}_{\text{EOQ}} = \frac{\text{EOQ}}{D}(365 \ \text{days/year}) = \frac{75}{936}(365) = 29.25 \ \text{days}$$

DECISION POINT

Using the EOQ, about 12 orders per year will be required. Using the current policy of 390 units per order, an average of 2.4 orders will be needed each year (every 5 months). The current policy saves on ordering costs but incurs a much higher cost for carrying the cycle inventory. Although it is easy to see which option is best on the basis of total ordering and holding costs, other factors may affect the final decision. For example, if the supplier would reduce the price per unit for large orders, it may be better to order the larger quantity.

Managerial Insights from the EOQ

Subjecting the EOQ formula to *sensitivity analysis* can yield valuable insights into the management of inventories. Sensitivity analysis is a technique for systematically changing crucial parameters to determine the effects of a change. Table 9.1 shows the effects on the EOQ when we substitute different values into the numerator or denominator of the formula.

TABLE 9.1 | **SENSITIVITY ANALYSIS OF THE EOQ**

Parameter	EOQ	Parameter Change	EOQ Change	Comments
Demand	$\sqrt{\frac{2DS}{H}}$	↑	↑	Increase in lot size is in proportion to the square root of *D*.
Order/Setup Costs	$\sqrt{\frac{2DS}{H}}$	↓	↓	Weeks of supply decreases and inventory turnover increases because the lot size decreases.
Holding Costs	$\sqrt{\frac{2DS}{H}}$	↓	↑	Larger lots are justified when holding costs decrease.

As Table 9.1 shows, the EOQ provides support for some of the intuition you may have about inventory management. However, the effect of ordering or setup cost changes on inventories is especially important for *lean systems*. This relationship explains why manufacturers are so concerned about reducing setup time and costs; it makes small lot production economic. Actually, lean systems provide an environment conducive to the use of the EOQ. For example, yearly, monthly, daily, or hourly demand rates are known with reasonable certainty in lean systems, and the rate of demand is relatively uniform. Lean systems (see Chapter 8, "Lean Systems") may have few process constraints if the firm practices *constraint management* (see Chapter 7, "Constraint Management"). In addition, lean systems strive for constant delivery lead times and dependable delivery quantities from suppliers, both of which are assumptions of the EOQ. Consequently, the EOQ as a lot sizing tool is quite compatible with the principles of lean systems.

Inventory Control Systems

The EOQ and other lot-sizing methods answer the important question: *How much* should we order? Another important question that needs an answer is: *When* should we place the order? An inventory control system responds to both questions. In selecting an inventory control system for a particular application, the nature of the demands imposed on the inventory items is crucial. An important distinction between types of inventory is whether an item is subject to dependent or independent demand. Retailers, such as JCPenney, and distributors must manage **independent demand items**—that is, items for which demand is influenced by market conditions and is not related to the inventory decisions for any other item held in stock or produced. Independent demand inventory includes

- Wholesale and retail merchandise
- Service support inventory, such as stamps and mailing labels for post offices, office supplies for law firms, and laboratory supplies for research universities
- Product and replacement-part distribution inventories
- Maintenance, repair, and operating (MRO) supplies—that is, items that do not become part of the final service or product, such as employee uniforms, fuel, paint, and machine repair parts

Managing independent demand inventory can be tricky because demand is influenced by external factors. For example, the owner of a bookstore may not be sure how many copies of the latest best-seller novel customers will purchase during the coming month. As a result, the manager may decide to stock extra copies as a safeguard. Independent demand, such as the demand for various book titles, must be *forecasted*.

In this chapter, we focus on inventory control systems for independent demand items, which is the type of demand the bookstore owner, other retailers, service providers, and distributors face. Even though demand from any one customer is difficult to predict, low demand from some customers for a particular item often is offset by high demand from others. Thus, total demand for any independent demand item may follow a relatively smooth pattern, with some random fluctuations. *Dependent demand items* are those required as components or inputs to a service or

independent demand items

Items for which demand is influenced by market conditions and is not related to the inventory decisions for any other item held in stock or produced.

Retailers typically face independent demands for their products. Here shoppers look for bargains at a JCPenney store in the Glendale Galleria in California.

product. Dependent demand exhibits a pattern very different from that of independent demand and must be managed with different techniques (see Chapter 16, "Resource Planning").

In this section, we discuss and compare two inventory control systems: (1) the continuous review system, called a *Q* system, and (2) the periodic review system, called a *P* system. We close with a look at hybrid systems, which incorporate features of both the *P* and *Q* systems.

Continuous Review System

A **continuous review (*Q*) system**, sometimes called a **reorder point (ROP) system** or *fixed order-quantity system*, tracks the remaining inventory of a SKU each time a withdrawal is made to determine whether it is time to reorder. In practice, these reviews are done frequently (e.g., daily) and often continuously (after each withdrawal). The advent of computers and electronic cash registers linked to inventory records has made continuous reviews easy. At each review, a decision is made about a SKU's inventory

continuous review (*Q*) system

A system designed to track the remaining inventory of a SKU each time a withdrawal is made to determine whether it is time to reorder.

reorder point (ROP) system

See continuous review (*Q*) system.

inventory position (IP)

The measurement of a SKU's ability to satisfy future demand.

scheduled receipts (SR)

Orders that have been placed but have not yet been received.

open orders

See scheduled receipts (SR).

reorder point (*R*)

The predetermined minimum level that an inventory position must reach before a fixed quantity *Q* of the SKU is ordered.

position. If it is judged to be too low, the system triggers a new order. The **inventory position (IP)** measures the SKU's ability to satisfy future demand. It includes **scheduled receipts (SR)**, which are orders that have been placed but have not yet been received, plus on-hand inventory (OH) minus backorders (BO). Sometimes. scheduled receipts are called **open orders**. More specifically,

$$\text{Inventory position} = \text{On-hand inventory} + \text{Scheduled receipts} - \text{Backorders}$$

$$\text{IP} = \text{OH} + \text{SR} - \text{BO}$$

When the inventory position reaches a predetermined minimum level, called the **reorder point (*R*)**, a fixed quantity *Q* of the SKU is ordered. In a continuous review system, although the order quantity *Q* is fixed, the time between orders can vary. Hence, *Q* can be based on the EOQ, a price break quantity (the minimum lot size that qualifies for a quantity discount), a container size (such as a truckload), or some other quantity selected by management.

Selecting the Reorder Point When Demand and Lead Time Are Constant   To demonstrate the concept of a reorder point, suppose that the demand for feeders at the museum gift shop in Example 9.3 is always 18 per week, the lead time is a constant 2 weeks, and the supplier always ships the exact number ordered on time. With both demand and lead time constant, the museum's buyer can wait until the inventory position drops to 36 units, or (18 units/week) (2 weeks), to place a new order. Thus, in this case, the reorder point, *R*, equals the *total demand during lead time*, with no added allowance for safety stock.

Figure 9.9 shows how the system operates when demand and lead time are constant. The downward-sloping line represents the on-hand inventory, which is being depleted at a constant rate. When it reaches reorder point *R* (the horizontal line), a new order for *Q* units is placed. The on-hand inventory continues to drop throughout lead time *L* until the order is received. At that time, which marks the end of the lead time, on-hand inventory jumps by *Q* units. A new order arrives just when inventory drops to 0. The TBO is the same for each cycle.

The inventory position, IP, shown in Figure 9.9 corresponds to the on-hand inventory, except during the lead time. Just after a new order is placed, at the start of the lead time, IP increases by *Q*, as shown by the dashed line. The IP exceeds OH by this same margin

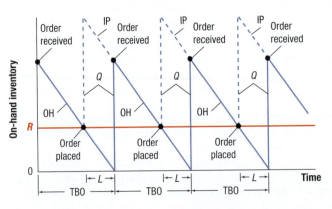

FIGURE 9.9 ▶
Q System When Demand and Lead Time Are Constant and Certain

throughout the lead time.[4] At the end of the lead time, when the scheduled receipts convert to on-hand inventory, IP = OH once again. The key point here is to compare IP, not OH, with R in deciding whether to reorder. A common error is to ignore scheduled receipts or backorders.

EXAMPLE 9.4	**Placing a New Order When Demand and Lead Time Are Constant**

Demand for chicken soup at a supermarket is always 25 cases a day and the lead time is always 4 days. The shelves were just restocked with chicken soup, leaving an on-hand inventory of only 10 cases. No backorders currently exist, but there is one open order in the pipeline for 200 cases. What is the inventory position? Should a new order be placed?

SOLUTION

$$R = \text{Total demand during lead time} = (25)(4) = 100 \text{ cases}$$
$$\text{IP} = \text{OH} + \text{SR} - \text{BO}$$
$$= 10 + 200 - 0 = 210 \text{ cases}$$

DECISION POINT

Because IP exceeds R (210 versus 100), do not reorder. Inventory is almost depleted, but a new order need not be placed because the scheduled receipt is in the pipeline.

Selecting the Reorder Point When Demand Is Variable and Lead Time Is Constant   In reality demand is not always predictable. For instance, the museum's buyer knows that *average* demand is 18 feeders per week. That is, a variable number of feeders may be purchased during the lead time, with an average demand during lead time of 36 feeders (assuming that each week's demand is identically distributed and lead time is a constant 2 weeks). This situation gives rise to the need for safety stocks. Suppose that the museum's buyer sets R at 46 units, thereby placing orders before they typically are needed. This approach will create a safety stock, or stock held in excess of expected demand, of 10 units $(46 - 36)$ to buffer against uncertain demand. In general

$$\text{Reorder point} = \text{Average demand during lead time} + \text{Safety stock}$$
$$= \overline{d}L + \text{safety stock}$$

where

$$\overline{d} = \text{average demand per week (or day or month)}$$
$$L = \text{constant lead tome in weeks (or days or months)}$$

Figure 9.10 shows how the Q system operates when demand is variable and lead time is constant. The wavy downward-sloping line indicates that demand varies from day to day. Its slope is steeper in the second cycle, which means that the demand rate is higher during this time period. The changing demand rate means that the time between orders changes, so $\text{TBO}_1 \neq \text{TBO}_2 \neq \text{TBO}_3$. Because of uncertain demand, sales during the lead time are unpredictable, and safety stock is added to hedge against lost sales. This addition is why R is higher in Figure 9.10 than in Figure 9.9. It also explains why the on-hand inventory usually does not drop to 0 by the time a replenishment order arrives. The greater the safety stock and thus the higher reorder point R, the less likely a stockout.

Because the average demand during lead time is variable, the real decision to be made when selecting R concerns the safety stock level. Deciding on a small or large safety stock is a trade-off between customer service and inventory holding costs. Cost minimization models can be used to find the best safety stock, but they require estimates of stockout and back-order costs, which are usually difficult to make with any precision because it is hard to estimate the effect of lost sales, lost customer confidence, future loyalty of customers, and market share because the customer went to a competitor. The usual approach for determining R is for management—based on

▼ **FIGURE 9.10**
Q System When Demand Is Uncertain

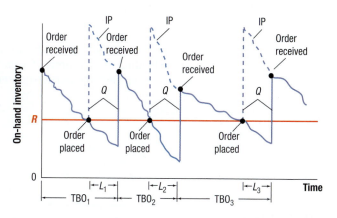

---

[4]A possible exception is the unlikely situation when more than one scheduled receipt is open at the same time because of long lead times.

judgment—to set a reasonable service-level policy for the inventory and then determine the safety stock level that satisfies this policy. There are three steps to arrive at a reorder point:

1. Choose an appropriate service-level policy.

2. Determine the distribution of demand during lead time.

3. Determine the safety stock and reorder point levels.

service level

The desired probability of not running out of stock in any one ordering cycle, which begins at the time an order is placed and ends when it arrives in stock.

cycle-service level

See service level.

protection interval

The period over which safety stock must protect the user from running out of stock.

Step 1: Service level policy   Select a **service level**, or **cycle-service level** (the desired probability of not running out of stock in any one ordering cycle), which begins at the time an order is placed and ends when it arrives in stock. The intent is to provide coverage over the **protection interval**, or the period over which safety stock must protect the user from running out of stock. For the Q system, the lead time is the protection interval. For example, in a bookstore the manager may select a 90 percent cycle-service level for a book. In other words, the probability is 90 percent that demand will not exceed the supply during the lead time. The probability of running short *during the protection interval*, creating a stockout or backorder, is only 10 percent $(100 - 90)$ in our example. This stockout risk, which occurs only during the lead time in the Q system, is greater than the overall risk of a stockout because the risk is nonexistent outside the ordering cycle.

Step 2: Distribution of demand during lead time   Determine the distribution of demand during lead time, which requires the specification of its mean and standard deviation. To translate a cycle-service level policy into a specific safety stock level, we must know how demand during the lead time is distributed. If demand and lead times vary little around their averages, the safety stock can be small. Conversely, if they vary greatly from one order cycle to the next, the safety stock must be large. Variability is measured by the distribution of demand during lead time. Sometimes, average demand during the lead time and the standard deviation of demand during the lead time are not directly available and must be calculated by combining information on the demand rate with information on the lead time. Suppose that lead time is constant and demand is variable, but records on demand are not collected for a time interval that is exactly the same as the lead time. The same inventory control system may be used to manage thousands of different SKUs, each with a different lead time. For example, if demand is reported *weekly*, these records can be used directly to compute the average and the standard deviation of demand during the lead time if the lead time is exactly 1 week. However, if the lead time is 3 weeks, the computation is more difficult.

We can determine the demand during the lead time distribution by making some reasonable assumptions. Suppose that the average demand, $\bar{d}$, is known along with the standard deviation of demand, σ_d, over some time interval such as days or weeks. Also, suppose that the probability distributions of demand for each time interval are identical and independent of each other. For example, if the time interval is a week, the probability distributions of demand are assumed to be the same each week (identical $\bar{d}$ and σ_d), and the total demand in 1 week does not affect the total demand in another week. Let L be the constant lead time, expressed in the same time units as the demand. Under these assumptions, average demand during the lead time will be the sum of the averages for each of the L identical and independent distributions of demand, or $\bar{d} + \bar{d} + \bar{d} + \ldots = \bar{d}L$. In addition, the variance of the distribution of demand during lead time will be the sum of the variances of the L identical and independent distributions of demand, or

$$\sigma_d^2 + \sigma_d^2 + \sigma_d^2 + \ldots = \sigma_d^2 L$$

Finally, the standard deviation of the distribution of demand during lead time is

$$\sigma_{dLT} = \sqrt{\sigma_d^2 L} = \sigma_d \sqrt{L}$$

Figure 9.11 shows how the demand distribution of the lead time is developed from the individual distributions of weekly demands, where $\bar{d} = 75$, $\sigma_d = 15$, and $L = 3$. In this example, average demand during the lead time is $(75)(3) = 225$ units and $\sigma_{dLT} = 15\sqrt{3} = 25.98$.

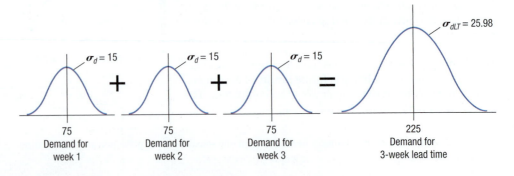

FIGURE 9.11 ▶
Development of Distribution of Demand During Lead Time

Step 3: Safety stock and reorder point   When selecting the safety stock, the inventory planner often assumes that demand during the lead time is normally distributed, as shown in Figure 9.12.

The average demand during the lead time is the centerline of the graph, with 50 percent of the area under the curve to the left and 50 percent to the right. Thus, if a cycle-service level of 50 percent were chosen, the reorder point R would be the quantity represented by this centerline. Because R equals the average demand during the lead time plus the safety stock, the safety stock is 0 when R equals this average demand. Demand is less than average 50 percent of the time and, thus, having no safety stock will be sufficient only 50 percent of the time.

To provide a service level above 50 percent, the reorder point must be higher than the average demand during the lead time. As Figure 9.12 shows, that requires moving the reorder point to the right of the centerline so that more than 50 percent of the area under the curve is to the left of R. An 85 percent cycle-service level is achieved in Figure 9.12 with 85 percent of the area under the curve to the left of R (in **blue**) and only 15 percent to the right (in **pink**). We compute the safety stock as follows:

$$\text{Safety stock} = z\sigma_{dLT}$$

where

z = the number of standard deviations needed to achieve the cycle-service level

σ_{dLT} = standard deviation of demand during the lead time

The reorder point becomes

$$R = \bar{d}L + \text{safety stock}$$

The higher the value of z, the higher the safety stock and the cycle-service level should be. If $z = 0$, there is no safety stock, and stockouts will occur during 50 percent of the order cycles. For a cycle-service level of 85 percent, $z = 1.04$. Example 9.5 shows how to use the Normal Distribution appendix to find the appropriate z value, safety stock, and reorder point.

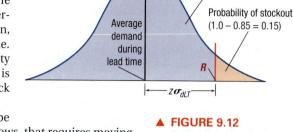

▲ **FIGURE 9.12**
Finding Safety Stock with Normal Probability Distribution for an 85 Percent Cycle-Service Level

---

EXAMPLE 9.5	**Reorder Point for Variable Demand and Constant Lead Time**

Let us return to the bird feeder in Example 9.3. The EOQ is 75 units. Suppose that the average demand is 18 units per week with a standard deviation of 5 units. The lead time is constant at 2 weeks. Determine the safety stock and reorder point if management wants a 90 percent cycle-service level.

MyOMLab

Tutor 9.3 in MyOMLab provides a new example to determine the safety stock and the reorder point for a Q system.

SOLUTION
In this case, $\sigma_d = 5$, $\bar{d} = 18$ units, and $L = 2$ weeks, so $\sigma_{dLT} = \sigma_d\sqrt{L} = 5\sqrt{2} = 7.07$. Consult the body of the table in the Normal Distribution appendix for 0.9000, which corresponds to a 90 percent cycle-service level. The closest number is 0.8997, which corresponds to 1.2 in the row heading and 0.08 in the column heading. Adding these values gives a z value of 1.28. With this information, we calculate the safety stock and reorder point as follows:

$$\text{Safety stock} = z\sigma_{dLT} = 1.28(7.07) = 9.05, \text{ or } 9 \text{ units}$$
$$\text{Reorder point} = \bar{d}L + \text{Safety stock}$$
$$= 2(18) + 9 = 45 \text{ units}$$

DECISION POINT
The Q system for the bird feeder operates as follows: Whenever the inventory position reaches 45 units, order the EOQ of 75 units. Various order quantities and safety stock levels can be used in a Q system. For example, management could specify a different order quantity (because of shipping constraints) or a different safety stock (because of storage limitations).

Selecting the Reorder Point When Both Demand and Lead Time Are Variable   In practice, it is often the case that both the demand and the lead time are variable. Unfortunately, the equations for the safety stock and reorder point become more complicated. In the model below we make two simplifying assumptions. First, the demand distribution and the lead time distribution are measured in the same time units. For example, both demand and lead time are measured in weeks. Second, demand and lead time are *independent*. That is, demand per week is not affected by the length of the lead time.

$$\text{Safety stock} = z\sigma_{dLT}$$

$$R = (\text{Average weekly demand} \times \text{Average lead time in weeks}) + \text{Safety stock}$$

$$= \overline{dL} + \text{Safety stock}$$

where

$$\overline{d} = \text{Average weekly (or daily or monthly) demand}$$

$$\overline{L} = \text{Average weekly (or daily or monthly) lead time}$$

$$\sigma_d = \text{Standard deviation of weekly (or daily or monthly) demand}$$

$$\sigma_{LT} = \text{Standard deviation of the lead time, and}$$

$$\sigma_{dLT} = \sqrt{\overline{L}\sigma_d^2 + \overline{d}^2\sigma_{LT}^2}$$

Now that we have determined the mean and standard deviation of the distribution of demand during lead time under these more complicated conditions, we can select the reorder point as we did before for the case where the lead time was constant.

EXAMPLE 9.6	**Reorder Point for Variable Demand and Variable Lead Time**

The Office Supply Shop estimates that the average demand for a popular ball-point pen is 12,000 pens per week with a standard deviation of 3,000 pens. The current inventory policy calls for replenishment orders of 156,000 pens. The average lead time from the distributor is 5 weeks, with a standard deviation of 2 weeks. If management wants a 95 percent cycle-service level, what should the reorder point be?

SOLUTION

We have $\overline{d} = 12{,}000$ pens, $\sigma_d = 3{,}000$ pens, $\overline{L} = 5$ weeks, and $\sigma_{LT} = 2$ weeks.

$$\sigma_{dLT} = \sqrt{\overline{L}\sigma_d^2 + \overline{d}^2\sigma_{LT}^2} = \sqrt{(5)(3{,}000)^2 + (12{,}000)^2(2)^2} = 24{,}919.87 \text{ pens}$$

Consult the body of the Normal Distribution appendix for 0.9500, which corresponds to a 95 percent cycle-service level. That value falls exactly in the middle of the tabular values of 0.9495 (for a z value of 1.64) and 0.9505 (for a z value of 1.65). Consequently, we will use the more conservative value of 1.65. We calculate the safety stock and reorder point as follows:

$$\text{Safety stock} = z\sigma_{dLT} = (1.65)(24{,}919.87) = 41{,}117.79, \text{ or } 41{,}118 \text{ pens}$$

$$\text{Reorder point} = \overline{dL} + \text{Safety stock} = (12{,}000)(5) + 41{,}118 = 101{,}118 \text{ pens}$$

DECISION POINT

Whenever the stock of ball-point pens drops to 101,118, management should place another replenishment order of 156,000 pens to the distributor.

Sometimes, the theoretical distributions for demand and lead time are not known. In those cases, we can use simulation to find the distribution of demand during lead time using discrete distributions for demand and lead times. Simulation can also be used to estimate the performance of an inventory system. More discussion, and an example, can be found in MyOMLab.

MyOMLab

visual system

A system that allows employees to place orders when inventory visibly reaches a certain marker.

Two-Bin System The concept of a Q system can be incorporated in a **visual system**, that is, a system that allows employees to place orders when inventory visibly reaches a certain marker. Visual systems are easy to administer because records are not kept on the current inventory position. The historical usage rate can simply be reconstructed from past purchase orders. Visual systems are intended for use with low-value SKUs that have a steady demand, such as nuts and bolts or office supplies. Overstocking is common, but the extra inventory holding cost is minimal because the items have relatively little value.

A visual system version of the Q system is the **two-bin system** in which a SKU's inventory is stored at two different locations. Inventory is first withdrawn from one bin. If the first bin is empty, the second bin provides backup to cover demand until a replenishment order arrives. An empty first bin signals the need to place a new order. Premade order forms placed near the bins let workers send one to purchasing or even directly to the supplier. When the new order arrives, the second bin is restored to its normal level and the rest is put in the first bin. The two-bin system operates like a Q system, with the normal level in the second bin being the reorder point R. The system also may be implemented with just one bin by marking the bin at the reorder point level.

Calculating Total Q System Costs   Total costs for the continuous review (Q) system is the sum of three cost components:

Total cost = Annual cycle inventory holding cost + annual ordering cost

+ annual safety stock holding cost

$$C = \frac{Q}{2}(H) + \frac{D}{Q}(S) + (H)(\text{Safety stock})$$

The annual cycle-inventory holding cost and annual ordering cost are the same equations we used for computing the total annual cycle-inventory cost in Example 9.2. The annual cost of holding the safety stock is computed under the assumption that the safety stock is on hand at all times. Referring to Figure 9.10 in each order cycle, we will sometimes experience a demand greater than the average demand during lead time, and sometimes we will experience less. On average over the year, we can assume the safety stock will be on hand. See Solved Problems 4 and 6 at the end of this chapter for an example of calculating the total costs for a Q system.

Periodic Review System

An alternative inventory control system is the **periodic review (P) system**, sometimes called a *fixed interval reorder system* or *periodic reorder system*, in which an item's inventory position is reviewed periodically rather than continuously. Such a system can simplify delivery scheduling because it establishes a routine. A new order is always placed at the end of each review, and the time between orders (TBO) is fixed at P. Demand is a random variable, so total demand between reviews varies. In a P system, the lot size, Q, may change from one order to the next, but the time between orders is fixed. An example of a periodic review system is that of a soft-drink supplier making weekly rounds of grocery stores. Each week, the supplier reviews the store's inventory of soft drinks and restocks the store with enough items to meet demand and safety stock requirements until the next week.

Under a P system, four of the original EOQ assumptions are maintained: (1) no constraints are placed on the size of the lot, (2) the relevant costs are holding and ordering costs, (3) decisions for one SKU are independent of decisions for other SKUs, and (4) lead times are certain and supply is known. However, demand uncertainty is again allowed for. Figure 9.13 shows the periodic review system under these assumptions. The downward-sloping line again represents on-hand inventory. When the predetermined time, P, has elapsed since the last review, an order is placed to bring the inventory position, represented by the dashed line, up to the target inventory level, T. The lot size for the first review is Q_1, or the difference between inventory position IP_1 and T. As with the continuous review system, IP and OH differ only during the lead time. When the order arrives at the end of the lead time, OH and IP again are identical. Figure 9.13 shows that lot sizes vary from one order cycle to the next. Because the inventory position is lower at the second review, a greater quantity is needed to achieve an inventory level of T.

Managerial Practice 9.1 shows how the use of periodic inventory review systems is important in supply chains in the chemical industry.

▼ **FIGURE 9.13**
P System When Demand Is Uncertain

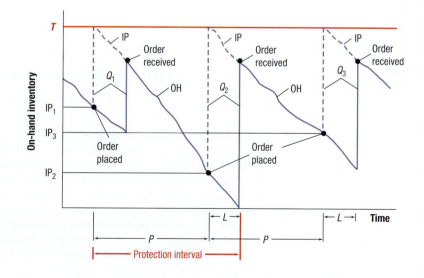

MANAGERIAL PRACTICE 9.1   The Supply Chain Implications of Periodic Review Inventory Systems at Celanese

What do products such as paints, adhesives, coatings, plastics, medicines, cosmetics, detergents, textiles, or fragrances have in common? All of these products use acetic acid as a major component. Celanese, a $6.5 billion chemical company with $640 million in total inventories, is a major supplier of acetic acid in the world. The large investment in inventory forces Celanese to take a hard look at inventory policies for all products, including acetic acid. The key to successful management of the inventories was to acknowledge the interaction between the inventory policies at each stage of the supply chain with the realities of material flows and logistics in the chemical industry.

The supply chain for acetic acid is complex, involving 90 stages. For example, the supply chain is comprised of stages such as vendors supplying a liner to transport the acid, manufacturing sites producing the acid, transportation modes moving acid, warehouses storing it, and customer demand locations to which the acid is finally shipped. There are four manufacturing facilities, three in the United States and one in Singapore, each of which supplies several storage locations worldwide. Transportation stages correspond to rail, barges, trucks, and ocean vessels. Material typically moves in large quantities because of economies of scale and transportation schedules. Storage facilities may be supplied by multiple upstream facilities as well as the manufacturing plant itself.

The use of periodic review inventory systems at the storage and demand locations in this supply chain scenario makes sense for several reasons. First, the transportation modes have defined schedules of operation. Review periods at storage facilities reflect the schedule of the supplying transportation mode. Second, customer orders are typically batched and timed with weekly, bi-weekly, or monthly frequencies. Celanese often assigns customers and storage facilities specific days to place orders so that their own production

Picture Contact BV/Alamy

Large, fixed capacity modes of transportation require defined schedules of operation. Such a situation supports the use of periodic inventory systems. Here ocean vessels await loads of petro-chemicals at the Vopak terminal in the Port of Rotterdam.

schedules can be coordinated. Finally, the cyclic ordering is often a function of the capital intensity of the industry. Long production runs are scheduled to gain production efficiency; it is costly to set up the equipment for another product.

Specifying the best review period and target inventory levels for the various stages of the supply chain takes sophisticated mathematical models. Regardless of the effort required, it is important to recognize the implications of the supply chain when determining inventory policies.

Source: John M. Bossert and Sean P. Williams, "A Periodic-Review Modeling Approach for Guaranteed Service Supply Chains," *Interfaces*, Vol. 37, No. 5 (September/October 2007), pp. 420–435; **http://finance.yahoo.com; www.Celanese.com**, 2008.

EXAMPLE 9.7   **Determining How Much to Order in a *P* System**

A distribution center has a backorder for five 46-inch LCD TV sets. No inventory is currently on hand, and now is the time to review. How many should be reordered if $T = 400$ and no receipts are scheduled?

SOLUTION

$$IP = OH + SR - BO$$

$$= 0 + 0 - 5 = -5 \text{ sets}$$

$$T - IP = 400 - (-5) = 405 \text{ sets}$$

That is, 405 sets must be ordered to bring the inventory position up to *T* sets.

Selecting the Time between Reviews   To run a *P* system, managers must make two decisions: the length of time between reviews, *P*, and the target inventory level, *T*. Let us first consider the time between reviews, *P*. It can be any convenient interval, such as each Friday or every other Friday. Another option is to base *P* on the cost trade-offs of the EOQ. In other words, *P* can be set equal to the average time between orders for the economic order quantity, or TBO_{EOQ}. Because demand is variable, some orders will be larger than the EOQ and some will be smaller. However, over an extended period of time, the average lot size should be close to the EOQ. If other models are used to determine the lot size (e.g., those described in Supplement C, "Special Inventory Models"), we divide the lot size chosen by the annual demand, *D*, and use this ratio as *P*. It will be expressed as the fraction of a year between orders, which can be converted into months, weeks, or days as needed.

Selecting the Target Inventory Level When Demand Is Variable and Lead Time Is Constant   Now, let us calculate the target inventory level, T, when demand is variable but the lead time is constant. Figure 9.13 reveals that an order must be large enough to make the inventory position, IP, last beyond the next review, which is P time periods away. The checker must wait P periods to revise, correct, and reestablish the inventory position. Then, a new order is placed, but it does not arrive until after the lead time, L. Therefore, as Figure 9.13 shows, a protection interval of $P + L$ periods is needed. A fundamental difference between the Q and P systems is the length of time needed for stockout protection. A Q system needs stockout protection only during the lead time because orders can be placed as soon as they are needed and will be received L periods later. A P system, however, needs stockout protection for the longer $P + L$ protection interval because orders are placed only at fixed intervals, and the inventory is not checked until the next designated review time.

As with the Q system, we need to develop the appropriate distribution of demand during the protection interval to specify the system fully. In a P system, we must develop the distribution of demand for $P + L$ time periods. The target inventory level T must equal the expected demand during the protection interval of $P + L$ periods, plus enough safety stock to protect against demand uncertainty over this same protection interval. We assume that lead time is constant and that demand in one period is independent of demand in the next period. Thus, the average demand during the protection interval is $\overline{d}(P + L)$, or

$$T = \overline{d}(P + L) + \text{Safety stock for the protection interval}$$

We compute safety stock for a P system much as we did for the Q system. However, the safety stock must cover demand uncertainty for a longer period of time. When using a normal probability distribution, we multiply the desired standard deviations to implement the cycle-service level, z, by the standard deviation of demand during the protection interval, σ_{P+L}. The value of z is the same as for a Q system with the same cycle-service level. Thus,

$$\text{Safety stock} = z\sigma_{P+L}$$

Based on our earlier logic for calculating σ_{dLT}, we know that the standard deviation of the distribution of demand during the protection interval is

$$\sigma_{P+L} = \sigma_d\sqrt{P + L}$$

Because a P system requires safety stock to cover demand uncertainty over a longer time period than a Q system, a P system requires more safety stock; that is, σ_{P+L} exceeds σ_{dLT}. Hence, to gain the convenience of a P system requires that overall inventory levels be somewhat higher than those for a Q system.

EXAMPLE 9.8   **Calculating P and T**

Again, let us return to the bird feeder example. Recall that demand for the bird feeder is normally distributed with a mean of 18 units per week and a standard deviation in weekly demand of 5 units. The lead time is 2 weeks, and the business operates 52 weeks per year. The Q system developed in Example 9.5 called for an EOQ of 75 units and a safety stock of 9 units for a cycle-service level of 90 percent. What is the equivalent P system? Answers are to be rounded to the nearest integer.

MyOMLab

Tutor 9.4 in MyOMLab provides a new example to determine the review interval and the target inventory for a P system.

SOLUTION

We first define D and then P. Here, P is the time between reviews, expressed in weeks because the data are expressed as demand *per week*:

$$D = (18 \text{ units/week})(52 \text{ weeks/year}) = 936 \text{ units}$$

$$P = \frac{\text{EOQ}}{D}(52) = \frac{75}{936}(52) = 4.2, \text{ or } 4 \text{ weeks}$$

With $\overline{d} = 18$ units per week, an alternative approach is to calculate P by dividing the EOQ by $\overline{d}$ to get $75/18 = 4.2$, or 4 weeks. Either way, we would review the bird feeder inventory every 4 weeks. We now find the standard deviation of demand over the protection interval $(P + L = 6)$:

$$\sigma_{P+L} = \sigma_d\sqrt{P + L} = 52\sqrt{6} = 12.25 \text{ units}$$

Before calculating T, we also need a z value. For a 90 percent cycle-service level, $z = 1.28$ (see the Normal Distribution appendix). The safety stock becomes

$$\text{Safety stock} = z\sigma_{P+L} = 1.28(12.25) = 15.68, \text{ or } 16 \text{ units}$$

We now solve for T:

$$T = \text{Average demand during the protection interval} + \text{Safety stock}$$

$$= \bar{d}(P + L) + \text{Safety stock}$$

$$= (18 \text{ units/week})(6 \text{ weeks}) + 16 \text{ units} = 124 \text{ units}$$

DECISION POINT

Every 4 weeks we would order the number of units needed to bring inventory position IP (counting the new order) up to the target inventory level of 124 units. The P system requires 16 units in safety stock, while the Q system only needs 9 units. If cost were the only criterion, the Q system would be the choice for the bird feeder. As we discuss later, other factors may sway the decision in favor of the P system.

Selecting the Target Inventory Level When Both Demand and Lead Time Are Variable   A useful approach for finding P and T in practice is simulation. Given discrete probability distributions for demand and lead time, simulation can be used to estimate the demand during the protection interval distribution. The *Demand During the Protection Interval Simulator* in OM Explorer can be used to determine the distribution. Once determined, the distribution can be used to select a value for T, given a desired cycle-service level. More discussion, and an example, can be found in MyOMLab.

MyOMLab

single-bin system

A system of inventory control in which a maximum level is marked on the storage shelf or bin, and the inventor is brought up to the mark periodically.

Single-Bin System   The concept of a P system can be translated into a simple visual system of inventory control. In the **single-bin system**, a maximum level is marked on the storage shelf or bin, and the inventory is brought up to the mark periodically—say, once a week. The single bin may be, for example, a gasoline storage tank at a service station or a storage bin for small parts at a manufacturing plant.

Calculating Total P System Costs   The total costs for the P system are the sum of the same three cost elements for the Q system. The differences are in the calculation of the order quantity and the safety stock. As shown in Figure 9.13, the average order quantity will be the average consumption of inventory during the P periods between orders. Consequently, $Q = \bar{d}P$. Total costs for the P system are

$$C = \frac{\bar{d}P}{2}(H) + \frac{D}{\bar{d}P}(S) + (H)(\text{Safety stock})$$

See Solved Problem 5 at the end of this chapter for an example of calculating total P system costs.

Comparative Advantages of the Q and P Systems

Neither the Q nor the P system is best for all situations. Three P-system advantages must be balanced against three Q-system advantages. The advantages of one system are implicitly disadvantages of the other system.

The primary advantages of P systems are the following:

1. The system is convenient because replenishments are made at fixed intervals. Fixed replenishment intervals allow for standardized pickup and delivery times.

2. Orders for multiple items from the same supplier can be combined into a single purchase order. This approach reduces ordering and transportation costs and can result in a price break from the supplier.

3. The inventory position, IP, needs to be known only when a review is made (not continuously, as in a Q system). However, this advantage is moot for firms using computerized record-keeping systems, in which a transaction is reported upon each receipt or withdrawal. When inventory records are always current, the system is called a **perpetual inventory system**.

perpetual inventory system

A system of inventory control in which the inventory records are always current.

The primary advantages of Q systems are the following:

1. The review frequency of each SKU may be individualized. Tailoring the review frequency to the SKU can reduce total ordering and holding costs.

2. Fixed lot sizes, if large enough, can result in quantity discounts. The firm's physical limitations, such as its truckload capacities, materials handling methods, and shelf space might also necessitate a fixed lot size.

3. Lower safety stocks result in savings.

In conclusion, the choice between Q and P systems is not clear cut. Which system is better depends on the relative importance of its advantages in various situations.

Hybrid Systems

Various hybrid inventory control systems merge some but not all the features of the P and Q systems. We briefly examine two such systems: (1) optional replenishment and (2) base stock.

Optional Replenishment System   Sometimes called the optional review, min–max, or (s, S) system, the **optional replenishment system** is much like the P system. It is used to review the inventory position at fixed time intervals and, if the position has dropped to (or below) a predetermined level, to place a variable-sized order to cover expected needs. The new order is large enough to bring the inventory position up to a target inventory, similar to T for the P system. However, orders are not placed after a review unless the inventory position has dropped to the predetermined minimum level. The minimum level acts as the reorder point R does in a Q system. If the target is 100 and the minimum level is 60, the minimum order size is 40 (or $100 - 60$). Because continuous reviews need not be made, this system is particularly attractive when both review and ordering costs are high.

Base-Stock System   In its simplest form, the **base-stock system** issues a replenishment order, Q, each time a withdrawal is made, for the same amount as the withdrawal. This one-for-one replacement policy maintains the inventory position at a base-stock level equal to expected demand during the lead time plus safety stock. The base-stock level, therefore, is equivalent to the reorder point in a Q system. However, order quantities now vary to keep the inventory position at R at all times. Because this position is the lowest IP possible that will maintain a specified service level, the base-stock system may be used to minimize cycle inventory. More orders are placed, but each order is smaller. This system is appropriate for expensive items, such as replacement engines for jet airplanes. No more inventory is held than the maximum demand expected until a replacement order can be received.

optional replenishment system

A system used to review the inventory position at fixed time intervals and, if the position has dropped to (or below) a predetermined level, to place a variable-sized order to cover expected needs.

base-stock system

An inventory control system that issues a replenishment order, Q, each time a withdrawal is made, for the same amount of the withdrawal.

LEARNING GOALS IN REVIEW

1 **Identify the advantages, disadvantages, and costs of holding inventory.** We cover these important aspects of inventories in the section "Inventory and Supply Chains," pp. 309–314. Focus on the pressures for small or large inventories and Figure 9.1.

2 **Define the different types of inventory and the roles they play in supply chains.** The section "Types of Inventory," p. 311–313, explains each type of inventory and provides an example in Figure 9.2. Example 9.1 and Solved Problem 1 show how to estimate inventory levels.

3 **Explain the basic tactics for reducing inventories in supply chains.** See the section "Inventory Reduction Tactics," pp. 313–314, for important approaches to managing inventory levels.

4 **Determine the inventory items deserving most attention and tightest inventory control.** The section "ABC Analysis," pp. 314–315, shows a simple approach to categorizing inventory items for ease of management oversight. Figure 9.4 has an example. Solved Problem 2 demonstrates the calculations.

5 **Calculate the economic order quantity and apply it to various situations.** See the section "Economic Order Quantity," pp. 315–319, for a complete discussion of EOQ model. Focus

on Figures 9.5 through 9.7 to see how the EOQ model affects inventory levels under the standard assumptions and how the EOQ provides the lowest cost solution. Review Examples 9.2 and 9.3 and Solved Problem 3 for help in calculating the total costs of various lot-size choices. Table 9.1 reveals important managerial insights from the EOQ.

6 **Determine the order quantity and reorder point for a continuous review inventory system.** The section "Continuous Review System," pp. 320–325, builds the essence of the Q system from basic principles to more-realistic assumptions. Be sure to understand Figures 9.10 and 9.12. Examples 9.4 through 9.6 and Solved Problems 4 and 6 show how to determine the parameters Q and R under various assumptions.

7 **Determine the review interval and target inventory level for a periodic review inventory control system.** We summarize the key concepts in the section "Periodic Review System," pp. 325–328. Figure 9.13 shows how a P system operates while Examples 9.7 and 9.8 and Solved Problem 5 demonstrate how to calculate the parameters P and T.

8 **Define the key factors that determine the appropriate choice of an inventory system.** See the section "Comparative Advantages of the Q and P Systems," pp. 328–329.

MyOMLab helps you develop analytical skills and assesses your progress with multiple problems on cycle inventory, pipeline inventory, safety stock, inventory turns, weeks of supply, aggregate inventory value, time between orders, optimal order quantity, EOQ, optimal interval between orders, reorder point, demand during lead time, cycle-service level, target inventory T, and P.

MyOMLab Resources	Titles	Link to the Book
Video	*Inventory and Textbooks*	Entire chapter
Active Model Exercise	9.1 Economic Order Quantity	Economic Order Quantity; Example 9.3 (p. 318); Active Model Exercise (p. 341)
OM Explorer Solvers	ABC Analysis	ABC Analysis (p. 314–315); Solved Problem 2 (p. 332–333)
	Inventory Systems Designer	Inventory Control Systems; Example 9.3 (p. 318); Example 9.8 (p. 327–328); Solved Problem 3 (p. 334); Solved Problem 4 (p. 334–335); Solved Problem 5 (p. 335); Figure 9.15 (p. 335)
	Demand During Protection Interval Simulator	Selecting the Reorder Point When Both Demand and Lead Time Are Variable; Selecting the Target Inventory Level When Both Demand and Lead Time Are Variable
	Q System Simulator	Continuous Review System
OM Explorer Tutors	9.1 Estimating Inventory Levels	Inventory and Supply Chains; Example 9.1 (p. 312–313); Figure 9.3 (p. 313); Solved Problem 1 (p. 332)
	9.2 ABC Analysis	ABC Analysis; Solved Problem 2 (p. 332–333); Figure 9.14 (p. 333)
	9.3 Finding EOQ and Total Cost	Economic Order Quantity; Example 9.3 (p. 318); Figure 9.8 (p. 318); Solved Problem 3 (p. 334)
	9.4 Finding the Safety Stock and R	Continuous Review System; Example 9.5 (p. 323); Solved Problem 4 (p. 334–335)
	9.5 Calculating P and T	Periodic Review System; Example 9.8 (p. 327–328); Solved Problem 5 (p. 335)
POM for Windows	Economic Order Quantity (EOQ) Model	Economic Order Quantity; Example 9.3 (p. 318); Solved Problem 3 (p. 334)
	ABC Analysis	ABC Analysis; Solved Problem 2 (p. 332–333)
Tutor Exercises	9.1 Finding EOQ, Safety Stock, R, P, T at Bison College Bookstore	Economic Order Quantity; Inventory Control Systems
Tutorial on Inventory Management Systems	Using Simulation to Develop Inventory Management Systems	Selecting the Reorder Point When Demand and Lead Time Are Variable; Selecting the Target Inventory Level When Both Demand and Lead Time Are Variable
Advanced Problems	1. Office Supply Shop Simulation	Continuous Review System; Selecting the Reorder Point When Both Demand and Lead Time Are Variable
	2. Grocery Store Simulation	Periodic Review System; Selecting the Target Inventory Level when Both Demand and Lead Time Are Variable
	3. Floral Shop Simulation	Continuous Review System; Selecting the Reorder Point When Demand and Lead Time Are Variable
Virtual Tours	Stickley	Inventory and Supply Chains
	United Wood Treating	Entire chapter
Internet Exercise	9.1 Round House	Inventory and Supply Chains
SimQuick Simulation Exercise	Inventory Control Systems	Continuous Review System; Periodic Review System
Key Equations		
Image Library		

Key Equations

1. Average cycle inventory: $\dfrac{Q}{2}$

2. Pipeline inventory: $\overline{D}_L = \overline{d}L$

3. Total annual cycle-inventory cost = Annual holding cost + Annual ordering or setup cost:

$$C = \frac{Q}{2}(H) + \frac{D}{Q}(S)$$

4. Economic order quantity:

$$\text{EOQ} = \sqrt{\frac{2DS}{H}}$$

5. Time between orders, expressed in weeks:

$$\text{TBO}_{\text{EOQ}} = \frac{\text{EOQ}}{D}(52 \text{ weeks/year})$$

6. Inventory position = On-hand inventory + Scheduled receipts − Backorders:

$$\text{IP} = \text{OH} + \text{SR} - \text{BO}$$

7. Continuous review system:

Protection interval = Lead time (L)

Standard deviation of demand during the lead time (constant L) $= \sigma_{dLT} = \sigma_d\sqrt{L}$

Standard deviation of demand during the lead time (variable L) =

$\sigma_{dLT} = \sqrt{\overline{L}\sigma_d^2 + \overline{d}^2\sigma_{LT}^2}$

Safety stock $= z\sigma_{dLT}$

Reorder point (R) for constant lead time $= \overline{d}L + $ Safety stock

Reorder point (R) for variable lead time $= \overline{d}\overline{L} + $ Safety stock

Order quantity $= \text{EOQ}$

Replenishment rule: Order EOQ units when IP $\leq R$

Total Q system cost: $C = \dfrac{Q}{2}(H) + \dfrac{D}{Q}(S) + (H)(\text{Safety stock})$

8. Periodic review system:

Review interval = Time between orders = P

Protection interval = Time between orders + Lead time = $P + L$

Standard deviation of demand during the protection interval $\sigma_{P+L} = \sigma_d\sqrt{P + L}$

Safety stock $= z\sigma_{P+L}$

Target inventory level (T) = Average demand during the protection interval + Safety stock

$$= \overline{d}(P + L) + \text{Safety stock}$$

Order quantity = Target inventory level − Inventory position = $T - \text{IP}$

Replenishment rule: Every P time periods, order $T - \text{IP}$ units

Total P system cost: $C = \dfrac{\overline{d}P}{2}(H) + \dfrac{D}{\overline{d}P}(S) + (H)(\text{Safety stock})$

Key Terms

ABC analysis 314
anticipation inventory 312
backorder 310
base-stock system 329
continuous review (Q) system 320
cycle counting 315
cycle inventory 311
cycle-service level 322
economic order quantity (EOQ) 315
finished goods (FG) 311
independent demand items 319
inventory 309
inventory holding cost 309
inventory management 308

inventory position (IP) 320
lot size 308
lot sizing 311
open orders 320
optional replenishment system 329
ordering cost 310
periodic review (P) system 325
perpetual inventory system 328
pipeline inventory 312
protection interval 322
quantity discount 311
raw materials (RM) 311
reorder point (R) 320
reorder point (ROP) system 320

repeatability 313
safety stock inventory 312
scheduled receipts (SR) 320
service level 322
setup cost 310
single-bin system 328
stock-keeping unit (SKU) 314
stockout 310
time between orders (TBO) 318
two-bin system 325
visual system 324
work-in-process (WIP) 311

Solved Problem 1

A distribution center experiences an average weekly demand of 50 units for one of its items. The product is valued at $650 per unit. Inbound shipments from the factory warehouse average 350 units. Average lead time (including ordering delays and transit time) is 2 weeks. The distribution center operates 52 weeks per year; it carries a 1-week supply of inventory as safety stock and no anticipation inventory. What is the value of the average aggregate inventory being held by the distribution center?

SOLUTION

Type of Inventory	Calculation of Aggregate Average Inventory	
Cycle	$\dfrac{Q}{2} = \dfrac{350}{2}$	= 175 units
Safety stock	1-week supply	= 50 units
Anticipation	None	
Pipeline	$\bar{d}L = (50 \text{ units/week})(2 \text{ weeks})$	= 100 units
	Average aggregate inventory	= 325 units
	Value of aggregate inventory	= $650(325)
		= $211,250

Solved Problem 2

Booker's Book Bindery divides SKUs into three classes, according to their dollar usage. Calculate the usage values of the following SKUs and determine which is most likely to be classified as class A.

SOLUTION

The annual dollar usage for each SKU is determined by multiplying the annual usage quantity by the value per unit. As shown in Figure 9.14, the SKUs are then sorted by annual dollar usage, in declining order. Finally, A–B and B–C class lines are drawn roughly, according to the guidelines presented in the text. Here, class A includes only one SKU (signatures), which represents only 1/7, or 14 percent, of the SKUs but accounts for 83 percent of annual dollar usage. Class B includes the next two SKUs, which taken together represent 28 percent of the SKUs and account for 13 percent of annual dollar usage. The final four SKUs, class C, represent over half the number of SKUs but only 4 percent of total annual dollar usage.

SKU Number	Description	Quantity Used per Year	Unit Value ($)
1	Boxes	500	3.00
2	Cardboard (square feet)	18,000	0.02
3	Cover stock	10,000	0.75
4	Glue (gallons)	75	40.00
5	Inside covers	20,000	0.05
6	Reinforcing tape (meters)	3,000	0.15
7	Signatures	150,000	0.45

SKU Number	Description	Quantity Used per Year		Unit Value ($)		Annual Dollar Usage ($)
1	Boxes	500	×	3.00	=	1,500
2	Cardboard (square feet)	18,000	×	0.02	=	360
3	Cover stock	10,000	×	0.75	=	7,500
4	Glue (gallons)	75	×	40.00	=	3,000
5	Inside covers	20,000	×	0.05	=	1,000
6	Reinforcing tape (meters)	3,000	×	0.15	=	450
7	Signatures	150,000	×	0.45	=	67,500
					Total	81,310

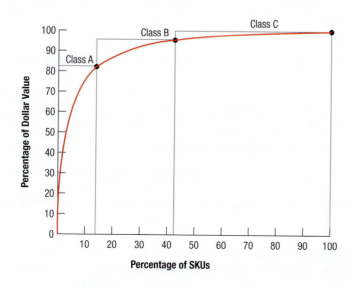

SKU #	Description	Qty Used/Year	Value	Dollar Usage	Pct of Total	Cumulative % of Dollar Value	Cumulative % of SKU	Class
7	Signatures	150,000	$0.45	$67,500	83.0%	83.0%	14.3%	A
3	Cover stock	10,000	$0.75	$7,500	9.2%	92.2%	28.6%	B
4	Glue	75	$40.00	$3,000	3.7%	95.9%	42.9%	B
1	Boxes	500	$3.00	$1,500	1.8%	97.8%	57.1%	C
5	Inside covers	20,000	$0.05	$1,000	1.2%	99.0%	71.4%	C
6	Reinforcing tape	3,000	$0.15	$450	0.6%	99.6%	85.7%	C
2	Cardboard	18,000	$0.02	$360	0.4%	100.0%	100.0%	C

Total $81,310

◄ **FIGURE 9.14**

Annual Dollar Usage for Class A, B, and C SKUs Using Tutor 9.2

Solved Problem 3

Nelson's Hardware Store stocks a 19.2 volt cordless drill that is a popular seller. Annual demand is 5,000 units, the ordering cost is $15, and the inventory holding cost is $4/unit/year.

a. What is the economic order quantity?

b. What is the total annual cost for this inventory item?

SOLUTION

a. The order quantity is

$$EOQ = \sqrt{\frac{2DS}{H}} = \sqrt{\frac{2(5,000)(\$15)}{\$4}} = \sqrt{37,500}$$

$$= 193.65, \text{ or } 194 \text{ drills}$$

b. The total annual cost is

$$C = \frac{Q}{2}(H) + \frac{D}{Q}(S) = \frac{194}{2}(\$4) + \frac{5,000}{194}(\$15) = \$774.60$$

Solved Problem 4

A regional distributor purchases discontinued appliances from various suppliers and then sells them on demand to retailers in the region. The distributor operates 5 days per week, 52 weeks per year. Only when it is open for business can orders be received. Management wants to reevaluate its current inventory policy, which calls for order quantities of 440 counter-top mixers. The following data are estimated for the mixer:

Average daily demand $(\bar{d})$ = 100 mixers

Standard deviation of daily demand (σ_d) = 30 mixers

Lead time (L) = 3 days

Holding cost (H) = $9.40/unit/year

Ordering cost (S) = $35/order

Cycle-service level = 92 percent

The distributor uses a continuous review (Q) system.

a. What order quantity Q, and reorder point, R, should be used?

b. What is the total annual cost of the system?

c. If on-hand inventory is 40 units, one open order for 440 mixers is pending, and no backorders exist, should a new order be placed?

SOLUTION

a. Annual demand is

$$D = (5 \text{ days/week})(52 \text{ weeks/year})(100 \text{ mixers/day}) = 26,000 \text{ mixers/year}$$

The order quantity is

$$EOQ = \sqrt{\frac{2DS}{H}} = \sqrt{\frac{2(26,000)(\$35)}{\$9.40}} = \sqrt{193,167} = 440.02, \text{ or } 440 \text{ mixers}$$

The standard deviation of the distribution of demand during lead time is

$$\sigma_{dLT} = \sigma_d\sqrt{L} = 30\sqrt{3} = 51.96$$

A 92 percent cycle-service level corresponds to z = 1.41 (see the Normal Distribution appendix). Therefore,

Safety stock = $z\sigma_{dLT}$ = 1.41(51.96 mixers) = 73.26, or 73 mixers

Average demand during the lead time = $\bar{d}L$ = 100(3) = 300 mixers

Reorder point (R) = Average demand during the lead time + Safety stock

= 300 mixers + 73 mixers = 373 mixers

With a continuous review system, Q = 440 and R = 373.

b. The total annual cost for the Q systems is

$$C = \frac{Q}{2}(H) + \frac{D}{Q}(S) + (H)\,(\text{Safety stock})$$

$$C = \frac{440}{2}(\$9.40) + \frac{26{,}000}{440}(35) + (\$9.40)(73) = \$4{,}822.38$$

c. Inventory position = On-hand inventory + Scheduled receipts − Backorders

$$\text{IP} = \text{OH} + \text{SR} - \text{BO} = 40 + 440 - 0 = 480 \text{ mixers}$$

Because IP (480) exceeds R (373), do not place a new order.

Solved Problem 5

Suppose that a periodic review (P) system is used at the distributor in Solved Problem 4, but otherwise the data are the same.

a. Calculate the P (in workdays, rounded to the nearest day) that gives approximately the same number of orders per year as the EOQ.

b. What is the target inventory level, T? Compare the P system to the Q system in Solved Problem 4.

c. What is the total annual cost of the P system?

d. It is time to review the item. On-hand inventory is 40 mixers; receipt of 440 mixers is scheduled, and no backorders exist. How much should be reordered?

SOLUTION

a. The time between orders is

$$P = \frac{\text{EOQ}}{D}(260 \text{ days/year}) = \frac{440}{26{,}000}(260) = 4.4, \text{ or } 4 \text{ days}$$

b. Figure 9.15 shows that $T = 812$ and safety stock $= (1.41)\,(79.37) = 111.91$, or about 112 mixers. The corresponding Q system for the counter-top mixer requires less safety stock.

c. The total annual cost of the P system is

$$C = \frac{\bar{d}P}{2}(H) + \frac{D}{\bar{d}P}(S) + (H)(\text{Safety stock})$$

$$C = \frac{(100)(4)}{2}(\$9.40) + \frac{26{,}000}{(100)(4)}(\$35) + (\$9.40)(1.41)(79.37)$$

$$= \$5{,}207.80$$

d. Inventory position is the amount on hand plus scheduled receipts minus backorders, or

$$\text{IP} = \text{OH} + \text{SR} - \text{BO} = 40 + 440 - 0 = 480 \text{ mixers}$$

The order quantity is the target inventory level minus the inventory position, or

$$Q = T - \text{IP} = 812 \text{ mixers} - 480 \text{ mixers} = 332 \text{ mixers}$$

An order for 332 mixers should be placed.

Continuous Review (Q) System		Periodic Review (P) System			
z	1.41	Time Between Reviews (P)	4.00	Days	
Safety Stock	73	☑ Enter manually			
Reorder Point	373	Standard Deviation of Demand During Protection Interval	79.37		
Annual Cost	$4,822.38	Safety Stock	112		
		Average Demand During Protection Interval	700		
		Target Inventory Level (T)	812		
		Annual Cost	$5,207.80		

◀ **FIGURE 9.15**
OM Explorer Solver for Inventory Systems

Solved Problem 6

Grey Wolf Lodge is a popular 500-room hotel in the North Woods. Managers need to keep close tabs on all room service items, including a special pine-scented bar soap. The daily demand for the soap is 275 bars, with a standard deviation of 30 bars. Ordering cost is $10 and the inventory holding cost is $0.30/bar/year. The lead time from the supplier is 5 days, with a standard deviation of 1 day. The lodge is open 365 days a year.

a.   What is the economic order quantity for the bar of soap?

b.   What should the reorder point be for the bar of soap if management wants to have a 99 percent cycle-service level?

c.   What is the total annual cost for the bar of soap, assuming a Q system will be used?

SOLUTION

a.   We have $D = (275)(365) = 100{,}375$ bars of soap; $S = \$10$; and $H = \$0.30$. The EOQ for the bar of soap is

$$\text{EOQ} = \sqrt{\frac{2DS}{H}} = \sqrt{\frac{2(100{,}375)(\$10)}{\$0.30}} = \sqrt{6{,}691{,}666.7}$$

$$= 2{,}586.83, \text{ or } 2{,}587 \text{ bars}$$

b.   We have $\bar{d} = 275$ bars/day, $\sigma_d = 30$ bars, $\bar{L} = 5$ days, and $\sigma_{LT} = 1$ day.

$$\sigma_{dLT} = \sqrt{\bar{L}\sigma_d^2 + \bar{d}^2\sigma_{LT}^2} = \sqrt{(5)(30)^2 + (275)^2(1)^2} = 283.06 \text{ bars}$$

Consult the body of the Normal Distribution appendix for 0.9900, which corresponds to a 99 percent cycle-service level. The closest value is 0.9901, which corresponds to a z value of 2.33. We calculate the safety stock and reorder point as follows:

$$\text{Safety stock } = z\sigma_{dLT} = (2.33)(283.06) = 659.53, \text{ or } 660 \text{ bars}$$

$$\text{Reorder point } = \bar{d}\bar{L} + \text{Safety stock} = (275)(5) + 600 = 2{,}035 \text{ bars}$$

c.   The total annual cost for the Q system is

$$C = \frac{Q}{2}(H) + \frac{D}{Q}(S) + (H)(\text{Safety stock})$$

$$= \frac{2{,}587}{2}(\$0.30) + \frac{100{,}375}{2{,}587}(\$10) + (\$0.30)(660) = \$974.05$$

Discussion Questions

1.   What is the relationship between inventory and the nine competitive priorities we discussed in Chapter 1, "Using Operations to Compete"? Suppose that two competing manufacturers, Company H and Company L, are similar except that Company H has much higher investments in raw materials, work-in-process, and finished goods inventory than Company L. In which of the nine competitive priorities will Company H have an advantage?

2.   Suppose that a large discount retailer with a lot of purchasing power in a supply chain requires that all suppliers incorporate a new information system that will reduce the cost of placing orders between the retailer and its suppliers as well as between the suppliers and their suppliers. Suppose also that order quantities and lead times are related; the smaller the order quantity the shorter the lead time from suppliers. Assume that all members of the supply chain use a continuous review system and EOQ order quantities. Explain the implications of the new information system for the supply chain in general and the inventory systems of the supply chain members in particular.

3.   Will organizations ever get to the point where they will no longer need inventories? Why or why not?

Problems

The OM Explorer and POM for Windows software is available to all students using the 10th edition of this textbook. Go to **www.pearsonhighered.com/krajewski** to download these computer packages. If you purchased MyOMLab, you also have access to Active Models software and significant help in doing the following problems. Check with your instructor on how best to use these resources. In many cases, the instructor wants you to understand how to do the calculations by hand. At the least, the software provides a check on your calculations. When calculations are particularly complex and the goal is interpreting the results in making decisions, the software replaces entirely the manual calculations.

1. A part is produced in lots of 1,000 units. It is assembled from 2 components worth $50 total. The value added in production (for labor and variable overhead) is $60 per unit, bringing total costs per completed unit to $110. The average lead time for the part is 6 weeks and annual demand is 3,800 units, based on 50 business weeks per year.

   a. How many units of the part are held, on average, in cycle inventory? What is the dollar value of this inventory?

   b. How many units of the part are held, on average, in pipeline inventory? What is the dollar value of this inventory? (*Hint:* Assume that the typical part in pipeline inventory is 50 percent completed. Thus, half the labor and variable overhead cost has been added, bringing the unit cost to $80, or $50 + $60/2.)

2. Prince Electronics, a manufacturer of consumer electronic goods, has five distribution centers in different regions of the country. For one of its products, a high-speed modem priced at $350 per unit, the average weekly demand at *each* distribution center is 75 units. Average shipment size to each distribution center is 400 units, and average lead time for delivery is 2 weeks. Each distribution center carries 2 weeks' supply as safety stock but holds no anticipation inventory.

   a. On average, how many dollars of pipeline inventory will be in transit to each distribution center?

   b. How much total inventory (cycle, safety, and pipeline) does Prince hold for all five distribution centers?

3. Terminator, Inc., manufactures a motorcycle part in lots of 250 units. The raw materials cost for the part is $150, and the value added in manufacturing 1 unit from its components is $300, for a total cost per completed unit of $450. The lead time to make the part is 3 weeks, and the annual demand is 4,000 units. Assume 50 working weeks per year.

   a. How many units of the part are held, on average, as cycle inventory? What is its value?

   b. How many units of the part are held, on average, as pipeline inventory? What is its value?

4. Oakwood Hospital is considering using ABC analysis to classify laboratory SKUs into three categories: those that will be delivered daily from their supplier (Class A items), those that will be controlled using a continuous review system (B items), and those that will be held in a two bin system (C items). The following table shows the annual dollar usage for a sample of eight SKUs. Rank the SKUs, and assign them to their appropriate category.

SKU	Dollar Value	Annual Usage
1	$0.01	1,200
2	$0.03	120,000
3	$0.45	100
4	$1.00	44,000
5	$4.50	900
6	$0.90	350
7	$0.30	70,000
8	$1.50	200

5. Southern Markets, Inc., is considering the use of ABC analysis to focus on the most critical SKUs in its inventory. Currently, there are approximately 20,000 different SKUs with a total dollar usage of $10,000,000 per year.

   a. What would you expect to be the number of SKUs and the total annual dollar usage for A items, B items and C items at Southern Markets, Inc.?

   b. The following table provides a random sample of the unit values and annual demands of eight SKUs. Categorize these SKUs as A, B, and C items.

SKU Code	Unit Value	Demand (Units)
A104	$2.10	2,500
D205	$2.50	30
X104	$0.85	350
U404	$0.25	250
L205	$4.75	20
S104	$0.02	4,000
X205	$0.35	1,020
L104	$4.25	50

6. Yellow Press, Inc., buys paper in 1,500-pound rolls for printing. Annual demand is 2,500 rolls. The cost per roll is $800, and the annual holding cost is 15 percent of the cost. Each order costs $50 to process.

   a. How many rolls should Yellow Press, Inc., order at a time?

   b. What is the time between orders?

7. Babble, Inc., buys 400 blank cassette tapes per month for use in producing foreign language courseware. The ordering cost is $12.50. Holding cost is $0.12 per cassette per year.

   a. How many tapes should Babble, Inc., order at a time?

   b. What is the time between orders?

8. At Dot Com, a large retailer of popular books, demand is constant at 32,000 books per year. The cost of placing an order to replenish stock is $10, and the annual cost of holding is $4 per book. Stock is received 5 working days after an order has been placed. No backordering is allowed. Assume 300 working days a year.

   a. What is Dot Com's optimal order quantity?

   b. What is the optimal number of orders per year?

   c. What is the optimal interval (in working days) between orders?

   d. What is demand during the lead time?

   e. What is the reorder point?

   f. What is the inventory position immediately after an order has been placed?

9. Leaky Pipe, a local retailer of plumbing supplies, faces demand for one of its SKUs at a constant rate of 30,000 units per year. It costs Leaky Pipe $10 to process an order to replenish stock and $1 per unit per year to carry the item in stock. Stock is received 4 working days after an order is placed. No backordering is allowed. Assume 300 working days a year.

a. What is Leaky Pipe's optimal order quantity?

b. What is the optimal number of orders per year?

c. What is the optimal interval (in working days) between orders?

d. What is the demand during the lead time?

e. What is the reorder point?

f. What is the inventory position immediately after an order has been placed?

10. Sam's Cat Hotel operates 52 weeks per year, 6 days per week, and uses a continuous review inventory system. It purchases kitty litter for $11.70 per bag. The following information is available about these bags.

Demand = 90 bags/week

Order cost = $54/order

Annual holding cost = 27 percent of cost

Desired cycle-service level = 80 percent

Lead time = 3 weeks (18 working days)

Standard deviation of *weekly* demand = 15 bags

Current on-hand inventory is 320 bags, with no open orders or backorders.

a. What is the EOQ? What would be the average time between orders (in weeks)?

b. What should R be?

c. An inventory withdrawal of 10 bags was just made. Is it time to reorder?

d. The store currently uses a lot size of 500 bags (i.e., $Q = 500$). What is the annual holding cost of this policy? Annual ordering cost? Without calculating the EOQ, how can you conclude from these two calculations that the current lot size is too large?

e. What would be the annual cost saved by shifting from the 500-bag lot size to the EOQ?

11. Consider again the kitty litter ordering policy for Sam's Cat Hotel in Problem 10.

a. Suppose that the weekly demand forecast of 90 bags is incorrect and actual demand averages only 60 bags per week. How much higher will total costs be, owing to the distorted EOQ caused by this forecast error?

b. Suppose that actual demand is 60 bags but that ordering costs are cut to only $6 by using the Internet to automate order placing. However, the buyer does not tell anyone, and the EOQ is not adjusted to reflect this reduction in S. How much higher will total costs be, compared to what they could be if the EOQ were adjusted?

12. In a Q system, the demand rate for strawberry ice cream is normally distributed, with an average of 300 pints *per week*. The lead time is 9 weeks. The standard deviation of *weekly* demand is 15 pints.

a. What is the standard deviation of demand during the 9-week lead time?

b. What is the average demand during the 9-week lead time?

c. What reorder point results in a cycle-service level of 99 percent?

13. Petromax Enterprises uses a continuous review inventory control system for one of its SKUs. The following information is available on the item. The firm operates 50 weeks in a year.

Demand = 50,000 units/year

Ordering cost = $35/order

Holding cost = $2/unit/year

Average lead time = 3 weeks

Standard deviation of weekly demand = 125 units

a. What is the economic order quantity for this item?

b. If Petromax wants to provide a 90 percent cycle-service level, what should be the safety stock and the reorder point?

14. In a continuous review inventory system, the lead time for door knobs is 5 weeks. The standard deviation of demand during the lead time is 85 units. The desired cycle-service level is 99 percent. The supplier of door knobs streamlined its operations and now quotes a one-week lead time. How much can safety stock be reduced without reducing the 99 percent cycle-service level?

15. In a two-bin inventory system, the demand for three-inch lag bolts during the 2-week lead time is normally distributed, with an average of 53 units per week. The standard deviation of weekly demand is 5 units.

a. What is the probability of demand exceeding the reorder point when the normal level in the second bin is set at 130 units?

b. What is the probability of demand exceeding the 130 units in the second bin if it takes 3 weeks to receive a replenishment order?

16. Nationwide Auto Parts uses a periodic review inventory control system for one of its stock items. The review interval is 6 weeks, and the lead time for receiving the materials ordered from its wholesaler is 3 weeks. Weekly demand is normally distributed, with a mean of 100 units and a standard deviation of 20 units.

a. What is the average and the standard deviation of demand during the protection interval?

b. What should be the target inventory level if the firm desires 97.5 percent stockout protection?

c. If 350 units were in stock at the time of a periodic review, how many units should be ordered?

17. In a P system, the lead time for a box of weed-killer is 2 weeks and the review period is 1 week. Demand during the protection interval averages 218 boxes, with a standard deviation of 40 boxes.

a. What is the cycle-service level when the target inventory is set at 300 boxes?

b. In the fall season, demand for weed-killer decreases but also becomes more highly variable. Assume that during the fall season, demand during the protection interval is expected to decrease to 180 boxes, but with a standard deviation of 50 boxes. What would be the cycle-service level if management keeps the target inventory level set at 300 boxes?

18. You are in charge of inventory control of a highly successful product retailed by your firm. Weekly demand for this item varies, with an average of 200 units and a standard deviation of 16 units. It is purchased from a wholesaler at a cost of $12.50 per unit. The supply lead time is 4 weeks. Placing an order costs $50, and the inventory carrying

rate per year is 20 percent of the item's cost. Your firm operates 5 days per week, 50 weeks per year.

a. What is the optimal ordering quantity for this item?

b. How many units of the item should be maintained as safety stock for 99 percent protection against stockouts during an order cycle?

c. If supply lead time can be reduced to 2 weeks, what is the percent reduction in the number of units maintained as safety stock for the same 99 percent stockout protection?

d. If through appropriate sales promotions, the demand variability is reduced so that the standard deviation of weekly demand is 8 units instead of 16, what is the percent reduction (compared to that in part [b]) in the number of units maintained as safety stock for the same 99 percent stockout protection?

19. Suppose that Sam's Cat Hotel in Problem 10 uses a P system instead of a Q system. The average daily demand is $\bar{d} = 90/6 = 15$ bags. and the standard deviation of

$daily$ demand is $\sigma_d = \dfrac{\sigma_{week}}{\sqrt{6}} = (15/\sqrt{6}) = 6.124$ bags.

a. What P (in working days) and T should be used to approximate the cost trade-offs of the EOQ?

b. How much more safety stock is needed than with a Q system?

c. It is time for the periodic review. How much kitty litter should be ordered?

20. Your firm uses a continuous review system and operates 52 weeks per year. One of the SKUs has the following characteristics.

Demand $(D) = 20,000$ units/year

Ordering cost $(S) = \$40/$order

Holding cost $(H) = \$2/$unit/year

Lead time $(L) = 2$ weeks

Cycle-service level $= 95$ percent

Demand is normally distributed, with a standard deviation of $weekly$ demand of 100 units.

Current on-hand inventory is 1,040 units, with no scheduled receipts and no backorders.

a. Calculate the item's EOQ. What is the average time, in weeks, between orders?

b. Find the safety stock and reorder point that provide a 95 percent cycle-service level.

c. For these policies, what are the annual costs of (i) holding the cycle inventory and (ii) placing orders?

d. A withdrawal of 15 units just occurred. Is it time to reorder? If so, how much should be ordered?

21. Your firm uses a periodic review system for all SKUs classified, using ABC analysis, as B or C items. Further, it uses a continuous review system for all SKUs classified as A items. The demand for a specific SKU, currently classified as an A item, has been dropping. You have been asked to evaluate the impact of moving the item from continuous review to periodic review. Assume your firm operates 52 weeks per year; the item's current characteristics are:

Demand $(D) = 15,080$ units/year

Ordering cost $(S) = \$125.00/$order

Holding cost $(H) = \$3.00/$unit/year

Lead time $(L) = 5$ weeks

Cycle-service level $= 95$ percent

Demand is normally distributed, with a standard deviation of weekly demand of 64 units.

a. Calculate the item's EOQ.

b. Use the EOQ to define the parameters of an appropriate continuous review and periodic review system for this item.

c. Which system requires more safety stock and by how much?

22. A company begins a review of ordering policies for its continuous review system by checking the current policies for a sample of SKUs. Following are the characteristics of one item.

Demand $(D) = 64$ units/week (Assume 52 weeks per year)

Ordering and setup cost $(S) = \$50/$order

Holding cost $(H) = \$13/$unit/year

Lead time $(L) = 2$ weeks

Standard deviation of $weekly$ demand $= 12$ units

Cycle-service level $= 88$ percent

a. What is the EOQ for this item?

b. What is the desired safety stock?

c. What is the reorder point?

d. What are the cost implications if the current policy for this item is $Q = 200$ and $R = 180$?

23. Using the same information as in Problem 22, develop the best policies for a periodic review system.

a. What value of P gives the same approximate number of orders per year as the EOQ? Round to the nearest week.

b. What safety stock and target inventory level provide an 88 percent cycle-service level?

24. Wood County Hospital consumes 1,000 boxes of bandages per week. The price of bandages is $35 per box, and the hospital operates 52 weeks per year. The cost of processing an order is $15, and the cost of holding one box for a year is 15 percent of the value of the material.

a. The hospital orders bandages in lot sizes of 900 boxes. What $extra$ cost does the hospital incur, which it could save by using the EOQ method?

b. Demand is normally distributed, with a standard deviation of weekly demand of 100 boxes. The lead time is 2 weeks. What safety stock is necessary if the hospital uses a continuous review system and a 97 percent cycle-service level is desired? What should be the reorder point?

c. If the hospital uses a periodic review system, with $P = 2$ weeks, what should be the target inventory level, T?

25. A golf specialty wholesaler operates 50 weeks per year. Management is trying to determine an inventory policy for its 1-irons, which have the following characteristics:

Demand $(D) = 2,000$ units/year

Demand is normally distributed

Standard deviation of $weekly$ demand $= 3$ units

Ordering cost = $40/order

Annual holding cost $(H) = $5/units

Desired cycle-service level = 90 percent

Lead time $(L) = 4$ weeks

a. If the company uses a periodic review system, what should P and T be? Round P to the nearest week.

b. If the company uses a continuous review system, what should R be?

26. Osprey Sports stocks everything that a musky fisherman could want in the Great North Woods. A particular musky lure has been very popular with local fishermen as well as those who buy lures on the Internet from Osprey Sports. The cost to place orders with the supplier is $30/order; the demand averages 4 lures per day, with a standard deviation of 1 lure; and the inventory holding cost is $1.00/lure/year. The lead time form the supplier is 10 days, with a standard deviation of 3 days. It is important to maintain a 97 percent cycle-service level to properly balance service with inventory holding costs. Osprey Sports is open 350 days a year to allow the owners the opportunity to fish for muskies during the prime season. The owners want to use a continuous review inventory system for this item.

a. What order quantity should be used?

b. What reorder point should be used?

c. What is the total annual cost for this inventory system?

27. The Farmer's Wife is a country store specializing in knickknacks suitable for a farm-house décor. One item experiencing a considerable buying frenzy is a miniature Holstein cow. Average weekly demand is 30 cows, with a standard deviation of 5 cows. The cost to place a replenishment order is $15 and the holding cost is $0.75/cow/year. The supplier, however, is in China. The lead time for new orders is 8 weeks, with a standard deviation of 2 weeks. The Farmer's Wife, which is open only 50 weeks a year, wants to develop a continuous review inventory system for this item with a cycle-service level of 90 percent.

a. Specify the continuous review system for the cows. Explain how it would work in practice.

b. What is the total annual cost for the system you developed?

Advanced Problems

It may be helpful to review MyOMLab Supplement E, "Simulation," before working Problem 29.

28. Muscle Bound is a chain of fitness stores located in many large shopping centers. Recently, an internal memo from the CEO to all operations personnel complained about the budget overruns at Muscle Bound's central warehouse. In particular, she said that inventories were too high and that the budget will be cut dramatically and proportionately equal for all items in stock. Consequently, warehouse management set up a pilot study to see what effect the budget cuts would have on customer service. They chose 5-pound barbells, which are a high volume SKU and consume considerable warehouse space. Daily demand for the barbells is 1,000 units, with a standard deviation of 150 units. Ordering costs are $40 per order. Holding costs are $2/unit/year. The supplier is located in the Philippines; consequently, the lead time is 35 days with a standard deviation of 5 days. Muscle Bound stores operate 313 days a year (no Sundays).

Suppose that the barbells are allocated a budget of $16,000 for total annual costs. If Muscle Bound uses a continuous review system for the barbells and cannot change the ordering costs and holding costs or the distributions of demand or lead time, what is the best cycle-service level management can expect from their system?

29. The Georgia Lighting Center stocks more than 3,000 lighting fixtures, including chandeliers, swags, wall lamps, and track lights. The store sells at retail, operates six days per week, and advertises itself as the "brightest spot in town." One expensive fixture is selling at an average rate of 5 units per day. The reorder policy is $Q = 40$ and $R = 15$. A new order is placed on the day the reorder point is reached. The lead time is 3 business days. For example, an order placed on Monday will be delivered on Thursday. Simulate the performance of this Q system for the next 3 weeks (18 workdays). Any stockouts result in lost sales (rather than backorders). The beginning inventory is 19 units, and no receipts are scheduled. Table 9.2 simulates the first week of operation. Extend Table 9.2 to simulate operations for the next 2 weeks if demand for the next 12 business days is 7, 4, 2, 7, 3, 6, 10, 0, 5, 10, 4, and 7.

a. What is the average daily ending inventory over the 18 days? How many stockouts occurred?

b. Simulate the inventory performance of the same item assuming a $Q = 30$, $R = 20$ system is used. Calculate the average inventory level and number of stockouts and compare with part (a).

TABLE 9.2 | **FIRST WEEK OF OPERATION**

Workday	Beginning Inventory	Orders Received	Daily Demand	Ending Inventory	Inventory Position	Order Quantity
1. Monday	19	—	5	14	14	40
2. Tuesday	14	—	3	11	51	—
3. Wednesday	11	—	4	7	47	—
4. Thursday	7	40	1	46	46	—
5. Friday	46	—	10	36	36	—
6. Saturday	36	—	9	27	27	—

Active Model Exercise

This Active Model appears in MyOMLab. It allows you to evaluate the sensitivity of the EOQ and associated costs to changes in the demand and cost parameters.

QUESTIONS

1. What is the EOQ and what is the lowest total cost?

2. What is the annual cost of holding inventory at the EOQ and the annual cost of ordering inventory at the EOQ?

3. From the graph, what can you conclude about the relationship between the lowest total cost and the costs of ordering and holding inventory?

4. How much does the total cost increase if the store manager orders twice as many bird feeders as the EOQ? How much does the total cost increase if the store manager orders half as many bird feeders as the EOQ?

5. What happens to the EOQ and the total cost when demand is doubled? What happens to the EOQ and the total cost when unit price is doubled?

6. Scroll through the lower order cost values and describe the changes to the graph. What happens to the EOQ?

7. Comment on the sensitivity of the EOQ model to errors in demand or cost estimates.

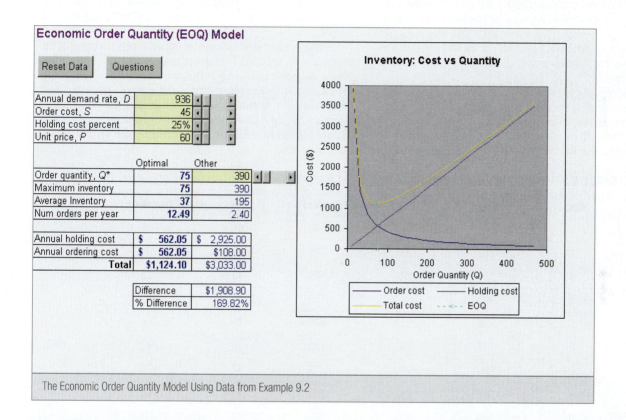

The Economic Order Quantity Model Using Data from Example 9.2

EXPERIENTIAL LEARNING   Swift Electronic Supply, Inc.

It was a typical fall afternoon in Southern California, with thousands of tourists headed to the beaches to have fun. About 40 miles away, however, Steven Holland, the CEO of the Swift Electronic Supply, Inc., faced a severe problem with Swift's inventory management.

An Intel veteran, Steven Holland worked in the electronic components distribution industry for more than 20 years. Seven years ago, he founded the Swift Electronic Supply, Inc., an electronic distributor. After several successful years, the company is now troubled with eroding profit margins. Recent economic downturns further worsened the situation. Factors such as the growth of B2B e-commerce, the globalization of markets, the increased popularity of value-added services, and ongoing consolidations among electronic distributors affect the future of Swift.

To reverse these influences, Holland talked to a prestigious local university. After consultation, Holland found the most effective way to increase profitability is to cut inventory costs. As a starting point, he studied in detail a representative product, dynamic random access memory (DRAM), as the basis for his plan.

Industry and Company Preview

Owing to a boom in the telecommunications industry and the information technology revolution, electronics distributors experienced double-digit annual growth over the last decade. To cut the cost of direct purchasing forces, large component manufacturers such as Intel, Cisco, and Texas Instruments decided to outsource their procurement so that they could focus on product development and manufacturing. Therefore, independent electronic distributors like Swift started offering procurement services to these companies.

Swift serves component manufacturers in California and Arizona. Working as the intermediary between its customers and overseas original equipment manufacturers (OEMs), Swift's business model is quite simple. Forecasting customer demand, Swift places orders to a number of OEMs, stocks those products, breaks the quantities down, and delivers the products to its end customers.

Recently, due to more intense competition and declines in demand, Swift offered more flexible delivery schedules and was willing to accommodate small order quantities. However, customers can always shift to Swift's competitors should Swift not fulfill their orders. Steven Holland was in a dilemma: The intangible costs of losing customers can be enormous; however, maintaining high levels of inventory can also be costly.

Dram

Holland turned his attention to DRAM as a representative product. Previously, the company ordered a large amount every time it felt it was necessary. Holland's assistant developed a table (Table 9.3) that has 2 months of demand history. From Holland's experience, the demand for DRAM is relatively stable in the company's product line and it had no sales seasonality. The sales staff agrees that conditions in the current year will not be different from those of past years, and historical demand will be a good indicator of what to expect in the future.

The primary manufacturers of DRAM are those in Southeast Asia. Currently, Swift can purchase one unit of 128M DRAM for $10. After negotiation with a reputable supplier, Holland managed to sign a long-term agreement, which kept the price at $10 and allowed Swift to place orders at any time. The supplier also supplies other items in Swift's inventory. In addition, it takes the supplier of the DRAM 2 days to deliver the goods to Swift's warehouse using air carriers.

When Swift does not have enough inventory to fill a customer's order, the sales are lost; that is, Swift is not able to backorder the shortage because its customers fill their requirements through competitors. The customers will accept partial shipments, however.

It costs Swift $200 to place an order with the suppliers. This amount covers the corresponding internal ordering costs and the costs of delivering the products to the company. Holland estimates that the cost of lost sales amounts to $2 per unit of DRAM. This rough estimate includes the loss of profits, as well as the intangible damage to customer goodwill.

To simplify its inventory management system, Swift has a policy of maintaining a cycle-service level of 95 percent. The holding cost per day per unit is estimated to be 0.5 percent of the cost of goods, regardless of the product. Inventory holding costs are calculated on the basis of the ending inventory each day. The current balance is 1,700 units of DRAM in stock.

The daily purchasing routine is as follows. Orders are placed at the *beginning* of the day, before Swift is open for customer business. The orders arrive at the beginning of the day, 2 days later, and can be used for sales that day. For example, an order placed at the beginning of day 1 will arrive at Swift before Swift is open for business on day 3. The actual daily demand is always recorded at the *end* of the day, after Swift has closed for customer business. All cost computations are done at the end of the day after the total demand has been recorded.

TABLE 9.3 | **HISTORICAL DEMAND DATA FOR THE DRAM (UNITS)**

Day	Demand	Day	Demand	Day	Demand
1	869	21	663	41	959
2	902	22	1,146	42	703
3	1,109	23	1,016	43	823
4	947	24	1,166	44	862
5	968	25	829	45	966
6	917	26	723	46	1,042
7	1,069	27	749	47	889
8	1,086	28	766	48	1,002
9	1,066	29	996	49	763
10	929	30	1,122	50	932
11	1,022	31	962	51	1,052
12	959	32	829	52	1,062
13	756	33	862	53	989
14	882	34	793	54	1,029
15	829	35	1,039	55	823
16	726	36	1,009	56	942
17	666	37	979	57	986
18	879	38	976	58	736
19	1,086	39	856	59	1,009
20	992	40	1,036	60	852

Simulation

Holland believes that simulation is a useful approach to assess various inventory control alternatives. The historical data from Table 9.3 could be used to develop attractive inventory policies. The table was developed to record various costs and evaluate different alternatives. An example showing some recent DRAM inventory decisions is shown in Table 9.4.

1. Design a new inventory system for Swift Electronic Supply, Inc., using the data provided.

2. Provide the rationale for your system, which should include the decision rules you would follow to determine how much to order and when.

3. Simulate the use of your inventory system and record the costs. Develop a table such as Table 9.4 to record your results. Your instructor will provide actual demands on a day-to-day basis during the simulation.

TABLE 9.4 | EXAMPLE SIMULATION

Day	1	2	3	4	5	6	7	8	9	10
Beginning inventory position	1,700	831	1,500	391	3,000	3,232	2,315			
Number ordered	1,500		3,000	1,200			1,900			
Daily demand	869	902	1,109	947	968	917	1,069			
Day-ending inventory	831	−71	391	−556	2,032	2,315	1,246			
Ordering costs ($200 per order)	200		200	200			200			
Holding costs ($0.05 per piece per day)	41.55	0.00	19.55	0.00	101.60	115.75	62.30			
Shortage costs ($2 per piece)	0	142	0	1,112	0	0	0			
Total cost for day	241.55	142.00	219.55	1,312.00	101.60	115.75	262.30			
Cumulative cost from last day	0.00	241.55	383.55	603.10	1,915.10	2,016.70	2,132.45			
Cumulative costs to date	241.55	383.55	603.10	1,915.10	2,016.70	2,132.45	2,394.75			

CASE   Parts Emporium

Parts Emporium, Inc., is a wholesale distributor of automobile parts formed by two disenchanted auto mechanics, Dan Block and Ed Spriggs. Originally located in Block's garage, the firm showed slow but steady growth for 7 years before it relocated to an old, abandoned meat-packing warehouse on Chicago's South Side. With increased space for inventory storage, the company was able to begin offering an expanded line of auto parts. This increased selection, combined with the trend toward longer car ownership, led to an explosive growth of the business. Fifteen years later, Parts Emporium was the largest independent distributor of auto parts in the north central region.

Recently, Parts Emporium relocated to a sparkling new office and warehouse complex off Interstate 55 in suburban Chicago. The warehouse space alone occupied more than 100,000 square feet. Although only a handful of new products have been added since the warehouse was constructed, its utilization increased from 65 percent to more than 90 percent of capacity. During this same period, however, sales growth stagnated. These conditions motivated Block and Spriggs to hire the first manager from outside the company in the firm's history.

It is June 6, Sue McCaskey's first day in the newly created position of materials manager for Parts Emporium. A recent graduate of a prominent business school, McCaskey is eagerly awaiting her first real-world problem. At approximately 8:30 A.M., it arrives in the form of status reports on inventory and orders shipped. At the top of an extensive computer printout is a handwritten note from Joe Donnell, the purchasing manager: "Attached you will find the inventory and customer service performance data. Rest assured that the individual inventory levels are accurate because we took a complete physical inventory count at the end of last week. Unfortunately, we do not keep compiled records in some of the areas as you requested. However, you are welcome to do so yourself. Welcome aboard!"

A little upset that aggregate information is not available, McCaskey decides to randomly select a small sample of approximately 100 items and compile inventory and customer service characteristics to get a feel for the "total picture." The results of this experiment reveal to her why Parts Emporium decided to create the position she now fills. It seems that the inventory is in all the wrong places. Although an *average* of approximately 60 days of inventory is on hand, the firm's customer service is inadequate. Parts Emporium tries to backorder the customer orders not immediately filled from stock, but some 10 percent of demand is being lost to competing distributorships. Because stockouts are costly, relative to inventory holding costs, McCaskey believes that a cycle-service level of at least 95 percent should be achieved.

McCaskey knows that although her influence to initiate changes will be limited, she must produce positive results immediately. Thus, she decides to concentrate on two products from the extensive product line: the EG151 exhaust gasket and the DB032 drive belt. If she can demonstrate significant gains from proper inventory management for just two products, perhaps Block and Spriggs will give her the backing needed to change the total inventory management system.

The EG151 exhaust gasket is purchased from an overseas supplier, Haipei, Inc. Actual demand for the first 21 weeks of this year is shown in the following table:

Week	Actual Demand	Week	Actual Demand
1	104	12	97
2	103	13	99
3	107	14	102
4	105	15	99
5	102	16	103
6	102	17	101
7	101	18	101
8	104	19	104
9	100	20	108
10	100	21	97
11	103		

A quick review of past orders, shown in another document, indicates that a lot size of 150 units is being used and that the lead time from Haipei is fairly constant at 2 weeks. Currently, at the end of week 21, no inventory is on hand, 11 units are backordered, and the company is awaiting a scheduled receipt of 150 units.

The DB032 drive belt is purchased from the Bendox Corporation of Grand Rapids, Michigan. Actual demand so far this year is shown in the following table:

Week	Actual Demand	Week	Actual Demand
11	18	17	50
12	33	18	53
13	53	19	54
14	54	20	49
15	51	21	52
16	53		

Because this product is new, data are available only since its introduction in week 11. Currently, 324 units are on hand, with no backorders and no scheduled receipts. A lot size of 1,000 units is being used, with the lead time fairly constant at 3 weeks.

The wholesale prices that Parts Emporium charges its customers are $12.99 for the EG151 exhaust gasket and $8.89 for the DB032 drive belt. Because no quantity discounts are offered on these two highly profitable items, gross margins based on current purchasing practices are 32 percent of the wholesale price for the exhaust gasket and 48 percent of the wholesale price for the drive belt.

Parts Emporium estimates its cost to hold inventory at 21 percent of its inventory investment. This percentage recognizes the opportunity cost of tying money up in inventory and the variable costs of taxes, insurance, and shrinkage. The annual report notes other warehousing expenditures for utilities and maintenance and debt service on the 100,000-square-foot warehouse, which was built for $1.5 million. However, McCaskey reasons that these warehousing costs can be ignored because they will not change for the range of inventory policies that she is considering.

Out-of-pocket costs for Parts Emporium to place an order with suppliers are estimated to be $20 per order for exhaust gaskets and $10 per order for drive belts. On the outbound side, the company can charge a delivery fee. Although most customers pick up their parts at Parts Emporium, some orders are delivered to customers. To provide this service, Parts Emporium contracts with a local company for a flat fee of $21.40 per order, which is added to the customer's bill. McCaskey is unsure whether to increase the ordering costs for Parts Emporium to include delivery charges.

QUESTIONS

1. Put yourself in Sue McCaskey's position and prepare a detailed report to Dan Block and Ed Spriggs on managing the inventory of the EG151 exhaust gasket and the DB032 drive belt. Be sure to present a proper inventory system and recognize all relevant costs.

2. By how much do your recommendations for these two items reduce annual cycle inventory, stockout, and ordering costs?

Selected References

Arnold, Tony J.R., Stephen Chapman, and Lloyd M. Clive. *Introduction to Materials Management*, 7th ed. Upper Saddle River, NJ: Prentice Hall, 2012.

Axsäter, Sven. *Inventory Control*, 2nd ed. New York: Springer Science + Business Media, LLC, 2006.

Bastow, B. J. "Metrics in the Material World." *APICS—The Performance Advantage* (May 2005), pp. 49–52.

Benton, W.C. *Purchasing and Supply Chain Management*, 2nd ed. New York: McGraw-Hill, 2010.

Callioni, Gianpaolo, Xavier de Montgros, Regine Slagmulder, Luk N. Van Wassenhove, and Linda Wright. "Inventory-Driven Costs." *Harvard Business Review* (March 2005), pp. 135–141.

Cannon, Alan R., and Richard E. Crandall. "The Way Things Never Were." *APICS—The Performance Advantage* (January 2004), pp. 32–35.

Hartvigsen, David. *SimQuick: Process Simulation with Excel*, 2nd ed. Upper Saddle River, NJ: Prentice Hall, 2004.

Operations Management Body of Knowledge. Falls Church, VA: American Production and Inventory Control Society, 2009.

Timme, Stephen G., and Christine Williams-Timme. "The Real Cost of Holding." *Supply Chain Management Review* (July/August 2003), pp. 30–37.

Walters, Donald. *Inventory Control and Management*, 2nd ed. West Sussex, England: John Wiley and Sons, Ltd, 2003.

Many real world problems require relaxation of certain assumptions on which the economic order quantity (EOQ) model is based. This supplement addresses three realistic situations that require going beyond the simple EOQ formulation.

1. Noninstantaneous Replenishment. Particularly in situations in which manufacturers use a continuous process to make a primary material, such as a liquid, gas, or powder, production is not instantaneous. Thus, inventory is replenished gradually, rather than in lots.

2. Quantity Discounts. Three annual costs are (1) the inventory holding cost, (2) the fixed cost for ordering and setup, and (3) the cost of materials. For service providers and for manufacturers alike, the unit cost of purchased materials sometimes depends on the order quantity.

3. One-Period Decisions. Retailers and manufacturers of fashion goods often face a situation in which demand is uncertain and occurs during just one period or season.

This supplement assumes you have read Chapter 9, "Supply Chain Inventory Management," and Supplement A, "Decision Making."

Noninstantaneous Replenishment

If an item is being produced internally rather than purchased, finished units may be used or sold as soon as they are completed, without waiting until a full lot is completed. For example, a restaurant that bakes its own dinner rolls begins to use some of the rolls from the first pan even before the baker finishes a five-pan batch. The inventory of rolls never reaches the full five-pan level, the way it would if the rolls all arrived at once on a truck sent by a supplier.

Figure C.1 depicts the usual case, in which the production rate, p, *exceeds* the demand rate, d. If demand and production were equal, manufacturing would be continuous with no buildup of cycle inventory. If the production rate is lower than the demand rate, sales opportunities are being missed on an ongoing basis. We assume that $p > d$ in this supplement.

LEARNING GOALS *After reading this supplement, you should be able to:*

1 Identify the situations where the economic lot size should be used rather than the economic order quantity.

2 Calculate the optimal lot size when replenishment is not instantaneous.

3 Define the relevant costs that should be considered to determine the order quantity when discounts are available.

4 Determine the optimal order quantity when materials are subject to quantity discounts.

5 Calculate the order quantity that maximizes the expected profits for a one-period inventory decision.

FIGURE C.1 ▶

Lot Sizing with Noninstantaneous
Replenishment

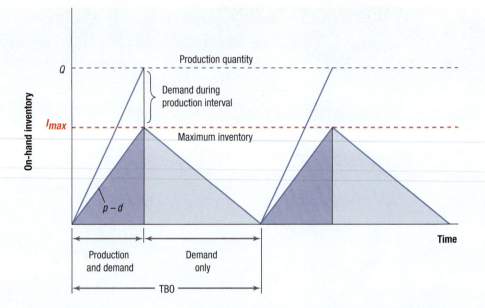

Ulrich Mueller/Shutterstock.com

This chemical plant stores its products in stainless steel silos. The production of each product is scheduled to start when its silo is nearly empty.

Cycle inventory accumulates faster than demand occurs; that is, a buildup of $p - d$ units occurs per time period. For example, if the production rate is 100 units per day and the demand is 5 units per day, the buildup is 95 (or 100 – 5) units each day. This buildup continues until the lot size, Q, has been produced, after which the inventory depletes at a rate of 5 units per day. Just as the inventory reaches 0, the next production interval begins. To be consistent, both p and d must be expressed in units of the same time period, such as units per day or units per week. Here, we assume that they are expressed in units per day.

The $p - d$ buildup continues for Q/p days because Q is the lot size and p units are produced each day. In our example, if the lot size is 300 units, the production interval is 3 days (300/100). For the given rate of buildup over the production interval, the maximum cycle inventory, I_{max}, is

$$I_{max} = \frac{Q}{p}(p - d) = Q\left(\frac{p - d}{p}\right)$$

Cycle inventory is no longer $Q/2$, as it was with the basic EOQ method; instead, it is $I_{max}/2$. Setting up the total annual cost equation for this production situation, where D is annual demand, as before, and d is daily demand, we get

Total annual cost = Annual holding cost + Annual ordering or setup cost

$$C = \frac{I_{max}}{2}(H) + \frac{D}{Q}(S) = \frac{Q}{2}\left(\frac{p - d}{p}\right)(H) + \frac{D}{Q}(S)$$

economic production lot size (ELS)

The optimal lot size in a situation in which replenishment is not instantaneous.

Based on this cost function, the optimal lot size, often called the **economic production lot size (ELS)**, is

$$ELS = \sqrt{\frac{2D\,S}{H}}\sqrt{\frac{p}{p - d}}$$

Because the second term is a ratio greater than 1, the ELS results in a larger lot size than the EOQ.

| **EXAMPLE C.1** | **Finding the Economic Production Lot Size** |

A plant manager of a chemical plant must determine the lot size for a particular chemical that has a steady demand of 30 barrels per day. The production rate is 190 barrels per day, annual demand is 10,500 barrels, setup cost is $200, annual holding cost is $0.21 per barrel, and the plant operates 350 days per year.

a. Determine the economic production lot size (ELS).

b. Determine the total annual setup and inventory holding cost for this item.

c. Determine the time between orders (TBO), or cycle length, for the ELS.

d. Determine the production time per lot.

What are the advantages of reducing the setup time by 10 percent?

MyOMLab

Tutor C.1 in MyOMLab provides a new example to determine the ELS.

MyOMLab

Active Model C.1 in MyOMLab provides additional insight on the ELS model and its uses.

SOLUTION

a. Solving first for the ELS, we get

$$ELS = \sqrt{\frac{2DS}{H}} \sqrt{\frac{p}{p-d}} = \sqrt{\frac{2(10,500)(\$200)}{\$0.21}} \sqrt{\frac{190}{190-30}}$$

$$= 4,873.4 \text{ barrels}$$

b. The total annual cost with the ELS is

$$C = \frac{Q}{2}\left(\frac{p-d}{p}\right)(H) + \frac{D}{Q}(S)$$

$$= \frac{4,873.4}{2}\left(\frac{190-30}{190}\right)(\$0.21) + \frac{10,500}{4,873.4}(\$200)$$

$$= \$430.91 + \$430.91 = \$861.82$$

c. Applying the TBO formula to the ELS, we get

$$TBO_{ELS} = \frac{ELS}{D}(350 \text{ days/year}) = \frac{4,873.4}{10,500}(350)$$

$$= 162.4, \quad \text{or} \quad 162 \text{ days}$$

d. The production time during each cycle is the lot size divided by the production rate:

$$\frac{ELS}{p} = \frac{4,873.4}{190} = 25.6, \quad \text{or} \quad 26 \text{ days}$$

DECISION POINT

As OM Explorer shows in Figure C.2, the net effect of reducing the setup cost by 10 percent is to reduce the lot size, the time between orders, and the production cycle time. Consequently, total annual costs are also reduced. This adds flexibility to the manufacturing process because items can be made more quickly with less expense. Management must decide whether the added cost of improving the setup process is worth the added flexibility and inventory cost reductions.

Period Used in Calculations	Day ▼
Demand per Day	30
Production Rate/Day	190
Annual Demand	10,500
Setup Cost	$180
Annual Holding Cost ($)	$0.21
Operating Days per Year	350

◉ Enter Holding Cost Manually   ○ Holding Cost As % of Value

Economic Lot Size (ELS)	4,623
Annual Total Cost	$817.60
Time Between Orders (days)	154.1
Production Time	24.3

◀ **FIGURE C.2**

OM Explorer Solver for the Economic Production Lot Size Showing the Effect of a 10 Percent Reduction in Setup Cost

Many hospitals join cooperatives (or co-ops) to gain the clout needed to garner price discounts from suppliers. Here a hospital pharmacist checks inventory records of supplies.

Quantity Discounts

Quantity discounts, which are price incentives to purchase large quantities, create pressure to maintain a large inventory. For example, a supplier may offer a price of $4.00 per unit for orders between 1 and 99 units, a price of $3.50 per unit for orders between 100 and 199 units, and a price of $3.00 per unit for orders of 200 or more units. The item's price is no longer fixed, as assumed in the EOQ derivation; instead, if the order quantity is increased enough, the price is discounted. Hence, a new approach is needed to find the best lot size—one that balances the advantages of lower prices for purchased materials and fewer orders (which are benefits of large order quantities) against the disadvantage of the increased cost of holding more inventory.

The total annual cost now includes not only the holding cost, $(Q/2)(H)$, and the ordering cost, $(D/Q)(S)$, but also the cost of purchased materials. For any per-unit price level, P, the total cost is

$$\text{Total annual cost} = \text{Annual holding cost} + \text{Annual ordering or setup cost}$$
$$+ \text{Annual cost of materials}$$

$$C = \frac{Q}{2}(H) + \frac{D}{Q}(S) + PD$$

The unit holding cost, H, usually is expressed as a percent of the unit price because the more valuable the item held in inventory, the higher the holding cost is. Thus, the lower the unit price, P, the lower H is. Conversely, the higher P is, the higher H is.

The total cost equation yields U-shaped total cost curves. Adding the annual cost of materials to the total cost equation raises each total cost curve by a fixed amount, as shown in Figure C.3 (a). The three cost curves illustrate each of the price levels. The top curve applies when no discounts are received; the lower curves reflect the discounted price levels. No single curve is relevant to all purchase quantities. The relevant, or feasible, total cost begins with the top curve, then drops down, curve by curve, at the price breaks. A price break is the minimum quantity needed to get a discount. In Figure C.3, two price breaks occur at $Q = 100$ and $Q = 200$. The result is a total cost curve, with steps at the price breaks.

▼ **FIGURE C.3**

Total Cost Curves with Quantity Discounts

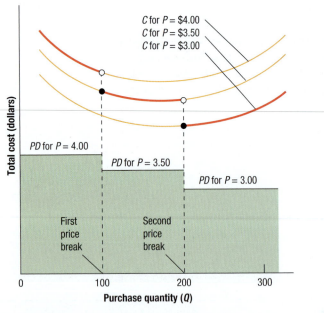

(a) Total cost curves with purchased materials added

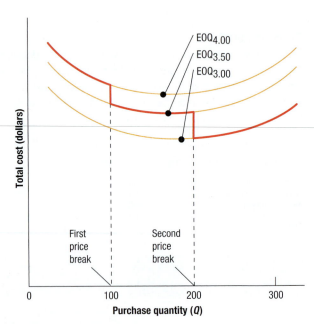

(b) EOQs and price break quantities

Figure C.3 (b) also shows three additional points—the minimum point on each curve—obtained with the EOQ formula at each price level. These EOQs do not necessarily produce the best lot size for two reasons.

1. The EOQ at a particular price level may not be feasible. The lot size may not lie in the range corresponding to its per-unit price. Figure C.3 (b) illustrates two instances of an infeasible EOQ. First, the minimum point for the $3.00 curve appears to be fewer than 200 units. However, the supplier's quantity discount schedule does not allow purchases of that small a quantity at the $3.00 unit price. Similarly, the EOQ for the $4.00 price level is greater than the first price break, so the price charged would be only $3.50.

2. The EOQ at a particular price level may be feasible but may not be the best lot size. The feasible EOQ may have a higher cost than is achieved by the EOQ or price break quantity on a lower price curve. In Figure C.3 (b), for example, the 200-unit price break quantity for the $3.00 price level has a lower total cost than the feasible EOQ for the $3.50 price level. A feasible EOQ always is better than any feasible point on cost curves with higher price levels, but not necessarily those with lower levels. Thus, the only time we can immediately conclude, without comparing total costs, that a feasible EOQ is the best order quantity is when it is on the curve for the lowest price level. This conclusion is not possible in Figure C.3 (b) because the only feasible EOQ is at the middle price level, $P = \$3.50$.

We must, therefore, pay attention only to feasible price–quantity combinations, shown as solid lines in Figure C.3 (b), as we search for the best lot size. The following two-step procedure may be used to find the best lot size.

Step 1. Beginning with the lowest price, calculate the EOQ for each price level until a feasible EOQ is found. It is feasible if it lies in the range corresponding to its price. Each subsequent EOQ is smaller than the previous one because P, and thus H, gets larger and because the larger H is in the denominator of the EOQ formula.

Step 2. If the first feasible EOQ found is for the lowest price level, this quantity is the best lot size. Otherwise, calculate the total cost for the first feasible EOQ and for the larger price break quantity at each lower price level. The quantity with the lowest total cost is optimal.

EXAMPLE C.2	**Finding Q with Quantity Discounts at St. LeRoy Hospital**

A supplier for St. LeRoy Hospital has introduced quantity discounts to encourage larger order quantities of a special catheter. The price schedule is

Order Quantity	Price per Unit
0 to 299	$60.00
300 to 499	$58.80
500 or more	$57.00

The hospital estimates that its annual demand for this item is 936 units, its ordering cost is $45.00 per order, and its annual holding cost is 25 percent of the catheter's unit price. What quantity of this catheter should the hospital order to minimize total costs? Suppose the price for quantities between 300 and 499 is reduced to $58.00. Should the order quantity change?

MyOMLab

Tutor C.2 in MyOMLab provides a new example for choosing the best order quantity when discounts are available.

MyOMLab

Active Model C.2 in MyOMLab provides additional insight on the quantity discount model and its uses.

SOLUTION

Step 1. Find the first feasible EOQ, starting with the lowest price level:

$$EOQ_{57.00} = \sqrt{\frac{2DS}{H}} = \sqrt{\frac{2(936)(\$45.00)}{0.25(\$57.00)}} = 77 \text{ units}$$

A 77-unit order actually costs $60.00 per unit, instead of the $57.00 per unit used in the EOQ calculation, so this EOQ is infeasible. Now, try the $58.80 level:

$$EOQ_{58.80} = \sqrt{\frac{2DS}{H}} = \sqrt{\frac{2(936)(\$45.00)}{0.25(\$58.80)}} = 76 \text{ units}$$

This quantity also is infeasible because a 76-unit order is too small to qualify for the $58.80 price. Try the highest price level:

$$EOQ_{60.00} = \sqrt{\frac{2DS}{H}} = \sqrt{\frac{2(936)(\$45.00)}{0.25(\$60.00)}} = 75 \text{ units}$$

This quantity is feasible because it lies in the range corresponding to its price, $P = \$60.00$.

Step 2. The first feasible EOQ of 75 does not correspond to the lowest price level. Hence, we must compare its total cost with the price break quantities (300 and 500 units) at the lower price levels ($58.80 and $57.00):

$$C = \frac{Q}{2}(H) + \frac{D}{Q}(S) + PD$$

$$C_{75} = \frac{75}{2}[(0.25)(\$60.00)] + \frac{936}{75}(\$45.00) + \$60.00(936) = \$57,284$$

$$C_{300} = \frac{300}{2}[(0.25)(\$58.80)] + \frac{936}{300}(\$45.00) + \$58.80(936) = \$57,382$$

$$C_{500} = \frac{500}{2}[(0.25)(\$57.00)] + \frac{936}{500}(\$45.00) + \$57.00(936) = \$56,999$$

The best purchase quantity is 500 units, which qualifies for the deepest discount.

DECISION POINT

If the price per unit for the range of 300 to 499 units is reduced to $58.00, the best decision is to order 300 catheters, as shown by OM Explorer in Figure C.4. This result shows that the decision is sensitive to the price schedule. A reduction of slightly more than 1 percent is enough to make the difference in this example. In general, however, it is not always the case that you should order more than the economic order quantity when given price discounts. When discounts are small, holding cost H is large, and demand D is small, small lot sizes are better even though price discounts are forgone.

FIGURE C.4 ▶

OM Explorer Solver for Quantity Discounts Showing the Best Order Quantity

	More	Fewer
Min. Amount Req'd for Price Point	Lot Sizes	Price/Unit
...	0–299	$60.00
300	300–499	$58.00
500	500 or more	$57.00

Annual Demand	936
Order Cost	$45
Holding Cost (% or price)	25%

Best Order Quantity	300

Price Point	EOQ or Req'd Order for Price Point	Inventory Cost	Order Cost	Purchase Cost	Total Cost
$60.00	75	$562.50	$561.60	$56,160	$57,284
$58.00	300	$2,175	$140.40	$54,288	$56,603
$57.00	500	$3,563	$84.24	$53,352	$56,999

One-Period Decisions

One of the dilemmas facing many retailers is how to handle seasonal goods, such as winter coats. Often, they cannot be sold at full markup the next year because of changes in styles. Furthermore, the lead time can be longer than the selling season, allowing no second chance to rush through another order to cover unexpectedly high demand. A similar problem exists for manufacturers of other fashion goods.

This type of situation is often called the *newsboy problem.* If the newspaper seller does not buy enough newspapers to resell on the street corner, sales opportunities are lost. If the seller buys too many newspapers, the overage cannot be sold because nobody wants yesterday's newspaper.

The following process is a straightforward way to analyze such problems and decide on the best order quantity.

1. List the different levels of demand that are possible, along with the estimated probability of each.

2. Develop a *payoff* table that shows the profit for each purchase quantity, Q, at each assumed demand level, D. Each row in the table represents a different order quantity, and each column represents a different demand level. The payoff for a given quantity–demand combination depends on whether all units are sold at the regular profit margin during the regular season, which results in two possible cases.

    a. If demand is high enough $(Q \le D)$, then all units are sold at the full profit margin, p, during the regular season,

    $$\text{Payoff} = (\text{Profit per unit}) (\text{Purchase quantity}) = pQ$$

    b. If the purchase quantity exceeds the eventual demand $(Q > D)$, only D units are sold at the full profit margin, and the remaining units purchased must be disposed of at a loss, l, after the season. In this case,

    $$\text{Payoff} = \left(\begin{array}{c}\text{Profit per unit sold}\\ \text{during season}\end{array}\right)(\text{Demand}) - \left(\begin{array}{c}\text{Loss per}\\ \text{unit}\end{array}\right)\left(\begin{array}{c}\text{Amount disposed of}\\ \text{after season}\end{array}\right)$$
    $$= pD - l(Q - D)$$

3. Calculate the expected payoff for each Q (or row in the payoff table) by using the *expected value* decision rule. For a specific Q, first multiply each payoff in the row by the demand probability associated with the payoff, and then add these products.

4. Choose the order quantity Q with the highest expected payoff.

Using this decision process for all such items over many selling seasons will maximize profits. However, it is not foolproof, and it can result in an occasional bad outcome.

EXAMPLE C.3   **Finding *Q* for One-Period Inventory Decisions**

One of many items sold at a museum of natural history is a Christmas ornament carved from wood. The gift shop makes a $10 profit per unit sold during the season, but it takes a $5 loss per unit after the season is over. The following discrete probability distribution for the season's demand has been identified:

Demand	10	20	30	40	50
Demand Probability	0.2	0.3	0.3	0.1	0.1

MyOMLab

Tutor C.3 in MyOMLab provides a new example to practice the one-period inventory decision.

MyOMLab

Active Model C.3 in MyOMLab provides additional insight on the one-period inventory decision model and its uses.

How many ornaments should the museum's buyer order?

SOLUTION

Each demand level is a candidate for best order quantity, so the payoff table should have five rows. For the first row, where $Q = 10$, demand is at least as great as the purchase quantity. Thus, all five payoffs in this row are

$$\text{Payoff} = pQ = (\$10)(10) = \$100$$

This formula can be used in other rows but only for those quantity–demand combinations where all units are sold during the season. These combinations lie in the upper-right portion of the payoff table, where $Q \le D$. For example, the payoff when $Q = 40$ and $D = 50$ is

$$\text{Payoff} = pQ = (\$10)(40) = \$400$$

The payoffs in the lower-left portion of the table represent quantity–demand combinations where some units must be disposed of after the season $(Q > D)$. For this case, the payoff must be calculated with the second formula. For example, when $Q = 40$ and $D = 30$,

$$\text{Payoff} = pD = l(Q - D) = (\$10)(30) - (\$5)(40 - 30) = \$250$$

OM Explorer or POM for Windows can be used to analyze this problem. Using OM Explorer, we obtain the payoff table in Figure C.5.

▼ **FIGURE C.5**

OM Explorer Solver for One-Period Inventory Decisions Showing the Payoff Table

Profit	$10.00	(if sold during preferred period)
Loss	$5.00	(if sold after preferred period)

Enter the possible demands along with the probability of each occuring. Use the buttons to increase or decrease the number of allowable demand forecasts. NOTE: Be sure to enter demand forecasts and probablities in all tinted cells, and be sure probabilities add up to 1.

	<	>			
Demand	10	20	30	40	50
Profitability	0.2	0.3	0.3	0.1	0.1

Payoff Table

		Demand			
Quantity	10	20	30	40	50
10	100	100	100	100	100
20	50	200	200	200	200
30	0	150	300	300	300
40	−50	100	250	400	400
50	−100	50	200	350	500

Now we calculate the expected payoff for each Q by multiplying the payoff for each demand quantity by the probability of that demand and then adding the results. For example, for $Q = 30$,

$$\text{Payoff} = 0.2(\$0) + 0.3(\$150) + 0.3(\$300) + 0.1(\$300) + 0.1(\$300) = \$195$$

Using OM Explorer, Figure C.6 shows the expected payoffs.

DECISION POINT

Because $Q = 30$ has the highest payoff at \$195, it is the best order quantity. Management can use OM Explorer or POM for Windows to do sensitivity analysis on the demands and their probabilities to see how confident they are with that decision.

FIGURE C.6 ▶

OM Explorer Solver Showing the Expected Payoffs

Weighted Payoffs

Order Quantity	Expected Payoff
10	100
20	170
30	195
40	175
50	140

Greatest Expected Payoff   195

Associated with Order Quantity   30

The need for one-time inventory decisions also can arise in manufacturing plants when (1) customized items are made (or purchased) to a single order, and (2) scrap quantities are high. A customized item produced for a single order is never intentionally held in stock because the demand for it is too unpredictable. In fact, it may never be ordered again so the manufacturer would like to make just the amount requested by the customer—no more, no less. The manufacturer also would like to satisfy an order in just one run to avoid an extra setup and a delay in delivering goods ordered. These two goals may conflict if the likelihood of some units being scrapped is high. Suppose that a customer places an order for 20 units. If the manager orders 20 units from the shop or from the supplier, one or two units may have to be scrapped. This shortage will force the manager to place a second (or even third) order to replace the defective units. Replacement can be costly if setup time is high and can also delay shipment to the customer. To avoid such problems, the manager could order more than 20 units the first time. If some units are left over, the customer might be willing to buy the extras or the manager might find an internal use for them. For example, some manufacturing companies set up a special account for obsolete materials. These materials can be "bought" by departments within the company at less than their normal cost, as an incentive to use them.

LEARNING GOALS IN REVIEW

1 Identify the situations where the economic lot size should be used rather than the economic order quantity. See the section "Noninstantaneous Replenishment," pp. 345–347. Figure C.1 shows the behavior of inventories when the ELS is appropriate.

2 Calculate the optimal lot size when replenishment is not instantaneous. Study Example C.1 and Solved Problem 1 for help on determining the ELS.

3 Define the relevant costs that should be considered to determine the order quantity when discounts are available. See the section "Quantity Discounts," pp. 348–350. Figure C.3 shows how the relevant costs affect the best lot size decision.

4 Determine the optimal order quantity when materials are subject to quantity discounts. Study Example C.2 and Solved Problem 2 for a step-by-step approach to determine the best order quantity.

5 Calculate the order quantity that maximizes the expected profits for a one-period inventory decision. See the section "One-Period Decisions," pp. 350–352. Be sure to understand Example C.3 and Solved Problem 3.

MyOMLab Resources	Titles	Link to the Book
Active Models Exercises	C.1 Economic Production Lot Size C.2 Quantity Discounts C.3 One-Time Inventory Decision	Noninstantaneous Replenishment; Example C.1 (p. 347); Solved Problem 1 (p. 354) Quantity Discounts; Example C.2 (pp. 349–350) One-Period Decisions; Example C.3 (pp. 351–352)
OM Explorer Solvers	Economic Production Lot Size Quantity Discounts One-Period Inventory	Noninstantaneous Replenishment; Example C.1 (p. 347); Figure C.2 (p. 347); Solved Problem 1 (p. 354) Quantity Discounts; Example C.2 (pp. 349–350); Figure C.4 (p. 350); Solved Problem 2 (pp. 354–355) One-Period Decisions; Example C.3 (pp. 351–352); Figure C.5 (p. 351); Figure C.6 (p. 352); Solved Problem 3 (p. 355)
OM Explorer Tutors	C.1 Economic Production Lot Size C.2 Finding Q with Quantity Discounts C.3 One-Period Inventory Decisions	Noninstantaneous Replenishment; Example C.1 (p. 347); Solved Problem 1 (p. 354) Quantity Discounts; Example C.2 (pp. 349–350); Solved Problem 2 (pp. 354–355) One-Period Decisions; Example C.3 (pp. 351–352); Solved Problem 3 (p. 355)
POM for Windows	Decision Tables Economic Production Lot Size Model Quantity Discount Model	One-Period Decisions; Example C.3 (pp. 351–352); Solved Problem 3 (p. 355) Noninstantaneous Replenishment; Example C.1 (p. 347); Solved Problem 1 (p. 354) Quantity Discounts; Example C.2 (pp. 349–350); Solved Problem 2 (pp. 354–355)
Virtual Tours	Sierra Nevada United Wood Treating Woot	Entire supplement Entire supplement Entire supplement
Internet Exercise	Continental Cement	Noninstantaneous Replenishment
Key Equations		
Image Library		

Key Equations

1. Noninstantaneous replenishment:

   Maximum inventory: $I_{\max} = Q\left(\dfrac{p - d}{p}\right)$

   Total annual cost = Annual holding cost + Annual ordering or setup cost

   $$C = \frac{Q}{2}\left(\frac{p - d}{p}\right)(H) + \frac{D}{Q}(S)$$

   Economic production lot size: $ELS = \sqrt{\dfrac{2DS}{H}}\sqrt{\dfrac{p}{p - d}}$

   Time between orders, expressed in years: $TBO_{ELS} = \dfrac{ELS}{D}$

2. Quantity discounts:

   Total annual cost = Annual holding cost + Annual ordering or setup cost + Annual cost of material

   $$C = \frac{Q}{2}(H) + \frac{D}{Q}(S) + PD$$

3. One-period decisions:

   Payoff matrix: $Payoff = \begin{cases} pQ & \text{if } Q \le D \\ pD - l(Q - D) & \text{if } Q > D \end{cases}$

Key Term

economic production lot size (ELS) 346

Solved Problem 1

Peachy Keen, Inc., makes mohair sweaters, blouses with Peter Pan collars, pedal pushers, poodle skirts, and other popular clothing styles of the 1950s. The average demand for mohair sweaters is 100 per week. Peachy's production facility has the capacity to sew 400 sweaters per week. Setup cost is $351. The value of finished goods inventory is $40 per sweater. The annual per-unit inventory holding cost is 20 percent of the item's value.

a. What is the economic production lot size (ELS)?

b. What is the average time between orders (TBO)?

c. What is the total of the annual holding cost and setup cost?

SOLUTION

a. The production lot size that minimizes total cost is

$$\text{ELS} = \sqrt{\frac{2DS}{H}} \sqrt{\frac{p}{p-d}} = \sqrt{\frac{2(100 \times 52)(\$351)}{0.20(\$40)}} \sqrt{\frac{400}{(400-100)}}$$

$$= \sqrt{456,300} \sqrt{\frac{4}{3}} = 780 \text{ sweaters}$$

b. The average time between orders is

$$\text{TBO}_{\text{ELS}} = \frac{\text{ELS}}{D} = \frac{780}{5,200} = 0.15 \text{ year}$$

Converting to weeks, we get

$$\text{TBO}_{\text{ELS}} = (0.15 \text{ year})(52 \text{ weeks/year}) = 7.8 \text{ weeks}$$

c. The minimum total of setup and holding costs is

$$C = \frac{Q}{2}\left(\frac{p-d}{p}\right)(H) + \frac{D}{Q}(S) = \frac{780}{2}\left(\frac{400-100}{400}\right)(0.20 \times \$40) + \frac{5,200}{780}(\$351)$$

$$= \$2,340/\text{year} + \$2,340/\text{year} = \$4,680/\text{year}$$

Solved Problem 2

A hospital buys disposable surgical packages from Pfisher, Inc. Pfisher's price schedule is $50.25 per package on orders of 1 to 199 packages and $49.00 per package on orders of 200 or more packages. Ordering cost is $64 per order, and annual holding cost is 20 percent of the per-unit purchase price. Annual demand is 490 packages. What is the best purchase quantity?

SOLUTION

We first calculate the EOQ at the *lowest* price:

$$\text{EOQ}_{49.00} = \sqrt{\frac{2DS}{H}} = \sqrt{\frac{2(490)(\$64.00)}{0.20(\$49.00)}} = \sqrt{6,400} = 80 \text{ packages}$$

This solution is infeasible because, according to the price schedule, we cannot purchase 80 packages at a price of $49.00 each. Therefore, we calculate the EOQ at the next lowest price ($50.25):

$$\text{EOQ}_{50.25} = \sqrt{\frac{2DS}{H}} = \sqrt{\frac{2(490)(\$64.00)}{0.20(\$50.25)}} = \sqrt{6,241} = 79 \text{ packages}$$

This EOQ is feasible, but $50.25 per package is not the lowest price. Hence, we have to determine whether total costs can be reduced by purchasing 200 units and thereby obtaining a quantity discount.

$$C = \frac{Q}{2}(H) + \frac{D}{Q}(S) + PD$$

$$C_{79} = \frac{79}{2}(0.20 \times \$50.25) + \frac{490}{79}(\$64.00) + \$50.25(490)$$

$$= \$396.98/\text{year} + \$396.68/\text{year} + \$24,622.50/\text{year} = \$25,416.44/\text{year}$$

$$C_{200} = \frac{200}{2}(0.20 \times \$49.00) + \frac{490}{200}(\$64.00) + \$49.00(490)$$

$$= \$980.00/\text{year} + \$156.80/\text{year} + \$24,010.00/\text{year} = \$25,146.80/\text{year}$$

Purchasing 200 units per order will save \$269.64/year, compared to buying 79 units at a time.

Solved Problem 3

Swell Productions is sponsoring an outdoor conclave for owners of collectible and classic Fords. The concession stand in the T-Bird area will sell clothing such as T-shirts and official Thunderbird racing jerseys. Jerseys are purchased from Columbia Products for \$40 each and are sold during the event for \$75 each. If any jerseys are left over, they can be returned to Columbia for a refund of \$30 each. Jersey sales depend on the weather, attendance, and other variables. The following table shows the probability of various sales quantities. How many jerseys should Swell Productions order from Columbia for this one-time event?

Sales Quantity	Probability	Quantity Sales	Probability
100	0.05	400	0.34
200	0.11	500	0.11
300	0.34	600	0.05

SOLUTION

Table C.1 is the payoff table that describes this one-period inventory decision. The upper-right portion of the table shows the payoffs when the demand, D, is greater than or equal to the order quantity, Q. The payoff is equal to the per-unit profit (the difference between price and

TABLE C.1 | PAYOFFS

	DEMAND, D						
Q	**100**	**200**	**300**	**400**	**500**	**600**	**Expected Payoff**
100	\$3,500	\$3,500	\$ 3,500	\$ 3,500	\$ 3,500	\$ 3,500	\$ 3,500
200	\$2,500	\$7,000	\$ 7,000	\$ 7,000	\$ 7,000	\$ 7,000	\$ 6,775
300	\$1,500	\$6,000	\$10,500	\$10,500	\$10,500	\$10,500	\$ 9,555
400	\$ 500	\$5,000	\$ 9,500	\$14,000	\$14,000	\$14,000	\$10,805
500	(\$ 500)	\$4,000	\$ 8,500	\$13,000	\$17,500	\$17,500	\$10,525
600	(\$1,500)	\$3,000	\$ 7,500	\$12,000	\$16,500	\$21,000	\$ 9,750

cost) multiplied by the order quantity. For example, when the order quantity is 100 and the demand is 200,

$$\text{Payoff} = (p-c)Q = (\$75 - \$40)100 = \$3,500$$

The lower-left portion of Table C.1 shows the payoffs when the order quantity exceeds the demand. Here the payoff is the profit from sales, pD, minus the loss associated with returning overstock, $l(Q - D)$, where l is the difference between the cost and the amount refunded for each jersey returned and $Q - D$ is the number of jerseys returned. For example, when the order quantity is 500 and the demand is 200,

$$\text{Payoff} = pD - l(Q - D) = (\$75 - \$40)200 - (\$40 - \$30)(500 - 200) = \$4,000$$

The highest expected payoff occurs when 400 jerseys are ordered:

$$\text{Expected payoff}_{400} = (\$500 \times 0.05) + (\$5,000 \times 0.11) + (\$9,500 \times 0.34)$$

$$+ (\$14,000 \times 0.34) + (\$14,000 \times 0.11) + (\$14,000 \times 0.05)$$

$$= \$10,805$$

Problems

The OM Explorer and POM for Windows software is available to all students using the 10th edition of this textbook. Go to **www.pearsonhighered.com/krajewski** to download these computer packages. If you purchased MyOMLab, you also have access to Active Models software and significant help in doing the following problems. Check with your instructor on how best to use these resources. In many cases, the instructor wants you to understand how to do the calculations by hand. At the least, the software provides a check on your calculations. When calculations are particularly complex and the goal is interpreting the results in making decisions, the software entirely replaces the manual calculations. The software also can be a valuable resource well after your course is completed.

1. Bold Vision, Inc., makes laser printer and photocopier toner cartridges. The demand rate is 625 EP cartridges per week. The production rate is 1,736 EP cartridges per week, and the setup cost is $100. The value of inventory is $130 per unit, and the holding cost is 20 percent of the inventory value. Bold Vision operates 52 weeks a year. What is the economic production lot size?

2. Sharpe Cutter is a small company that produces specialty knives for paper cutting machinery. The annual demand for a particular type of knife is 100,000 units. The demand is uniform over the 250 working days in a year. Sharpe Cutter produces this type of knife in lots and, on average, can produce 450 knives a day. The cost to set up a production lot is $300, and the annual holding cost is $1.20 per knife.

   a. Determine the economic production lot size (ELS).

   b. Determine the total annual setup and inventory holding cost for this item.

   c. Determine the TBO, or cycle length, for the ELS.

   d. Determine the production time per lot.

3. Suds's Bottling Company does bottling, labeling, and distribution work for several local microbreweries. The demand rate for Wortman's beer is 600 cases (24 bottles each) per week. Suds's bottling production rate is 2,400 cases per week, and the setup cost is $800. The value of inventory is $12.50 per case, and the annual holding cost is 30 percent of the inventory value. Suds's facilities operate 52 weeks each year. What is the economic production lot size?

4. The Bucks Grande exhibition baseball team plays 50 weeks each year and uses an average of 350 baseballs per week. The team orders baseballs from Coopers-Town, Inc., a ball manufacturer noted for six-sigma–level consistency and high product quality. The cost to order baseballs is $100 per order and the annual holding cost per ball is 38 percent of the purchase price. Coopers-Town's price structure is:

Order Quantity	Price per Unit
1–999	$7.50
1,000–4999	$7.25
5,000 or more	$6.50

   a. How many baseballs should the team buy per order?

   b. What is the total annual cost associated with the best order quantity?

   c. Coopers-Town, Inc., discovers that, owing to special manufacturing processes required for the Buck's baseballs, it has underestimated the setup time required on a capacity-constrained piece of machinery. Coopers-Town adds another category to the price structure to provide an incentive for larger orders and thereby hopes to reduce the number of setups required. If the Bucks buy 15,000 baseballs or more, the price will drop to $6.25 each. Should the Bucks revise their order quantity?

5. To boost sales, Pfisher (refer to Solved Problem 2) announces a new price structure for disposable surgical packages. Although the price break no longer is available at 200 units, Pfisher now offers an even greater discount if larger quantities are purchased. On orders of 1 to 499 packages, the price is $50.25 per package. For orders of 500 or more, the price per unit is $47.80. Ordering costs, annual holding costs, and annual demand remain at $64 per order, 20 percent of the per-unit cost, and 490 packages per year, respectively. What is the new lot size?

6. The University Bookstore at a prestigious private university buys mechanical pencils from a wholesaler. The wholesaler offers discounts for large orders according to the following price schedule:

Order Quantity	Price per Unit
0 to 200	$4.00
201 to 2,000	$3.50
2,001 or more	$3.25

The bookstore expects an annual demand of 2,500 units. It costs $10 to place an order, and the annual cost of holding a unit in stock is 30 percent of the unit's price. Determine the best order quantity.

7. Mac-in-the-Box, Inc., sells computer equipment by mail and telephone order. Mac sells 1,200 flat-bed scanners per year. Ordering cost is $300, and annual holding cost is 16 percent of the item's price. The scanner manufacturer offers the following price structure to Mac-in-the-Box:

Order Quantity	Price per Unit
0 to 11	$520
12 to 143	$500
144 or more	$400

What order quantity minimizes total annual costs?

8. As inventory manager, you must decide on the order quantity for an item that has an annual demand of 2,000 units. Placing an order costs you $20 each time. Your annual holding cost, expressed as a percentage of

average inventory value, is 20 percent. Your supplier has provided the following price schedule:

Minimum Order Quantity	Price per Unit
1	$2.50
200	$2.40
300	$2.25
1,000	$2.00

What ordering policy do you recommend?

9. Downtown Health Clinic needs to order influenza vaccines for the next flu season. The Clinic charges its patients $15.00 per vaccination and each dose of vaccine costs the clinic $4.00 to purchase. The Center for Disease Control has a long standing policy of buying back unused vaccines for $1.00 per dose. The Clinic estimates the following probability distribution for the season's demand:

Demand	Probability
2,000	0.05
3,000	0.20
4,000	0.25
5,000	0.40
6,000	0.10

  a. How many vaccines should the Clinic order to maximize its expected profit?

  b. The Clinic is trying to determine if they should participate in a new Federal program in which the cost of each dose is reduced to $2.00. However, to participate in the program, they can charge no more than $10.00 per vaccine. On strictly a profit maximizing basis, should the Clinic agree to participate?

10. Dorothy's pastries are freshly baked and sold at several specialty shops throughout Perth. When they are a day

old, they must be sold at reduced prices. Daily demand is distributed as follows:

Demand	Probability
50	0.25
150	0.50
200	0.25

Each pastry sells for $1.00 and costs $0.60 to make. Each one not sold at the end of the day can be sold the next day for $0.30 as day-old merchandise. How many pastries should be baked each day?

11. The Aggies will host Tech in this year's homecoming football game. Based on advance ticket sales, the athletic department has forecast hot dog sales as shown in the following table. The school buys premium hot dogs for $1.50 and sells them during the game at $3.00 each. Hot dogs left over after the game will be sold for $0.50 each to the Aggie student cafeteria, where they will be used in making hotdog casserole.

Sales Quantity	Probability
2,000	0.10
3,000	0.30
4,000	0.30
5,000	0.20
6,000	0.10

Use a payoff matrix to determine the number of hot dogs to buy for the game.

12. Bold Vision, Inc. (from Problem 1), must purchase toner from a local supplier. The company does not wish to carry raw material inventory and therefore only purchases enough toner to satisfy the demand of each individual batch of cartridges. Each toner cartridge requires one pound of toner. The raw material supplier offers Bold Vision a purchase discount of $2.00 per pound if the company orders at least 2,000 pounds at a time. Should Bold Vision accept this offer and alter its toner purchase quantity?

Selected References

Arnold, Tony J.R., Stephen Chapman, and Lloyd M. Clive. *Introduction to Materials Management*, 7th ed. Upper Saddle River, NJ: Prentice Hall, 2012.

Axsäter, Sven. *Inventory Control*, 2nd ed. New York: Springer Science + Business Media, LLC, 2006.

Bastow, B.J. "Metrics in the Material World." *APICS—The Performance Advantage* (May 2005), pp. 49–52.

Benton, W.C. *Purchasing and Supply Chain Management*, 2nd ed. New York: McGraw-Hill, 2010.

Callioni, Gianpaolo, Xavier de Montgros, Regine Slagmulder, Luk N. Van Wassenhove, and Linda Wright. "Inventory-Driven Costs." *Harvard Business Review* (March 2005), pp. 135–141.

Cannon, Alan R., and Richard E. Crandall. "The Way Things Never Were." *APICS—The Performance Advantage* (January 2004), pp. 32–35.

Hartvigsen, David. *SimQuick: Process Simulation with Excel*, 2nd ed. Upper Saddle River, NJ: Prentice Hall, 2004.

Manikas, Andrew. "Fighting Pests with the EOQ," *APICS Magazine* (April 2007), pp. 34–37.

Operations Management Body of Knowledge. Falls Church, VA: American Production and Inventory Control Society, 2009.

Timme, Stephen G., and Christine Williams-Timme. "The Real Cost of Holding." *Supply Chain Management Review* (July/August 2003), pp. 30–37.

Walters, Donald. *Inventory Control and Management*, 2nd ed. West Sussex, England: John Wiley and Sons, Ltd, 2003.

John Sommers II/Reuters

SUPPLY CHAIN DESIGN

A UPS employee loads packages into air cargo containers at the UPS Worldport All Points International Hub in Louisville, Kentucky. Many companies outsource their distribution processes to UPS because of its expertise in that critical function.

Nikon

Nikon is a $10 billion-a-year manufacturer of precision optical instruments in four product lines. Imaging Products includes a host of digital SLR cameras and the new Coolpix compact camera. Sports Optics includes binoculars and the new golfer's laser rangefinder. Precision Equipment covers scanners used for the manufacturing of semiconductors and Instrument Products, which includes a variety of microscopes. The product lines have their own supply chains because suppliers and their requirements differ. Imaging Products, because of its volatility due to technological innovation, is a case where the existing supply chain had to be redesigned to be more flexible and faster. As new digital products were developed, Nikon wanted to have the capability to deliver those products to retailers quickly so that they could keep up with demand from tech-savvy consumers and professional photographers. However, creating a new supply chain with those characteristics would place a burden on its existing infrastructure and put customer service in jeopardy. Nikon took the rare step of outsourcing the distribution of an entire consumer electronics product line to a third-party logistics provider (3PL), in this case United Parcel Service (UPS).

UPS helped Nikon create an entirely new distribution strategy that moves digital products to retail stores throughout the United States, Latin America, and the Caribbean. The journey for a shipment begins at manufacturing centers in Korea, Japan, and Indonesia, continues as air or ocean freight, passes through customs, and is directed to Louisville, Kentucky, which is UPS's all-points connection for global operations. At this point, UPS can perform some of the final assembly operations for Nikon. For example, accessories such as batteries and chargers can be added, or the products can be repackaged to meet the requirements for retailer displays. Finally, the products can be shipped to thousands of retail outlets in the United States, Latin America, and the Caribbean.

The redesigned supply chain not only improves the flow of products from origin to destination, it also provides Nikon with timely information of the status of shipments and advance notices of their delivery throughout the extent of the supply chain, including the retailers. Armed with this information, Nikon can make adjustments to delivery times to accommodate sales opportunities that otherwise would be missed. Despite the complexity of the supply chain, products leaving Nikon manufacturing facilities can now be on a retailer's shelves in as few as two days.

Sources: UPS Supply Chain Solutions: Nikon Focuses on Supply Chain Innovation—and Makes New Product Distribution a Snap, **www.ups-scs.com,** 2005; **www.Nikon.com**, 2011

LEARNING GOALS *After reading this chapter, you should be able to:*

1. Explain the strategic importance of supply chain design.
2. Identify the nature of supply chains for service providers, as well as for manufacturers.
3. Define the critical supply chain performance measures.
4. Explain the strategy of mass customization and its implications for supply chain design.
5. Define the important decision factors to be considered when employing an outsourcing or offshoring strategy.
6. Explain how efficient supply chains differ from responsive supply chains and the environments best suited for each type of supply chain.

Nikon, partnering with third-party logistics provider UPS, is an excellent example of how a supply chain can be successfully tailored to the market needs of a dynamic product. A *supply chain* is the interrelated series of processes within a firm and across different firms that produces a service or product to the satisfaction of customers. More specifically, it is a network of service, material, monetary, and information flows that link a firm's customer relationship, order fulfillment, and supplier relationship processes to those of its suppliers and customers. It is important to note, however, that a firm such as Nikon may have multiple supply chains, depending on the mix of services or products it produces. A supplier in one supply chain may not be a supplier in another supply chain because the service or product may be different or the supplier may simply be unable to negotiate a successful contract.

supply chain design

Designing a firm's supply chain to meet the competitive priorities of the firm's operations strategy.

The firm's operations strategy and competitive priorities guide its supply chain choices. Figure 10.1 shows the three major areas of focus in creating an effective supply chain.

1. *Link Service/Products with Internal Processes.* Parts 1 and 2 of this text have shown how firms coordinate internal process decisions with the competitive priorities of the services or products covered in the operations strategy.

▼ **FIGURE 10.1**
Creating an Effective Supply Chain

2. *Link Services/Products with the External Supply Chain.* The competitive priorities assigned to the firm's services or products must be reflected in the design of the network of suppliers.

3. *Link Services/Products with Customers, Suppliers and Supply Chain Processes.* The firm's processes that enable it to develop what customers want, interact with suppliers, deliver services or products, interact with customers, address environmental and ethical issues, and provide the information and planning tools needed to execute the operations strategy are the glue that binds the effective supply chain.

Service/Product

Link Services/Products with Internal Processes

Link Services/Products with External Supply Chain

Processes

Supply Chain

Link Services/Products with Customers, Suppliers, and Supply Chain Processes

Supply chain management, the synchronization of a firm's processes with those of its suppliers and customers to match the flow of materials, services, and information with demand, is a critical skill in most organizations. A key part of supply chain management is **supply chain design**, which seeks to design a firm's supply chain to meet the competitive priorities of the firm's operations strategy. To get a better understanding of the importance of supply chain design, consider Figure 10.2, which conceptually shows the challenges facing supply chain managers. The blue line is an *efficiency curve,* which shows the trade-off between costs and performance for the current supply chain design if the supply chain is operated as efficiently as it can be. Now, suppose that your firm plots its actual costs and performance, as indicated by the red dot. It is far off of the efficiency curve, which is not an uncommon occurrence. The challenge is to move operations into the tinted area, as close to the curve as possible, which can be accomplished by better forecasting, inventory management, operations planning and scheduling, and resource planning. We have already discussed inventory management; the other topics will follow. However, quantum steps in improvement can be obtained by improving the design of the supply chain in accordance with a sound operations strategy, which moves the curve as shown by the dashed red line. The goal is to reduce costs as well as increase performance. Design issues include placement of inventories, mass customization, outsourcing, supply chain collaboration, supplier selection, closed-loop supply chains, and facility location, which are just some of the topics we discuss in this and the next two chapters. In this chapter, we will also discuss the important measures of supply chain performance and show how sound supply chain design can improve key financial measures.

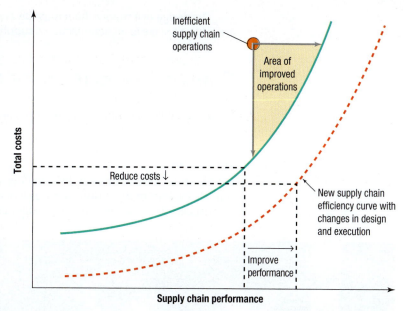

▲ **FIGURE 10.2**
Supply Chain Efficiency Curve

Supply Chain Design across the Organization

Creating a global supply chain involves more than supply chain managers designing the infrastructure or finding the best suppliers. There are also internal organizational pressures from groups such as sales, marketing, and product development that need to be recognized. These pressures are (1) dynamic sales volumes, (2) customer service levels, and (3) service/product proliferation.

Dynamic sales volumes One of the most costly operating aspects of supply chains is trying to meet the needs of volatile sales volumes. Often this involves excessive inventories, underutilized personnel, or more expensive delivery options to meet customer demands on time. While sometimes these volatile demands are caused by external sources such as the customers themselves, they are often caused internally by end-of-month sales promotions. Supply chain design should involve close collaboration between top level managers across the organization so that unnecessary costly supply chain options are avoided. We will discuss the implications of supply chain dynamics in more depth in Chapter 12, "Supply Chain Integration."

Customer service levels We have discussed customer service levels as they relate to an organization's internal inventories in Chapter 9, "Supply Chain Inventory Management." Here we focus on the organizational pressures emanating from the sales and marketing groups for superior service levels for the organization's customers. Questions such as "What service level should be guaranteed?" or "How speedy must our deliveries be?" need collaborative discussion from the sales, marketing, and finance groups. The answers to these questions impinge on the design of the supply chain, particularly its points of supply and the choice of suppliers.

Service/product proliferation The sales and marketing groups provide the momentum to create new services or products because they are closely in touch with customers and their needs. The survival of any organization depends on the development of new markets. However, adding more services or products often adds complexity to the supply chain. It is not an unusual circumstance to find that a relatively large proportion of SKUs contribute only a small percentage of the revenues. Generally, these niche services or products have low volumes and therefore cost more to produce, market, and deliver. A thoughtful balance needs to be struck between the cost of operating the supply chain and the need to market new services and products.

Creating Value through Operations Management

Using Operations to Compete
Project Management

Managing Processes

Process Strategy
Process Analysis
Quality and Performance
Capacity Planning
Constraint Management
Lean Systems

Managing Supply Chains

Supply Chain Inventory Management
Supply Chain Design
Supply Chain Location Decisions
Supply Chain Integration
Supply Chain and the Environment
Forecasting
Operations Planning and Scheduling
Resource Planning

The design of a supply chain must be a collaborative effort from the CEO down. All functional areas have a stake in an organization's supply chains.

Supply Chains for Services and Manufacturing

Every firm or organization is a member of some supply chain. In this section, we show the similarities and differences between supply chains for services and manufacturing.

Services

Supply chain design for a service provider is driven by the need to provide support for the essential elements of the various services it delivers. Consider the example of Flowers-on-Demand, a florist with 27 retail stores in the greater Boston metropolitan area.[1] Customers can place orders for customized floral arrangements by visiting one of the stores, using a toll-free number, or going to the florist's Web page. The 800 number and the Web page are operated by a local Internet services company, which takes orders and relays them to the florist. The arrangements are produced at a distribution center, and deliveries are made using either local couriers, or FedEx, if the delivery is outside of the Boston area. Fresh flowers, flown in from all over the world, are used in the arrangements.

Chuck Pefley/Alamy

An employee at a warehouse for a commercial flower farm packages flowers for local delivery. The farm acts as a supplier to grocery stores and retail florists.

What differentiates Flowers-on-Demand from floral wire services, such as Teleflora or FTD, is that it assembles all the arrangements and can ship out-of-area orders for next-day delivery anywhere in the country. To do business, the florist must have a supply chain that provides retail stores, a delivery center, computers, point-of-sale equipment, and employees. It must purchase flowers that are sourced globally as well as arrangement materials, such as pots, baskets, greeting cards, and packing materials. The florist must arrange the flowers per the customer's order and ensure that the arrangement is delivered as specified by the customer, using local services or FedEx. The design of its supply chain must provide convenience, which is facilitated by the location of the retail outlets and the opportunity to place orders via the Internet or the toll-free number.

Figure 10.3 illustrates a simplified supply chain for the florist. Each of the suppliers, of course, has its own supply chain (not shown). For example, the supplier for the arrangement materials may get baskets from one supplier and pots from another. The suppliers in the florist's supply chain play an integral role in its ability to meet its competitive priorities, such as top quality, delivery speed, and customization.

Manufacturing

A fundamental purpose of supply chain design for manufacturers is to control inventory by managing the flow of materials. The typical manufacturer spends more than 60 percent of its total income from sales on purchased services and materials, whereas the typical service provider spends only 30 to 40 percent. Because materials comprise such a large component of the sales dollar, manufacturers can reap large profits with a small reduction in the cost of materials, which makes supply chain management a key competitive weapon.

The supply chain for a manufacturing firm can be complicated, as Figure 10.4 illustrates. However, the supply chain depicted is an oversimplification because many companies have hundreds, if not thousands, of suppliers. In this example, the firm is in Ireland and deals with an international supply chain. In addition, it owns its distribution and transportation services. Suppliers are often identified by their position in the supply chain. Here, tier 1 suppliers provide

---

[1]The florist depicted is real; however, the name has been changed.

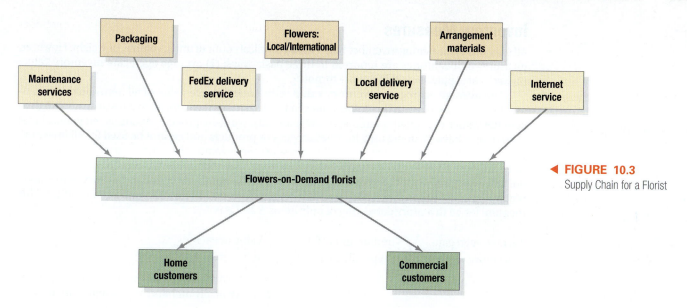

◀ FIGURE 10.3
Supply Chain for a Florist

major subassemblies that are assembled by the manufacturing firm, tier 2 suppliers provide tier 1 suppliers with components, and so on. Not all companies have the same number of levels in their supply chains. For example, companies that engineer products to customer specifications normally do not have distribution centers as part of their supply chains. Such companies often ship products directly to their customers.

▼ FIGURE 10.4
Supply Chain for a Manufacturing Firm

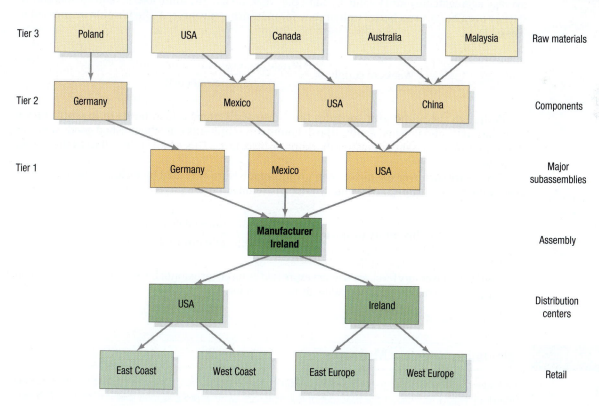

Measures of Supply Chain Performance

Managers need performance measures to assess the implications of changes to a supply chain. Before discussing the major supply chain design decisions, we define the typical inventory measures and financial measures used to monitor supply chain performance and evaluate alternative supply chain designs.

Inventory Measures

All methods of measuring inventory begin with a physical count of units, volume, or weight. However, measures of inventories are reported in three basic ways: (1) average aggregate inventory value, (2) weeks of supply, and (3) inventory turnover.

The **average aggregate inventory value** is the total average value of all items held in inventory by a firm. We express the dollar values in this inventory measure at cost because we can then sum the values of individual items in raw materials, work-in-process, and finished goods. Final sales dollars have meaning only for final services or products and cannot be used for all inventory items. It is an average because it usually represents the inventory investment over some period of time. Suppose a retailer holds items A and B in stock. One unit of item A may be worth only a few dollars, whereas one unit of item B may be valued in the hundreds of dollars because of the labor, technology, and other value-added operations performed in manufacturing the product. This measure for an inventory consisting of only items A and B is

$$\begin{aligned} \text{Average aggregate} \atop \text{inventory value} = &\left(\begin{array}{c}\text{Number units of item A} \\ \text{typically on hand}\end{array}\right)\left(\begin{array}{c}\text{Value of each} \\ \text{unit of item A}\end{array}\right) + \\ &\left(\begin{array}{c}\text{Number of units of item B} \\ \text{typically on hand}\end{array}\right)\left(\begin{array}{c}\text{Value of each} \\ \text{unit of item B}\end{array}\right) \end{aligned}$$

Summed over all items in an inventory, this total value tells managers how much of a firm's assets are tied up in inventory. Manufacturing firms typically have about 25 percent of their total assets in inventory, whereas wholesalers and retailers average about 75 percent.

To some extent, managers can decide whether the aggregate inventory value is too low or too high by historical or industry comparisons or by managerial judgment. However, a better performance measure would take demand into account because it would show how long the inventory resides in the firm. **Weeks of supply** is an inventory measure obtained by dividing the average aggregate inventory value by sales per week at cost. (In some low-inventory operations, days or even hours are a better unit of time for measuring inventory.) The formula (expressed in weeks) is

$$\text{Weeks of supply} = \frac{\text{Average aggregate inventory value}}{\text{Weekly sales (at cost)}}$$

Although the numerator includes the value of all items a firm holds in inventory (raw materials, WIP, and finished goods), the denominator represents only the finished goods sold—at cost rather than the sale price after markups or discounts. This cost is referred to as the *cost of goods sold*.

Inventory turnover (or *turns*) is an inventory measure obtained by dividing annual sales at cost by the average aggregate inventory value maintained during the year, or

$$\text{Inventory turnover} = \frac{\text{Annual sales (at cost)}}{\text{Average aggregate inventory value}}$$

The "best" inventory level, even when expressed as turnover, cannot be determined easily. A good starting point is to benchmark the leading firms in an industry.

EXAMPLE 10.1	**Calculating Inventory Measures**

The Eagle Machine Company averaged $2 million in inventory last year, and the cost of goods sold was $10 million. Figure 10.5 shows the breakout of raw materials, work-in-process, and finished goods inventories. The best inventory turnover in the company's industry is six turns per year. If the company has 52 business weeks per year, how many weeks of supply were held in inventory? What was the inventory turnover? What should the company do?

Cost of Goods Sold	$10,000,000
Weeks of Operation	52

	Item Number	Average Level	Unit Value	Total Value
Raw Materials	1	1,400	$50.00	$70,000
	2	1,000	$32.00	$32,000
	3	400	$60.00	$24,000
	4	2,400	$10.00	$24,000
	5	800	$15.00	$12,000
Work in Process	6	320	$700.00	$224,000
	7	160	$900.00	$144,000
	8	280	$750.00	$210,000
	9	240	$800.00	$192,000
	10	400	$1,000.00	$400,000
Finished Goods	11	60	$2,000.00	$120,000
	12	40	$3,500.00	$140,000
	13	50	$2,800.00	$140,000
	14	20	$5,000.00	$100,000
	15	40	$4,200.00	$168,000
Total				$2,000,000

Average Weekly Sales at Cost	$192,308
Weeks of Supply	10.4
Inventory Turnover	5.0

◀ **FIGURE 10.5**
Calculating Inventory Measures Using *Inventory Estimator* Solver

SOLUTION

The average aggregate inventory value of $2 million translates into 10.4 weeks of supply and 5 turns per year, calculated as follows:

$$\text{Weeks of supply} = \frac{\$2 \text{ million}}{(\$10 \text{ million})/(52 \text{ weeks})} = 10.4 \text{ weeks}$$

$$\text{Inventory turns} = \frac{\$10 \text{ million}}{\$2 \text{ million}} = 5 \text{ turns/year}$$

DECISION POINT

The analysis indicates that management must improve the inventory turns by 20 percent. Management should improve its order fulfillment process to reduce finished goods inventory. Supply chain operations can also be improved to reduce the need to have so much raw materials and work-in-process inventory stock. It will take an inventory reduction of about 16 percent to achieve the target of 6 turns per year. However, inventories would not have to be reduced as much if sales increased. If the sales department targets an increase in sales of 8 percent ($10.8 million), inventories need only be reduced by 10 percent ($1.8 million) to get 6 turns a year. Management can now do sensitivity analyses to see what effect reductions in the inventory of specific items or increases in the annual sales have on weeks of supply or inventory turns.

Financial Measures

How the supply chain is designed and managed has a huge financial impact on the firm. Inventory is an investment because it is needed for future use. However, inventory ties up funds that might be used more profitably in other operations. Figure 10.6 shows how supply chain decisions can affect financial measures.

Total Revenue Supply chain performance measures related to time, which is a critical dimension of supply chain operations, have financial implications. Many service providers and manufacturers measure the percent of on-time deliveries of their services or products to their customers, as well as services and materials from their suppliers. Increasing the percent of on-time deliveries to customers, for example, will increase *total revenue* because satisfied customers will buy more services and products from the firm.

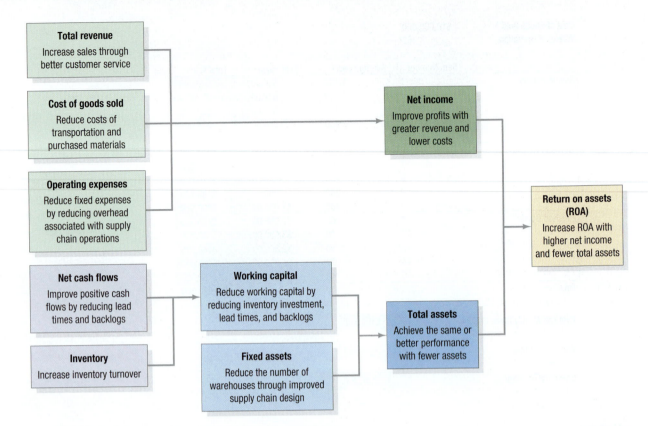

FIGURE 10.6 ▲
How Supply Chain Decisions Can
Affect ROA

Cost of Goods Sold Being able to buy materials or services at a better price and transform them more efficiently into services or products will improve a firm's *cost of goods sold* measure and ultimately its *net income*. These improvements will also have an effect on *contribution margin*, which is the difference between price and the variable costs to produce a service or good. Reducing production, material, transportation, and poor quality costs increases the contribution margin, allowing for greater profits. Contribution margins are often used as inputs to decisions regarding the portfolio of services or products the firm offers.

Operating Expenses Selling expenses, fixed expenses, and depreciation are considered operating expenses. Designing a supply chain with minimal capital investment can reduce depreciation charges. Changes to the supply chain infrastructure can have an effect on overhead, which is considered a fixed expense.

Cash Flow The supply chain design can improve positive net cash flows by focusing on reducing lead times and backlogs of orders. The Internet brings another financial measure related to cash flows to the forefront: *Cash-to-cash* is the time lag between paying for the services and materials needed to produce a service or product and receiving payment for it. The shorter the time lag, the better the *cash flow* position of the firm because it needs less working capital. The firm can then use the freed-up funds for other projects or investments. Redesigning the order placement process, so that payment for the service or product by the customer is made at the time the order is placed, can reduce the time lag. By contrast, billing the customer after the service is performed or the order is shipped increases the need for working capital. The goal is to have a negative cash-to-cash situation, which is possible when the customer pays for the service or product before the firm has to pay for the resources and materials needed to produce it. In such a case, the firm must have supplier inventories on consignment, which allows it to pay for materials as it uses them.

Working Capital Weeks of inventory and inventory turns are reflected in another financial measure, *working capital*, which is money used to finance ongoing operations. Decreasing weeks of supply or increasing inventory turns reduces the working capital needed to finance inventories. Reductions in working capital can be accomplished by improving the customer relationship, order fulfillment, or supplier relationship processes. For example, reducing supplier lead times has the effect of reducing weeks of supply and increasing inventory turns. Matching the input and output flows of materials is easier because shorter-range, more reliable forecasts of demand can be used.

Return on Assets Designing and managing the supply chain so as to reduce the aggregate inventory investment or fixed investments such as warehouses will reduce the *total assets* portion of the firm's balance sheet. An important financial measure is *return on assets* (*ROA*), which is net income divided by total assets. Consequently, reducing aggregate inventory investment and fixed investments, or increasing net income by better cost management, will increase ROA. Techniques for reducing inventory, transportation, and operating costs related to resource usage and scheduling are discussed in the chapters to follow.

We now turn to a discussion of several major supply chain design decisions and their implications for a firm's performance.

Inventory Placement

A fundamental supply chain design decision that affects performance is where to locate an inventory of finished goods. Placing inventories can have strategic implications, as in the case of international companies locating *distribution centers* (DCs) in foreign countries to preempt local competition by reducing customer delivery times. However, the issue for any firm producing standardized products is where to position the inventory in the supply chain. At one extreme, the firm could use **centralized placement**, which means keeping all the inventory of a product at a single location, such as a firm's manufacturing plant or a warehouse, and shipping directly to each of its customers. The advantage would come from what is referred to as **inventory pooling**, which is a reduction in inventory and safety stock because of the merging of uncertain and variable demands from the customers. A higher-than-expected demand from one customer can be offset by a lower-than-expected demand from another so that the total demand remains fairly stable. A disadvantage of placing inventory at a central location, however, is the added cost of shipping smaller, uneconomical quantities directly to customers over long distances.

Another approach is to use **forward placement**, which means locating stock closer to customers at a warehouse, DC, wholesaler, or retailer. Forward placement can have two advantages—faster delivery times and reduced transportation costs—that can stimulate sales. However, as inventory is placed closer to the customer, such as at a DC, the pooling effect of the inventories is reduced because safety stocks for the item must increase to take care of uncertain demands at each DC, rather than just a single location. Nonetheless, the time to get the product to the customer is reduced. Consequently, service to the customer is quicker, and the firm can take advantage of larger, less costly shipments to the DCs from the manufacturing plant, at the expense of larger overall inventories.

centralized placement

Keeping all the inventory of a product at a single location such as a firm's manufacturing plant or a warehouse and shipping directly to each of its customers.

inventory pooling

A reduction in inventory and safety stock because of the merging of variable demands from customers.

forward placement

Locating stock closer to customers at a warehouse, DC, wholesaler, or retailer.

Mass Customization

A firm's supply chain must be capable of addressing certain competitive priorities that will win orders from customers. Often customers want more than a wide selection of standard services or products; they want a personalized service or product and they want it fast. For example, suppose you want to paint your living room a new color. You need to complement all of the existing furnishings, wall decorations, and carpet. You go to your local paint retail store and select a color from a stack of books that spans every color of the rainbow. The store can give you all the paint you need in your selected color while you wait. How can the store provide that service economically? Certainly the store cannot stock thousands of colors in sufficient quantities for any job. The store stocks the base colors and pigments separately and mixes them as needed, thereby supplying an unlimited variety of colors without maintaining the inventory required to match each customer's particular color needs. The paint retailer is practicing a strategy known as *mass customization*, whereby a firm's highly divergent processes generate a wide variety of customized services or products at reasonably low costs. Essentially, the firm allows customers to select from a variety of standard options to create the service or product of their choice.

Competitive Advantages

A mass customization strategy has three important competitive advantages.

- *Managing Customer Relationships.* Mass customization requires detailed inputs from customers so that the ideal service or product can be produced. The firm can learn a lot about its customers from the data it receives. Once customers are in the database, the firm can keep track of them over time. A significant competitive advantage is realized through these close customer relationships based on a strategy of mass customization.

- *Eliminating Finished Goods Inventory.* Producing to a customer's order is more efficient than producing to a forecast because forecasts are not perfect. The trick is to have everything

ROBERTO GONZALEZ KRT/Newscom

At My Twinn doll company, kids can create their own doll online. Artisans can even match the doll face to the child's face from a photo you supply. My Twinn continues to market clothing accessories as the doll "grows up" with its human twin.

you need to produce the order quickly. A technology some firms use for their order placement process is a software system called a *configurator*, which gives firms and customers easy access to data relevant to the options available for the service or product. Dell uses a configurator that allows customers to design their own computer from a set of standard components that are in stock. Once the order is placed, the product is assembled and then delivered. Using sales promotions, the firm can exercise some control over the requirements for the inventory of components by steering customers away from options that are out of stock in favor of options that are in stock. This capability takes pressure off the supply chain while keeping the customer satisfied.

Service providers also take advantage of mass customization to reduce the level of inventory. British Airways is trying to personalize the service of customers once they are on board. It has a software system that tracks the preferences of its most favored customers down to the magazines they read. This information allows the airline to more accurately plan what to pack on each flight. This information saves the airline a significant amount of money because it does not pack amenities passengers do not want.

■ *Increasing Perceived Value of Services or Products.* With mass customization, customers can have it their way. In general, mass customization often has a higher value in the mind of the customer than it actually costs to produce. This perception allows firms to charge prices that provide a nice margin.

Supply Chain Design for Mass Customization

How does mass customization affect the design of supply chains? We address three major considerations.

Assemble-to-Order Strategy   The underlying process design is an assemble-to-order strategy. This strategy involves two stages in the provision of the service or product. Initially, standardized components are produced or purchased and held in stock. This stage is important because it enables the firm to produce or purchase these standard items in large volumes to keep the costs low. In the second stage, the firm assembles these standard components to a specific customer order. In mass customization, this stage must be flexible to handle a large number of potential combinations, and be capable of producing the order quickly and accurately. For example, for My Twinn customized dolls, customers can choose from more than 325,000 different doll combinations. To ensure accuracy, the Web site takes the customer through the required choices and allows the customer to see the doll as the various options are chosen. Figure 10.7 shows how the customer

FIGURE 10.7 ▼

Supply Chain Design for
Assemble-to-Order Strategy

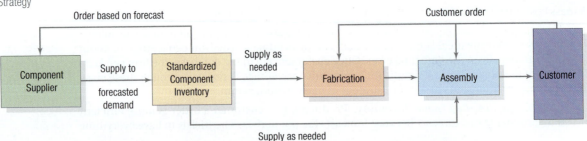

order is transmitted to the fabrication and assembly operations and that standardized purchased components are taken to the point of fabrication or assembly as needed for the order. Notice that there is no finished goods inventory.

Modular Design   The service or product must have a modular design that enables the "customization" the customer desires. This approach requires careful attention to service/product designs so that the final service or product can be assembled from a set of standardized modules economically and fast in response to a customer order.

Postponement   Finally, successful mass customizers postpone the task of differentiating a service or product for a specific customer until the last possible moment. *Postponement* is a concept whereby some of the final activities in the provision of a service or product are delayed until the orders are received. Doing so allows the greatest application of standard modules before specific customization is done. Postponement is a key decision because it specifies where in the supply chain volume-oriented, standardized operations are separated from custom-oriented, assembly operations. Sometimes the final customization occurs in the last step.

The assemble-to-order strategy and postponement can be extended to supply chains. The costs of inventory and transportation often determine the extent to which a manufacturer uses postponement in the supply chain. With postponement, manufacturers can avoid inventory buildup. Some firms take advantage of a process called **channel assembly**, whereby members of the distribution channel act as if they were assembly stations in the factory. Distribution centers or warehouses can perform the last-minute customizing operations after specific orders have been received. That was the case with Nikon in the chapter opener. UPS added batteries and chargers or repackaged the digital cameras to specific retailer requests just before delivering the orders. Channel assembly is particularly useful when the required customizing has some geographical rationale, such as language differences or technical requirements. In general, beyond the inventory advantages, the advantage of postponement in the distribution channel is that the firm's plants can focus on the standardized aspects of the product, while the distributor can focus on customizing a product that may require additional components from local suppliers.

channel assembly

The process of using members of the distribution channel as if they were assembly stations in the factory.

Outsourcing Processes

All businesses buy at least some inputs to their processes (such as professional services, raw materials, or manufactured parts) from other producers. Most businesses also purchase services to get their products to their customers. How many of the processes that produce those purchased items and services should a firm own and operate instead? The answer to that question determines the extent of the firm's vertical integration. The more processes in the supply chain that the organization performs itself, the more vertically integrated it is. If it does not perform some processes itself, it must rely on **outsourcing**, or paying suppliers and distributors to perform those processes and provide needed services and materials. When managers opt for more vertical integration, by definition less outsourcing occurs. These decisions are sometimes called **make-or-buy decisions**, with a *make* decision meaning more vertical integration and a *buy* decision meaning more outsourcing. After deciding what to outsource and what to do in-house, management must find ways to coordinate and integrate the various processes and suppliers involved. Example 10.2 shows how break-even analysis, which can be found in Supplement A, "Decision Making," can be used for the make-or-buy decision.

outsourcing

Paying suppliers and distributors to perform processes and provide needed services and materials.

make-or-buy decision

A managerial choice between whether to outsource a process or do it in-house.

EXAMPLE 10.2	**Using Break-Even Analysis for the Outsourcing Decision**

Thompson manufacturing produces industrial scales for the electronics industry. Management is considering outsourcing the shipping operation to a logistics provider experienced in the electronics industry. Thompson's annual fixed costs of the shipping operation are $1,500,000, which includes costs of the equipment and infrastructure for the operation. The estimated variable cost of shipping the scales with the in-house operation is $4.50 per ton-mile. If Thompson outsourced the operation to Carter Trucking, the annual fixed costs of the infrastructure and management time needed to manage the contract would be $250,000. Carter would charge $8.50 per ton-mile. What is the break-even quantity?

MyOMLab

Active Model A.2 in MyOMLab provides additional insight on the make-or-buy decision and its extensions.

MyOMLab

Tutor A.2 in MyOMLab provides a new example to practice break-even analysis on make-or-buy decisions.

SOLUTION

From Supplement A, "Decision Making," the formula for the break-even quantity yields

$$Q = \frac{F_m - F_b}{c_b - c_m}$$

$$= \frac{1,500,000 - 250,000}{8.50 - 4.50} = 312,500 \text{ ton-miles}$$

DECISION POINT

Thompson management must now assess how many ton-miles of product will likely be shipped now and in the future. If that estimate is less than 312,500 ton-miles, the best option is to outsource the operation to Carter Trucking.

Vertical Integration

backward integration

A firm's movement upstream toward the sources of raw materials, parts, and services through acquisitions.

forward integration

Acquiring more channels of distribution, such as distribution centers (warehouses) and retail stores, or even business customers.

Vertical integration can be in two directions. **Backward integration** represents a firm's movement upstream in the supply chain toward the sources of raw materials, parts, and services through acquisitions, such as a major grocery chain having its own plants to produce house brands of ice cream, frozen pizza dough, and peanut butter. Backward integration has the effect of reducing the risk of supply. **Forward integration** means that the firm acquires more channels of distribution, such as its own distribution centers (warehouses) and retail stores. It can also mean that the firm goes even farther by acquiring its business customers. A firm chooses vertical integration when it has the skills, volume, and resources to hit the competitive priorities better than outsiders can. Doing the work within its organizational structure may mean better quality and more timely delivery, as well as taking better advantage of the firm's human resources, equipment, and space. Extensive vertical integration is generally attractive when input volumes are high because high volumes allow task specialization and greater efficiency. It is also attractive if the firm has the relevant skills and views the processes that it is integrating as particularly important to its future success. However, care must be exercised that excessive vertical integration does not lead to a loss of focus for the firm in delivering value in its core business.

Management must identify, cultivate, and exploit its core competencies to prevail in global competition. Recall that core competencies reflect the collective learning of the organization, especially its ability for coordinating diverse processes and integrating multiple technologies. They define the firm and provide its reason for existence. Management must be constantly attentive to bolstering core competencies, perhaps by looking upstream toward its suppliers and downstream toward its customers and acquiring those processes that support its core competencies—those that allow the firm to organize work and deliver value better than its competitors. To do otherwise poses a risk that the firm will lose control over critical areas of its business.

Outsourcing

offshoring

A supply chain strategy that involves moving processes to another country.

Notwithstanding the arguments in favor of increased vertical integration, some firms outsource important processes such as accounting, marketing, or manufacturing. Many firms outsource payroll, security, cleaning, and other types of services, rather than employ personnel to provide these services. Outsourcing is a particularly attractive option to those firms that have low volumes. What prompts a firm to outsource rather than vertically integrate? An outsourcing firm realizes that another firm can perform the outsourced process more efficiently and with better quality than it can. They opt to add external suppliers to their supply chains rather than to keep internal suppliers. However, the outsourcing decision is a serious one because the firm can lose the skills and knowledge needed to conduct the process. All learning about process advancements is left to the outsourcing partner, which makes it difficult to ever bring that process back into the firm.

The strategy of globalizing a firm adds a new dimension to the development of supply chains and the use of outsourcing. **Offshoring** is a supply chain strategy that involves moving processes to another country. As such, offshoring is more encompassing than outsourcing because it also includes vertical integration by locating internal processes in other countries. Firms are motivated to initiate operations offshore by the market potential and the cost advantages it provides. The firm may be able to create new markets because of its presence in other countries and its ability to offer competitive prices due to its cost efficiencies. Competitive priorities other than low costs, such as delivery speed to distant customers, can drive the decision, too.

Decision Factors The decision to outsource or offshore a process is complex and involves a number of factors.

- *Comparative Labor Costs.* Some countries such as China and India have traditionally held a huge edge when it comes to labor costs. In India, the salary for a computer programmer is much less than that of a programmer in the United States with comparable skills. In China, the average monthly wages are much less than those in Japan. However, the advantage of doing business in these and other low-wage countries is eroding as wages in those countries rise due to increased demands. In some cases, the labor-cost advantage may only be a short-term one because of local economic conditions, such as the lowered wage structures in the United States because of the recession of 2008.

■ *Rework and Product Returns.* While labor wage rates may be low in a particular location, the quality of workmanship must also be considered. Internal rework costs and the cost of product returns may offset the advantage in wage rates.

■ *Logistics Costs.* Even if labor costs are not favorable, it may still be less costly to outsource or offshore final assembly processes to other countries to reduce the logistical costs of delivering products to international customers. Moving processes closer to the customer and using more local suppliers reduces the cost of transporting the final product to its ultimate destination. Using shipping or air transportation can be costly because of their dependence on oil. The savings in logistical costs can offset the higher labor costs in those countries.

■ *Tariffs and Taxes.* Some countries offer tax incentives to firms that do business within their borders. Tariffs can also be a stumbling block for firms looking to do business in a country. Sometimes they are high enough that the firm decides to assemble the products in that country rather than export the products in.

■ *Market Effects.* Not to be overlooked is the potential advantage of offshoring a process in a location where the presence of the firm can have a positive effect on local sales.

■ *Labor Laws and Unions.* Some countries have fewer unions or restrictions on the flexible use of labor. The ability to use workers to perform a number of different tasks without restrictions can be important to firms trying to achieve flexibility in operations and reduce costs. Nonetheless, firms must be cognizant of local labor laws and customs and strive to achieve a high level of ethical behavior when doing business in other countries.

■ *Internet.* The Internet reduces the transaction costs of managing distant partners or operations.

Potential Pitfalls Even though outsourcing may appear to offer some big advantages, it also has some pitfalls that should be carefully explored before using this strategy.

■ *Pulling the Plug Too Quickly.* A major mistake is to decide to outsource a process before making a good-faith effort to fix the existing one. We discussed many ways to improve processes in parts 1 and 2 of this text; these methods should be explored first. It is not always the case that outsourcing is the answer, even if local labor wages far exceed those of other countries. Make sure you really need to outsource in order to accomplish your operations strategy.

■ *Technology Transfer.* Often an outsourcing strategy involves creating a *joint venture* with a company in another country. With a joint venture, two firms agree to jointly produce a service or product together. Typically, a transfer of technology takes place to bring one partner up to speed regarding the service or product. The danger is that the firm with the technology advantage will essentially be setting up the other firm to be a future competitor.

■ *Process Integration.* Despite the power of the Internet, it is difficult to fully integrate outsourced processes with the firm's other processes. Time, distance, and communication can be formidable hurdles, especially if the supplier is on the other side of the world. Managing offshore processes will not be the same as managing processes located next door. Often considerable managerial time must be expended to coordinate offshore processes. Managerial Practice 10.1 reveals outsourcing on a global scale can be challenging for management.

Employees work on the Wuling minivan engine assembly line at the SAIC GM Wuling Automobile Co., Ltd. factory in Liuzhou, Guangxi Province, China. SAIC got a license to use GM's technical knowledge for the design and manufacture of automobiles.

Qilai Shen/In Pictures/Corbis

MANAGERIAL PRACTICE 10.1   Building a Supply Chain for the Dreamliner

Suppose that you had the freedom to totally design the supply chain for one of the most highly anticipated airliners of modern times. The airliner, the Boeing 787 Dreamliner, is a super-efficient commercial airplane that can carry up to 330 passengers on routes as long as 8,000 nautical miles at speeds up to 850 miles per hour. It is constructed with carbon-fiber composite materials, which are lightweight and not susceptible to corrosion or fatigue like aluminum. This plane uses 50 percent composite materials; Boeing used only 10 to 12 percent in the 777. Boeing's goal was to bring the most complex machine in mass production to market in just over 4 years, or 2 years less time than other projects. Boeing had two options for the design of the supply chain: (1) Produce about 50 percent of the plane in-house, including the wing and fuselage as in existing Boeing planes, and run the risk that production lead times will suffer because of capacity constraints; or (2) outsource about 85 percent of the plane, essentially only constructing the vertical fin in-house, and manage the global suppliers responsible for design as well as production of major components. Boeing selected option 2.

There are some good reasons for this choice. First, a number of big customers for the 787, such as India and Japan, require that significant portions of the aircraft must be manufactured in their countries. Using major contractors within those countries satisfies the requirement. Second, a shortage of high-quality engineering talent also puts pressure on outsourcing. Third, the sheer complexity of the airplane makes it necessary to share the load. Boeing, even with all of its resources, could not build all of the components and pieces in one facility or region. Finally, work on the plane can proceed concurrently, rather than sequentially, thereby saving time and money. For example, the modular design of the plane allows Boeing to utilize flexible tooling to move planes through the factory much more quickly. The suppliers design and deliver the subsystems on a just-in-time basis where they are "snapped" together by a smaller number of factory workers in a matter of days rather than a month, the typical time for a plane of that complexity.

Boeing chose to design its supply chain with 43 top-tier suppliers on three continents. Outsourcing so much responsibility requires a lot of managerial attention; you have to know what is going on in each factory at all times. As expected with something so complex, major glitches popped up like gremlins. The first Dreamliner to show up at Boeing's factory was missing tens

The first Boeing 787 Dreamliner takes shape in the final assembly plant in Everett, Washington. The new commercial airplane is assembled with major components produced worldwide.

AP Photo/John Froschauer

of thousands of parts. Supplier problems ranged from language barriers to problems caused by some contractors who outsourced major portions of their assigned work and then experienced problems with their suppliers. The first fuselage section, the big multi-part cylindrical barrel that encompasses the passenger seating area, failed in company testing, causing Boeing to make more sections than planned and to reexamine quality and safety concerns. Software programs designed by a variety of manufacturers had trouble talking to one another and the overall weight of the airplane was too high, especially the carbon-fiber wing. These and many other glitches caused major delays in the promised deliveries of the first 787s. The in-service date for the 787 is now set at late-2011 or even 2012, more than three years behind schedule.

Did the advantages of collaboration on such a large scale outweigh the loss of logistical and design control? The jury is still out on that question; however, Boeing's customers are not happy with all of the delays. Nonetheless, Boeing has more than 843 orders for the Dreamliner.

Source: Elizabeth Rennie, "Beyond Borders," *APICS Magazine* (March 2007), pp. 34–38; Stanley Holmes, "The 787 Encounters Turbulence," *Business Week* (June 19, 2006), pp. 38–40; J. Lynn Lunsford, "Boeing Scrambles to Repair Problems With New Plane," *The Wall Street Journal* (December 7, 2007), p. A1. "Boeing 787 Dreamliner," http://en.wikipedia.org/wiki/Boeing_787_Dreamliner

Strategic Implications

A supply chain is, of course, a network of firms. Thus, each firm in the chain should design its own supply chains to support the competitive priorities of its services or products. Even though extensive technologies such as the Internet, computer-assisted design, flexible manufacturing, and automated warehousing have been applied to all stages of the supply chain, the performance of many supply chains remains dismal. A study of the U.S. food industry estimated that poor coordination among supply chain partners wastes $30 billion annually. One possible cause for failures is that managers do not understand the nature of the demand for their services or products and, therefore, cannot design supply chains to satisfy those demands. Two distinct designs used to competitive advantage are *efficient supply chains* and *responsive supply chains*. Table 10.1 shows the environments that best suit each design.

Efficient Supply Chains

The nature of demand for the firm's services or products is a key factor in the best choice of supply chain strategy. Efficient supply chains work best in environments where demand is highly predictable, such as demand for staple items purchased at grocery stores or demand for a package delivery service.

TABLE 10.1 | ENVIRONMENTS BEST SUITED FOR EFFICIENT AND RESPONSIVE SUPPLY CHAINS

Factor	Efficient Supply Chains	Responsive Supply Chains
Demand	Predictable, low forecast errors	Unpredictable, high forecast errors
Competitive priorities	Low cost, consistent quality, on-time delivery	Development speed, fast delivery times, customization, volume flexibility, variety, top quality
New-service/product introduction	Infrequent	Frequent
Contribution margins	Low	High
Product variety	Low	High

Common Designs There is one popular design for efficient supply chains.

- *Build-to-stock* (BTS): The product is built to a sales forecast and sold to the customer from a finished goods stock. The end customer has no individual inputs into the configuration of the product and typically purchases the product from a retailer. Examples include groceries, books, appliances, and housewares.

The focus of the BTS supply chain is on efficient service, material, monetary, and information flows; and keeping inventories to a minimum. Because of the markets the firms serve, service or product designs last a long time, new introductions are infrequent, and variety is small. Such firms typically produce for markets in which price is crucial to winning an order. Contribution margins are low and efficiency is important. Consequently, efficient supply chains have competitive priorities of low-cost operations, consistent quality, and on-time delivery.

Responsive Supply Chains

Responsive supply chains are designed to react quickly in order to hedge against uncertainties in demand. They work best when firms offer a great variety of services or products and demand predictability is low.

Common Designs There are three popular designs for responsive supply chains.

- *Assemble-to-order* (ATO): The product is built to customer specifications from a stock of existing components. Customers can choose among various standard components in arriving at their own products; however they have no control over the design of the components. Assembly is delayed until the order is received. This is the design mass customizers use as depicted in Figure 10.7. Examples include Dell's approach to customizing desktops and laptops and automobile manufacturers who offer a selection of options with each model.

- *Make-to-order* (MTO): The product is based on a standard design; however, component production and manufacture of the final product is linked to the customer's specifications. Examples include custom made clothing, such as that offered by Land's End and Tommy Hilfiger, pre-designed houses, and commercial aircraft, such as Boeing as depicted in Managerial Practice 10.1.

- *Design-to-order* (DTO): The product is designed and built entirely to the customer's specifications. This supply chain allows customers

Efficient supply chains need to keep logistical costs to a minimum. Here vessels loaded with containers berth at Singapore's Keppel Port, one of the world's most efficient, and busiest, sea ports.

Roslan Rahman/AFP/Getty Images

to design the product to fit their specific needs. Examples include large construction projects, women's designer dresses, custom made men's suits, and original architecture house construction.

To stay competitive, firms in a responsive supply chain frequently introduce new services or products. Nonetheless, because of the innovativeness of their services or products, they enjoy high contribution margins. Typical competitive priorities for responsive supply chains are development speed, fast delivery times, customization, variety, volume flexibility, and top quality. The firms may not even know what services or products they need to provide until customers place orders. In addition, demand may be short-lived, as in the case of fashion goods. The focus of responsive supply chains is reaction time, which helps avoid keeping costly inventories that ultimately must be sold at deep discounts.

A firm may need to utilize both types of supply chains, especially when it focuses its operations on specific market segments or it can segment the supply chain to achieve two different requirements. For example, the supply chain for a standard product, such as an oil tanker, has different requirements than that for a customized product, such as a luxury liner, even though both are ocean-going vessels and both may be manufactured by the same company. You might also see elements of efficiency and responsiveness in the same supply chain. For example, Gillette uses an efficient supply chain to manufacture its products so that it can utilize a capital-intensive manufacturing process, and then it uses a responsive supply chain for the packaging and delivery processes to be responsive to retailers. The packaging operation involves customization in the form of printing in different languages. Just as processes can be broken into parts, with different process structures for each, supply chain processes can be segmented to achieve optimal performance.

The Design of Efficient and Responsive Supply Chains

Table 10.2 contains the basic design features for efficient and responsive supply chains. The more downstream in an efficient supply chain that a firm is, the more likely it is to have a line-flow strategy that supports high volumes of standardized services or products. Consequently, suppliers in efficient supply chains should have low capacity cushions because high utilization keeps the cost per unit low. High inventory turns are desired because inventory investment must be kept low to achieve low costs. Firms should work with their suppliers to shorten lead times, but care must be taken to use tactics that do not appreciably increase costs. For example, lead times for a supplier could be shortened by switching from rail to air transportation; however, the added cost may offset the savings obtained from the shorter lead times. Suppliers should be selected with emphasis on low prices, consistent quality, and on-time delivery. Because of low capacity cushions, disruptions in an efficient supply chain can be costly and must be avoided. Figure 10.8 shows that firms with large batch, line, or continuous processes are more likely to be part of an efficient supply chain.

By contrast, firms in a responsive supply chain should be flexible and have high capacity cushions. WIP inventories should be positioned in the chain to support delivery speed, but inventories of expensive finished goods should be avoided. Firms should aggressively work with their suppliers to shorten lead times because it allows them to wait longer before committing to a customer order—in other words, it gives them greater flexibility. Firms should select suppliers to support the competitive priorities of the services or products provided, which in this case would include the ability to provide quick deliveries, customize services or components, adjust volumes

TABLE 10.2 | **DESIGN FEATURES FOR EFFICIENT AND RESPONSIVE SUPPLY CHAINS**

Factor	Efficient Supply Chains	Responsive Supply Chains
Operation strategy	Make-to-stock standardized services or products; emphasize high volumes	Assemble-to-order, make-to-order, or design-to-order customized services or products; emphasize variety
Capacity cushion	Low	High
Inventory investment	Low; enable high inventory turns	As needed to enable fast delivery time
Lead time	Shorten, but do not increase costs	Shorten aggressively
Supplier selection	Emphasize low prices, consistent quality, on-time delivery	Emphasize fast delivery time, customization, variety, volume flexibility, top quality

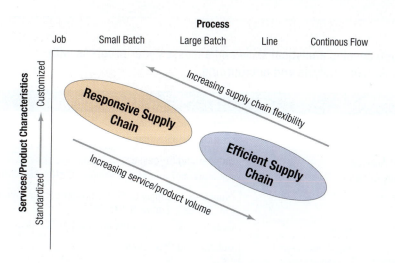

Process

Job   Small Batch   Large Batch   Line   Continous Flow

Services/Product Characteristics (Standardized → Customized)

Responsive Supply Chain

Efficient Supply Chain

Increasing supply chain flexibility

Increasing service/product volume

◄ **FIGURE 10.8**

Linking Supply Chain Design to Processes and Service/Product Characteristics

quickly to match demand cycles, offer variety, and provide top quality. Figure 10.8 shows that firms with job or small batch processes are more likely to be a part of a responsive supply chain.

Poor supply chain performance often is the result of using the wrong supply chain design for the services or products provided. A common mistake is to use an efficient supply chain in an environment that calls for a responsive supply chain. Over time, a firm may add options to its basic service or product, or introduce variations, so that the variety of its offerings increases dramatically and demand for any given service or product predictability drops. Yet, the firm continues to measure the performance of its supply chain as it always has, emphasizing efficiency, even when contribution margins would allow a responsive supply chain design. Clearly, aligning supply chain operations to the firm's competitive priorities has strategic implications.

Gillette uses both supply chain designs: efficient and responsive. The capital intensive processes in its Boston factory support an efficient supply chain to keep costs down. Gillette uses a responsive supply chain for packaging and delivery to service its retail customers.

Brooks Kraft/Sygma/Corbis

LEARNING GOALS IN REVIEW

1 Explain the strategic importance of supply chain design. Review Figures 10.1 and 10.2 for the big picture of supply chain design. The section "Supply Chain Design across the Organization," pp. 361–362, reveals the internal functional pressures that impinge on supply chain design.

2 Identify the nature of supply chains for service providers, as well as for manufacturers. See the section "Supply Chain for Services and Manufacturing," pp. 362–363, and Figures 10.3 and 10.4 for examples.

3 Define the critical supply chain performance measures. The section "Measures of Supply Chain Performance," pp. 363–367, discusses the important inventory and financial measures. Be sure to understand Example 10.1 and the Solved Problem.

4 Explain the strategy of mass customization and its implications for supply chain design. The section "Mass Customization," pp. 367–369, describes the competitive advantages and the implications for supply chain design.

5 Define the important decision factors to be considered when employing an outsourcing or offshoring strategy. Outsourcing and offshoring are discussed in detail in the section "Outsourcing Processes," pp. 369–372. Be sure to review Example 10.2, which uses break-even analysis for the make-or-buy decision. Managerial Practice 10.1 shows how complex the outsourcing decision can become.

6 Explain how efficient supply chains differ from responsive supply chains and the environments best suited for each type of supply chain. Review the section "Strategic Implications," pp. 372–375. Be sure to understand Tables 10.1 and 10.2.

MyOMLab helps you develop analytical skills and assesses your progress with multiple problems on break-even analysis and outsourcing.

MyOMLab Resources	Titles	Link to the Book
Video	*Inventory and Textbooks*	Strategic Implications
	Clif Bar: Supply Chain	Strategic Implications
Active Models Exercises	A.1 Break-Even Analysis	Outsourcing Processes
	A.2 Make-or-Buy Decision	Outsourcing Processes; Example 10.2 (pp. 369–370)
OM Explorer Solvers	Inventory Estimator	Measures of Supply Chain Performance; Example 10.1 (pp. 364–365); Figure 10.5 (p. 365)
	Financial Measures Analyzer	Case: Brunswick Distribution, Inc. (pp. 381–383)
OM Explorer Tutors	A.1 Break-Even, Evaluating Products and Services	Outsourcing Processes
	A.2 Break-Even Evaluating Processes	Outsourcing Processes; Example 10.2 (pp. 369–370)
	10.1 Calculating Inventory Measures	Measures of Supply Chain Performance; Example 10.1 (pp. 364–365)
	F.4 NPV, IRR, Payback	Measures of Supply Chain Performance; Case: Brunswick Distribution, Inc. (pp. 381–383)
POM for Windows	Break-Even Analysis	Outsourcing Processes; Example 10.2 (pp. 369–370)
	Financial Analysis	Measures of Supply Chain Performance; Case: Brunswick Distributors, Inc. (pp. 381–383)
Tutor Exercise	10.1 Calculating inventory measures under different scenarios	Inventory Measures; Example 10.1 (pp. 364–365)
Virtual Tours	10.1 Yamaha Construction	Strategic Implications
	10.2 Jagger Yarn	Supply Chains for Services and Manufacturing; Figures 10.3 and 10.4 (p. 363)
MyOMLab Supplement	F. Financial Analysis	Measures of Supply Chain Performance; Case: Brunswick Distributors, Inc. (pp. 381–383)
Internet Exercise	10.1 Walmart	Strategic Implications
Key Equations		
Image Library		

Key Equations

1. Average aggregate inventory value = average inventory of each SKU multiplied by its value, summed over all SKUs held in stock.

2. Weeks of supply $= \dfrac{\text{Average aggregate inventory value}}{\text{Weekly sales (at cost)}}$

3. Inventory turnover $= \dfrac{\text{Annual sales (at cost)}}{\text{Average aggregate inventory value}}$

4. Make-or-buy break-even quantity: $Q = \dfrac{F_m - F_b}{c_b - c_m}$

Key Terms

average aggregate inventory value 364
backward integration 370
centralized placement 367
channel assembly 369
forward integration 370

forward placement 367
inventory pooling 367
inventory turnover 364
make-or-buy decision 369
offshoring 370

outsourcing 369
supply chain design 361
weeks of supply 364

Solved Problem

A firm's cost of goods sold last year was $3,410,000, and the firm operates 52 weeks per year. It carries seven items in inventory: three raw materials, two work-in-process items, and two finished goods. The following table contains last year's average inventory level for each item, along with its value.

a. What is the average aggregate inventory value?

b. How many weeks of supply does the firm maintain?

c. What was the inventory turnover last year?

Category	Part Number	Average Level	Unit Value
Raw materials	1	15,000	$3.00
	2	2,500	5.00
	3	3,000	1.00
Work-in-process	4	5,000	14.00
	5	4,000	18.00
Finished goods	6	2,000	48.00
	7	1,000	62.00

SOLUTION

a.

Part Number	Average Level		Unit Value		Total Value
1	15,000	×	$3.00	=	$ 45,000
2	2,500	×	5.00	=	12,500
3	3,000	×	1.00	=	3,000
4	5,000	×	14.00	=	70,000
5	4,000	×	18.00	=	72,000
6	2,000	×	48.00	=	96,000
7	1,000	×	62.00	=	62,000
			Average aggregate inventory value	=	$360,500

b. Average weekly sales at cost = $3,410,000/52 weeks = $65,577/week

$$\text{Weeks of supply} = \frac{\text{Average aggregate inventory value}}{\text{Weekly sales (at cost)}} = \frac{\$360,500}{\$65,577} = 5.5 \text{ weeks}$$

c. $$\text{Inventory turnover} = \frac{\text{Annual sales (at cost)}}{\text{Average aggregate inventory value}} = \frac{\$3,410,000}{\$360,500} = 9.5 \text{ turns}$$

Discussion Questions

1. Explain how a firm can reduce costs while improving the performance of its supply chain.

2. The Walmart retail chain sells standardized items and enjoys great purchasing clout with its suppliers, none of which it owns. The Limited retail chain sells fashion goods and owns Mast Industries, which is responsible for producing many of the items sold in The Limited stores. The Limited boasts that it can go from the concept for a new garment to the store shelf in 1,000 hours. Compare and contrast the implications for supply chain design for these two retail systems.

3. Canon, a Japanese manufacturer of photographic equipment, decided against offshoring and kept its manufacturing and new product development processes in Japan, which has relatively high labor costs. In contrast, GM, headquartered in the United States, has a joint venture with Shanghai Auto Industry Corporation (SAIC) to produce cars in China. Given our discussion of outsourcing, offshoring, and supply chain design, discuss how these two seemingly diverse decisions could be supportive of each company's operations strategy.

Problems

The OM Explorer and POM for Windows software is available to all students using the 10th edition of this textbook. Go to **www.pearsonhighered.com/krajewski** to download these computer packages. If you purchased MyOMLab, you also have access to Active Models software and significant help in doing the following problems. Check with your instructor on how best to use these resources. In many cases, the instructor wants you to understand how to do the calculations by hand. At the least, the software provides a check on your calculations. When calculations are particularly complex and the goal is interpreting the results in making decisions, the software replaces entirely the manual calculations.

1. EBI Solar uses a high-tech process to turn silicon wafers into tiny solar panels. These efficient and inexpensive panels are used to power low-energy hand-held electronic devices. Last year, EBI Solar turned their inventory 4.5 times and had a cost of goods sold of $2.5 million. Assuming 52 business weeks per year:

   a. Express last year's average inventory in weeks of supply.

   b. After several supply chain improvement initiatives, inventory investment has dropped across all inventory categories. While EBI's cost of goods sold is not expected to change from last year's level, the value of raw materials has dropped to $100,500; work-in-process to $25,800; and finished goods to $16,200. Assuming 52 business weeks per year, express EBI's current total inventory level in weeks of supply and inventory turns.

2. Buzzrite, a retailer of casual clothes, ended the current year with annual sales (at cost) of $48 million. During the year, the inventory of apparel turned over six times. For the next year, Buzzrite plans to increase annual sales (at cost) by 25 percent.

   a. What is the increase in the average aggregate inventory value required if Buzzrite maintains the same inventory turnover during the next year?

   b. What change in inventory turns must Buzzrite achieve if, through better supply chain management, it wants to support next year's sales with no increase in the average aggregate inventory value?

3. Jack Jones, the materials manager at Precision Enterprises, is beginning to look for ways to reduce inventories. A recent accounting statement shows the following inventory investment by category: raw materials, $3,129,500; work-in-process, $6,237,000; and finished goods, $2,686,500. This year's cost of goods sold will be about $32.5 million. Assuming 52 business weeks per year, express total inventory as

   a. Weeks of supply

   b. Inventory turns

4. One product line has 10 turns per year and an annual sales volume (at cost) of $985,000. How much inventory is being held, on average?

5. The Bawl Corporation supplies alloy ball bearings to auto manufacturers in Detroit. Because of its specialized manufacturing process, considerable work-in-process and raw materials are needed. The current inventory levels are $2,470,000 and $1,566,000, respectively. In addition, finished goods inventory is $1,200,000 and sales (at cost) for the current year are expected to be about $48 million. Express total inventory as

   a. Weeks of supply

   b. Inventory turns

6. The following data were collected for a retailer:

Cost of goods sold	$3,500,000
Gross profit	$700,000
Operating costs	$500,000
Operating profit	$200,000
Total inventory	$1,200,000
Fixed assets	$750,000
Long-term debt	$300,000

Assuming 52 business weeks per year, express total inventory as

   a. Weeks of supply

   b. Inventory turns

Advanced Problems

Problems 8, 10, 11, and 12 require prior reading of Supplement A, "Decision Making."

7. Sterling, Inc., operates 52 weeks per year, and its cost of goods sold last year was $6,500,000. The firm carries eight items in inventory: four raw materials, two work-in-process items, and two finished goods. Table 10.3 shows last year's average inventory levels for these items, along with their unit values.

   a. What is the average aggregate inventory value?

   b. How many weeks of supply does the firm have?

   c. What was the inventory turnover last year?

TABLE 10.3 | STERLING INVENTORY ITEMS

Category	Part Number	Average Inventory Units	Value per Unit
Raw materials	RM-1	20,000	$1
	RM-2	5,000	5
	RM-3	3,000	6
	RM-4	1,000	8
Work-in-process	WIP-1	6,000	10
	WIP-2	8,000	12
Finished goods	FG-1	1,000	65
	FG-2	500	88

8. A large global automobile manufacturer is considering outsourcing the manufacturing of a solenoid used in the transmission of its SUVs. The company estimates that annual fixed costs of manufacturing the part in-house, which include equipment, maintenance, and management, amounts to $6 million. The variable costs of labor and material are $5.00 per unit. The company has an offer from a major subcontractor to produce the part for $8.00 per unit. However, the subcontractor wants the company to share in the costs of the equipment. The automobile company estimates that the total cost would be $4 million, which also includes management oversight for the new supply contact.

   a. How many solenoids would the automobile company need per year to make the in-house option least costly?

   b. What other factors, besides costs, should the automobile company consider before revising its supply chain for SUVs?

9. Dogs-R-Us and K-9, Inc. are two retail stores that cater to the needs of dog owners in the greater Charleston area. There is healthy competition between these two establishments. Both operate 52 weeks a year and both sell approximately the same type and dollar value of items. Table 10.4 provides the cost of goods sold, the average inventory level, and unit value of each item sold in the two stores.

   a. Compare the two retail stores in terms of average aggregate inventory value.

   b. Compare the two retail stores in terms of weeks of supply.

   c. Compare the two retail stores in terms of inventory turnover.

10. Black Bear Outfitters is an international supplier of outdoor gear for families. Currently, the company uses a logistical provider to provide warehouse services and handle packages destined for ground delivery. The contract calls for $9 million in annual fixed charges, which covers the provider's overhead and warehouse costs, and variable costs of $15 per package shipped. Recently, Black Bear Outfitters found a warehouse it could lease at a cost of $16 million per year, which includes lease costs, labor, and management oversight. Furthermore, the company found another provider who would deliver packages from the warehouse for $6.00 per package. Considering only costs, how many packages must Black Bear Outfitters ship to make the vertical integration into warehouse operations beneficial?

11. At the BlueFin Bank corporate headquarters, management was discussing the potential of outsourcing the processing of credit card transactions to DataEase, an international provider of banking operational services. Processing of the transactions at BlueFin has been a costly element of the annual profit and loss statement and the continual investment in equipment to keep up to date has been draining capital reserves. Based upon initial study and negotiations, DataEase will charge $0.02 more per transaction than BlueFin's cost per transaction, and DataEase will want $12 million per year to cover equipment and overhead costs associated with the contract. BlueFin has yet to develop an estimate for the annual overhead and fixed costs associated with processing the transactions. These costs include supervision, administrative support, maintenance, equipment depreciation, and overhead. If BlueFin must process 20 million transactions per year, how high must those fixed costs be before it would pay to use DataEase?

12. A global manufacturer of electrical switching equipment (ESE) is considering outsourcing the manufacturing of an electrical breaker used in the manufacturing of switch boards. The company estimates that the annual fixed cost of manufacturing the part in-house, which includes equipment, maintenance, and management, amounts to $8 million. The variable cost of labor and materials are $11.00 per breaker. The company has an offer from a major subcontractor to produce the part for $16.00 per breaker.

TABLE 10.4 | INVENTORY DATA FOR DOGS-R-US AND K-9, INC. STORES

	DOGS-R-US		K-9, INC.	
Cost of Goods Sold	$560,000.00		$640,000.00	
Category	Average Inventory in Units	Value per Unit	Average Inventory in Units	Value per Unit
Dog Beds	200	$55.00	140	$55.00
Dog Bones & Treats	1,200	$2.50	250	$2.50
Pet Feeders	50	$12.50	20	$12.50
Flea & Tick	350	$7.50	75	$7.50
Dog Kennels	10	$65.00	2	$65.00
Dog Pens	10	$220.00	3	$220.00
Patio Pet Doors	5	$120.00	2	$120.00
Dog Ramps	5	$150.00	2	$150.00
Pet Strollers	10	$40.00	2	$40.00
Pet Supplements	1,400	$4.50	150	$4.50
Dog Toys	250	$2.20	100	$2.20

a. How many breakers would the electrical switching equipment company need per year to make the in-house option the least costly?

b. Assume the subcontractor wants the company to share in the costs of the equipment. The ESE company estimates that the total annual cost would be $5 million, which also includes management oversight for the new supply contract. For this concession, the subcontractor will drop the per unit price to $12.00. Under this assumption, how many breakers would the ESE company need per year to make the in-house option least costly?

c. If the ESE manufacturer is expecting to use 1,500,000 breakers per year, which option (make in-house, use subcontractor without sharing in the cost of equipment, use subcontractor with sharing in the cost of equipment) is the least costly?

EXPERIENTIAL LEARNING   Sonic Distributors

Scenario

Sonic Distributors produces and sells music CDs. The CDs are pressed at a single facility (factory), issued through the company's distribution center, and sold to the public from various retail stores. The goal is to operate the distribution chain at the lowest total cost.

Materials (available from instructor)

Retail and distributor purchase order forms

Factory work order forms

Factory and distributor materials delivery forms

Inventory position worksheets

A means of generating random demand (typically a pair of dice)

Setup

Each team is in the business of manufacturing music CDs and distributing them to retail stores where they are sold. Two or more people play the role of retail outlet buyers. Their task is to determine the demand for the CDs and order replenishment stock from the distributor. The distributor carries forward-placed stock obtained from the factory. The factory produces in lot sizes either to customer order or to stock.

Tasks

Divide into teams of four or five.

Two or three people operate the retail stores.

One person operates the distribution center.

One person schedules production at the factory.

Every day, as play progresses, the participants at each level of the supply chain estimate demand, fill customer orders, record inventory levels, and decide how much to order or produce and when to place orders with their supplier.

Costs and Conditions

Unless your instructor indicates otherwise, the following costs and conditions hold.

Costs

Holding cost per unit per day	Retail outlets: $1.00/CD/day
	Distribution Center: $0.50/CD/day
	Factory: $0.25/CD/day
Pipeline inventory cost	Assume that pipeline cost can be ignored for this exercise (consider it zero).
Ordering cost (retailers and distributors)	$20/order
Factory setup cost (to run an order)	$50 (Note: Cost is per order, not per day, because even though successive orders from distributors are for the same item, the factory is busy fabricating other items between orders.)
Stockout (lost margin) cost	Retail Store: $8 per CD sale lost in a period
	$0 for backorders for shortages from the factory or shipping new orders
Shipping cost	Because other products are already being distributed through this chain and because CDs are light and take up little volume, consider the cost to be zero.

Conditions

Starting inventory	Retail stores each have 15 CDs
	Distribution center has 25 CDs
	Factory has 100 CDs
Lot-sizing restrictions	Retail outlets and distribution centers—no minimum order. Any amount may be stored. Factory production lot sizes and capacity—produce in minimum lots of 20. Maximum capacity: 200/day.
Outstanding orders	None

Delays

Ordering Delay. One day to send an order from a retail store to the distributor or from the distributor to the factory (that is, 1 day is lost between placing an order and the recipient acting on it).

No delay occurs in starting up production once an order has been received (but 1 day is needed for delivery of an order from the distributor to the factory).

Delivery Delay. One-day shipping time between the distributor and a retail store or between the factory and the distributor (that is, 1 day is lost between shipping an order and receiving it).

Run the Exercise

For simplicity's sake, assume all transactions take place simultaneously at the middle of the day. For every simulated day, the sequence of play goes as follows.

Retailers

a. Each retailer receives any shipment due in from its distributor (1 day after shipment) and places it in sales inventory (adds the quantity indicated on any incoming Material Delivery Form from the distributor—after its 1-day delay—to the previous day's ending inventory level on the Retailer's Inventory Position Worksheet). (*Note:* For the first day of the exercise, no order will come in.)

b. The retailers each determine the day's retail demand (the quantity of CDs requested) by rolling a pair of dice. The roll determines the number demanded.

c. Retailers fill demand from available stock, if possible. Demand is filled by subtracting it from the current inventory level to develop the ending inventory level, which is recorded. If demand exceeds supply, sales are lost. Record all lost sales on the worksheet.

d. Retailers determine whether an order should be placed. If an order is required, the desired quantity of CDs is written on a Retail Store Purchase Order, which is forwarded to the distributor (who receives it after a 1-day delay). If an order is made, it should be noted on the worksheet. Retailers may also desire to keep track of outstanding orders separately.

Distributor

a. The distributor receives any shipment due in from the factory and places the CDs in available inventory (adds the quantity indicated on any incoming Material Delivery Form from the factory—after its 1-day delay—to the previous day's ending inventory level on the distributor's inventory position worksheet).

b. All outstanding backorders are filled (the quantity is subtracted from the current inventory level indicated on the worksheet) and prepared for shipment. CDs are shipped by filling out a Distribution Center Material Delivery Form indicating the quantity of CDs to be delivered.

c. The distributor uses the purchase orders received from the retail stores (after the designated 1-day delay) to prepare shipments for delivery from available inventory. Quantities shipped are subtracted from the current level to develop the ending inventory level, which is recorded. If insufficient supply exists, backorders are generated.

d. The distributor determines whether a replenishment order should be placed. If an order is required, the quantity of CDs is written on a Distribution Center Purchase Order, which is forwarded to the factory (after a 1-day delay). If an order is made, it should be noted on the worksheet. The distributor may also desire to keep track of outstanding orders separately.

Factory

a. The factory places any available new production into inventory (adds the items produced the previous day to the previous day's ending inventory level on the Factory Inventory Position Worksheet).

b. All outstanding backorders are filled (the quantity is subtracted from the current inventory level indicated on the worksheet) and prepared for shipment. CDs are shipped by filling out a Factory Material Delivery Form, indicating the quantity of CDs to be delivered.

c. The factory obtains the incoming distributor's purchase orders (after the designated 1-day delay) and ships them from stock, if it can. These amounts are subtracted from the current values on the inventory worksheet. Any unfilled orders become backorders for the next day.

d. The factory decides whether to issue a work order to produce CDs either to stock or to order. If production is required, a Factory Work Order is issued, and the order is noted on the inventory worksheet. Remember that a setup cost applies to each *production* order. It is important to keep careful track of all production in process.

Remember, once an order has been placed, it cannot be changed and no partial shipments can be made. For each day, record your ending inventory position, backorder or lost sales amount, and whether an order was made (or a production run initiated). After everyone completes the transactions for the day, the sequence repeats, beginning at retailer step (a). Your instructor will tell you how many simulated days to run the exercise.

When the play is stopped, find the cumulative amount of inventory and other costs. You can do so by summing up the numbers in each column and then multiplying these totals by the costs previously listed. Use the total of these costs to assess how well your team operated the distribution chain.

Source: This exercise was developed by Larry Meile, Carroll School of Management, Boston College. By permission of Larry Meile.

CASE   Brunswick Distribution, Inc.

Alex Brunswick, CEO of Brunswick Distribution, Inc. (BDI), looked out his office window at another sweltering day and wondered what could have gone wrong at his company. He just finished reviewing his company's recent financial performance and noticed something that worried him. BDI had experienced a period of robust growth over the last 4 years. "What could be going wrong?" he thought to himself. "Our sales have been growing at an average rate of 8 percent over the last 4 years but we still appear to be worse off than before." He sat back in his chair with a heavy sigh and continued reviewing the report on his desk.

Sales had risen consistently over the past 4 years but the future was uncertain. Alex Brunswick was aware that part of the past growth had largely been the result of a few competitors in the region going out of business, a situation that was unlikely to continue. Net earnings, however, had been declining for the last 3 years and were expected to decline next year.

Brunswick was determined to turn his company around within the next 3 years. He sat back from his desk and buzzed his personal assistant: "Carla, could you ask Marianna and Bradley to come up?"

Background

The distribution business, in its simplest form, involves the purchase of inventory from a variety of manufacturers and its resale to retailers. Over the last 3 to 5 years, demands on inventory changed considerably; neither manufacturers nor retailers want to handle inventory, leaving distributors to pick up the slack. In addition, an increased tendency of retailers to order directly from manufacturers placed further strain on the profitability of distributorships in general.

After humble beginnings in a shed behind the house of Brunswick's grandmother, the company moved to a 10,000 square-foot leased facility. Ten years ago, BDI began distributing high-end appliance products to supplement its low-margin products. BDI entered into an agreement with KitchenHelper Corp., a large manufacturer of high-end kitchen appliances, located 35 miles from Moline, Illinois, to distribute KitchenHelper appliances to customers in the region. Over the years BDI enjoyed steady growth and expanded its area of coverage. Currently, Brunswick was covering an area with a radius of 200 miles from the company's main facility. Given the rapid growth, BDI purchased the leased facility and made additions to bring its capacity to 30,000 square feet.

The demise of several of its competitors resulted in the acquisition of new retailer customers and some new product lines. Traditional ordering in the retailer-distributor-manufacturer chain took place via fax or telephone. Brunswick considered implementing an Internet-based ordering system but was unsure of the potential operational and marketing benefits that it could provide.

Concerns

Market

Direct competition from distributors increased over the past 5 years. As a result, the most successful distributors adopted a value-added strategy in order to remain competitive. Retailers want dependable delivery to support sales promotions and promises to customers. They also want the freedom to hold sales promotions at any time as competitive conditions dictate and with only short notice to distributors. They also want the opportunity to choose from a wide variety of appliances. Nonetheless, many orders are won on the basis of price and lost on the basis of delivery problems.

Financial

Manufacturers commonly demand payment in 30 to 45 days and provide no financing considerations. Retailers, on the other hand, pay in 50 to 60 days. This difference often leaves BDI in a cash-poor situation that puts an unnecessary strain on its current operating loan. The company's borrowing capacity has almost been exhausted. Any additional financing will have to be sought from alternative sources. Given BDI's financial situation, any additional financing will be issued at a higher charge than the company's existing debt.

Operations

Inventory turnover also presented a problem for the past 5 years. In the past 2 years, however, a significant downturn in turnover occurred. This trend seems likely to continue.

Orders from retailers come in as their customers near completion of construction or renovations. Even though historical information provided a good benchmark of future sales, the changing market lessened the reliability of the information. The changes also affect BDI's ordering. Manufacturers require projections 60, 90, and 120 days out in order to budget their production. Sometimes penalties are assessed when BDI changes an order after it is placed with a manufacturer.

Strategic Issues

As Marianna and Bradley walked into Brunswick's office, he was still pondering the report. "Grab a seat," he grunted. They knew they were going to have a long day. Brunswick quickly briefed them on why he had summoned them, and they all immediately dove into a spirited discussion. Brunswick pointed out that BDI would need to be properly structured to deal with the recession and the reality of today's market. "We need to be well-positioned for growth

as the market stabilizes," he said. In order to meet this challenge, BDI must evaluate a number of alternative options. Some of the possible options might include expanding current systems and, when necessary, developing new systems that interface with suppliers, customers, and commercial transportation resources to gain total asset visibility.

Before making any investment decision, Brunswick reminded them that BDI would have to evaluate any new capital requirements, as well as the expected contribution to the company's bottom line and market share, that any option might provide. Exhibit 1 shows the income statement for the current year.

Investing in New Infrastructure

Bradley Pulaski, vice president of operations, said, "Since Associated Business Distribution Corp. ceased operations 4 years ago, we have been inundated with phone calls and e-mails from potential customers across the Midwest looking for an alternative to ABD's services. These requests come not only from former ABD customers, but also from potential customers that have not dealt with either ABD or us in the past. We cannot adequately service this market from our current warehouse because the customers do not want to wait for lengthy deliveries. We are currently servicing some customers in that region; however, I do not think we can keep them much longer because of delayed deliveries. In order to take advantage of this opportunity, we would have to construct a new storage facility to complement our already strained resources and 'forward position' inventory to shorten our delivery times to customers on short notice. We are challenged by an inadequate infrastructure far too small for our requirements. We only have the Moline warehouse at this time." The addition of new facilities would provide BDI with an opportunity for increased penetration in key industrial markets in the upper Midwest where the company has had a limited presence.

EXHIBIT 1 ▼

Company Income Statement ($000's)

Revenue		33,074
Cost of Goods Sold		
Shipping costs	8,931	
Direct materials	5,963	
Direct labor and other	6,726	
Total	21,620	
Gross Profit		11,454
Operating Expenses		
Selling expenses	2,232	
Fixed expenses	2,641	
Depreciation	1,794	
Total	6,667	
Earnings before Interest and Taxes		4,787
Interest expense		838
Earnings before Taxes		3,949
Taxes @ 35%		1,382
Net Income		2,567

The financing resources for this option would be a challenge, given that BDI was approaching its credit limit with its principal bank. Additional financing from larger banks in Chicago, however, was not ruled out. It would be expensive (with current interest rates for long-term loans starting at 11 percent). According to Bradley, this option would cost $2 million for property and $10 million for plant and equipment. The new warehouse facilities would be depreciated over 20 years. The 20-year loan would be repaid with a single balloon payment at the end of the loan. With the additional infrastructure, BDI would be able to increase its annual sales by $4,426,000. In addition, delivery lead times to customers in the region would be reduced from 5 days to 2 days, which would be very competitive. Because of the added warehouse capacity, BDI could also increase the number of brands and models of appliances to better serve the retailers' needs for more variety. However, certain categories in the costs of goods sold would also increase. Total annual shipping costs, which include supplier deliveries to the warehouse as well as deliveries to the customer, would increase by $955,000. Annual materials costs (for the sold appliances) and labor costs would each increase by 6 percent. Total assets would increase from $30,170,000 to $43,551,000. This increase takes into account changes to inventory investment, which would become $7,200,000, accounts receivable, property, and plant and equipment.

Streamlining the Distribution System

Marianna Jackson, the vice president of logistics, stated, "I believe there is an opportunity to capitalize on the void left by our fallen rivals by utilizing a cost-efficient distribution system. We do not need a new facility; we can continue to serve the customers in the Midwest as best we can. However, what we do need is an efficient distribution system. We are holding a considerable amount of stock that has not moved simply because of our inefficient inventory systems. One of our top priorities is working diligently with the inventory control department to keep what we need and dispose of what we do not need. This approach will allow us to use the space recovered from the unneeded items for automated warehouse equipment that will enable us to become more efficient. Everything we do and every dollar we spend affects our customers. We need to keep our prices competitive. Our cost of operations is our customers' cost. Our goal is to enable customers to spend their resources on readiness and the tools of their trade, not logistics. This option will not help us much with product variety or delivery speed; however, it will increase our on-time delivery performance and improve our flexibility to respond to changes in retailer orders to support their sales programs."

The option of having an integrated center, comprised of sophisticated automation systems, advanced materials handling equipment, and specially developed information technology, would provide BDI with both the versatility and capacity to offer improved products and services to Brunswick's customers. The system would support real-time ordering, logistics planning and scheduling, and after-sales service. When an order is received through a call center at Brunswick's offices in Moline, it will be forwarded to a logistics center for processing. The customer is given a delivery date based on truck availability. Orders would be grouped by destination so that trucks could be efficiently loaded to maximize the truck capacity. The order would then be scheduled for delivery and the customer notified of the estimated arrival. This new information technology would improve BDI's reliability in delivering the products when promised. The system also includes an automatic storage and retrieval system (AS/RS). The AS/RS selects a customer order and moves it to a dock for loading on a truck headed for the customer's location. The capital costs for this system would be $7 million, which would be depreciated over a 10-year period. The operating costs, including training, would run at $0.5 million each year. These costs would be considered fixed expenses by Brunswick. The improved system, however, would have tremendous cost savings. Marianna estimated that the system would save up to 16 percent in shipping expenses and 16 percent in labor expenses annually. Total assets would increase from $30,170,000 to $35,932,000 to account for changes in accounts receivables and equipment. Aggregate inventories would be only $4,500,000 because of the reduced need for safety stock inventories. BDI could finance this option using a 10-year loan at a 10 percent rate of interest. The loan would be repaid with a balloon payment at the end of the loan.

These savings would come from more efficient handling of customers' orders by the call center, better planning and scheduling of shipments, and improved communication with the warehouse and the customer, resulting in a dramatic reduction in the shipping costs in the supply chain. Additional savings would result from the reduction in personnel costs; fewer operators would be required. Marianna Jackson thought that BDI could maintain its current level of service with her option while becoming much more efficient.

The Decision

Alex Brunswick pondered the two options posed by Bradley Pulaski and Marianna Jackson. Bradley's option enabled the firm to increase its revenues by serving more customers. The capital outlay was sizable, however. Marianna's option focused on serving the firm's existing customers more efficiently. The value of that option was its dramatic reduction in costs; however, it was uncertain whether BDI could hold onto its current upper Midwest customers. Brunswick realized that he could not undertake both options, given the company's current financial position. Brunswick uses a 12 percent cost of capital as the discount rate when making financial decisions. How will each option affect the firm's operational and financial performance measures, which investors watch closely? Which supply chain design option would be better for the company?

Selected References

Aron, Ravi, and Jitendra V. Singh. "Getting Offshoring Right." *Harvard Business Review* (December 2005), pp. 135–143.

de Waart, Dick, and Steve Kemper. "5 Steps to Service Supply Chain Excellence." *Supply Chain Management Review* (January/February 2004), pp. 28–35.

Duray, Rebecca. "Mass Customization Origins: Mass or Custom Manufacturing?" *International Journal of Operations and Production Management*, vol. 22, no. 3 (2002), pp. 314–328.

Ellram, Lisa M., and Baohong Liu. "The Financial Impact of Supply Management." *Supply Chain Management Review* (November/December 2002), pp. 30–37.

Fisher, Marshall L. "What Is the Right Supply Chain for Your Product?" *Harvard Business Review* (March/April 1997), pp. 105–116.

Flynn, Laurie J. "Built to Order." *Knowledge Management.* (C=December 11, 2000), www.destinationkm.com

Garber, Randy, and Suman Sarkar. "Want a More Flexible Supply Chain?" *Supply Chain Management Review* (January/February 2007), pp. 28–34.

Glatzel, Christoph, Jocher Großpietsch, and Ildefonso Silva. "Is Your Top Team Undermining Your Supply Chain?" *McKinsey Quarterly*, (January 2011), pp. 1–6.

Goel, Ajay K., Nazgol Moussavi, and Vats N. Srivastan. "Time to Rethink Offshoring?" *McKinsey on Business Technology: Innovations in IT Management*, No. 14 (Winter 2008), pp. 32–35.

Grey, William, Kaan Katircioglu, Dailun Shi, Sugato Bagchi, Guillermo Gallego, Mark Adelhelm, Dave Seybold, and Stavros Stefanis. "Beyond ROI." *Supply Chain Management Review* (March/April 2003), pp. 20–27.

Hartly-Urquhart, Roland. "Managing the Financial Supply Chain." *Supply Chain Management Review* (September 2006), pp. 18–25.

Hartvigsen, David. *SimQuick: Process Simulation with Excel*, 2nd ed. Upper Saddle River, NJ: Prentice Hall, 2004.

Hofman, Debra. "Supply Chain Measurement: Turning Data Into Action." *Supply Chain Management Review* (November 2007), pp. 20–26.

Lee, Hau L. "The Triple-A Supply Chain." *Harvard Business Review* (October 2004), pp. 102–112.

Reeve, James M., and Mandyam M. Srinivasan. "Which Supply Chain Design is Right for You?" *Supply Chain Management Review* (May/June 2005), pp. 50–57.

Roberts, Dexter. "China's Factory Blues." *Businessweek* (April 7, 2008), pp. 78–82.

Slone, Reuben E., John T. Mentzer, and J. Paul Dittmann. "Are You the Weakest Link in Your Supply Chain?" *Harvard Business Review* (September 2007), pp. 116–127.

Stavrulaki, Euthemia, and Mark Davis. "Aligning Products with Supply Chain Processes and Strategy." *The International Journal of Logistics Management*, vol. 21, no. 1 (2010), pp. 127–151.

Tiede, Tom, and Kay Ree Lee. "What is an Optimal Distribution Network Strategy?" *Supply Chain Management Review* (November 2005), pp. 32–39.

SUPPLY CHAIN LOCATION DECISIONS

A completed BMW X6 crossover vehicle awaits final inspection at the company's plant in Spartanburg, South Carolina, U.S. This plant location was carefully chosen after an extensive study and numerous government concessions. The X6 and X5 'Sport Activity Vehicles,' as well as the Z4 roadster have been designed and built at this plant.

Bavarian Motor Works (BMW)

Bavarian Motor Works (BMW), founded in 1917 and headquartered in Munich, Germany, is a manufacturer of select premium segment brands such as BMW, MINI, and Rolls-Royce Motor Cars in the international automobile market. When faced with fluctuating exchange rates and increasing production costs in the late 1980s, BMW decided that it was time to consider operating a new production facility outside the European borders. A "blank page" approach was used to compile a list of 250 potential worldwide plant sites. Further analysis pared the list down to 10 viable options; a plant location in the United States was preferred due to its proximity to a large market segment for BMW's automobiles.

The selection of the plant site involved many factors that had to be analyzed prior to its construction. BMW considered the labor climate in each country, geographical requirements and constraints, and its relations with the governments of the countries in which the prospective sites were located. In terms of the labor climate, a technologically capable workforce was needed due to the complex nature of the automotive manufacturing process. Because the cost to train a single worker in the automotive industry is between $10,000 and $20,000, this factor was especially critical. Geographical factors had to be examined because thousands of automobile parts needed to be delivered from both domestic and foreign suppliers. In order to keep the supply chain costs down, it was decided that the new location should have ample highway/interstate access and be reasonably close to a port from which both supplies and finished automobiles could be easily transported. Another consideration was easy access to an airport for BMW's executives traveling back and forth to its headquarters in Germany. The final location factor was government related. BMW wanted to move to a location that was "business friendly" in terms of making concessions on issues such as infrastructural improvements, tax abatements, and employee screening and education programs. The overall goal was to make the relationship between BMW and the local community as mutually beneficial as possible through a coordinated improvement effort.

After a 3 1/2-year search process that stringently evaluated the 10 viable options across these location factors, BMW finally decided to build a new 2 million square-foot production facility in Spartanburg, South Carolina. The final decision was made based on a good match between the aforementioned selection criteria and the environment in Spartanburg. South Carolina lawmakers proved flexible and open as to how the state would address the needs set forth by BMW. For instance, they agreed to acquire the 500 acres necessary to build the plant (requiring a $25 million bond package be passed), improve the highway system around the facility (requiring $10 million), and lengthen the runway and modernize the terminal at the Spartanburg airport ($40 million expenditure). The legislature also agreed to provide tax incentives, property tax relief, and establish an employee screening and training program to ensure the right mix of workers were available. (Processing the applications alone proved to be a daunting task because more than 50,000 applications were received.) South Carolina may not have scored the highest on each decision criterion, but taken as a whole, the Spartanburg location was best for BMW.

This location proved to be a good one. The plant, which opened in July 1994, subsequently underwent a $200 million expansion in 1996, a $50 million expansion in 1999, a $300 million expansion in 2000, and another $750 million dollar expansion in March 2008. BMW Manufacturing Corporation in South Carolina today is part of BMW Group's global manufacturing network, and currently employs over 7,000 people to produce the X3 and X5 Sports Activity Vehicle and the X6 Sports Activity Coupe at its 1,150-acre, 4-million-square-foot campus. Apart from a nearly $5 billion dollar investment by BMW, South Carolina also reaped rewards in the form of business growth (BMW has 170 North American suppliers with 40 in South Carolina alone), employment, and community improvements—a success story all around.

Source: "Manager's Journal: Why BMW Cruised into Spartanburg," *Wall Street Journal* (July 6, 1992), p. A10; "BMW Announces Its Plans for a Plant in South Carolina," *Wall Street Journal* (June 24, 1992), p. B2; P. Galuszka, "The South Shall Rise Again," *Chief Executive* (November 2004), pp. 50–54; Southern Business & Development, **www.sb-d.com** (June 2005); **www.bmwusfactory.com**, May 2011.

LEARNING GOALS *After reading this chapter, you should be able to:*

1. Explain how location decisions relate to the design of supply chains.

2. Identify factors affecting location choices.

3. Understand the role of geographical information systems in making location decisions.

4. Understand single facility location techniques.

5. Understand multiple facilities location methods.

Firms like BMW evaluate their supply chain network in its entirety when deciding where to locate a new facility. **Facility location** is the process of determining geographic sites for a firm's operations, which could include a manufacturing plant, a distribution center, and a customer service center. **Distribution center** is a warehouse or a stocking point where goods are stored for subsequent distribution to manufacturers, wholesalers, retailers, and customers. Location choices can be critically important for firms, and have a profound impact on its supply chains. For example, they can affect the supplier relationship process. The expanding global economy gives firms greater access to suppliers around the world, many of whom can offer lower input costs or better-quality services and products. Nonetheless, when manufacturing facilities are offshored, locating far from one's suppliers can lead to higher transportation costs and coordination difficulties. The customer relationship process can also be affected by the firm's location decisions. If the customer must be physically present at the process, it is unlikely that a location will be acceptable if the time or distance between the service provider and customer is great. If, on the other hand, customer contact is more passive and impersonal, or if materials or information is processed rather than people, then location may be less of an issue. Information technology and the Internet can sometimes help overcome the disadvantages related to a company's location. Still, one thing is clear: The location of a business's facilities has a significant impact on the company's operating costs, the prices it charges for services and goods, and its ability to compete in the marketplace and penetrate new customer segments.

Analyzing location patterns to discover a firm's underlying strategy is fascinating. Recognizing the strategic impact location decisions have on implementing a firm's strategy and supply chain design, we first consider the qualitative factors that influence location choices and their implications across the organization. Subsequently we examine an important trend in location patterns: the use of geographical information systems (GIS) to identify market segments and how serving each segment can profitably affect the firm's location decisions. We end by presenting some analytic techniques for making single- and multiple-facility location decisions and understanding their impact across the supply chain.

Location Decisions across the Organization

Location decisions affect processes and departments throughout the organization. When locating new retail facilities, such as Wendy's stores, marketing must carefully assess how the location will appeal to customers and possibly open up new markets. Relocating all or part of an organization can significantly affect the attitudes of the firm's workforce and the organization's ability to operate effectively across departmental lines. Location also has implications for a firm's human resources department, which must be attuned to the firm's hiring and training needs. Locating new facilities or relocating existing facilities is usually costly; therefore, these decisions must be carefully evaluated by the organization's accounting and finance departments. For instance, when BMW located its manufacturing plant in South Carolina, the economic environment of the state and the monetary incentives offered by its legislators played a role in the financial payoff associated with the proposed new plant. Finally, operations also has an important stake in location decisions because the location needs to be able to meet current customer demand and provide the right amount of customer contact (for both external and internal customers). When their manufacturing plants are far away, firms like Gillette create active involvement by locating distribution centers in foreign countries where employees know the local culture and the language, and offer "one face to the customer." International operations, like those of McDonald's, Starbucks, Toyota, and Walmart, introduce a new set of challenges because setting up and managing facilities and employees in foreign countries can be extremely time-consuming and difficult. Yet, it is an important part of a firm's growth. For instance, Starbucks has over 17,000 locations, with 6,000 of these sites being located in 50 countries spread across the globe.

Factors Affecting Location Decisions

Managers of both service and manufacturing organizations must weigh many factors when assessing the desirability of particular locations, including their proximity to customers and suppliers, labor costs, and transportation costs. Managers generally can disregard factors that fail to meet at least one of the following two conditions:

1. *The Factor Must Be Sensitive to Location.* In other words, managers should not consider a factor not affected by the location decision. For example, if community attitudes are uniformly good at all the locations under consideration, community attitudes should not be considered as a factor.

---

facility location

The process of determining geographic sites for a firm's operations.

distribution center

A warehouse or stocking point where goods are stored for subsequent distribution to manufacturers, wholesalers, retailers, and customers.

Creating Value through Operations Management

Using Operations to Compete
Project Management

Managing Processes

Process Strategy
Process Analysis
Quality and Performance
Capacity Planning
Constraint Management
Lean Systems

Managing Supply Chains

Supply Chain Inventory Management
Supply Chain Design
Supply Chain Location Decisions
Supply Chain Integration
Supply Chain Sustainability and Humanitarian Logistics
Forecasting
Operations Planning and Scheduling
Resource Planning

Starbucks within malls are a given, but only in Dubai will you find such a fancy mall as the massive (and themed!) Ibn Battuta Mall. While the Starbucks store is nothing special to gawk at, the Persia Court where it's located is a stunning tiled dome. The mall is essentially a homage to the Arab traveller and adventurer Ibn Battuta and each part of the mall represents a different land he traveled to. This Starbucks is located in Persia and the dome is made up of Persian blue art and tiles. But we don't think Battuta actually encountered any Starbucks on his travels.

Typhoonsk/Dreamstime.com

2. *The Factor Must Have a High Impact on the Company's Ability to Meet Its Goals.* For example, although different facilities will be located at different distances from suppliers, if the shipments from them can take place overnight and the communication with them is done via fax, e-mail, or teleconferencing, the distance is not likely to have a large impact on the firm's ability to meet its goals. It should therefore not be considered as a factor.

Managers can divide location factors into dominant and secondary factors. Dominant factors are derived from competitive priorities (cost, quality, time, and flexibility) and have a particularly strong impact on sales or costs. For example, a favorable labor climate and monetary incentives were dominant factors affecting the decision to locate the BMW plant in Spartanburg, South Carolina. Secondary factors also are important, but management may downplay or even ignore some of these secondary factors if other factors are more important. Thus, for GM's Saturn plant, which makes many of its parts onsite, inbound transportation costs were considered less important and therefore were a secondary factor.

Dominant Factors in Manufacturing

The following seven groups of factors dominate the decisions firms, including BMW, make about the location of new manufacturing plants or distribution centers. Often there is a trade-off among factors. Locating facilities in, say, a location with high labor costs might make sense if other factors such as logistics, taxes, and proximity to customers are favorable. Lowering the total costs of designing, developing, manufacturing, and distributing a product to its market becomes especially important in developing international supply chains and finding locations for plants, distribution centers, software design studios, and the like.

Favorable Labor Climate   A favorable labor climate may well be the most important factor for labor-intensive firms in industries such as textiles, furniture, and consumer electronics. Labor climate is a function of wage rates, training requirements, attitudes toward work, worker productivity, and union strength. Many executives perceive weak unions or a low probability of union organizing efforts as a distinct advantage. Having a favorable climate applies not only to the workforce already on site, but also to the employees that a firm hopes will transfer to or will be attracted to the new site. Boeing made a decision in 2009 to locate its assembly lines for the Dreamliner planes in Charleston, South Carolina, because of the favorable labor climate, as well as presence of other Boeing facilities and suppliers in the area. It was a very carefully thought and crafted decision because there are only three sites worldwide at which commercial wide body jets are assembled—Everett, Washington; Charleston, South Carolina; and Toulouse, France (Airbus plants). The 1.2 million square foot plant was formally inaugurated in June 2011 despite a complaint filed by the National Labor Relations Board on behalf of the labor unions in the state of Washington. Boeing would maintain assembly of the Dreamliner planes in both locations, and actually added 2000 new jobs in Washington State to support that effort.

Proximity to Markets   After determining where the demand for services and goods is greatest, management must select a location for the facility that will supply that demand. Locating near markets is particularly important when the final goods are bulky or heavy and *outbound* transportation rates are high. For example, manufacturers of products such as plastic pipe and heavy metals require proximity to their markets.

Impact on Environment   As the focus on sustainability has increased, firms are looking to recognize the impact of the location decisions on the environment. Along with minimizing the carbon footprint of the new facility and its accompanying facilities in the supply chain, consideration must also be given to reducing overall energy costs. These and related issues are covered in greater detail in Chapter 13, "Supply Chain Sustainability and Humanitarian Logistics."

Quality of Life   Good schools, recreational facilities, cultural events, and an attractive lifestyle contribute to **quality of life**. This factor can make the difference in location decisions. In the United States during the past 2 decades, more than 50 percent of new industrial jobs went to nonurban regions. A similar shift is taking place in Japan and Europe. Reasons for this movement include high cost of living, high crime rate, and general decline in the quality of life in many large cities.

Proximity to Suppliers and Resources   Firms dependent on inputs of bulky, perishable, or heavy raw materials emphasize proximity to their suppliers and resources. In such cases, *inbound* transportation costs become a dominant factor, encouraging such firms to locate facilities near suppliers. For example, locating paper mills near forests and food-processing facilities near farms is practical. Another advantage of locating near suppliers is the ability to maintain lower inventories (see Chapter 8, "Lean Systems," and Chapter 10, "Supply Chain Design").

Proximity to the Parent Company's Facilities   In many companies, plants supply parts to other facilities or rely on other facilities for management and staff support. These ties require frequent communication and coordination, which can become more difficult as distance increases.

Utilities, Taxes, and Real Estate Costs   Other location decision factors include utility costs (telephone, energy, and water), local and state taxes, financing incentives offered by local or state governments, relocation costs, and land costs. For example, the location of the Daimler plant in Alabama for manufacturing its "M series" vehicles, the BMW plant in South Carolina in the opening vignette, and a Toyota plant in Georgetown, Kentucky, were all attractive to these companies in part due to the incentives from local governments.

Other Factors   Still other secondary factors may need to be considered, including room for expansion, construction costs, accessibility to multiple modes of transportation, the cost of shuffling people and materials between plants, insurance costs, competition from other firms for the workforce, local ordinances (such as pollution or noise control regulations), community attitudes, and many others. For global operations, firms need a good local infrastructure and local employees who are educated and have good skills. Many firms are concluding that large, centralized manufacturing facilities in low-cost countries with poorly trained workers are not sustainable. Smaller, flexible facilities located in the countries that the firm serves allow it to avoid problems related to trade barriers like tariffs and quotas and the risk that changing exchange rates will adversely affect its sales and profits.

Dominant Factors in Services

The factors mentioned for manufacturers also apply to service providers with one important addition: the impact of location on sales and customer satisfaction. Customers usually care about how close a service facility is, particularly if the process requires considerable customer contact.

Proximity to Customers   Location is a key factor in determining how conveniently customers can carry on business with a firm. For example, few people will patronize a remotely located dry cleaner or supermarket if another is more convenient. Thus, the influence of location on revenues tends to be the dominant factor. In addition, customer proximity by itself is not enough—the key is proximity to customers who will patronize the facility and seek its services. Being close to customers who match a firm's target market and service offerings is thus important for profitability.

Transportation Costs and Proximity to Markets   For warehousing and distribution operations, transportation costs and proximity to markets are extremely important. With a warehouse nearby, many firms can hold inventory closer to the customer, thus reducing delivery time and promoting sales. For example, Invacare Corporation of Elyria, Ohio, gained a competitive edge in the distribution of home health care products by decentralizing inventory into 32 warehouses across the country. Invacare sells wheelchairs, hospital beds, and other patient aids—some of which it produces and some of which it buys from other firms—to small dealers who sell to consumers. Previously the dealers, often small mom-and-pop operations, had to wait three weeks for deliveries, which meant their cash was tied up in excess inventory. With Invacare's new distribution

This Scottrade office in Indianapolis, Indiana, is among 363 local brokerage branches located near where their customers live and work.

network, the dealers get daily deliveries of products from one source. Invacare's location strategy shows how timely delivery can be a competitive advantage, and helped it to be ranked among the 2009 *Industry Week's* U.S. 500 manufacturing firms.

Location of Competitors   One complication related to estimating the sales potential of different locations is the impact of competitors. Management must not only consider the current location of competitors, but also try to anticipate their reaction to the firm's new location. Avoiding areas where competitors are already well-established often pays off. However, in some industries, such as new-car sales showrooms and fast-food chains, locating near competitors is actually advantageous. The strategy is to create a **critical mass**, whereby several competing firms clustered in one location attract more customers than the total number who would shop at the same stores at scattered locations. Recognizing this effect, some firms use a follow-the-leader strategy when selecting new sites.

Site-Specific Factors   Retailers also must consider the level of retail activity, residential density, traffic flow, and site visibility. Retail activity in the area is important because shoppers often decide on impulse to go shopping or to eat in a restaurant. Traffic flows and visibility are important because customers arrive in cars. Management considers possible traffic tie-ups, traffic volume and direction by time of day, traffic signals, intersections, and the position of traffic medians. Visibility involves distance from the street and the size of nearby buildings and signs. A high residential density increases nighttime and weekend business if the population in the area fits the firm's competitive priorities and target market segment.

Geographical Information Systems and Location Decisions

A **geographical information system (GIS)** is a system of computer software, hardware, and data that the firm's personnel can use to manipulate, analyze, and present information relevant to a location decision. A GIS can also integrate different systems to create a visual representation of a firm's location choices. Among other things, it can be used to (1) store databases, (2) display maps, and (3) create models that can take information from existing datasets, apply analytic functions, and write results into newly derived datasets. Together, these three functionalities of data storage, map displays, and modeling are critical parts of an intelligent GIS, and are used to a varying extent in all GIS applications.

A GIS system can be a really useful decision-making tool because many of the decisions made by businesses today have a geographical aspect. A GIS stores information in several databases that can be naturally linked to places, such as customer sales and locations, or a census tract, or the percentage of residents in the tract that make a certain amount of money a year. The demographics of an area include the number of people in the metropolitan statistical area, city, or ZIP code; average income; number of families with children; and so forth. These demographics may all be important variables in the decision of how best to reach the target market. Similarly, the road system, including bridges and highways, location of nearby airports and seaports, and the terrain (mountains, forests, lakes, and so forth), play an important role in facility location decisions. As such, a GIS can have a diverse set of location-related applications in different industries such as the retail, real estate, government, transportation, and logistics industries.

Managerial Practice 11.1 illustrates how fast-food chains use GIS to select sites. Governmental data can provide a statistical mother lode of information used to make better GIS-based location decisions. Internet sites on Yahoo!, MapQuest, and Expedia, among others, allow people to pull up maps, distances and travel times, and routes between locations, such as between Toronto, Ontario, and San Diego, California. In addition, search engines such as Google can be integrated with population demographics to create information of interest in social and business domains. Web sites are using Google maps to display high crime areas, the location of cheap gas, and apartments for rent.

Many different types of GIS packages are available, such as ArcInfo (from ESRI), MapInfo (from MapInfo), SAS/GIS (from SAS Institute, Inc.), and Microsoft's MapPoint. Many of these systems are tailored to a specific application such as locating retail stores, redistricting legislative districts, analyzing logistics and marketing data, environmental management, and so forth. Because of its wide spread availability and ease of use, MapPoint by Microsoft is an easy-to-use and fairly inexpensive GIS that mainly focuses on everyday business use by nontechnical analysts. Its ability to display information on maps can be a powerful decision-making tool especially since the maps and much of the census data comes with the software itself instead of having to be purchased separately from the GIS vendor in many other systems. MyOMLab has three videos on how MapPoint can be used to make location decisions.

GIS can be useful for identifying locations that relate well to a firm's target market based on customer demographics. When coupled with other location models, sales forecasting models, and geo-demographic systems, it can give a firm a formidable array of decision-making tools for its location decisions.

---

critical mass

A situation whereby several competing firms clustered in one location attract more customers than the total number who would shop at the same stores at scattered locations.

geographical information system (GIS)

A system of computer software, hardware, and data that the firm's personnel can use to manipulate, analyze, and present information relevant to a location decision.

MyOMLab

MANAGERIAL PRACTICE 11.1   How Fast-Food Chains Use GIS to Select Their Sites

Until recently, fast-food chains used consultants to analyze geo-demographic data (description of different characteristics about people based upon the location where they live or work) for strategic planning and making franchise location and marketing decisions. Now with the availability of easy-to-use GIS systems that cost less than $5,000 and can be operated on a regular PC, small and large fast-food chains are doing it on their own. For instance, Marco's Franchising, headquartered in Toledo, Ohio, uses MapInfo's (a Windows based mapping and geographical analysis application) GIS solutions to identify new markets where the customer and competitor landscape are best for new sites. MapInfo's Smart Site Solutions and AnySite Online technologies supply interactive mapping and reporting functionality to examine market level deployment strategies and individual site opportunities. These programs can estimate the total dollars up for grabs in a market by analyzing local age and income data from the U.S. Census Bureau as well as sales data from stores in an area—numbers that are commonly available through third-party vendors. The programs can also tell the optimal number and locations of stores in a market, and how much in sales a store can expect. Analyses can be run for any U.S. market and can rank markets in order of viability. A list of realistic sites with high sales potential can be put together at times in less than a minute. Other small fast-food chains in the United States, like Cousins Subs and 99 Restaurants and Pubs, are using in-house GIS and getting a handsome return on the investments. For instance, 99 Restaurants and Pubs found that it was able to recoup its GIS-related investment in a single week.

Bigger nationwide fast-food chains such as Domino's Pizza use GIS software to screen alternative sites for new franchises, determine how moving a store a few blocks away can affect sales, and decide when they should relocate or remodel existing stores. They can also use GIS to identify

A Domino's Pizza store in the NYC neighborhood of Chelsea

overlapping delivery zones and zones that are not being covered. AFC Enterprises, which owns and franchises the Popeye's and Church's chains of restaurants, uses GIS to help it sell franchises. The level of detailed information that it can provide to prospective franchisees can make all the difference when it comes to closing the deal.

Because of its ability to provide these insights, GIS is a useful tool for expanding fast-food chains that need to quickly master the demographic details of competitive terrains in thousands of locations across the country.

Source: www.gis.com/whatisgis/index.html; Ed Rubinstein, "Chains Chart Their Course of Actions with Geographic Information Systems," *Nation's Restaurant News*, vol. 32, no. 6 (1998), p. 49; "MapInfo Delivers Location Intelligence for Marco's Pizza," *Directions Magazine* (December 14, 2004), www.directionsmag.com/press.releases/?duty=Show&id=10790; Ryan Chittum, "Location, Location, and Technology: Where to Put That New Store? Site-Selection Software May Be Able to Help," *Wall Street Journal* (July 18, 2005), p. R7; http://www.pbinsight.com/welcome/mapinfo/ (May 27, 2011).

Locating a Single Facility

Having examined trends and important factors in location, we now consider more specifically how a firm can make location decisions. In this section, we consider the case of locating only one new facility. Managers must first decide whether to expand onsite, build another facility, or relocate to another site. Onsite expansion has the advantage of keeping people together, reducing construction time and costs, and avoiding splitting up operations. However, as a firm expands a facility, at some point diseconomies of scale set in. Poor materials handling, increasingly complex production control, and simple lack of space are reasons for building a new plant or relocating the existing plant.

The advantages of building a new plant or moving to a new retail or office space are that the firm does not have to rely on the production from a single plant. A new plant allows it to hire more employees, install newer, more productive machinery and better technology, and reduce transportation costs. Most firms that choose to relocate are small (comprised of fewer than 10 employees). They tend to be single-location companies cramped for space and needing to redesign their production processes and layouts. More than 80 percent of all relocations are made within 20 miles of companies' original locations, which enables the firms to retain their current employees.

On an average, it is less costly to relocate a service-oriented business than a manufacturing business. For instance, relocating a major check or credit card transaction processing facility for a bank may not be as difficult as moving a manufacturing plant. Because proximity to customers is important, the location of service facilities must constantly be reevaluated in the context of shifting populations and their changing needs. At times, a combination of all three options—staying at the same location, relocating, and opening a new facility—might simultaneously be considered.

When the facility is part of a firm's larger network of facilities, we assume that it is not interdependent; that is, a decision to open a restaurant in Tampa, Florida, is independent of whether the chain has a restaurant in Austin, Texas. Let us begin by considering how to decide whether a new location is needed, and then examine a systematic selection process aided by what is known as the *load–distance method* to deal with proximity.

Comparing Several Sites

A systematic selection process begins after perception or evidence indicates that opening a retail outlet, warehouse, office, or plant in a new location will improve performance. The process of selecting a new facility location involves a series of steps.

1. Identify the important location factors and categorize them as dominant or secondary.

2. Consider alternative regions; then narrow the choices to alternative communities and finally to specific sites.

3. Collect data on the alternatives from location consultants, state development agencies, city and county planning departments, chambers of commerce, land developers, electric power companies, banks, and onsite visits. Some of these data and information may also be contained inside the GIS.

4. Analyze the data collected, beginning with the *quantitative* factors—factors that can be measured in dollars, such as annual transportation costs or taxes. The quantitative factors can also be measured in terms other than dollars, such as driving time and miles. These values may be broken into separate cost categories (for example, inbound and outbound transportation, labor, construction, and utilities) and separate revenue sources (say sales, stock or bond issues, and interest income). These financial factors can then be converted to a single measure of financial merit such as total costs, return on investment (ROI), or net present value (NPV), and used to compare two or more sites, especially if capital costs for the new facility are also considered.

5. Bring the qualitative factors pertaining to each site into the evaluation. A *qualitative* factor is one that cannot be evaluated in dollar terms, such as community attitudes, environmental factors, or quality of life. To merge quantitative and qualitative factors, some managers review the expected performance of each factor, while others assign each factor a weight of relative importance and calculate a weighted score for each site, using a preference matrix (see Supplement A, "Decision Making"). What is important in one situation may be unimportant or less important in another. The site with the highest weighted score is best.

After thoroughly evaluating all potential sites, those making the study prepare a final report containing site recommendations, along with a summary of the data and analyses on which they are based. An audiovisual presentation of the key findings usually is delivered to top management in large firms.

EXAMPLE 11.1	**Calculating Weighted Scores in a Preference Matrix**

MyOMLab

Tutor 11.1 in MyOMLab provides another example to practice with a preference matrix for location decisions.

A new medical facility, Health-Watch, is to be located in Erie, Pennsylvania. The following table shows the location factors, weights, and scores (1 = poor, 5 = excellent) for one potential site. The weights in this case add up to 100 percent. A weighted score (*WS*) will be calculated for each site. What is the *WS* for this site?

Location Factor	Weight	Score
Total patient miles per month	25	4
Facility utilization	20	3
Average time per emergency trip	20	3
Expressway accessibility	15	4
Land and construction costs	10	1
Employee preferences	10	5

SOLUTION

The *WS* for this particular site is calculated by multiplying each factor's weight by its score and adding the results:

$$WS = (25 \times 4) + (20 \times 3) + (20 \times 3) + (15 \times 4) + (10 \times 1) + (10 \times 5)$$
$$= 100 + 60 + 60 + 60 + 10 + 50$$
$$= 340$$

The total *WS* of 340 can be compared with the total weighted scores for other sites being evaluated.

Applying the Load–Distance Method

In the systematic selection process, the analyst must identify attractive candidate locations and compare them on the basis of quantitative factors. The load–distance method is one way to facilitate this step. It works much like the weighted-distance method does for designing layouts for divergent processes (see Chapter 3, "Process Strategy"), whereby loads represent the weights. Several location factors relate directly to distance: proximity to markets, average distance to target customers, proximity to suppliers and resources, and proximity to other company facilities. The **load–distance method** is a mathematical model used to evaluate locations based on proximity factors. The objective is to select a location that minimizes the sum of the loads multiplied by the distance the load travels. Time may be used instead of distance if so desired.

load–distance method
A mathematical model used to evaluate locations based on proximity factors.

Calculating a Load–Distance Score Suppose that a firm planning a new location wants to select a site that minimizes the distances that loads, particularly the larger ones, must travel to and from the site. Depending on the industry, a *load* may be shipments from suppliers, shipments between plants or to customers, or it may be customers or employees traveling to and from the facility. The firm seeks to minimize its load–distance (*ld*) score, generally by choosing a location that ensures loads go short distances.

To calculate the *ld* score for any potential location, we use the actual distance between any two points using a GIS system and simply multiply the loads flowing to and from the facility by the distances traveled. Alternately, rectilinear or Euclidean distance as calculated in Chapter 3, "Process Strategy," can also be used as an approximation for distance using the *x* coordinate and *y* coordinate. Travel time, actual miles, or rectilinear distances when using a grid approach are all appropriate measures for distance. The formula for the *ld* score is

$$ld = \sum_i l_i d_i$$

These loads may be expressed as the number of potential customers needing physical presence for a service facility; loads may be tons or number of trips per week for a manufacturing facility. The score is the sum of these load–distance products. By selecting a new location based on *ld* scores, customer service is improved or transportation costs reduced.

The goal is to find one acceptable facility location that minimizes the score, where the location is defined by its *x* coordinate and *y* coordinate or the longitude and the latitude. Practical considerations rarely allow managers to select the exact location with the lowest possible score. For example, land might not be available there at a reasonable price, or other location factors may make the site undesirable.

Center of Gravity Testing different locations with the load–distance model is relatively simple if some systematic search process is followed. **Center of gravity** is a good starting point to evaluate locations in the target area using the load–distance method. The first step is to determine the *x* and *y* coordinates of different locations either in the form of the longitude and latitude of the locations, or by creating an (*x*, *y*) grid as was used in constructing layouts in Chapter 3, "Process Strategy." The center of gravity's *x* coordinate, denoted x^*, is found by multiplying each point's *x* coordinate (either the longitude of the location or the *x* coordinate on a grid), by its load (l_i), summing these products ($\sum_i l_i x_i$), and then dividing by the sum of the loads ($\sum_i l_i$). The center of gravity's *y* coordinate (either the latitude or the *y* coordinate on a grid), denoted y^*, is found the same way. The formulas are as follows:

center of gravity
A good starting point to evaluate locations in the target area using the load–distance model.

$$x^* = \frac{\sum_i l_i x_i}{\sum_i l_i} \quad \text{and} \quad y^* = \frac{\sum_i l_i y_i}{\sum_i l_i}$$

This location generally is not the optimal one for the distance measures, but it still is an excellent starting point. The load–distance scores for locations in its vicinity can be calculated until the solution is near optimal.

EXAMPLE 11.2	Finding the Center of Gravity for an Electric Utilities Supplier

MyOMLab

Tutor 11.2 in MyOMLab
provides another example on
how to calculate the center of
gravity.

A supplier to the electric utility industry produces power generators; the transportation costs are high. One market area includes the lower part of the Great Lakes region and the upper portion of the southeastern region. More than 600,000 tons are to be shipped to eight major customer locations as shown below:

Customer Location	Tons Shipped	x, y Coordinates
Three Rivers, MI	5,000	(7, 13)
Fort Wayne, IN	92,000	(8, 12)
Columbus, OH	70,000	(11, 10)
Ashland, KY	35,000	(11, 7)
Kingsport, TN	9,000	(12, 4)
Akron, OH	227,000	(13, 11)
Wheeling, WV	16,000	(14, 10)
Roanoke, VA	153,000	(15, 5)

What is the center of gravity for the electric utilities supplier? Using rectilinear distance, what is the resulting load–distance score for this location?

SOLUTION

The center of gravity is calculated (with tons-shipped values in thousands) as shown below:

$$\sum_i l_i = 5 + 92 + 70 + 35 + 9 + 227 + 16 + 153 = 607$$

$$\sum_i l_i x_i = 5(7) + 92(8) + 70(11) + 35(11) + 9(12) + 227(13) + 16(14) + 153(15)$$

$$= 7,504$$

$$x^* = \frac{\sum_i l_i x_i}{\sum_i l_i} = \frac{7,504}{607} = 12.4$$

$$\sum_i l_i y_i = 5(13) + 92(12) + 70(10) + 35(7) + 9(4) + 227(11) + 16(10) + 153(5) = 5,572$$

$$y^* = \frac{\sum_i l_i y_i}{\sum_i l_i} = \frac{5,572}{607} = 9.2$$

The resulting load–distance score is

$$ld = \sum_i l_i d_i = 5(5.4 + 3.8) + 92(4.4 + 2.8) + 70(1.4 + 0.8) + 35(1.4 + 2.2)$$

$$+ 9(0.4 + 5.2) + 227(0.6 + 1.8) + 16(1.6 + 0.8) + 153(2.6 + 4.2)$$

$$= 2,662.4$$

where

$$d_i = |x_i - x^*| + |y_i - y^*|$$

DECISION POINT

The center of gravity is (12.4, 9.2) and the load–distance score is 2,662,400. Solved Problem 3 at the end of this chapter illustrates an example of using latitude and longitude rather than grid coordinates for finding center of gravity.

Using Break-Even Analysis

Break-even analysis can help a manager compare location alternatives on the basis of quantitative factors that can be expressed in terms of total cost (See Supplement A, "Decision Making."). It is particularly useful when the manager wants to define the ranges over which each alternative is best. The basic steps for graphic and algebraic solutions are as follows:

1. Determine the variable costs and fixed costs for each site. Recall that *variable costs* are the portion of the total cost that varies directly with the volume of output. Recall that *fixed costs* are the portion of the total cost that remains constant regardless of output levels.

2. Plot the total cost lines—the sum of variable and fixed costs—for all the sites on a single graph (for assistance, see Tutors A.1 and A.2 in OM Explorer).

3. Identify the approximate ranges for which each location has the lowest cost.

4. Solve algebraically for the break-even points over the relevant ranges.

The Ridge Golf Course in Sedona, Arizona. Amazon.com and Target announced they would both open fulfillment centers in Arizona, due in part to the availability of affordable real estate.

Chad Ehlers/Alamy

| EXAMPLE 11.3 | **Break-Even Analysis for Location** |

An operations manager narrowed the search for a new facility location to four communities. The annual fixed costs (land, property taxes, insurance, equipment, and buildings) and the variable costs (labor, materials, transportation, and variable overhead) are as follows:

Community	Fixed Costs per Year	Variable Costs per Unit
A	$150,000	$62
B	$300,000	$38
C	$500,000	$24
D	$600,000	$30

MyOMLab

Active Model 11.1 in MyOMLab provides insight on defining the three relevant ranges for this example.

MyOMLab

Tutor 11.3 in MyOMLab provides another example to practice break-even analysis for location decisions.

Step 1: Plot the total cost curves for all the communities on a single graph. Identify on the graph the approximate range over which each community provides the lowest cost.

Step 2: Using break-even analysis, calculate the break-even quantities over the relevant ranges. If the expected demand is 15,000 units per year, what is the best location?

SOLUTION

Step 1: To plot a community's total cost line, let us first compute the total cost for two output levels: $Q = 0$ and $Q = 20,000$ units per year. For the $Q = 0$ level, the total cost is simply the fixed costs. For the $Q = 20,000$ level, the total cost (fixed plus variable costs) is as follows:

Community	Fixed Costs	VARIABLE COSTS (Cost per Unit) (No. of Units)	TOTAL COST (Fixed + Variable)
A	$150,000	$62(20,000) = $1,240,000	$1,390,000
B	$300,000	$38(20,000) = $ 760,000	$1,060,000
C	$500,000	$24(20,000) = $ 480,000	$ 980,000
D	$600,000	$30(20,000) = $ 600,000	$1,200,000

Figure 11.1 shows the graph of the total cost lines. The line for community A goes from (0, 150) to (20, 1,390). The graph indicates that community A is best for low volumes, B for intermediate volumes, and C for high volumes. We should no longer consider community D, because both its fixed *and* its variable costs are higher than community C's.

Step 2: The break-even quantity between A and B lies at the end of the first range, where A is best, and the beginning of the second range, where B is best. We find it by setting both communities' total cost equations equal to each other and solving:

(A)	(B)
$150,000 + $62Q = $300,000 + $38Q	
Q = 6,250 units	

The break-even quantity between B and C lies at the end of the range over which B is best and the beginning of the final range where C is best. It is

(B)	(C)
$300,000 + $38Q = $500,000 + $24Q	
Q = 14,286 units	

No other break-even quantities are needed. The break-even point between A and C lies above the shaded area, which does not mark either the start or the end of one of the three relevant ranges.

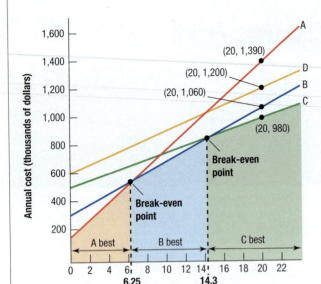

FIGURE 11.1 ▲
Break-Even Analysis of Four
Candidate Locations

DECISION POINT

Management located the new facility at community C, because the 15,000 units-per-year demand forecast lies in the high-volume range. These results can also be used as an input for a final decision using a preference matrix, where other non-quantitative factors could also be incorporated into the decision-making process.

Locating a Facility Within a Supply Chain Network

When a firm with a network of existing facilities plans a new facility, one of two conditions exists: (1) Either the facilities operate independently (examples include a chain of restaurants, health clinics, banks, or retail establishments) or (2) the facilities interact (examples include component manufacturing plants, assembly plants, and warehouses). Independently operating units can be located by treating each as a separate single facility, as described in the preceding section. Managerial Practice 11.2 shows how General Electric's industrial expansion in India requires it to think about locating multiple plants in different industries in a geographically diverse country, as well as having a single plant serve different businesses.

Locating interacting facilities introduces new issues, such as how to allocate work between the facilities and how to determine the best capacity for each. In addition, the facilities may be scattered internationally around the globe with suppliers in Asia, manufacturing plants in Latin America, and distribution warehouses in Europe and United States. Changing the work allocations in turn affects the size (or capacity utilization) of the various facilities. Thus, the multiple-facility location problem has three dimensions—location, allocation, and capacity—that must be solved simultaneously. In many cases, the analyst can identify a workable solution merely by looking for patterns in the cost, demand, and capacity data and using trial-and-error calculations. In other cases, more formal approaches are needed.

Moe's Grill

On March 12, 2007, Raving Brands, owners of Moe's Southwest Grill, announced it inked development agreements for the Middle East, Singapore, and Canada. According to the announcement, these three independent deals resulted in 150 restaurants. Moe's subsequently determined specific locations for these multiple facilities.

MANAGERIAL PRACTICE 11.2 — General Electric's Expansion in India

General Electric (GE) is a diversified global infrastructure, finance, and media conglomerate with presence in appliances, aviation, consumer products, electrical distribution, energy, lighting, and rail industries, among others. With revenues of about 150 billion and a worldwide workforce of 280,000 employees, GE is one of the largest firms in the world. India's infrastructure sector in industries such as energy, aviation, and healthcare is particularly attractive to GE, especially since it plans to generate over 60 percent of its revenue from outside the United States with a focus on emerging markets such as China, Russia, India, and Brazil.

In October 2010, GE received a contract to supply engines for India's lightweight jetfighter, and in the same month won a contract from India's Reliance Power, Ltd. to supply $750 million of turbine equipment for a 2400 megawatt expansion of the Samalkot power plant in the southern state of Andhra Pradesh. It also formed a joint venture with Triveni Engineering and Industries, Ltd. to make steam turbines, as well as formed a health care joint venture with software firm Wipro. To enable the strategic expansion in India for its products and services, GE announced plans in March 2011 to invest $50 million to build a factory in India. This plant would begin at 400,000 square feet in size, and eventually expand to 70,000 square feet. Mr. John Flannery, President and Chief Executive of GE India, characterized this proposed facility as a "multimodal multibusiness facility that several GE businesses will share." Determining the location of such a multibusiness plant is quite a complex task, and akin to locating a facility within a network of facilities. It must take into consideration different facilities in different businesses

Workers putting together an ultra scanner machine at General Electric Wipro in Bangalore, India

that this plant will serve, and the diverse supply chains with which it must interact. Future GE plants in the sub-continent would have to similarly determine their locations to strategically support a multibusiness manufacturing capability in India, particularly as the domestic production expands across a range of industries and GE's investment in India potentially grows to 200 million and beyond.

Source: Santanu Chowdhury. "GE Plans Multiuse Factory in India." *Wall Street Journal* (March 14, 2011); **http://www.ge.com/products_services/index.html**, May 23, 2011.

The GIS Method for Locating Multiple Facilities

GIS tools help visualize customer locations and data, as well as the transportation structure of roads and interstate highways. These capabilities allow the analyst to quickly arrive at a reasonable solution to the multiple-facility location problems. Load–distance score and center of gravity data can be merged with customer databases in Excel to arrive at trial locations for facilities, which can then be evaluated for annual driving time or distance using a GIS such as MapPoint and Excel. A five-step framework that captures the use of GIS for locating multiple facilities is outlined here.

1. Map the data for existing customers and facilities in the GIS.

2. Visually split the entire operating area into the number of parts or subregions that equal the number of facilities to be located.

3. Assign a facility location for each region based on the visual density of customer concentration or other factors. Alternately, determine the center of gravity for each part or subregion identified in step 2 as the starting location point for the facility in that subregion.

4. Search for alternate sites around the center of gravity to pick a feasible location that meets management's criteria such as environmental issues, availability to major metropolitan areas, or proximity to highways.

5. Compute total load–distance scores and perform capacity checks before finalizing the locations for each region.

Such an approach can have many applications, including the design of supply chain distribution networks as illustrated in the Witherspoon Automotive video in MyOMLab.

MyOMLab

The Transportation Method

The **transportation method for location problems** is a quantitative approach that can help solve multiple-facility location problems. We use it here to determine the allocation pattern that minimizes the cost of shipping products from two or more plants, or *sources of supply*, to two or more warehouses, or *destinations*. We focus on the setup and interpretation of the

transportation method for location problems

A quantitative approach that can help solve multiple-facility location problems.

problem, leaving the rest of the solution process to a software package on a computer such as POM for Windows. A fuller development of this problem can be found in Supplement D, "Linear Programming," and textbooks covering quantitative methods and management science.

The transportation method does not solve *all* facets of the multiple-facility location problem. It only finds the *best* shipping pattern between plants and warehouses for a particular set of plant locations, each with a given capacity. The analyst must try a variety of location–capacity combinations and use the transportation method to find the optimal distribution for each one. Distribution costs (variable shipping and possibly variable production costs) are but one important input in evaluating a particular location–allocation combination. Investment costs and other fixed costs also must be considered, along with various qualitative factors. This complete analysis must be made for each reasonable location–capacity combination. Because of the importance of making a good decision, this extra effort is well worth its cost.

Setting Up the Initial Tableau   The first step in solving a transportation problem is to format it in a standard matrix, sometimes called a *tableau*. The basic steps in setting up an initial tableau are as follows:

1. Create a row for each plant (existing or new) being considered and a column for each warehouse.

2. Add a column for plant capacities and a row for warehouse demands and insert their specific numerical values.

3. Each cell not in the requirements row or capacity column represents a shipping route from a plant to a warehouse. Insert the unit costs in the upper right-hand corner of each of these cells.

The Sunbelt Pool Company is considering building a new 500-unit plant because business is booming. One possible location is Atlanta. Figure 11.2 shows a tableau with its plant capacity, warehouse requirements, and shipping costs. The tableau shows, for example, that shipping one unit from the existing Phoenix plant to warehouse 1 in San Antonio, Texas, costs $5.00. Costs are assumed to increase linearly with the size of the shipment; that is, the cost is the same *per unit* regardless of the size of the total shipment.

Plant	Warehouse			Capacity
	San Antonio, TX (1)	Hot Springs, AR (2)	Sioux Falls, SD (3)	
Phoenix	5.00	6.00	5.40	400
Atlanta	7.00	4.60	6.60	500
Requirements	200	400	300	900 / 900

FIGURE 11.2 ▲
Initial Tableau

In the transportation method, the sum of the shipments in a row must equal the corresponding plant's capacity. For example, in Figure 11.2, the total shipments from the Atlanta plant to warehouses 1, 2, and 3 located in San Antonio, Texas; Hot Springs, Arkansas; and Sioux Falls, South Dakota, respectively must add up to 500. Similarly, the sum of shipments to a column must equal the corresponding warehouse's demand requirements. Thus, shipments to warehouse 1 in San Antonio, Texas, from Phoenix and Atlanta must total 200 units.

Dummy Plants or Warehouses   The transportation method also requires that the sum of capacities equal the sum of demands, which happens to be the case at 900 units (see Figure 11.2). In many real problems, total capacity exceeds requirements, or vice versa. If capacity exceeds requirements by *r* units, we add an extra column (a *dummy warehouse*) with a demand of *r* units and make the shipping costs $0 in the newly created cells. Shipments are not actually made, so they represent unused plant capacity. Similarly, if requirements exceed capacity by *r* units, we add an extra row (a *dummy plant*) with a capacity of *r* units. We assign shipping costs equal to the stockout costs of the new cells. If stockout costs are unknown or are the same for all warehouses, we simply assign shipping costs of $0 per unit to each cell in the dummy row. The optimal solution will not be affected because the shortage of *r* units is required in all cases. Adding a dummy warehouse or dummy plant ensures that the sum of capacities equals the sum of demands. Some software packages, such as POM for Windows, automatically add them when we make the data inputs.

Finding a Solution   After the initial tableau has been set up, the goal is to find the least-cost allocation pattern that satisfies all demands and exhausts all capacities. This pattern can be found by using the transportation method, which guarantees the optimal solution. The initial tableau is filled in with a feasible solution that satisfies all warehouse demands and exhausts all plant capacities. Then a new tableau is created, defining a new solution that has a lower total cost. This iterative process continues until no improvements can be made in the current solution, signaling that the optimal solution has been found. When using a computer package, all that you have to input is the information for the initial tableau.

Another procedure is the simplex method (see Supplement D, "Linear Programming"), although more inputs are required. The transportation problem is actually a special case of linear programming, which can be modeled with a decision variable for each cell in the tableau, a constraint for each row in the tableau (requiring that each plant's capacity be fully utilized), and a constraint for each column in the tableau (requiring that each warehouse's demand be satisfied).

Whichever method is used, the number of nonzero shipments in the optimal solution will never exceed the sum of the numbers of plants and warehouses minus 1. The Sunbelt Pool Company has two plants and three warehouses, so there need not be more than 4 (or 3 + 2 − 1) shipments in the optimal solution.

EXAMPLE 11.4	**Interpreting the Optimal Solution**

The optimal solution for the Sunbelt Pool Company, found with POM for Windows, is shown in Figure 11.3. Figure 11.3(a) displays the data inputs, with the cells showing the unit costs, the bottom row showing the demands, and the last column showing the supply capacities. Figure 11.3(b) shows how the existing network of plants supplies the three warehouses to minimize costs for a total of $4,580. Verify that each plant's capacity is exhausted and that each warehouse's demand is filled. Finally, Figure 11.3(c) shows the total quantity and cost of each shipment. The total optimal cost reported in the upper-left corner of Figure 11.3(b) is $4,580, or 200($5.00) + 200($5.40) + 400($4.60) + 100($6.60) = $4,580.

▼ **FIGURE 11.3**
POM for Windows Screens for Sunbelt Pool Company

Figure 11.3(a) Input Data

	San Antonio	Hot Springs	Sioux Falls	SUPPLY
Phoenix	5	6	5.4	400
Atlanta	7	4.6	6.6	500
DEMAND	200	400	300	

Figure 11.3(b) Optimal Shipping Pattern

Optimal cost = $4580	San Antonio	Hot Springs	Sioux Falls
Phoenix	200		200
Atlanta		400	100

Figure 11.3(c) Cost Breakdown

	San Antonio	Hot Springs	Sioux Falls
Phoenix	200/$1000		200/$1080
Atlanta		400/$1840	100/$660

▼ **FIGURE 11.4**
Optimal Transportation Solution for Sunbelt Pool Company

SOLUTION

Figure 11.4 is a map created with the MapPoint software that shows how the plants supply the three warehouses. The Phoenix plant and its shipments are represented in red and the Atlanta plant and its shipments are represented in yellow. The size of the circles for the three warehouses represents their capacities and how much of that capacity is being supplied from which plant. For example, Phoenix ships 200 units to warehouse 1 in San Antonio, Texas, and 200 units to warehouse 3 in Sioux Falls, South Dakota, exhausting its 400-unit capacity. Atlanta ships 400 units of its 500-unit capacity to warehouse 2 in Hot Springs, Arkansas, and the remaining 100 units to warehouse 3 in Sioux Falls, South Dakota. All warehouse demand

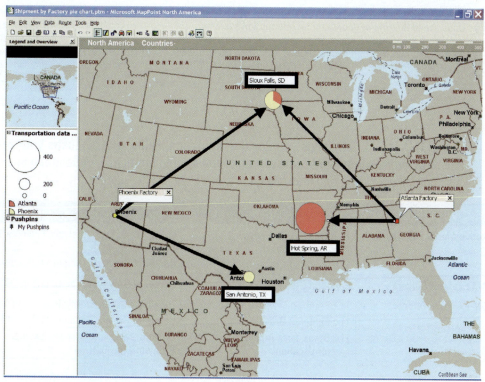

is satisfied: warehouse 1 in San Antonio, Texas is fully supplied by Phoenix and warehouse 2 in Hot Springs, Arkansas by Atlanta. Warehouse 3 in Sioux Falls, South Dakota receives 200 units from Phoenix and 100 units from Atlanta, satisfying its 300-unit demand. The total transportation cost is 200($5.00) + 200($5.40) + 400($4.60) + 100($6.60) = $4,580.

DECISION POINT

Management must evaluate other plant locations before deciding on the best one. The optimal solution does not necessarily mean that the best choice is to open an Atlanta plant. It just means that the best allocation pattern for the current choices on the other two dimensions of this multiple-facility location problem (that is, a capacity of 400 units at Phoenix and the new plant's location at Atlanta) results in total *transportation* costs of $4,580.

The Larger Solution Process   Other costs and various qualitative factors also must be considered as additional parts of a complete evaluation. For example, the annual profits earned from the expansion must be balanced against the land and construction costs of a new plant in Atlanta. Thus, management might use the preference matrix approach (see Example 11.1) to account for the full set of location factors.

The analyst should also evaluate other capacity and location combinations. For example, one possibility is to expand in Phoenix and build a smaller plant at Atlanta. Alternatively, a new plant could be built at another location, or several new plants could be built. The analyst must repeat the analysis for each such likely location strategy.

LEARNING GOALS IN REVIEW

1 **Explain how location decisions relate to the design of supply chains.** Review the opening vignette on BMW, its subsequent discussion on how location decisions impact the core business and supply chain processes, and the section "Location Decisions across the Organization," p. 387.

2 **Identify factors affecting location choices.** See the section "Factors Affecting Location Decisions," pp. 387–390. Focus on understanding the key differences between locating manufacturing versus service facilities.

3 **Understand the role of geographical information systems in making location decisions.** The section "Geographical Information Systems and Location Decisions," pp. 390–391, and Managerial Practice 11.1 shows you how firms are using GIS

software packages to make demographic data driven location decisions that are cheap as well as effective in simultaneously considering several location decision variables.

4 **Understand single facility location techniques.** The section "Locating a Single Facility," pp. 391–396, shows how firms can use the load-distance method, center of gravity, and break-even analysis to find single locations in a simple and easy to use manner.

5 **Understand multiple facilities location methods.** The section "Locating a Facility within a Supply Chain Network," pp. 396–400, shows how the transportation method can be used to minimize production and distribution costs in a supply network characterized by multiple plants and warehouses.

MyOMLab helps you develop analytical skills and assesses your progress with multiple problems on weighted scores, preference matrix, break even, Euclidean and rectilinear distances, center of gravity, and load-distance model.

MyOMLab Resources	Titles	Link to the Book
Active Model Exercise	11.1 Break-Even Analysis for Location	Using Break-Even Analysis; Example 11.3 (p. 395); Solved Problem 2 (p. 403)
	11.2 Center of Gravity	Center of Gravity; Example 11.2 (p. 394); Solved Problem 3 (p. 403)
OM Explorer Solvers	Center of Gravity	Center of Gravity; Example 11.2 (p. 394); Solved Problem 3 (p. 403)
OM Explorer Tutors	11.1 Preference Matrix for Location	Locating a Single Facility; Example 11.1 (p. 392); Solved Problem 1 (p. 402)
	11.2 Center of Gravity	Center of Gravity; Example 11.2 (p. 394); Solved Problem 3 (p. 403)
	11.3 Break-Even Analysis for Location	Using Break-Even Analysis; Example 11.3 (p. 395); Solved Problem 2 (p. 403)

MyOMLab Resources	Titles	Link to the Book
POM for Windows	Transportation Method (Location)	Factors Affecting Location Decisions; The GIS Method for Locating Multiple Facilities; Example 11.4 (p. 399); Solved Problem 4 (p. 405)
	Weighting Method	Factors Affecting Location Decisions; Locating a Facility within a Supply Chain Network
	Two-Dimensional Siting	Factors Affecting Location Decisions; Locating a Facility within a Supply Chain Network
	Cost-Volume Analysis	Factors Affecting Location Decisions
Tutor Exercises	11.1 Center of Gravity	Center of Gravity; Example 11.2 (p. 394); Solved Problem 3 (p. 403)
	11.2 Cost Changes at Able and Charlie	Using Break-Even Analysis; Example 11.3 (p. 395); Solved Problem 2 (p. 403)
MapPoint Videos	Starbucks	Factors Affecting Location Decisions
	Witherspoon Automotive	Locating Facilities within a Supply Chain Network
	Tyler EMS	Factors Affecting Location Decisions
Virtual Tours	McCadam Cheese Company	Factors Affecting Location Decisions
	Jack Daniels Distillery Tour	Factors Affecting Location Decisions
MyOMLab Supplements	F. Financial Analysis	MyOMLab Supplement F
Internet Exercise	Dunkin Donuts	Factors Affecting Location Decisions
	MapQuest	Factors Affecting Location Decisions
	Expedia	Expedia Factors Affecting Location Decisions
	Yahoo	Factors Affecting Location Decisions
	U.S. Census Bureau	Factors Affecting Location Decisions
Key Equations		
Image Library		

Key Equations

1. Load–distance score:

$$ld = \sum_i l_i d_i$$

2. Center of gravity:

$$x^* = \frac{\sum_i l_i x_i}{\sum_i l_i} \quad \text{and} \quad y^* = \frac{\sum_i l_i y_i}{\sum_i l_i}$$

Key Terms

center of gravity 393
critical mass 390
distribution center 387

facility location 387
geographical information system (GIS) 390
load–distance method 393

quality of life 389
transportation method for location
    problems 397

Solved Problem 1

An electronics manufacturer must expand by building a second facility. The search is narrowed to four locations, all of which are acceptable to management in terms of dominant factors. Assessment of these sites in terms of seven location factors is shown in Table 11.1. For example, location A has a factor score of 5 (excellent) for labor climate; the weight for this factor (20) is the highest of any.

Calculate the weighted score for each location. Which location should be recommended?

SOLUTION
Based on the weighted scores shown in Table 11.2, location C is the preferred site, although location B is a close second.

TABLE 11.1 | FACTOR INFORMATION FOR ELECTRONICS MANUFACTURER

Location Factor	Factor Weight	Factor Score for Each Location			
		A	B	C	D
1. Labor climate	20	5	4	4	5
2. Quality of life	16	2	3	4	1
3. Transportation system	16	3	4	3	2
4. Proximity to markets	14	5	3	4	4
5. Proximity to materials	12	2	3	3	4
6. Taxes	12	2	5	5	4
7. Utilities	10	5	4	3	3

TABLE 11.2 | CALCULATING WEIGHTED SCORES FOR ELECTRONICS MANUFACTURER

Location Factor	Factor Weight	Weighted Score for Each Location			
		A	B	C	D
1. Labor climate	20	100	80	80	100
2. Quality of life	16	32	48	64	16
3. Transportation system	16	48	64	48	32
4. Proximity to markets	14	70	42	56	56
5. Proximity to materials	12	24	36	36	48
6. Taxes	12	24	60	60	48
7. Utilities	10	50	40	30	30
Totals	100	348	370	374	330

Solved Problem 2

The operations manager for Mile-High Lemonade narrowed the search for a new facility location to seven communities. Annual fixed costs (land, property taxes, insurance, equipment, and buildings) and variable costs (labor, materials, transportation, and variable overhead) are shown in Table 11.3.

a. Which of the communities can be eliminated from further consideration because they are dominated (both variable and fixed costs are higher) by another community?

b. Plot the total cost curves for all remaining communities on a single graph. Identify on the graph the approximate range over which each community provides the lowest cost.

c. Using break-even analysis, calculate the break-even quantities to determine the range over which each community provides the lowest cost.

TABLE 11.3 | FIXED AND VARIABLE COSTS FOR MILE-HIGH LEMONADE

Community	Fixed Costs per Year	Variable Costs per Barrel
Aurora	$1,600,000	$17.00
Boulder	$2,000,000	$12.00
Colorado Springs	$1,500,000	$16.00
Denver	$3,000,000	$10.00
Englewood	$1,800,000	$15.00
Fort Collins	$1,200,000	$15.00
Golden	$1,700,000	$14.00

SOLUTION

a. Aurora and Colorado Springs are dominated by Fort Collins, because both fixed and variable costs are higher for those communities than for Fort Collins. Englewood is dominated by Golden.

b. Figure 11.5 shows that Fort Collins is best for low volumes, Boulder for intermediate volumes, and Denver for high volumes. Although Golden is not dominated by any community, it is the second or third choice over the entire range. Golden does not become the lowest-cost choice at any volume.

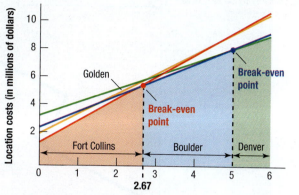

◀ **FIGURE 11.5**
Break-Even Analysis for Four Candidate Locations

c. The break-even point between Fort Collins and Boulder is

$$\$1,200,000 + \$15Q = \$2,000,000 + \$12Q$$
$$Q = 266,667 \text{ barrels per year}$$

The break-even point between Denver and Boulder is

$$\$3,000,000 + 10Q = \$2,000,000 \ \$12Q$$
$$Q = 500,000 \text{ barrels per year}$$

Solved Problem 3

The new Health-Watch facility is targeted to serve seven census tracts in Erie, Pennsylvania, whose latitudes and longitudes are shown in Table 11.4, along with the population in each census tract (in thousands). Customers will travel from the seven census-tract centers to the new facility when they need health care. What is the target area's center of gravity for the Health-Watch medical facility?

TABLE 11.4 | LOCATION DATA AND CALCULATIONS FOR HEALTH-WATCH

Census Tract	Population	Latitude	Longitude	Population × Latitude	Population × Longitude
15	2,711	42.134	−80.041	114,225.27	−216,991.15
16	4,161	42.129	−80.023	175,298.77	−332,975.70
17	2,988	42.122	−80.055	125,860.54	−239,204.34
25	2,512	42.112	−80.066	105,785.34	−201,125.79
26	4,342	42.117	−80.052	182,872.01	−347,585.78
27	6,687	42.116	−80.023	281,629.69	−535,113.80
28	6,789	42.107	−80.051	285,864.42	−543,466.24
Total	**30,190**			**1,271,536.04**	**−2,416,462.80**

SOLUTION

We solve for the center of gravity x^* and y^*. Because the coordinates are given as longitude and latitude, x^* is the longitude and y^* is the latitude for the center of gravity.

$$x^* = \frac{1,271,536.05}{30,190} = 42.1178$$

$$y^* = \frac{-2,416,462.81}{30,190} = -80.0418$$

The center of gravity is (42.12 North, 80.04 West), and is shown in Figure 11.6 to be fairly central to the target area.

Active Model 11.2 in MyOMLab confirms these calculations for the center of gravity, and allows us to explore other alternative locations as well.

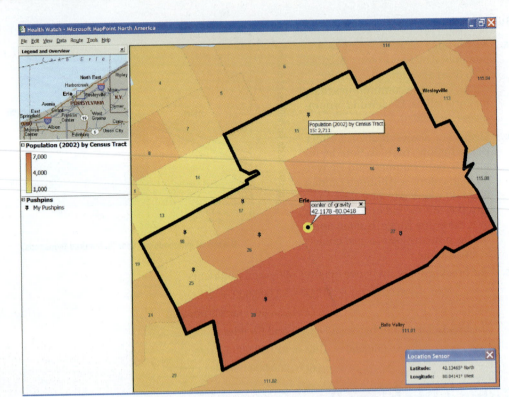

FIGURE 11.6 ▶
Center of Gravity for Health-Watch

Solved Problem 4

The Arid Company makes canoe paddles to serve distribution centers in Worchester, Rochester, and Dorchester from existing plants in Battle Creek and Cherry Creek. Arid is considering locating a plant near the headwaters of Dee Creek. Annual capacity for each plant is shown in the right-hand column, while annual demand is shown in the bottom row of the tableau in Figure 11.7. Transportation costs per paddle are shown in the tableau in the small boxes. For example, the cost to ship one paddle from Battle Creek to Worchester is $4.37. The optimal allocations are also shown in Figure 11.7. For example, Battle Creek ships 12,000 units to Rochester. What are the estimated transportation costs associated with this allocation pattern?

Source	Destination			Capacity
	Worchester	Rochester	Dorchester	
Battle Creek	$4.37	$4.25	$4.89	12,000
		12,000		
Cherry Creek	$4.00	$5.00	$5.27	10,000
	6,000	4,000		
Dee Creek	$4.13	$4.50	$3.75	18,000
		6,000	12,000	
Demand	6,000	22,000	12,000	40,000

▲ FIGURE 11.7
Optimal Solution for Arid Company

SOLUTION

The total cost is $167,000.

Ship 12,000 units from Battle Creek to Rochester @ $4.25.	Cost =	$51,000
Ship 6,000 units from Cherry Creek to Worchester @ $4.00.	Cost =	$24,000
Ship 4,000 units from Cherry Creek to Rochester @ $5.00.	Cost =	$20,000
Ship 6,000 units from Dee Creek to Rochester @ $4.50.	Cost =	$27,000
Ship 12,000 units from Dee Creek to Dorchester @ $3.75.	Cost =	$45,000
	Total =	$167,000

Discussion Questions

1. Break into teams. Select two organizations, one in services and one in manufacturing, which are known to some of your team members. What are the key factors that each organization would consider in locating a new facility? What data would you want to collect before evaluating the location options, and how would you collect the data? Explain.

2. The owner of a Major League Baseball team is considering moving his team from its current city in the upper Midwest to a city in the Southeast that offers a larger television market, a new stadium, and holds the potential for greater fan support. What other factors should the owner consider before actually making the decision to relocate?

3. A firm in Ohio is thinking of buying a plant from a regional business group located in a Southeast Asian country. The business group selling the plant has questionable labor and management practices, some of which are also in conflict with the OSHA (Occupational Safety and Health Administration) and EPA (Environmental Protection Agency) regulations. What ethical and environmental factors should the U.S. firm consider before finalizing the plant location decision?

Problems

The OM Explorer and POM for Windows software is available to all students using the 10th edition of this textbook. Go to **www.pearsonhighered.com/krajewski** to download these computer packages. If you purchased MyOMLab, you also have access to Active Models software and significant help in doing the following problems. Check with your instructor on how best to use these resources. In many cases, the instructor wants you to understand how to do the calculations by hand. At the least, the software provides a check on your calculations. When calculations are particularly complex and the goal is interpreting the results in making decisions, the software entirely replaces the manual calculations.

1. Calculate the weighted score for each location (A, B, C, and D) shown in Table 11.5. Which location would you recommend?

2. John and Jane Darling are newlyweds trying to decide among several available rentals. Alternatives were scored on a scale of 1 to 5 (5 = best) against weighted performance criteria, as shown in Table 11.6. The criteria included rent, proximity to work and recreational opportunities, security, and other neighborhood characteristics associated with the couple's values and lifestyle. Alternative A is an apartment, B is a bungalow, C is a condo, and D is a downstairs apartment in Jane's parents' home.

Which location is indicated by the preference matrix? What qualitative factors might cause this preference to change?

3. Two alternative locations are under consideration for a new plant: Jackson, Mississippi, and Dayton, Ohio.

TABLE 11.5 | FACTORS FOR LOCATIONS A–D

Location Factor	Factor Weight	Factor Score for Each Location			
		A	B	C	D
1. Labor climate	5	5	4	3	5
2. Quality of life	30	2	3	5	1
3. Transportation system	5	3	4	3	5
4. Proximity to markets	25	5	3	4	4
5. Proximity to materials	5	3	2	3	5
6. Taxes	15	2	5	5	4
7. Utilities	15	5	4	2	1
Total	100				

TABLE 11.6 | FACTORS FOR NEWLYWEDS

Location Factor	Factor Weight	Factor Score for Each Location			
		A	B	C	D
1. Rent	25	3	1	2	5
2. Quality of life	20	2	5	5	4
3. Schools	5	3	5	3	1
4. Proximity to work	10	5	3	4	3
5. Proximity to recreation	15	4	4	5	2
6. Neighborhood security	15	2	4	4	4
7. Utilities	10	4	2	3	5
Total	100				

The Jackson location is superior in terms of costs. However, management believes that sales volume would decline if this location were chosen because it is farther from the market, and the firm's customers prefer local suppliers. The selling price of the product is $250 per unit in either case. Use the following information to determine which location yields the higher total profit per year:

Location	Annual Fixed Cost	Variable Cost per Unit	Forecast Demand per Year
Jackson	$1,500,000	$50	30,000 units
Dayton	$2,800,000	$85	40,000 units

4. Fall-Line, Inc., is a Great Falls, Montana, manufacturer of a variety of downhill skis. Fall-Line is considering four locations for a new plant: Aspen, Colorado; Medicine Lodge, Kansas; Broken Bow, Nebraska; and Wounded Knee, South Dakota. Annual fixed costs and variable costs per pair of skis are shown in the following table:

Location	Annual Fixed Costs	Variable Cost per Pair
Aspen	$8,000,000	$250
Medicine Lodge	$2,400,000	$130
Broken Bow	$3,400,000	$90
Wounded Knee	$4,500,000	$65

a. Plot the total cost curves for all the communities on a single graph (see Solved Problem 2). Identify on the graph the range in volume over which each location would be best.

b. What break-even quantity defines each range?

Although Aspen's fixed and variable costs are dominated by those of the other communities, Fall-Line believes that both the demand and the price would be higher for skis made in Aspen than for skis made in the other locations. The following table shows those projections:

Location	Price per Pair	Forecast Demand per Year
Aspen	$500	60,000 pairs
Medicine Lodge	$350	45,000 pairs
Broken Bow	$350	43,000 pairs
Wounded Knee	$350	40,000 pairs

c. Determine which location yields the highest total profit per year.

d. Is this location decision sensitive to forecast accuracy? At what minimum sales volume does Aspen become the location of choice?

5. Wiebe Trucking, Inc., is planning a new warehouse to serve the western United States. Denver, Santa Fe, and Salt Lake City are under consideration. For each location, annual

fixed costs (rent, equipment, and insurance) and average variable costs per shipment (labor, transportation, and utilities) are listed in the following table. Sales projections range from 550,000 to 600,000 shipments per year.

Location	Annual Fixed Costs	Variable Costs per Shipment
Denver	$5,000,000	$4.65
Santa Fe	$4,200,000	$6.25
Salt Lake City	$3,500,000	$7.25

a. Plot the total cost curves for all the locations on a single graph.

b. Which city provides the lowest overall costs?

6. Sam Hutchins is planning to operate a specialty bagel sandwich kiosk, but is undecided about whether to locate in the downtown shopping plaza or in a suburban shopping mall. Based on the following data, which location would you recommend?

Location	Downtown	Suburban
Annual rent, including utilities	$12,000	$8,000
Expected annual demand (sandwiches)	30,000	25,000
Average variable costs per sandwich	$1.50	$1.00
Average selling price per sandwich	$3.25	$2.85

7. The following three points are the locations of important facilities in a transportation network: (20, 20), (50, 10), and (50, 60). The coordinates are in miles.

a. Calculate the Euclidean distances (in miles) between each of the three pairs of facilities.

b. Calculate these distances using rectilinear distances.

8. West Gorham High School is to be located at the population center of gravity of three communities: Westbrook, population 16,000; Scarborough, population 22,000; and Gorham, population 36,500. Westbrook is located at 43.6769°N, 70.3717°W; Scarborough is located at 43.5781°N, 70.3222°W; and Gorham is located at 43.6795°N, 70.4447°W.

a. Where should West Gorham High School be located?

b. If only two pieces of adequate land are available for sale: Baker's Field at 43.6784°N, 70.3827°W; or Lonesome Acres at 43.5119°N, 70.3856°W, using rectilinear distances, which is closer to the site located in part (a)?

9. Prescott Industries transports sand and stone extracted from its open-pit mines located in Odessa and Bryan to its concrete block manufacturing facilities in Abilene, Tyler, and San Angelo. For the capacities, locations, and shipment costs per truckload shown in the Figure 11.8, determine the shipping pattern that will minimize transportation costs. What are the estimated transportation costs associated with this optimal allocation pattern?

Source	Destination			Capacity
	Abilene	Tyler	San Angelo	
Odessa	$60	$50	$40	12,000
Bryan	$70	$30	$90	10,000
Demand	8,000	10,000	4,000	22,000

▲ **FIGURE 11.8**
Transportation Tableau for Prescott Industries

10. The Winston Company has four distribution centers (A, B, C, and D) that require 40,000, 60,000, 30,000, and 50,000 gallons of de-ionized water, respectively, per month for cleaning their long-haul trucks. Three de-ionized water wholesalers (1, 2, and 3) indicated their willingness to supply as many as 50,000, 70,000, and 60,000 gallons, respectively. The total cost (shipping plus price) of delivering 1,000 gallons of de-ionized water from each wholesaler to each distribution center is shown in the following table:

Wholesaler	Distribution Center			
	A	B	C	D
1	$1.30	$1.40	$1.80	$1.60
2	$1.30	$1.50	$1.80	$1.60
3	$1.60	$1.40	$1.70	$1.50

a. Determine the optimal solution. Show that all capacities have been exhausted and that all demands can be met with this solution.
b. What is the total cost of the solution?

11. Val's Pizza is looking for a single central location to make pizza for delivery only. This college town is arranged on a grid with arterial streets, as shown in Figure 11.9. The main campus (A), located at 14th and R, is the source of 4,000 pizza orders per week. Three smaller campuses (B, C, and D) are located at 52nd

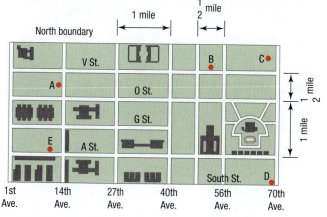

▲ **FIGURE 11.9**
Map of Campus Area

and V, at 67th and Z, and at 70th and South. Orders from the smaller campuses average 1,000 pizzas per week. In addition, the State Patrol headquarters (E) at 10th and A orders 500 pizzas per week.

a. At about what intersection should Val start looking for a suitable site? (Estimate coordinates for the major demands accurate to the nearest one-quarter mile, and then find the center of gravity.)
b. What is the rectilinear weekly load–distance score for this location?
c. If the delivery person can travel 1 mile in 2 minutes on arterial streets and ¼ mile per minute on residential streets, going from the center of gravity location to the farthest demand location will take how long?

12. A larger and more modern main post office is to be constructed at a new location in Davis, California. Growing suburbs caused a shift in the population density from where it was 40 years ago, when the current facility was built. Annette Werk, the postmaster, asked her assistants to draw a grid map of the seven points where mail is picked up and delivered in bulk. The coordinates and trips per day to and from the seven mail source points and the current main post office, M, are shown in the following table. M will continue to act as a mail source point after relocation.

Mail Source Point	Round Trips Per Day	x, y Coordinates (miles)
1	6	(2, 8)
2	3	(6, 1)
3	3	(8, 5)
4	3	(13, 3)
5	2	(15, 10)
6	7	(6, 14)
7	5	(18, 1)
M	3	(10, 3)

a. Calculate the center of gravity as a possible location for the new facility (round to the nearest whole number).
b. Compare the load–distance scores for the location in part (a) and the current location, using rectilinear distance.

13. Rauschenberg Manufacturing is investigating which locations would best position its new plant relative to three important customers (located in cities A, B, and C). As shown in the table below, all three customers require multiple daily deliveries. Management limited the search for this plant to those three locations and compiled the following information:

Location	Coordinates (miles)	Deliveries per day
A	(100, 200)	8
B	(400, 100)	4
C	(100, 100)	3

a. Which of these three locations yields the smallest total travel distance, based on Euclidean distances?

b. Which of these locations is best, based on rectilinear distances?

c. What are the coordinates of the center of gravity?

14. A personal computer manufacturer plans to locate its assembly plant in Taiwan and to ship its computers back to the United States through either Los Angeles or San Francisco. It has distribution centers in Atlanta, New York, and Chicago and will ship to them from whichever city is chosen as the port of entry on the west coast. Overall transportation cost is the only criterion for choosing the port. Use the load–distance model and the information in Table 11.7 to select the more cost-effective city.

TABLE 11.7 | DISTANCES AND COSTS FOR PC MANUFACTURER

		Distribution Center (units/year)		
		Chicago (10,000)	Atlanta (7,500)	New York (12,500)
	Los Angeles			
	Distance (miles)	1,800	2,600	3,200
	Shipping cost ($/unit)	0.0017/mile	0.0017/mile	0.0017/mile
PORT OF ENTRY				
	San Francisco			
	Distance (miles)	1,700	2,800	3,000
	Shipping cost ($/unit)	0.0020/mile	0.0020/mile	0.0020/mile

Advanced Problems

15. Oscar's Bowling, Inc., wants to break into the Phoenix metropolitan market with one of its super-sized 200 lane, 24-hour bowling alleys. It, however, only has enough capital to build one facility. Oscar wants it to be centered by population, as determined by the center of gravity method. The following information is given:

City	Population	x coordinate	y coordinate
Tempe	250,000	5	5
Scottsdale	400,000	5	10
Chandler	300,000	5	0
Mesa	700,000	10	1
Glendale	350,000	1	10

a. Where should Oscar build?

b. If Oscar wanted to relocate to the closest city near his new facility—where would he live?

16. The Acme Company operates four factories that ship products to five warehouses. The shipping costs, requirements capacities, and optimal allocations are shown in Figure 11.10. What is the total cost of the optimal solution?

Factory	Shipping Cost per Case to Warehouse					Capacity
	W1	W2	W3	W4	W5	
F1	$1 / 60,000	$3 / 20,000	$4	$5	$6	80,000
F2	$2	$2	$1 / 50,000	$4 / 10,000	$5	60,000
F3	$1	$5	$1	$3 / 20,000	$1 / 40,000	60,000
F4	$5	$2 / 50,000	$4	$5	$4	50,000
Demand	60,000	70,000	50,000	30,000	40,000	250,000

▲ **FIGURE 11.10**
Optimal Solution for Acme Company

17. Dennison Manufacturing makes large helical springs used in aircraft landing gear. The company has narrowed its potential choices for its new manufacturing facility to four cities. The following information is known about the manufacturing and shipping costs of locating in each of these four cities:

	Fixed Costs per Year	Variable Manufacturing Costs per Unit	Variable Shipping Costs per Unit
Phoenix	$300,000	$70.00	$5.00
Buffalo	$600,000	$56.00	$4.00
Seattle	$1,500,000	$36.00	$2.00
Atlanta	$1,750,000	$42.00	$5.00

a. Use break-even point analysis to determine where Dennison should locate.

b. Based solely on break-even quantity, if Dennison's manufacturing forecast for the foreseeable future is 40,000 units annually, where should he locate?

18. The Giant Farmer Company processes food for sale in discount food stores. It has two plants: one in Chicago and one in Houston. The company also operates warehouses in Miami, Florida; Denver, Colorado; Lincoln, Nebraska; and Jackson, Mississippi. Forecasts indicate that demand soon will exceed supply and that a new plant with a capacity of 8,000 cases per week is needed. The question is where to locate the new plant. Two potential sites are Buffalo, New York, and Atlanta. The two tables at the bottom of this page give data on capacities, forecasted demand, and shipping costs that have been gathered.

For each alternative new plant location, determine the shipping pattern that will minimize total transportation costs. Where should the new plant be located?

Plant	Capacity (Cases per Week)	Warehouse	Demand (Cases per Week)
Chicago	10,000	Miami	7,000
Houston	7,500	Denver	9,000
New plant	8,000	Lincoln	4,500
	Total 25,500	Jackson	5,000
			Total 25,500

	Shipping Cost to Warehouse (per Case)			
Plant	Miami	Denver	Lincoln	Jackson
Chicago	$7.00	$2.00	$4.00	$5.00
Houston	$3.00	$1.00	$5.00	$2.00
Buffalo (alternative 1)	$6.00	$9.00	$7.00	$4.00
Atlanta (alternative 2)	$2.00	$10.00	$8.00	$3.00

19. The Thor International Company operates four factories that ship products to five warehouses. The shipping costs, requirements, and capacities are shown in Figure 11.11. Use the transportation method to find the shipping schedule that minimizes shipping cost.

Factory	Shipping Cost per Case to each Warehouse						Capacity
	W1	W2	W3	W4	W5	Dummy	
F1	$2	$3	$3	$2	$6	$0	50,000
F2	$2	$3	$2	$4	$5	$0	80,000
F3	$4	$2	$4	$2	$3	$0	80,000
F4	$3	$4	$4	$5	$2	$0	40,000
Demand	45,000	30,000	30,000	35,000	50,000	60,000	250,000

▲ **FIGURE 11.11**

Transportation Tableau for Thor International

20. Consider further the Thor International Company situation described in Problem 19. Thor decides to close F4 because of high operating costs. The logistics manager is worried about the effect of this move on transportation costs. Currently, F4 is shipping 40,000 units to W5 at cost of $80,000 [or 40,000($2)]. If this warehouse were to be served by F1 (currently not being used), the cost would increase to $240,000 [or 40,000(6)]. As a result, the Ajax logistics manager requests a budget increase of $160,000 (or $240,000 − $80,000).

a. Should the logistics manager get the budget increase?

b. If not, how much would you budget for the increase in shipping costs?

21. Consider the facility location problem at the Giant Farmer Company described in Problem 18. Management is considering a third site, at Memphis. The shipping costs per case from Memphis are $3 to Miami, $11 to Denver, $6 to Lincoln, and $5 to Jackson. Find the minimum cost plan for an alternative plant in Memphis. Would this result change the decision in Problem 18?

22. The Chambers Corporation produces and markets an automotive theft-deterrent product, which it stocks in various warehouses throughout the country. Recently, its market research group compiled a forecast indicating that a significant increase in demand will occur in the near future, after which demand will level off for the foreseeable future. The company decides to satisfy this demand by constructing new plant capacity. Chambers already has plants in Baltimore and Milwaukee and has no desire to relocate those facilities. Each plant is capable of producing 600,000 units per year.

After a thorough search, the company developed three site and capacity alternatives. Alternative 1 is to build a 600,000-unit plant in Portland. Alternative 2 is to build a 600,000-unit plant in San Antonio. Alternative 3 is to build a 300,000-unit plant in Portland and a 300,000-unit plant in San Antonio. The company's four warehouses distribute the product to retailers. The market research study provided the following data.

Warehouse	Expected Annual Demand
Atlanta (AT)	500,000
Columbus (CO)	300,000
Los Angeles (LA)	600,000
Seattle (SE)	400,000

The logistics department compiled the following cost table specifying the cost per unit to ship the product from each plant to each warehouse in the most economical manner, subject to the reliability of the various carriers involved.

Plant	Warehouse			
	AT	CO	LA	SE
Baltimore	$0.35	$0.20	$0.85	$0.75
Milwaukee	$0.55	$0.15	$0.70	$0.65
Portland	$0.85	$0.60	$0.30	$0.10
San Antonio	$0.55	$0.40	$0.40	$0.55

As one part of the location decision, management wants an estimate of the total distribution cost for each alternative. Use the transportation method to calculate these estimates.

Active Model Exercise

Active Model 11.2 appears in MyOMLab. It allows you to find the location that minimizes the total load–distance score.

QUESTIONS

1. What is the total load–distance score to the new Health-Watch medical facility if it is located at the center of gravity?

2. Fix the y coordinate, and use the scroll bar to modify the x coordinate. Can you reduce the total load–distance score?

3. Fix the x coordinate, and use the scroll bar to modify the y coordinate. Can you reduce the total load–distance score?

4. The center of gravity does not necessarily find the site with the minimum total load–distance score. Use both scroll bars to move the trial location, and see whether you can improve (lower) the total load–distance score.

Selected References

"BMW Announces Its Plans for a Plant in South Carolina." *Wall Street Journal* (June 24, 1992), p. B2.

Chittum, Ryan. "Location, Location, and Technology: Where to Put That New Store? Site-Selection Software May Be Able to Help." *Wall Street Journal* (July 18, 2005), p. R7.

Chowdhury, Santanu. "GE Plans Multiuse Factory in India." *Wall Street Journal* (March 14, 2011).

"Doing Well by Doing Good." *The Economist* (April 22, 2000), pp. 65–67.

Deeds, David. "Increasing the Rate of New Venture Creation: Does Location Matter?" *Academy of Management Executive*, vol. 18, no. 2 (2004), pp. 152–154.

Galuszka, P. "The South Shall Rise Again." *Chief Executive*, November 2004, pp. 50–54.

Lovelock, Christopher H., and George S, Yip. "Developing Global Strategies for Service Businesses." *California Management Review*, vol. 38, no. 2 (1996), pp. 64–86.

Hahn, E.D., and K. Bunyaratavej. "Services cultural alignment in offshoring: The impact of cultural dimensions on offshoring location decisions." *Journal of Operations Management*, vol. 28, no. 3 (2010), pp. 186–193.

"Manager's Journal: Why BMW Cruised into Spartanburg." *Wall Street Journal* (July 6, 1992), p. A10.

"MapInfo Delivers Location Intelligence for Marco's Pizza." *Directions Magazine* (December 14, 2004), www.directions mag.com/press .releases/?duty=Show&id=10790

Melo, M.T., S. Nickel, and F. Saldanha-da-Gama. "Facility Location and Supply Chain Management a Review." *European Journal of Operational Research*, vol. 196, no. 2 (2009), pp. 401–412.

Rubinstein, Ed. "Chain Chart Their Course of Actions with Geographic Information Systems." *Nation's Restaurant News*, vol. 32, no. 6 (1998), p. 49.

"The Science of Site Selection." *National Real Estate Investor* (October 11, 2002).

12

SUPPLY CHAIN INTEGRATION

An Eastman Kodak employee assembles a Kodak NexPress digital production printer at the graphic communication plant in Rochester, New York.

Eastman Kodak

Customer service after the sale, which requires having the right repair part in the right location in the service parts supply chain, is a key competitive dimension of business today for service providers as well as manufacturers. Overstocking the parts throughout the supply chain is not an option because of the expense. Eastman Kodak, a $7.2 billion manufacturer of digital imaging products, found this out for its digital after-market parts. These parts, which become obsolete quickly and are very expensive, included circuit boards, print head controllers, CPUs, optical drives, and monitors. Kodak faced a typical conflict: the product group managers wanted low inventories and the field engineers, who did the repair in the field and had to face the customer, wanted high inventories. Facing a need to reduce the cost of inventories while maintaining or improving service, Kodak needed to redesign its supply chain and integrate the entities for smooth operations.

Rather than allowing the expensive parts to reside with the field engineers, Kodak managers realized that centralizing expensive parts at strategic field locations and delivering them on demand to field engineers was an option worth pursuing. Kodak made three major decisions to implement the new supply chain design. First, the number of forward stock locations (FSLs) to stock the expensive parts and the inventory levels at each stock location had to be determined using software from Baxter Planning Systems. Second, UPS Supply Chain Solutions was chosen as the logistics provider to operate the more than 100 FSLs and deliver the parts as needed to the field engineers. Finally, and perhaps most importantly, Kodak fully incorporated the field engineers and service parts supply chain personnel employees in the design process and executed a four-month pilot program to help employees accept the new design.

Kodak showed that fully integrating the key elements of the new supply chain, including the software provider, logistics provider, and internal employees, can lead to significant results. During the first year of operation, the new program has reduced FSL inventory items by 66 percent, central inventory by 32 percent, priority shipments by 22 percent, and had no negative impact on call duration.

Source: Mark Brienzi and Dr. Sham Kekre, "How Kodak Transformed Its Service Parts Supply Chain," *Supply Chain Management Review* (October 2005), pp. 25–32; **www.bybaxter.com/news_archive.html**; **www.kodak.com**.

LEARNING GOALS *After reading this chapter, you should be able to:*

1 Identify the major causes of dynamics in a supply chain.

2 Explain how integrated supply chains can mitigate supply chain dynamics.

3 Explain the nature and purpose of the key nested processes within the new service or product development, supplier relationship, order fulfillment, and customer relationship processes.

4 Use a total annual cost analysis and a preference matrix to select suppliers and the expected value decision rule to determine logistics capacity.

5 Identify key levers and performance measures useful for improving supply chain performance.

supply chain integration

The effective coordination of supply chain processes through the seamless flow of information up and down the supply chain.

The development and delivery of services and products has become increasingly complex in today's global economy. The Kodak example shows how a firm can redesign its supply chain for competitive advantage by changing the location of critical stocks, adding a new supplier in the form of a logistics provider, and changing the way field engineers access that inventory. To be effective, accurate inventory and demand information had to be available along with considerable collaboration between the parties involved. Kodak was successful at **supply chain integration**, which is the effective coordination of supply chain processes through the seamless flow of information up and down the supply chain. Supply chain integration provides each member of the supply chain visibility into the capacities and inventories of other members of the supply chain to aid in planning and scheduling. It facilitates collaboration between firms in a supply chain; in effect, it is an enabler of supply chain management. In this chapter, we explore the processes involved in supply chain integration, the benefits to be had, and the role supply chain integration can play in addressing environmental concerns.

Supply Chain Integration across the Organization

As the Kodak example portrayed, supply chain integration involves internal as well as external processes. Figure 12.1 shows just how interconnected processes and firms in a supply chain can be. Think of a supply chain as a river that flows from raw material suppliers to consumers. For example, a ketchup factory gets its major supply from the tomato paste factories, which for the ketchup factory is a tier 1 supplier. In turn, the tomato paste factories get their major sup-

FIGURE 12.1 ▶
Supply Chain for a Ketchup Factory

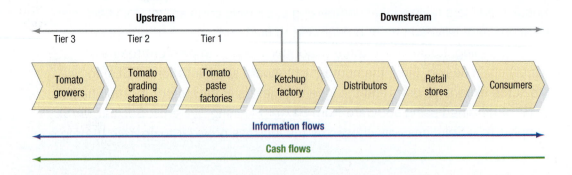

plies from the tomato grading stations, which are tier 2 suppliers of the ketchup factory. Finally, the tomato growers ship their product directly to the tomato grading stations. The tier 1, tier 2, and tier 3 suppliers are all *upstream* from the ketchup factory, which means that they control the flow of supply to the ketchup factory. Suppose the tomato paste factory has a major process failure. The flow of tomato paste to the ketchup factory would dwindle to a trickle, as if someone built a dam across a river. Indeed, even those entities *downstream* from the ketchup factory could feel the effects after inventories of ketchup have been consumed. When a link in the supply chain fails, whether it is an internal process or one at a supplier, the rest of the chain feels the effects.

Mitigating the effects of supply chain disruptions is an important benefit of supply chain integration. Information flows, both upstream and downstream, provide visibility to supply chain members regarding supplies, capacities, and plans. Cash flows move upstream and are affected by pricing, promotional programs, and supply contracts. Understanding the implications of material, information, and cash flows in a supply chain is important for all employees in an organization.

Supply Chain Dynamics

Supply chain dynamics can wreak havoc on supply chain performance. Each firm in a supply chain depends on other firms for services, materials, or the information needed to supply its immediate external customer in the chain. Because firms are typically owned and managed independently, the actions of downstream supply chain members (positioned nearer the end user of the service or product) can affect the operations of upstream members. The reason is that upstream members of a supply chain must react to the demands placed on them by downstream members of the chain. These demands are a function of the policies downstream firms have for replenishing their inventories, the actual levels of those inventories, the demands of their customers, and the accuracy of the information they have to work with. As you examine the order patterns of firms in a supply chain, you will frequently see the variability in order quantities increase as you proceed upstream. This increase in variability is referred to as the **bullwhip effect**, which gets its name from the action of a bullwhip—the handle of the whip initiates the action; however, the tip of the whip experiences the wildest action. The slightest change in customer demands can ripple through the entire chain, with each member receiving more variability in demands from the member immediately downstream. A firm contributes to the bullwhip effect if the variability of the orders to its suppliers exceeds the variability of the orders from its immediate customers.

Figure 12.2 shows the bullwhip effect in a supply chain for facial tissue. The variability in the orders increases as you go upstream in the supply chain. Because supply patterns do not match demand patterns, inventories accumulate in some firms and shortages occur in others. The firms with too much inventory stop ordering, and those that have shortages place expedited orders. The culprits are unexpected changes in demands or supplies that are based on a number of causes.

Creating Value through Operations Management

Using Operations to Compete
Project Management

Managing Processes

Process Strategy
Process Analysis
Quality and Performance
Capacity Planning
Constraint Management
Lean Systems

Managing Supply Chains

Supply Chain Inventory Management
Supply Chain Design
Supply Chain Location Decisions
Supply Chain Integration
Supply Chain Sustainability and Humanitarian Logistics
Forecasting
Operations Planning and Scheduling
Resource Planning

▼ **FIGURE 12.2**
Supply Chain Dynamics for Facial Tissue

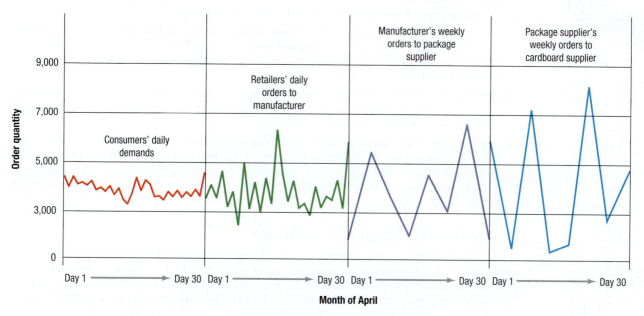

bullwhip effect

The phenomenon in supply chains whereby ordering patterns experience increasing variance as you proceed upstream in the chain.

External Causes

A firm has the least amount of control over its external customers and suppliers, who can periodically cause disruptions. Typical external disruptions include the following:

- *Volume Changes.* Customers may change the quantity of the service or product they had ordered for a specific date or unexpectedly demand more of a standard service or product. If the market demands short lead times, the firm needs a quick reaction from its suppliers.

- *Service and Product Mix Changes.* Customers may change the mix of items in an order and cause a ripple effect throughout the supply chain. For example, a major-appliance store chain may change the mix of washing machines in its orders from 60 percent Whirlpool brand and 40 percent Kitchen Aid brand to 40 percent Whirlpool and 60 percent Kitchen Aid. This decision changes the production schedule of the Whirlpool plant that makes both brands, causing imbalances in its inventories. In addition, the tier 1 supplier that makes the face plates for the washing machines must change its schedules, thereby affecting *its* suppliers.

- *Late Deliveries.* Late deliveries of materials or delays in essential services can force a firm to switch its schedule from production of one product model to another. Firms that supply model-specific items may have their schedules disrupted. For example, the Whirlpool plant may find that a component supplier for its Model A washing machine could not supply the part on-time. To avoid shutting down the assembly line, which is an expensive action, Whirlpool may decide to switch to Model B production. Suddenly, the demand on the suppliers of Model B-specific parts increases.

- *Underfilled Shipments.* Suppliers that send partial shipments do so because of disruptions at their own plants. The effects of underfilled shipments are similar to those of late shipments unless the underfilled shipment contains enough materials to allow the firm to operate until the next shipment.

Internal Causes

A famous line from a *Pogo* cartoon is "We have seen the enemy, and it is us!" Unfortunately, this statement is true for many firms when it comes to disruptions in the supply chain. A firm's own operations can be the culprit in what becomes the source of constant dynamics in the supply chain. Typical internal disruptions include the following:

- *Internally Generated Shortages.* A shortage of parts manufactured by a firm may occur because of machine breakdowns or inexperienced workers. This shortage can cause a change in the firm's production schedule, thus affecting suppliers.

- *Engineering Changes.* Changes to the design of services or products can have a direct impact on suppliers. For example, changing cable TV feed lines to fiber-optic technology increases the benefits to the cable company's customers but affects the demand for co-axial cable.

- *Order Batching.* Suppliers may offer a quantity discount, which gives an incentive to firms to purchase large quantities of an item less frequently, thereby raising the variability in orders to the supplier. Order batching may also result in transportation economies; larger orders may enable full-truckload shipments and thus create more variability.

- *New Service or Product Introductions.* A firm decides on the number of new service or product introductions, as well as their timing, and hence introduces a dynamic in the supply chain. New services or products may even require a new supply chain or the addition of new members to an existing supply chain.

- *Service or Product Promotions.* A common practice of firms producing standardized services or products is to use price discounts to promote sales. Price discounting creates a spike in demand that is felt throughout the supply chain.

- *Information Errors.* Demand forecast errors can cause a firm to order too many or too few services and materials or can precipitate expedited orders that force suppliers to react more quickly to avoid shortages in the supply chain. In addition, errors in the physical count of items in stock can cause shortages (leading to panic purchases) or too much inventory (leading to a slowdown in purchases). Finally, communication links between buyers and suppliers can be faulty.

Implications for Supply Chain Design

Sometimes the dynamics in a supply chain can be reduced by redesigning the supply chain. Consider Figure 12.3, which shows a generalized relationship between the annual volumes of a firm's stock-keeping units (SKUs) and their weekly demand variability. For this firm, some SKUs have high volume and low variability while some SKUs have low volume and high variability. To the extent that the volatility in demands is outside the control of the firm, hanging on to the firm's legacy supply chain may be too costly because of supply chain dynamics. Mapping the portfolio of products along the dimensions of annual volume and weekly demand variability may reveal better configurations of the supply chain. A firm may therefore have two distinctly different supply chain designs serving two different product groups. For example, SKUs with lower volumes and higher weekly variability may be best served with a responsive supply chain design, such as assemble-to-order (ATO) or make-to-order (MTO). Doing so reduces the need for finished goods inventories where customer demands are very unpredictable. Alternatively, serving the high volume, low weekly demand variability SKUs with an efficient supply chain design such as build-to-stock would be better. Forecasts are more accurate and using a finished goods inventory is an effective strategy. Redesigning a supply chain is a costly endeavor and should only be contemplated when the dynamics cannot be sufficiently reduced by other means.

It is true that many disruptions are simply caused by ineffective coordination in the supply chain because so many firms and separate operations are involved. It is therefore unrealistic to think that all disruptions can be eliminated. Nonetheless, the challenge for supply chain managers is to remove as many disruptions as possible and minimize the impact of those that cannot be eliminated.

▲ **FIGURE 12.3**

Annual Volume versus Variability in Weekly Demands for a Firm's SKUs

Integrated Supply Chains

Regardless of the supply chain design, minimizing supply chain disruptions begins with a high degree of functional and organizational integration. Such integration does not happen overnight; it must include linkages between the firm and its suppliers and customers, as shown in Figure 12.4. The new service or product development, supplier relationship, order fulfillment, and customer relationship processes, as well as their internal and external linkages, are integrated into the normal business routine. The firm takes on a customer orientation. However, rather than merely reacting to customer demand, the firm strives to work with its customers and suppliers so that everyone benefits from improved flows of services and materials. The firm must also develop a better understanding of its suppliers' organizations, capacities, strengths, and weaknesses—and include its suppliers earlier into the design of new services or products.

▼ **FIGURE 12.4**

External Supply Chain Linkages

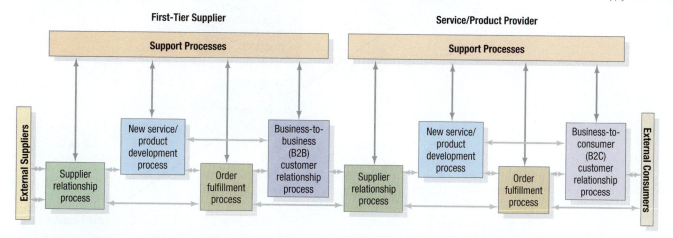

Another integrative frame of reference is the *supply chain operations reference model*, known as *SCOR*, developed by the Supply Chain Council with the assistance of 70 of the world's leading manufacturing companies. Figure 12.5 shows that the **SCOR model** focuses on a basic supply chain of *plan, source, make, deliver,* and *return* processes, repeated again and again along the supply chain. The return processes handle the return of recyclable materials and defective products, which we will discuss in more detail in Chapter 13, "Supply Chain Sustainability and Humanitarian Logistics." Much like our model shown in Figure 12.4, the SCOR model emphasizes that the design of an integrated supply chain is complex and requires a *process view*. We already provided some key insights into process design decisions in Part 1 and Part 2 of the text. These insights must be applied to the new service or product development, supplier relationship, order fulfillment,

SCOR model

A framework that focuses on a basic supply chain of plan, source, make, deliver, and return processes, repeated again and again along the supply chain.

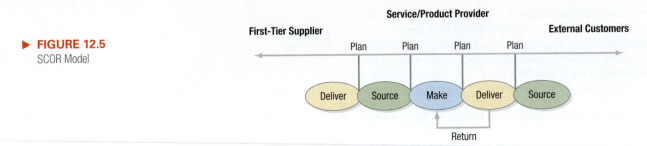

▶ **FIGURE 12.5**
SCOR Model

and customer relationship processes. Beyond that, these processes need to be integrated both within a firm and across the supply chain. It is important to know that an integrated supply chain, implied by Figure 12.4 and the SCOR model in Figure 12.5, provides a framework for the operating decisions in a firm and that these processes play a major role.

New Service or Product Development Process

Competitive priorities help managers develop services and products that customers want. New services or products are essential to the long-term survival of the firm. *New* refers to both brand new services or products or major changes to existing services or products. The new service/product development process is an integral element in a firm's supply chain because it defines the nature of the materials, services, and information flows the supply chain must support. As shown in Figure 12.6, it begins with the consideration of the development strategy and ends with the launch of the new offering. Here are some considerations for supply chain managers.

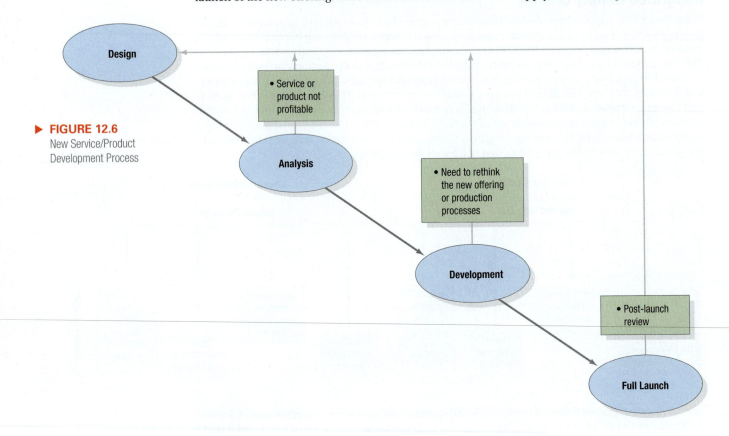

▶ **FIGURE 12.6**
New Service/Product
Development Process

Design

The *design* stage is critical because it links the creation of new services or products to the corporate strategy of the firm and defines the requirements for the firm's supply chain. As we have already noted, the corporate strategy specifies the long-term objectives and the markets in which the firm wishes to compete. In the design stage, ideas for new offerings are proposed and screened for feasibility and market worthiness. These ideas specify how the customer connects with the service or manufacturing firm, the benefits and outcomes for the customer, and the value of the service or product. Proposals also specify how the new offering will be produced and delivered—an important consideration for the supply chain. Often critical choices must be made, such as the raw materials, degree of modularity in the design, or the nature of the logistical services needed to get the service or product to market. Even though many of the detailed specifics of the service or product and its processes have not yet been developed, the inputs of designers, engineers, suppliers, supply chain managers, and potential customers in this stage can avoid costly mistakes.

Analysis

The second stage, *analysis*, involves a critical review of the new offering and how it will be produced to make sure that it fits the corporate strategy, is compatible with regulatory standards, presents an acceptable market risk, and satisfies the needs of the intended customers. The resource requirements for the new offering must be examined from the perspective of the core capabilities of the firm and the need to acquire additional resources. The existing supply chain must be evaluated for its appropriateness for the new offering. If change is needed, the design might have to be revised (efficient or responsive) or new capabilities acquired by forming strategic partnerships with new firms. If the analysis reveals that the new offering has good market potential and that the firm has the capability (or can acquire it), the authorization is given to proceed to the next stage.

Development

The third stage, *development*, brings more specificity to the new offering. The required competitive priorities are used as inputs to the design (or redesign) of the processes that will be involved in delivering the new offering. The processes are analyzed, including those of suppliers; each activity is designed to meet its required competitive priorities and to add value to the service or product. Once the new offering is specified and the capability of the processes verified, the market program can be designed. Finally, personnel are trained and some pilot runs can be conducted to iron out the kinks in production and supply. At this stage it is possible that some unforeseen problems may arise, forcing a reconsideration of the service or product or the processes required to produce it. The supply chain might have to be redesigned as well.

To avoid costly mismatches between the design of a new offering and the capability of the processes and supply chain required to produce it, many firms engage in a concept called **concurrent engineering**, which brings product engineers, process engineers, marketers, buyers, information specialists, quality specialists, and suppliers together to design a product and the processes that will meet customer expectations. Changes are much simpler and less costly at this stage. However, problems with the design of the new offering or the capability to deliver it may be discovered during this stage. The proposal for the new offering may have to be scrapped or completely rethought.

concurrent engineering

A concept that brings product engineers, process engineers, marketers, buyers, information specialists, quality specialists, and suppliers together to design a product and the processes that will meet customer expectations.

Full Launch

The final stage, *full launch*, involves the coordination of many internal processes as well as those both upstream and downstream in the supply chain. Promotions for the new offering must be initiated, sales personnel briefed, distribution processes activated, and old services or products that the new offering is to replace withdrawn. A particular strain is placed on the supply chain during a period referred to as *ramp-up*, when the production processes must increase volume to meet demands while coping with quality problems and last-minute design changes. The more integrated the supply chain, the easier the ramp-up period. Flexibility in the supply chain is a desirable attribute during the ramp-up period. Later, as the service or product matures and volume increases sufficiently, a supply chain based on efficiency may have to be developed.

Regardless of the service or product, a post launch review should compare the competitive priorities of the supply chain to its competitive capabilities, perhaps signaling a need to rethink the original service or product idea or the supply chain. The review should also get inputs from customers, who may divulge their experiences and may share ideas for change.

Supplier Relationship Process

The nature of the service or product determines the design requirements for the upstream supply chain. The supplier relationship process, which focuses on the interaction of the firm with upstream suppliers, includes five major nested processes: (1) sourcing, (2) design collaboration, (3) negotiation, (4) buying, and (5) information exchange. For many firms, these processes are the organizational responsibility of **purchasing**, which is the activity that decides the suppliers to use, negotiates contracts, maintains information flows, and determines whether to buy locally.

purchasing

The activity that decides which suppliers to use, negotiates contracts, and determines whether to buy locally.

Sourcing

The sourcing process is involved in the selection, certification, and evaluation of suppliers and, in general, the management of supply contracts.

Supplier Selection   A starting point for selecting suppliers is to perform a total cost analysis. There are four key costs to consider for each supplier.

- *Material costs.* Negotiating with suppliers for the provision of a service or product results in a price per unit (or application of the service). Material costs equal annual requirements (D) multiplied by the price per unit, p.

$$\text{Annual material costs} = pD$$

- *Freight costs.* The costs of transporting the product or the equipment and personnel who will perform the service can vary greatly depending on the location of the supplier, the size of the shipments (full truckload shipments [TL] are cheaper per pound than less-than-truckload shipments [LTL]), the number of shipments per year, and the mode of transportation (air transportation is more expensive than truck or rail transportation).

- *Inventory costs.* Buyers interested in purchasing products must consider the shipping quantity and the lead time from the supplier. The shipping quantity, Q, will determine the cycle inventory the buyer must maintain until the next shipment of the product.

$$\text{Cycle inventory} = Q/2$$

The lead time, L, and the average requirements per day (or week) $\bar{d}$, will determine the level of the pipeline inventory, which also may be the responsibility of the buyer. Assuming a constant lead time,

$$\text{Pipeline inventory} = \bar{d}L$$

The buyer must pay inventory holding costs on the cycle and pipeline inventories. Annual inventory costs equal the sum of the cycle and pipeline inventories multiplied by the annual holding cost per unit, H. See Chapter 9, "Supply Chain Inventory Management" for a review of inventory holding costs and cycle and pipeline inventories.

$$\text{Annual inventory costs} = (Q/2 + \bar{d}L)H$$

- *Administrative costs.* Supply contacts must be monitored and frequent interactions with the supplier may be required. Administrative costs include the managerial time, travel, and other variable costs associated with interacting with a supplier. These costs may vary greatly depending on the location of the supplier; more distant suppliers typically require more administrative attention.

The total annual cost for a supplier is the sum of these costs:

$$\text{Total Annual Cost} = pD + \text{Freight costs} + (Q/2 + \bar{d}L)H + \text{Administrative costs}$$

EXAMPLE 12.1	**Total Cost Analysis for Supplier Selection**

Compton Electronics manufactures laptops for major computer manufacturers. A key element of the laptop is the keyboard. Compton has identified three potential suppliers for the keyboard, each located in a different part of the world. Important cost considerations are the price per keyboard, freight costs, inventory costs, and contract administrative costs. The annual requirements for the keyboard are 300,000 units. Assume Compton has 250 business days a year. Managers have acquired the following data for each supplier.

	ANNUAL FREIGHT COSTS		
	Shipping Quantity (Units/Shipment)		
Supplier	**10,000**	**20,000**	**30,000**
Belfast	$380,000	$260,000	$237,000
Hong Kong	$615,000	$547,000	$470,000
Shreveport	$285,000	$240,000	$200,000

		KEYBOARD COSTS AND SHIPPING LEAD TIMES		
Supplier	Price/Unit	Annual Inventory Carrying Cost/Unit	Shipping Lead Time (Days)	Administrative Costs
Belfast	$100	$20.00	15	$180,000
Hong Kong	$96	$19.20	25	$300,000
Shreveport	$99	$19.80	5	$150,000

Which supplier provides the lowest annual total cost to Compton?

SOLUTION

The average requirements per day are

$$\bar{d} = 300{,}000/250 = 1{,}200 \text{ keyboards.}$$

Each option must be evaluated with consideration for the shipping quantity using the following equation:

Total Annual Cost = Material costs + Freight costs + Inventory costs + Administrative costs

$= pD + $ Freight costs $ + (Q/2 + \bar{d}L)H + $ Administrative costs

For example, consider the Belfast option for a shipping quantity of $Q = 10{,}000$ units. The costs are

Material costs $= pD = $ ($100/unit)(300,000 units) $= \$30{,}000{,}000$

Freight costs $= \$380{,}000$

Inventory costs $= $ (cycle inventory + pipeline inventory)$H = Q/2 + \bar{d}L)H$

$= $ (10,000 units/2 + 1,200 units/day (15 days))$20/unit/year $= \$460{,}000$

Administrative costs $= \$180{,}000$

Total Annual Cost $= \$30{,}000{,}000 + \$380{,}000 + \$460{,}000 + \$180{,}000 = \$31{,}020{,}000$

The total costs for all three shipping quantity options are similarly calculated and are contained in the following table.

	TOTAL ANNUAL COSTS FOR THE KEYBOARD SUPPLIERS		
	Shipping Quantity		
Supplier	10,000	20,000	30,000
Belfast	$31,020,000	$31,000,000	$31,077,000
Hong Kong	$30,387,000	$30,415,000	$30,434,000
Shreveport	$30,352,800	$30,406,800	$30,465,800

DECISION POINT

Notice that the shipping quantity plays an important role; the lowest cost for the Belfast supplier comes with a shipping quantity of 20,000 keyboards. Nonetheless, based on the total cost analysis, the Shreveport supplier will provide the lowest cost to Compton. Compton should choose a shipping quantity of 10,000 keyboards, which implies that there will be 30 shipments a year (or 300,000/10,000). While the Hong Kong supplier had the lowest price per keyboard, the Shreveport supplier could deliver the keyboards to Compton with the lowest overall cost, which includes logistics, inventory, and administrative costs.

While total cost is an important consideration in the selection of suppliers, other performance dimensions may also be important. The quality of a supplier's materials may be critical since hidden costs of poor quality may be high. Similarly, shorter lead times and on-time delivery may allow the buying firm to maintain acceptable customer service with less inventory. So management must review the market segments it wants to serve and accordingly relate its supplier selection needs to the supply chain. Competitive priorities and order winners for the firm are a good starting point in developing a list of performance criteria to be used. For example, if you were a manager of a food-service firm, in addition to total costs, you would likely use on-time delivery and quality as top criteria for selecting suppliers. These criteria reflect the requirements that food service supply chains need to meet.

Another criterion that is becoming important in the selection of suppliers is environmental impact. Many firms are engaging in **green purchasing**, which involves identifying, assessing, and managing the flow of environmental waste and finding ways to reduce it and minimize its impact on the environment. Suppliers are being asked to be environmentally conscious when designing and producing their services or products. Claims such as *green, biodegradable, natural,* and *recycled* must be substantiated when bidding on a contract. In the not-too-distant future, this criterion could become an important one in the selection of suppliers. We will have more to say about this topic in Chapter 13, "Supply Chain Sustainability and Humanitarian Logistics."

When faced with multiple criteria in the supplier selection problem, management can use a preference matrix as shown in Example 12.2. See Supplement A, "Decision Making" for a review of this approach.

| EXAMPLE 12.2 | **Using a Preference Matrix for Selecting Suppliers** |

The management of Compton Electronics has done a total cost analysis for three international suppliers of keyboards (see Example 12.1). Compton also considers on-time delivery, consistent quality, and environmental stewardship in its selection process. Each criterion is given a weight (total of 100 points), and each supplier is given a score (1 = poor, 10 = excellent) on each criterion. The data are shown in the following table.

Criterion	Weight	SCORE Belfast	SCORE Hong Kong	SCORE Shreveport
Total Cost	25	5	8	9
On-Time Delivery	30	9	6	7
Consistent Quality	30	8	9	6
Environment	15	9	6	8

SOLUTION

The weighted score for each supplier is calculated by multiplying the weight by the score for each criterion and arriving at a total. For example, the Belfast weighted score is

$$WS = (25 \times 5) + (30 \times 9) + (30 \times 8) + (15 \times 9) = 770$$

Similarly, the weighted score for Hong Kong is 740, and for Shreveport, 735. Consequently, Belfast is the preferred supplier.

DECISION POINT

Even though Belfast had a higher total cost based on the calculations in Example 12.1, it significantly outperformed the other suppliers on the criteria Compton considered very important. Given the weights placed on the criteria, it is clear that Compton is willing to pay extra for better delivery performance, quality, and environmental stewardship.

Supplier Certification and Evaluation Supplier certification programs verify that potential suppliers have the capability to provide the services or materials the buying firm requires. ISO 9001:2008 is one such program (see Chapter 5, "Quality and Performance," for more details). Nonetheless, certification typically involves site visits by a cross-functional team from the buying firm, which does an in-depth evaluation of the supplier's capability to meet cost, quality, delivery, and flexibility targets from process and information system perspectives. The team may consist of members from operations, purchasing, engineering, information systems, and accounting. Every aspect of producing the services or materials is explored. The team observes the supplier's processes in action and reviews the documentation for completeness and accuracy. Once certified, the supplier can be used by the purchasing department without its having to make background checks.

Certification does not give the supplier a free pass on future evaluation. Performance is regularly monitored and performance records are kept. Periodic visits by the certification team may take place. Recertification may be required after a certain period of time or if performance declines.

Design Collaboration

The design collaboration process focuses on jointly designing new services or products with key suppliers; it facilitates concurrent engineering by drawing key suppliers into the new service/product development process, particularly in the design and development stages. This process seeks to eliminate costly delays and mistakes incurred when many suppliers concurrently design service packages or manufactured components.

An approach that many firms are using in their design collaboration process is called **early supplier involvement**, which is a program that includes suppliers in the design phase of a service or product. Suppliers provide suggestions for design changes and materials choices that will result in more efficient operations and higher quality. In the automotive industry, an even higher level of early supplier involvement is known as **presourcing**, whereby suppliers are selected early in a product's concept development stage and are given significant, if not total, responsibility for the design of certain components or systems of the product. Presourced suppliers also take responsibility for the cost, quality, and on-time delivery of the items they produce.

Firms can also improve performance by engaging in **value analysis**, which is a systematic effort to reduce the cost or improve the performance of services or products, either purchased or produced. It is an intensive examination of the services, materials, processes, information systems, and flows of material involved in the production of a service or an item. Benefits include reduced production, materials, and distribution costs; improved profit margins; and increased customer satisfaction.

Negotiation

The negotiation process focuses on obtaining an effective contract that meets the price, quality, and delivery requirements of the supplier relationship process's internal customers. The nature of the relations maintained with suppliers can affect the quality, timeliness, and price of a firm's services and products. The firm's orientation toward supplier relations will affect the negotiation and design collaboration processes.

BIOTA's premium natural spring water originates from one of the world's highest protected alpine springs perched above Ouray, Colorado, and is packaged in the world's first biodegradable bottle. Made from corn, a 100 percent renewable resource, BIOTA bottles break down in approximately 80 days in a commercial composting environment.

PRNewsFoto/BIOTA Brands of America, Inc./AP Photos

Competitive Orientation   The **competitive orientation** sees negotiations between buyer and seller as a zero-sum game: Whatever one side loses, the other side gains. Short-term advantages are prized over long-term commitments. The buyer may try to beat the supplier's price down to the lowest survival level or push demand to high levels during boom times and order almost nothing during recessions. In contrast, the supplier presses for higher prices for specific levels of quality, customer service, and volume flexibility. Which party wins depends largely on who has the most clout.

Purchasing power determines the clout that a firm has. A buyer has purchasing power when the purchasing volume represents a significant share of the supplier's sales or the purchased service or item is standardized and many substitutes are available. We refer to this condition as *economic dependency*. However, firms may have other sources of power in a relationship with suppliers. These sources include:

- *Referent*–the supplier values identification with the buyer. For example, the fact that a firm is supplying IBM may open the door for business with other customers.

- *Expert*–the buyer has access to knowledge, information, and skills desired by the supplier. For example, a supplier to UPS may get access to logistics planning skills possessed by UPS.

value analysis

A systematic effort to reduce the cost or improve the performance of services or products, either purchased or produced.

competitive orientation

A supplier relation that views negotiations between buyer and seller as a zero-sum game: Whatever one side loses, the other side gains, and short-term advantages are prized over long-term commitments.

- *Reward*–the buyer has the ability to give rewards to the supplier. Often, the rewards may involve the promise for future business or the opportunity to become a partner sometime in the future.

- *Legal*–the buyer has the legal right to prescribe behavior for the supplier. For example, the buyer may demand strict compliance to the negotiated contract or else the matter will be taken to court.

- *Coercive*–the buyer has the ability to punish the supplier. For example, the buyer may threaten to cancel future business with the supplier unless the supplier adheres to the buyer's demands.

It should be noted that the buyer does not always possess the power in a relationship. Sometimes the supplier holds the power; consequently, the supplier could exercise these sources of power as well.

cooperative orientation

A supplier relation in which the buyer and seller are partners, each helping the other as much as possible.

Cooperative Orientation   The **cooperative orientation** emphasizes that the buyer and the seller are partners, each helping the other as much as possible. A cooperative orientation means long-term commitment, joint work on quality and service or product designs, and support by the buyer of the supplier's managerial, technological, and capacity development. A cooperative orientation favors fewer suppliers of a particular service or item, with just one or two suppliers being the ideal number. As order volumes increase, the supplier gains economies of scale, which lowers costs. When contracts are large and a long-term relationship is ensured, the supplier might even build a new facility and hire a new workforce, perhaps relocating close to the buyer's plant. Suppliers become almost an extension of the buyer.

A cooperative orientation means that the buyer shares more information with the supplier about its future buying intentions. This forward visibility allows suppliers to make better, more reliable forecasts of future demand. The buyer visits suppliers' plants and cultivates cooperative attitudes. The buyer may even suggest ways to improve the suppliers' operations. This close cooperation with suppliers could even mean that the buyer does not need to inspect incoming materials. It also could mean involving the supplier more in designing services or products, implementing cost-reduction ideas, and sharing in savings.

sole sourcing

The awarding of a contract for a service or item to only one supplier.

One advantage of a cooperative orientation is that of reducing the number of suppliers in the supply chain, which reduces the complexity of managing them. However, reducing the number of suppliers for a service or item may increase the risk of an interruption in supply. It also means less opportunity to drive a good bargain unless the buyer has a lot of clout. **Sole sourcing**, which is the awarding of a contract for a service or item to only one supplier, can amplify any problems with the supplier that may crop up.

Both the competitive and cooperative orientations have their advantages and disadvantages. The key is to use the approach that serves the firm's competitive priorities best. Some companies utilize a mixed strategy, applying a competitive approach for its commodity-like supplies and a cooperative approach for its complex, high-valued, or high-volume services and materials.

Buying

The buying process relates to the actual procurement of the service or material from the supplier. This process includes the creation, management, and approval of purchase orders and determines the locus of control for purchasing decisions. Although not all purchasing opportunities involve the Internet, virtual marketplaces have provided firms with many opportunities to improve their buying and information exchange processes. Here we discuss four approaches to e-purchasing: (1) electronic data interchange, (2) catalog hubs, (3) exchanges, and (4) auctions, and close with the implications of choosing a locus of control.

electronic data interchange (EDI)

A technology that enables the transmission of routine business documents having a standard format from computer to computer over telephone or direct leased lines.

Electronic Data Interchange   A traditional form of e-purchasing is **electronic data interchange (EDI)**, a technology that enables the transmission of routine, standardized business documents from computer to computer over telephone or direct leased lines. Special communications software translates documents into and out of a generic form, allowing organizations to exchange information even if they have different hardware and software components. Invoices, purchase orders, and payments information are some of the routine documents that EDI can handle—it replaces the phone call or mailed document.

catalog hubs

A system whereby suppliers post their catalog of items on the Internet and buyers select what they need and purchase them electronically.

Catalog Hubs   **Catalog hubs** can be used to reduce the costs of placing orders to suppliers as well as the costs of the services or goods themselves. Suppliers post their catalog of items on the hub, and buyers select what they need and purchase them electronically. The hub connects the firm to potentially hundreds of suppliers through the Internet, saving the costs of EDI, which requires one-to-one connections to individual suppliers. Moreover, the buying firm can negotiate prices with individual suppliers for items such as office supplies, technical equipment, services, and so forth. The catalog that the buying firm's employees see consists only of the approved items and the prices the buyer has prenegotiated with its suppliers. Employees use their PCs to select the items they need, and the system generates the purchase orders, which are electronically dispatched to the suppliers.

Exchanges   An **exchange** is an electronic marketplace where buying firms and selling firms come together to do business. The exchange maintains relationships with buyers and sellers, making it easy to do business without the aspect of contract negotiations or other types of long-term conditions. Exchanges are often used for "spot" purchases to satisfy an immediate need at the lowest possible cost. Commodity items such as oil, steel, or energy fit this category. However, exchanges can also be used for most any item, such as hotel or hospital supplies.

Auctions   An extension of the exchange is the **auction**, where firms place competitive bids to buy something. For example, a site may be formed for a particular industry, and firms with excess capacity or materials can offer them for sale to the highest bidder. Bids can either be closed or open to the competition. The Autodaq Private Auction service provides dealers and wholesalers of used automobiles complete condition and guidebook price information and a Web site where they can search for vehicles. Dealers may bid in real time or submit "proxy bids" by entering the highest amount they would pay for a given vehicle and allowing the Autodaq system to automatically bid for them in $50 increments up to that proxy amount. Other industries where auctions have value include steel and chemicals. Examples involving consumers include eBay and Priceline.com.

An approach that has received considerable attention is the so-called *reverse auction*, where suppliers bid for contracts with buyers. Each bid is posted, so suppliers can see how much lower their next bid must be to remain in the running for the contract. Each contract has an electronic prospectus that provides all the specifications, conditions, and other requirements that are non-negotiable. The only thing left to determine is the cost to the buyer. Savings to the buyer can be dramatic, sometime as much as 20 to 30 percent over typical contract prices. Ford, GM, and Chrysler put together a reverse auction marketplace to procure parts from suppliers. Similar marketplaces are forming around the buying and selling of paper, plastic, steel, bandwidth, chemicals, and the like.

Our discussion of these electronic purchasing approaches should not leave you with the impression that cost is the only consideration firms make. Exchanges and auctions are more useful for commodities, near-commodities, or infrequently needed items that require only short-term relationships with suppliers. Nonetheless, suppliers should be thought of as partners when the needed supply is significant and steady over extended periods of time. Supplier involvement in service or product design and supply chain performance improvement requires long-term relationships not found with competitive pricing on the Internet.

Locus of Control   When an organization has several facilities (stores, hospitals, or plants, for example), management must decide whether to buy locally or centrally. This decision has implications for the control of supply chain flows.

Centralized buying has the advantage of increasing purchasing power by creating a situation where suppliers are economically dependent on the buyer. Savings can be significant, often on the order of 10 percent or more. Increased buying power can mean getting better service, ensuring long-term supply availability, or developing new supplier capability. Companies with overseas suppliers favor centralization because the specialized skills (for example, understanding of foreign languages and cultures) needed to buy from foreign sources can be centralized in one location. Buyers also need to understand international commercial and contract law regarding the transfer of services and goods. Another trend that favors centralization is the growth of computer-based information systems and the Internet, which give specialists at headquarters access to data previously available only at the local level.

Probably the biggest disadvantage of centralized buying is loss of control at the local level. Centralized buying is undesirable for items unique to a particular facility. These items should be purchased locally whenever possible. The same holds for purchases that must be closely meshed with production schedules. Localized buying is also an advantage when the firm has major facilities in foreign countries because the managers there, often foreign nationals, have a much better understanding of the local culture than staff members at the home office. Also, centralized buying often means longer lead times.

exchange

An electronic marketplace where buying firms and selling firms come together to do business.

auction

A marketplace where firms place competitive bids to buy something.

Perhaps the best solution is a compromise strategy, whereby both local autonomy and centralized buying are possible. For example, the corporate purchasing group at IBM negotiates contracts on a centralized basis only at the request of local plants. Management at one of the plants then monitors the contract for all the participating plants.

Information Exchange

The information exchange process facilitates the exchange of pertinent operating information, such as forecasts, schedules, and inventory levels between the firm and its suppliers. New technology in the form of radio frequency identification facilitates the flow of inventory information. Beyond inventory information, the exchange of forecasts and other demand-related data facilitates integrating activities such as vendor managed inventories.

Radio Frequency Identification   An important requirement for any information exchange process in a supply chain context is accurate information regarding the quantity and location of inventories. A new application of an old technology presents some tantalizing benefits. **Radio frequency identification (RFID)** is a method for identifying items through the use of radio signals from a tag attached to an item. The tag has information about the item and sends signals to a device that can read the information and even write new information on the tag. Data from the tags can be transmitted wirelessly from one place to another through electronic product code (EPC) networks and the Internet, making it theoretically possible to uniquely identify every item a company produces and track it until the tag is destroyed.

Walmart, Target, Intel, Gillette, and the Department of Defense, among a number of large retailers, manufacturers, government agencies, and suppliers, are implementing RFID in their supply chains. The use of RFID data can increase a supplier's service level and reduce theft. Gillette is using RFID to reduce the amount of razor-blade theft, which amounts to as much as 30 percent of sales.

Individual firms can use RFID within their own operations and avoid costly coordination with other firms in the supply chain. Benefits of using RFID internal to the firm are limited, however, and so is the investment. The larger potential gains come with application to the supply chain. To be successful, all members of the supply chain must benefit from the investment in RFID, not just the firm pushing the project. This is particularly true for global operations. Global data synchronization using industry standards is critical to ensure that accurate and consistent product information is exchanged between trading partners, a very challenging task.

Vendor Managed Inventories   Reliable information regarding inventories up and down the supply chain allows firms to collaborate on effective ways to improve material flows. A tactic that requires a reliable information exchange process is **vendor-managed inventories (VMI)**, a system in which the supplier has access to the customer's inventory data and is responsible for maintaining the inventory level required by the customer. The inventory is on the customer's site, and often the supplier retains possession of the inventory until it is used by the customer. Companies such as Walmart and Dell leverage their market position to mandate VMI. Vendor-managed inventories have several key elements.

- *Collaborative Effort.* For VMI to succeed, the customers must be willing to allow the supplier access to their inventory information, which is facilitated by RFID but must be bolstered by information on forecasts, sales promotions, and other demand related data. The implication is that the supplier assumes an important administrative role in the management of the inventory. Thus, an atmosphere of trust and accountability is required.

- *Cost Savings.* Suppliers and customers eliminate the need for excess inventory through better operational planning. VMI lowers costs by reducing administrative and inventory costs. Order placement costs are also reduced.

- *Customer Service.* The supplier is frequently on the customer's site and better understands the operations of the customer, improving response times and reducing stockouts.

- *Written Agreement.* It is important that both parties fully understand the responsibilities of each partner. Areas such as billing procedures, forecast methods, and replenishment schedules should be clearly specified. Further, the responsibility for obsolete inventory resulting from forecast revisions and changes in contract lengths should be included.

VMI can be used both by service providers as well as manufacturers. AT&T, Roadway Express, Walmart, Dell, Westinghouse, and Bose are among the companies that use it.

radio frequency identification (RFID)

A method for identifying items through the use of radio signals from a tag attached to an item.

vendor-managed inventories (VMI)

A system in which the supplier has access to the customer's inventory data and is responsible for maintaining the inventory on the customer's site.

The Order Fulfillment Process

The order fulfillment process produces and delivers the service or product to the firm's customers. There are four key nested processes: (1) customer demand planning, (2) supply planning, (3) production, and (4) logistics.

Customer Demand Planning

The customer demand planning (CDP) process facilitates the collaboration of a supplier and its customers for the purpose of forecasting customer requirements for a service or product. CDP is a business-planning process that enables sales teams (and customers) to develop demand forecasts as input to service-planning processes, production and inventory planning, and revenue planning. Forecasts must generally precede plans: It is not possible to make decisions on staffing levels, purchasing commitments, and inventory levels until forecasts are developed that give reasonably accurate views of demand over the forecasting time horizon. Chapter 14, "Forecasting," contains many practical tools for forecasting customer demands.

Supply Planning

The supply planning process takes the demand forecasts produced by the customer demand planning process, the customer service levels and inventory targets provided by inventory management, and the resources provided by aggregate and detailed capacity planning to generate a plan to meet the demand. Regardless of whether the firm is producing services or a product, this process is critical for effective execution in the supply chain.

An important supply planning activity for a firm is planning its aggregate resource levels so that supply is in balance with demands. Usually stated at the firm or departmental level for several months to a year in advance, these aggregate plans specify output rates, workforce levels, and inventory levels that are consistent with demand forecasts, capacity constraints, and inventory plans. Such plans amount to planning staffing levels for service firms and production rates and finished goods inventory levels for manufacturing firms. Once the aggregate resource levels are determined, they must be scheduled to achieve the level of performance envisioned by management. In Chapter 15, "Operations Planning and Scheduling," we show how firms plan and schedule productive resources to provide an appropriate level of supply for services or products.

Finally, as the saying goes, "The devil is in the details." Planning for resources at aggregate levels is useful for making hiring decisions or acquiring capital goods so as to meet aggregate levels of demand. It ultimately becomes necessary to take each individual service or product and plan for its specific supply. With the aggregate operations plans and schedules as a basis, information such as time standards, routings, and other specific information as to how a service or product is produced is used to plan the specific inputs required. These plans are very detailed but very important for the efficient flow of services and products. In Chapter 16, "Resource Planning," we show how this detailed level of planning is done and how the information from this planning effort is used by other functional areas in the firm.

Production

The production process executes the supply plan to produce the service or product. Nonetheless, the production process must be integrated with the processes that supply the inputs, establish the demands, and deliver the product to the customers. For example, while customers can always shop for standardized Dell computer packages at retail stores like Best Buy and Walmart, order placement, buying, production, and logistics processes are tightly linked at Dell when a customized machine is being ordered directly from the computer manufacturer. Dell's supply chain is designed to support an assemble-to-order strategy, thereby providing speedy service with minimal inventories.

Integrating the supply-facing and customer-facing processes to the production process is as important to service firms as it is to manufacturing firms. The best firms tightly link their production process to suppliers as well as customers.

Logistics

A key aspect of order fulfillment is the logistics process, which delivers the product or service to the customer. Five important decisions determine the design and implementation of logistics processes: (1) degree of ownership, (2) facility location, (3) mode selection, (4) capacity level, and (5) amount of cross docking.

■ *Ownership.* The firm has the most control over the logistics process if it owns and operates it, thereby becoming a *private carrier.* Although this approach may help to better achieve the firm's competitive priorities, the cost of equipment, labor, facilities, and maintenance could be high. The firm could instead leave the distribution to a *third-party logistics provider* (3PL), negotiating with the carrier for specific services. Those services could involve taking over a major portion of the order fulfillment process. 3PLs typically provide integrated services, from transportation and packaging services to warehousing and inventory management, for corporate clients that need to get their products to market. They can help with the design of a client's supply chain and facilitate the flow of information up and down the supply chain.

■ *Facility Location.* A critical decision affecting the effectiveness of supply chains is the location of facilities that serve as points of service, storage, or manufacture. As the opener to this chapter showed, Kodak went though considerable analysis to arrive at its forward stocking locations. Chapter 11, "Supply Chain Location Decisions," provides a complete discussion of facility location choices.

■ *Mode Selection.* The five basic modes of transportation are (1) truck, (2) train, (3) ship, (4) pipeline, and (5) airplane. The drivers for the selection should be the firm's competitive priorities. Trucks provide the greatest flexibility because they can go wherever roads go. Transit times are good, and rates are usually better than trains for small quantities and short distances. Rail transportation can move large quantities cheaply; however, the transit times are long and often variable. Water transportation provides high capacity and low costs, and is necessary for overseas shipments of bulky items; however, the transit times are slow and highway or rail transportation is often needed to get the product to its ultimate destination. Pipeline transportation is highly specialized and is used for liquids, gases, or solids in slurry form. Although it has limited geographical flexibility, transporting via pipeline requires no packaging, and the operating costs per mile are low. Finally, air transportation is the fastest and most costly mode per mile. Nonetheless, getting a product to the customer fast using air transportation may actually reduce total costs when the costs of inventory and warehouse handling are considered for alternative modes. The cost of the funds tied up in some in-transit inventories can be considerable. Firms can also use mixed modal transportation, whereby a given shipment may combine two or more different modes. For example, containers can be carried by trucks, trains, or ships over different portions of their transit, and can often give the best tradeoffs between cost and delivery times.

The Trans Alaska Pipeline System traverses 800 miles from the North Slope of Alaska to the northern most ice-free port of Valdez, Alaska. The pipeline is 48 inches in diameter, has 11 pump stations, and has a maximum throughput of 2 million barrels per day.

Bkp/Shutterstock

■ *Capacity.* The performance of a logistics process is directly linked to its capacity. The ownership decision and the modal selection decision are often intertwined because the question of how much capacity is needed must be resolved. If ownership of the equipment and facilities is under consideration, capital costs as well as variable operating costs must be weighed against the costs of obtaining the logistics services from a supplier. Making things more difficult is the fact that the requirements for the logistics process are rarely known with certainty. In such cases, management can use the *expected value decision rule* to evaluate capacity alternatives. The expected value of an alternative is calculated as follows:

> *Expected value of an alternative* = (probability of a level of demand occurring)(payoff for using the alternative if that level of demand materialized) summed over all possible levels of demand.

See Supplement A, "Decision Making," for details on this approach. Example 12.3 demonstrates the use of the expected value decision rule for analyzing truck capacity.

EXAMPLE 12.3	Using the Expected Value Decision Rule for Truck Capacity

Tower Distributors provides logistical services to local manufacturers. Tower picks up products from the manufacturers, takes them to its distribution center, and then assembles shipments to retailers in the region. Tower needs to build a new distribution center; consequently, it needs to make a decision on how many trucks to use. The monthly amortized capital cost of ownership is $2,100 per truck. Operating variable costs are $1 per mile for each truck owned by Tower. If capacity is exceeded in any month, Tower can rent trucks at $2 per mile. Each truck Tower owns can be used 10,000 miles per month. The requirements for the trucks, however, are uncertain. Managers have estimated the following probabilities for several possible demand levels and corresponding fleet sizes.

Requirements (miles/month)	100,000	150,000	200,000	250,000
Fleet Size (trucks)	10	15	20	25
Probability	0.2	0.3	0.4	0.1

Notice that the sum of the probabilities must equal 1.0. If Tower Distributors wants to minimize the expected cost of operations, how many trucks should it use?

SOLUTION

We use the expected value decision rule to evaluate the alternative fleet sizes where we want to minimize the expected monthly cost. To begin, the monthly cost, C, must be determined for each possible combination of fleet size and requirements. The cost will depend on whether additional capacity must be rented for the month. For example, consider the *10 truck fleet size* alternative, which represents a capacity of 100,000 miles per month. C = monthly capital cost of ownership + variable operating cost per month + rental costs if needed:

C (100,000 miles/month) = ($2,100/truck)(10 trucks) + ($1/mile)(100,000 miles) = $121,000

C (150,000 miles/month) = ($2,100/truck)(10 trucks) + ($1/mile)(100,000 miles) + ($2 rent/mile)(150,000 miles
$\qquad$ − 100,000 miles) = $221,000

C (200,000 miles/month) = ($2,100/truck)(10 trucks) + ($1/mile)(100,000 miles) + ($2 rent/mile)(200,000 miles
$\qquad$ − 100,000 miles) = $321,000

C (250,000 miles/month) = ($2,100/truck)(10 trucks) + ($1/mile)(100,000 miles) + ($2 rent/mile)(250,000 miles
$\qquad$ − 100,000 miles) = $421,000

Next, calculate the expected value for the *10 truck fleet size* alternative as follows:

Expected Value (10 trucks) = 0.2($121,000) + 0.3($221,000) + 0.4($321,000) + 0.1($421,000) = $261,000

Using similar logic, we can calculate the expected costs for each of the other fleet-size options:

Expected Value (15 trucks) = 0.2($131,500) + 0.3($181,500) + 0.4($281,500) + 0.1($381,500) = $231,500

Expected Value (20 trucks) = 0.2($142,000) + 0.3($192,000) + 0.4($242,000) + 0.1($342,000) = $217,000

Expected Value (25 trucks) = 0.2($152,500) + 0.3($202,500) + 0.4($252,500) + 0.1($302,500) = $222,500

Using the expected value decision rule, Tower Distributors should use a fleet of 20 trucks.

DECISION POINT

The fleet size of 20 trucks means that Tower will have enough capacity to handle 90 percent of its requirements (sum of the probabilities for 100,000 miles, 150,000 miles, and 200,000 miles). Further, there will be a 50 percent chance that it will have excess capacity (sum of the probabilities for 100,000 miles and 150,000 miles). While the decision to invest in 20 trucks will minimize expected costs, it will also provide slack capacity 50 percent of the time, which is reflective of the relatively high cost for being short of capacity.

■ *Cross-Docking.* Low-cost operations and delivery speed can be enhanced with a technique called **cross-docking**, which is the packing of products on incoming shipments so that they can be easily sorted at intermediate warehouses for outgoing shipments based on their final destinations; the items are carried from the incoming-vehicle docking point to the outgoing-vehicle docking point without being stored in inventory at the warehouse. The warehouse becomes a short-term staging area for organizing efficient shipments to customers. The benefits

cross-docking

The packing of products on incoming shipments so that they can be easily sorted at intermediate warehouses for outgoing shipments based on their final destinations.

of cross-docking include reductions in inventory investment, storage space requirements, handling costs, and lead times, as well as increased inventory turnover and accelerated cash flow. Management must decide where cross-docking operations are best placed, given the overall flow of items and their destinations.

Managerial Practice 12.1 shows the complexity of the the order fulfillment process aboard a cruise ship.

MANAGERIAL PRACTICE 12.1   Order Fulfillment aboard the Coral Princess

Regardless of where you are right now, or what the weather is like in your home town, think of lounging on the deck of the Coral Princess somewhere in the Caribbean just after passing through the Panama Canal. The view is gorgeous, the breezes soft and cool; you find it difficult to contemplate that you and 1974 other guests are in a top-rated hotel with amenities that range from an outdoor movie theater to a casino with every game of chance a gambler ever wanted. If you want a wedding, there is a chapel. There is a swimming pool with a retractable glass dome, a cigar bar, and a complete TV studio.

While all those amenities sound good, there is one thing that all cruise liner guests look forward to and that is eating. There are five high-quality restaurants on board the Coral Princess. When the meals arrive in front of you, have you ever wondered how they got there? The supply of food, and its preparation, puts a tremendous strain on the order fulfillment process and the supply chain for the Coral Princess. Here, four days at sea, the hotel manager just cannot call the suppliers and say that he forgot the carrots, the butter, and the sugar. Running out of stock is not an acceptable option. Pure volume is a complicating factor. On an average 15-day cruise, the Coral Princess will use 175 tons of food, including 43,200 eggs, 2,425 pounds of pasta, 7,245 pounds of rice, 84,000 pounds of vegetables, 13,000 pounds of chicken, 8,800 pounds of fish, and 10,500 pounds of beef. Two hundred galley employees prepare over 10,000 meals daily; all washed down with 3,800 bottles of wine and 12,000 bottles/cans of beer a cruise. Not everything is purchased, nor can all 175 tons of food be stored on board at one time. Certain items such as ice cream, dessert pastries, and breads are produced onboard to ensure their freshness. Other items must be restocked at selected ports on a schedule put in place long in advance of the cruise. While the guests are enjoying a port of call, the ship personnel are restocking the hold with food from suppliers in that area that have been proven to supply the best quality. Cruise ships must be very careful not to accept food that will cause sickness onboard.

Even a well-oiled wagon can run off the road, and so it can with the order fulfillment process of a cruise ship. Good planning, including contingency plans, and quick reactions are desirable attributes. Natural disasters, such as hurricanes or earthquakes, cause cruise ships to reschedule their ports of call, which can disrupt the scheduled supply of food, boutique items,

The cruise liner Coral Princess awaits food and supplies at the Cruise Terminal in the Port of Los Angeles, San Pedro, Los Angeles, California.

hotel supplies, and maintenance items. For example, the massive Japanese earthquake and tsunami of March 11, 2011 not only caused cruise ships to revise their itineraries, but it will also affect the supply chain for a very long time, given the scare from radiation getting into the food supply. Another example of a disruptive event is a strike at a port of supply for a cruise ship. A labor strike at the Ports of Los Angeles and Long Beach caused a massive congestion of 70 vessels waiting to be unloaded, thereby putting pressure on the Princess Cruises schedule for replenishing a cruise ship headed for Mexico. The solution was to use trucks to move part of the shipment in ocean freight containers to go by sea using a vessel at the Port of Oakland, 450 miles away, and the rest of the shipment by land using 53-foot trailers, timed so that both shipments would meet the cruise ship in Mexico. The two loads required considerable coordination for health inspections, security, customs clearance, and refrigerated equipment.

Cruise ships strive to provide the best possible experience for their guests. The order fulfillment process is at the heart of that effort.

*Source: Handout, The Coral Princess Food & Beverage Department, August 16, 2011; News archive at **www.princess.com/news/article**; Case Study Cruise Line Logistics-Princess Cruises, Agility Logistics, **www.agilitylogistics.com***

The Customer Relationship Process

The customer relationship process addresses the interface between the firm and its customers downstream in the supply chain. The purpose of the customer relationship process, which supports *customer relationship management (CRM)* programs, is to identify, attract, and build relationships with customers and to facilitate the transmission and tracking of orders. Key nested processes include the marketing, order placement, and customer service processes.

Marketing

The marketing process focuses on such issues as determining the customers to target, how to target them, what services or products to offer and how to price them, and how to manage promotional campaigns. In this regard, **electronic commerce (e-commerce)**, which is the application of

electronic commerce (e-commerce)

The application of information and communication technology anywhere along the supply chain of business processes.

information and communication technology anywhere along the supply chain of business processes, has had a huge impact up and down the supply chain. There are two e-commerce technologies that relate to the marketing process: (1) business-to-consumer and (2) business-to-business systems.

Business-to-Consumer Systems   Business-to-consumer (B2C) systems, which allow customers to transact business over the Internet, are commonplace. B2C e-commerce offers a new distribution channel for businesses, and consumers can avoid crowded department stores with long checkout lines and parking-space shortages. Many of the advantages of e-commerce were first exploited by retail "e-businesses," such as Amazon.com, E*TRADE, and Autobytel. These three companies created Internet versions of traditional bookstores, brokerage firms, and auto dealerships. The Internet is changing operations, processes, and cost structures for even traditional retailers, and the overall growth in its usage has been dramatic. Today, anyone with an Internet connection can open a store in cyberspace. Even well-established retailers like Walmart are increasing their web-based shopping presence through their site-to-store program, where customers can actually shop on the Internet and pick up their order at the closest store within a few days.

Business-to-Business Systems   The biggest growth, however, has been in business-to-business (B2B) e-commerce systems, or commerce between firms. In fact, business-to-business e-commerce outpaces business-to-consumer transactions, with trade between businesses making up more than 70 percent of the regular economy. These systems facilitate trade up and down the supply chain, making it easier to purchase or sell services or products. B2B systems can also help manage the flow of materials. For example, if a distributor is out of stock, the firm's central warehouse can be notified to immediately ship replenishment stock directly to the customer.

Order Placement

The order placement process involves the activities required to execute a sale, register the specifics of the order request, confirm the acceptance of the order, and track the progress of the order until it is completed. Often the firm has a sales force that visits prospective and current customers to encourage a sale.

The Internet enables firms to reengineer their order placement process to benefit both the customer and the firm. The Internet provides the following advantages for a firm's order placement process.

- *Cost Reduction.* Using the Internet can reduce the costs of processing orders because it allows for greater participation by the customer. Customers can select the services or products they want and place an order with the firm without actually talking to anyone. This approach reduces the need for call centers, which are labor intensive and often take longer to place orders.

- *Revenue Flow Increase.* A firm's Web page can allow customers to enter credit card information or purchase-order numbers as part of the order placement process. This approach reduces the time lags often associated with billing the customer or waiting for checks sent in the mail.

- *Global Access.* Another advantage the Internet provides to firms is the opportunity to accept orders 24-hours a day. Traditional brick-and-mortar firms only take orders during their normal business hours. Firms with Internet access can reduce the time it takes to satisfy customers, who can shop and purchase at any time. This access gives these firms a competitive advantage over brick-and-mortar firms.

- *Pricing Flexibility.* Firms with their services and products posted on the Web can easily change prices as the need arises, thereby avoiding the cost and delay of publishing new catalogs. Customers placing orders have current prices to consider when making their choices. From the perspective of supply chains, Dell uses this capability to control for component shortages. Because of its direct-sales approach and promotional pricing, Dell can steer customers to certain configurations of computers for which ample supplies exist.

Customer Service

The customer service process helps customers with answers to questions regarding the service or product, resolves problems, and, in general, provides information to assist customers. It is an important point of contact between the firm and its customers, who may judge the firm on the basis of their experiences with this process. The age-old tradeoff between cost and quality, however, enters the picture, especially for call centers. In an effort to reduce the cost of their customer service process, many firms have opted to replace human service agents with automated systems, which often require customers to wade through an exhausting sequence of options that sometimes only lead to frustration. Other firms are using Verbots®, or "verbal robots," which are supported by sophisticated artificial intelligence. They have personalities, ask and respond to questions, and in

Sherwin Crasto/Reuters/Corbis

Many firms have outsourced their customer service processes, particularly if the service can be transacted over the phone. Here Indian employees at a call center in the southern city of Bangalore, India, provide service support to international customers.

some cases are almost indistinguishable from humans over the phone. Nonetheless, most customers and others seeking information about a service or product prefer humans. Consequently, in consideration for the cost involved, many companies have expanded their supply chain by outsourcing the customer service process to an off-shore site where labor costs are low. In this regard, India has responded in a big way to the international need for low-cost call centers. Of course, the big risk in outsourcing the customer service process, or a part of it, is that the firm loses some control over a process that has direct interface with its customers. This consideration should be carefully weighed in the final analysis.

Levers for Improved Supply Chain Performance

Now that we have discussed a framework for integrated supply chains, we can return to the problems caused by supply chain dynamics and how integrated supply chains can mitigate the poor performance from dynamics such as the bullwhip effect. In this section, we pose some levers that can be utilized with integrated supply chains to improve performance and reveal important performance measures for tracking supply chain operations.

The Levers

Integrated supply chains facilitate the application of the following options for overall improvement in performance:

- *Sharing data*–one source of dynamics in supply chains is the lack of visibility of end-user demand by suppliers upstream in the supply chain. To facilitate planning at all levels in the supply chain, point-of-sale (POS) data, which records actual customer purchases of the final service or product, can be shared with all suppliers. RFID can also be used to track quantities of inventory throughout the supply chain.

- *Collaborative activities*–working closely with customers and suppliers in customer demand planning (CDP) and environmental health and safety programs as well as the design collaboration process improves information flows, improves environmental stewardship, and reduces surprises from demand spikes due to promotions or supply hang-ups because of poorly designed services or products.

- *Reduce replenishment lead times*–improving internal processes and working with suppliers to reduce lead times allows the firm to wait longer before reacting to a change in demand levels, thereby mitigating the bullwhip effect. In addition, shorter lead times leads to smaller pipeline inventories.

- *Reduce order lot sizes*–working on ways to reduce the costs associated with ordering, transporting and receiving inventory throughout the supply chain will reduce order lot sizes and thereby decrease the amount of fluctuation in the size of orders in the supply chain.

- *Ration short supplies*–when a shortage exists, customers sometimes artificially inflate their orders to protect themselves, only to cancel them later when the shortage is relieved. To counteract this behavior, suppliers can ration short supplies to customers on the basis of their past sales, rather than their current orders.

- *Use every day low pricing (EDLP)*–promotional or discount pricing encourages spikes in demand. Using a stable pricing program such as EDLP, as is done by Walmart, discourages customers from buying excess stock at discounted prices so they can offer price promotions, a practice called *forward buying*. EDLP levels the demand.

- *Be cooperative and trustworthy*–being cooperative in solving supply issues and providing information that can be trusted serves to reduce costs for all members of the supply chain and mitigates environmental problems and the deleterious effects of supply chain dynamics.

Performance Measures

It is important to monitor the performance of supply chains to see where improvements can be made or to measure the impact of applying the levers. Supply chain managers monitor performance by measuring costs, time, quality, and environmental impact. Table 12.1 contains examples of commonly used performance measures for three supply chain processes. Managers periodically collect data on these measures and track them to note changes in level or direction. Statistical process control charts can be used to determine whether the changes are statistically significant.

Integrated supply chains are powerful tools for achieving competitiveness along many performance measures. Currently, concerns about the environment are prompting supply chain managers to take a careful look at their operations and those of their suppliers. In Chapter 13, "Supply Chain Sustainability and Humanitarian Logistics", we take a look at the impact of environmental concerns on supply chains and how integrated supply chains can be used to relieve some of those concerns and still be profitable.

TABLE 12.1 | **SUPPLY CHAIN PROCESS MEASURES**

Customer Relationship	Order Fulfillment	Supplier Relationship
■ Percent of orders taken accurately ■ Time to complete the order placement process ■ Customer satisfaction with the order placement process ■ Customer's evaluation of firm's environmental stewardship	■ Percent of incomplete orders shipped ■ Percent of orders shipped on-time ■ Time to fulfill the order ■ Percent of botched services or returned items ■ Cost to produce the service or item ■ Customer satisfaction with the order fulfillment process ■ Inventory levels of work-in-process and finished goods ■ Amount of greenhouse gasses emitted into the air	■ Percent of suppliers' deliveries on-time ■ Suppliers' lead times ■ Percent defects in services and purchased materials ■ Cost of services and purchased materials ■ Inventory levels of supplies and purchased components ■ Evaluation of suppliers' collaboration on streamlining and waste conversion ■ Amount of transfer of environmental technologies to suppliers

LEARNING GOALS IN REVIEW

1 **Identify the major causes of dynamics in a supply chain.** Figure 12.1 shows the interconnectedness of supply chains, which is the engine for supply chain dynamics. The section "Supply Chain Dynamics," pp. 413–416, explains the major internal and external causes. Figure 12.3 shows how SKU volume and weekly demand volatility can be key factors in supply chain design.

2 **Explain how integrated supply chains can mitigate supply chain dynamics.** See the section "Integrated Supply Chains," pp. 415–416, and Figures 12.4 and 12.5, which show integrative frameworks.

3 **Explain the nature and purpose of the key nested processes within the new service or product development, supplier relationship, order fulfillment, and customer relationship**

processes. Because of their importance, each of these processes has a major section in this chapter.

4 **Use a total annual cost analysis and a preference matrix to select suppliers and the expected value decision rule to determine logistics capacity.** See the "Supplier Relationship Process," pp. 417–424. Study Example 12.1 on total cost analysis, Example 12.2 on the use of preference matrices, and Example 12.3 on the use of the expected value model. See the three Solved Problems for additional help.

5 **Identify key levers and performance measures useful for improving supply chain performance.** See the section "Levers for Improved Supply Chain Performance," pp. 430–431. Table 12.1 contains examples of supply chain performance measures.

MyOMLab helps you develop analytical skills and assesses your progress with multiple problems on supplier selection, preference matrix, and expected annual cost.

MyOMLab Resources	Titles	Link to the Book
Video	*Sourcing Strategy at Starwood*	Supplier Relationship Process; Levers for Improved Supply Chain Performance
	Pearson Education: Information Technology	Customer Relationship Process
OM Explorer Solvers	Preference Matrix	Supplier Relationship Process; Example 12.2 (p. 420); Solved Problem 2 (p. 433)
	Decision Theory	The Order Fulfillment Process; Example 12.3 (p. 427); Solved Problem 3 (pp. 433–434)

MyOMLab Resources	Titles	Link to the Book
OM Explorer Tutors	A.3 Preference Matrix	Supplier Relationship Process; Example 12.2 (p. 420); Solved Problem 2 (p. 433)
	A.5 Decisions Under Risk	The Order Fulfillment Process; Example 12.3 (p. 427); Solved Problem 3 (pp. 433–434)
POM for Windows	Preference Matrix	Supplier Relationship Process; Example 12.2 (p. 420); Solved Problem 2 (p. 433)
	Decision Tables	The Order Fulfillment Process; Example 12.3 (p. 427); Solved Problem 3 (pp. 433–434)
Virtual Tours	Monaco Coach Corporation	Entire chapter
Internet Exercise	1. Dell	Supplier Relationship Process; Customer Relationship Process
	2. Federal Mogul	Supplier Relationship Process; Customer Relationship Process
Key Equations		
Image Library		

Key Equations

1. Total Annual Cost $= pD +$ Freight costs $+ (Q/2 + \bar{d}L)H +$ Administrative costs

2. *Expected value of an alternative* = (probability of a level of demand occurring)(payoff for using the alternative if that demand materialized) summed over all possible levels of demand.

Key Terms

auction 423
bullwhip effect 413
catalog hubs 422
competitive orientation 421
concurrent engineering 417
cooperative orientation 422
cross-docking 427

early supplier involvement 421
electronic commerce (e-commerce) 428
electronic data interchange (EDI) 422
exchange 423
green purchasing 420
presourcing 421
purchasing 417

radio frequency identification (RFID) 424
SCOR model 415
sole sourcing 422
supply chain integration 412
value analysis 421
vendor-managed inventories (VMI) 424

Solved Problem 1

ABC Electric Repair is a repair facility for several major electric appliance manufacturers. ABC wants to find a low-cost supplier for an electric relay switch used in many appliances. The annual requirements for the relay switch (D) are 100,000 units. ABC operates 250 days a year. The following data are available for two suppliers, Kramer and Sunrise, for the part:

Supplier	Freight Costs Shipping Quantity (Q)		Price/Unit (p)	Carrying Cost/Unit (H)	Lead Time (L) (days)	Administrative Costs
	2,000	10,000				
Kramer	$30,000	$20,000	$5.00	$1.00	5	$10,000
Sunrise	$28,000	$18,000	$4.90	$0.98	9	$11,000

Which supplier will provide the lowest annual total costs?

SOLUTION

The daily requirements for the relay switch are:

$$\bar{d} = 100,000/250 = 400 \text{ units.}$$

We must calculate the total annual costs for each alternative:

Total annual cost = Material costs + Freight costs + Inventory costs + Administrative costs

$$= pD + \text{Freight costs} + (Q/2 + \bar{d}L)H + \text{Administrative costs}$$

Kramer

$Q = 2,000$:   ($5.00)(100,000) + $30,000 + (2,000/2 + 400(5))($1) + $10,000 = $543,000

$Q = 10,000$:   ($5.00)(100,000) + $20,000 + (10,000/2 + 400(5))($1) + $10,000 = $537,000

Sunrise

$Q = 2,000$:   ($4.90)(100,000) + $28,000 + (2,000/2 + 400(9))($0.98) + $11,000 = $533,508

$Q = 10,000$:   ($4.90)(100,000) + $18,000 + (10,000/2 + 400(9))($0.98) + $11,000 = $527,428

The analysis reveals that using Sunrise and a shipping quantity of 10,000 units will yield the lowest annual total costs.

Solved Problem 2

ABC Electric Repair wants to select a supplier based on total annual cost, consistent quality, and delivery speed. The following table shows the weights management assigned to each criterion (total of 100 points) and the scores assigned to each supplier (Excellent = 5, Poor = 1).

Criterion	Weight	Scores	
		Kramer	Sunrise
Total annual cost	30	4	5
Consistent quality	40	3	4
Delivery speed	30	5	3

Which supplier should ABC select, given these criteria and scores?

SOLUTION

Using the preference matrix approach, the weighted scores for each supplier are:

$$\textit{Kramer:}\ WS = (30 \times 4) + (40 \times 3) + (30 \times 5) = 390$$

$$\textit{Sunrise:}\ WS = (30 \times 5) + (40 \times 4) + (30 \times 3) = 400$$

Based on the weighted scores, ABC should select Sunrise even though delivery speed performance would be better with Kramer.

Solved Problem 3

Schneider Logistics Company has built a new warehouse in Columbus, Ohio, to facilitate the consolidation of freight shipments to customers in the region. George Schneider must determine how many teams of dock workers he should hire to handle the cross docking operations and the other warehouse activities. Each team costs $5,000 a week in wages and overhead. Extra capacity can be subcontracted at a cost of $8,000 a team per week. Each team, whether in-house or subcontracted, can satisfy 200 labor hours of work a week. The labor hour requirements for the new facility are uncertain. Management has estimated the following probabilities for the requirements:

Requirements (hours/wk)	200	400	600
Number of teams	1	2	3
Probability	0.20	0.50	0.30

How many teams should Schneider hire?

SOLUTION

We use the expected value decision rule by first computing the cost for each option for each possible level of requirements and then using the probabilities to determine the expected value for each option. The option with the lowest expected cost is the one Schneider will implement. We demonstrate the approach using the "one team" in-house option.

One Team In-House

$$C(200) = \$5,000$$

$$C(400) = \$5,000 + \$8,000 = \$13,000$$

$$C(600) = \$5,000 + \$8,000 + \$8,000 = \$21,000$$

Expected Value

$$(\text{One Team}) = 0.20(\$5,000) + 0.50(\$13,000) + 0.30(\$21,000) = \$13,800.$$

A table of the complete results is below.

| In-House | Weekly Labor Requirements | | | Expected Value |
	200 hrs	400 hrs	600 hrs	
One team	$5,000	$13,000	$21,000	$13,800
Two teams	$10,000	$10,000	$18,000	$12,400
Three teams	$15,000	$15,000	$15,000	$15,000

Based on the expected value decision rule, Schneider should employ two teams at the warehouse.

Discussion Questions

1. Supply chain dynamics can cause excessive costs and poor customer service. Explain how the redesign of a supply chain can help to mitigate the effects of supply chain dynamics.

2. Chrysler and General Motors vigorously compete with each other in many automobile and truck markets. When Jose Ignacio Lopez was vice president of purchasing for GM, he made it clear that his buyers were not to accept luncheon invitations from suppliers. Thomas Stalcamp, head of purchasing for Chrysler at the time, instructed his buyers to take suppliers to lunch. Rationalize these two directives in light of supplier relations and the impact on supply chain management.

3. Firms such as Walmart, General Electric, Chase Manhattan, and Boeing have a lot of influence in their respective supply chains because of the power they have. Explain how firms with a lot of power can influence supply chain integration.

4. We discussed the inventory and supply chain considerations such as small lot sizes, close supplier ties, and quality at the source in Chapter 8, "Lean Systems." What are the implications of these principles for supply chain integration?

Problems

The OM Explorer and POM for Windows software is available to all students using the 10th edition of this textbook. Go to **www.pearsonhighered.com/krajewski** to download these computer packages. If you purchased MyOMLab, you also have access to Active Models software and significant help in doing the following problems. Check with your instructor on how best to use these resources. In many cases, the instructor wants you to understand how to do the calculations by hand. At the least, the software provides a check on your calculations. When calculations are particularly complex and the goal is interpreting the results in making decisions, the software entirely replaces the manual calculations.

1. Horizon Cellular manufactures cell phones for exclusive use in its communication network. Management must select a circuit board supplier for a new phone soon to be introduced to the market. The annual requirements are 50,000 units and Horizon's plant operates 250 days per year. The data for three suppliers are in Table 12.2.

   Which supplier and shipping quantity will provide the lowest total cost for Horizon Cellular?

2. Eight Flags operates several amusement parks in the Midwest. The company stocks machine oil to service the machinery for the many rides at the parks. Eight Flags needs 30,000 gallons of oil annually; the parks operate 50 weeks a year. Management is unsatisfied with the current supplier of oil and has obtained two bids from other suppliers. The data are contained in Table 12.3.

   Which supplier and which shipping quantity will provide the lowest costs for Eight Flags?

TABLE 12.2 | **DATA FOR SUPPLIERS TO HORIZON CELLULAR**

Supplier	Annual Freight Costs Shipping Quantity		Price/Unit	Annual Holding Cost/Unit	Lead Time (Days)	Annual Administrative Cost
	10,000	20,000				
Abbott	$10,000	$7,000	$30	$6.00	4	$10,000
Baker	$12,000	$9,000	$28	$5.60	7	$12,000
Carpenter	$9,000	$6,500	$31	$6.20	3	$9,000

TABLE 12.3 | **DATA FOR SUPPLIERS TO EIGHT FLAGS**

Supplier	Annual Freight Costs Shipping Quantity			Price/Unit	Annual Holding Cost/Unit	Lead Time (wks)	Annual Administrative Cost
	5,000	10,000	15,000				
Sharps	$5,000	$2,600	$2,000	$4.00	$0.80	4	$4,000
Winkler	$5,500	$3,200	$2,900	$3.80	$0.76	6	$5,000

3. The Bennet Company purchases one of its essential raw materials from three suppliers. Bennet's current policy is to distribute purchases equally among the three. The owner's son, Benjamin Bennet, just graduated from a business college. He proposes that these suppliers be rated (high numbers mean a good performance) on six performance criteria weighted as shown in the table. A total score hurdle of 0.60 is proposed to screen suppliers. Purchasing policy would be revised to order raw materials from suppliers with performance scores greater than the total score hurdle, in proportion to their performance rating scores.

Performance Criterion	Weight	Rating		
		Supplier A	Supplier B	Supplier C
1. Price	0.2	0.6	0.5	0.9
2. Quality	0.2	0.6	0.4	0.8
3. Delivery	0.3	0.6	0.3	0.8
4. Production facilities	0.1	0.5	0.9	0.6
5. Environmental protection	0.1	0.7	0.8	0.6
6. Financial position	0.1	0.9	0.9	0.7

a. Use a preference matrix to calculate the total weighted score for each supplier.

b. Which supplier(s) survived the total score hurdle? Under the younger Bennet's proposed policy, what proportion of orders would each supplier receive?

c. What advantages does the proposed policy have over the current policy?

4. Beagle Clothiers uses a weighted score for the evaluation and selection of its suppliers of trendy fashion garments. Each supplier is rated on a 10-point scale (10 = highest) for four different criteria: price, quality, delivery, and flexibility (to accommodate changes in quantity and timing). Because of the volatility of the business in which Beagle operates, flexibility is given twice the weight of each of the other three criteria, which are equally weighted. The table below shows the scores for three potential suppliers for the four performance criteria. Based on the highest weighted score, which supplier should be selected?

Criteria	Supplier A	Supplier B	Supplier C
Price	8	6	6
Quality	9	7	7
Delivery	7	9	6
Flexibility	5	8	9

5. Wingman Distributing Company is expanding its supply chain to include a new distribution hub in South Bend. A key decision involves the number of trucks for the facility. The particular model of truck Wingman is considering can be used 8,000 miles a month and will cost $1,500 a month in capital costs. In addition, each mile a truck is used costs $0.90 for maintenance. A local truck rental firm will rent trucks at a cost of $1.40 per mile. Given the distribution of likely requirements for trucks, management has come up with three alternatives to consider as shown in the table:

Monthly requirements (miles)	40,000	80,000	120,000
Probability	0.30	0.40	0.30
Fleet size (trucks)	5	10	15

Which fleet size will yield the lowest expected monthly costs for Wingman?

6. Sanchez Trucking has been experiencing delays at its warehouse operations. Management hired a consultant to find out why service deliveries to local businesses have taken longer than they should. The consultant narrowed down the problem to the number of work crews loading and unloading trucks. Each crew consists of 6 employees

who work as a team on a variety of tasks; each employee works a full 40 hours a week. However, costs are also a concern. The consultant advised management that they could supplement work crews with short-term employees, at a higher cost, to cover unexpected needs on a weekly basis. Each work crew permanently hired by Sanchez costs $3,200 per week in wages and benefits, while a crew of short-term employees costs $5,000 per week. Complicating the decision is the fact that the weekly hourly requirements for work crews is uncertain because of the volatility in the number of deliveries to be made.

Deliberating with management, the consultant arrived at the following data:

Requirements (labor-hours)	720	960	1,200	1,440
Probability	0.2	0.4	0.3	0.1
Number of Crews	3	4	5	6

If the consultant wants to offer a solution that minimizes the expected weekly costs for Sanchez, how many work crews should Sanchez have on its permanent payroll?

Advanced Problems

7. Bradley Solutions and Alexander Limited are two well-established suppliers of inexpensive tools. Weekend Projects is a national chain of retail outlets that caters to the occasional fixer-upper who would prefer to get the job done fast rather that investing in a well-appointed tool box. Weekend Projects wants to find a supplier for a particular tool set that promises to be a big seller. Expected annual sales are 100,000 units. Weekend's warehouses operate 50 weeks a year. Management collected data on the two suppliers, which are contained in the first table.

   a. Which of the two suppliers would provide the lowest annual cost to Weekend Projects? What shipping quantity would you suggest?

   b. Before management could make a decision, another option became available. Zelda Tools offered the tool set for only $8.00; however, the lead time is longer than the other two suppliers. Zelda is a new supplier and has not been in the industry very long. Additional data for Zelda are in the second table.

   Management has begun to assess the administrative costs to manage the contract with Zelda. What is the lowest level of administrative costs at which Weekend Projects would be indifferent between using Zelda versus the option you chose in part (a)?

Supplier	Freight Costs Shipping Quantity 10,000	25,000	50,000	Price/Unit	Annual Holding Cost/Unit	Lead Time (wks)	Annual Administrative Cost
Bradley	$35,000	$25,000	$18,000	$8.10	$1.62	6	$10,000
Alexander	$40,000	$28,000	$19,000	$8.10	$1.62	4	$15,000

Supplier	Freight Costs Shipping Quantity 10,000	25,000	50,000	Price/Unit	Annual Holding Cost/Unit	Lead Time (wks)
Zelda	$45,000	$25,000	$17,000	$8.00	$1.60	7

8. Wanda Lux must select a supplier for a plastic bottle and proprietary dispenser for its new hair shampoo. Three suppliers have placed bids; at Wanda's request, all bids are for a shipping quantity of 20,000 bottles with annual requirements of 40,000 units. Wanda's factory operates 250 days a year. The first table (below) shows each supplier's price, estimated annual freight costs, and current lead times; management has added estimates for holding costs and administrative oversight costs for each supplier.

   Beyond costs, however, Wanda has three other criteria considered important in the selection of a supplier. The second table (top of next page) shows all the criteria, their weights,

and the scores for all of them except total costs, where a score of 1.0 indicates "poor" and 10 indicates "superior." Because all three suppliers have done business with Wanda Lux before, management will assign a score of "10" to the supplier with the lowest total annual cost, a score of "8.5" for the next lowest cost, and a score of "7.0" for the worst cost of the three.

   a. Which of the three suppliers will provide the lowest annual cost to Wanda Lux?

   b. Given Wanda's criteria and weighting system, which supplier should Wanda award the contract to?

Suppliers	Freight Costs	Price/Unit	Annual Holding Cost Per Unit	Lead Time (days)	Annual Administrative Costs
Dover Plastics	$3,500	$5.10	$1.02	15	$4,000
Evan & Sons	$3,000	$5.05	$1.01	12	$6,000
Farley, Inc.	$4,500	$5.00	$1.00	20	$3,000

Criterion	Weight	Score		
		Dover	Evan	Farley
Total Cost	30	?	?	?
Consistent Quality	30	9	9	7
On-Time Delivery	20	8	9	9
Environment	20	8	7	7

9. Adelie Enterprises is exploring a new service to provide weekly delivery of grocery items to homes in the greater Greenwood area. The company's customers place Web-based orders and Adelie's team assembles and delivers the orders in specially designed cardboard boxes. Management, interested in locating a supplier that can provide boxes cheaply and efficiently, has discovered that each potential supplier's ability to satisfy the company's requirements is influenced by the level of demand. The following table provides Adelie's vendor selection criteria, criterion weights, and rankings (1–10 with 10 being the highest) under the assumption that low, medium, or high demand is generated for their service.

SUPPLIER RATING UNDER LOW, MODERATE, AND HIGH LEVELS OF DEMAND

	Weight	Local Supplier			National Supplier			International Supplier		
		Low	Moderate	High	Low	Moderate	High	Low	Moderate	High
Product Quality	0.35	8	6	5	7	7	7	6	6	6
Delivery Speed	0.15	9	7	3	6	6	6	4	5	7
Product Price	0.25	5	5	3	5	7	9	7	7	9
Environmental Impact	0.25	9	9	9	7	7	7	8	8	8

a. Which supplier should be selected if there is low demand for Adelie's new service; which supplier should be selected under moderate demand assumptions; under high demand assumptions?

b. Which supplier is selected if Adelie evaluates each alternative using a Maximin decision criterion (see Supplement A, "Decision Making")?

c. Which supplier achieves the highest expected ranking if the probability of low demand is 35 percent, moderate demand is 45 percent, and high demand is 20 percent?

10. Adelie Enterprises (from Problem 9) has decided to drop the International Supplier from consideration. Furthermore, Adelie has decided to order boxes in lots of 10,000 in order to optimize the use of available storage space at its distribution facility. In order to more completely consider the cost/volume tradeoffs associated with selecting the local or national supplier, management has collected the following data. Adelie services its customers 250 days per year.

	Demand Level	Demand	Price/unit	Freight Cost/1,000	Carrying Cost/unit	Lead Time (days)	Administrative costs
Local Supplier	Low	50,000	$1.25	$20.00	$0.10	1	$15,000.00
	Moderate	100,000	$1.25	$20.00	$0.10	1	$15,000.00
	High	250,000	$1.25	$20.00	$0.10	1	$15,000.00
National Supplier	Low	50,000	$1.35	$120.00	$0.10	15	$12,500.00
	Moderate	100,000	$1.25	$120.00	$0.10	15	$12,500.00
	High	250,000	$1.00	$120.00	$0.10	15	$12,500.00

a. On purely a total cost basis, which supplier should be selected if there is low demand for Adelie's new service; which supplier should be selected under moderate demand assumptions; under high demand?

b. Which supplier achieves the lowest expected cost if the probability of low demand is 35 percent, moderate demand is 45 percent, and high demand is 20 percent?

VIDEO CASE   Sourcing Strategy at Starwood

Bath towels. Televisions. Fresh produce. Uniforms. On the surface, these items may not appear to have any relationship to each other. Sure, they exist in most households, even though they were probably bought independently of one another. Yet to the supply chain manager employed in the hospitality industry, they not only have a relationship, but their purchase can be critical to gaining a competitive advantage.

Just ask Paul Davis, vice president of strategic sourcing for Starwood's North American operations. With hundreds of hotels and resorts in the United States, Canada, and the Caribbean, Davis's goal is to create the hospitality industry's best supply chain organization. The items procured within his organization not only include replenishable goods such as fresh produce and food items, but also extend to the sourcing of national contracts for nonperishable goods such as bath towels, electronics, staff apparel, energy, and contract services.

It is easy to confuse supply chain processes with the routine procurement of goods and services. Starwood's supply chain certainly does include contracting, but it is much more: It consists of the customer relationship, order fulfillment, and supplier relationship processes. Strong linkages exist among the company's upstream suppliers of services, materials, and information, and the customers of Starwood's hotels and resorts. If the upstream relationships are not carefully managed, downstream delivery of consistency, quality, and value to Starwood's guests may suffer. As a result, significant effort is placed on the nested processes within the supplier relationship process such as design collaboration, sourcing, negotiation, contracting, and information exchange.

Any number of events will trigger the involvement of Paul Davis's supply chain team:

- Existing contracts expire.
- Individual hotel brands seek new products.
- Hotel property design teams generate ideas.
- New categories of products emerge and need evaluation.
- A particular hotel needs help with a local service contract.

When a product or service needs to be sourced, the specifications are driven by internal customers such as restaurant chefs, housekeeping, and maintenance. If the product or service does not already exist, domestic and international suppliers that might be able to create the item are researched, as are regional and local vendors. Sometimes, sourcing an existing item simply means renewing an agreement with a current supplier. Still other situations demand creating a new category that has not been sourced before or using a third party to help locate sources.

Due diligence is always performed by sending potential suppliers a "request for information" in either paper or electronic form. The responses returned by the suppliers are entered into a database and help Starwood to prequalify the suppliers. A good match is sought, requiring the suppliers to meet minimum requirements for financial viability, quality, scope of operations, references, and legal risk avoidance. With a suitable potential supplier candidate pool, Starwood then takes one of two paths. The first one is to conduct a reverse auction where preselected vendors bid against each other. This method is used with shorter-term contracts on commodity items that have low external customer visibility. Kitchen uniforms, hotel room door keys, and paint are sourced this way. The second option is to send out a request for proposal (RFP), which requires the vendor to put its best terms forward at the outset for consideration.

After review by the supply chain team, the vendor winning the auction or emerging from the RFP review activity as the best fit is engaged in negotiations. Throughout the supplier relationship building process, Starwood gets to know the vendors, but it becomes much more personal at this point as both parties move toward concluding their contract negotiations.

Starwood maintains a cooperative orientation toward its supplier relationships, building a partnership to maximize value for each party to ensure that each side is comfortable with the price, quality, and delivery requirements it has agreed upon in the contract negotiation process. When contract negotiations are complete, the different brands are notified and the buying and information exchange processes begin.

At this point, you might think the job of the supply chain team is done. Yet managing the existing supplier relationship after the contract ink dries is perhaps the most challenging task of all. The contract involving sourcing of bed linens and terrycloth items is a perfect example. Not long after the contract was finalized, an alternate supplier approached Starwood with an offer to supply comparable quality goods at a much lower cost. Supply chain managers had a choice to make: continue to work with the existing supplier or buy out the current supplier's contract and begin sourcing with the new one.

QUESTIONS

1. Should Starwood maintain a cooperative orientation or a competitive orientation with its suppliers for the kind of items described here?

2. What types of information should Starwood exchange with its bed linens and terrycloth supplier? What does Starwood risk by sharing too much information?

3. How would you approach the sourcing of bed linens and terrycloth items? That is, would you use a reverse auction or request for proposal? Under what circumstances would you change suppliers?

4. In addition to performing value analysis on the services its properties offer, Starwood evaluates the performance of its suppliers against contract metrics. Using the bed linens and terrycloth supplier as an example, describe some of the metrics Starwood should use.

CASE | Wolf Motors

John Wolf, president of Wolf Motors, just returned to his office after visiting the company's newly acquired automotive dealership. It was the fourth Wolf Motors' dealership in a network that served a metropolitan area of 400,000 people. Beyond the metropolitan area, but within a 45-minute drive, were another 500,000 people. Each of the dealerships in the network marketed a different brand of automobile and historically had operated autonomously.

Wolf was particularly excited about this new dealership because it was the first "auto supermarket" in the network. Auto supermarkets differ from traditional auto dealerships in that they sell multiple brands of automobiles at the same location. The new dealership sold a full line of Chevrolets, Nissans, and Volkswagens.

Starting 15 years ago with the purchase of a bankrupt Dodge dealership, Wolf Motors had grown steadily in size and in reputation. Wolf attributed this success to three highly interdependent factors. The first was volume. By maintaining a high volume of sales and turning over inventory rapidly, economies of scale could be achieved, which reduced costs and provided customers with a large selection. The second factor was a marketing approach called the "hassle-free buying experience." Listed on each automobile was the "one price–lowest price." Customers came in, browsed, and compared prices without being approached by pushy salespeople. If they had questions or were ready to buy, a walk to a customer service desk produced a knowledgeable salesperson to assist them. Finally, and Wolf thought perhaps most importantly, was the after-sale service. Wolf Motors established a solid reputation for servicing, diagnosing, and repairing vehicles correctly and in a timely manner—the first time.

High-quality service after the sale depended on three essential components. First was the presence of a highly qualified, well-trained staff of service technicians. Second was the use of the latest tools and technologies to support diagnosis and repair activities. Third was the availability of the full range of parts and materials necessary to complete the service and repairs without delay. Wolf invested in training and equipment to ensure that the trained personnel and technology were provided. What he worried about, as Wolf Motors grew, was the continued availability of the right parts and materials. This concern caused him to focus on the supplier relationship process and management of the service parts and materials flows in the supply chain.

Wolf thought back to the stories in the newspaper's business pages describing the failure of companies that had not planned appropriately for growth. These companies outgrew their existing policies, procedures, and control systems. Lacking a plan to update their systems, the companies experienced myriad problems that led to inefficiencies and an inability to compete effectively. He did not want that to happen to Wolf Motors.

Each of the four dealerships purchased its own service parts and materials. Purchases were based on forecasts derived from historical demand data, which accounted for factors such as seasonality. Batteries and alternators had a high failure rate in the winter, and air-conditioner parts were in great demand during the summer. Similarly, coolant was needed in the spring to service air conditioners for the summer months, whereas antifreeze was needed in the fall to winterize automobiles. Forecasts also were adjusted for special vehicle sales and service promotions, which increased the need for materials used to prep new cars and service other cars.

One thing that made the purchase of service parts and materials so difficult was the tremendous number of different parts that had to be kept on hand. Some of these parts would be used to service customer automobiles, and others would be sold over the counter. Some had to be purchased from the automobile manufacturers or their certified wholesalers, and to support, for example, the "guaranteed GM parts" promotion. Still, other parts and materials such as oils, lubricants, and fan belts could be purchased from any number of suppliers. The purchasing department had to remember that the success of the dealership depended on (1) lowering costs to support the hassle-free, one price–lowest price concept and (2) providing the right parts at the right time to support fast, reliable after-sale service.

As Wolf thought about the purchasing of parts and materials, two things kept going through his mind: the amount of space available for parts storage and the level of financial resources available to invest in parts and materials. The acquisition of the auto supermarket dealership put an increased strain on both finances and space, with the need to support three different automobile lines at the same facility. Investment dollars were becoming scarce, and space was at a premium. Wolf wondered what could be done in the purchasing area to address some of these concerns and alleviate some of the pressures.

QUESTIONS

1. What recommendations would you make to John Wolf with respect to structuring the supplier relationship process for the Wolf Motors dealership network?

2. How might purchasing policies and procedures differ as the dealerships purchase different types of service parts and materials (for example, lubricants versus genuine GM parts)?

3. How can supply chain design and integration help John Wolf reduce investment and space requirements while maintaining adequate service levels?

Source: This case was prepared by Dr. Brooke Saladin, Wake Forest University, as a basis for classroom discussion. Copyright © Brooke Saladin. Used with permission.

Selected References

Benton, W.C. *Purchasing and Supply Management*, 2nd ed. New York: McGraw-Hill, 2010.

Bowersox, Donald. *Supply Chain Logistics Management*, 3rd ed. New York: McGraw-Hill, 2010.

Brienzi, Mark, and Dr. Sham Kekre. "How Kodak Transformed its Service Parts Supply Chain." *Supply Chain Management Review* (October 2005), pp. 25–32.

Chopra, Sunil, and Peter Meindl. *Supply Chain Management*, 4th ed. Upper Saddle River, NJ: Prentice Hall, 2010.

Chopra, Sunil, and ManMohan S. Sodhi. "Looking for the Bang from the RFID Buck." *Supply Chain Management Review* (May/June 2007), pp. 34–41.

Cook, Robert L., Brian Gibson, and Douglas MacCurdy. "A Lean Approach to Cross-Docking." *Supply Chain Management Review* (March 2005), pp. 54–59.

Fleck, Thomas. "Supplier Collaboration in Action at IBM." *Supply Chain Management Review* (March 2008), pp. 30–37.

Freund, Brian C., and June M. Freund. "Hands-On VMI." *APICS—The Performance Advantage* (March 2003), pp. 34–39.

Fugate, Brian S., and John T. Mentzer. "Dell's Supply Chain DNA." *Supply Chain Management Review* (October 2004), pp. 20–24.

Handfield, Robert B., and Kevin McCormack. "What you Need to Know About Sourcing From China." *Supply Chain Management Review* (September 2005), pp. 28–33.

Hartvigsen, David. *SimQuick: Process Simulation with Excel*, 2nd ed. Upper Saddle River, NJ: Prentice Hall, 2004.

Lee, Hau L. "The Triple-A Supply Chain." *Harvard Business Review* (October 2004), pp. 102–112.

Liker, Jeffrey K., and Thomas Y. Choi. "Building Deep Supplier Relationships." *Harvard Business Review* (December 2004), pp. 104–113.

Malik, Yogesh, Alex Niemeyer, and Brian Ruwadi. "Building the Supply Chain of the Future." *McKinsey Quarterly* (January 2011), pp. 1–10.

Maloni, M. J., and W.C. Benton. "Power Influences in the Supply Chain." *Journal of Business Logistics*, vol. 21 (2000), pp. 49–73.

Melnyk, Steven, Robert Sroufe, and Roger Calantone. "Assessing the Impact of Environmental Management Systems on Corporate and Environmental Performance." *Journal of Operations Management*, vol. 21, no. 3 (2003).

Miller, Jamey. "Shared Success: Working Together to Find the Value of VMI." *APICS Magazine* (November/December 2007), pp. 37–39.

Murphy-Hoye, Mary, Hau L. Lee, and James B. Rice, Jr. "A Real-World Look at RFID." *Supply Chain Management Review* (July/August 2005), pp. 18–26.

Randall, Taylor, Serguei Netessine, and Nils Rudi. "Should You Take the Virtual Fulfillment Path?" *Supply Chain Management Review* (November/December 2002), pp. 54–58.

Sullivan, Laurie. "Walmart's Way." *Informationweek.com* (September 27, 2004), pp. 37–50.

Trent, Robert J. "What Everyone Needs to Know About SCM." *Supply Chain Management Review* (March 2004), pp. 52–59.

Wagner, Richard H. "An Interview with Dirk Brand: Hotel Manager of Queen Mary 2." *Beyondships.com* (December 2007), pp. 1–4.

AP Photo/Shawano Cleary

FedEx employees prepare a shipment of medical supplies from International Aid of Spring Lake, Michigan to aid disaster victims.

13

SUPPLY CHAIN SUSTAINABILITY AND HUMANITARIAN LOGISTICS

FedEx

FedEx is a $33 billion-a-year delivery service company that is used to dealing with crises. Hardly a day goes by without disruptive events such as social unrest, major storms, or unanticipated labor strikes somewhere in the world. FedEx has designed a supply chain that is flexible and responsive to unpredictable catastrophic events. This capability has made FedEx an important resource in disaster relief supply chains, which experience major mismatches of supply and demand that require a global response as quickly as possible. A wait-and-see attitude will not work; advanced planning is necessary. Such was the case with Hurricane Katrina, the costliest natural disaster in the history of the United States. Five days before Katrina hit the southeast Louisiana coast, the chief of FedEx's Global Operations Control in Memphis, Tennessee, conducted a twice-daily conference call for more than 100 people. Before the storm hit, FedEx positioned 30,000 bags of ice, 30,000 gallons of water, and 85 home generators outside Baton Rouge, Louisiana, and Tallahassee, Florida, for quick deployment to relieve employees, and 60 tons of American Red Cross equipment to strategic locations to await the storm. While the high winds were expected, the massive flooding took the company by surprise. FedEx returned 10,000 packages to their senders and had to reprogram 100,000 devices to avoid sending new shipments to closed areas. Since the New Orleans airport was

441

closed, FedEx had to shift its hub to Lafayette, Louisiana, in only a few days, a move of 135 miles that would have normally taken 6 months to complete. Nonetheless, within a week after the devastation, FedEx used its vast logistics network to move more than 900 tons of relief supplies to affected areas in Louisiana, Mississippi, and the Houston Astrodome.

Source: Ellen Florian Kratz, "For FedEx, It Was Time to Deliver," *Fortune* (October 3, 2005), pp. 83–84; *Business Wire,* **www.businesswire.com**, November 14, 2006; LN Van Wassenhove, "Humanitarian Aid Logistics: Supply Chain Management in High Gear," *Journal of the Operational Research Society,* 2006, vol. 57, pp. 475–489; **www.fedex.com/us/about/responsibility/katrina.html**

LEARNING GOALS *After reading this chapter, you should be able to:*

1 Define the three elements of supply chain sustainability.

2 Explain the reverse logistics process and its implications for supply chain design.

3 Explain how firms can improve the energy efficiency of their supply chains.

4 Explain how supply chains can be designed to support the response and recovery operations of disaster relief efforts.

5 Explain how ethical issues confront supply chain managers.

sustainability

A characteristic of processes that are meeting humanity's needs without harming future generations.

financial responsibility

An element of sustainability that addresses the financial needs of the shareholders, employees, customers, business partners, financial institutions, and any other entity that supplies the capital for the production of services or products or relies on the firm for wages or reimbursements.

environmental responsibility

An element of sustainability that addresses the ecological needs of the planet and the firm's stewardship of the natural resources used in the production of services and products.

social responsibility

An element of sustainability that addresses the moral, ethical, and philanthropic expectations that society has of an organization.

humanitarian logistics

The process of planning, implementing and controlling the efficient, cost-effective flow and storage of goods and materials, as well as related information, from the point of origin to the point of consumption for the purpose of alleviating the suffering of vulnerable people.

FedEx is an excellent example of how a firm designs its own supply chain to be reactive to major disruptions and natural disasters, a capability that also comes in handy when called upon to assist others in disaster relief causes. While effective supply chain design and integration is a capability that an organization can use to be more competitive in its industry, it can also be used to become a better corporate citizen. A growing theme among many corporations is that of responsible stewardship of the capital, ecological, and human resources that they and their suppliers use in the production of their services or products. Translated into a goal, these corporations strive to have services, products, and processes with the characteristic of **sustainability**, which means that they are meeting humanity's needs without harming future generations.[1] The topic of sustainability has many implications for a firm; however, in this chapter we focus only on supply chains and how they can be used to achieve sustainability.

As Figure 13.1 shows, supply chain sustainability has three elements. First, **financial responsibility** addresses the financial needs of the shareholders, employees, customers, business partners, financial institutions, and any other entity that supplies the capital for the production of services or products or relies on the firm for wages or reimbursements. Supply chains, for their part, support the financial responsibility of a firm by influencing elements that contribute to the return on assets as explained in Chapter 10, "Supply Chain Design." Furthermore, because supply chains are essentially linked processes, any improvement to processes or their management, as explained in Parts 1 and 2 of the text, improves the financial well-being of the firm and increases its chances of survival in a competitive world. Second, **environmental responsibility** addresses the ecological needs of the planet and the firm's stewardship of the natural resources used in the production of services and products. The goal is to leave as small an environmental footprint as possible so that future generations can make use of abundant natural resources. The design and integration of supply chains can play a major role in preserving resources. We shall examine how supply chains can be designed to produce a product and then reprocess them at the end of their lives to yield value in the form of remanufactured products or recycled materials. We will also examine how supply routes can be planned to reduce the amount of energy consumed in delivering materials or products to customers.

Finally, **social responsibility** addresses the moral, ethical, and philanthropic expectations that society has of an organization. While this responsibility covers a broad range of activities, supply chains can be used to meet such expectations. Firms can engage in **humanitarian logistics**, which is the process of planning, implementing, and controlling the efficient, cost-effective flow and storage of goods and materials, as well as related information, from the point of origin to the

---

[1] For a more complete discussion of sustainability and what major corporations are doing, see Pete Engardio, "Beyond the Green Corporation," *Business Week* (January 29, 2007), pp. 50–64.

point of consumption for the purpose of alleviating the suffering of vulnerable people.[2] As such, firms can use their expertise in supply chain management to design supply chains that provide disaster relief, or to supply much needed drugs and food to undeveloped areas of the world. Ethical considerations also arise in the choice of suppliers regarding their practices in the use of labor and natural resources, the relationship between buyers and sellers, the location of facilities, and inventory management. Table 13.1 provides some examples of well-known firms addressing sustainability.

Sustainability across the Organization

Integrated supply chains can facilitate the implementation of sustainable operations because of their established communication and material flows. Nonetheless, achieving sustainability throughout the supply chain is no easy task. It requires cross-functional and inter-firm cooperation to address challenges such as:

- *Environmental protection*–firms should monitor their own processes and those of their suppliers to improve waste elimination methods, to reduce the pollution of the air, streams, and rivers, and to increase efforts at ecological stewardship for the protection of flora and fauna.

- *Productivity improvement*–firms should examine processes up and down the supply chain to increase material conservation, to increase energy efficiency, and to look for ways to convert waste into useful by-products.

- *Risk minimization*–as the supply chain grows, particularly on a global basis, firms should take great care to ensure that the materials that go into their services, products, or processes do not pose health or safety hazards to customers.

- *Innovation*–as new services, products, or technologies are designed and developed, firms should strive to make sure that they support their financial, environmental, and social responsibilities while serving the needs of customers.

A survey of 766 CEOs of major corporations conducted by the Accenture Institute for High Performance revealed that 93 percent of the CEOs believe that sustainability issues are critical to the future success of their companies, and 91 percent of them will employ new technologies (renewable energy, energy efficiency, information and communication) over the next five years.[3] Further, 88 percent of the CEOs believe that they should be integrating sustainability through their supply chains; however only 54 percent believe that it has been achieved. Why is there a gap? While intuitively appealing, and the opinions of the CEOs notwithstanding, sustainability is not the easiest strategy for which approval can be gained from most boards of directors. Directors often need some sort of demonstrable return on investment to justify the costs. For sustainability efforts, however, managers must have a long-term view that accepts lower returns in the near future to improve the chances of survival and better returns in the future.

As we have already devoted most of the text to the financial implications of processes, supply chains, and their management, we now turn to the role of supply chains in achieving a firm's environmental and social responsibilities.

Supply Chains and Environmental Responsibility

Environmental concerns regarding business are voiced every day in the popular media. Service providers are examining ways to increase efficiency and reduce the impact of their operations on the environment. Manufacturers are feeling pressure to take responsibility for their products from birth to death. In this section, we will discuss some environmental concerns and how integrated supply chains can address them. We will discuss the supply chain implications for implementing an approach called "reverse logistics," which responds to the need to salvage products at the end of their life cycles. We will also show how the need for energy efficiency impacts decision making in supply chains.

---

[2]A. Thomas. "Humanitarian Logistics: Enabling Disaster Response," White paper: The Fritz Institute, San Francisco, California, 2003.

[3]"A New Era of Sustainability: UN Global Compact–Accenture," Accenture Institute for High Performance. (June 2010), pp. 1–66.

FIGURE 13.1 ▲
Supply Chains and Sustainability

Creating Value through Operations Management

Using Operations to Compete
Project Management

Managing Processes

Process Strategy
Process Analysis
Quality and Performance
Capacity Planning
Constraint Management
Lean Systems

Managing Supply Chains

Supply Chain Inventory Management
Supply Chain Design
Supply Chain Location Decisions
Supply Chain Integration
Supply Chain Sustainability and Humanitarian Logistics
Forecasting
Operations Planning and Scheduling
Resource Planning

TABLE 13.1 | EXAMPLES OF SUPPLY CHAIN SUSTAINABILITY EFFORTS

Financial Responsibility		Example
		NCR: In moving from a country-centric model to a global low-cost manufacturing model, where the majority of product is produced in Asia, import duties, tariffs, distances, fuel, and lack of capacity in the logistical carrier base must be considered along with the occasional need to create the logistics infrastructure to move the product.
		Nike: In analyzing the tradeoff in the costs of manufacturing products on-shore versus offshore, Nike has found that for plants in Europe it could pay 13 percent more in price for local sourcing and still be as well off as sourcing the production from offshore manufacturers.
Environmental Responsibility		
	Reverse Logistics	*Caterpillar:* Its remanufacturing facilities recycle over two million pieces and 140 million pounds of materials annually in state-of-the-art factories worldwide.
		IBM: Its Global Asset Recovery division in one year collected over one million units of used information technology equipment that was converted to billions of dollars in revenues on the second-hand equipment, parts, and materials markets.
	Efficiency	*RR Donnelley:* Proactively worked with suppliers to help find ways to use less packaging and to reuse or recycle what cannot be eliminated.
		UPS: Will add 48 heavy-duty trucks, powered by liquefied natural gas (LPG), to cut diesel use by 95 percent while emitting 25 percent fewer greenhouse gasses.
Social Responsibility		
	Disaster Relief Supply Chains	*Intel and Solectron:* In collaboration with the International Rescue Committee, these firms use their corporate expertise on disaster relief to significantly streamline procurement and create processes to substantially reduce response times.
		DHL: Uses its comprehensive logistics network and worldwide presence to help people and communities affected by major sudden-onset natural disasters.
	Ethics	*Airbus:* Places its highest priority on environmental performance, and supports green economy and technology transfers in developing countries.
		The Body Shop: Produces environmentally and ethically-focused cosmetics and requires all suppliers to sign on to its corporate code of conduct before trading with them.

reverse logistics

The process of planning, implementing, and controlling the efficient, cost effective flow of products, materials, and information from the point of consumption back to the point of origin for returns, repair, remanufacture, or recycling.

closed-loop supply chain

A supply chain that integrates forward logistics with reverse logistics, thereby focusing on the complete chain of operations from the birth to the death of a product.

Reverse Logistics

To address environmental concerns and to manage their products throughout their life cycles, firms such as Coors Brewing Company, Dell, and Caterpillar are turning to **reverse logistics**, which is the process of planning, implementing, and controlling the efficient, cost effective flow of products, materials, and information from the point of consumption back to the point of origin for returns, repair, remanufacture, or recycling.

Supply Chain Design How are supply chains designed to be environmentally responsible for the entire life cycle of a product? A supply chain that integrates forward logistics with reverse logistics is called a **closed-loop supply chain** because it focuses on the complete chain of operations from

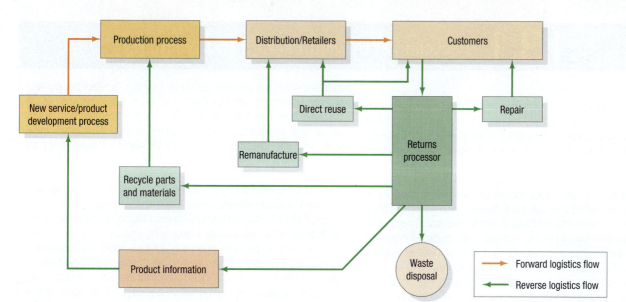

FIGURE 13.2 ▲
Flows in a Closed-Loop Supply Chain

the birth to the death of a product. Figure 13.2 shows how a product starts its journey at the new service/product development process, makes its way to the customer, and then enters the reverse logistics chain that attempts to maximize the value of the item at the end of its useful life.

It is clear that the reverse logistics operations are considerably different from the forward logistics flows, and considerably more expensive. A firm must establish convenient collection points to receive the used goods from the final customer and transport the goods to a *returns processor*, which is a facility owned by the manufacturer or outsourced to a supplier that is proficient at disassembling products and gleaning any remaining value from them. Several options exist. If the item is inoperable, it could be repaired and returned to the customer. Another option is that it could be cleaned and refurbished for direct use and returned either to the distribution channel, which is the case with leased products, or back to customers, which is the case with a maintenance warranty. The product could be remanufactured by tearing it down, rebuilding it with new parts as needed, and returning it to the distribution channel. Finally, the product could be completely disassembled and the usable parts and materials cleaned, tested, and returned to the production process. There are two important by-products of the reverse logistics process: waste, which must be properly disposed of, and product information, which is transmitted to the new service/product development process so that improvements can be made to future generations of the product.

Reverse logistics processes can be particularly important in the electronics industry. Have you ever wondered what happens to your old computer after you purchase a new one? You may have given it to the store where you purchased your new one, or merely slipped it into the household trash. More retired computers in the United States are dumped in landfills than are recycled. Old computers contain electronics components with materials that can be recycled. However, they also have toxins that leach into the soil if these components are left unprocessed: lead and cadmium in computer circuit boards, lead oxide and barium in computer monitors and cathode-ray tubes (CRT), mercury in switches and flat screens, and fire retardant on printed circuit boards and plastic casings.

Many recycling processors are located in developing countries and are typically low tech. Recyclable material is collected and processed. Workers, who usually do not wear protective gear, often toss the chemicals that come out during the processing into nearby streams and rivers. Other materials not processed often are left in the dumps, allowing the toxins to leak out. This disregard for the environment prompted a backlash against the improper disposal of electronic equipment in developed countries. The European Union (EU) passed a law requiring electronics manufacturers to take back and recycle 75 percent of the products they sell in the EU. Some states in the United States banned *e-waste*, electronic components and the chemical by-products of the recycling process, from landfills and are considering making electronics manufacturers responsible for managing e-waste.

Recycling is a major aspect of reverse logistics. Managerial Practice 13.1 provides examples of how two companies use recycling to be environmentally responsible.

Hewlett-Packard

Each year the United States generates about 2 million tons of e-waste, which can be environmentally destructive if not properly disposed of. The United States alone is responsible for more than 2 billion kilograms of technology trash annually, a 300 percent increase in the past decade. The growing need to be environmentally responsible in the electronics manufacturing industry provides an opportunity for firms to not only recover scarce resources, but also raise their public image. Hewlett-Packard (HP), in conjunction with a Canadian metals and mining company called Noranda, built a recycling plant in Roseville, near Sacramento, California. A football field-sized loading dock is stacked high with mobile phones, copy machines, computer monitors, printers, PCs and servers—a one-day supply. Technicians recover reusable parts before the machines are put on conveyors, chopped up by powerful shredders, smashed to bits by a granulator, and sorted by magnets and air currents. All that remains are mounds of plastic, steel, and aluminum in two-centimeter chunks, destined for a smelter. Precious metals go to Noranda; aluminum, glass, and plastic are sold to recyclers. Nothing goes to landfills. HP has recycling plants worldwide.

HP Recycling plant operations in Roseville, California.

Robyn Beck/AFP/Getty Images/Newscom

Walmart

All companies can save money by reducing the amount of waste they must dispose. Walmart, owing to its size, is certainly no exception. Many trash items, such as loose plastic, plastic hangers, office paper, and aluminum cans, are unruly and difficult to collect for recycling. To attack this problem, Walmart initiated its "super sandwich bale" at all of its stores and clubs in the United States. Associates place 10 to 20 inches of cardboard at the bottom of large trash compactors. Commodities, such as loose plastic bags, aluminum cans, plastic hangers, and plastic water and soda bottles are loaded in, and another layer of cardboard is placed on top. The compactor then presses the bale into a "sandwich" with 9 to 18 inches of recyclables in the middle. The bales are then loaded onto a truck to be recycled into various raw materials to ultimately become products once again. For example, in one of its sustainability programs, Walmart directs recycled plastics and cardboard to Worldwise, a leader in developing, manufacturing, and marketing sustainable pet products, where they are transformed into a stylish and durable line of dog beds. Plastic hangers are turned into litter pans, plastic bags into litter liners, and corrugated cardboard into cat scratchers. To get a sense of the value involved, Walmart used to pay trash companies to haul more than one billion plastic hangers from its stores and clubs each year. Now, it gets paid 15 to 20 cents a pound for them. The money adds up in a hurry. Who said that reverse logistics supply chains are not profitable? It is clear that environmentally conscious supply chain operations can literally turn "trash" into "cash."

Walmart Sandwich Recycling Process.

Najlah Feanny/News Archive/Corbis

Source: Marc Gunther. "The End of Garbage," *FORTUNE,* (March 19, 2007), pp. 158–166; Zachary Slobig. "Hewlett-Packard E-Cyclers are Gold Miners of the Internet-Age," Terradaily, (June 15, 2007), pp. 1–3; Oliver Ryan. "10 Green Giants," CNNMoney.com, 2011; "Waste,"walmartstores.com, 2008; "Walmart Rolls Out the Plastic Sandwich Bale,"walmartstores.com, 2005.

Financial Implications   Some firms participate in reverse supply chains by owning and operating processes such as remanufacturing, recycling, or repairing the used products and materials. These firms benefit from reclaiming usable parts for their manufacturing operations or by selling remanufactured products at competitive prices. If an original equipment manufacturer (OEM) participates in remanufacturing, however, there is a fear that the remanufactured products will cannibalize the sales of the firm's new products. Often these fears give rise to restrictions such as floors on the price that can be charged for remanufactured products, limits on the markets where they can be sold, limits on the distribution channels the products can be sold through, and reduced warranties that can be offered on them. Of course, there are also opportunity costs for not remanufacturing. There is the danger that environmentally irresponsible product disposal practices on the part of the firm's customers may result in the firm facing costly regulatory restrictions in the future. Also, third-party manufacturers may participate in the reverse logistics supply chain by collecting or purchasing the unclaimed old products, remanufacturing them, and then selling them in competition with the firm's new products. That is the case with refilled laser printer and inkjet cartridges, which erodes profit margins for original printer manufacturers like Hewlett-Packard, Dell, and Epson.

Other firms, as well as individuals, participate in the reverse logistics supply chain by supplying their used products and materials for processing. A continuous supply of these unused products and materials is needed to make the reverse supply chain financially viable. Various incentives may be used to influence the quantity, the quality (condition), and the timing of supply. Examples of incentives include:

- *Fee.* A fee is paid to the user when a used product or recyclable material is delivered for recovery. Usually the fee depends on the condition of the product or material because this may determine the possibilities for its reuse. Of course, we have seen how some firms such as Walmart can garner substantial revenues from the recycling of useful materials they previously discarded.

- *Deposit fee.* Such fees provide incentive for the user to return the product or containers of the product to get the reimbursement of the deposit fee. This fee may relate to the product itself, such as a rental trailer, which must be refurbished (cleaned, maintained) before allowing the next customer to use it. Alternatively, the fee may apply to the distribution of the product, such as the deposit fee on beer bottles in some states. Harris Teeter, a grocery chain in North Carolina and other states, charges a fee on glass bottles of organic milk. This fee, which can be as much as $1.50 for milk that costs $3.49, is refunded when the customers bring back the empty glass bottles to the store.

- *Take back.* A company may offer to collect its products from its customers for no charge when those customers want to dispose of them. Dell, for example, charges no fees to recycle old computers from customers. Dell has designed its computers to make them easier to disassemble and recycle.

- *Trade-in.* One can get a new copy of a product if another copy is returned. For example, purchasing a refurbished engine for an automobile often requires the owner to turn in the old one, which might be disassembled for its parts or refurbished for sale to another customer.

- *Community programs.* Often communities or groups will set aside special days for the disposal of various items that are difficult to dispose of, such as automobile tires, paint, metal, and other things that the trash collectors do not normally pick up.

Many firms and individuals submit their used products and materials to be recycled for no other reason than it is the environmentally correct thing to do. Nonetheless, without the incentives, many reverse logistics supply chains would dry up.

Energy Efficiency

Supply chains involve the flow of materials and services from their origination to their ultimate destination. As such, supply chains consume energy. Energy consumption not only is expensive from a business perspective, but it can also have negative effects on the environment. Increasingly more firms are measuring their **carbon footprint**, which is the total amount of greenhouse gasses produced to support their operations, usually expressed in equivalent tons of carbon dioxide (CO_2). Major contributors to carbon footprints are fossil fuels, in particular oil, diesel, and gasoline, which are used extensively in supply chain logistical operations. Therefore, it behooves firms to be efficient with respect to energy consumption throughout their supply chain. In this section, we will discuss four levers supply chain managers can use to increase the energy efficiency of their operations: (1) transportation distance, (2) freight density, (3) transportation mode, and (4) transportation technology.

carbon footprint

The total amount of greenhouse gasses produced to support operations, usually expressed in equivalent tons of carbon dioxide (CO_2).

Transportation Distance   Supply chain managers can decrease the amount of energy consumed in moving materials or supplying services by reducing the distance traveled. There are two ways this can be accomplished. The first involves the design of the supply chain itself. Locating service facilities or manufacturing plants in close proximity to customer populations reduces the distance required to supply the service or product. Furthermore, selecting suppliers that are in close proximity to the service facilities or manufacturing plants reduces the amount of fuel needed to procure those materials. Of course, these suppliers must meet the firm's quality and performance needs.

A second way to improve energy efficiency involving transportation distances is **route planning**, which seeks to find the shortest route to deliver a service or product. Once the design of the supply chain has been determined, attention turns to minimizing the distance traveled to supply the service or product to customers on a daily basis. There are two traditional versions of

route planning

An activity that seeks to find the shortest route to deliver a service or product.

shortest route problem

A problem whose objective is to find the shortest distance between two cities in a network or map.

traveling salesman problem

A problem whose objective is to find the shortest possible route that visits each city exactly once and returns to the starting city.

this problem. The first is the **shortest route problem**, which seeks to find the shortest distance between two cities in a network or map. While elegant mathematical methods have been developed for solving this problem, today we are fortunate to have GPS systems for vehicles and Web sites such as Mapquest™ where very good, energy efficient, routes can be obtained quickly. Manufacturers with their own delivery fleets or third party logistics providers (3PLs) can use these routes to minimize their costs of making the deliveries.

The second version is known as the **traveling salesman problem**, which seeks to find the shortest possible route that visits each location or city exactly once and returns to the starting location. This problem is a much more difficult one to solve, yet one which delivery services face every day. Starting from a central location, such as a warehouse, distribution center, or hub, orders headed for multiple destinations are loaded into a truck that must make the deliveries, or alternately pick up supplies, and return to the central location. The problem is to find the sequence of cities the truck must visit so that the total miles traveled is minimized. Figure 13.3 shows a four-city traveling salesman problem with the driving miles between each city shown on the arc connecting them. How many different routes are there in Figure 13.3? Because we are dealing with driving distances between the cities, and there are no anomalies such as one-way streets or road blockages, the route Central Hub –A–C–B– Central Hub, for example, has the same total distance as the reverse of that route, Central Hub –B–C–A– Central Hub. Consequently, there are only three different routes:

Central Hub –A–B–C– Central Hub, which is $90 + 100 + 120 + 80 = 390$ miles.
Central Hub –B–C–A– Central Hub, which is $85 + 120 + 130 + 90 = 425$ miles.
Central Hub –C–A–B– Central Hub, which is $80 + 130 + 100 + 85 = 395$ miles.

The optimal route for the example is Central Hub –A–B–C– Central Hub.

It may look easy to find the best route; simply evaluate each possible route as we just did. However, if the truck must visit n cities there are $(n-1)!/2$ different routes to consider. For example, if there are only 8 cities to visit, there are 2,520 possible routes to consider.[4] While complete enumeration of all the feasible routes is one way to solve this problem, the computational effort becomes onerous. If you were faced with the problem in Figure 13.3, how might you proceed to solve it in the absence of a brute force approach? You might do the following: Start with the Central Hub and go to the closest unvisited city; from that city find the next closest unvisited city, and repeat until you get back to the Central Hub. This approach is called the **nearest neighbor (NN) heuristic**, and has the following steps:

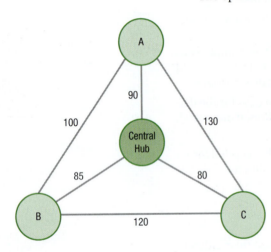

FIGURE 13.3 ▲
Four-City Traveling Salesman Problem

nearest neighbor (NN) heuristic

A technique that creates a route by deciding the next city to visit on the basis of its proximity.

1. Start with the city that is designated as the central location. Call this city the *start* city. Place all other cities in an *unvisited* set.

2. Choose the city in the unvisited set that is closest to the start city. Remove that city from the unvisited set.

3. Repeat the procedure with the latest visited city as the start city.

4. Conclude when all the cities have been visited, and return back to the central location.

5. Compute the total distance traveled along the selected route.

Using the NN heuristic for the problem in Figure 13.3 yields the following route: Central Hub –C–B–A– Central Hub, for a total distance of 390 miles. Notice that this is the optimal solution to the problem. The NN heuristic does not always yield the optimal solution; however its main advantages are that it is fast and that it generally provides reasonable solutions to a very complex problem. Example 13.1 shows the application of the NN heuristic for the delivery of natural food products.

Freight Density   Truck vans, containers, and rail cars all have limits with respect to cargo volume and weight. By reducing the volume that a product displaces while staying within the weight limits of the conveyance, the firm can use fewer trucks, containers, or rail cars to ship the same number of units. Freight density, measured in pounds per cubic foot, determines the freight class and the cost a shipper must pay. The lower the freight density, the higher the freight class because the volume of the conveyance will be maxed out before the weight limit is reached. For example, 1,000 pounds of ping pong balls occupy much more room in a trailer than 1,000 pounds of

---

[4]We assume that there are roads connecting each pair of cities. The factorial is divided by 2 because the sequence of cities visited in one tour is assumed to have the same distance as a tour with the cities in reverse sequence.

| EXAMPLE 13.1 | Finding an Energy-Efficient Route Using the Nearest Neighbor Heuristic |

Hillary and Adams, Inc., is a privately owned firm located in Atlanta that serves as the regional distributor of natural food products for Georgia, Kentucky, North Carolina, South Carolina, and Tennessee. They are particularly well known for their unique blend of fiery hot Habanera sauces. Every week, a truck leaves the large distribution center in Atlanta to stock local warehouses located in Charlotte NC, Charleston SC, Columbia SC, Knoxville TN, Lexington KY, and Raleigh NC. The truck visits each local warehouse only once, and returns to Atlanta after all the deliveries have been completed. The distance between any two cities in miles is given below.

From/To	Atlanta	Charleston	Charlotte	Columbia	Knoxville	Lexington	Raleigh
Atlanta	0	319	244	225	214	375	435
Charleston	319	0	209	116	373	540	279
Charlotte	244	209	0	93	231	398	169
Columbia	225	116	93	0	264	430	225
Knoxville	214	373	231	264	0	170	351
Lexington	375	540	398	430	170	0	498
Raleigh	435	279	169	225	351	498	0

John Jensen, vice president of logistics at Hillary and Adams, Inc., is worried about the rising fuel costs. With a reduction in operating budgets, he is interested in finding a route that would minimize the distance traveled by the truck.

Use the Nearest Neighbor heuristic to identify a route for the truck and compute the total distance traveled.

SOLUTION

The application of the NN heuristic results in the following steps:

1. Start with Atlanta and place all other cities in the unvisited set : Charlston, Charlotte, Columbia, Knoxville, Lexington, Raleigh.

2. Select the closest city to Atlanta in the unvisited set, which is Knoxville. Remove Knoxville from the unvisited set. The partial route is now Atlanta – Knoxville, which is 214 miles.

3. Scan the unvisited set for the city closest to Knoxville, which is Lexington. Remove Lexington from the unvisited set. The partial route is now Atlanta – Knoxville – Lexington, which is 214 + 170 = 384 miles.

4. Repeat the procedure until all cities have been removed from the unvisited city set. Connect the last city to Atlanta to complete the route.

5. Compute the total distance traveled along the selected route. The route using the nearest neighbor heuristic is Atlanta – Knoxville – Lexington – Charlotte – Columbia – Charleston – Raleigh – Atlanta. The total distance traveled is (214 + 170 + 398 + 93 + 116 + 279 + 435) = 1705 miles.

Note that using the same sequence of cities we could start the route at any one of them and travel the same total distance. For example, the route Lexington – Charlotte – Columbia – Charleston – Raleigh – Atlanta – Knoxville – Lexington will also be 1705 miles. This fact allows us to use the NN heuristic again to see if a better solution exists; we repeat the heuristic six more times using each city as the starting point. This approach results in the following routes:

- Charleston – Columbia – Charlotte – Raleigh – Knoxville – Lexington – Atlanta – Charleston (116 + 93 + 169 + 351 + 170 + 375 + 319) = 1,593 miles.

- Charlotte – Columbia – Charleston – Raleigh – Knoxville – Lexington – Atlanta – Charlotte (93 + 116 + 279 + 351 + 170 + 375 + 244) = 1,628 miles.

- Columbia – Charlotte – Raleigh – Charleston – Atlanta – Knoxville – Lexington – Columbia (93 + 169 + 279 + 319 + 214 + 170 + 430) = 1,674 miles.

- Knoxville – Lexington – Atlanta – Columbia – Charlotte – Raleigh – Charleston – Knoxville (170 + 375 + 225 + 93 + 169 + 279 + 373) = 1,684 miles.

- Lexington – Knoxville – Atlanta – Columbia – Charlotte – Raleigh – Charleston – Lexington (170 + 214 + 225 + 93 + 169 + 279 + 540) = 1,690 miles.

- Raleigh – Charlotte – Columbia – Charleston – Atlanta – Knoxville – Lexington – Raleigh (169 + 93 + 116 + 319 + 214 + 170 + 498) = 1,579 miles.

Of the seven routes produced with the NN heuristic, the best is the last one with a travel distance of 1,579 miles.

DECISION POINT

Minimizing the number of miles to complete a route reduces the amount of fuel consumed by Hilary and Adams for the delivery process. Since each route is a loop, the truck driver would be instructed to go from Atlanta to Knoxville to Lexington to Raleigh to Charlotte to Columbia to Charleston and back to Atlanta. Alternatively, the reverse sequence could be taken; go to Charleston first and so on. The sequence of cities dictates how the truck is loaded. The travel distances would still be 1,579 miles. While the NN heuristic cannot guarantee an optimal solution, it can help John Jensen avoid a costly mistake. For example, the route Atlanta – Raleigh – Lexington – Charleston – Knoxville – Columbia – Charlotte – Atlanta is 2,447 miles, a 55 percent increase in mileage over the best NN solution. Consequently, in addition to being environmentally responsible, the NN solution supports the competitive priority of low cost operations. Minimizing the distance traveled to complete the route also shortens the time required to make the deliveries, which supports the competitive priority of delivery speed.

The shipper of these wrapped pallets of cardboard boxes will be charged a freight rate based on six factors, including density, weight and distance.

bowling balls. Firms can increase the freight density by reducing the volume of packaging, redesigning the product to take less volume, or postponing the assembly of the product until the customer takes possession.

Firms using third-party logistics providers (3PLs) to get their materials or products to their customers must pay a freight rate based on six factors:

1. The freight density

2. The shipment's weight

3. The distance the shipment is moving

4. The commodity's susceptibility to damage

5. The value of the commodity

6. The commodity's loadability and handling characteristics (Loadability refers to how efficiently the items being shipped fit into a standard container or truck van. Also, in some cases special care must be taken in handling during the loading and unloading processes.)

Table 13.2 shows example freight rates and weight breaks based on various freight classifications for a shipment scheduled between two specific zip codes. The rates are given in dollars per hundredweight (cwt). The freight classification for a shipment is determined by the National Motor Freight Classification (NMFC) tariff and is based upon the last four factors in the list above. There are eighteen possible freight classes ranging from 50 to 500. The tariffs in the table are based on the distance to be traveled and are modified by the weight and freight class. Notice how the rates increase as the class goes up, and decrease as the weight goes up.

Using Table 13.2, a shipment weighing 2,000 pounds and freight class 85 would cost 20(34.87) = $690.74. A 5,000 pound shipment of the same commodity would cost 50(26.60) = $1,330,

TABLE 13.2 | **EXAMPLE MATRIX OF WEIGHT BREAKS AND FREIGHT CLASS ($/CWT)**

Class	< 500 (lbs)	500 (lbs)	1,000 (lbs)	2,000 (lbs)	5,000 (lbs)	10,000 (lbs)	≥ 20,000 (lbs)
50	34.30	28.32	24.25	23.04	17.58	15.74	10.47
55	36.94	30.50	26.12	24.82	18.93	17.41	11.58
60	39.59	32.69	27.99	26.60	20.29	19.08	12.69
65	41.94	34.64	29.66	28.18	21.49	20.27	13.48
70	44.64	36.86	31.56	29.99	22.88	21.94	14.59
77.5	48.10	39.72	34.01	32.32	24.65	23.85	15.86
85	51.90	42.86	36.70	34.87	26.60	26.24	17.45
92.5	55.89	46.15	39.52	37.56	28.64	28.38	18.87
100	60.27	49.77	42.61	40.50	30.89	30.77	20.46

James Hardy/PhotoAlto/Alamy

which is more costly than the 2,000 pound shipment as we would expect. What rate will be charged if the shipper has a 4,000 pound shipment? Because the shipping weight falls between two weight breaks, we must see which of the two rates applies. Using the 2,000 pound rate the total charge would be 40(34.87) = $1,394.80. At the 5,000 pound rate, the total charge is 40(26.60) = $1,064.00. In this case, the shipper would only be charged $1,064.00 even though the actual shipment does not meet the minimum for that rate. To determine the break-even weight between two adjacent weight breaks, we define the following variables:

x = break-even weight
A = lower weight bracket
B = next highest weight bracket
C = freight rate relative to A
D = freight rate relative to B

The break-even weight is given by

$$x = (BD)/C$$

In our example, the break-even weight is (50)(26.60)/(34.87) = 38.14, or 3,814 pounds. Any shipment greater than 3,814 pounds would enjoy the lower freight rate. Example 13.2 shows how a firm can find out if increasing freight density is in its best interests.

EXAMPLE 13.2	**Evaluating an Increase in Freight Density**

One of the products produced by Kitchen Tidy is Squeaky Kleen, a tile cleaner used by restaurants and hospitals. Squeaky Kleen comes in 5-gallon containers, each weighing 48 pounds. Currently, Kitchen Tidy ships four pallets of 25 units each week to a distribution center. The freight classification for this commodity is 100. Table 13.2 has the freight rates governing this shipment.

In an effort to be more environmentally responsible, Kitchen Tidy asked their product engineers to evaluate a plan to convert Squeaky Kleen into a concentrated liquid by removing some water from the product. The product would be essentially the same; however the customer would have to add water to the concentrate before using it. This would allow engineers to design a smaller container so that 50 units can be loaded on each pallet. Each container of Squeaky Kleen would weigh only 42 pounds. The product redesign would increase the product's freight density and therefore reduce the freight class, which now would be 92.5.

Use Table 13.2 to determine the savings in freight costs Kitchen Tidy might expect from the new product design.

SOLUTION

Current product design

- The weekly shipment in pounds is (number of pallets)(units per pallet)(pounds per unit) = (4)(25)(48) = 4,800 pounds.

- The freight class is 100. The shipping weight falls between two weight break brackets. The break-even weight between these two weight breaks is (50)(30.89)/(40.50) = 38.14, or 3,814 pounds. Therefore, the shipment qualifies for the lower freight rate.

- The total weekly shipping cost is 48(30.89) = $1,482.72.

New product design

- The weekly shipment in pounds is (number of pallets)(units per pallet)(pounds per unit) = (2)(50)(42) = 4,200 pounds.

- The freight class is 92.5. The shipping weight falls between two weight break brackets. The break-even weight between these two weight breaks is (50)(28.64)/(37.56) = 38.126, or 3,813 pounds. The shipment of 4,200 pounds will get the lower rate.

- The total weekly shipping cost is 42(28.64) = $1,202.88.

The new product design will save Kitchen Tidy $1,482.72 − $1,202.88 = $279.84 each week.

DECISION POINT

There are other potential benefits to the decision to move forward on the new product design. The smaller container for the product means less expensive packaging, which increases the profit margin of Kitchen Tidy. A smaller cubic volume also implies less warehouse space that must be devoted to this product. The environmental impact of this effort could be used to raise the image of Kitchen Tidy in the industry. Of course, care must be taken to educate the customers of Squeaky Kleen about the new product design so that they do not think they are getting less value for their money due to the smaller containers.

Transtock/SuperStock

Cargo containers arrive by truck and get loaded on railroad cars at Logistics Park in Joliet, Illinois.

intermodal shipments

Mixing the modes of transportation for a given shipment, such as moving shipping containers or truck trailers on rail cars.

Transportation Mode The four major modes of transportation are (1) air freight, (2) trucking, (3) shipping by water, and (4) rail. From an energy perspective, air freight and trucking are much less efficient than shipping or rail. According to the Association of American Railroads, on average, railroads are three times more fuel efficient than trucks, capable of moving a ton of freight 436 miles per gallon of fuel. The EPA also considers railroads best when it comes to noxious emissions per ton-mile. Further, freight railroads help relieve congestion on the highways; a typical train takes the freight equivalent of several hundred trucks. Trucks, nonetheless, are more flexible and can make deliveries right to the customer's door. Shippers can have the door-to-door convenience of trucks with the long-haul economy of railroads or ocean containers by employing **intermodal shipments**, which involves mixing the modes of transportation for a given shipment, such as moving shipping containers or truck trailers on rail cars. A huge range of consumer goods, from bicycles and lawn mowers to greeting cards and clothing, and an increasing amount of industrial and agricultural products, are being transported by intermodal shipments. All of these factors should be considered when designing an environmentally responsible supply chain.

There are other considerations, however, that enter the mode selection. Air freight is the fastest, but costly. Trucks, being the most flexible, can reach destinations where air freight, shipping by water along rivers or oceans, or rail are not economical or feasible. Shipping by water, a preferred mode for intercontinental shipments, typically can handle containers of greater weight, thereby minimizing the number of necessary shipments. Designing a supply chain that addresses environmental concerns, while meeting a firm's competitive priorities, can be a challenging task.

Transportation Technology Each of the transportation modes offers opportunities for improving energy efficiency through improved designs. Design factors include:

- Relative drag—the energy needed for propulsion of a vehicle of a given size at a given speed
- Payload ratio—the cargo-carrying capacity of the vehicle relative to the vehicle's weight when fully loaded
- Propulsion systems—the technology used to move the vehicle

Walmart, for example, purchased diesel-electric and refrigerated trucks with a power unit that could keep cargo cold without the engine running, saving nearly $75 million in fuel costs and eliminating an estimated 400,000 tons of CO_2 pollution in one year alone. Manufacturers and transportation services companies such as FedEx, UPS, and DHL actively replace old equipment with newer, energy efficient equipment and greatly reduce their carbon footprints. In New Delhi, India, all commercial vehicles were mandated to use liquefied natural gas (LNG) in order to reduce the air pollution levels in the capital city.

Supply Chains, Social Responsibility, and Humanitarian Logistics

Beyond the financial and environmental responsibilities, firms and organizations are also recognizing that there are social responsibilities that must be recognized if they are to be considered good corporate citizens. Supply chain managers are in a unique position to be catalysts for social responsibility activities because they are boundary spanners: They interact with other key functional areas of the organization as well as externally with suppliers and customers. Nonetheless, supply chain managers cannot do it alone. Social responsibility should be the focus of the entire organization, including the top management. In this section, we discuss humanitarian logistics operations in the form of disaster relief supply chains and supply chain ethics, two areas in which supply chain managers can make a major contribution.

Disaster Relief Supply Chains

According to the United Nations, a *disaster* is a serious disruption of the functioning of society, causing widespread human, material, or environmental losses which exceed the ability of the affected people to cope using only its own resources. Disasters can be human-related (epidemic, war, genocide, insurgency, arson, or terrorism), or natural (earthquake, tsunami, hurricane, tornado, flood, or volcanic activity). Some disasters allow more planning time than others; all disasters put pressure on relief operations. Recent disasters such as the earthquakes in Haiti and Chile in 2010 and the earthquake and tsunami in Japan in 2011 are cases in point. Between 400 and 500 natural disasters strike per year affecting more than 250 million people, and 80 percent of all relief operations for all types of disasters require supply chains of some sort. Needless to say, supply chain managers play a vital role.

UPS hybrid electric vehicle making deliveries.

Disaster relief operations for major disasters often involve many organizations, typically led by the United Nations. Agencies such as the International Federation of Red Cross and Red Crescent Societies and many other philanthropic and faith-based organizations assist in the relief efforts under the leadership of the United Nations through programs such as the World Food Program or the UN Development Program. In addition, private third-party logistics providers, such as Agility, TNT, and UPS, have partnered with the UN to provide additional transportation capacity in the event of large-scale international disasters and to provide warehousing services in Italy, United Arab Emirates, Panama, and Ghana. The warehouses, referred to as strategic hubs, stockpile vital supplies in anticipation of major disasters in that area of the world. In a similar vein, FedEx has partnered with Heart to Heart International, a global humanitarian organization that works to improve health and to respond to the needs of disaster victims worldwide. Together, they have established four Forward Response Centers located in Kansas City, and internationally in Mexico, the Philippines, and the United Arab Emirates. Each center is stocked with 60–80 pallets containing basic relief supplies, which are ready to be moved to the affected area on a moment's notice. As these examples show, private corporations can be socially responsible by transferring, acquiring, and sharing their expertise and access to needed resources through formalized collaborations such as these.

Figure 13.4 shows the three major humanitarian supply chain operations relating to disaster relief—preparation, response, and recovery—and the temporal relationship between them.

- *Preparation.* Forecasts and early warning systems can sometimes provide enough lead time to assemble the resources and organize the relief efforts. Often, however, disasters happen with little or no warning. Nonetheless, relief agencies can do some advance planning to reduce the response time. Communication protocols and the information technology infrastructure can be prepared. Strategic partnerships with other agencies and private companies can be formalized and training of agency personnel can be undertaken before the next disaster. Kits of standardized, nonperishable items can be pre-assembled and stocked, and some items can be placed in strategic hubs to reduce the delivery time when the need arises.

- *Response.* After the disaster strikes, the resources are mobilized and sent to the disaster location as soon as possible. The initial procurement of food, water, materials, and medicines are made, and personnel to provide assistance and humanitarian aid are dispatched to the region. A preliminary needs assessment is made by small multidisciplinary teams of experienced humanitarian workers and logisticians. Communication lines often are lacking. Coordination between international suppliers of needed items and the local authorities is often difficult because ports of entry are inaccessible, or the authorities are overtaken by events and inexperienced in events of this magnitude, or lacking in resources.

▼ **FIGURE 13.4**
Humanitarian Supply Chain Operations

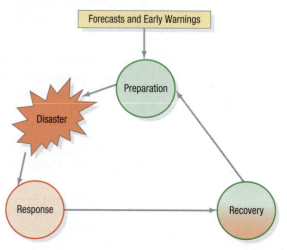

■ *Recovery.* As time progresses, the crisis mode of response operations gives way to a focus on rebuilding the information and logistical infrastructure and rehabilitating the affected population. Supplies, food, and medicines can be ordered with more normal lead times as quantities can be better estimated. More focus is placed on the cost of these items. Collaboration and cooperation improves; however, supplies from all over the world sent during the response phase, often without consultation with teams on location, cause oversupply of some commodities.

Pacific Press Service/Alamy

Police officers, fire fighters, and Self-Defense Force soldiers from all over Japan work as a rescue party after the earthquake and tsunami in Japan.

Role of Supply Chain Management   Regarding the structure of supply chains and disaster relief operations, a firm's main supply chain processes of Supplier Relationship Process, Order Fulfillment Process, and Customer Relationship Process remain essentially intact. The difference when it comes to disaster relief is that the timetable and ultimate customer for a supplier changes rapidly. Nonetheless, supply chain managers can provide the glue between the disaster relief operations. From the perspective of the disaster relief agencies, which oversee the relief operations relative to their organizations, supply chains must be designed to link the preparation activities to the initial response activities and ultimately the recovery operations. The procurement of materials, food, and medicines must be matched with their distribution to the affected areas, often involving trade-offs in delivery speed, cost, and consistent quality with regard to the type of goods and their quantities. Supply chain managers can do this; they can also link disaster relief headquarters with the operations in the field, whether those operations are response or recovery operations. For this reason, disaster relief organizations need supply chain managers on their staffs, something that few of them have.

In disaster relief operations, suppliers must understand that the morphology of supply chains changes over time. The life cycle of disaster relief supply chains has five stages: (1) brief needs assessment; (2) development of the initial supply chains for flexibility; (3) speedy distribution of the supplies to the affected region based on forecasted needs; (4) increased structuring of the supply chain as time progresses, whereby supplies arrive by a fixed schedule or on request; and (5) dismantling or turning over of the supply chain to local agencies. Suppliers, and their suppliers, need to be on the same page as this life cycle plays out; the timeframe and requirements can be different for each disaster. Supply chain managers who understand the implications of the disaster relief life cycle can mitigate its effects on their firm's operations while doing their best to support the relief agency's goals.

Supply Chain Management Challenges   The unpredictability and severity of disasters pose unique challenges to supply chain managers.

■ *Design implications.* Many disaster relief supply chains exist for only short times. At the onset of a disaster, the supply chain could require a new design from scratch featuring quick response capabilities involving innovative suppliers. Response operations are all about speed and agility in the supply of needed items. Risk-taking is encouraged because the priority is quick access. Recovery operations, however, require a more scheduled program, favoring an efficient supply chain design.

■ *Command and control.* In major international disasters, the United Nations typically has the leadership role. Disaster relief agencies work to supply the items and services they have access to. However, the national and local government of the affected region must be recognized and included. Sometimes the national government will not grant access to the area until it can ensure its security or it will not allow access to relief agencies from some countries because of political reasons. In other words, critical supplies may not be deployed as soon as they are available because of national or local roadblocks.

■ *Cargo security.* Shipping containers of goods in some Third World countries may experience theft or extra delays. Delays can be due to numerous police checks or weight checks. Sometimes, bribes are needed to move the cargo through the check points, all of which cause delays when speed is of the essence.

- *Donor independence.* There are many disaster relief agencies, all of which having the best of intentions to help relieve pain and suffering at the point of the disaster. If they are not coordinated with a list of required supplies, each sends what they think is needed. The result can be confusion, congestion, over stocking of some items, and under stocking of other items.

- *Change in work flow.* During the response operations, supplies are sent without waiting for demand to be accurately determined. Supplies are provided according to forecasts, however imperfect, using a *push flow* from suppliers. Once relief efforts reach the recovery operations, actual needs should dictate the required volume of supplies and the supply chains should switch to a *pull flow.*

- *Local infrastructure.* Because disasters often cause major damage to the infrastructure, roads, ports, railways, and airports may be compromised, thereby limiting the logistical movement of needed supplies. Local transportation capability may be limited. Often novel approaches must be used, such as helicopters moving supplies from ships at sea.

- *High employee turnover.* The needs of the disaster relief and the availability of qualified local labor are highly unpredictable. The manual processes often needed in disasters are poorly defined. Coupled with the uncertainty of funding from the relief agency, employee turnover tends to be high in disaster relief operations.

- *Poor communication.* Information technology is fragmented: telephone lines are disrupted, cell service is limited, and Internet access is unreliable.

Consequently, each disaster requires a unique supply chain solution.

Supply Chain Ethics

Supply chains, by virtue of their interconnectedness with other firms and their intense internal and external human interactions, often pose ethical issues for their managers. In this section, we will explore some of the ethical issues associated with buyer-supplier relationships, location of facilities, and inventory management.

Buyer-Supplier Relationships   When an ethical issue arises, customers often blame the firm that sold them the service or product when in fact a supplier farther upstream was at fault. Selecting suppliers that adhere to ethical codes of conduct is a critical aspect of designing a supply chain. This is a difficult task; however, socially responsible firms have some guidance for selecting ethical suppliers. Social Accountability International, an organization dedicated to defining and verifying the implementation of ethical workplaces, has compiled **SA8000:2008**, which is a list of standards covering nine dimensions of ethical workforce management:

SA8000:2008

A list of standards covering nine dimensions of ethical workforce management.

1 *Child Labor:* Employ no underage workers, usually taken to be under 15 years of age.

2 *Forced Labor:* Prohibit the use of forced labor, including prison or debt bondage labor.

3 *Health and Safety:* Provide a safe and healthy work environment.

4 *Freedom of Association and Right to Collective Bargaining:* Respect the right to form and join trade unions and bargain collectively.

5 *Discrimination:* Avoid discrimination based on race, caste, origin, religion, disability, gender, sexual orientation, union or political affiliation, or age; no sexual harassment.

6 *Discipline:* Use no corporal punishment, mental or physical coercion, or verbal abuse.

7 *Working Hours:* Require no more than 48 hours per week with at least one day off for every 7-day period with overtime hours and pay subject to the collective bargaining agreement.

8 *Compensation:* Wages paid for a standard work week must meet the legal and industry standards and be sufficient to meet the basic need of workers and their families.

9 *Management Systems:* Facilities seeking to gain and maintain certification must go beyond simple compliance to integrate the standard into their management systems and practices.

Once certified, firms need to be recertified every 3 years. Nonetheless, standards such as SA8000:2008 go a long way toward building a supply chain that is socially responsible.

Beyond identifying suppliers with ethical workforce practices, firms should strive to select suppliers impartially, guided strictly by market criteria and competitive priorities. Preferential treatment of suppliers because of friendships, family ties, or investment in the supplier should be avoided. Buyers should be candid when negotiating contracts and have respect for the supplier's cost structure and any special efforts for performance improvements. Gratuities to the buyer should be limited or excluded. Other unethical activities include:

- Revealing confidential bids and allowing certain suppliers to rebid
- Making reciprocal arrangements whereby the firm purchases from a supplier who in turn purchases from the firm
- Exaggerating situations to get better deals
- Using company resources for personal gain

One final buyer-supplier relationship deserves special mention. In most every supply chain, purchasing power plays a role in the relationship between a buyer and supplier. Buyers who represent a large portion of the revenue for a supplier can exact concessions from the supplier that may not be in its best interests. For example, should a powerful buyer force suppliers to take a loss, even for only the short term, on the premise that they, and everyone else, will benefit in the long run? Some would say that was the case when Walmart required all suppliers to invest in RFID (radio frequency identification) technology to track inventories and shipments. Some suppliers, especially those selling commodities with small profit margins, were forced to take a loss to remain a supplier to Walmart. Others would say that in the long run everyone in the supply chain will be better off than before the switch to RFID. Nonetheless, the initial attempts at establishing RFID in Walmart were not well received by suppliers and their push-back caused the program to falter. Today, Walmart shares in the supplier costs and focuses more on RFID at the item level, especially in apparel, where benefits can be more clearly defined. Privacy watchers, however, are still concerned over the use of RFID because they think the privacy of individual consumers may be compromised as they take products home with the tags attached.

It should be pointed out that ethical issues such as this also confront powerful suppliers, particularly in their contracts with retailers in regard to the exclusivity of their products or control over the ordering policy.

Facility Location We have already discussed the cost implications of the decision to locate facilities in Chapter 11, "Supply Chain Location Decisions." However, major decisions such as facility location have ethical considerations as well. The construction and operation of new facilities may affect the natural environment by disrupting ecosystems, primarily through habitat destruction and increased air, water, and noise pollution. Energy efficiency is also a concern. These considerations raise an ethical dilemma: Should a location based on traditional construction and logistical costs be changed to a more environmentally responsible location if it will increase costs? The location of a facility to avoid the disruption of natural habitats or taking steps to include noise-abatement and air pollution reduction technologies because of local ordinances may increase both start-up as well as operating costs. A balancing act between financial responsibility and environmental responsibility takes place. Locating a facility in a Third-World country to avoid some of the environmental laws in place in more developed countries may be less costly, but is it ethical? However, in some ways, identifying the least costly location may also help the environment. For example, minimizing the total material and personnel travel distances to and from a facility reduces operating costs and increases energy efficiency.

Inventory Management Inventory policies for independent demand inventories are discussed in Chapter 9, "Supply Chain Inventory Management." Reducing the order quantity for an item reduces the cycle inventory held in storage while increasing the number of orders per year. The ultimate is a *just-in-time system* (JIT), discussed in Chapter 8, "Lean Systems," where orders of small quantities are placed as they are needed. Imagine a large metropolitan area where most businesses are using JIT systems and the traffic congestion that results from delivery trucks carrying the small order quantities. For example, much of the congestion in Tokyo is attributed to JIT deliveries. While the cost of the inventory system for a given company is minimized in the traditional sense, noise pollution, energy consumption, air pollution, and travel time have increased for the community at large. Once again, there is a balancing

Rush hour traffic jams in Tokyo, Japan, are common, as in most large cities. A high proportion of delivery vehicles exacerbates the problem.

act between financial responsibility and environmental responsibility. What can a firm do? Steps can be taken to minimize the material movement to and from the firm by (1) consolidating shipments of items using *a periodic review system* (P-system with the same review period for a group of items from the same supplier), or (2) increasing inventory levels by either adjusting the timing of deliveries to avoid rush hours or reducing the total number of shipments. Of course, these remedies increase inventory costs to the firm.

Managing Sustainable Supply Chains

How can a firm manage its supply chains to ensure that they are sustainable? Firms might consider the following steps:

1. Develop a *sustainable supply chain framework*. Define what "sustainability" means for the firm in clear terms. Use SA8000:2008 as a guideline for workplace issues.

2. Gather data on the performance of current suppliers and use the same questionnaire to screen potential new suppliers. Use the supply chain sustainability framework as a foundation.

3. Require compliance to the sustainable supply chain framework across all business units, including their dealings with current suppliers and the selection of future suppliers.

4. Engage in active supplier management and utilize all available ethical means to influence their behavior.

5. Provide periodic reports on the impact the supply chains have on sustainability.

Designing and managing a sustainable supply chain is not an easy task. Nonetheless, many firms are making sustainability a major goal of their operations.

LEARNING GOALS IN REVIEW

❶ **Define the three elements of supply chain sustainability.** See the opening paragraphs of this chapter, Figure 13.1, and Table 13.1, which contains examples of financial, environmental, and social responsibility involving major companies. Also see the section "Sustainability across the Organization," pp. 443, for the challenges organizations face.

❷ **Explain the reverse logistics process and its implications for supply chain design.** See the sections "Supply Chains and Environmental Responsibility," pp. 443–452, and study Figure 13.2. Managerial Practice 13.1 shows how Walmart and Hewlett-Packard recycle used products and materials.

❸ **Explain how firms can improve the energy efficiency of their supply chains.** See the section "Supply Chains and Environmental Responsibility," pp. 443–452. Be sure you understand the nearest neighbor heuristic, Example 13.1,

and Solved Problem 1. Study the break-even approach to identifying break-even weights for the shipment of goods, Table 13.2, Example 13.2, and Solved Problem 2.

❹ **Explain how supply chains can be designed to support the response and recovery operations of disaster relief efforts.** The section "Supply Chains, Social Responsibility, and Humanitarian Logistics" pp. 452–457, contains a discussion of the role of supply chain managers and the challenges they face in disaster relief efforts. Figure 13.4 shows the three major disaster relief operations that supply chains must support.

❺ **Explain how ethical issues confront supply chain managers.** The section "Supply Chains, Social Responsibility, and Humanitarian Logistics" pp. 452–457, discusses the nature of the ethical issues that supply chain managers face and how they can deal with them.

MyOMLab helps you develop analytical skills and assess your progress with multiple problems on the traveling salesman problem using the nearest neighbor heuristic and break-even analysis to find the break-even rates for shipping costs.

MyOMLab Resources	Titles	Link to the Book
Video	Clif Bar: Supply Chain	Supply Chains and Environmental Responsibility
Virtual Tours	1. Caterpillar	Reverse Logistics
	2. Patagonia	Supply Chains and Environmental Responsibility; Supply Chain Ethics
Internet Exercises	13.1 Newsweek Green Rankings	Entire chapter
	13.2 Steelcase	Entire chapter
Key Equations		
Image Library		

Key Equations

Break-even weight

A = lower weight bracket

B = next highest weight bracket

C = freight rate relative to A

D = freight rate relative to B

$x = (BD)/C$

Key Terms

carbon footprint 447
closed-loop supply chain 444
environmental responsibility 442
financial responsibility 442
humanitarian logistics 442

intermodal shipments 452
nearest neighbor (NN) heuristic 448
reverse logistics 444
route planning 447
SA8000:2008 455

shortest route problem 448
social responsibility 442
sustainability 442
traveling salesman problem 448

Solved Problem 1

Greenstreets Recycling, Inc., collects used motor oil from several collection sites around the Greater Stanford area. In order to minimize the use, and thereby the cost, of its labor, vehicle, and energy resources, the company is interested in locating the shortest route that will allow its collection vehicle to visit each collection site exactly once. The following table provides the travel distances in miles between each site. Note that the company's recycling facility is located at site A.

To/From	A	B	C	D	E	F
A (depot)	–	25	50	48	41	60
B	25	–	35	22	23	43
C	50	35	–	25	47	65
D	48	22	25	–	24	40
E	41	23	47	24	–	21
F	60	43	65	40	21	–

Provide an efficient route for the collection vehicle.

SOLUTION

a. Begin at the recycling facility (site A) and proceed to its nearest neighbor (site B) which is 25 miles away.

b. From site B proceed to its nearest unvisited neighbor. Proceed from B to D—distance 22 miles.

c. From site D proceed to site E—distance 24 miles.

d. From site E proceed to site F—distance 21 miles.

e. From site F proceed to site C (the only remaining unvisited site)—distance 65 miles.

f. From site C return to A—distance 50 miles.

The completed route is A–B–D–E–F–C–A with a total distance traveled of 207 miles $(25 + 22 + 24 + 21 + 65 + 50)$.

To see if a better solution exists, the nearest neighbor heuristic should repeated using each city in turn as the starting point.

City B: B–D–E–F–A–C–B with a total distance of $(22 + 24 + 21 + 60 + 50 + 35) = 212$ miles
City C: C–D–B–E–F–A–C with a total distance of $(25 + 22 + 23 + 21 + 60 + 50) = 201$ miles
City D: D–B–E–F–A–C–D with a total distance of $(22 + 23 + 21 + 60 + 50 + 25) = 201$ miles
City E: E–F–D–B–A–C–E with a total distance of $(21 + 40 + 22 + 25 + 50 + 47) = 205$ miles
City F: F–E–B–D–C–A–F with a total distance of $(21 + 23 + 22 + 25 + 50 + 60) = 201$ miles

Note that the solutions located by using cities C, D, and F as the starting point all provide the equally short route. Thus, with the recycling facility at site A, the collection vehicle should proceed to F then E then B then D then C and finally back to A with a total distance traveled of 201 miles. It should also be noted that since the distances are symmetric, a route of reverse order, A–C–D–B–E–F–A, provides the same total distance traveled.

Solved Problem 2

Kayco Stamping in Ft. Worth, Texas, ships sheet metal components to a switch box assembly plant in Waterford, Virginia. Each component weighs approximately 25 pounds, and 50 components fit on a standard pallet. A complete pallet ships as freight class classification 92.5. Use Table 13.2 to calculate the weekly shipment cost for the following demand quantities and calculate the shipping cost per component.

a. 3 pallets

b. 13 pallets

SOLUTION

a. At 3 pallets, or 150 pieces, the shipping weight $= 150 \times (25$ pounds$) = 3{,}750$ pounds

At a freight classification of 92.5, using Table 13.2, the break-even weight $= 50(28.64)/37.56 = 38.13$ or 3,813 pounds, thus the shipment does not qualify for the lower rate.

Total weekly shipping cost $= 37.5(37.56) = \$1{,}408.50$

The per-component shipping charge is $\$1{,}408.50/150 = \$9.39$

b. At 13 pallets, or 650 pieces, the shipping weight $= 650 \times (25$ pounds$) = 16{,}250$ pounds

At a freight classification of 92.5, using Table 13.2, the break-even weight $= 200(18.87)/28.38 = 132.98$ or 13,298 pounds, thus the shipment qualifies for the lower rate.

Total weekly shipping cost $= 162.5(18.87) = \$3{,}066.38$

The per-component shipping charge is $\$3{,}066.38/650 = \$4.72$.

Discussion Questions

1. In the chapter opener, we get a glimpse of how a major corporation can make a significant social contribution by assisting in a disaster relief supply chain. Disasters often occur with little warning; humanitarian aid needs to get to the affected areas very quickly. From the perspective of supply chain design, what capabilities should disaster relief supply chains have?

2. Designing supply chains that are energy efficient and environmentally responsible helps a firm achieve a competitive priority of low cost operations, which supports the firm's financial responsibility to its shareholders. Explain how focusing on energy efficiency can pose some ethical dilemmas for supply chain managers.

3. As more firms entertain the option of developing reverse logistics supply chains, explain the financial implications they should consider.

4. Traditional buyer-supplier relationships revolve around financial considerations. Buyers and suppliers both try to get the best deals for their companies; often one's gain is the other's loss. Many firms today are valuing sustainable operations, which involves financial, environmental, and social responsibility. Explain how buyer-supplier relationships are evolving in this new environment.

Problems

The OM Explorer and POM for Windows software is available to all students using the 10th edition of this textbook. Go to **www.pearsonhighered.com/krajewski** to download these computer packages. If you purchased MyOMLab, you also have access to Active Models software and significant help in doing the following problems. Check with your instructor on how best to use these resources. In many cases, the instructor wants you to understand how to do the calculations by hand. At the least, the software provides a check on your calculations. When calculations are particularly complex and the goal is interpreting the results in making decisions, the software entirely replaces the manual calculations.

1. On a daily basis, the Vampire Van is dispatched from Maplewood Hospital to pickup blood and platelet donations made at its local donation centers. The distances in miles between all locations may be found in Table 13.3.

   a. The Vampire Van travels from Maplewood Hospital (A) to (B) to (C) to (D) to (E) and then returns to the hospital (A). What is the total number of miles that the van must travel using this route?

   b. Using Maplewood Hospital as the beginning location, create a route using the Nearest Neighbor heuristic. What is the total number of miles that the van must travel using this route?

   c. Using Valley Hills (E) as the beginning location, create a route using the Nearest Neighbor heuristic. What is the total number of miles that the van must travel using this route?

2. Royal Seafood delivers fresh fin and shellfish to specialty grocery stores in the state of Oregon. The company packs a delivery truck in Corvallis and then drives in one single route to its five customers spread throughout the state. The distances in miles between all locations may be found in the table below.

	Corvallis [A]	Roseburg [B]	Bend [C]	Baker [D]	Lakeview [E]	Burns [F]
Corvallis [A]	–	93	102	268	219	220
Roseburg [B]	93	–	116	296	167	216
Bend [C]	102	116	–	181	138	117
Baker [D]	268	296	181	–	223	106
Lakeview [E]	219	167	138	223	–	118
Burns [F]	220	216	117	106	118	–

   a. Propose an efficient route by using the Nearest Neighbor heuristic with Corvallis as the starting city. What is the total distance traveled?

   b. Use the Nearest Neighbor heuristic to calculate five routes, each starting from one of Royal Seafood's customer's location. What is the best route for Royal Seafood?

3. On Thursdays, Traxis Consolidated delivers liquid oxygen to its industrial customers in northern Michigan. The table below provides the driving time in minutes among all customers and the Traxis liquid oxygen depot location.

	A [depot]	B	C	D	E	F	G	H
A [depot]	–	26	38	31	49	33	40	52
B	26	–	53	54	75	35	56	73
C	38	53	–	46	45	68	70	77
D	31	54	46	–	25	41	30	32
E	49	75	45	25	–	69	55	44
F	33	35	68	41	69	–	27	50
G	40	56	70	30	55	27	–	21
H	52	73	77	32	44	50	21	–

   a. Currently, Traxis travels from the depot (A) to (F) to (G) to (D) to (E) to (H) to (B) to (C), then returns to (A). What is the total driving time using this route?

   b. Using the depot (A) as the beginning location, create a route using the Nearest Neighbor heuristic. What is the total driving time using this route?

   c. Use the Nearest Neighbor heuristic to calculate seven routes, each starting from one of Traxis's customer's locations. What are your conclusions?

4. Big Jim plows snow for five residential customers in northern New Hampshire. Placing a Cartesian coordinate system on a map of his service area, with his home at the origin (A), Big Jim located his five customers at coordinates: B (10,40); C (22,20); D (35,37); E (40,25); F (50,40). He has two ways to measure distances between his customers: Euclidian and rectilinear (see Chapter 3, "Process Strategy" and the section on layouts for the definitions of these measures). He is wondering if his method of measuring distances will affect the sequence

TABLE 13.3 | MILEAGE DATA FOR VAMPIRE VAN

	Maplewood Hospital (A)	City Center Donation Site (B)	Westbrook Donation Site (C)	Municipal Park Donation Site (D)	Valley Hills Donation Site (E)
Maplewood Hospital (A)	–	3.0	3.5	4.0	4.1
City Center Donation Site (B)	3.0	–	6.1	7.0	4.3
Westbrook Donation Site (C)	3.5	6.1	–	4.2	3.6
Municipal Park Donation Site (D)	4.0	7.0	4.2	–	7.2
Valley Hills Donation Site (E)	4.1	4.3	3.6	7.2	–

of customer locations he must visit to minimize his energy costs.

a. Use the Nearest Neighbor heuristic to locate the best route for Big Jim, assuming that he is interested in minimizing Euclidean distances.

b. Use the Nearest Neighbor heuristic to locate the best route for Big Jim, assuming that he is interested in minimizing rectilinear distances.

5. Arts N Crafts Industries manufacturers high-end light fixtures which it sells internationally. The company is responsible for paying for shipping its product to a distributor located in Atlanta, Georgia. Due to the high cost of shipping, the company is considering shipping its products unassembled, but will include detailed assembly instructions in each package. By shipping unassembled products, Arts N Crafts will be able to shrink the size of its cartons and thereby increase the number of products per pallet from 16 (4 rows of 4) to 25 (5 rows of 5). Note that this change will not appreciably increase each product's 8 pound shipping weight. Furthermore, since individual fixtures will be packed more tightly, they will be less susceptible to damage and easier to handle. Thus, the company expects the freight class to decrease from 85 to 70. Currently the company ships 400 units per week.

a. Using Table 13.2, assess the impact of the proposed change in packaging on weekly shipping cost.

b. How will your analysis change if demand for the product increases to 500 units per week?

6. Microtech Incorporated has decided to package its cell phone in a smaller, recyclable package. Additionally, the company will discontinue the practice of shipping each phone with a 250-page user manual and instead will make the manual available on line. These changes will result in a lighter, but more difficult package for shippers to handle. The weight of each packaged phone has dropped from 1.2 pounds to 0.5 pounds as a result of these changes and the freight classification has worsened from 55 to 70. Use Table 13.2 to calculate the difference in Microtech's monthly shipping charges if the company ships 10,000 phones per month.

VIDEO CASE   Supply Chain Sustainability at Clif Bar & Company

When Gary Erickson started out on one of his typical long-distance bike treks in 1990, he expected to have a great ride. What he did not expect was to come home with the idea for a sports nutrition energy bar that became the genesis of Clif Bar & Company.

After consuming most of his supply of Power Bars part way through the 175-mile ride, Erickson could not face the prospect of consuming another one. Instead, he figured he could come up with something better. Something sports enthusiasts would not mind eating in quantities while out on the road or the trail. The result? The CLIF® BAR, an all-natural combination of grains, nuts, and fruit, with far more taste appeal than the standard energy bar of the time.

Today, the Berkeley, California-based company has over 100 stock-keeping units, or SKUs, in its product mix, including numerous flavorful versions of the original CLIF BAR plus brands such as LUNA®, The Whole Nutrition Bar for Women™; CLIF Kid Organic ZBaR™, The Whole Grain Energy bar for kids; CLIF Mojo™, The Sweet & Salty Trail Mix Bar; and CLIF Nectar®, an organic fruit and nut bar made with just five ingredients. CLIF BARs were originally distributed through cycling shops and other niche retail outlets, but can now be found in a wide variety of retail outlets in the United States, such as Whole Foods, Trader Joe's, REI, and even your local grocery store.

From the beginning, CLIF BARs were made from wholesome ingredients. Yet as Erickson looked at the ingredients being sourced for CLIF BAR, he realized that making a healthy food product and sourcing ingredients from farmers, ranchers, and cooperatives using organic growing techniques was a "natural" fit. The company made a commitment to both sustainable growing techniques and using only organic raw ingredients back then, and by 2003, had made CLIF BAR 70 percent certified organic. Since then, six of Clif Bar & Company's CLIF and LUNA brands are now made with 70 percent organic ingredients or more.

The impact on the supply chain for sourcing organic ingredients is tremendous. First, there is a limited—but growing—number of organic growers for the ingredients Clif Bar uses. Second, growers who do not use pesticides, herbicides, and genetically modified plants are sometimes at risk of producing lower crop yields. Third, it can be more costly to store the ingredients. And fourth, as more companies commit to environmentally responsible programs and organic ingredients, competition is great for the available global supply.

The company's forecasters and planners work hard to manage both the raw ingredient inventory flows from upstream suppliers, and the finished goods flows to downstream customers, to be sure products are available in the right quantities and right locations. Ordering on raw materials and packaging materials is aggregated to provide efficient sourcing. Production is planned, based on both input and output forecasts, to maximize customer service and minimize inventory. Several times a year, production plans are shared with business partners at all points of the supply chain to make sure the flow of ingredients and products is smooth, and that inventories do not accumulate at any point in the supply chain beyond planned volumes. Monthly forecasts and changes to plans also are communicated to all supply chain partners.

Clif Bar & Company managers know that consumers' tastes for products change regularly, so new flavors and brands are periodically introduced into the various brands. Likewise, flavors are sometimes retired to make room for new ones. As the company's research and development team prepares to move a new product idea from the test kitchen to the manufacturing plant, supply chain managers must get to work assuring any new ingredients can be procured.

As a smaller and privately owned company, Clif Bar does not own its manufacturing plants and distribution centers, and relies on contractual agreements with outsourcers in the United States. These supply chain business partners are carefully chosen for their ability to manufacture and distribute Clif Bar's products, their commitment to quality, and their alignment with Clif Bar's own value system. This value system, referred to as the company's "Five Aspirations," holds that Clif Bar & Company will work toward sustaining its people, brands, business, community, and the planet. Greg Ginsburg, Vice President of Supply Chain, wants to be sure all parts of the supply chain—owned or not—are in agreement with these aspirations. "We look at their energy sourcing, labor practices, and workplace environments. And where we source products from small cooperatives, we'll go as far back as possible to assure those tiny growers know about our expectations," says Ginsburg.

QUESTIONS

1. In what ways does Clif Bar have a sustainable supply chain?

2. Regarding financial responsibility, what business risks does Clif Bar & Company face with so many parts of its supply chain outsourced?

3. What issues or risks to sustainability could Clif Bar & Company encounter if it chose to expand to international markets?

Selected References

Beamon, Benita M. "Environmental and Sustainability Ethics in Supply Chain Management." *Science and Engineering Ethics*, vol. 11 (2005), pp. 221–234.

Beschorner, Thomas, and Martin Muller. "Social Standards: Toward an Active Ethical Involvement of Business in Developing Countries." *Journal of Business Ethics*, vol. 73 (2006), pp. 11–23.

Bonini, Sheila, Steven Gorner, and Alissa Jones. "How Companies Manage Sustainability." *McKinsey & Company Global Survey Results*, (February 2010), pp. 1–8.

Carter, C.R., and M.M. Jennings. "The Role of Purchasing in Corporate Social Responsibility." *Journal of Business Logistics*, vol. 25, no. 1 (2004), pp. 145–186.

Conner, Martin P. "The Supply Chain's Role in Leveraging Product Life Cycle Management." *Supply Chain Management Review* (March 2004), pp. 36–43.

Curkovic, Sime, and Robert Sroufe. "Using ISO 14001 to Promote a Sustainable Supply Chain." *Business Strategy and the Environment*, vol. 20 (2011), pp. 71–93.

Day, J.M., I. Junglas, and L. Silva. "Information Flow Impediments in Disaster Relief Supply Chains." *Journal of the Association for Information Systems*, vol. 10, no. 8 (August 2009), pp. 637–660.

Drake, Matthew J., and John Teepen Schlachter. "A Virtue-Ethics Analysis of Supply Chain Collaboration." *Journal of Business Ethics*, vol. 82 (2007), pp. 851–864.

Ferguson, Mark. "Making Your Supply Chain More Sustainable by Closing the Loop." *The European Business Review* (November–December 2010), pp. 28–31.

Fiksel, Joseph, Douglas Lambert, Les B. Artman, John A. Harris, and Hugh M. Share. "Environmental Excellence: The New Supply Chain Edge." *Supply Chain Management Review* (July/August 2004), pp. 50–57.

"Global Responsibility," *RR Donnelley*, (2010), pp. 1–21.

Handfield, Robert B., and David L. Baumer. "Managing Conflicts of Interest in Purchasing." *Journal of Supply Chain Management*, (Summer 2006), pp. 41–50.

Handfield, Robert, S. Walton, Robert Sroufe, and Steven Melnyk. "Applying Environmental Criteria to Supplier Assessment: A Study of the Application of the Analytical Hierarchy Process." *European Journal of Operational Research*, vol. 41, no. 1 (2002), pp. 70–87.

Hartvigsen, David. *SimQuick: Process Simulation with Excel*, 2nd ed. Upper Saddle River, NJ: Prentice Hall, 2004.

Hindo, Brian. "Everything Old is New Again." *BusinessWeek* (September 25, 2006), pp. 64–70.

"How the Calculate the Weight Break," http://freightlogistics.wordpress.com, (December 19, 2007).

Keating, B., A. Quazi, A. Kriz, and T. Coltman. "In Pursuit of a Sustainable Supply Chain: Insights from Weatpac Banking Corporation." *Research Online*, http://ro.uow.edu.au/infopapers/688, (2008).

Kulwiec, Ray. "Reverse Logistics Provides Green Benefits." *Target*, vol. 22, no. 3 (Third Issue 2006), pp. 11–20.

Lacy, Peter, Tim Cooper, Rob Hayward, and Lisa Neuberger. "A New Era of Sustainability," *UN Global Compact-Accenture CEO Study 2010*, (June 2010), pp. 1–60.

Lillywhite, Serena. "Responsible Supply Chain Management: Ethical Purchasing in Practice." *Brotherhood of St. Laurence* (October 2004), pp. 1–5.

Maon, Francois, Adam Lindgreen, and Joelle VanHamme. "Supply Chains in Disaster Relief Operations: Cross-Sector Socially Oriented Collaborations." *Hull University Business School*, Research Memorandum 80, (April 2009), pp. 1–35.

Martha, Joseph, and Sunil Subbakrishna. "Targeting a Just-In-Case Supply Chain for the Inevitable Next Disaster." *Supply Chain Management Review* (September/October 2002), pp. 18–23.

Melnyk, Steven, E.W. Davis, R.E. Speckman, and J. Sandor. "Outcome Driven Supply Chains." *Sloan Management Review*, vol. 51, no. 2 (2010), pp. 33–38.

Meyer, Tobias A. "Increasing the Energy Efficiency of Supply Chains." *McKinsey Quarterly*, (August 2009), pp. 1–2.

Mollenkopf, Diane A., and David J. Closs. "The Hidden Value in Reverse Logistics." *Supply Chain Management Review* (July/August 2005), pp. 34–43.

Plambeck, Erica L. "The Greening of Walmart's Supply Chain." *Supply Chain Management Review* (July/August 2007), pp. 18–25.

"Social Accountability International – SA8000," http://www.sa-intl.org, (2011).

Thomas, A. and L. Kopczak. "Life-Saving Supply Chains and the Path Forward." In Lee, H. and C.Y. Lee (Eds.) *Building Supply Chain Excellence in Emerging Economies*, London, Springer Science and Business Media LLC, 2007.

Van Wassenhove, L. "Humanitarian Aid Logistics: Supply Chain Management in High Gear." *Journal of the Operational Research Society*, vol. 57, no. 5, (2006), pp. 475–489.

Whybark, D.C. "Issues in Managing Disaster Relief Inventories." *International Journal of Production Economics*, vol. 108, no. 1 (July 2007), pp. 228–235.

Whybark, D.C., Steven A. Melnyk, Jamison Day, and Ed Davis. "Disaster Relief Supply Chain Management: New Realities, Management Challenges, Emerging Opportunities." *Decision Line* (May 2010), pp. 4–7.

A Motorola Droid phone displayed Google's homepage in Washington, D.C. on August 15, 2011. Google Inc. bought the phone manufacturer Motorola Mobility for $12.5 billion. Motorola considerably improved its demand forecasting process, with payoffs in how it managed its supply chain.

Motorola Mobility

Motorola Mobility makes mobile phone handsets, smartphones, tablets, and cable set-top box assets. In the early 2000s, Motorola's leadership and market share were eroding. Motorola realized that it must transform its supply chain, and embarked on a major initiative to tighten communications and collaboration along its supply chain. It put collaborative planning, forecasting, and replenishment (*CPFR*) into action in 2002. The payoff has been significant.

Motorola sells over 120 handset models globally. Forecasting how many of which models to make and sell is difficult, and accurate replenishment of retailers' shelves is critical. If a customer's favorite handset is not in stock, there is a real risk that Motorola loses that customer for life, and not just for the next service contract. Approximately one half of all stockouts result in lost sales. To make matters worse, a phone model can have multiple SKUs, life cycles average little more than a year, and new product introductions are rapid.

Prior to adopting CPFR, Motorola Mobile's sales were highly variable and were not synchronized with customer demand. Motorola had visibility only for its shipments to retailers' distribution centers, but not for shipments from the retailers' distribution centers to the stores. Knowing what retailers are selling is much more valuable information in forecasting future demand than knowing what retailers are buying. Without this information, forecast errors were very high, resulting in excessive stockouts. CPFR enabled Motorola to collaborate with its retailers' distribution centers' customers and increase its ability to forecast effectively. Motorola launched an organization-wide shift to customer-focused operations teams. They shared with their retailers their real-time data and plans, including forecasts, inventories, sales to retailers' shelves, promotions, product plans, and exceptions. Traditionally, suppliers and buyers in most supply chains prepare independent demand forecasts.

Before CPFR, the retailers' forecasts were developed at the end of the second week of each month while Motorola's assembled its sales and operations plan earlier in the second week. Motorola convinced the retailer to move up its planning cycle by just two or three days, which eliminated a seven-week forecast lag resulting from the forecast not being incorporated until the next month's planning cycle. Now, the retailer loads its forecasts for the next month on Monday. On Tuesday, Motorola loads its forecast. During the weekly call on Wednesday, the two teams jointly resolve discrepancies line-by-line. The inclusion of a forecasting analyst means they can immediately resolve issues arising from the discrepancies.

The real key to a successful implementation of CPFR is the forging of a cultural alliance that involves peer-to peer relations and cross-functional teams. Prior to CPFR, retailers sometimes gave Motorola "C," "D," and "F" rating on metrics such as on-time delivery, ease of doing business, and stockouts. After CPFR, they give Motorola "A" ratings. Motorola's CPFR initiative reduced forecast error to a fraction of its previous level, allowed quick reductions in safety stock, cut transportation costs in half because of fewer less-than-truckload shipments, and cut stockouts to less than a third of previous levels. Such success is one reason Google paid big ($12.5 billion) to buy Motorola's cellphone business in August 2011.

Source: Jerold P. Cederlund, Rajiv Kohli, Susan A. Sherer, and Yuliang Yao, "How Motorola Put CPFR into Action," *Supply Chain Management Review* (October 2007), pp. 28–35; Sharyn Leaver, Patrick Connaughton, and Elisse Gaynor, "Case Study: Motorola's Quest for Supply Chain Excellence," *Forrester Research, Inc.* (October, 2006), pp. 1–12, www .motorola.com, April 29, 2011; Amir Efrati and Spencer A. Ante, "Google's $12.5 Billion Gamble," *The Wall Street Journal*, August 12, 2011.

LEARNING GOALS *After reading this chapter, you should be able to:*

1. Identify the five basic patterns of most demand time series.
2. Identify the various measures of forecast errors.
3. Use regression to make forecasts with one or more independent variables.
4. Make forecasts using the most common approaches for time-series analysis.
5. Make forecasts using trend projection with regression.
6. Describe a typical forecasting process used by businesses.
7. Explain collaborative planning, forecasting, and replenishment (CPFR).

forecast

A prediction of future events used for planning purposes.

Balancing supply and demand begins with making accurate forecasts, and then reconciling them across the supply chain as shown by Motorola Mobility. A **forecast** is a prediction of future events used for planning purposes. Planning, on the other hand, is the process of making management decisions on how to deploy resources to best respond to the demand forecasts. Forecasting methods may be based on mathematical models that use available historical data, or on qualitative methods that draw on managerial experience and judgments, or on a combination of both.

In this chapter, our focus is on demand forecasts. We begin with different types of demand patterns. We examine forecasting methods in three basic categories: (1) judgment, (2) causal, and (3) time-series methods. Forecast errors are defined, providing important clues for making better forecasts. We next consider the forecasting techniques themselves, and then how they can be combined to bring together insights from several sources. We conclude with overall processes for making forecasts and designing the forecasting system.

Forecasts are useful for both managing processes and managing supply chains. At the supply chain level, a firm needs forecasts to coordinate with its customers and suppliers. At the process level, output forecasts are needed to design the various processes throughout the organization, including identifying and dealing with in-house bottlenecks.

Forecasting across the Organization

The organization-wide forecasting process cuts across functional areas. Forecasting overall demand typically originates with marketing, but internal customers throughout the organization depend on forecasts to formulate and execute their plans as well. Forecasts are critical inputs to business plans, annual plans, and budgets. Finance needs forecasts to project cash flows and capital requirements. Human resources uses forecasts to anticipate hiring and training needs. Marketing is an important source for sales forecast information because it is closest to external customers. Operations and supply chain managers need forecasts to plan output levels, purchases of services and materials, workforce and output schedules, inventories, and long-term capacities.

Managers throughout the organization make forecasts on many variables other than future demand, such as competitor strategies, regulatory changes, technological changes, processing times, supplier lead times, and quality losses. Tools for making these forecasts are basically the same tools covered here for demand forecasting: judgment, opinions of knowledgeable people, averages of experience, regression, and time-series techniques. Using these tools, forecasting can be improved. Still, forecasts are rarely perfect. As Samuel Clemens (Mark Twain) said in *Following the Equator*, "Prophesy is a good line of business, but it is full of risks." Smart managers recognize this reality and find ways to update their plans when the inevitable forecast error or unexpected event occurs.

Demand Patterns

Forecasting customer demand is a difficult task because the demand for services and goods can vary greatly. For example, demand for lawn fertilizer predictably increases in the spring and summer months; however, the particular weekends when demand is heaviest may depend on uncontrollable factors such as the weather. Sometimes, patterns are more predictable. Thus, the peak hours of the day for a large bank's call center are from 9:00 A.M. to 12:00 P.M., and the peak day of the week is Monday. For its statement-rendering processes, the peak months are January, April, July, and October, which is when the quarterly statements are sent out. Forecasting demand in such situations requires uncovering the underlying patterns from available information. In this section, we discuss the basic patterns of demand.

The repeated observations of demand for a service or product in their order of occurrence form a pattern known as a **time series**. There are five basic patterns of most demand time series:

1. *Horizontal.* The fluctuation of data around a constant mean.
2. *Trend.* The systematic increase or decrease in the mean of the series over time.
3. *Seasonal.* A repeatable pattern of increases or decreases in demand, depending on the time of day, week, month, or season.
4. *Cyclical.* The less predictable gradual increases or decreases in demand over longer periods of time (years or decades).
5. *Random.* The unforecastable variation in demand.

Cyclical patterns arise from two influences. The first is the business cycle, which includes factors that cause the economy to go from recession to expansion over a number of years. The other influence is the service or product life cycle, which reflects the stages of demand from development through decline. Business cycle demand is difficult to predict because it is affected by national or international events.

The four patterns of demand—horizontal, trend, seasonal, and cyclical—combine in varying degrees to define the underlying time pattern of demand for a service or product. The fifth pattern, random variation, results from chance causes and thus, cannot be predicted. Random variation is an aspect of demand that makes every forecast ultimately inaccurate. Figure 14.1 shows the first four patterns of a demand time series, all of which contain random variations.

Creating Value through
Operations Management

↓

Using Operations to Compete
Project Management

Managing Processes

↓

Process Strategy
Process Analysis
Quality and Performance
Capacity Planning
Constraint Management
Lean Systems

Managing Supply Chains

↓

Supply Chain Inventory
Management
Supply Chain Design
Supply Chain Location
Decisions
Supply Chain Integration
Supply Chain Sustainability
and Humanitarian Logistics
Forecasting
Operations Planning and
Scheduling
Resource Planning

---

time series

The repeated observations of demand for a service or product in their order of occurrence.

FIGURE 14.1 ▶
Patterns of Demand

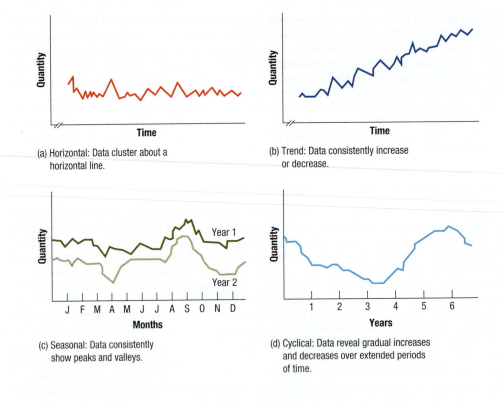

(a) Horizontal: Data cluster about a
    horizontal line.

(b) Trend: Data consistently increase
    or decrease.

(c) Seasonal: Data consistently
    show peaks and valleys.

(d) Cyclical: Data reveal gradual increases
    and decreases over extended periods
    of time.

Key Decisions on Making Forecasts

Before using forecasting techniques, a manager must make two decisions: (1) what to forecast, and (2) what type of forecasting technique to select for different items.

Deciding What to Forecast

Although some sort of demand estimate is needed for the individual services or goods produced by a company, forecasting total demand for groups or clusters and then deriving individual service or product forecasts may be easiest. Also, selecting the correct unit of measurement (e.g., service or product units or machine-hours) for forecasting may be as important as choosing the best method.

Level of Aggregation   Few companies err by more than 5 percent when forecasting the annual total demand for all their services or products. However, errors in forecasts for individual items and shorter time periods may be much higher. Recognizing this reality, many companies use a two-tier forecasting system. They first cluster (or "roll up") several similar services or products in a process called **aggregation**, making forecasts for families of services or goods that have similar demand requirements and common processing, labor, and materials requirements. Next, they derive forecasts for individual items, which are sometimes called stock-keeping units. A *stock-keeping unit (SKU)* is an individual item or product that has an identifying code and is held in inventory somewhere along the supply chain, such as in a distribution center.

Units of Measurement   Rather than using dollars as the initial unit of measurement, forecasts often begin with service or product units, such as SKUs, express packages to deliver, or customers needing maintenance service or repairs for their cars. Forecasted units can then be translated to dollars by multiplying them by the unit price. If accurately forecasting demand for a service or product is not possible in terms of number of units, forecast the standard labor or machine-hours required of each of the critical resources.

Choosing the Type of Forecasting Technique

Forecasting systems offer a variety of techniques, and no one of them is best for all items and situations. The forecaster's objective is to develop a useful forecast from the information at hand with the technique that is appropriate for the different patterns of demand. Two general types of forecasting techniques are used: judgment methods and quantitative methods. **Judgment methods** translate the opinions of managers, expert opinions, consumer surveys, and salesforce estimates

aggregation

The act of clustering several similar services or products so that forecasts and plans can be made for whole families.

judgment methods

A forecasting method that translates the opinions of managers, expert opinions, consumer surveys, and salesforce estimates into quantitative estimates.

into quantitative estimates. Quantitative methods include causal methods, time-series analysis, and trend projection with regression. **Causal methods** use historical data on independent variables, such as promotional campaigns, economic conditions, and competitors' actions, to predict demand. **Time-series analysis** is a statistical approach that relies heavily on historical demand data to project the future size of demand and recognizes trends and seasonal patterns. **Trend projection using regression** is a hybrid between a time-series technique and the causal method.

Forecast Error

For any forecasting technique, it is important to measure the accuracy of its forecasts. Forecasts almost always contain errors. Random error results from unpredictable factors that cause the forecast to deviate from the actual demand. Forecasting analysts try to minimize forecast errors by selecting appropriate forecasting models, but eliminating all forms of errors is impossible.

Forecast error for a given period t is simply the difference found by subtracting the forecast from actual demand, or

$$E_t = D_t - F_t$$

where

$$E_t = \text{forecast error for period } t$$

$$D_t = \text{actual demand for period } t$$

$$F_t = \text{forecast for period } t$$

This equation (notice the alphabetical order with D_t coming before F_t) is the starting point for creating several measures of forecast error that cover longer periods of time. Figure 14.2 shows the output from the *Error Analysis* routine in Forecasting's dropdown menu of POM for Windows. Part (a) gives a big picture view of how well the forecast has been tracking the actual demand. Part (b) shows the detailed calculations needed to obtain the summary error terms. Finally, Part (c) gives the summary error measures summarized across all 10 time periods, as derived from Part (b).

The **cumulative sum of forecast errors (CFE)** measures the total forecast error:

$$CFE = \sum E_t$$

CFE is a cumulative sum. Figure 14.3(b) shows that it is the sum of the errors for all 10 periods. For any given period, it would be the sum of errors up through that period. For example, it would be −8 (or −2 −6) for period 2. CFE is also called the *bias error* and results from consistent mistakes—the forecast is always too high or too low. This type of error typically causes the greatest disruption to planning efforts. For example, if a forecast is consistently lower than actual demand, the value of CFE will gradually get larger and larger. This increasingly large error indicates some systematic deficiency in the forecasting approach. The average forecast error, sometimes called the *mean bias*, is simply

$$\bar{E} = \frac{CFE}{n}$$

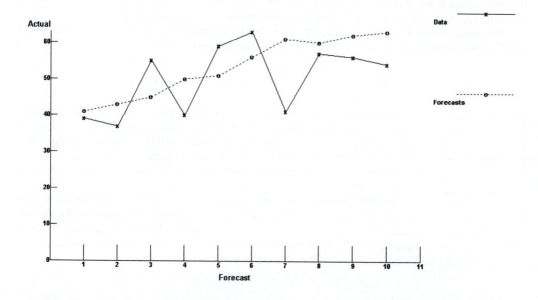

◄ **FIGURE 14.2(a)**
Graph of Actual and Forecast Demand Using *Error Analysis* of Forecasting in POM for Windows

▶ **FIGURE 14.2(b)**
Detailed Calculations of
Forecast Errors

		Forecast	Error	|Error|	Error^2	|Pct Error|
Past period 1	39	41	-2	2	4	5.128%
Past period 2	37	43	-6	6	36	16.216%
Past period 3	55	45	10	10	100	18.182%
Past period 4	40	50	-10	10	100	25%
Past period 5	59	51	8	8	64	13.559%
Past period 6	63	56	7	7	49	11.111%
Past period 7	41	61	-20	20	400	48.78%
Past period 8	57	60	-3	3	9	5.263%
Past period 9	56	62	-6	6	36	10.714%
Past period 10	54	63	-9	9	81	16.667%
TOTALS	501		-31	81	879	170.621%
AVERAGE	50.1		-3.1	8.1	87.9	17.062%
			(Bias)	(MAD)	(MSE)	(MAPE)
				Std dev	29.648	

mean squared error (MSE)

A measurement of the dispersion of forecast errors.

standard deviation (σ)

A measurement of the dispersion of forecast errors.

mean absolute deviation (MAD)

A measurement of the dispersion of forecast errors.

The **mean squared error (MSE)**, **standard deviation of the errors** (σ), and **mean absolute deviation (MAD)** measure the dispersion of forecast errors attributed to trend, seasonal, cyclical, or random effects:

$$MSE = \frac{\sum E_t^2}{n}$$

$$\sigma = \sqrt{\frac{\sum (E_t - \overline{E})^2}{n-1}}$$

$$MAD = \frac{\sum |E_t|}{n}$$

Figure 14.2(b) shows the squared error in period 1 is 4, and MSE is 87.9 for the whole sample. The standard deviation of the errors is calculated using one of the functions available in Excel and is not shown in Figure 14.2(b). The absolute value of the error in period 2 is 6, and MAD is 8.1 across the whole sample.

The mathematical symbol | | is used to indicate the absolute value—that is, it tells you to disregard positive or negative signs. If MSE, σ, or MAD is small, the forecast is typically close to actual demand; by contrast, a large value indicates the possibility of large forecast errors. The measures do differ in the way they emphasize errors. Large errors get far more weight in MSE and σ because the errors are squared. MAD is a widely used measure of forecast error and is easily understood; it is merely the mean of the absolute forecast errors over a series of time periods, without regard to whether the error was an overestimate or an underestimate.

mean absolute percent error (MAPE)

A measurement that relates the forecast error to the level of demand and is useful for putting forecast performance in the proper perspective.

The **mean absolute percent error (MAPE)** relates the forecast error to the level of demand and is useful for putting forecast performance in the proper perspective:

$$MAPE = \frac{(\sum |E_t| / D_t)\,(100)}{n} \text{ (expressed as a percentage)}$$

▼ **FIGURE 14.2(c)**
Error Measures

Measure	Value
Error Measures	
CFE (Cumulative Forecast Error)	-31
MAD (Mean Absolute Deviation)	8.1
MSE (Mean Squared Error)	87.9
Standard Deviation of Errors	29.648
MAPE (Mean Absolute Percent	17.062%

For example, an absolute forecast error of 100 results in a larger percentage error when the demand is 200 units than when the demand is 10,000 units. MAPE is the best error measure to use when making comparisons between time series for different SKUs. Looking again at Figure 14.2(b), the percent error in period 2 is 16.22 percent, and MAPE, the average over all 10 periods, is 17.06 percent.

Finally, Figure 14.2(c) summarizes the key error terms across all 10 time periods. They are actually found in selected portions of Figure 14.2(b). For example, CFE is -31, which is in the error column of Figure 14.2(b) in the TOTALS row. MAD is 8.1, found in the |Error| column and AVERAGE row. Finally, | | is 17.06%, which is in the |Pct Error| column and AVERAGE row.

EXAMPLE 14.1	**Calculating Forecast Error Measures**

The following table shows the actual sales of upholstered chairs for a furniture manufacturer and the forecasts made for each of the last 8 months. Calculate CFE, MSE, σ, MAD, and MAPE for this product.

Month, t	Demand, D_t	Forecast, F_t	Error, E_t	Error, Squared, E_t^2	Absolute Error $|E_t|$	Absolute Percent Error, $(|E_t|/D_t)(100)$
1	200	225	−25	625	25	12.5%
2	240	220	20	400	20	8.3
3	300	285	15	225	15	5.0
4	270	290	−20	400	20	7.4
5	230	250	−20	400	20	8.7
6	260	240	20	400	20	7.7
7	210	250	−40	1,600	40	19.0
8	275	240	35	1,225	35	12.7
		Total	−15	5,275	195	81.3%

SOLUTION
Using the formulas for the measures, we get

Cumulative forecast error (bias):

CFE $= -15$ (the bias, or the sum of the errors for all time periods in the time series)

Average forecast error (mean bias):

$$\bar{E} = \frac{CFE}{n} = \frac{-15}{8} = -1.875$$

Mean squared error:

$$MSE = \frac{\sum E_t^2}{n} = \frac{5,275}{8} = 659.4$$

Standard deviation of the errors:

$$\sigma = \sqrt{\frac{\sum [E_t - (-1.875)]^2}{7}} = 27.4$$

Mean absolute deviation:

$$MAD = \frac{\sum |E_t|}{n} = \frac{195}{8} = 24.4$$

Mean absolute percent error:

$$MAPE = \frac{[\sum |E_t|/D_t]100}{n} = \frac{81.3\%}{8} = 10.2\%$$

A CFE of −15 indicates that the forecast has a slight bias to overestimate demand. The MSE, σ, and MAD statistics provide measures of forecast error variability. A MAD of 24.4 means that the average forecast error was 24.4 units in absolute value. The value of σ, 27.4, indicates that the sample distribution of forecast errors has a standard deviation of 27.4 units. A MAPE of 10.2 percent implies that, on average, the forecast error was about 10 percent of actual demand. These measures become more reliable as the number of periods of data increases.

DECISION POINT
Although reasonably satisfied with these forecast performance results, the analyst decided to test out a few more forecasting methods before reaching a final forecasting method to use for the future.

Computer Support

Computer support, such as from OM Explorer or POM for Windows, makes error calculations easy when evaluating how well forecasting models fit with past data. Errors are measured across past data, often called the *history file* in practice. They show the various error measures across the entire history file for each forecasting method evaluated. They also make forecasts into the future, based on the method selected.

Judgment Methods

Forecasts from quantitative methods are possible only when there is adequate historical data, (i.e., the *history file*). However, the history file may be nonexistent when a new product is introduced or when technology is expected to change. The history file might exist but be less useful when certain events (such as rollouts or special packages) are reflected in the past data, or when certain events are expected to occur in the future. In some cases, judgment methods are the only practical way to make a forecast. In other cases, judgment methods can also be used to modify forecasts that are generated by quantitative methods. They may recognize that one or two quantitative models have been performing particularly well in recent periods. Adjustments certainly would be called for if the forecaster has important contextual knowledge. *Contextual knowledge* is knowledge that practitioners gain through experience, such as cause-and-effect relationships, environmental cues, and organizational information that may have an effect on the variable being forecast. Adjustments also could account for unusual circumstances, such as a new sales promotion or unexpected international events. They could also have been used to remove the effect of special one-time events in the history file before quantitative methods are applied. Four of the more successful judgment methods are as follows: (1) salesforce estimates, (2) executive opinion, (3) market research, and (4) the Delphi method.

Salesforce estimates are forecasts compiled from estimates made periodically by members of a company's salesforce. The salesforce is the group most likely to know which services or products customers will be buying in the near future and in what quantities. Forecasts of individual salesforce members can be combined easily to get regional or national sales estimates. However, individual biases of the salespeople may taint the forecast. For example, some people are naturally optimistic, whereas others are more cautious. Adjustments in forecasts may need to be made to account for these individual biases.

Executive opinion is a forecasting method in which the opinions, experience, and technical knowledge of one or more managers or customers are summarized to arrive at a single forecast. All of the factors going into judgmental forecasts would fall into the category of executive opinion. Executive opinion can also be used for **technological forecasting**. The quick pace of technological change makes keeping abreast of the latest advances difficult.

Market research is a systematic approach to determine external consumer interest in a service or product by creating and testing hypotheses through data-gathering surveys. Conducting a market research study includes designing a questionnaire, deciding how to administer it, selecting a representative sample, and analyzing the information using judgment and statistical tools to interpret the responses. Although market research yields important information, it typically includes numerous qualifications and hedges in the findings.

The **Delphi method** is a process of gaining consensus from a group of experts while maintaining their anonymity. This form of forecasting is useful when no historical data are available from which to develop statistical models and when managers inside the firm have no experience on which to base informed projections. A coordinator sends questions to each member of the group of outside experts, who may not even know who else is participating. The coordinator prepares a statistical summary of the responses along with a summary of arguments for particular responses. The report is sent to the same group for another round, and the participants may choose to modify their previous responses. These rounds continue until consensus is obtained.

In the remainder of this chapter, we turn to the commonly used quantitative forecasting approaches.

Causal Methods: Linear Regression

Causal methods are used when historical data are available and the relationship between the factor to be forecasted and other external or internal factors (e.g., government actions or advertising promotions) can be identified. These relationships are expressed in mathematical terms and can be complex. Causal methods are good for predicting turning points in demand and for preparing long-range forecasts. We focus on linear regression, one of the best known and most commonly used causal methods.

In **linear regression**, one variable, called a dependent variable, is related to one or more independent variables by a linear equation. The **dependent variable** (such as demand for door

---

salesforce estimates

The forecasts that are compiled from estimates of future demands made periodically by members of a company's salesforce.

executive opinion

A forecasting method in which the opinions, experience, and technical knowledge of one or more managers are summarized to arrive at a single forecast.

technological forecasting

An application of executive opinion to keep abreast of the latest advances in technology.

market research

A systematic approach to determine external consumer interest in a service or product by creating and testing hypotheses through data-gathering surveys.

Delphi method

A process of gaining consensus from a group of experts while maintaining their anonymity.

linear regression

A causal method in which one variable (the dependent variable) is related to one or more independent variables by a linear equation.

dependent variable

The variable that one wants to forecast.

hinges) is the one the manager wants to forecast. The **independent variables** (such as advertising expenditures and new housing starts) are assumed to affect the dependent variable and thereby "cause" the results observed in the past. Figure 14.3 shows how a linear regression line relates to the data. In technical terms, the regression line minimizes the squared deviations from the actual data.

In the simplest linear regression models, the dependent variable is a function of only one independent variable and, therefore, the theoretical relationship is a straight line:

$$Y = a + bX$$

where

$Y =$ dependent variable

$X =$ independent variable

$a =$ Y-intercept of the line

$b =$ slope of the line

The objective of linear regression analysis is to find values of a and b that minimize the sum of the squared deviations of the actual data points from the graphed line. Computer programs are used for this purpose. For any set of matched observations for Y and X, the program computes the values of a and b and provides measures of forecast accuracy. Three measures commonly reported are (1) the sample correlation coefficient, (2) the sample coefficient of determination, and (3) the standard error of the estimate.

The *sample correlation coefficient, r,* measures the direction and strength of the relationship between the independent variable and the dependent variable. The value of r can range from -1.00 to $+1.00$. A correlation coefficient of $+1.00$ implies that period-by-period changes in direction (increases or decreases) of the independent variable are always accompanied by changes in the same direction by the dependent variable. An r of -1.00 means that decreases in the independent variable are always accompanied by increases in the dependent variable, and vice versa. A zero value of r means no linear relationship exists between the variables. The closer the value of r is to ± 1.00, the better the regression line fits the points.

The *sample coefficient of determination* measures the amount of variation in the dependent variable about its mean that is explained by the regression line. The coefficient of determination is the square of the correlation coefficient, or r^2. The value of r^2 ranges from 0.00 to 1.00. Regression equations with a value of r^2 close to 1.00 mean a close fit.

The *standard error of the estimate, s_{xy},* measures how closely the data on the dependent variable cluster around the regression line. Although it is similar to the sample standard deviation, it measures the error from the dependent variable, Y, to the regression line, rather than to the mean. Thus, it is the standard deviation of the difference between the actual demand and the estimate provided by the regression equation.

independent variables

Variables that are assumed to affect the dependent variable and thereby "cause" the results observed in the past.

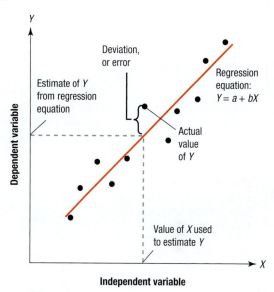

FIGURE 14.3 ▲
Linear Regression Line Relative to Actual Demand

EXAMPLE 14.2 | **Using Linear Regression to Forecast Product Demand**

The supply chain manager seeks a better way to forecast the demand for door hinges and believes that the demand is related to advertising expenditures. The following are sales and advertising data for the past 5 months:

Month	Sales (Thousands of Units)	Advertising (Thousands of $)
1	264	2.5
2	116	1.3
3	165	1.4
4	101	1.0
5	209	2.0

MyOMLab

Active Model 14.1 in MyOMLab provides insight on varying the intercept and slope of the model.

The company will spend $1,750 next month on advertising for the product. Use linear regression to develop an equation and a forecast for this product.

SOLUTION

We used POM for Windows to determine the best values of *a*, *b*, the correlation coefficient, the coefficient of determination, and the standard error of the estimate.

$$a = -8.135$$
$$b = 109.229X$$
$$r = 0.980$$
$$r^2 = 0.960$$
$$s_{yx} = 15.603$$

The regression equation is

$$Y = -8.135 + 109.229X$$

and the regression line is shown in Figure 14.4. The sample correlation coefficient, *r*, is 0.98, which is unusually close to 1.00 and suggests an unusually strong positive relationship exists between sales and advertising expenditures. The sample coefficient of determination, r^2, implies that 96 percent of the variation in sales is explained by advertising expenditures.

FIGURE 14.4 ▶
Linear Regression Line for the Sales and Advertising Data Using POM for Windows

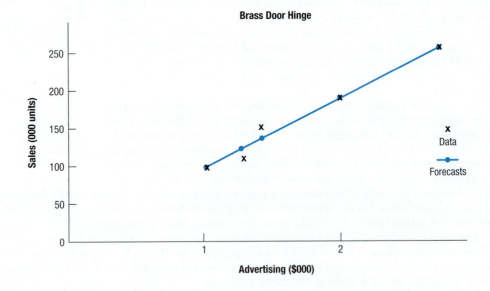

DECISION POINT

The supply chain manager decided to use the regression model as input to planning production levels for month 6. As the advertising expenditure will be $1,750, the forecast for month 6 is $Y = -8.135 + 109.229(1.75) = 183.016$, or 183,016 units.

Often several independent variables may affect the dependent variable. For example, advertising expenditures, new corporation start-ups, and residential building contracts all may be important for estimating the demand for door hinges. In such cases, *multiple regression analysis* is helpful in determining a forecasting equation for the dependent variable as a function of several independent variables. Such models can be analyzed with POM for Windows or OM Explorer and can be quite useful for predicting turning points and solving many planning problems.

Time-Series Methods

Rather than using independent variables for the forecast as regression models do, time-series methods use historical information regarding only the dependent variable. These methods are based on the assumption that the dependent variable's past pattern will continue in the future. Time-series analysis identifies the underlying patterns of demand that combine to produce an observed historical pattern of the dependent variable and then develops a model to replicate it. In this section, we focus on time-series methods that address the horizontal, trend, and seasonal patterns of demand. Before we discuss statistical methods, let us take a look at the simplest time-series method for addressing all patterns of demand—the naïve forecast.

Naïve Forecast

A method often used in practice is the **naïve forecast**, whereby the forecast for the next period (F_{t+1}) equals the demand for the current period (D_t). So if the actual demand for Wednesday is 35 customers, the forecasted demand for Thursday is 35 customers. Despite its name, the naïve forecast can perform well.

The naïve forecast method may be adapted to take into account a demand trend. The increase (or decrease) in demand observed between the last two periods is used to adjust the current demand to arrive at a forecast. Suppose that last week the demand was 120 units and the week before it was 108 units. Demand increased 12 units in 1 week, so the forecast for next week would be $120 + 12 = 132$ units. The naïve forecast method also may be used to account for seasonal patterns. If the demand last July was 50,000 units, and assuming no underlying trend from one year to the next, the forecast for this July would be 50,000 units. The method works best when the horizontal, trend, or seasonal patterns are stable and random variation is small.

naïve forecast

A time-series method whereby the forecast for the next period equals the demand for the current period, or Forecast $= D_t$.

Estimating the Average

We begin our discussion of statistical methods of time-series forecasting with demand that has no apparent trend, seasonal, or cyclical patterns. The horizontal pattern in a time series is based on the mean of the demands, so we focus on forecasting methods that estimate the average of a time series of data. The forecast of demand for *any* period in the future is the average of the time series computed in the current period. For example, if the average of past demand calculated on Tuesday is 65 customers, the forecasts for Wednesday, Thursday, and Friday are 65 customers each day.

Consider Figure 14.5, which shows patient arrivals at a medical clinic over the past 28 weeks. Assuming that the time series has only a horizontal and random pattern, one approach is simply to calculate the average of the data. However, this approach has no adaptive quality if there is a trend, seasonal, or cyclical pattern. The statistical techniques that do have an adaptive quality in estimating the average in a time series are (1) simple moving averages, (2) weighted moving averages, and (3) exponential smoothing. Another option is the simple average, but it has no adaptive capability.

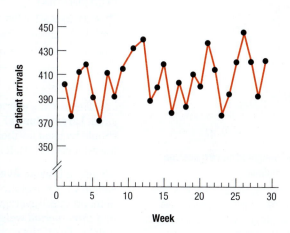

FIGURE 14.5 ▲
Weekly Patient Arrivals at a Medical Clinic

Simple Moving Averages The **simple moving average method** simply involves calculating the average demand for the n most recent time periods and using it as the forecast for future time periods. For the next period, after the demand is known, the oldest demand from the previous average is replaced with the most recent demand and the average is recalculated. In this way, the n most recent demands are used, and the average "moves" from period to period.

Specifically, the forecast for period $t + 1$ can be calculated at the end of period t (after the actual demand for period t is known) as

simple moving average method

A time-series method used to estimate the average of a demand time series by averaging the demand for the n most recent time periods.

$$F_{t+1} = \frac{\text{Sum of last } n \text{ demands}}{n} = \frac{D_t + D_{t-1} + D_{t-2} + \cdots + D_{t-n+1}}{n}$$

where

$D_t =$ actual demand in period t

$n =$ total number of periods in the average

$F_{t+1} =$ forecast for period $t + 1$

EXAMPLE 14.3	Using the Moving Average Method to Estimate Average Demand

a. Compute a *three-week* moving average forecast for the arrival of medical clinic patients in week 4. The numbers of arrivals for the past 3 weeks were as follows:

Week	Patient Arrivals
1	400
2	380
3	411

b. If the actual number of patient arrivals in week 4 is 415, what is the forecast error for week 4?

c. What is the forecast for week 5?

MyOMLab

Active Model 14.2 in MyOMLab provides insight on the impact of varying n using the example in Figure 14.5.

MyOMLab

Tutor 14.1 in MyOMLab provides another example to practice making forecasts with the moving average method.

SOLUTION

a. The moving average forecast at the end of week 3 is

$$F_4 = \frac{411 + 380 + 400}{3} = 397.0$$

b. The forecast error for week 4 is

$$E_4 = D_4 - F_4 = 415 - 397 = 18$$

c. The forecast for week 5 requires the actual arrivals from weeks 2 through 4, the 3 most recent weeks of data.

$$F_5 = \frac{415 + 411 + 380}{3} = 402.0$$

DECISION POINT

Thus, the forecast at the end of week 3 would have been 397 patients for week 4, which fell short of actual demand by 18 patients. The forecast for week 5, made at the end of week 4, would be 402 patients. If a forecast is needed now for week 6 and beyond, it would also be for 402 patients.

The moving average method may involve the use of as many periods of past demand as desired. Large values of n should be used for demand series that are stable, and small values of n should be used for those that are susceptible to changes in the underlying average. If n is set to its lowest level (i.e., 1), it becomes the naïve method.

weighted moving average method

A time-series method in which each historical demand in the average can have its own weight; the sum of the weights equals 1.0.

Weighted Moving Averages   In the simple moving average method, each demand has the same weight in the average—namely, $1/n$. In the **weighted moving average method**, each historical demand in the average can have its own weight. The sum of the weights equals 1.0. For example, in a *three-period* weighted moving average model, the most recent period might be assigned a weight of 0.50, the second most recent might be weighted 0.30, and the third most recent might be weighted 0.20. The average is obtained by multiplying the weight of each period by the value for that period and adding the products together:

$$F_{t+1} = 0.50D_t + 0.30D_{t-1} + 0.20D_{t-2}$$

For a numerical example of using the weighted moving average method to estimate average demand, see Solved Problem 2 and Tutor 14.2 of OM Explorer in MyOMLab.

The advantage of a weighted moving average method is that it allows you to emphasize recent demand over earlier demand. (It can even handle seasonal effects by putting higher weights on prior years in the same season.) The forecast will be more responsive to changes in the underlying average of the demand series than the simple moving average forecast.

exponential smoothing method

A weighted moving average method that calculates the average of a time series by implicitly giving recent demands more weight than earlier demands.

Exponential Smoothing   The **exponential smoothing method** is a sophisticated weighted moving average method that calculates the average of a time series by implicitly giving recent demands more weight than earlier demands, all the way back to the first period in the history file. It is the most frequently used formal forecasting method because of its simplicity and the small amount of data needed to support it. Unlike the weighted moving average method, which requires n periods of past demand and n weights, exponential smoothing requires only three items of data: (1) the last period's forecast; (2) the actual demand for this period; and (3) a smoothing parameter, alpha (α), which has a value between 0 and 1.0. The equation for the exponentially smoothed forecast for period $t + 1$ is calculated

$$F_{t+1} = \alpha D_t + (1 - \alpha)F_t$$

Unilever—the purveyor of Lipton Tea, Dove, Hellmann's, and hundreds of other brands, must forecast demand around the world. It has a state-of-the-art forecasting system. Using software from Manugistics, the system blends forecasts from time series techniques with judgmental adjustments for planned promotions from its sales teams. Unilever compares point-of-sales data with its own forecasts. The forecasts are reviewed and judgmentally adjusted as needed.

Art Directors & TRIP/Alamy

The emphasis given to the most recent demand levels can be adjusted by changing the smoothing parameter. Larger α values emphasize recent levels of demand and result in forecasts more responsive to changes in the underlying average. Smaller α values treat past demand more uniformly and result in more stable forecasts. Smaller α values are analogous to increasing the value of n in the moving average method and giving greater weight to past demand. In practice, various values of α are tried and the one producing the best forecasts is chosen.

Exponential smoothing requires an initial forecast to get started. There are several ways to get this initial forecast. OM Explorer and POM for Windows use as a default setting the actual demand in the first period, which becomes the forecast for the second period. Forecasts and forecast errors then are calculated beginning with period 2. If some historical data are available, the initial forecast can be found by calculating the average of several recent periods of demand. The effect of the initial estimate of the average on successive estimates of the average diminishes over time.

EXAMPLE 14.4   Using Exponential Smoothing to Estimate Average Demand

a. Reconsider the patient arrival data in Example 14.3. It is now the end of week 3, so the actual number of arrivals is known to be 411 patients. Using $\alpha = 0.10$, calculate the exponential smoothing forecast for week 4.

b. What was the forecast error for week 4 if the actual demand turned out to be 415?

c. What is the forecast for week 5?

MyOMLab

Active Model 14.3 in MyOMLab provides insight on the impact of varying α in Figure 14.5.

SOLUTION

MyOMLab

a. The exponential smoothing method requires an initial forecast. Suppose that we take the demand data for the first 2 weeks and average them, obtaining $(400 + 380)/2 = 390$ as an initial forecast. (POM for Windows and OM Explorer simply use the actual demand for the first week as a default setting for the initial forecast for period 1, and do not begin tracking forecast errors until the second period). To obtain the forecast for week 4, using exponential smoothing with $D_3 = 411$, $\alpha = 0.10$, and $F_3 = 390$, we calculate the forecast for week 4 as

Tutor 14.3 in MyOMLab provides a new practice example of how to make forecasts with the exponential smoothing method.

$$F_4 = 0.10(411) + 0.90(390) = 392.1$$

Thus, the forecast for week 4 would be 392 patients.

b. The forecast error for week 4 is

$$E_4 = 415 - 392 = 23$$

c. The new forecast for week 5 would be

$$F_5 = 0.10(415) + 0.90(392.1) = 394.4$$

or 394 patients. Note that we used F_4, not the integer-value forecast for week 4, in the computation for F_5. In general, we round off (when it is appropriate) only the final result to maintain as much accuracy as possible in the calculations.

DECISION POINT

Using this exponential smoothing model, the analyst's forecasts would have been 392 patients for week 4 and then 394 patients for week 5 and beyond. As soon as the actual demand for week 5 is known, then the forecast for week 6 will be updated.

Because exponential smoothing is simple and requires minimal data, it is inexpensive and attractive to firms that make thousands of forecasts for each time period. However, its simplicity also is a disadvantage when the underlying average is changing, as in the case of a demand series with a trend. Like any method geared solely to the assumption of a stable average, exponential smoothing results will lag behind changes in the underlying average of demand. Higher α values may help reduce forecast errors when there is a change in the average; however, the lags will still occur if the average is changing systematically. Typically, if large α values (e.g., > 0.50) are required for an exponential smoothing application, chances are good that another model is needed because of a significant trend or seasonal influence in the demand series.

Trend Projection with Regression

Let us now consider a demand time series that has a trend. A *trend* in a time series is a systematic increase or decrease in the average of the series over time. Where a significant trend is present, forecasts from naïve, moving average, and exponential smoothing approaches are adaptive, but still lag behind actual demand and tend to be below or above the actual demand.

Trend projection with regression is a forecasting model that accounts for the trend with simple regression analysis. To develop a regression model for forecasting the trend, let the dependent variable, *Y*, be a period's demand and the independent variable, *t*, be the time period. For the first period, let $t = 1$; for the second period, let $t = 2$; and so on. The regression equation is

$$F_t = a + bt$$

One advantage of the trend projection with regression model is that it can forecast demand well into the future. The previous models project demand just one period ahead, and assume that demand beyond that will remain at that same level. Of course, all of the models (including the trend projection with regression model) can be updated each period to stay current. One *apparent* disadvantage of the trend with regression model is that it is not adaptive. The solution to this problem comes when you answer the following question. If you had the past sales of Ford automobiles since 1920, would you include each year in your regression analysis, giving equal weight to each year's sales, or include just the sales for more recent years? You most likely would decide to include just the more recent years, making your regression model more adaptive. The trend projection with regression model can thus be made more or less adaptive by the selection of historical data periods to include in the same way that moving average (changing *n*) or exponential smoothing (changing α) models do.

The trend projection with regression model can be solved with either the *Trend Projection with Regression* Solver or the *Time Series Forecasting* Solver in OM Explorer. Both solvers provide the regression coefficients, coefficient of determination r^2, error measures, and forecasts into the future. POM for Windows has an alternative model (we do not cover in the textbook, although a description is provided in MyOMLab) that includes the trend, called the *Trend-Adjusted Smoothing* model.

MyOMLab

The *Trend Projection with Regression* Solver focuses exclusively on trend analysis. Its graph gives a big-picture view of how well the model fits the actual demand. Its sliders allow you to control when the regression begins, how many periods are included in the regression analysis, and how many periods you want forecasted into the future. The *Time Series Forecasting* Solver, on the other hand, covers all time series models, including the trend projection with regression. It also computes a combination forecast, which we cover in a subsequent section on using multiple techniques.

| EXAMPLE 14.5 | **Using Trend Projection with Regression to Forecast a Demand Series with a Trend** |

MyOMLab

Active Model 14.4 in MyOMLab provides insight on the behavior of the Trend Projetion with Regression model on the Medanalysis data.

Medanalysis, Inc., provides medical laboratory services to patients of Health Providers, a group of 10 family-practice doctors associated with a new health maintenance program. Managers are interested in forecasting the number of blood analysis requests per week. Recent publicity about the damaging effects of cholesterol on the heart has caused a national increase in requests for standard blood tests. The arrivals over the last 16 weeks are given in Table 14.1. What is the forecasted demand for the next three periods?

TABLE 14.1   ARRIVALS AT MEDANALYSIS FOR LAST 16 WEEKS

Week	Arrivals	Week	Arrivals
1	28	9	61
2	27	10	39
3	44	11	55
4	37	12	54
5	35	13	52
6	53	14	60
7	38	15	60
8	57	16	75

SOLUTION

Figure 14.6(a) shows the results using the *Trend Projection with Regression* Solver when all 16 weeks are included in the regression analysis, with Figure 14.6(b) showing the worksheet that goes with it.

Solver - Trend Projection with Regression

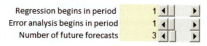

Regression begins in period   1
Error analysis begins in period   1
Number of future forecasts   3

▼ **FIGURE 14.6(a)**
First Model

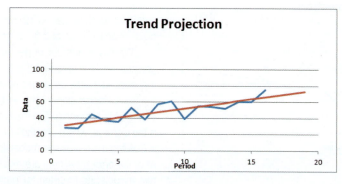

a (Y intercept)	28.50
b (slope or trend)	2.35
r2	0.69
CFE	0.00
MAD	6.21
MSE	52.96
MAPE	13.53%
Forecast for period 17	68.375
Forecast for period 18	70.72059
Forecast for period 19	73.06618

▼ **FIGURE 14.6(b)**
Detailed Calculations of Forecast Errors for First Model

					Averages		
				CFE	MSE	MAD	MAPE
				0.000	**52.958**	**6.210**	**13.53%**
	Actual				Error	Absolute	Abs %
Period #	Demand	Forecast	Error	Running CFE	Squared	Error	error
1	28	31	-2.846	-2.846	8.097	2.846	10.16%
2	27	33	-6.191	-9.037	38.331	6.191	22.93%
3	44	36	8.463	-0.574	71.626	8.463	19.23%
4	37	38	-0.882	-1.456	0.779	0.882	2.38%
5	35	40	-5.228	-6.684	27.331	5.228	14.94%
6	53	43	10.426	3.743	108.711	10.426	19.67%
7	38	45	-6.919	-3.176	47.874	6.919	18.21%
8	57	47	9.735	6.559	94.776	9.735	17.08%
9	61	50	11.390	17.949	129.725	11.390	18.67%
10	39	52	-12.956	4.993	167.855	12.956	33.22%
11	55	54	0.699	5.691	0.488	0.699	1.27%
12	54	57	-2.647	3.044	7.007	2.647	4.90%
13	52	59	-6.993	-3.949	48.897	6.993	13.45%
14	60	61	-1.338	-5.287	1.791	1.338	2.23%
15	60	64	-3.684	-8.971	13.571	3.684	6.14%
16	75	66	8.971	0.000	80.471	8.971	11.96%

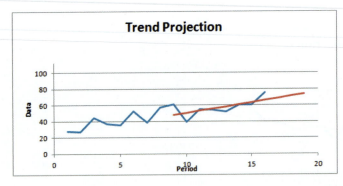

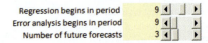

Solver - Trend Projection with Regression

Regression begins in period	9
Error analysis begins in period	9
Number of future forecasts	3

Trend Projection

a (Y intercept)	24.86
b (slope or trend)	2.57
r2	0.39

CFE	0.00
MAD	5.96
MSE	55.29
MAPE	11.10%

Forecast for period 17	68.57143
Forecast for period 18	71.14286
Forecast for period 19	73.71429

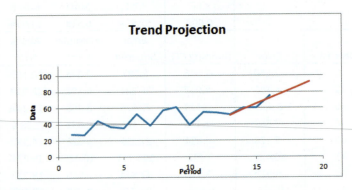

Solver - Trend Projection with Regression

Regression begins in period	13
Error analysis begins in period	13
Number of future forecasts	3

FIGURE 14.6(d) ▶
Third Model

Trend Projection

a (Y intercept)	-38.30
b (slope or trend)	6.90
r2	0.86

CFE	0.00
MAD	2.60
MSE	9.67
MAPE	4.13%

Forecast for period 17	79
Forecast for period 18	85.9
Forecast for period 19	92.8

Looking at the Results sheet of Figure 14.6(a), we see that the Y intercept of the trend line (a) is 28.50 and the slope of the line (b) is 2.35. Thus, the trend equation is $F_t = a + bt$, where t is the time period for which you are forecasting. The forecast for period 19 is $28.5 + 2.35(19) = 73$. The error terms are CFE = 0 (which is to be expected when the regression begins at the same time that error analysis begins), MAD = 6.21, MSE = 52.96, and MAPE = 13.53 percent. The coefficient of determination r^2 is decent at 0.69. The trend line is rising gently and reaches 73 for period 19. Each period the forecast predicts an increase of 2.35 arrivals per week.

When the number of periods included in the regression analysis is reduced to 9, Figure 14.6(c) shows this second model produces mixed results. The trend line has a steeper slope. MAD and MAPE are better, but r^2 and MSE are worse. The third model in Figure 14.6(d) is the extreme, where only the last four periods are used in building the regression model. It has the best r^2, and all of the error measures are much better than the first two models. Its forecast for period 19 is 93 arrivals. However, this model is based only on the last 4 weeks of data, ignoring all previous data in the history file. For that reason, management decided to split the difference with a forecast of 83 arrivals. It is halfway between the more conservative forecast of 73 in Figure 14.6(a) and Figure 14.6(c), and the optimistic forecast of 93 in Figure 14.6(d).

Seasonal Patterns

Seasonal patterns are regularly repeating upward or downward movements in demand measured in periods of less than one year (hours, days, weeks, months, or quarters). In this context, the time periods are called *seasons*. For example, customer arrivals at a fast-food shop on any day may peak between 11 A.M. and 1 P.M. and again from 5 P.M. to 7 P.M.

An easy way to account for seasonal effects is to use one of the techniques already described, but to limit the data in the time series to those time periods in the same season. For example, for a day-of-the-week seasonal effect, one time series would be for Mondays, one for Tuesdays, and so on. Such an approach accounts for seasonal effects, but has the disadvantage of discarding considerable information on past demand.

Other methods are available that analyze all past data, using one model to forecast demand for all of the seasons. We describe only the **multiplicative seasonal method**, whereby an estimate of average demand is multiplied by seasonal factors to arrive at a seasonal forecast. The four-step procedure presented here involves the use of simple averages of past demand, although more sophisticated methods for calculating averages, such as a moving average or exponential smoothing approach, could be used. The following description is based on a seasonal pattern lasting one year and seasons of one month, although the procedure can be used for any seasonal pattern and season of any length.

multiplicative seasonal method

A method whereby seasonal factors are multiplied by an estimate of average demand to arrive at a seasonal forecast.

1. For each year, calculate the average demand per season by dividing annual demand by the number of seasons per year.

2. For each year, divide the actual demand for a season by the average demand per season. The result is a *seasonal index* for each season in the year, which indicates the level of demand relative to the average demand. For example, a seasonal index of 1.14 calculated for April implies that April's demand is 14 percent greater than the average demand per month.

3. Calculate the average seasonal index for each season, using the results from step 2. Add the seasonal indices for a season and divide by the number of years of data.

4. Calculate each season's forecast for next year. Begin by forecasting next year's annual demand using the naïve method, moving averages, exponential smoothing, or trend projection with regression. Then, divide annual demand by the number of seasons per year to get the average demand per season. Finally, make the seasonal forecast by multiplying the average demand per season by the appropriate seasonal index found in step 3.

EXAMPLE 14.6	Using the Multiplicative Seasonal Method to Forecast the Number of Customers

The manager of the Stanley Steemer carpet cleaning company needs a quarterly forecast of the number of customers expected next year. The carpet cleaning business is seasonal, with a peak in the third quarter and a trough in the first quarter. The manager wants to forecast customer demand for each quarter of year 5, based on an estimate of total year 5 demand of 2,600 customers.

SOLUTION

The following table calculates the seasonal factor for each week.

It shows the quarterly demand data from the past 4 years, as well as the calculations performed to get the average seasonal factor for each quarter.

	YEAR 1		YEAR 2		YEAR 3		YEAR 4		
Quarter	Demand	Seasonal Factor (1)	Demand	Seasonal Factor (2)	Demand	Seasonal Factor (3)	Demand	Seasonal Factor (4)	Average Seasonal Factor [(1+2+3+4)/4]
1	45	45/250 = 0.18	70	70/300 = 0.23333	100	100/450 = 0.22222	100	100/550 = 0.18182	0.2043
2	335	335/250 = 1.34	370	370/300 = 1.23333	585	585/450 = 1.30	725	725/550 = 1.31818	1.2979
3	520	520/250 = 2.08	590	590/300 = 1.96667	830	830/450 = 1.84444	1160	1160/550 = 2.10909	2.0001
4	100	100/250 = 0.40	170	170/300 = 0.56667	285	285/450 = 0.63333	215	215/550 = 0.39091	0.4977
Total	1,000		1,200		1,800		2,200		
Average	1,000/4 = 250		1,200/4 = 300		1,800 = 450		2,200/4 = 550		

For example, the seasonal factor for quarter 1 in year 1 is calculated by dividing the actual demand (45) by the average demand for the whole year (1000/4 = 250). When this is done for all 4 years, we then can average the seasonal factors for quarter 1 over all 4 years. The result is a seasonal factor of 0.2043 for quarter 1.

Once seasonal factors are calculated for all four seasons (see last column in the table on the previous page), we then turn to making the forecasts for year 5. The manager suggests a forecast of 2,600 customers for the whole year, which seems reasonable given that the annual demand has been increasing by an average of 400 customers each year (from 1,000 in year 1 to 2,200 in year 4, or 1,200/3 = 400. The computed forecast demand is found by extending that trend, and projecting an annual demand in year 5 of 2,200 + 400 = 2,600 customers. (This same result is confirmed using the *Trend Projection with Regression* Solver of OM Explorer.) The quarterly forecasts are straight-forward. First, find the average demand forecast for year 5, which is 2,600/4 = 650. Then multiple this average demand by the average seasonal index, giving us

Quarter	Forecast
1	650 × 0.2043 = 132.795
2	650 × 1.2979 = 843.635
3	650 × 2.0001 = 1,300.065
4	650 × 0.4977 = 323.505

Figure 14.7 shows the computer solution using the *Seasonal Forecasting* Solver in OM Explorer. Figure 14.7(b) confirms all of the calculations made above. Notice in Figure 14.7(a) that a computer demand forecast is provided as a default for year 5. However, there is an option for user-supplied demand forecast that overrides the computer-supplied forecast if the manager wishes to make a judgmental forecast based on additional information.

FIGURE 14.7 ▶

Demand Forecasts Using the *Seasonal Forecasting* Solver of *OM Explorer*

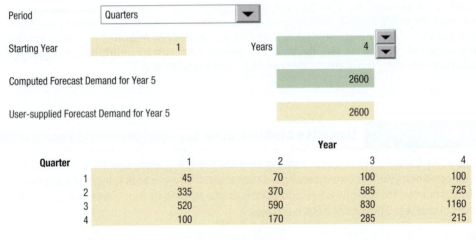

Period	Quarters			
Starting Year	1	Years	4	
Computed Forecast Demand for Year 5			2600	
User-supplied Forecast Demand for Year 5			2600	

		Year		
Quarter	**1**	**2**	**3**	**4**
1	45	70	100	100
2	335	370	585	725
3	520	590	830	1160
4	100	170	285	215

(a) Inputs sheet

Quarter	Seasonal Index	Forecast
1	0.2043	132.795
2	1.2979	843.635
3	2.0001	1300.065
4	0.4977	323.505

(b) Results

DECISION POINT

Using this seasonal method, the analyst makes a demand forecast as low as 133 customers in the first quarter and as high as 1,300 customers in the third quarter. The season of the year clearly makes a difference.

An alternative to the multiplicative seasonal method is the **additive seasonal method**, whereby seasonal forecasts are generated by adding or subtracting a seasonal constant (say, 50 units) to the estimate of average demand per season. This approach is based on the assumption that the seasonal pattern is constant, regardless of average demand. The amplitude of the seasonal adjustment remains the same regardless of the level of demand.

Choosing a Quantitative Forecasting Method

Criteria for Selecting Time-Series Methods

Forecast error measures provide important information for choosing the best forecasting method for a service or product. They also guide managers in selecting the best values for the parameters needed for the method: n for the moving average method, the weights for the weighted moving average method, α for the exponential smoothing method, and when regression data begins for the trend projection with regression method. The criteria to use in making forecast method and parameter choices include (1) minimizing bias (CFE); (2) minimizing MAPE, MAD, or MSE; (3) maximizing r^2; (4) meeting managerial expectations of changes in the components of demand; and (5) minimizing the forecast errors in recent periods. The first three criteria relate to statistical measures based on historical performance, the fourth reflects expectations of the future that may not be rooted in the past, and the fifth is a way to use whatever method seems to be working best at the time a forecast must be made.

Using Statistical Criteria   Statistical performance measures can be used in the selection of which forecasting method to use. The following guidelines will help when searching for the best time-series models:

1. For projections of more stable demand patterns, use lower α values or larger n values to emphasize historical experience.

2. For projections of more dynamic demand patterns using the models covered in this chapter, try higher α values or smaller n values. When historical demand patterns are changing, recent history should be emphasized.

Often, the forecaster must make trade-offs between bias (CFE) and the measures of forecast error dispersion (MAPE, MAD, and MSE). Managers also must recognize that the best technique in explaining the past data is not necessarily the best technique to predict the future, and that "overfitting" past data can be deceptive. Such was the case in Example 14.5. All of the forecast error measures suggested that the regression model in Figure 14.5(d) was best, but management was hesitant because it used so little of the time series. A forecasting method may have small errors relative to the history file, but may generate high errors for future time periods. For this reason, some analysts prefer to use a **holdout sample** as a final test (see the two Experiential Learning Exercises at the end of this chapter). To do so, they set aside some of the more recent periods from the time series and use only the earlier time periods to develop and test different models. Once the final models have been selected in the first phase, they are tested again with the holdout sample. Performance measures, such as MAD and CFE, would still be used but they would be applied to the holdout sample. Whether this idea is used or not, managers should monitor future forecast errors, and modify their forecasting approaches as needed. Maintaining data on forecast performance is the ultimate test of forecasting power—rather than how well a model fits past data or holdout samples.

Tracking Signals

A **tracking signal** is a measure that indicates whether a method of forecasting is accurately predicting actual changes in demand. The tracking signal measures the number of MADs represented by the cumulative sum of forecast errors, the CFE. The CFE tends to be close to 0 when a correct forecasting system is being used. At any time, however, random errors can cause the CFE to be a nonzero number. The tracking signal formula is

$$\text{Tracking signal} = \frac{\text{CFE}}{\text{MAD}} \quad \text{or} \quad \frac{\text{CFE}}{\text{MAD}_t}$$

Each period, the CFE and MAD are updated to reflect current error, and the tracking signal is compared to some predetermined limits. The MAD can be calculated in one of two ways: (1) as the simple average of all absolute errors (as demonstrated in Example 14.1) or (2) as a weighted average determined by the exponential smoothing method:

$$\text{MAD}_t = \alpha |E_t| + (1 - \alpha)\text{MAD}_{t-1}$$

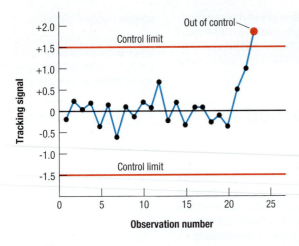

FIGURE 14.8 ▲
Tracking Signal

If forecast errors are normally distributed with a mean of 0, the relationship between σ and MAD is simple:

$$\sigma = (\sqrt{\pi/2})(\text{MAD}) \cong 1.25(\text{MAD})$$

$$\text{MAD} = 0.7978\sigma \cong 0.8\sigma$$

where

$$\pi = 3.1416$$

This relationship allows use of the normal probability tables to specify limits for the tracking signal. If the tracking signal falls outside those limits, the forecasting model no longer is tracking demand adequately. A tracking system is useful when forecasting systems are computerized because it alerts analysts when forecasts are getting far from desirable limits. Figure 14.8 shows tracking signal results for 23 periods plotted on a *control chart*. The control chart is useful for determining whether any action needs to be taken to improve the forecasting model. In the example, the first 20 points cluster around 0, as we would expect if the forecasts are not biased. The CFE will tend toward 0. When the underlying characteristics of demand change but the forecasting model does not, the tracking signal eventually goes out of control. The steady increase after the 20th point in Figure 14.8 indicates that the process is going out of control. The 21st and 22nd points are acceptable, but the 23rd point is not.

Using Multiple Techniques

We described several individual forecasting methods and showed how to assess their forecast performance. However, we need not rely on a single forecasting method. Several different forecasts can be used to arrive at a final forecast. Initial statistical forecasts using several time-series methods and regression are distributed to knowledgeable individuals, such as marketing directors and sales teams, (and sometimes even suppliers and customers) for their adjustments. They can account for current market and customer conditions that are not necessarily reflected in past data. Multiple forecasts may come from different sales teams, and some teams may have a better record on forecast errors than others.

combination forecasts

Forecasts that are produced by averaging independent forecasts based on different methods, different sources, or different data.

Research during the last two decades suggests that combining forecasts from multiple sources often produces more accurate forecasts. **Combination forecasts** are forecasts that are produced by averaging independent forecasts based on different methods, different sources, or different data. It is intriguing that combination forecasts often perform better over time than even the *best* single forecasting procedure. For example, suppose that the forecast for the next period is 100 units from technique 1 and 120 units from technique 2 and that technique 1 has provided more accurate forecasts to date. The combination forecast for next period, giving equal weight to each technique, is 110 units (or $0.5 \times 100 + 0.5 \times 120$). When this averaging technique is used consistently into the future, its combination forecasts often will be much more accurate than those of any single best forecasting technique (in this example, technique 1). Combining is most effective when the individual forecasts bring different kinds of information into the forecasting process. Forecasters have achieved excellent results by weighting forecasts equally, and this is a good starting point. However, unequal weights may provide better results under some conditions.

OM Explorer and POM for Windows allow you to evaluate several forecasting models, and then you can create combination forecasts from them. In fact, the *Time-Series Forecasting* Solver of OM Explorer automatically computes a combination forecast as a weighted average, using the weights that you supply for the various models that it evaluates. The models include the naïve, moving average, exponential smoothing, and regression projector methods. Alternately, you can create a simple Excel spreadsheet that combines forecasts generated by POM for Windows to create combination forecasts. The *Time Series Forecasting* Solver also allows you evaluate your forecasting process with a holdout sample. The forecaster makes a forecast just one period ahead, and learns of given actual demand. Next the solver computes forecasts and forecast errors for the period. The process continues to the next period in the holdout sample with the forecaster committing to a forecast for the next period. To be informed, the forecaster should also be aware of how well the other forecasting methods have been performing, particularly in the recent past.

MANAGERIAL PRACTICE 14.1 — Combination Forecasts and the Forecasting Process

Fiskars Brands, Inc., totally overhauled its forecasting process. It serves 2,000 customers ranging from large discounters to local craft stores providing about 2,300 finished SKUs. Its parent company, Fiskars Corporation, is the second oldest incorporated entity in the world and produces a variety of high-quality products such as garden shears, pruners, hand tools, scissors for preschoolers, ratchet tools, screwdrivers, and the like. Business is highly seasonal and prices quite variable. About 10 percent to 15 percent of the annual revenue comes from one-time promotions, and 25 percent to 35 percent of its products are new every year.

It introduced a statistical-based analysis along with a Web-based business intelligence tool for reporting. It put much more emphasis on combination forecasts. Instead of asking members of the sales staff to provide their own forecasts, forecasts were sent to them, and they were asked for their validation and refinement. Their inputs are most useful relative to additions, deletions, and promotions. Converting multiple forecasts into one number (forecasts from time-series techniques, sales input, and customer input) creates more accurate forecasts by SKU. Fiskars's software has the ability to weigh each input. It gives more weight to a statistical forecast for in-line items, and inputs from the sales staff get much more weight for promoted products and new items.

It also segments SKUs by value and forecastability so as to focus forecasting efforts on SKUs that have the biggest impact on the business. High-value items ("A" items identified with ABC analysis in Chapter 9, "Supply Chain Inventory Management") that also have high forecastability (stable demand with low forecast errors to date) tend to do well with the time-series techniques, and **judgmental adjustments** are made with caution. High-value items with low forecastability get top priority in the forecasting effort, such as with CPFR. Much less attention is given to improving forecasts for "C" items for which there is some history and fairly steady demand.

Finally, Fiskars instituted a Web-based program that gives the entire company visibility to forecast information in whatever form it needs. For example, Finance wants monthly, quarterly, and yearly projections in dollars, whereas Operations wants projections in units as well as accuracy measures. Everybody can track updated forecast information by customer, brand, and SKU.

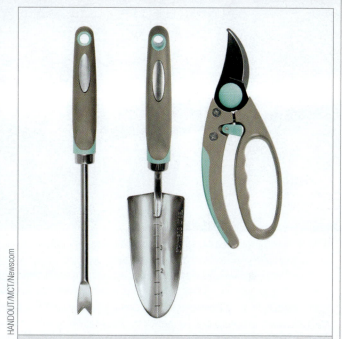

HANDOUT/MCT/Newscom

Fiskars Brands, Inc., totally overhauled its forecasting process. Its products include garden shears, pruners, hand tools, scissors, ratchet tools, and the like. It introduced time-series tools, with much emphasis placed on combination forecasts. Instead of asking members of the sales staff to provide their own forecasts, forecasts were sent to them, and they were asked for validation and refinement. Their judgmental inputs provide valuable information relative to additions, deletions, and promotions. Combining multiple forecasts (forecasts from several time-series techniques and judgment inputs) into one number creates more accurate forecasts by SKU.

Source: David Montgomery, "Flashpoints for Changing Your Forecasting Process," *The Journal of Business Forecasting*, (Winter 2006–2007), pp. 35–37; **http://www.fiskars.com**, May 21, 2011.

Another way to take advantage of multiple techniques is **focus forecasting**, which selects the best forecast (based on past error measures) from a group of forecasts generated by individual techniques. Every period, all techniques are used to make forecasts for each item. The forecasts are made with a computer because there can be 100,000 SKUs at a company, each needing to be forecast. Using the history file as the starting point for each method, the computer generates forecasts for the current period. The forecasts are compared to actual demand, and the method that produces the forecast with the least error is used to make the forecast for the next period. The method used for each item may change from period to period.

Putting It All Together: Forecasting as a Process

Often companies must prepare forecasts for hundreds or even thousands of services or products repeatedly. For example, a large network of health care facilities must calculate demand forecasts for each of its services for every department. This undertaking involves voluminous data that must be manipulated frequently. However, software such as Motorola Mobility's system can ease the burden of making these forecasts and coordinating the forecasts between customers

judgmental adjustment

An adjustment made to forecasts from one or more quantitative models that accounts for recognizing which models are performing particularly well in recent past, or take into account contextual information.

focus forecasting

A method of forecasting that selects the best forecast from a group of forecasts generated by individual techniques.

FRED GREAVES Feature Photo Service/Newscom

West Marine acquired its East Coast competitor, E&B Marine, in 1997. The consequences were quickly apparent. Peak-season out-of-stock levels rose more than 12 percent compared to the prior year. After six years of steady growth, net income dropped from $15 million in 1997 to not much more than $1 million the next year. Fast-forward six years. They had no supply problems in any of their warehouses or stores. What changed? Two words: supply chain. Managers recognized that they needed to make a significant shift in managing its supply chain. A crucial element was greater collaboration with their suppliers. It is not enough to coordinate the supply chain within the boundaries of a single organization.

and suppliers. Many forecasting software packages are available, including Manugistics, Forecast Pro, and SAS. The forecasting routines in OM Explorer and POM for Windows give some hint of their capabilities. Forecasting is not just a set of techniques, but instead a process that must be designed and managed. While there is no one process that works for everyone, here we describe two comprehensive processes that can be quite effective in managing operations and the supply chain.

A Typical Forecasting Process

Many *inputs* to the forecasting process are informational, beginning with the *history file* on past demand. The history file is kept up-to-date with the actual demands. Clarifying notes and adjustments are made to the database to explain unusual demand behavior, such as the impact of special promotions and closeouts. Often the database is separated into two parts: *base* data and *nonbase* data. The second category reflects irregular demands. Final forecasts just made at the end of the prior cycle are entered in the history file, so as to track forecast errors. Other information sources are from salesforce estimates, outstanding bids on new orders, booked orders, market research studies, competitor behavior, economic outlook, new product introductions, pricing, and promotions. If CPFR is used, as is done by Motorola Mobility in our opening vignette, then considerable information sharing will take place with customers and suppliers. For new products, a history database is fabricated based on the firm's experience with prior products and the judgment of personnel.

Outputs of the process are forecasts for multiple time periods into the future. Typically, they are on a monthly basis and are projected out from six months to two years. Most software packages have the ability to "roll up" or "aggregate" forecasts for individual stock-keeping units (SKUs) into forecasts for whole product families. Forecasts can also be "blown down" or "disaggregated" into smaller pieces. In a make-to-stock environment, forecasts tend to be more detailed and can get down to specific individual products. In a make-to-order environment, the forecasts tend to be for groups of products. Similarly, if the lead times to buy raw materials and manufacture a product or provide a service are long, the forecasts go farther out into the future.

The forecast process itself, typically done on a monthly basis, consists of structured steps. These steps often are facilitated by someone who might be called a demand manager, forecast analyst, or demand/supply planner. However, many other people are typically involved before the plan for the month is authorized.

Step 1. The cycle begins mid-month just after the forecasts have been finalized and communicated to the stakeholders. Now is the time to update the history file and review forecast accuracy. At the end of the month, enter actual demand and review forecast accuracy.

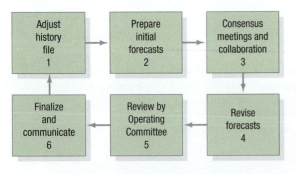

Step 2. Prepare initial forecasts using some forecasting software package and judgment. Adjust the parameters of the software to find models that fit the past demand well and yet reflect the demand manager's judgment on irregular events and information about future sales pulled from various sources and business units.

Step 3. Hold consensus meetings with the stakeholders, such as marketing, sales, supply chain planners, and finance. Make it easy for business unit and field sales personnel to make inputs. Use the Internet to get collaborative information from key customers and suppliers. The goal is to arrive at consensus forecasts from all of the important players.

Step 4. Revise the forecasts using judgment, considering the inputs from the consensus meetings and collaborative sources.

Step 5. Present the forecasts to the operating committee for review and to reach a final set of forecasts. It is important to have a set of forecasts that everybody agrees upon and will work to support.

Step 6. Finalize the forecasts based on the decisions of the operating committee and communicate them to the important stakeholders. Supply chain planners are usually the biggest users.

As with all work activity, forecasting is a process and should be continually reviewed for improvements. A better process will foster better relationships between departments such as marketing, sales, and operations. It will also produce better forecasts. This principle is the first one in Table 14.2 to guide process improvements.

Adding Collaboration to the System

This process is similar to the first one, except that it adds considerable collaboration with the company's customers and suppliers, particularly in step 3. **Collaborative planning, forecasting, and replenishment (CPFR)** is a specific nine-step process for supply chain integration that allows a supplier and its customers to collaborate on making the forecast by using the Internet. Many other firms, including Motorola as described in the opening vignette, are turning to CPFR to coordinate up and down the supply chain.

Forecasting as a Nested Process

Forecasting is not a stand-alone activity, but instead part of a larger process that encompasses the remaining chapters. After all, demand is only half of the equation—the other half is supply. Future plans must be developed to supply the resources needed to meet the forecasted demand. Resources include the workforce, materials, inventories, dollars, and equipment capacity. Making sure that demand and supply plans are in balance begins in the next chapter, Chapter 15, "Operations Planning and Scheduling" and continues with Chapter 16, "Resource Planning."

Mark Peterson/CORBIS

Walmart has long been known for its careful analysis of cash register receipts and for working with suppliers to reduce inventories. In the past, like many other retailers, Walmart did not share its forecasts with its suppliers. The result was forecast errors as much as 60 percent of actual demand. Retailers ordered more than they needed, and suppliers produced more than they could sell. To combat the ill effects of forecast errors on inventories, Benchmarking Partners, Inc., was funded in the mid-1990s by Walmart, IBM, SAP, and Manugistics to develop a software package. Walmart initiated this new approach with Listerine, a primary product of Warner-Lambert (now produced and distributed by Johnson & Johnson). The system worked in the following way during this pilot period. Walmart and Warner-Lambert independently calculated the demand they expected for Listerine six months into the future, taking into consideration factors such as past sales trends and promotion plans. They then exchanged their forecasts over the Internet. If the forecasts differed by more than a predetermined percentage, the retailer and the manufacturer used the Internet to exchange written comments and supporting data. The parties went through as many cycles as needed to converge on an acceptable forecast. They passed the pilot period in flying colors. Benefits to Walmart included a reduction in stockouts from 15 percent to 2 percent as well as significant increases in sales and reductions in inventory costs. Likewise, Warner-Lambert benefited by having a smoother production plan and lower average costs. This system was later generalized and dubbed CPFR, which stands for collaborative planning, forecasting, and replenishment.

TABLE 14.2 | **SOME PRINCIPLES FOR THE FORECASTING PROCESS**

- Better processes yield better forecasts.
- Demand forecasting is being done in virtually every company, either formally or informally. The challenge is to do it well—better than the competition.
- Better forecasts result in better customer service and lower costs, as well as better relationships with suppliers and customers.
- The forecast can and must make sense based on the big picture, economic outlook, market share, and so on.
- The best way to improve forecast accuracy is to focus on reducing forecast error.
- Bias is the worst kind of forecast error; strive for zero bias.
- Whenever possible, forecast at more aggregate levels. Forecast in detail only where necessary.
- Far more can be gained by people collaborating and communicating well than by using the most advanced forecasting technique or model.

Source: Based on Thomas F. Wallace and Robert A. Stahl, *Sales Forecasting: A New Approach* (Cincinnati, OH: T. E. Wallace & Company, 2002), p. 112. Copyright © 2002 T.E. Wallace & Company. Used with permission.

collaborative planning, forecasting, and replenishment (CPFR)

A nine-step process for supply chain integration that allows a supplier and its customers to collaborate on making the forecast by using the Internet.

LEARNING GOALS IN REVIEW

1 **Identify the various forecasting methods available to forecasting systems.** The section "Choosing the Type of Forecasting Technique," on pp. 466–467, gives a quick introduction to the four groups of forecasting techniques. Each is described fully in subsequent sections.

2 **Identify the various measures of forecast errors.** Review the "Forecast Error," pp. 467–469, and "Choosing a Quantitative Forecasting Method," pp. 481–482, to understand CFE, MSE, σ, MAD, MAPE, and the tracking signals.

3 **Use regression to make forecasts with one or more independent variables.** The "Causal Methods: Linear Regression" section and Example 14.2, pp. 470–472, describe how linear regression, when historical data is available, can express demand as a linear function of one or more independent variables. Example 14.2 on pp. 471–472 and Solved Problem 1 illustrate the computer output, including the various statistics on how well the regression equation fits the data.

4 **Make forecasts using the most common approaches for time-series analysis.** The "Time-Series Methods" on pp. 472–481

explain the naïve method, simple moving average, weighted moving average, and exponential smoothing techniques that are used. Examples 14.3 and 14.4 demonstrate some of the methods, as do Solved Problems 2 and 3.

5 **Make forecasts using Trend Projection with Regression.** We cover this technique on pp. 476–478, with four figures illustrating the computer output and how varying number of periods included in the regression analysis can impact the results.

6 **Describe a typical forecasting process used by businesses.** See the section "A Typical Forecasting Process,"pp. 484–485, and the six steps involved. There is much more complexity when you realize the number of SKUs involved and the need to update the history file.

7 **Explain collaborative planning, forecasting, and replenishment (CPFR).** The "Adding Collaboration to the System" section, p. 485, is a big step in increasing the coordination up and down the supply chain. In the chapter opener, we see what Motorola Mobility has done with its customers to improve its demand forecasts.

MyOMLab helps you develop analytical skills and assesses your progress with multiple problems on moving average, mean absolute deviation, mean absolute percent error, mean squared error (MSE), exponential smoothing, MAD, MAPE, multiplicative seasonal method, least squares regression model, and trend projection with regression.

MyOMLab Resources	Titles	Link to the Book
Video	*Forecasting and Supply Chain Management at Deckers Outdoor Corporation*	Using Multiple Techniques; Putting It All Together: Forecasting as a Process
Active Model Exercises	14.1 Linear Regression	Casual Methods: Linear Regression; Example 14.2 (pp. 471–472); Solved Problem 1
	14.2 Simple Moving Averages	Estimating the Average; Example 14.3 (pp. 473–474); Solved Problem 2
	14.3 Exponential Smoothing	Exponential Smoothing; Example 14.4 (p. 475); Solved Problem 3
OM Explorer Solvers	Regression Analysis	Casual Methods: Linear Regression; Example 14.2 (pp. 471–472); Solved Problem 1
	Seasonal Forecasting	Seasonal Patterns; Example 14.6 (pp. 479–480); Solved Problem 4
	Time Series Forecasting	Time Series Methods Examples 14.3 – 14.5
	Trend Projection with Regression	Trend Projection with Regression; Example 14.5 (pp. 476–478)
OM Explorer Tutors	14.1 Moving Average Method	Estimating the Average; Example 14.3 (pp. 473–474); Solved Problem 2
	14.2 Weighted Moving Average Method	
	14.3 Exponential Smoothing	Weighted Moving Average and Solved Problem 2
		Exponential Smoothing; Example 14.4 (p. 475); Solved Problem 3
Virtual Tours	Ferrara Pan	Key Decisions on Making Forecasts
	Cape Cod Chips	Demand Patterns; Judgment Methods
POM for Windows	Time Series Analysis	Time Series Methods; Examples 14.3 – 14.5; Seasonal Patterns; Example 14.6 (pp. 479–480); Solved Problem 4
	Regression Projector	Casual Methods: Linear Regression; Example 14.2 (pp. 471–472); Solved Problem 1
	Least Squares – Simple and Multiple Regression	Casual Methods: Linear Regression; Example 14.2 (pp. 471–472); Solved Problem 1
	Error Analysis	Judgment Methods; Forecast Error, Example 14.1 (p. 469), Choosing a Quantitative Forecasting Method; Solved Problem 3

MyOMLab Resources	Titles	Link to the Book
Student Data File	Experiential Exercise Two	Time-Series Methods; Choosing a Time-Series Method; Using Multiple Techniques
Tutorial	Trend-Adjusted Exponential Smoothing	Trend Projection with Regression (pp. 476–478)
Internet Exercise	National Climate Data Center	Casual Methods: Linear Regression; Example 14.2 (pp. 471–472); Trend Projection with Regression; Example 14.5 (pp. 476–478)
Key Equations		
Image Library		

Key Equations

1. Forecast error:

$$E_t = D_t - F_t$$

$$\text{CFE} = \sum E_t$$

$$\bar{E} = \frac{\text{CFE}}{n}$$

$$\text{MSE} = \frac{\sum E_t^2}{n}$$

$$\sigma = \sqrt{\frac{\sum (E_t - \bar{E})^2}{n - 1}}$$

$$\text{MAD} = \frac{\sum |E_t|}{n}$$

$$\text{MAPE} = \frac{(\sum |E_t|/D_t)(100\%)}{n}$$

2. Linear regression:

$$Y = a + bX$$

3. Naïve forecasting:

$$\text{Forecast} = D_t$$

4. Simple moving average:

$$F_{t+1} = \frac{D_t + D_{t-1} + D_{t-2} + \cdots + D_{t-n+1}}{n}$$

5. Weighted moving average:

$$F_{t+1} = \text{Weight}_1(D_t) + \text{Weight}_2(D_{t-1}) + \text{Weight}_3(D_{t-2}) + \cdots + \text{Weight}_n(D_{t-n+1})$$

6. Exponential smoothing:

$$F_{t+1} = \alpha D_t + (1 - \alpha)F_t$$

7. Trend Projection using Regression

$$F_t = a + bt$$

8. Tracking signal:

$$\frac{\text{CFE}}{\text{MAD}} \quad \text{or} \quad \frac{\text{CFE}}{\text{MAD}_t}$$

9. Exponentially smoothed error:

$$\text{MAD}_t = \alpha |E_t| + (1 - \alpha)\text{MAD}_{t-1}$$

Key Terms

additive seasonal method 481

aggregation 466

causal methods 467

collaborative planning, forecasting, and replenishment (CPFR) 485

combination forecasts 482

cumulative sum of forecast errors (CFE) 467

dependent variable 470

Delphi method 470

executive opinion 470

exponential smoothing method 474

focus forecasting 483

forecast 464

forecast error 467

holdout sample 481

independent variables 471

judgment methods 466

judgmental adjustment 483

linear regression 470

market research 470

mean absolute deviation (MAD) 468

mean absolute percent error (MAPE) 468

mean squared error (MSE) 468

multiplicative seasonal method 479

naïve forecast 473

salesforce estimates 470

simple moving average method 473

standard deviation (σ) 468

technological forecasting 470

time series 465

time-series analysis 467

tracking signal 481

trend projection with regression 467

weighted moving average method 474

Solved Problem 1

Chicken Palace periodically offers carryout five-piece chicken dinners at special prices. Let Y be the number of dinners sold and X be the price. Based on the historical observations and calculations in the following table, determine the regression equation, correlation coefficient, and coefficient of determination. How many dinners can Chicken Palace expect to sell at $3.00 each?

Observation	Price (X)	Dinners Sold (Y)
1	$ 2.70	760
2	$ 3.50	510
3	$ 2.00	980
4	$ 4.20	250
5	$ 3.10	320
6	$ 4.05	480
Total	$19.55	3,300
Average	$ 3.258	550

SOLUTION

We use the computer (*Regression Analysis* Solver of OM Explorer or *Regression Projector* module of POM for Windows) to calculate the best values of a, b, the correlation coefficient, and the coefficient of determination.

$$a = 1,454.60$$
$$b = -277.63$$
$$r = -0.84$$
$$r^2 = 0.71$$

The regression line is

$$Y = a + bX = 1,454.60 - 277.63X$$

The correlation coefficient ($r = -0.84$) shows a negative correlation between the variables. The coefficient of determination ($r^2 = 0.71$) is not too large, which suggests that other variables (in addition to price) might appreciably affect sales.

If the regression equation is satisfactory to the manager, estimated sales at a price of $3.00 per dinner may be calculated as follows:

$$Y = a + bX = 1,454.60 - 277.63(3.00)$$
$$= 621.71 \text{ or } 622 \text{ dinners}$$

Solved Problem 2

The Polish General's Pizza Parlor is a small restaurant catering to patrons with a taste for European pizza. One of its specialties is Polish Prize pizza. The manager must forecast weekly demand for these special pizzas so that he can order pizza shells weekly. Recently, demand has been as follows:

Week	Pizzas	Week	Pizzas
June 2	50	June 23	56
June 9	65	June 30	55
June 16	52	July 7	60

a. Forecast the demand for pizza for June 23 to July 14 by using the simple moving average method with $n = 3$. Then, repeat the forecast by using the weighted moving average method with $n = 3$ and weights of 0.50, 0.30, and 0.20, with 0.50 applying to the most recent demand.

b. Calculate the MAD for each method.

SOLUTION

a. The simple moving average method and the weighted moving average method give the following results:

Current Week	Simple Moving Average Forecast for Next Week	Weighted Moving Average Forecast for Next Week
June 16	$\dfrac{52 + 65 + 50}{3} = 55.7$ or 56	$[(0.5 \times 52) + (0.3 \times 65) + (0.2 \times 50)] = 55.5$ or 56
June 23	$\dfrac{56 + 52 + 65}{3} = 55.7$ or 58	$[(0.5 \times 56) + (0.3 \times 52) + (0.2 \times 65)] = 56.6$ or 57
June 30	$\dfrac{55 + 56 + 52}{3} = 54.3$ or 54	$[(0.5 \times 55) + (0.3 \times 56) + (0.2 \times 52)] = 54.7$ or 55
July 7	$\dfrac{60 + 55 + 56}{3} = 57.0$ or 57	$[(0.5 \times 60) + (0.3 \times 55) + (0.2 \times 56)] = 57.7$ or 58

Forecasts in each row are for the next week's demand. For example, the simple moving average and weighted moving average forecasts (both are 56 units) calculated after learning the demand on June 16 apply to June 23's demand forecast.

b. The mean absolute deviation is calculated as follows:

Week	Actual Demand	SIMPLE MOVING AVERAGE Forecast for This Week	Absolute Errors $\lvert E_t \rvert$	WEIGHTED MOVING AVERAGE Forecast for This Week	Absolute Errors $\lvert E_t \rvert$
June 23	56	56	$\lvert 56 - 56 \rvert = 0$	56	$\lvert 56 - 56 \rvert = 0$
June 30	55	58	$\lvert 55 - 58 \rvert = 3$	57	$\lvert 55 - 57 \rvert = 2$
July 7	60	54	$\lvert 60 - 54 \rvert = 6$	55	$\lvert 60 - 55 \rvert = 5$
			$\text{MAD} = \dfrac{0 + 3 + 6}{3} = 3.0$		$\text{MAD} = \dfrac{0 + 2 + 5}{3} = 2.3$

For this limited set of data, the weighted moving average method resulted in a slightly lower mean absolute deviation. However, final conclusions can be made only after analyzing much more data.

Solved Problem 3

The monthly demand for units manufactured by the Acme Rocket Company has been as follows:

Month	Units	Month	Units
May	100	September	105
June	80	October	110
July	110	November	125
August	115	December	120

a. Use the exponential smoothing method to forecast the number of units for June to January. The initial forecast for May was 105 units; $\alpha = 0.2$.

b. Calculate the absolute percentage error for each month from June through December and the MAD and MAPE of forecast error as of the end of December.

c. Calculate the tracking signal as of the end of December. What can you say about the performance of your forecasting method?

SOLUTION

a.

Current Month, t	Calculating Forecast for Next Month $F_{t+1} = \alpha D_t + (1 - \alpha)F_t$	Forecast for Month $t + 1$
May	$0.2(100) + 0.8(105) = 104.0$ or 104	June
June	$0.2(80) + 0.8(104.0) = 99.2$ or 99	July
July	$0.2(110) + 0.8(99.2) = 101.4$ or 101	August
August	$0.2(115) + 0.8(101.4) = 104.1$ or 104	September
September	$0.2(105) + 0.8(104.1) = 104.3$ or 104	October
October	$0.2(110) + 0.8(104.3) = 105.4$ or 105	November
November	$0.2(125) + 0.8(105.4) = 109.3$ or 109	December
December	$0.2(120) + 0.8(109.3) = 111.4$ or 111	January

b.

Month, t	Actual Demand, D_t	Forecast, F_t	Error, $E_t = D_t - F_t$	Absolute Error, $\lvert E_t \rvert$	Absolute Percentage Error, $(\lvert E_t \rvert / D_t)(100\%)$
June	80	104	−24	24	30.0%
July	110	99	11	11	10.0
August	115	101	14	14	12.0
September	105	104	1	1	1.0
October	110	104	6	6	5.5
November	125	105	20	0	16.0
December	120	109	11	11	9.2
Total	765		39	87	83.7%

$$\text{MAD} = \frac{\Sigma \lvert E_t \rvert}{n} = \frac{87}{7} = 12.4 \text{ and MAPE} = \frac{(\Sigma \lvert E_t \rvert / D_t)(100)}{n} = \frac{83.7\%}{7} = 11.96\%$$

c. As of the end of December, the cumulative sum of forecast errors (CFE) is 39. Using the mean absolute deviation calculated in part (b), we calculate the tracking signal:

$$\text{Tracking signal} = \frac{\text{CFE}}{\text{MAD}} = \frac{39}{12.4} = 3.14$$

The probability that a tracking signal value of 3.14 could be generated completely by chance is small. Consequently, we should revise our approach. The long string of forecasts lower than actual demand suggests use of a trend method.

Solved Problem 4

The Northville Post Office experiences a seasonal pattern of daily mail volume every week. The following data for two representative weeks are expressed in thousands of pieces of mail:

Day	Week 1	Week 2
Sunday	5	8
Monday	20	15
Tuesday	30	32
Wednesday	35	30
Thursday	49	45
Friday	70	70
Saturday	15	10
Total	224	210

a. Calculate a seasonal factor for each day of the week.

b. If the postmaster estimates 230,000 pieces of mail to be sorted next week, forecast the volume for each day of the week.

SOLUTION

a. Calculate the average daily mail volume for each week. Then, for each day of the week, divide the mail volume by the week's average to get the seasonal factor. Finally, for each day, add the two seasonal factors and divide by 2 to obtain the average seasonal factor to use in the forecast (see part [b]).

Day	WEEK 1		WEEK 2		Average Seasonal Factor [(1) + (2)]/2
	Mail Volume	Seasonal Factor (1)	Mail Volume	Seasonal Factor (2)	
Sunday	5	5/32 = 0.15625	8	8/30 = 0.26667	0.21146
Monday	20	20/32 = 0.62500	15	15/30 = 0.50000	0.56250
Tuesday	30	30/32 = 0.93750	32	32/30 = 1.06667	1.00209
Wednesday	35	35/32 = 1.09375	30	30/30 = 1.00000	1.04688
Thursday	49	49/32 = 1.53125	45	45/30 = 1.50000	1.51563
Friday	70	70/32 = 2.18750	70	70/30 = 2.33333	2.26042
Saturday	15	15/32 = 0.46875	10	10/30 = 0.33333	0.40104
Total	224		210		
Average	224/7 = 32		210/7 = 30		

b. The average daily mail volume is expected to be 230,000/7 = 32,857 pieces of mail. Using the average seasonal factors calculated in part (a), we obtain the following forecasts:

Day	Calculation		Forecast
Sunday	0.21146(32,857) =		6,948
Monday	0.56250(32,857) =		18,482
Tuesday	1.00209(32,857) =		32,926
Wednesday	1.04688(32,857) =		34,397
Thursday	1.51563(32,857) =		49,799
Friday	2.26042(32,857) =		74,271
Saturday	0.40104(32,857) =		13,177
		Total	230,000

Discussion Questions

1. Figure 14.9 shows summer air visibility measurements for Denver, Colorado. The acceptable visibility standard is 100, with readings above 100 indicating clean air and good visibility, and readings below 100 indicating temperature inversions caused by forest fires, volcanic eruptions, or collisions with comets.

   a. Is a trend evident in the data? Which time-series techniques might be appropriate for estimating the average of these data?

   b. A medical center for asthma and respiratory diseases located in Denver has great demand for its services when air quality is poor. If you were in charge of developing a short-term (say, 3-day) forecast of visibility, which causal factor(s) would you analyze? In other words, which external factors hold the potential to significantly affect visibility in the *short term*?

   c. Tourism, an important factor in Denver's economy, is affected by the city's image. Air quality, as measured by visibility, affects the city's image. If you were responsible for development of tourism, which causal factor(s) would you analyze to forecast visibility for the *medium term* (say, the next two summers)?

   d. The federal government threatens to withhold several hundred million dollars in Department of Transportation funds unless Denver meets visibility standards within 8 years. How would you proceed to generate a *long-term* judgment forecast of technologies that will be available to improve visibility in the next 10 years?

2. Kay and Michael Passe publish *What's Happening?*—a biweekly newspaper to publicize local events. *What's Happening?* has few subscribers; it typically is sold at checkout stands. Much of the revenue comes from advertisers of garage sales and supermarket specials. In an effort to reduce costs associated with printing too many papers or delivering them to the wrong location, Michael implemented a computerized system to collect sales data. Sales-counter scanners accurately record sales data for each location. Since the system was implemented, total sales volume has steadily declined. Selling advertising space and maintaining shelf space at supermarkets are getting more difficult.

Reduced revenue makes controlling costs all the more important. For each issue, Michael carefully makes a forecast based on sales data collected at each location. Then, he orders papers to be printed and distributed in quantities matching the forecast. Michael's forecast reflects a downward trend, which *is* present in the sales data. Now only a few papers are left over at only a few locations. Although the sales forecast accurately predicts the actual sales at most locations, *What's Happening?* is spiraling toward oblivion. Kay suspects that Michael is doing something wrong in preparing the forecast but can find no mathematical errors. Tell her what is happening.

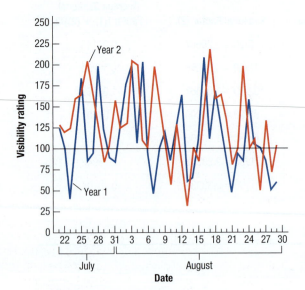

▲ **FIGURE 14.9**
Summer Air Visibility Measurements

Problems

The OM Explorer and POM for Windows software is available to all students using the 10th edition of this textbook. Go to **www.pearsonhighered.com/krajewski** to download these computer packages. If you purchased MyOMLab, you also have access to Active Models software and significant help in doing the following problems. Check with your instructor on how best to use these resources. In many cases, the instructor wants you to understand how to do the calculations by hand. At the least, the software provides a check on your calculations. When calculations are particularly complex and the goal is interpreting the results in making decisions, the software entirely replaces the manual calculations.

1. The owner of a computer store rents printers to some of her preferred customers. She is interested in arriving at a forecast of rentals so that she can order the correct quantities of supplies that go with the printers. Data for the last 10 weeks are shown here.

Week	Rentals	Week	Rentals
1	23	6	28
2	24	7	32
3	32	8	35
4	26	9	26
5	31	10	24

a. Prepare a forecast for weeks 6 through 10 by using a 5-week moving average. What is the forecast for week 11?

b. Calculate the mean absolute deviation as of the end of week 10.

2. Sales for the past 12 months at Dalworth Company are given here.

Month	Sales ($ millions)	Month	Sales ($ millions)
January	20	July	53
February	24	August	62
March	27	September	54
April	31	October	36
May	37	November	32
June	47	December	29

a. Use a three-month moving average to forecast the sales for the months May through December.

b. Use a four-month moving average to forecast the sales for the months May through December.

c. Compare the performance of the two methods by using the mean absolute deviation as the performance criterion. Which method would you recommend?

d. Compare the performance of the two methods by using the mean absolute percent error as the performance criterion. Which method would you recommend?

e. Compare the performance of the two methods by using the mean squared error as the performance criterion. Which method would you recommend?

3. Karl's Copiers sells and repairs photocopy machines. The manager needs weekly forecasts of service calls so that he can schedule service personnel. Use the actual demand in the first period for the forecast for the first week so error measurement begins in the second week. The manager uses exponential smoothing with $\alpha = 0.20$. Forecast the number of calls for week 6, which is next week.

Week	Actual Service Calls
1	24
2	32
3	36
4	23
5	25

4. Consider the sales data for Dalworth Company given in Problem 2.

a. Use a 3-month weighted moving average to forecast the sales for the months April through December. Use weights of (3/6), (2/6), and (1/6), giving more weight to more recent data.

b. Use exponential smoothing with $\alpha = 0.6$ to forecast the sales for the months April through December. Assume that the initial forecast for January was $22 million. Start error measurement in April.

c. Compare the performance of the two methods by using the mean absolute deviation as the performance criterion, with error measurement beginning in April. Which method would you recommend?

d. Compare the performance of the two methods by using the mean absolute percent error as the performance criterion, with error measurement beginning in April. Which method would you recommend?

e. Compare the performance of the two methods by using the mean squared error as the performance criterion, with error measurement beginning in April. Which method would you recommend?

5. A convenience store recently started to carry a new brand of soft drink. Management is interested in estimating future sales volume to determine whether it should continue to carry the new brand or replace it with another brand. The table at the top of the next page provides the number of cans sold per week. Use both the trend projection with regression and the exponential smoothing (let $\alpha = 0.4$ with an initial forecast for week 1 of 617) methods to forecast demand for week 13. Compare these methods by using the mean absolute deviation and mean absolute percent error performance criteria. Does your analysis suggest that sales are trending and if so, by how much?

Week	1	2	3	4	5	6	7	8	9	10	11	12
Sales	617	617	648	739	659	623	742	704	724	715	668	740

6. Community Federal Bank in Dothan, Alabama, recently increased its fees to customers who use employees as tellers. Management is interested in whether its new fee policy has increased the number of customers now using its automatic teller machines to that point that more machines are required. The following table provides the number of automatic teller transactions by week. Use trend projection with regression to forecast usage for weeks 13–16.

Week	1	2	3	4	5	6	7	8	9	10	11	12
Transactions	716	721	833	639	689	736	779	711	723	835	829	667

7. The number of heart surgeries performed at Heartville General Hospital has increased steadily over the past several years. The hospital's administration is seeking the best method to forecast the demand for such surgeries in year 6. The data for the past 5 years are shown.

Year	Demand
1	45
2	50
3	52
4	56
5	58

The hospital's administration is considering the following forecasting methods. Begin error measurement in year 3, so all methods are compared for the same years.

i. Exponential smoothing, with $\alpha = 0.6$. Let the initial forecast for year 1 be 45, the same as the actual demand.

ii. Exponential smoothing, with $\alpha = 0.9$. Let the initial forecast for year 1 be 45, the same as the actual demand.

iii. Trend projection with regression.

iv. Two-year moving average.

v. Two-year weighted moving average, using weights 0.6 and 0.4, with more recent data given more weight.

vi. If MAD is the performance criterion chosen by the administration, which forecasting method should it choose?

vii. If MSE is the performance criterion chosen by the administration, which forecasting method should it choose?

viii. If MAPE is the performance criterion chosen by the administration, which forecasting method should it choose?

8. The following data are for calculator sales in units at an electronics store over the past 9 weeks:

Week	Sales	Week	Sales
1	46	6	58
2	49	7	62
3	43	8	56
4	50	9	63
5	53		

Use trend projection with regression to forecast sales for weeks 10–13. What are the error measures (CFE, MSE, σ, MAD, and MAPE) for this forecasting procedure? How about r^2?

9. The demand for Krispee Crunchies, a favorite breakfast cereal of people born in the 1940s, is experiencing a decline. The company wants to monitor demand for this product closely as it nears the end of its life cycle. The following table shows the actual sales history for January–October. Generate forecasts for November–December, using the trend projection by regression method. Looking at the accuracy of its forecasts over the history file, as well as the other statistics provided, how confident are you in these forecasts for November–December?

Month	Sales	Month	Sales
January	890,000	July	710,000
February	800,000	August	730,000
March	825,000	September	680,000
April	840,000	October	670,000
May	730,000	November	
June	780,000	December	

10. Forrest and Dan make boxes of chocolates for which the demand is uncertain. Forrest says, "That's life." On the other hand, Dan believes that some demand patterns exist that could be useful for planning the purchase of sugar, chocolate, and shrimp. Forrest insists on placing a surprise chocolate-covered shrimp in some boxes so that "You never know what you'll get." Quarterly demand (in boxes of chocolates) for the last 3 years follows:

Quarter	Year 1	Year 2	Year 3
1	3,000	3,300	3,502
2	1,700	2,100	2,448
3	900	1,500	1,768
4	4,400	5,100	5,882
Total	10,000	12,000	13,600

a. Use intuition and judgment to estimate quarterly demand for the fourth year.

b. If the expected sales for chocolates are 14,800 cases for year 4, use the multiplicative seasonal method to prepare a forecast for each quarter of the year. Are any of the quarterly forecasts different from what you thought you would get in part (a)?

11. The manager of Snyder's Garden Center must make the annual purchasing plans for rakes, gloves, and other gardening items. One of the items the company stocks is Fast-Grow, a liquid fertilizer. The sales of this item are seasonal, with peaks in the spring, summer, and fall months. Quarterly demand (in cases) for the past 2 years follows:

Quarter	Year 1	Year 2
1	40	60
2	350	440
3	290	320
4	210	280
Total	890	1,100

If the expected sales for Fast-Grow are 1,150 cases for year 3, use the multiplicative seasonal method to prepare a forecast for each quarter of the year.

12. The manager of a utility company in the Texas panhandle wants to develop quarterly forecasts of power loads for the next year. The power loads are seasonal, and the data on the quarterly loads in megawatts (MW) for the last 4 years are as follows:

Quarter	Year 1	Year 2	Year 3	Year 4
1	103.5	94.7	118.6	109.3
2	126.1	116.0	141.2	131.6
3	144.5	137.1	159.0	149.5
4	166.1	152.5	178.2	169.0

The manager estimates the total demand for the next year at 600 MW. Use the multiplicative seasonal method to develop the forecast for each quarter.

13. Demand for oil changes at Garcia's Garage has been as follows:

Month	Number of Oil Changes
January	41
February	46
March	57
April	52
May	59
June	51
July	60
August	62

a. Use simple linear regression analysis to develop a forecasting model for monthly demand. In this application, the dependent variable, Y, is monthly demand and the independent variable, X, is the month. For January, let $X = 1$; for February, let $X = 2$; and so on.

b. Use the model to forecast demand for September, October, and November. Here, $X = 9$, 10, and 11, respectively.

14. At a hydrocarbon processing factory, process control involves periodic analysis of samples for a certain process quality parameter. The analytic procedure currently used is costly and time consuming. A faster and more economical alternative procedure has been proposed. However, the numbers for the quality parameter given by the alternative procedure are somewhat different from those given by the current procedure, not because of any inherent errors but because of changes in the nature of the chemical analysis.

Management believes that if the numbers from the new procedure can be used to forecast reliably the corresponding numbers from the current procedure, switching to the new procedure would be reasonable and cost effective. The following data were obtained for the quality parameter by analyzing samples using both procedures:

Current (Y)	Proposed (X)	Current (Y)	Proposed (X)
3.0	3.1	3.1	3.1
3.1	3.9	2.7	2.9
3.0	3.4	3.3	3.6
3.6	4.0	3.2	4.1
3.8	3.6	2.1	2.6
2.7	3.6	3.0	3.1
2.7	3.6	2.6	2.8

a. Use linear regression to find a relation to forecast Y, which is the quality parameter from the current procedure, using the values from the proposed procedure, X.

b. Is there a strong relationship between Y and X? Explain.

15. Ohio Swiss Milk Products manufactures and distributes ice cream in Ohio, Kentucky, and West Virginia. The company wants to expand operations by locating another plant in northern Ohio. The size of the new plant will be a function of the expected demand for ice cream within the area served by the plant. A market survey is currently under way to determine that demand.

Ohio Swiss wants to estimate the relationship between the manufacturing cost per gallon and the number of gallons sold in a year to determine the demand for ice cream and, thus, the size of the new plant. The following data have been collected:

a. Develop a regression equation to forecast the cost per gallon as a function of the number of gallons produced.

Plant	Cost per Thousand Gallons (Y)	Thousands of Gallons Sold (X)
1	$ 1,015	416.9
2	973	472.5
3	1,046	250.0
4	1,006	372.1
5	1,058	238.1
6	1,068	258.6
7	967	597.0
8	997	414.0
9	1,044	263.2
10	1,008	372.0
Total	$10,182	3,654.4

b. What are the correlation coefficient and the coefficient of determination? Comment on your regression equation in light of these measures.

c. Suppose that the market survey indicates a demand of 325,000 gallons in the Bucyrus, Ohio, area. Estimate the manufacturing cost per gallon for a plant producing 325,000 gallons per year.

Advanced Problems

16. Franklin Tooling, Inc., manufactures specialty tooling for firms in the paper-making industry. All of their products are engineer-to-order and so the company never knows exactly what components to purchase for a tool until a customer places an order. However, the company believes that weekly demand for a few components is fairly stable. Component 135.AG is one such item. The last 26 weeks of historical use of component 135.AG is recorded below.

Week	Demand	Week	Demand
1	137	14	131
2	136	15	132
3	143	16	124
4	136	17	121
5	141	18	127
6	128	19	118
7	149	20	120
8	136	21	115
9	134	22	106
10	142	23	120
11	125	24	113
12	134	25	121
13	118	26	119

Use OM Explorer's *Time Series Forecasting* Solver to evaluate the following forecasting methods. Start error measurement in the fifth week, so all methods are evaluated over the same time interval. Use the default settings for initial forecasts.

i. Naïve (1-Period Moving Average)

ii. 3-Period Moving Average

iii. Exponential Smoothing, with $\alpha = .28$

iv. Trend Projection with Regression

v. Which forecasting method should management use, if the performance criterion it chooses is:

- CFE?
- MSE?
- MAD?
- MAPE?

17. Create an Excel spreadsheet on your own that can create combination forecasts for Problem 16. Create a combination forecast using all four techniques from problem 16. Give each technique an equal weight. Create a second combination forecast by using the three techniques that seem best based on MAD. Give equal weight to each technique. Finally, create a third forecast by equally weighting the two best techniques. Calculate CFE, MAD, MSE, and MAPE for the combination forecast. Are these forecasts better or worse than the forecasting techniques identified in Problem 16?

18. The director of a large public library must schedule employees to reshelf books and periodicals checked out of the library. The number of items checked out will determine the labor requirements. The following data reflect

the number of items checked out of the library for the past 3 years:

Month	Year 1	Year 2	Year 3
January	1,847	2,045	1,986
February	2,669	2,321	2,564
March	2,467	2,419	2,635
April	2,432	2,088	2,150
May	2,464	2,667	2,201
June	2,378	2,122	2,663
July	2,217	2,206	2,055
August	2,445	1,869	1,678
September	1,894	2,441	1,845
October	1,922	2,291	2,065
November	2,431	2,364	2,147
December	2,274	2,189	2,451

The director needs a time-series method for forecasting the number of items to be checked out during the next month. Find the best simple moving average forecast you can. Decide what is meant by "best" and justify your decision.

19. Using the data in Problem 18, find the best exponential smoothing solution you can. Justify your choice.

20. Using the data in Problem 18, find the best trend projection with regression solution you can. Compare the performance of this method with those of the best moving average method (from Problem 18) and the exponential smoothing method (from Problem 19). Which of these three methods would you choose?

21. Cannister, Inc., specializes in the manufacture of plastic containers. The data on the monthly sales of 10-ounce shampoo bottles for the past 5 years are as follows:

Year	1	2	3	4	5
January	742	741	896	951	1,030
February	697	700	793	861	1,032
March	776	774	885	938	1,126
April	898	932	1,055	1,109	1,285
May	1,030	1,099	1,204	1,274	1,468
June	1,107	1,223	1,326	1,422	1,637
July	1,165	1,290	1,303	1,486	1,611
August	1,216	1,349	1,436	1,555	1,608
September	1,208	1,341	1,473	1,604	1,528
October	1,131	1,296	1,453	1,600	1,420
November	971	1,066	1,170	1,403	1,119
December	783	901	1,023	1,209	1,013

a. Using the multiplicative seasonal method, calculate the monthly seasonal indices.

b. Develop a simple linear regression equation to forecast annual sales. For this regression, the dependent variable, Y, is the demand in each year and the independent variable, X, is the index for the year (i.e., $X = 1$ for year 1, $X = 2$ for year 2, and so on until $X = 5$ for year 5).

c. Forecast the annual sales for year 6 by using the regression model you developed in part (b).

d. Prepare the seasonal forecast for each month by using the monthly seasonal indices calculated in part (a).

22. The Midwest Computer Company serves a large number of businesses in the Great Lakes region. The company sells supplies and replacements and performs service on all computers sold through seven sales offices. Many items are stocked, so close inventory control is necessary to assure customers of efficient service. Recently, business has been increasing, and management is concerned about stockouts. A forecasting method is needed to estimate requirements several months in advance so that adequate replenishment quantities can be purchased. An example of the sales growth experienced during the last 50 months is the growth in demand for item EP-37, a laser printer cartridge, shown in Table 14.3.

a. Develop a trend projection with regression solution using OM Explorer. Forecast demand for month 51.

b. A consultant to Midwest's management suggested that new office building leases would be a good leading indicator for company sales. The consultant quoted a recent university study finding that new office building leases precede office equipment and supply sales by 3 months. According to the study findings, leases in month 1 would affect sales in month 4, leases in month 2 would affect sales in month 5, and so on. Use POM for Windows' linear regression module to develop a forecasting model for sales, with leases as the independent variable. Forecast sales for month 51.

c. Which of the two models provides better forecasts? Explain.

23. A certain food item at P&Q Supermarkets has the demand pattern shown in the table at the bottom of the next page. There are 5 periods per cycle. Find the "best" forecast you can for month 25 and justify your methodology. If you wish to explore the Seasonal Forecasting method as one of the techniques tested, you will find that OM Explorer's *Seasonal Forecasting* Solver does not cover the case where there are 5 periods in a cycle (or seasons in a year). You must do some manual calculations or write an Excel spreadsheet on your own.

TABLE 14.3 | EP-37 SALES AND LEASE DATA

Month	EP-37 Sales	Leases	Month	EP-37 Sales	Leases
1	80	32	26	1,296	281
2	132	29	27	1,199	298
3	143	32	28	1,267	314
4	180	54	29	1,300	323
5	200	53	30	1,370	309
6	168	89	31	1,489	343
7	212	74	32	1,499	357
8	254	93	33	1,669	353
9	397	120	34	1,716	360
10	385	113	35	1,603	370
11	472	147	36	1,812	386
12	397	126	37	1,817	389
13	476	138	38	1,798	399
14	699	145	39	1,873	409
15	545	160	40	1,923	410
16	837	196	41	2,028	413
17	743	180	42	2,049	439
18	722	197	43	2,084	454
19	735	203	44	2,083	441
20	838	223	45	2,121	470
21	1,057	247	46	2,072	469
22	930	242	47	2,262	490
23	1,085	234	48	2,371	496
24	1,090	254	49	2,309	509
25	1,218	271	50	2,422	522

Period	Demand	Period	Demand
1	33	13	37
2	37	14	43
3	31	15	56
4	39	16	41
5	54	17	36
6	38	18	39
7	42	19	41
8	40	20	58
9	41	21	42
10	54	22	45
11	43	23	41
12	39	24	38

24. The data for the visibility chart in Discussion Question 1 are shown in Table 14.4. The visibility standard is set at 100. Readings below 100 indicate that air pollution has reduced visibility, and readings above 100 indicate that the air is clearer.

   a. Use several methods to generate a visibility forecast for August 31 of the second year. Which method seems to produce the best forecast?

   b. Use several methods to forecast the visibility index for the summer of the third year. Which method seems to produce the best forecast? Support your choice.

25. Tom Glass forecasts electrical demand for the Flatlands Public Power District (FPPD). The FPPD wants to take its Comstock power plant out of service for maintenance when demand is expected to be low. After shutdown, performing maintenance and getting the plant back on line takes two weeks. The utility has enough other generating capacity to satisfy 1,550 megawatts (MW) of demand while Comstock is out of service. Table 14.5 at the end of

TABLE 14.4 | VISIBILITY DATA

Date	Year 1	Year 2	Date	Year 1	Year 2	Date	Year 1	Year 2
July 22	125	130	Aug 5	105	200	Aug 19	170	160
23	100	120	6	205	110	20	125	165
24	40	125	7	90	100	21	85	135
25	100	160	8	45	200	22	45	80
26	185	165	9	100	160	23	95	100
27	85	205	10	120	100	24	85	200
28	95	165	11	85	55	25	160	100
29	200	125	12	125	130	26	105	110
30	125	85	13	165	75	27	100	50
31	90	105	14	60	30	28	95	135
Aug 1	85	160	15	65	100	29	50	70
2	135	125	16	110	85	30	60	105
3	175	130	17	210	150			
4	200	205	18	110	220			

the Advanced Problems shows weekly peak demands (in MW) for the past several autumns. When next fall should the Comstock plant be scheduled for maintenance?

26. A manufacturing firm seeks to develop a better forecast for an important product, and believes that there is a trend to the data. OM Explorer's *Trend Projection with Regression* Solver has been set up with the 47 demands in the history file. Note the "Load Problem 26 Data" button in the *Trend Projection with Regression* Solver that when clicked will automatically input the demand data. Otherwise, you can enter the demand data directly into the Inputs sheet.

Yr	1	2	3	4
Jan	4507	4589	4084	4535
Feb	4400	4688	4158	4477
Mar	4099	4566	4174	4601
Apr	4064	4485	4225	4648
May	4002	4385	4324	4860
Jun	3963	4377	4220	4998
Jul	4037	4309	4267	5003
Aug	4162	4276	4187	4960
Sep	4312	4280	4239	4943
Oct	4395	4144	4352	5052
Nov	4540	4219	4331	5107
Dec	4471	4052	4371	

a. What is your forecast for December of Year 4, making period 1 as the starting period for the regression?

b. The actual demand for period 48 was just learned to be 5,100. Add this demand to the Inputs file and change the starting period for the regression to period 2 so that the number of periods in the regression remains unchanged. How much or little does the forecast for period 49 change from the one for period 48? The error measures? Are you surprised?

c. Now change the time when the regression starts to period 25 and repeat the process. What differences do you note now? What forecast will you make for period 49?

27. A manufacturing firm has developed a skills test, the scores from which can be used to predict workers' production rating factors. Data on the test scores of various workers and their subsequent production ratings are shown.

Worker	Test Score	Production Rating	Worker	Test Score	Production Rating
A	53	45	K	54	59
B	36	43	L	73	77
C	88	89	M	65	56
D	84	79	N	29	28
E	86	84	O	52	51
F	64	66	P	22	27
G	45	49	Q	76	76
H	48	48	R	32	34
I	39	43	S	51	60
J	67	76	T	37	32

a. Using POM for Windows' least squares-linear regression module, develop a relationship to forecast production ratings from test scores.

b. If a worker's test score was 80, what would be your forecast of the worker's production rating?

c. Comment on the strength of the relationship between the test scores and production ratings.

28. The materials handling manager of a manufacturing company is trying to forecast the cost of maintenance for the company's fleet of over-the-road tractors. The manager believes that the cost of maintaining the tractors increases with their age. The following data was collected:

Age (years)	Yearly Maintenance Cost ($)	Age (years)	Yearly Maintenance Cost ($)
4.5	619	5.0	1,194
4.5	1,049	0.5	163
4.5	1,033	0.5	182
4.0	495	6.0	764
4.0	723	6.0	1,373
4.0	681	1.0	978
5.0	890	1.0	466
5.0	1,522	1.0	549
5.5	987		

a. Use POM for Windows' least squares-linear regression module to develop a relationship to forecast the yearly maintenance cost based on the age of a tractor.

b. If a section has 20 three-year-old tractors, what is the forecast for the annual maintenance cost?

TABLE 14.5 | **WEEKLY PEAK POWER DEMANDS**

Year	AUGUST			SEPTEMBER				OCTOBER				NOVEMBER	
	1	2	3	4	5	6	7	8	9	10	11	12	13
1	2,050	1,925	1,825	1,525	1,050	1,300	1,200	1,175	1,350	1,525	1,725	1,575	1,925
2	2,000	2,075	2,225	1,800	1,175	1,050	1,250	1,025	1,300	1,425	1,625	1,950	1,950
3	1,950	1,800	2,150	1,725	1,575	1,275	1,325	1,100	1,500	1,550	1,375	1,825	2,000
4	2,100	2,400	1,975	1,675	1,350	1,525	1,500	1,150	1,350	1,225	1,225	1,475	1,850
5	2,275	2,300	2,150	1,525	1,350	1,475	1,475	1,175	1,375	1,400	1,425	1,550	1,900

VIDEO CASE | **Forecasting and Supply Chain Management at Deckers Outdoor Corporation**

Deckers Outdoor Corporation's footwear products are among some of the most well-known brands in the world. From UGG sheepskin boots and Teva sport sandals to Simple shoes, Deckers flip-flops, and Tsubo footwear, Deckers is committed to building niche footwear brands into global brands with market leadership positions. Net sales for fiscal year 2007 were close to $449 million. In addition to traditional retail store outlets for Deckers' footwear styles, the company maintains an active and growing "direct to consumer" e-commerce business. Since most retail stores cannot carry every style in every color and size, the company offers the full line for each of its brands directly to consumers through the brands' individual Web sites. Online sales at its virtual store are handled by its e-commerce group. Customers who want a pair of shoes not available at the retail store can always buy from the virtual store.

Founded in 1973, the company manufactured a single line of sandals in a small factory in Southern California. The challenges of managing the raw materials and finished goods inventories were small compared to today's global sourcing and sales challenges for the company's various brands. Today, each brand has its own development team and brand managers who generate, develop, and test-market the seasonal styles that appear on the shelves of retailers such as Nordstrom, Lord & Taylor, REI, the Walking Company, and the company's own UGG brand retail stores in the United States and Japan.

At Deckers, forecasting is the starting point for inventory management, sales and operations planning, resource planning, and scheduling—in short, managing its supply chain. It carries a considerable amount of seasonal stock. Shoes with seasonal demand that are left over at the end of their season must be sold at heavily discounted prices. Its products fall into three categories: (1) carry-over items that were sold in prior years, (2) new items that look similar to past models, and (3) completely new designs that are fashionable with no past history.

Pearson

Twice a year, the brand development teams work on the fall and spring product lines. They come up with new designs about one year in advance of each season. Each brand (UGG, Teva, Simple, Tsubo, and Deckers) contains numerous products, called stock keeping units (SKUs). The materials for new designs are selected and tested in prototypes. Approved designs are put into the seasonal line-up. Forecasts must be made at both the SKU and aggregate levels months before the season begins. "Bottoms-up" forecasts for each SKU begin by analyzing any available history files of past demand. Judgment forecasts are also important inputs, particularly for the second and third categories of shoes that are not carry-overs. For example, Char Nicanor-Kimball is an expert in spotting trends in shoe sales and makes forecasts for the virtual store. For new designs, historical sales on similar items are used to make a best guess on demand for those items. This process is facilitated by a forecasting and inventory system on the company's Intranet. At the same time, the sales teams for each brand call on their retail accounts and secure customer orders of approved designs for the coming season. Then, the virtual store forecasts are merged with orders from the retail store orders to get the total seasonal demand forecasted by SKU. Next, the SKU forecasts are "rolled up" by category and "top down" forecasts are also made.

These forecasts then go to top management where some adjustments may be made to account for financial market conditions, consumer credit, weather, demographic factors, and customer confidence. The impact of public relations and advertising must also be considered.

Actually, forecasting continues on throughout the year on a daily and weekly basis to "get a handle" on demand. Comparing actual demand with what was forecasted for different parts of the season also helps the forecasters make better forecasts for the future and better control inventories.

Based on initial demand forecasts, the company must begin sourcing the materials needed to produce the footwear. The company makes most of its products in China and sources many of the raw materials there as well. For UGG products sheepskin sourcing occurs in Australia with top grade producers, but the rawhide tanning still takes places in China. With potential suppliers identified and assurance from internal engineering that the footwear can be successfully made, the engineering and material data are handed over to the manufacturing department to determine how best to make the footwear in mass quantities. At this point, Deckers places a seasonal "buy" with its suppliers.

The orders for each SKU are fed into the manufacturing schedules at the Chinese factories. All the SKUs for a given brand are manufactured at the same factory. While Deckers agents negotiate the raw materials contracts early in the development process, the factories only place the orders for the raw materials when the company sends in the actual orders for the finished goods. No footwear is made by the factories until orders are received.

At the factories, finished goods footwear is inspected and packaged for the month-long ocean voyage from Hong Kong to ports in the United States. Deckers ships fifty containers a week from its Chinese manufacturing sources, each holding approximately 5,000 pairs of shoes. Ownership of the finished goods transfers from the factories to Deckers in Hong Kong.

When the shipping containers arrive in the United States, the footwear is transferred to Deckers' distribution centers in Southern California. Teva products are warehoused in Ventura, California; all other products are handled by the company's state-of-the-art facility in Camarillo, California. Typically, Deckers brings product into the distribution centers two to three months in advance of expected needs so that the production at the suppliers' factories and the labor activities at the distribution centers are leveled. There are definitive spikes in the demand for footwear, with Teva spiking in Quarter 1 and UGG spiking in Quarter 4. The leveling approach works to keep costs low in the supply chain. However, it also means that Deckers must maintain sizeable inventories. Most shipments from suppliers come in to the distribution centers and are stored in inventory for one to two months awaiting a customer order. By the time the footwear is stocked in the distribution center, the company knows which retail customers will be getting the various products, based on the orders booked months earlier. Then, according to delivery schedules negotiated with the customers, the company begins filling orders and shipping products to retail locations. The warehouse tracks incoming shipments, goods placed on the shelves for customers, and outgoing orders. The inventory system helps manage the customer order filling process.

Because the booked orders are a relatively large proportion of the total orders from retailers, and the number of unanticipated orders is very small, only small safety stocks are needed to service the retailers. Occasionally, the purchase order from Deckers to one of its suppliers matches the sales order from the customer. In such a case, Deckers uses a "cross-dock" system. When the shipment is received at the distribution center, it is immediately checked in and loaded on another truck for delivery to customers. Cross docking reduces the need to store vast quantities of product for long periods of time and cuts down on warehousing expenses for Deckers. The company has been successful in turning its inventory over about four times a year, which is in line with footwear industry standards.

The online sales traffic is all managed centrally. In fact, for ordering and inventory management purposes, the online side of the business is treated just like another major retail store account. As forecasted seasonal orders are generated by each brand's sales team, a manufacturing order for the online business is placed by the e-commerce sales team at the same time. However, unlike the retail outlets that take delivery of products on a regular schedule, the inventory pledged to the online business is held in the distribution center until a Web site order is received. Only then is it shipped directly to the consumer who placed the online order. If actual demand exceeds expected demand, Char Nicanor-Kimball checks if more inventory can be secured from other customer orders that have scaled back.

The forecasting and supply chain management challenges now facing Deckers are two-fold. First, the company plans to grow the brands that have enjoyed seasonal sales activity into year-round footwear options for consumers by expanding the number of SKUs for those brands. For example, most sales for UGG footwear occur in the fall/winter season. Sales for Teva historically have been in the spring and summer. Product managers are now working to develop styles that will allow the brands to cross over the seasons. Second, the company plans to expand internationally, and will have retail outlets in Europe, China, and other Asian locations in the very near future. Company managers are well aware of the challenges and opportunities such global growth will bring, and are taking steps now to assure that the entire supply chain is prepared to forecast and handle the demand when the time comes.

QUESTIONS

1. How much does the forecasting process at Deckers correspond with the "typical forecasting process" described at the end of this chapter?

2. Based on what you see in the video, what kinds of information technology are used to make forecasts, maintain accurate inventory records, and project future inventory levels?

3. What factors make forecasting at Deckers particularly challenging? How can forecasts be made for seasonal, fashionable products for which there is no history file? What are the costs of over-forecasting demand for such items? Under-forecasting?

4. How does the concept of *postponement* get implemented at Deckers by having online sales and positioning inventory at the DCs for every model, color, and size?

5. Where in the supply chain are cycle, pipeline, safety stock, and anticipation inventories being created?

6. What are the benefits of leveling aggregate demand by having a portfolio of SKUs that create 365-day demand?

7. Deckers plans to expand internationally, thereby increasing the volume of shoes it must manage in the supply chain and the pattern of material flows. What implications does this strategy have on forecasting, order quantities, logistics, and relationships with its suppliers and customers?

CASE   Yankee Fork and Hoe Company

The Yankee Fork and Hoe Company is a leading producer of garden tools ranging from wheelbarrows, mortar pans, and hand trucks to shovels, rakes, and trowels. The tools are sold in four different product lines ranging from the top-of-the-line Hercules products, which are rugged tools for the toughest jobs, to the Garden Helper products, which are economy tools for the occasional user. The market for garden tools is extremely competitive because of the simple design of the products and the large number of competing producers. In addition, more people are using power tools, such as lawn edgers, hedge trimmers, and thatchers, reducing demand for their manual counterparts. These factors compel Yankee to maintain low prices while retaining high quality and dependable delivery.

Garden tools represent a mature industry. Unless new manual products can be developed or a sudden resurgence occurs in home gardening, the prospects for large increases in sales are not bright. Keeping ahead of the competition is a constant battle. No one knows this better than Alan Roberts, president of Yankee.

The types of tools sold today are, by and large, the same ones sold 30 years ago. The only way to generate new sales and retain old customers is to provide superior customer service and produce a product with high customer value. This approach puts pressure on the manufacturing system, which has been having difficulties lately. Recently, Roberts has been receiving calls from long-time customers, such as Sears and True Value Hardware Stores, complaining about late shipments. These customers advertise promotions for garden tools and require on-time delivery.

Roberts knows that losing customers like Sears and True Value would be disastrous. He decides to ask consultant Sharon Place to look into the matter and report to him in one week. Roberts suggests that she focus on the bow rake as a case in point because it is a high-volume product and has been a major source of customer complaints of late.

Planning Bow Rake Production

A bow rake consists of a head with 12 teeth spaced 1 inch apart, a hardwood handle, a bow that attaches the head to the handle, and a metal ferrule that reinforces the area where the bow inserts into the handle. The bow is a metal strip that is welded to the ends of the rake head and bent in the middle to form a flat tab for insertion into the handle. The rake is about 64 inches long.

Place decides to find out how Yankee plans bow rake production. She goes straight to Phil Stanton, who gives the following account:

Planning is informal around here. To begin, marketing determines the forecast for bow rakes by month for the next year. Then they pass it along to me. Quite frankly, the forecasts are usually inflated—must be their big egos over there. I have to be careful because we enter into long-term purchasing agreements for steel, and having it just sitting around is expensive. So I usually reduce the forecast by 10 percent or so. I use the modified forecast to generate a monthly final-assembly schedule, which determines what I need to have from the forging and woodworking areas. The system works well if the forecasts are good. But when marketing comes to me and says they are behind on customer orders, as they often do near the end of the year, it wreaks havoc with the schedules. Forging gets hit the hardest. For example, the presses that stamp the rake heads from blanks of steel can handle only 7,000 heads per day, and the bow rolling machine can do only 5,000 per day. Both operations are also required for many other products.

Because the marketing department provides crucial information to Stanton, Place decides to see the marketing manager, Ron Adams. Adams explains how he arrives at the bow rake forecasts.

Things do not change much from year to year. Sure, sometimes we put on a sales promotion of some kind, but we try to give Phil enough warning before the demand kicks in—usually a month or so. I meet with several managers from the various sales regions to go over shipping data from last year and discuss anticipated promotions, changes in the economy, and shortages we experienced last year. Based on these meetings, I generate a monthly forecast for the next year. Even though we take a lot of time getting the forecast, it never seems to help us avoid customer problems.

The Problem

Place ponders the comments from Stanton and Adams. She understands Stanton's concerns about costs and keeping inventory low and Adams's concern about having enough rakes on hand to make timely shipments. Both are also somewhat concerned about capacity. Yet she decides to check actual customer demand for the bow rake over the past 4 years (in Table 14.6) before making her final report to Roberts.

QUESTIONS

1. Comment on the forecasting system being used by Yankee. Suggest changes or improvements that you believe are justified.

2. Develop your own forecast for bow rakes for each month of the next year (year 5). Justify your forecast and the method you used.

TABLE 14.6 | **FOUR-YEAR DEMAND HISTORY FOR THE BOW RAKE**

Month	DEMAND			
	Year 1	Year 2	Year 3	Year 4
1	55,220	39,875	32,180	62,377
2	57,350	64,128	38,600	66,501
3	15,445	47,653	25,020	31,404
4	27,776	43,050	51,300	36,504
5	21,408	39,359	31,790	16,888
6	17,118	10,317	32,100	18,909
7	18,028	45,194	59,832	35,500
8	19,883	46,530	30,740	51,250
9	15,796	22,105	47,800	34,443
10	53,665	41,350	73,890	68,088
11	83,269	46,024	60,202	68,175
12	72,991	41,856	55,200	61,100

Note: The demand figures shown in the table are the number of units promised for delivery each month. Actual delivery quantities differed because of capacity or shortages of materials.

EXPERIENTIAL LEARNING 14.1 | Forecasting with Holdout Sample

A company's history file, as shown in the following table, gives monthly sales in thousands of dollars "rolled up" into aggregated totals for one of its major product lines.

Your team should use the *Time Series Forecasting* Solver to make forecasts into the future. Note the "Load EL1 Data" button in this solver when clicked will automatically input the demand data. Otherwise, you can enter the demand data directly into the Inputs sheet. Seek out which models you wish to use in making in-class forecasts of monthly sales for the last two months of year 8 and several months into year 9. Perhaps you might want to know the forecasts of all of them, or alternately focus on just two or three of them. If one of the models is the combination forecast, you must decide on the weights to give the models going into its forecast. The weights should add up to 1.0.

Bring to class a one-page document that

- characterizes the monthly sales of the product line in terms of its forecastability.

- identifies the relative importance of four demand patterns: horizontal, trend, seasonal, and cyclical.

- identifies the forecasting models that you will use to make the forecasts for the last of year 8, and future months into year 9, and the extent that judgmental adjustments might be used during the holdout sample exercise. Explain why you made this selection, given that MAD will be used as your error measure.

- makes the November forecast for year 8.

At the start of the in-class portion of the experiential exercise, hand in your one-page document and open the final *Time Series Forecasting* Solver file that you used in modeling the history file. Do not change any of the final parameters chosen for your various forecasting models using the history file

(*n* for moving average, weights for weighted moving average, α for exponential smoothing, and weights for combination).

To start the Holdout Sample session, click on the Worksheet tab and set the time when error analysis begins to be period 95 (November, Year 8). In doing so, error analysis will be tracked only for the holdout periods. Now click on the "Holdout Sample" tab to begin the session. You will initially be presented with the November forecasts for all of the techniques used during your analysis of the history file (including the combination model if you used it). Your next step is to input your team's forecast for November. It can be the forecast from any of the techniques shown, or one of your own if you believe that judgmental adjustment is appropriate. You have no contextual information, but may observe that one model has been performing particularly well in the last few months. You team might have different opinions, but you must reach a consensus. Your instructor will then provide November's actual sales from the holdout sample. After you input that additional information, forecast errors are computed for each model and for your team's November forecast. In addition, computer forecasts (naïve, moving average, weighted moving average, exponential smoothing, trend projection, and combination) are posted for December.

Begin December by inputting your team's forecast. The instructor then provides December's actual sales, and so forth. Continue this process until all errors are calculated for the last period in the holdout sample, which will be announced by your instructor. At the end of this exercise, create a second one-page document that reports your forecasts for the holdout sample, the corresponding average MAD, and CFE whether (and how) you modified your forecasting process as the exercise progressed, and what you learned from this exercise. You will need to write an Excel spreadsheet to calculate the MAD and CFE statistics for the holdout sample. Its output can be attached to your second one-page document. Submit your report to your instructor at the end of the class session.

Your grade on this exercise will be based on (1) the insights provided in the two documents (50 percent of grade) (2) the average MAD for the history file (25 percent of grade), and (3) the average MAD for the holdout sample (25 percent of grade).

Yr	Jan	Feb	Mar	Apr	May	Jun	Jul	Aug	Sep	Oct	Nov	Dec
1	3,255	3,420	3,482	3,740	3,713	3,785	3,817	3,900	3,878	3,949	4,004	4,035
2	3,892	3,730	4,115	4,054	4,184	4,321	4,307	4,481	4,411	4,443	4,395	4,403
3	4,507	4,400	4,099	4,064	4,002	3,963	4,037	4,162	4,312	4,395	4,540	4,471
4	4,589	4,688	4,566	4,485	4,385	4,377	4,309	4,276	4,280	4,144	4,219	4,052
5	4,084	4,158	4,174	4,225	4,324	4,220	4,267	4,187	4,239	4,352	4,331	4,371
6	4,535	4,477	4,601	4,648	4,860	4,998	5,003	4,960	4,943	5,052	5,107	5,100
7	5,303	5,550	5,348	5,391	5,519	5,602	5,557	5,608	5,663	5,497	5,719	5,679
8	5,688	5,604	5,703	5,899	5,816	5,745	5,921	5,900	5,911	5,987		

Sources: This experiential exercise was adapted from an in-class exercise prepared by Dr. Richard J. Penlesky, Carroll University, as a basis for classroom discussion. By permission of Richard J. Penlesky.

EXPERIENTIAL LEARNING 14.2   Forecasting a Vital Energy Statistic

The following time series data captures the weekly average of East Coast crude oil imports in thousands of barrels per day.

QUARTER 2 2010		QUARTER 3 2010		QUARTER 4 2010		QUARTER 1 2011	
Time Period	Data	Time Period	Data	Time Period	Data	Time Period	Data
Apr 02, 2010	1,160	Jul 02, 2010	1,116	Oct 01, 2010	1,073	Dec 31, 2010	994
Apr 09, 2010	779	Jul 09, 2010	1,328	Oct 08, 2010	857	Jan 07, 2011	1,307
Apr 16, 2010	1,134	Jul 16, 2010	1,183	Oct 15, 2010	1,197	Jan 14, 2011	997
Apr 23, 2010	1,275	Jul 23, 2010	1,219	Oct 22, 2010	718	Jan 21, 2011	1,082
Apr 30, 2010	1,355	Jul 30, 2010	1,132	Oct 29, 2010	817	Jan 28, 2011	887
May 07, 2010	1,513	Aug 06, 2010	1,094	Nov 05, 2010	946	Feb 04, 2011	1,067
May 14, 2010	1,394	Aug 13, 2010	1,040	Nov 12, 2010	725	Feb 11, 2011	890
May 21, 2010	1,097	Aug 20, 2010	1,053	Nov 19, 2010	748	Feb 18, 2011	865
May 28, 2010	1,206	Aug 27, 2010	1,232	Nov 26, 2010	1,031	Feb 25, 2011	858
Jun 04, 2010	1,264	Sep 03, 2010	1,073	Dec 03, 2010	1,061	Mar 04, 2011	814
Jun 11, 2010	1,153	Sep 10, 2010	1,329	Dec 10, 2010	1,074	Mar 11, 2011	871
Jun 18, 2010	1,424	Sep 17, 2010	1,096	Dec 17, 2010	941	Mar 18, 2011	1,255
Jun 25, 2010	1,274	Sep 24, 2010	1,125	Dec 24, 2010	994	Mar 25, 2011	980

Your instructor has a "holdout" sample representing the values for April 1, 2011 and beyond. Your task is to use the POM for Windows *Time Series Forecasting* module and the history file to project this statistic into the future. If you have MyOMLab, the demand data is available in the *Exercise 2* Excel file. It can be pasted into the Data Table of the *Time Series Forecasting*

module. Otherwise, you can enter the demand data directly into the Data Table. Prior to your next class meeting:

a. Use the POM for Windows *Time Series Forecasting* module to locate the best naïve, moving average, weighted moving average, and trend projection with regression models that you think will most accurately forecast demand during the holdout sample. *Begin your error calculations with period 5 (April 30, 2010).*

b. Create an Excel spreadsheet that begins with inputs of the four forecasts from the *Time Series Forecasting* module. Its purpose is to develop a combination forecast that will serve as your team's forecasts for each period. Assign a weight to each forecast model (the sum of all four forecast weights for one period should equal 1.0) and develop a "combination forecast" by multiplying each forecast by its weight. Keep the weights constant for the whole history file as you search for the best set of weights. If you do not like a particular model, give it a weight of 0. Calculate appropriate forecast error measures for your combination forecast in your Excel spreadsheet.

c. Create a management report that shows your period-by-period forecasts and their overall historical CFE and MAPE performance for each model and your combination forecast.

In-Class Exercise—Part 1

a. Input into your Excel spreadsheet the forecasts from the POM for Windows *Time Series Forecasting* module to get the combination forecast for the first period (the week of April 1, 2011) in the holdout sample. The combination forecast is considered your team's forecast.

b. Enter the actual data announced by your instructor, and have Excel compute appropriate forecast error measures for your four models and the combination forecast. Decide on any revisions of weights for the combination forecast.

c. Update the POM for Windows *Time Series Forecasting* module with the actual demand for the new period and get the new forecasts.

In-Class Exercise—Part 2

a. Input the forecasts from the POM for Windows *Time Series Forecasting* module into your Excel spreadsheet to get the final combination forecast for the next period (the week of April 8, 2011). At this point, you may change this period's weights on each forecasting technique going into the combination forecast. You have no contextual information, but may observe that one model has been performing particularly well in the last few periods. Your team might have different opinions, but you must reach a consensus.

b. Enter the actual data announced by your instructor, with Excel computing appropriate forecast error measures for your four models and the combination forecast.

c. Update the POM for Windows *Time Series Forecasting* module with the actual demand for the new period and get the new forecasts.

In-Class Exercise—Parts 3 and beyond

Continue in the fashion of Parts 1 and 2 to produce forecasts as directed by your instructor. At the end of the exercise, create a second management report that shows for the holdout sample your period-by-period forecasts, their individual forecast errors and percent deviations for each model and your combination forecast. Explain your logic regarding any changes made to your combination forecast weights over the holdout period.

Source: This experiential exercise was prepared as an in-class exercise prepared by Dr. John Jensen, University of South Carolina, as a basis for classroom discussion. By permission of John B. Jensen.

Selected References

Armstrong, J. Scott. "Findings from Evidence-based Forecasting: Methods for Reducing Forecast Error." *International Journal of Forecasting,* vol. 22, no. 3 (2006), pp. 583–598.

Attaran, Mohsen, and Sharmin Attaran. "Collaborative Supply Chain Management." *Business Process Management Journal Management Journal,* vol. 13, no. 13 (June 2007), pp. 390–404.

Cederlund, Jerold P., Rajiv Kohli, Susan A. Sherer, and Yuliang Yao. "How Motorola Put CPFR into Action." *Supply Chain Management Review* (October 2007), pp. 28–35.

Daugherty, Patricia J., R. Glenn Richey, Anthony S. Roath, Soonhong Min, Haozhe Chen, Aaron D. Arndt, and Stefan E. Genchev. "Is Collaboration Paying Off for Firms?" *Business Horizons* (2006), pp. 61–70.

Fildes, Robert, Paul Goodwin, Michael Lawrence, and Konstantinos Nikolopoulos. "Effective Forecasting and Judgmental Adjustments: An Empirical Evaluation and Strategies for Improvement in Supply-Chain Planning." *International Journal of Forecasting,* vol. 25, no. 1 (2009), pp. 3–23.

Lawrence, Michael, Paul Goodwin, Marcus O'Connor, and Dilek Onkal. "Judgmental Forecasting: A Review of Progress over the Last 25 Years." *International Journal of Forecasting* (June 2006), pp. 493–518.

McCarthy, Teresa, Donna F. Davis, Susan L. Golicic, and John T. Mentzer. "The Evolution of Sales Forecasting Management: A 20-Year Longitudinal Study of Forecasting Practices." *Journal of Forecasting,* vol. 25 (2006), pp. 303–324.

Min, Hokey, and Wen-Bin Vincent Yu. "Collaborative Planning, Forecasting and Replenishment: Demand Planning in Supply Chain Management." *International Journal of Information Technology and Management,* vol. 7, no. 1 (2008), pp. 4–20.

Montgomery, David. "Flashpoints for Changing Your Forecasting Process." *The Journal of Business Forecasting* (Winter 2006–2007), pp. 35–42.

Principles of Forecasting: A Handbook for Researchers and Practitioners. J. Scott Armstrong (ed.). Norwell, MA: Kluwer Academic Publishers, 2001. Also visit **http://www.forecastingprinciples.com** for valuable information on forecasting, including frequently asked questions, a forecasting methodology tree, and a dictionary.

Saffo, Paul. "Six Rules for Effective Forecasting." *Harvard Business Review* (July–August 2007), pp. 1–30.

Smaros, Johanna. "Forecasting Collaboration in the European Grocery Sector: Observations from a Case Study." *Journal of Operations Management,* vol. 25, no. 3 (April 2007), pp. 702–716.

Smith, Larry. "West Marine: A CPFR Success Story." *Supply Chain Management Review* (March 2006), pp. 29–36.

Syntetos, Aris Konstantinos Nikolopoulos, John Boylan, Robert Fildes, and Paul Goodwin. "The Effects of Integrating Management Judgement into Intermittent Demand Forecasts." *International Journal of Production Economics,* vol. 118, no. 1 (March, 2009), pp. 72–81.

Wikipedia, "Collaborative Planning, Forecasting, and Replenishment," http:en.wikipedia.org/wiki/Collaborative Planning Forecasting and Replenishment, (April, 2011).

15

OPERATIONS PLANNING AND SCHEDULING

Operations planning and scheduling at an airline like Air New Zealand goes through several stages to match supply with demand, from aggregate plans to short-term schedules. Even after finalizing flights and crew roster schedules, severe weather conditions or mechanical failures can cause last-minute changes. Long-term competitive strength depends on how well it performs this process.

Air New Zealand

How important is scheduling to an airline company? Certainly, customer satisfaction regarding on-time schedule performance is critical in a highly competitive industry such as air transportation. In addition, airlines lose a lot of money when expensive equipment, such as an aircraft, is idle. Flight and crew scheduling, however, is a complex process. For example, Air New Zealand is a group of five airlines with a combined fleet of 96 aircraft, with another 22 more on order. The average utilization is 8:44 hours per day. It has undergone an $800 million upgrade to its long-haul service, refitting its Boeing 747 fleet and adding eight new Boeing 777-200 aircraft for flights to North America. It directly serves 50 ports—26 domestic and 24 international within 15 countries. It carries 11.7 million passengers annually, and its network incorporates flight times ranging from 15 minutes to 13 hours. Operations planning and scheduling at the aggregate level begins with a market plan that identifies the new and existing flight segments that are needed to remain competitive. This general plan is further refined to a three-year plan, and then is put into an annual budget in which flight segments have specific departure and arrival times.

Next, crew availability must be matched to the flight schedules. The two types of crews—pilots and attendants—each comes with its own set of constraints. Pilots, for example, cannot be scheduled for more than 35 hours in a 7-day week and no more than 100 hours in a 28-day cycle. They also must have a 36-hour break every 7 days and 30 days off in an 84-day cycle. Each pilot's tour of duty begins and ends at a crew base and consists of an alternating sequence of duty periods and rest periods, with duty periods including one or more flights. The schedule must ensure that each flight has a qualified crew and that each crew member has a feasible tour of duty over the roster period. From the crew's point of view, it is also important to satisfy as many crew requests and preferences as possible.

Source: "Service Scheduling at Air New Zealand," *Operations Management 10e Video Library* (Upper Saddle River, NJ: Prentice Hall, 2010); **www.airnewzealand.com** (August, 2011). See video in MyOMLab.

LEARNING GOALS *After reading this chapter, you should be able to:*

1. Describe the operations planning and scheduling process.
2. Explain why the process of matching supply with demand begins with aggregation.
3. Identify the different demand and supply options.
4. Explain how operations plans and schedules relate to other plans.
5. Use spreadsheets to create sales and operations plans.
6. Develop employee schedules.
7. Develop schedules for single workstations.

Creating Value through Operations Management

Using Operations to Compete
Project Management

Managing Processes

Process Strategy
Process Analysis
Quality and Performance
Capacity Planning
Constraint Management
Lean Systems

Managing Supply Chains

Supply Chain Inventory Management
Supply Chain Design
Supply Chain Location Decisions
Supply Chain Integration
Supply Chain Sustainability and Humanitarian Logistics
Forecasting
Operations Planning and Scheduling
Resource Planning

Managing supply chains effectively requires more than just good demand forecasts. Demand is the first half of the equation, and the other half is supply. The firm must develop plans to supply the resources needed to meet the forecasted demand. These resources include the workforce, materials, inventories, dollars, and equipment capacity.

Operations planning and scheduling is the process of making sure that demand and supply plans are in balance, from the aggregate level down to the short-term scheduling level. As we saw with Air New Zealand, the process begins at the aggregate level and gets progressively more specific until all crew members know their tour of duty. Operations planning and scheduling lies at the core of supply chain integration, around which plans are made up and down the supply chain, from supplier deliveries to customer due dates and services. Table 15.1 defines several types of plans related to operations planning and scheduling.

In this chapter, we focus on two major parts of the overall process: (1) sales and operations planning and (2) scheduling. We begin with the purpose of aggregation in sales and operations planning. We examine how S&OP relates with other plans and functional areas within the firm. We describe a typical planning process, and various strategies to cope with uneven demand. We show how spreadsheets can help find good solutions. Then, we conclude with scheduling, including performance measures and some basic techniques for creating schedules. MyOMLab Supplement J, "Operations Scheduling", provides additional help with scheduling problems.

Operations Planning and Scheduling across the Organization

Operations planning and scheduling is meaningful for each organization along the supply chain. First, it requires managerial inputs from all of the firm's functions. Marketing provides inputs on demand and accounting provides important cost data and a firm's financial condition. Second, each function is affected by the plan. A plan that calls for expanding the workforce has a direct impact on the hiring and training requirements for the human resources function. As the plan is implemented, it creates revenue and cost streams that finance must deal with as it manages the firm's cash flows. Third, each department and group in a firm has its own workforce. Managers of these departments must make choices on hiring, overtime, and vacations.

Scheduling is important for both service and manufacturing processes. Whether the business is an airline, hotel, computer manufacturer, or university, schedules are a part of everyday life. Schedules involve an enormous amount of detail and affect every process in a firm. For example, service, product, and employee schedules determine specific cash flow requirements, trigger the firm's billing process, and initiate requirements for the employee training process. Firms use the scheduling process to lower their costs and improve their responsiveness, affecting operations up and down the supply chain worldwide.

TABLE 15.1 | TYPES OF PLANS WITH OPERATIONS PLANNING AND SCHEDULING

Key Term	Definition
Sales and operations plan (S&OP)	A plan of future aggregate resource levels so that supply is in balance with demand. It states a company's or department's production rates, workforce levels, and inventory holdings that are consistent with demand forecasts and capacity constraints. The S&OP is time-phased, meaning that it is projected for several time periods (such as months or quarters) into the future.
Aggregate plan	Another term for the sales and operations plan.
Production plan	A sales and operations plan for a *manufacturing firm* that centers on production rates and inventory holdings.
Staffing plan	A sales and operations plan for a *service firm*, which centers on staffing and on other human resource-related factors.
Resource plan	An intermediate step in the planning process that lies between S&OP and scheduling. It determines requirements for materials and other resources on a more detailed level than the S&OP. It is covered in the next chapter.
Schedule	A detailed plan that allocates resources over shorter time horizons to accomplish specific tasks.

Stages in Operations Planning and Scheduling

In this section, we explain why companies begin with plans that take a macro, or big picture, view of their business. We also describe how these plans relate to their other plans, and how the long-term plans ultimately are translated into detailed schedules ready for immediate action.

Aggregation

The sales and operations plan is useful because it focuses on a general course of action, consistent with the company's strategic goals and objectives, without getting bogged down in details. We must first aggregate, and then use the targets and resources from the plan to create effective, coordinated schedules. A company's managers must determine whether they can satisfy budgetary goals without having to schedule each of the company's thousands of products and employees individually. While schedules with such detail are the goal, the operations planning and scheduling process begins at the aggregate level.

In general, companies perform aggregation along three dimensions: (1) services or products, (2) workforce, and (3) time.

Product Families   A group of customers, services, or products that have similar demand requirements and common process, workforce, and materials requirements is called a **product family**. Sometimes, product families relate to market groupings or to specific processes. A firm can aggregate its services or products into a set of relatively broad families, avoiding too much detail at this stage of the planning process. For instance, a manufacturer of bicycles that produces 12 different models of bikes might divide them into two groups, mountain bikes and road bikes, for the purpose of preparing the sales and operations plan. Common and relevant measurements should be used.

Workforce   A company can aggregate its workforce in various ways as well, depending on its flexibility. For example, if workers at the bicycle manufacturer are trained to work on either mountain bikes or road bikes, for planning purposes management can consider its workforce to be a single aggregate group, even though the skills of individual workers may differ.

Time   The planning horizon covered by a sales and operations plan typically is one year, although it can differ in various situations. To avoid the expense and disruptive effect of frequent changes in output rates and the workforce, adjustments usually are made monthly or quarterly. In other words, the company looks at time in the aggregate—months, quarters, or seasons—rather than in weeks, days, or hours.

The Relationship of Operations Plans and Schedules to Other Plans

A financial assessment of the organization's near future—that is, for 1 or 2 years ahead—is called either a business plan (in for-profit firms) or an annual plan (in nonprofit service organizations). A **business plan** is a projected statement of income, costs, and profits. It usually is accompanied by budgets, a projected (pro forma) balance sheet, and a projected cash flow statement, showing sources and allocations of funds. The business plan unifies the plans and expectations of

operations planning and scheduling

The process of balancing supply with demand, from the aggregate level down to the short-term scheduling level.

product family

A group of services or products that have similar demand requirements and common process, labor, and materials requirements.

business plan

A projected statement of income, costs, and profits.

a firm's operations, finance, sales, and marketing managers. In particular, it reflects plans for market penetration, new product introduction, and capital investment. Manufacturing firms and for-profit service organizations, such as a retail store, a firm of attorneys, or a hospital, prepare such plans. A nonprofit service organization, such as the United Way or a municipal government, prepares a different type of plan for financial assessment, called an **annual plan or financial plan**.

annual plan or financial plan

A plan for financial assessment used by a nonprofit service organization.

Figure 15.1 illustrates the relationships among the business or annual plan, sales and operations plan, and detailed plans and schedules derived from it. For **service providers** in the supply chain, top management sets the organization's direction and objectives in the business plan (in a for-profit organization) or annual plan (in a not-for-profit organization). This plan then provides the framework for developing the sales and operations plan, which typically focuses on staffing and other human resource-related factors at a more aggregate level. It presents the number and types of employees needed to meet the objectives of the business or annual plan.

FIGURE 15.1 ▶

The Relationship of Sales and Operations Plans and Schedules to Other Plans

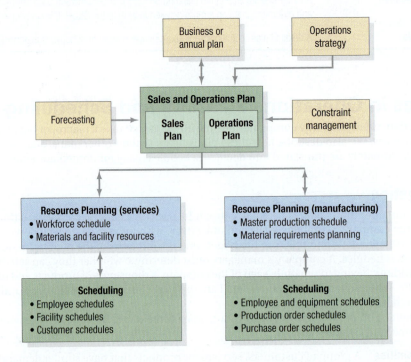

Based on the sales and operations plan for a service provider, the next planning level is *resource planning* to determine the firm's workforce schedules and other resource requirements, such as materials and facilities, on a more detailed level. The *workforce schedule* details the specific work schedule for each category of employee. For example, a sales and operations plan might allocate 10 police officers for the day shift in a particular district; the workforce schedule might assign 5 of them to work Monday through Friday and the other 5 to work Wednesday through Sunday to meet the varying daily needs for police protection in that district. The lowest planning level is *scheduling*, which puts together day-to-day schedules for individual employees and customers.

For **manufacturing firms** in the supply chain, top management sets the company's strategic objectives for at least the next year in the business plan. It provides the overall framework, along with inputs coming from operations strategy, forecasting, and capacity constraint management. The sales and operations plan specifies product family production rates, inventory levels, and workforce levels. The next planning level beneath the sales and operations plan is resource planning, which we cover in the next chapter. Resource planning gets specific as to individual products within each product family, purchased materials, and resources on a detailed level. The *master production schedule* specifies the timing and size of production quantities for each product in the product families. The *material requirements planning* process then derives plans for components, purchased materials, and workstations. As with service providers, the lowest and most detailed planning level is scheduling. It puts together day-to-day schedules or priorities for employees, equipment, and production or purchase orders. Thus, the sales and operations plan plays a key role in translating the strategies of the business plan into an operational plan for the manufacturing process.

As the arrows in Figure 15.1 indicate, information flows in two directions: from the top down (broad to detailed) and from the bottom up (detailed to broad). If a sales and operations plan cannot be developed to satisfy the objectives of the business or annual plan with the existing resources, the business or annual plan might need some adjustment. Similarly, if a feasible master production schedule or workforce schedule cannot be developed, the sales and operations plan might need some adjustment. The planning process is dynamic, with periodic plan revisions or adjustments based on two-way information flows, typically on a monthly basis.

Managing Demand

Matching supply with demand becomes a challenge when forecasts call for uneven demand patterns—and uneven demand is more the rule than the exception. Demand swings can be from one month to the next, one week to the next, or even one hour to the next. Peaks and valleys in demand are costly or can cause poor customer service. Air New Zealand can lose sales because capacity is exceeded for one of its flights, while another of its flights to the same destination at about the same time has many empty seats. If nothing is done to even out demand, sales are lost or greater capacity cushions might be needed. For other companies, the supply options for handing uneven demand could be overtime, hiring and curtailing the workforce, and anticipation inventories. All come at an extra cost. Here we deal with **demand management**, the process of changing demand patterns using one or more demand options.

> **demand management**
> The process of changing demand patterns using one or more demand options.

Demand Options

Various options are available in managing demand, including complementary products, promotional pricing, prescheduled appointments, reservations, revenue management, backlogs, backorders, and stockouts. The manager may select one or more of them, as we illustrate below.

Complementary Products   One demand option for a company to even out the load on resources is to produce **complementary products**, or services that have similar resource requirements but different demand cycles. For example, manufacturers of matzoh balls for the Jewish Passover holiday are in a seasonal business. The B. Manischewitz Company, a kosher foods manufacturer in Jersey City, New Jersey, previously experienced 40 percent of its annual sales for the 8-day Passover holiday alone. It expanded toward markets with year-round appeal such as low-carb, low-fat foods, including canned soups and crackers, borscht, cake mixes, dressing and spreads, juices, and condiments.

> **complementary products**
> Services or products that have similar resource requirements but different demand cycles.

For service providers, a city parks and recreation department can counterbalance seasonal staffing requirements for summer activities by offering ice skating, tobogganing, or indoor activities during the winter months. The key is to find services and products that can be produced with the existing resources and can level off the need for resources over the year.

Promotional Pricing   Promotional campaigns are designed to increase sales with creative pricing. Examples include automobile rebate programs, price reductions for winter clothing in the late summer months, reduced prices for hotel rooms during off-peak periods, and "two-for-the-price-of-one" automobile tire sales. Lower prices can increase demand for the product or service from new and existing customers, take sales from competitors, or encourage customers to move up future buying. The first two outcomes increase overall demand, while the third shifts demand to the current period.

Prescheduled Appointments   Service providers often can schedule customers for definite periods of order fulfillment. With this approach, demand is leveled to not exceed supply capacity. An appointment system assigns specific times for service to customers. The advantages of this method are timely customer service and the high utilization of service personnel.

Doctors, dentists, lawyers, and automobile repair shops are examples of service providers that use appointment systems. Doctors can use the system to schedule parts of their day to visit hospital patients, and lawyers can set aside time to prepare cases. Care must be taken to tailor the length of appointments to individual customer needs rather than merely scheduling customers at equal time intervals.

This doctor's office uses the appointment system to schedule patients for definite periods of order fulfillment. Times are selected that are agreeable with the patient and fit with the doctor's schedule. Demand is leveled and does not exceed capacity.

Thinkstock Images/Getty Images

Reservations   Reservation systems, although quite similar to appointment systems, are used when the customer actually occupies or uses facilities associated with the service. For example, customers reserve hotel rooms, automobiles, airline seats, and concert seats. The major advantage of reservation systems is the lead time they give service managers and the ability to level demand. Managers can deal with no-shows with a blend of overbooking, deposits, and cancellation penalties. Sometimes overbooking means that a customer with reservations cannot be served as promised. In such cases, bonuses can be offered for compensation. For example, an airline passenger might not only get on the next available flight, but also may be given a free ticket for a second flight sometime in the future.

Revenue Management   A specialized combination of the pricing and reservation options for service providers is revenue management. **Revenue management** (sometimes called *yield management*) is the process of varying price at the right time for different customer segments to maximize revenues generated from existing supply capacity. It works best if customers can be segmented, prices can be varied by segment, fixed costs are high, variable costs are low, service duration is predictable, and capacity is lost if not used (sometimes called *perishable capacity*). Airlines, hotels, cruise lines, restaurants (early-bird specials), and rental cars are good examples. Computerized reservation systems can make hour-by-hour updates, using decision rules for opening or closing price classes depending on the difference between supply and continually updated demand forecasts. In the airlines industry, prices are lowered if a particular airline flight is not selling as fast as expected, until more seats are booked. Alternately, if larger than expected demand is developing, prices for the remaining seats may be increased. Last-minute business travelers pay the higher prices, whereas leisure travelers making reservations well in advance and staying over the weekend get the bargain prices. Southwest Airlines now segments its customers by creating a "Business Select" ticket class that rewards more perks to frequent fliers willing to pay higher prices. Managerial Practice 15.1 describes an unusual way of segmenting customers and setting prices at casinos.

revenue management

Varying price at the right time for different customer segments to maximize revenues yielded by existing supply capacity.

MANAGERIAL PRACTICE 15.1   Harrah's Cherokee Casino & Hotel

Decision rules in revenue management systems are entirely different at casinos. The payoff from revenue management, when combined with its customer relationship process, is particularly pronounced at Harrah's Cherokee Casino & Hotel even though it serves no alcohol and has no traditional gaming tables. The difference with casinos is that their biggest source of revenue is from gambling, not their hotel rooms and restaurants. Customers are segmented not by when they make the reservation or whether their booking includes a weekend, but by how much they are likely to gamble. Top gamblers may lose thousands of dollars per day, whereas risk-adverse customers may lose only $25 per day. Information on the willingness to gamble comes from Cherokee's customer relationship management system (see customer relationship process in Chapter 12, "Supply Chain Integration") that uses a "Total Rewards" card program. Customers present this card each time they make a bet, and in return receive perks and gifts based on the number of points accumulated. When a customer calls in to make reservations, the state-of-the-art revenue management system takes over. The customer is asked for his or her Total Rewards number, which the system uses to estimate how much the customer bets per night, based on past history. The number can be translated probabilistically into the casino's expected net profit per night from the customer. Considerable room capacity is reserved up to the last minute for the big betters. They not only get a room, but one at no cost. With Cherokee's revenue

Harrah's Cherokee Casino & Hotel has a unique approach to revenue management. Using a state-of-the-art revenue management system, considerable room capacity is reserved for big betters, and their rooms are free. Everyone self prices, depending on their gambling history.

management system, everyone self-prices, depending on how much they want to gamble.

DAVID T. FOSTER III KRT/Newscom

Source: Richard Metters, C. Queenan, M. Ferguson, L. Harrison, J. Higbie, S. Ward, B. Barfield, T. Farley, A. Kuyumcu, and A. Duggasani, "The 'Killer Application' of Revenue Management: Harrah's Cherokee Casino & Hotel," *Interfaces,* vol. 38 (2008); **www.harrascherokee.com** (2011)

Backlogs   Much like the appointments or reservations of service providers, a **backlog** is an accumulation of customer orders that a manufacturer has promised for delivery at some future date. Manufacturers in the supply chain that maintain a backlog of orders as a normal business practice can allow the backlog to grow during periods of high demand and then reduce it during periods of low demand. Airplane manufacturers do not promise instantaneous delivery, as do wholesalers or retailers farther forward in the supply chain. Instead, they impose a lead time between when the order is placed and when it is delivered. For example, an automotive parts manufacturer may agree to deliver to the repair department of a car dealership a batch of 100 door latches for a particular car model next Tuesday. The parts manufacturer uses that due date to plan its production of door latches within its capacity limits. Firms that are most likely to use backlogs—and increase the size of them during periods of heavy demand—make customized products and tend to have a make-to-order strategy. Backlogs reduce the uncertainty of future production requirements and also can be used to level demand. However, they become a competitive disadvantage if they get too big.

Backorders and Stockouts   A last resort in demand management is to set lower standards for customer service, either in the form of backorders or stockouts (see Chapter 10, "Supply Chain Design"). Not to be confused with a backlog, a *backorder* is a customer order that cannot be filled immediately but is filled as soon as possible. Demand may be too unpredictable or the item may be too costly to hold it in inventory. Although the customer is not pleased with the delay, the customer order is not lost and it is filled at a later date. In contrast, a *stockout* is much the same, except that the order is lost and the customer goes elsewhere. A backorder adds to the next period's demand requirement, whereas a stockout does not. Backorders and stockouts can lead dissatisfied customers to do their future business with another firm. Generally, backorders and stockouts are to be avoided.

Combinations of demand options can also be used. For example, a manufacturer of lighting equipment had several products characterized as "slow movers with spikes," where only 2 or 3 units were sold for several weeks, and then suddenly there was a huge order for 10,000 units the next week. The reason is that their product was purchased by commercial property managers who might be upgrading the lighting in a large office building. The result was a forecasting nightmare and having to resort to high cost supply options to meet the demand spikes. The breakthrough in solving this problem was to combine the pricing and backlog options. Contractors are now offered a 3 percent discount (the pricing option) on any order in excess of 10,000 units that are placed five or more weeks before they are needed (the backlog option). The advanced warning allows the manufacturer to smooth out its production processes, saving millions of dollars annually.

The left side of Table 15.2 summarizes the demand options for operations planning and scheduling, and the right side lists the supply options for balancing supply with demand. The following two sections on sales and operations planning and scheduling cover the supply options.

backlog

An accumulation of customer orders that a manufacturer has promised for delivery at some future date.

TABLE 15.2 | **DEMAND AND SUPPLY OPTIONS FOR OPERATIONS PLANNING AND SCHEDULING**

Demand Options	Supply Options
Complementary products	Anticipation inventory
Promotional pricing	Workforce adjustment (hiring and layoffs)
Prescheduled appointments	Workforce utilization (overtime and undertime)
Reservations	Part-time workers and subcontractors
Revenue management	Vacation schedules
Backlogs	Workforce schedules
Backorders	Job and customer sequences
Stockouts	Expediting

Sales and Operations Plans

Developing sales and operations plans means making decisions. In this section, we concentrate on the information inputs, the supply options themselves, and strategies that go into the sales and operations planning (S&OP) decisions.

Information Inputs

Just as it is needed to manage the demand side, consensus is needed among the firm's departments when decisions for the supply side are made. Information inputs are sought to create a plan that works for all. Figure 15.2 lists inputs from each functional area. They must be accounted for to make sure that the plan is a good one and also doable. Such coordination helps synchronize the flow of services, materials, and information through the supply chain to best balance supply with customer demand.

FIGURE 15.2 ▶

Managerial Inputs from
Functional Areas to Sales
and Operations Plans

Operations
• Current machine capacities
• Plans for future capacities
• Workforce capacities
• Current staffing level

Distribution and marketing
• Customer needs
• Demand forecasts
• Competition behavior

Materials
• Supplier capabilities
• Storage capacity
• Materials availability

Sales and Operations Plan

Accounting and finance
• Cost data
• Financial condition of firm

Engineering
• New products
• Product design changes
• Machine standards

Human resources
• Labor market conditions
• Training capacity

Supply Options

Given demand forecasts, as modified by demand management choices, operations managers must develop a plan to meet the demand, drawing from the supply options listed in Table 15.2.

Anticipation Inventory   *Anticipation inventory* can be used to absorb uneven rates of demand or supply. For example, a plant facing seasonal demand can stock anticipation inventory during light demand periods and use it during heavy demand periods. Manufacturers of air conditioners, such as Whirlpool, can experience 90 percent of their annual demand during just three months of a year. Extra, or anticipation inventory, also can help when supply, rather than demand, is uneven. For example, a company can stock up on a certain purchased item if the company's suppliers expect severe capacity limitations. Despite its advantages, anticipation inventory can be costly to hold, particularly if stocked in its finished state. Moreover, when services or products are customized, anticipation inventory is not usually an option. Service providers in the supply chain generally cannot use anticipation inventory because services cannot be stocked.

An employee stocks a Whirlpool air conditioner at a Lowe's store in Westborough, Massachusetts. The demand for window units is highly seasonal and also depends on variations in the weather. Typically, Whirlpool begins production of room air conditioners in the fall and holds them as inventory until they are shipped in the spring. Building anticipation inventory in the slack season allows the company to even out production rates over much of the year and still satisfy demand in the peak periods (spring and summer) when retailers are placing most of their orders.

Workforce Adjustment   Management can adjust workforce levels by hiring or laying off employees. The use of this alternative can be attractive if the workforce is largely unskilled or semiskilled and the labor pool is large. These conditions are more likely found in some countries than in others. However, for a particular company, the size of the qualified labor pool may limit the number of new employees that can be hired at any one time. Also, new employees must be trained, and the capacity of the training facilities themselves might limit the number of new hires at any one time. In some industries, laying off employees is difficult or unusual for contractual reasons (unions); in other industries, such as tourism and agriculture, seasonal layoffs and hirings are the norm.

overtime

The time that employees work that is longer than the regular workday or workweek for which they receive additional pay.

Workforce Utilization   An alternative to a workforce adjustment is a change in workforce utilization involving overtime and undertime. **Overtime** means that employees work longer than the regular workday or workweek and receive additional pay for the extra hours. It can be used to

Photo by Michael Springer/Bloomberg via Getty Images

satisfy output requirements that cannot be completed on regular time. Overtime is expensive (typically 150 percent of the regular-time pay rate) and workers often do not want to work a lot of overtime for an extended period of time. Excessive overtime also can result in declining quality and productivity. On the other hand, it helps avoid the costly fringe benefits (such as health insurance, dental care, Social Security, retirement funds, paid vacations, and holidays) that come with hiring a new full-time employee,

Undertime means that employees do not have enough work for the regular-time workday or workweek. For example, they cannot be fully utilized for eight hours per day or for five days per week. Undertime occurs when labor capacity exceeds demand requirements (net of anticipation inventory), and this excess capacity cannot or should not be used productively to build up inventory or to satisfy customer orders earlier than the delivery dates already promised.

Undertime can either be paid or unpaid. An example of *paid undertime* is when employees are kept on the payroll rather than being laid off. In this scenario, employees work a full day and receive their full salary but are not as busy because of the light workload. Some companies use paid undertime (though they do not call it that) during slack periods, particularly with highly skilled, hard-to-replace employees or when there are obstacles to laying off workers. The disadvantages of paid undertime include the cost of paying for work not performed and lowered productivity.

Part-Time Workers Another option apart from undertime is to hire part-time workers, who are paid only for the hours and days worked. Perhaps they only work during the peak times of the day or peak days of the week. Sometimes, part-time arrangements provide predictable work schedules, but in other cases workers are not called in if the workload is light. Such arrangements are more common in low-skill positions or when the supply of workers seeking such an arrangement is sufficient. Part-time workers typically do not receive fringe benefits.

Subcontractors Subcontractors can be used to overcome short-term capacity shortages, such as during peaks of the season or business cycle. Subcontractors can supply services, make components and subassemblies, or even assemble an entire product.

Vacation Schedules A manufacturer can shut down during an annual lull in sales, leaving a skeleton crew to cover operations and perform maintenance. Hospital employees might be encouraged to take all or part of their allowed vacation time during slack periods. The use of this alternative depends on whether the employer can mandate the vacation schedules of its employees. In any case, employees may be strongly discouraged from taking vacations during peak periods or encouraged to take vacations during slack periods.

Planning Strategies

Here we focus on supply options that define output rates and workforce levels. Two basic strategies are useful starting points in searching for the best plan.

1. *Chase Strategy.* The **chase strategy** involves hiring and laying off employees to match the demand forecast over the planning horizon. Varying the workforce's regular-time capacity to equate supply to demand requires no inventory investment, overtime, or undertime. The drawbacks are the expense of continually adjusting workforce levels, the potential alienation of the workforce, and the loss of productivity and quality because of constant changes in the workforce.

2. *Level Strategy.* The **level strategy** involves keeping the workforce constant (except possibly at the beginning of the planning horizon). It can vary its utilization to match the demand forecast via overtime, undertime (paid or unpaid), and vacation planning (i.e., paid vacations when demand is low). A constant workforce can be sized at many levels: Managers can choose to maintain a large workforce so as to minimize the planned use of overtime during peak periods (which, unfortunately, also maximizes the need for undertime during slack periods). Alternately, they can choose to maintain a smaller workforce and rely heavily on overtime during the peak periods (which places a strain on the workforce and endangers quality).

undertime
The situation that occurs when employees do not have enough work for the regular-time workday or workweek.

chase strategy
A strategy that involves hiring and laying off employees to match the demand forecast.

level strategy
A strategy that keeps the workforce constant, but varies its utilization via overtime, undertime, and vacation planning to match the demand forecast.

Even though Hallmark's business is seasonal, the company has never laid off employees. Instead the company's employees are trained to do different jobs at different times, and at different plants, if need be. Because they know they have greater job security, they work hard to keep setup times short and Hallmark's costs low.

These two "pure" strategies used alone usually do not produce the best sales and operations plan. It might not be best to keep the workforce exactly level, or to vary it to exactly match forecasted demand on a period-by-period basis. The best strategy, therefore, usually is a **mixed strategy** that considers the full range of supply options. The chase strategy is limited to just hiring and laying off employees. The level strategy is limited to overtime, undertime, and vacation schedules. The mixed strategy opens things up to all options, including anticipation inventory, part-time workers, subcontractors, backorders, and stockouts.

Constraints and Costs

An acceptable sales and operations plan must recognize relevant constraints or costs. Constraints can be either physical limitations or related to managerial policies. Examples of physical constraints might be machine capacities that limit maximum output or inadequate inventory storage space. Policy constraints might include limitations on the number of backorders or the use of subcontractors or overtime, as well as the minimum inventory levels needed to achieve desired safety stocks. Ethical issues may also be involved, such as excessive layoffs or required overtime.

Typically, many plans can contain a number of constraints. Table 15.3 lists the costs that the planner considers when preparing sales and operations plans.

Sales and Operations Planning as a Process

Sales and operations planning is a decision-making process, involving both planners and management. It is dynamic and continuing, as aspects of the plan are updated periodically when new information becomes available and new opportunities emerge. It is a cross-functional process that seeks a set of plans that all of a firm's functions can support. For each product family, decisions are made based on cost trade-offs, recent history, recommendations by planners and middle management, and the executive team's judgment.

Figure 15.3 shows a typical plan for a manufacturer. The plan is for one of the manufacturer's make-to-stock product families expressed in aggregate units. This simple spreadsheet shows the interplay between demand and supply. The history on the left for January through March shows how forecasts are tracking actual sales, and how well actual production conforms to the plan. The inventory projections are of particular interest to finance because they significantly affect the manufacturer's cash requirements. The last two columns on the top right show how current fiscal year sales projections match up with the current business plan.

This particular plan is projected out for 18 months, beginning with April. The forecast, operations, and inventory sections for the first 6 months are shown on a month-by-month basis. They then are shown on a quarterly basis for the second 6 months. Finally, the totals for the last 6 months in the time horizon are given in just one column. This display gives more precision to the short term and yet gives coverage well into the future—all with a limited number of columns.

This particular make-to-stock family experiences highly seasonal demand. The operations plan is to build up seasonal inventory in the slack season, schedule vacations as much as

TABLE 15.3 | TYPES OF COSTS WITH SALES AND OPERATIONS PLANNING

Cost	Definition
Regular time	Regular-time wages paid to employees plus contributions to benefits, such as health insurance, dental care, Social Security, retirement funds, and pay for vacations, holidays, and certain other types of absences.
Overtime	Wages paid for work beyond the normal workweek, typically 150 percent of regular-time wages (sometimes up to 200 percent for Sundays and holidays), exclusive of fringe benefits. Overtime can help avoid the extra cost of fringe benefits that come with hiring another full-time employee.
Hiring and layoff	Costs of advertising jobs, interviews, training programs for new employees, scrap caused by the inexperience of new employees, loss of productivity, and initial paperwork. Layoff costs include the costs of exit interviews, severance pay, retaining and retraining remaining workers and managers, and lost productivity.
Inventory holding	Costs that vary with the level of inventory investment: the costs of capital tied up in inventory, variable storage and warehousing costs, pilferage and obsolescence costs, insurance costs, and taxes.
Backorder and stockout	Additional costs to expedite past-due orders, the costs of lost sales, and the potential cost of losing a customer to a competitor (sometimes called loss of goodwill).

Artic Air Company—April Sales and Operations Plan

Family: Medium window units (make-to-stock)   *Unit of measure:* 100 units

	HISTORY			A*	M	J	J	A	S	3rd 3 Mos**	4th 3 Mos	Mos 13–18	Fiscal Year Projection ($000)	Business Plan ($000)
SALES	J	F	M											
New forecast	45	55	60	70	85	95	130	110	70	150	176	275	$8,700	$8,560
Actual sales	52	40	63											
Diff for month	7	–15	3											
Cum		–8	–5											
OPERATIONS														
New Plan	75	75	75	75	75	85	85	85	75	177	225			
Actual	75	78	76											
Diff for month	0	3	1											
Cum		3	4											
INVENTORY														
Plan	85	105	120	125	115	105	60	35	40	198	321			
Actual	92	130	143											

DEMAND ISSUES AND ASSUMPTIONS
1. New product design to be launched in January of next year.

SUPPLY ISSUES
1. Vacations primarily in November and December.
2. Overtime in July–August.

*   April is the first month of the planning horizon for this current plan. When next month's plan is developed, its first month in the planning
horizon will be May, and the most recent month of the history will be April (with January no longer shown in the history).

** This column provides the sales, operations, and inventory totals for October through December. For example, the forecast of 150 units
translates into an average of 50 units per month (or 150/3 = 50).

possible in November and December, and use overtime in the peak season of June, July, and
August. For example, the Operations plan increases monthly production from 75 to 85 for June
through August, returns to 75 for September, and then drops to an average of only 59 (or 177/3)
for October through December. Plan spreadsheets use different formats depending on production and inventory strategy. For an assemble-to-order strategy, the inventory does not consist
of finished goods. Instead, it is inventory of standardized components and subassemblies built
for the finishing and assembly operations. For the make-to-order strategy, the inventory section
in the plan of Figure 15.3 is replaced by a section showing the planned and actual order backlog
quantities.

Plans for service providers are quite different. For one thing, their plan does not contain an
inventory section, but focuses instead on the demand and supply of human resources. Forecasts
are typically expressed in terms of employees required, with separate rows for regular time, overtime, vacations, part-time workers, and so on. Different departments or worker classifications replace product families.

The process itself, typically done on a monthly basis, consists of six basic steps. They are
much like the steps we discussed in Chapter 14, "Forecasting."

Step 1. Begin to "roll forward" the plan for the new planning horizon. Start preliminary work
right after the month's end. Update files with actual sales, production, inventory, costs, and
constraints.

Step 2. Participate in the forecasting and demand planning to create the authorized demand forecasts. For service providers, the forecasts are staff requirements for each workforce group. For example, a director of nursing in a hospital can develop a workload index for a nursing staff and
translate a projection of the month-to-month patient load into an equivalent total amount of
nursing care time—and thus the number of nurses—required for each month of the year.

Step 3. Update the sales and operations planning spreadsheet for each
family, recognizing relevant constraints and costs including availability of materials from suppliers, training facilities capable of handling
only so many new hires at a time, machine capacities, or limited storage space. Policy constraints might include limitations on the number
of backorders, or the use of subcontractors or overtime, as well as the
minimum inventory levels needed to achieve desired safety stocks. Typically, many plans can satisfy a specific set of constraints. The planner
searches for a plan that best balances costs, customer service, workforce
stability, and the like. This process may necessitate revising the plan
several times.

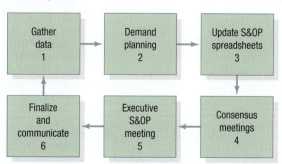

FUJIFILM Imaging Colorants makes inks and dyes, primarily for inkjet printer cartridges, and operates an effective S&OP process. It must coordinate between a U.S. finishing plant and a UK bulk manufacturing plant. Managers from all functions teleconference at the U.S. site with seven other UK managers. At this Partnership meeting (step 4 in the S&OP process), they review the demand, production, and inventory plans, as well as the projected working capital plan.

Step 4. Have one or more consensus meetings with the stakeholders on how best to balance supply with demand. Participants could include the supply chain manager, plant manager, controller, purchasing manager, production control manager, or logistics manager. The goal is one set of recommendations to present at the firm's executive sales and operations planning (S&OP) meeting. Where agreement cannot be reached, prepare scenarios of alternative plans. Also prepare an updated financial view of business by rolling up the plans for all product families into a spreadsheet expressed in total dollars.

Step 5. Present recommendations by product family at the executive S&OP meeting, which typically includes the firm's president and the vice presidents of functional areas. The plan is reviewed relative to the business plan, new product issues, special projects, and other relevant factors. The executives may ask for final changes to the plan, such as to balance conflicting objectives better. Acceptance of this authorized plan does not necessarily mean that everyone is in total agreement, but it does imply that everyone will work to achieve the plan.

Step 6. Update the spreadsheets to reflect the authorized plan, and communicate the plans to the important stakeholders for implementation. Important recipients include those who do resource planning, covered in the next chapter.

Using Spreadsheets

The sales and operations plan in Figure 15.3 does not show much on the supply options used in the operations plan or their cost implications. Here we discuss using spreadsheets that do just that. Supplement D, "Linear Programming," describes using the transportation method for production planning. Both techniques could be used on the side as a planner develops prospective plans in step 3 of the planning process.

Various spreadsheets can be used, including ones that you develop on your own. Here we work with the *Sales and Operations Planning with Spreadsheets* Solver in OM Explorer. Figure 15.4 shows a plan for a manufacturer, which uses all supply options except overtime.

FIGURE 15.4 ▶

Manufacturer's Plan Using a Spreadsheet and Mixed Strategy

	1	2	3	4	5	6	Total
Inputs							
Forecasted demand	24	142	220	180	136	168	870
Workforce level	120	158	158	158	158	158	910
Undertime	6	0	0	0	0	0	6
Overtime	0	0	0	0	0	0	0
Vacation time	20	6	0	0	4	10	40
Subcontracting time	0	0	0	0	0	6	6
Backorders	0	0	0	4	0	0	4
Derived							
Utilized time	94	152	158	158	154	148	864
Inventory	70	80	18	0	14	0	182
Hires	0	38	0	0	0	0	38
Layoffs	0	0	0	0	0	0	0
Calculated							
Utilized time cost	$376,000	$608,000	$632,000	$632,000	$616,000	$592,000	$3,456,000
Undertime cost	$24,000	$0	$0	$0	$0	$0	$24,000
Overtime cost	$0	$0	$0	$0	$0	$0	$0
Vacation time cost	$80,000	$24,000	$0	$0	$16,000	$40,000	$160,000
Inventory cost	$2,800	$3,200	$720	$0	$560	$0	$7,280
Backorders cost	$0	$0	$0	$4,000	$0	$0	$4,000
Hiring cost	$0	$91,200	$0	$0	$0	$0	$91,200
Layoff cost	$0	$0	$0	$0	$0	$0	$0
Subcontracting cost	$0	$0	$0	$0	$0	$43,200	$43,200
Total cost	$482,800	726,400	632,720	636,000	632,560	675,200	$3,785,680

Spreadsheets for a Manufacturer   The top part of the spreadsheet shows the *input values* that give the forecasted demand requirements and the supply option choices period by period. Vary these "levers" as you search for better plans.

The next part of the spreadsheet (in green) shows the *derived values* that must follow from the input values. The first row of derived values is called *utilized time*, which is that portion of the workforce's regular time that is paid for and productively used. In any period, the utilized time equals the workforce level minus undertime and vacation time. For example, in period 1 the utilized time is 94 (or 120 – 6 – 20). The hires and layoffs rows can be derived from the workforce levels. In this example, the workforce is increased for period 2 from its initial size of 120 employees to 158, which means that 38 employees are hired. Because the workforce size remains constant at 158 throughout the rest of the planning horizon, no other hirings or layoffs happen. When additional alternatives, such as vacations, inventory, and backorders are all possible, the overtime and undertime cannot be derived just from information on forecasted demand and workforce levels. Thus, undertime and overtime are shown as input values (rather than derived values) in the spreadsheet, and the user must be careful to specify consistent input values.

The final part of the spreadsheet, the *calculated values* of the plan, shows the plan's cost consequences. Along with qualitative considerations, the cost of each plan determines whether the plan is satisfactory or whether a revised plan should be considered. When seeking clues about how to improve a plan already evaluated, we identify its highest cost elements. Revisions that would reduce these specific costs might produce a new plan with lower overall costs. Spreadsheet programs make analyzing these plans easy, and they present a whole new set of possibilities for developing sound sales and operations plans.

The plan in Figure 15.4 definitely is for a manufacturer because it uses inventory to advantage, particularly in the first two periods. It is a mixed strategy, and not just because it uses anticipation inventory, backorders, and subcontracting. The workforce level changes in period 2, but it does not exactly match the forecasted demand as with a chase strategy. It has some elements of the level strategy, because undertime and vacation time are part of the plan, but it does not rely exclusively on these supply options.

Care must be taken to recognize differences in how inputs are measured. The workforce level might be expressed as the number of employees, but the forecasted demand and inventory are expressed as units of the product. The OM Explorer spreadsheets require a common unit of measure, so we must translate some of the data prior to entering the input values. Perhaps the easiest approach is to express the forecasted demand and supply options as *employee-period equivalents*. If demand forecasts are given as units of product, we can convert them to employee-period equivalents by dividing them by the productivity of a worker. For example, if the demand is for 1,500 units of product and the average employee produces 100 units in one period, the demand requirement is 15 employee-period equivalents.

Spreadsheets for a Service Provider   The same spreadsheets can be used by service providers, except anticipation inventory is not an option. You can unprotect the sheet and then hide the rows that are not relevant. It is useful not to hide the inventory row until the end, however, because positive or negative values signal an inconsistency in your plan. Whereas Figure 15.4 shows a good plan found after several revisions, here we illustrate with Example 15.1 how to find a good plan for a service provider beginning with the chase and level (ignoring vacations) strategies. These plans can provide insights that lead to even better mixed strategy plans.

EXAMPLE 15.1	Using the Chase and Level Strategies as Starting Points

The manager of a large distribution center must determine how many part-time stockpickers to maintain on the payroll. She wants to develop a staffing plan that minimizes total costs, and wants to begin with the chase strategy and level strategy. For the level strategy, she wants to first try the workforce level that meets demand with the minimum use of undertime and not consider vacation scheduling.

First, the manager divides the next year into six time periods, each one 2 months long. Each part-time employee can work a maximum of 20 hours per week on regular time, but the actual number can be less. Instead of paying undertime, each worker's day is shortened during slack periods. Once on the payroll, each worker is used each day, but they may work only a few hours. Overtime can be used during peak periods.

The distribution center's forecasted demand is shown as the number of part-time employees required for each time period at the maximum regular time of 20 hours per week. For example, in period 3, an estimated 18 part-time employees working 20 hours per week on regular time will be needed.

	1	2	3	4	5	6	Total
Forecasted demand*	6	12	18	15	13	14	78

*Number of part-time employees

Currently, 10 part-time clerks are employed. They have not been subtracted from the forecasted demand shown. Constraints and cost information are as follows:

a.   The size of training facilities limits the number of new hires in any period to no more than 10.

b.   No backorders are permitted; demand must be met each period.

c.   Overtime cannot exceed 20 percent of the regular-time capacity (that is, 4 hours) in any period. Therefore, the most that any part-time employee can work is 1.20(20) = 24 hours per week.

d.   The following costs can be assigned:

Regular-time wage rate	$2,000 per time period at 20 hours per week
Overtime wages	150 percent of the regular-time rate
Hires	$1,000 per person
Layoffs	$500 per person

Framed by thousands of ski poles, a part-time worker sorts and inventories new products in the receiving department of REI's distribution center in Sumner, Washington. REI employs a high percentage of part-time workers, many of whom are college students. They tend to be young people who participate in outdoor sports and are familiar with the equipment that REI sells.

AP Photos

MyOMLab

Tutor 15.1 in MyOMLab provides a new example for planning using the chase strategy with hiring and layoffs.

SOLUTION

a.   **Chase Strategy**

This strategy simply involves adjusting the workforce as needed to meet demand, as shown in Figure 15.5. Rows in the spreadsheet that do not apply (such as inventory and vacations) are hidden. The workforce level row is identical to the forecasted demand row. A large number of hirings and layoffs begin with laying off four part-time employees immediately because the current staff is 10 and the staff level required in period 1 is only six. However, many employees, such as college students, prefer part-time work. The total cost is $173,500, and most of the cost increase comes from frequent hiring and layoffs, which add $17,500 to the cost of utilized regular-time costs.

FIGURE 15.5 ▶

Spreadsheet for Chase Strategy

	1	2	3	4	5	6	Total
Inputs							
Forecasted demand	6	12	18	15	13	14	78
Workforce level	6	12	18	15	13	14	78
Undertime	0	0	0	0	0	0	0
Overtime	0	0	0	0	0	0	0
Derived							
Utilized time	6	12	18	15	13	14	78
Hires	0	6	6	0	0	1	13
Layoffs	4	0	0	3	2	0	9
Calculated							
Utilized time cost	$12,000	$24,000	$36,000	$30,000	$26,000	$28,000	$156,000
Undertime cost	$0	$0	$0	$0	$0	$0	$0
Hiring cost	$0	$6,000	$6,000	$0	$0	$1,000	$13,000
Layoff cost	$2,000	$0	$0	$1,500	$1,000	$0	$4,500
Total cost	$14,000	30,000	42,000	31,500	27,000	29,000	$173,500

b. Level Strategy

In order to minimize undertime, the maximum use of overtime possible must occur in the peak period. For this particular level strategy (other workforce options are possible), the most overtime that the manager can use is 20 percent of the regular-time capacity, w, so

$$1.20w = 18 \text{ employees required in peak period (period 3)}$$

$$w = \frac{18}{1.20} = 15 \text{ employees}$$

A 15-employee staff size minimizes the amount of undertime for this level strategy. Because the staff already includes 10 part-time employees, the manager should immediately hire five more. The complete plan is shown in Figure 15.6. The total cost is $164,000, which seems reasonable because the minimum conceivable cost is only $156,000 (78 periods × $2,000/period). This cost could be achieved only if the manager found a way to cover the forecasted demand for all 78 periods with regular time. The plan seems reasonable primarily because it involves the use of large amounts of undertime (15 periods), which in this example are unpaid.

◀ **FIGURE 15.6**
Spreadsheet for Level Strategy

MyOMLab

Tutor 15.2 in MyOMLab provides a new example for planning using the level strategy with overtime and undertime.

	1	2	3	4	5	6	Total
Inputs							
Forecasted demand	6	12	18	15	13	14	78
Workforce level	15	15	15	15	15	15	90
Undertime	9	3	0	0	2	1	15
Overtime	0	0	3	0	0	0	3
Derived							
Utilized time	6	12	15	15	13	14	75
Hires	5	0	0	0	0	0	5
Layoffs	0	0	0	0	0	0	0
Calculated							
Utilized time cost	$12,000	$24,000	$30,000	$30,000	$26,000	$28,000	$150,000
Undertime cost	$0	$0	$0	$0	$0	$0	$0
Overtime cost	$0	$0	$9,000	$0	$0	$0	$9,000
Hiring cost	$5,000	$0	$0	$0	$0	$0	$5,000
Layoff cost	$0	$0	$0	$0	$0	$0	$0
Total cost	$17,000	24,000	39,000	30,000	26,000	28,000	$164,000

DECISION POINT

The manager, now having a point of reference with which to compare other plans, decided to evaluate some other plans before making a final choice, beginning with the chase strategy. The only way to reduce costs is somehow to reduce the premium for three overtime employee periods (3 periods × $3,000/period) or to reduce the hiring cost of five employees (5 hires × $1,000/person). Nonetheless, better solutions may be possible. For example, undertime can be reduced by delaying the hiring until period 2 because the current workforce is sufficient until then. This delay would decrease the amount of unpaid undertime, which is a qualitative improvement. See Active Model 15.1 for additional insights.

MyOMLab

Active Model 15.1 in MyOMLab shows the impact of changing the workforce level, the cost structure, and overtime capacity.

Scheduling

Scheduling is the last step in Figure 15.1. It takes the operations and scheduling process from planning to execution, and is where the "rubber meets the road." This important aspect of supply chain management is itself a process. It requires gathering data from sources such as demand forecasts or specific customer orders, resource availability from the sales and operations plan, and specific constraints to be reckoned with from employees and customers. It then involves generating a work schedule for employees or sequences of jobs or customers at workstations. The schedule has to be coordinated with the employees and suppliers to make sure that all constraints are satisfied. Here we cover Gantt charts, employee schedules, job sequencing at workstations, and software support.

Gantt Charts

Schedules can be displayed in various ways. For different jobs or activities they can simply list their due dates, show in a table their start and finish times, or show in a graph their start and finish times. The *Gantt chart* uses the third approach. Figure 2.5 back in Chapter 2 demonstrates how a "picture can be worth a thousand words" in managing projects. Associates not familiar with scheduling techniques can still grasp the essence of the plan. This tool can be used to monitor the progress of work and to view the load on workstations. The chart takes two basic forms: (1) the job or activity progress chart and (2) the workstation chart. The *Gantt progress chart* graphically displays the current status of each job or activity relative to its scheduled completion date. For example, suppose that an automobile parts manufacturer has three jobs under way, one each for Ford, Nissan, and Buick. The actual status of these orders is shown by the colored bars in Figure 15.7; the red lines indicate the desired schedule for the start and finish of each job. For the current date, April 21, this Gantt chart shows that the Ford order is behind schedule because operations has completed only the work scheduled through April 18. The Nissan order is exactly on schedule, and the Buick order is ahead of schedule.

FIGURE 15.7 ▶
Gantt Progress Chart for an Auto Parts Company

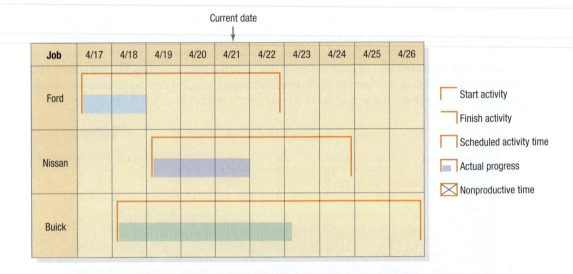

Figure 15.8 shows a *Gantt workstation chart* of the operating rooms at a hospital for a particular day. Using the same notation as in Figure 15.7, the chart shows the load on the operating rooms and the nonproductive time. The time slots assigned to each doctor include the time needed to clean the room prior to the next surgery. The chart can be used to identify time slots for unscheduled emergency surgeries. It can also be used to accommodate requests to change the time of surgeries. For example, Dr. Flowers may be able to change the start of her surgery to 2 P.M. by swapping time slots with Dr. Gillespie in operating room C or by asking Dr. Brothers to start her surgery one hour earlier in operating room A and asking Dr. Bright to schedule her surgery for the morning in operating room C. In any event, the hospital administrator would have to get involved in rescheduling the surgeries.

FIGURE 15.8 ▼
Gantt Workstation Chart for Operating Rooms at a Hospital

Scheduling Employees

workforce scheduling

A type of scheduling that determines when employees work.

Another way to manage capacity is **workforce scheduling**, which is a type of scheduling that determines when employees work. Of particular interest are situations when not all employees work the same five days a week, and same eight hours per day. The schedule specifies the on-duty and off-duty periods for each employee over a certain time period, as in assigning postal clerks, nurses, pilots, attendants, or police officers to specific workdays and shifts. This approach is used when customers demand quick response and total demand can be forecasted with reasonable accuracy. In these instances, capacity is adjusted to meet the expected loads on the service system.

Workforce schedules translate the staffing plan into specific schedules of work for each employee. Determining the workdays for each employee in itself does not make the staffing plan operational. Daily workforce requirements, stated in aggregate terms in the staffing plan, must be satisfied. The workforce capacity available each day must meet or exceed daily workforce requirements. If it does not, the scheduler must try to rearrange days off until the requirements are met. If no such schedule can be found, management might have to change the staffing plan and hire more employees, authorize overtime hours, or allow for larger backlogs.

Constraints   The technical constraints imposed on the workforce schedule are the resources provided by the staffing plan and the requirements placed on the operating system. However, other constraints, including legal and behavioral considerations, also can be imposed. For example, Air New Zealand is required to have at least a minimum number of flight attendants on duty at all times. Similarly, a minimum number of fire and safety personnel must be on duty at a fire station at all times. Such constraints limit management's flexibility in developing workforce schedules.

An air hostess serves coffee to the pilots in the cockpit. Airlines provide service for more than eight hours per day and five days per week, and there are many technical, behavioral, and legal constraints to be met. Such conditions make scheduling a challenging task.

The constraints imposed by the psychological needs of workers complicate scheduling even more. Some of these constraints are written into labor agreements. For example, an employer might agree to give employees a certain number of consecutive days off per week or to limit employees' consecutive workdays to a certain maximum. Other provisions might govern the allocation of vacations, days off for holidays, or rotating shift assignments. In addition, the preferences of the employees themselves need to be considered.

One way that managers deal with certain undesirable aspects of scheduling is to use a **rotating schedule**, which rotates employees through a series of workdays or hours. Thus, over a period of time, each person has the same opportunity to have weekends and holidays off and to work days, as well as evenings and nights. A rotating schedule gives each employee the next employee's schedule the following week. In contrast, a **fixed schedule** calls for each employee to work the same days and hours each week.

rotating schedule

A schedule that rotates employees through a series of workdays or hours.

fixed schedule

A schedule that calls for each employee to work the same days and hours each week.

Developing a Workforce Schedule   Suppose that we are interested in developing a workforce schedule for a company that operates seven days a week and provides each employee with two consecutive days off. In this section, we demonstrate a method that recognizes this constraint. The objective is to identify the two consecutive days off for each employee that will minimize the amount of total slack capacity, thereby maximizing the utilization of the workforce. The work schedule for each employee, then, is the five days that remain after the two days off have been determined. The procedure involves the following steps.

Step 1.   From the schedule of net requirements for the week, find all the pairs of consecutive days, excluding the day (or days) with the maximum daily requirement. Select the unique pair that has the lowest total requirements for the two days. In some unusual situations, all pairs may contain a day with the maximum requirements. If so, select the pair with the lowest total requirements. Suppose that the numbers of employees required are

Monday: 8	Thursday: 12	Saturday: 4
Tuesday: 9	Friday: 7	Sunday: 2
Wednesday: 2		

The maximum daily requirement is 12 employees, on Thursday. The consecutive pair with the lowest total requirements is Saturday and Sunday, with $4 + 2 = 6$.

Step 2.   If a tie occurs, choose one of the tied pairs, consistent with the provisions written into the labor agreement, if any. Alternatively, the tie could be broken by asking the employee being scheduled to make the choice. As a last resort, the tie could be broken arbitrarily. For example, preference could be given to Saturday–Sunday pairs.

Step 3. Assign the employee the selected pair of days off. Subtract the requirements satisfied by the employee from the net requirements for each day the employee is to work. In this example, the employee is assigned Saturday and Sunday off. After requirements are subtracted, Monday's requirement is 7, Tuesday's is 8, Wednesday's is 1, Thursday's is 11, and Friday's is 6. Saturday's and Sunday's requirements do not change because no employee is yet scheduled to work those days.

Step 4. Repeat steps 1 through 3 until all the requirements have been satisfied or a certain number of employees have been scheduled.

This method reduces the amount of slack capacity assigned to days with low requirements and forces the days with high requirements to be scheduled first. It also recognizes some of the behavioral and contractual aspects of workforce scheduling in the tie-breaking rules.

EXAMPLE 15.2	Developing a Workforce Schedule

The Amalgamated Parcel Service is open seven days a week. The schedule of requirements is

Day	M	T	W	Th	F	S	Su
Required number of employees	6	4	8	9	10	3	2

The manager needs a workforce schedule that provides two consecutive days off and minimizes the amount of total slack capacity. To break ties in the selection of off days, the scheduler gives preference to Saturday and Sunday if it is one of the tied pairs. If not, she selects one of the tied pairs arbitrarily.

MyOMLab

Tutor 15.3 in MyOMLab provides a new example to practice workforce scheduling.

SOLUTION

Friday contains the maximum requirements, and the pair S–Su has the lowest total requirements. Therefore, Employee 1 is scheduled to work Monday through Friday.

Note that Friday still has the maximum requirements and that the requirements for the S–Su pair are carried forward because these are Employee 1's days off. These updated requirements are the ones the scheduler uses for the next employee.

The day-off assignments for the employees are shown in the following table.

SCHEDULING DAYS OFF

M	T	W	Th	F	S	Su	Employee	Comments
6	4	8	9	10	3	2	1	The S–Su pair has the lowest total requirements. Assign Employee **1** to a Monday through Friday schedule and update the requirements.
5	3	7	8	9	3	2	2	The S–Su pair has the lowest total requirements. Assign Employee **2** to a Monday through Friday schedule and update the requirements.
4	2	6	7	8	3	2	3	The S–Su pair has the lowest total requirements. Assign Employee **3** to a Monday through Friday schedule and update the requirements.
3	1	5	6	7	3	2	4	The M–T pair has the lowest total requirements. Assign Employee **4** to a Wednesday through Sunday schedule and update the requirements.
3	1	4	5	6	2	1	5	The S–Su pair has the lowest total requirements. Assign Employee **5** to a Monday through Friday schedule and update the requirements.
2	0	3	4	5	2	1	6	The M–T pair has the lowest total requirements. Assign Employee **6** to a Wednesday through Sunday schedule and update the requirements.
2	0	2	3	4	1	0	7	The S–Su pair has the lowest total requirements. Assign Employee **7** to a Monday through Friday schedule and update the requirements.
1	0	1	2	3	1	0	8	Four pairs have the minimum requirement and the lowest total: S–Su, Su–M, M–T, and T–W. Choose the S–Su pair according to the tie-breaking rule. Assign Employee **8** to a Monday through Friday schedule and update the requirements.
0	0	0	1	2	1	0	9	Arbitrarily choose the Su–M pair to break ties because the S–Su pair does not have the lowest total requirements. Assign Employee **9** to a Tuesday through Saturday schedule and update the requirements.
0	0	0	0	1	0	0	10	Choose the S–Su pair according to the tie-breaking rule. Assign Employee **10** to a Monday through Friday schedule.

In this example, Friday always has the maximum requirements and should be avoided as a day off. The final schedule for the employees is shown in the following table.

FINAL SCHEDULE

Employee	M	T	W	Th	F	S	Su	Total
1	X	X	X	X	X	off	off	
2	X	X	X	X	X	off	off	
3	X	X	X	X	X	off	off	
4	off	off	X	X	X	X	X	
5	X	X	X	X	X	off	off	
6	off	off	X	X	X	X	X	
7	X	X	X	X	X	off	off	
8	X	X	X	X	X	off	off	
9	off	X	X	X	X	X	off	
10	X	X	X	X	X	off	off	
Capacity, C	7	8	10	10	10	3	2	50
Requirements, R	6	4	8	9	10	3	2	42
Slack, C − R	1	4	2	1	0	0	0	8

DECISION POINT

With its substantial amount of slack capacity, the schedule is not unique. Employee 9, for example, could have Sunday and Monday, Monday and Tuesday, or Tuesday and Wednesday off without causing a capacity shortage. Indeed, the company might be able to get by with one fewer employee because of the total of eight slack days of capacity. However, all 10 employees are needed on Fridays. If the manager were willing to get by with only nine employees on Fridays or if someone could work one day of overtime on a rotating basis, he would not need Employee 10. As indicated in the table, the net requirement left for Employee 10 to satisfy amounts to only one day, Friday. Thus, Employee 10 can be used to fill in for vacationing or sick employees.

Managers try to staff their call centers to meet certain performance measures. One such measure is the percentage of calls answered (PCA) within a specified time interval, termed the service objective (SO). Typically in a call center, the PCA is in the range of 80 to 90 percent, and the SO is 15 to 30 seconds. The problem is that the requirements for agents change over time, depending on the time of day and day of the year. Also, the callers are likely to speak different languages. Determining how many of each group of agents to have on hand at all times is not easy. Fortunately, employee scheduling software is available to estimate call volumes, project skill requirements, and identify employee start and end times or preferred days off.

Sequencing Jobs at a Workstation

Another aspect of scheduling is sequencing work at workstations. **Sequencing** determines the order in which jobs or customers are processed in the waiting line at a workstation. When combined with the expected processing time, the sequence allows you to estimate the start and finish times of each job.

sequencing

Determining the order in which jobs or customers are processed in the waiting line at a workstation.

priority sequencing rule

A rule that specifies the job or customer processing sequence when several jobs are waiting in line at a workstation.

first-come, first served (FCFS)

A priority sequencing rule that specifies that the job or customer arriving at the workstation first has the highest priority.

earliest due date (EDD)

A priority sequencing rule that specifies that the job or customer with the earliest due date is the next job to be processed.

expediting

The process of completing a job or finishing with a customer sooner than would otherwise be done.

flow time

The amount of time a job spends in the service or manufacturing system.

past due

The amount of time by which a job missed its due date.

tardiness

See past due.

Priority Sequencing Rules   One way to determine what job or customer to process next is with the help of a **priority sequencing rule**. The following two priority sequencing rules are commonly used in practice.

- *First-Come, First-Served.* The job or customer arriving at the workstation first has the highest priority under a **first-come, first-served (FCFS)** rule. This rule is the most "democratic" in that each job is treated equally, with no one stepping ahead of others already in line. It is commonly used at service facilities, and is the rule that was assumed in Supplement B, "Waiting Lines."

- *Earliest Due Date.* The job or customer with the **earliest due date (EDD)** is the next one to be processed. The *due date* specifies when work on a job or customer should be finished. Due dates are commonly used by manufacturers and suppliers in the supply chain. For example, a product cannot be assembled until all of its purchased and produced components are available. If these components were not already in inventory, they must be ordered prior to when the product assembly can begin. Their due date is the start date for assembling the product to be assembled. This simple relationship is fundamental to coordinating with suppliers and with the manufacturer's own shops in the supply chain. It is also the key to **expediting**, which is the process of completing a job sooner than would otherwise be done. Expediting can be done by revising the due date, moving the job to the front of the waiting line, making a special appeal by phone or e-mail to the supplier, adding extra capacity, or even putting a red tag on the job that says the job is urgent.

Neither rule guarantees finding an optimal solution. Different sequences found by trial and error can produce better schedules. In fact, there are multiple performance measures for judging a schedule. A schedule that does well on one measure may do poorly on another.

Performance Measures   The quality of a schedule can be judged in various ways. Two commonly used performance measures are flow time and past due.

- *Flow Time.* The amount of time a job spends in the service or manufacturing system is called **flow time**. It is the sum of the waiting time for servers or machines; the process time, including setups; the time spent moving between operations; and delays resulting from machine breakdowns, unavailability of facilitating goods or components, and the like. Flow time is sometimes referred to as *throughput time* or *time spent in the system, including service.* For a set of jobs to be processed at a single workstation. a job's flow time is

$$\text{Flow time} = \text{Finish time} + \text{Time since job arrived at workstation}$$

When using this equation, we assume for convenience that the first job scheduled starts at time zero (0). At time 0, all the jobs were available for processing at the workstation.

- *Past Due.* The measure **past due** can be expressed as the amount of time by which a job missed its due date (also referred to as **tardiness**) or as the percentage of total jobs processed over some period of time that missed their due dates. Minimizing these past due measures supports the competitive priorities of cost (penalties for missing due dates), quality (perceptions of poor service), and time (on-time delivery).

EXAMPLE 15.3	**Using the Two Priority Sequencing Rules**

Currently a consulting company has five jobs in its backlog. The time since the order was placed, processing time, and promised due dates are given in the following table. Determine the schedule by using the FCFS rule, and calculate the average days past due and flow time. How can the schedule be improved, if average flow time is the most critical?

Customer	Time Since Order Arrived (days ago)	Processing Time (days)	Due Date (days from now)
A	15	25	29
B	12	16	27
C	5	14	68
D	10	10	48
E	0	12	80

SOLUTION

a. The FCFS rule states that Customer A should be the first one in the sequence, because that order arrived earliest—15 days ago. Customer E's order arrived today, so it is processed last. The sequence is shown in the following table, along with the days past due and flow times.

Customer Sequence	Start Time (days)		Processing Time (days)		Finish Time (days)	Due Date	Days Past Due	Days Ago Since Order Arrived	Flow Time (days)
A	0	+	25	=	25	29	0	15	40
B	25	+	16	=	41	27	14	12	53
D	41	+	10	=	51	48	3	10	61
C	51	+	14	=	65	68	0	5	70
E	65	+	12	=	77	80	0	0	77

The *finish time* for a job is its start time plus the processing time. Its finish time becomes the start time for the next job in the sequence, assuming that the next job is available for immediate processing. The days past due for a job is zero (0) if its due date is equal to or exceeds the finish time. Otherwise it equals the shortfall. The flow time for each job equals its finish time plus the number of days ago since the order first arrived at the workstation. For example, Customer C's flow time is its scheduled finish time of 65 days plus the 5 days since the order arrived, or 70 days. The days past due and average flow time performance measures for the FCFS schedule are

$$\text{Average days past due} = \frac{0 + 14 + 3 + 0 + 0}{5} = 3.4 \text{ days}$$

$$\text{Average flow time} = \frac{40 + 53 + 61 + 70 + 77}{5} = 60.2 \text{ days}$$

b. The average flow time can be reduced. One possibility is the sequence shown in the following table, which uses the Shortest Processing Time (SPT) rule, which is one of several rules developed more fully in MyOMLab Supplement J, "Operations Scheduling." (For still another possibility, see Solved Problem 3, which applies the EDD rule.)

MyOMLab

Customer Sequence	Start Time (days)		Processing Time (days)		Finish Time (days)	Due Date	Days Past Due	Days Ago Since Order Arrived	Flow Time (days)
D	0	+	10	=	10	48	0	10	20
E	10	+	12	=	22	80	0	0	22
C	22	+	14	=	36	68	0	5	41
B	36	+	16	=	52	27	25	12	64
A	52	+	25	=	77	29	48	15	92

$$\text{Average days past due} = \frac{0 + 0 + 0 + 25 + 48}{5} = 14.6 \text{ days}$$

$$\text{Average flow time} = \frac{20 + 22 + 41 + 64 + 92}{5} = 47.8 \text{ days}$$

This schedule reduces the average flow time from 60.2 to 47.8 days—a 21 percent improvement. However, the past due times for jobs A and B have increased.

DECISION POINT

Management decided to use a modified version of the second schedule, adding overtime when Customer B is processed. Further, Customer A agreed to extend its due date to 77 days, because in this case the advanced warning allowed it to reschedule its own operations with little problem.

When Nissan introduced the Almera to the European market, it decided to produce it at the most efficient Sunderland plant in the United Kingdom. It already manufactured the Micra and Primera models there. Multi-vehicle scheduling is quite complex because of the many constraints involved. The painting portion of the job is particularly time consuming, but is only one of thousands of tasks the plant must efficiently schedule. A sophisticated software package, called ILOG Solver, assists the scheduling process for a coordinated flow.

Software Support

Computerized scheduling systems are available to cope with the complexity of workforce scheduling, such as the myriad constraints and concerns at Air New Zealand. In some types of firms, such as telephone companies, mail-order catalog houses, or emergency hotline agencies, employees must be on duty 24 hours a day, 7 days a week.

Sometimes a portion of the staff is part time, which allows management a great deal of flexibility but adds considerable complexity to the scheduling requirements. The flexibility comes from the opportunity to match anticipated loads closely through the use of overlapping shifts or odd shift lengths; the complexity comes from the need to evaluate the numerous possible alternatives. Management also must consider the timing of lunch breaks and rest periods, the number and starting times of shift schedules, and the days off for each employee. The programs select the schedule that minimizes the sum of expected costs of over- and understaffing.

Software is also available for sequencing jobs at workstations. They help firms design and manage the linkages between customers and suppliers in the supply chain. True integration requires the manipulation of large amounts of complex data in real time because the customer order work flow must be synchronized with the required material, manufacturing, and distribution activity. Coupled with the Internet and improved data storage and manipulation methods, such computer software has given rise to **advanced planning and scheduling (APS) systems**, which seek to optimize resources across the supply chain and align daily operations with strategic goals. A firm's ability to change its schedules quickly and still keep the goods and services flowing smoothly through the supply chain provides a competitive edge.

advanced planning and scheduling (APS) systems

Computer software systems that seek to optimize resources across the supply chain and align daily operations with strategic goals.

LEARNING GOALS IN REVIEW

① **Describe the operations planning and scheduling process.** The section "Stages in Operations Planning and Scheduling," pp. 509–511, shows its various stages and how it relates with other plans in the organization. Pay particular attention to Figure 15.1.

② **Explain why the process of matching supply with demand begins with aggregation.** Aggregation allows the general course of action to be established, without being bogged down in details. The section on "Aggregation," p. 509, shows that aggregation is performed across three dimensions: product families, workforce, and time.

③ **Identify the different demand and supply options.** Sections "Managing Demand" and "Sales and Operations Plans," pp. 511–521, describe the various options, along with their positive and negative aspects. Focus on Table 15.2 for the complete list.

④ **Explain how operations plans and schedules relate to other plans.** Figure 15.2 shows how Sales and Operations Plans are related with other functional areas. Also, check out the section "The Relationship of Operations Plans and Schedules to Other Plans," pp. 509–511.

⑤ **Use spreadsheets to create sales and operations plans.** Turn to "Using Spreadsheets" on pp. 518–521 to better understand how to use this tool. Solved Problem 1 is also helpful.

⑥ **Develop employee schedules.** The section on "Scheduling Employees," pp. 522–525, describes how to create a workforce schedule. Also see Example 15.2, pp. 524–525, and Solved Problem 2, pp. 531–532.

⑦ **Develop schedules for single workstations.** The FCFS and EDD rules, and the flow time and past due performance measures, are described in the "Sequencing Jobs at a Workstation," pp. 525–527. Example 15.3 illustrates their use, along with Solved Problem 3.

MyOMLab helps you develop analytical skills and assesses your progress with multiple problems on level, chase, and mixed strategy plans, total costs, anticipation inventory, break-even analysis, FCFS and EDD scheduling rules, average flow time, and average days past due.

MyOMLab Resources	Titles	Link to the Book
Video	Air New Zealand: Service Scheduling	Operations Planning and Scheduling Across the Organization; Stages in Operations Planning and Scheduling;
	Sales and Operations Planning at Starwood	Stages in Operations Planning and Scheduling; Demand Options
Active Model Exercise	15.1 Level Strategy	Planning Strategies; Example 15.1 (pp. 519–521)
OM Explorer Solvers	Sales and Operations Planning with Spreadsheets	Using Spreadsheets; Figure 15.3 (p. 510); Figure 15.4 (p. 518); Solved Problem 1 (pp. 530–531)
	Workforce Scheduler	Scheduling Employees; Example 15.2 (pp. 524–525); Solved Problem 2
	Single Work-Station Scheduler	Sequencing Jobs at a Workstation; Example 15.3 (pp. 526–527); Solved Problem 3 (p. 533)
OM Explorer Tutors	15.1 Chase Strategy	Planning Strategies; Example 15.1 (pp. 519–521); Solved Problem 1
	15.2 Level Strategy	Planning Strategies; Example 15.1 (pp. 519–521); Solved Problem 1
	15.3 Developing a Workforce Schedule	Scheduling Employees; Example 15.2 (pp. 524–525); Solved Problem 2
	15.4 Staffing Strategies with Spreadsheets (4 pds)	Scheduling Employees; Example 15.2 (pp. 524–525); Solved Problem 2
Tutor Exercises	15.1 Results of Different Scenarios with a Level Strategy	Planning Strategies; Example 15.1 (pp. 519–521); Solved Problem 2
	15.2 Staffing for the Newest MBA Class	Scheduling Employees; Example 15.2 (pp. 524–525)
POM for Windows	Scheduling	Sequencing Jobs at a Workstation; Example 15.3 (pp. 526–527); Solved Problem 3 (p. 533)
Virtual Tours	Stratton Furniture Company	Stages in Operations Planning and Scheduling; Demand Options; Supply Options; Planning Strategies
	Stihl Chain Saws	
	Stihl Factory Tour	
MyOMLab Supplements	H. Measuring Output Rates	MyOMLab Supplement H
	I. Learning Curve Analysis	MyOMLab Supplement I
Internet Exercise	United Parcel Service	Managing Demand; Supply Options; Planning Strategies
	United Parcel Service Centers	
	Internal Revenue Service	Managing Demand; Supply Options; Planning Strategies
	H&R Block	
	Hoovers	
Additional Case	Food King	Scheduling Employees; Example 15.2 (pp. 524–525)
Key Equations		
Image Library		

Key Terms

advanced planning and scheduling (APS) systems 528
aggregate plan 509
annual plan or financial plan 510
backlog 513
backorder and stockout 516
business plan 509
chase strategy 515
complementary products 511
demand management 511
earliest due date (EDD) 526
expediting 526
first-come, first-served (FCFS) 526

fixed schedule 523
flow time 526
hiring and layoff 516
inventory holding 516
level strategy 515
mixed strategy 516
operations planning and scheduling 508
overtime 514
past due 526
priority sequencing rule 526
product family 509
production plan 509

regular time 516
resource plan 509
revenue management 512
rotating schedule 523
sales and operations planning (S&OP) 509
schedule 509
sequencing 525
staffing plan 509
tardiness 526
undertime 515
workforce scheduling 522

Solved Problem 1

MyOMLab

Tutor 15.4 in MyOMLab provides another example for practicing sales and operations planning using a variety of strategies.

The Cranston Telephone Company employs workers who lay telephone cables and perform various other construction tasks. The company prides itself on good service and strives to complete all service orders within the planning period in which they are received.

Each worker puts in 600 hours of regular time per planning period and can work as many as an additional 100 hours of overtime. The operations department has estimated the following workforce requirements for such services over the next four planning periods:

Planning Period	1	2	3	4
Demand (hours)	21,000	18,000	30,000	12,000

Cranston pays regular-time wages of $6,000 per employee per period for any time worked up to 600 hours (including undertime). The overtime pay rate is $15 per hour over 600 hours. Hiring, training, and outfitting a new employee costs $8,000. Layoff costs are $2,000 per employee. Currently, 40 employees work for Cranston in this capacity. No delays in service, or backorders, are allowed. Use the spreadsheet approach to answer the following questions:

a. Prepare a chase strategy using only hiring and layoffs. What are the total numbers of employees hired and laid off?

b. Develop a workforce plan that uses the level strategy, relaying only on overtime and undertime. Maximize the use of overtime during the peak period so as to minimize the workforce level and amount of undertime.

c. Propose an effective mixed-strategy plan.

d. Compare the total costs of the three plans.

SOLUTION

a. The chase strategy workforce is calculated by dividing the demand for each period by 600 hours, or the amount or regular-time work for one employee during one period. This strategy calls for a total of 20 workers to be hired and 40 to be laid off during the four-period plan. Figure 15.9 shows the "chase strategy" solution that OM Explorer's *Sales and Operations Planning with Spreadsheets* Solver produces. We simply hide any unneeded columns and rows in this general-purpose solver.

FIGURE 15.9 ▶

Spreadsheet for Chase Strategy

	1	2	3	4	Total
Inputs					
Forecasted demand	35	30	50	20	135
Workforce level	35	30	50	20	135
Undertime	0	0	0	0	0
Overtime	0	0	0	0	0
Derived					
Utilized time	35	30	50	20	135
Hires	0	0	20	0	20
Layoffs	5	5	0	30	40
Calculated					
Utilized time cost	$210,000	$180,000	$300,000	$120,000	$810,000
Undertime cost	$0	$0	$0	$0	$0
Overtime cost	$0	$0	$0	$0	$0
Hiring cost	$0	$0	$160,000	$0	$160,000
Layoff cost	$10,000	$10,000	$0	$60,000	$80,000
Total cost	$220,000	190,000	460,000	180,000	$1,050,000

b. The peak demand is 30,000 hours in period 3. As each employee can work 700 hours per period (600 on regular time and 100 on overtime), the workforce level of the level strategy that minimizes undertime is $30,000/700 = 42.86$, or 43 employees. This strategy calls for three employees to be hired in the first quarter and for none to be laid off. To convert the demand requirements into employee-period equivalents, divide the demand in hours by 600. For example, the demand of 21,000 hours in period 1 translates into 35 employee-period equivalents (21,000/600) and demand in period 3 translates into 50 employee-period equivalents (30,000/600). Figure 15.10 shows OM Explorer's spreadsheet for this level strategy that minimizes undertime.

	1	2	3	4	Total
Inputs					
Forecasted demand	35	30	50	20	135
Workforce level	43	43	43	43	172
Undertime	8	13	0	23	44
Overtime	0	0	7	0	7
Derived					
Utilized time	35	30	43	20	128
Hires	3	0	0	0	3
Layoffs	0	0	0	0	0
Calculated					
Utilized time cost	$210,000	$180,000	$258,000	$120,000	$768,000
Undertime cost	$48,000	$78,000	$0	$138,000	$264,000
Overtime cost	$0	$0	$63,000	$0	$63,000
Hiring cost	$24,000	$0	$0	$0	$24,000
Layoff cost	$0	$0	$0	$0	$0
Total cost	$282,000	258,000	321,000	258,000	$1,119,000

◀ **FIGURE 15.10**
Spreadsheet for Level Strategy

c. The mixed-strategy plan that we propose uses a combination of hires, layoffs, and overtime to reduce total costs. The workforce is reduced by 5 at the beginning of the first period, increased by 8 in the third period, and reduced by 13 in the fourth period. Figure 15.11 shows the results.

	1	2	3	4	Total
Inputs					
Forecasted demand	35	30	50	20	135
Workforce level	35	35	43	30	143
Undertime	0	5	0	10	15
Overtime	0	0	7	0	7
Derived					
Utilized time	35	30	43	20	128
Hires	0	0	8	0	8
Layoffs	5	0	0	13	18
Calculated					
Utilized time cost	$210,000	$180,000	$258,000	$120,000	$768,000
Undertime cost	$0	$30,000	$0	$60,000	$90,000
Overtime cost	$0	$0	$63,000	$0	$63,000
Hiring cost	$0	$0	$64,000	$0	$64,000
Layoff cost	$10,000	$0	$0	$26,000	$36,000
Total cost	$220.000	210,000	385,000	206,000	$1,021,000

◀ **FIGURE 15.11**
Spreadsheet for Mixed Strategy

d. The total cost of the chase strategy is $1,050,000. The level strategy results in a total cost of $1,119,000. The mixed-strategy plan was developed by trial and error and results in a total cost of $1,021,000. Further improvements are possible.

Solved Problem 2

The Food Bin grocery store operates 24 hours per day, 7 days per week. Fred Bulger, the store manager, has been analyzing the efficiency and productivity of store operations recently. Bulger decided to observe the need for checkout clerks on the first shift for a one-month period. At the end of the month, he calculated the average number of checkout registers that should be open during the first shift each day. His results showed peak needs on Saturdays and Sundays.

Day	M	T	W	Th	F	S	Su
Number of Clerks Required	3	4	5	5	4	7	8

Bulger now has to come up with a workforce schedule that guarantees each checkout clerk two consecutive days off, but still covers all requirements.

a. Develop a workforce schedule that covers all requirements while giving two consecutive days off to each clerk. How many clerks are needed? Assume that the clerks have no preference regarding which days they have off.

b. Plans can be made to use the clerks for other duties if slack or idle time resulting from this schedule can be determined. How much idle time will result from this schedule, and on what days?

SOLUTION

a. We use the method demonstrated in Example 15.2 to determine the number of clerks needed. The minimum number of clerks is eight.

	DAY						
	M	**T**	**W**	**Th**	**F**	**S**	**Su**
Requirements	3	4	5	5	4	7	8*
Clerk 1	off	off	X	X	X	X	X
Requirements	3	4	4	4	3	6	7*
Clerk 2	off	off	X	X	X	X	X
Requirements	3	4	3	3	2	5	6*
Clerk 3	X	X	X	off	off	X	X
Requirements	2	3	2	3	2	4	5*
Clerk 4	X	X	X	off	off	X	X
Requirements	1	2	1	3	2	3	4*
Clerk 5	X	off	off	X	X	X	X
Requirements	0	2	1	2	1	2	3*
Clerk 6	off	off	X	X	X	X	X
Requirements	0	2*	0	1	0	1	2*
Clerk 7	X	X	off	off	X	X	X
Requirements	0	1*	0	1*	0	0	1*
Clerk 8	X	X	X	X	off	off	X
Requirements	0	0	0	0	0	0	0

*Maximum requirements

b. Based on the results in part (a), the number of clerks on duty minus the requirements is the number of idle clerks available for other duties:

	M	**T**	**W**	**Th**	**F**	**S**	**Su**
Number on duty	5	4	6	5	5	7	8
Requirements	3	4	5	5	4	7	8
Idle clerks	2	0	1	0	1	0	0

The slack in this schedule would indicate to Bulger the number of employees he might ask to work part time (fewer than 5 days per week). For example, Clerk 7 might work Tuesday, Saturday, and Sunday and Clerk 8 might work Tuesday, Thursday, and Sunday. That would eliminate slack from the schedule.

Solved Problem 3

Revisit Example 15.3, where the consulting company has five jobs in its backlog. Create a schedule using the EDD rule, calculating the average days past due and flow time. In this case, does EDD outperform the FCFS rule?

SOLUTION

Customer Sequence	Start Time (days)		Processing Time (days)		Finish Time (days)	Due Date	Days Past Due	Days Ago Since Order Arrived	Flow Time (days)
B	0	+	16	=	16	27	**0**	12	**28**
A	16	+	25	=	41	29	**12**	15	**56**
D	41	+	10	=	51	48	**3**	10	**61**
C	51	+	14	=	65	68	**0**	5	**70**
E	65	+	12	=	77	80	**0**	0	**77**

The days past due and average flow time performance measures for the EDD schedule are

$$\text{Average days past due} = \frac{0 + 12 + 3 + 0 + 0}{5} = 3.0 \text{ days}$$

$$\text{Average flow time} = \frac{28 + 56 + 61 + 70 + 77}{5} = 58.4 \text{ days}$$

By both measures, EDD outperforms the FCFS (3.0 versus 3.4 past due and 58.4 versus 60.2 flow time). However, the solution found in part (b) of Example 15.3 still has the best average flow time of only 47.8 days.

Discussion Questions

1. Quantitative methods can help managers evaluate alternative sales and operations plans on the basis of cost. These methods require cost estimates for each of the controllable variables, such as overtime, subcontracting, hiring, firing, and inventory investment. Say that the existing workforce is made up of 10,000 workers, each having skills valued at $40,000 per year. The plan calls for "creating alternative career opportunities"—in other words, laying off 500 employees. List the types of costs incurred when employees are laid off, and make a rough estimate of the length of time required for payroll savings to recover restructuring costs. If business is expected to improve in one year, are layoffs financially justified? What costs are incurred in a layoff that are difficult to estimate in monetary terms?

2. In your community, some employers maintain stable workforces at all costs, and others furlough and recall workers seemingly at the drop of a hat. What are the differences in markets, management, products, financial position, skills, costs, and competition that could explain these two extremes in personnel policy?

3. Consider the revenue management policies used by Harrah's Cherokee Casino & Hotel, as described in Managerial Practice 15.1. From a business ethics perspective, argue for the policies. Now argue against them.

4. Explain why management should be concerned about priority systems in service and manufacturing organizations.

Problems

The OM Explorer and POM for Windows software is available to all students using the 10th edition of this textbook. Go to **www.pearsonhighered.com/krajewski** to download these computer packages. If you purchased MyOMLab, you also have access to Active Models software and significant help in doing the following problems. Check with your instructor on how best to use these resources. In many cases, the instructor wants you to understand how to do the calculations by hand. At the least, the software provides a check on your calculations. When calculations are particularly complex and the goal is interpreting the results in making decision, the software entirely replaces the manual calculations.

1. The Barberton Municipal Division of Road Maintenance is charged with road repair in the city of Barberton and the surrounding area. Cindy Kramer, road maintenance director, must submit a staffing plan for the next year based on a set schedule for repairs and on the city budget. Kramer estimates that the labor hours required for the next four quarters are 6,000, 12,000, 19,000, and 9,000, respectively. Each of the 11 workers on the workforce can contribute 500 hours per quarter. Payroll costs are $6,000 in wages per worker for regular time worked up to 500 hours, with an overtime pay rate of $18 for each overtime hour. Overtime is limited to 20 percent of the

regular-time capacity in any quarter. Although unused overtime capacity has no cost, unused regular time is paid at $12 per hour. The cost of hiring a worker is $3,000, and the cost of laying off a worker is $2,000. Subcontracting is not permitted.

a. Find a level workforce plan that relies just on overtime and the minimum amount of undertime possible. Overtime can be used to its limits in any quarter. What is the total cost of the plan and how many undertime hours does it call for?

b. Use a chase strategy that varies the workforce level without using overtime or undertime. What is the total cost of this plan?

c. Propose a plan of your own. Compare your plan with those in part (a) and part (b) and discuss its comparative merits.

2. Bob Carlton's golf camp estimates the following workforce requirements for its services over the next 2 years.

Quarter	1	2	3	4
Demand (hours)	4,200	6,400	3,000	4,800
Quarter	5	6	7	8
Demand (hours)	4,400	6,240	3,600	4,800

Each certified instructor puts in 480 hours per quarter regular time and can work an additional 120 hours overtime. Regular-time wages and benefits cost Carlton $7,200 per employee per quarter for regular time worked up to 480 hours, with an overtime cost of $20 per hour. Unused regular time for certified instructors is paid at $15 per hour. There is no cost for unused overtime capacity. The cost of hiring, training, and certifying a new employee is $10,000. Layoff costs are $4,000 per employee. Currently, eight employees work in this capacity.

a. Find a workforce plan using the level strategy that allows for no delay in service. It should rely only on overtime and the minimum amount of undertime necessary. What is the total cost of this plan?

b. Use a chase strategy that varies the workforce level without using overtime or undertime. What is the total cost of this plan?

c. Propose a better plan and calculate its total cost.

3. Continuing Problem 2, now assume that Carlton is permitted to employ some uncertified, part-time instructors, provided they represent no more than 15 percent of the total workforce hours in any quarter. Each part-time instructor can work up to 240 hours per quarter, with no overtime or undertime cost. Labor costs for part-time instructors are $12 per hour. Hiring and training costs are $2,000 per uncertified instructor, and there are no layoff costs.

a. Propose a low-cost, mixed-strategy plan and calculate its total cost.

b. What are the primary advantages and disadvantages of having a workforce consisting of both regular and temporary employees?

4. The Donald Fertilizer Company produces industrial chemical fertilizers. The projected manufacturing requirements (in thousands of gallons) for the next four quarters are 80, 50, 80, and 130, respectively. A level workforce is desired, relying only on anticipation inventory as a supply option. Stockouts and backorders are to be avoided, as are overtime and undertime.

a. Determine the quarterly production rate required to meet total demand for the year, and minimize the anticipation inventory that would be left over at the end of the year. Beginning inventory is zero.

b. Specify the anticipation inventory that will be produced.

c. Suppose that the requirements for the next four quarters are revised to 80, 130, 50, and 80, respectively. If total demand is the same, what level of production rate is needed now, using the same strategy as part (a)?

5. Management at the Kerby Corporation has determined the following aggregated demand schedule (in units):

Month	1	2	3	4
Demand	500	800	1,000	1,400
Month	5	6	7	8
Demand	2,000	3,000	2,700	1,500
Month	9	10	11	12
Demand	1,400	1,500	2,000	1,200

An employee can produce an average of 10 units per month. Each worker on the payroll costs $2,000 in regular-time wages per month. Undertime is paid at the same rate as regular time. In accordance with the labor contract in force, Kerby Corporation does not work overtime or use subcontracting. Kerby can hire and train a new employee for $2,000 and lay off one for $500. Inventory costs $32 per unit on hand at the end of each month. At present, 140 employees are on the payroll and anticipation inventory is zero.

a. Prepare a production plan that only uses a level workforce and anticipation inventory as its supply options. Minimize the inventory left over at the end of the year. Layoffs, undertime, vacations, subcontracting, backorders, and stockouts are not options. The plan may call for a one-time adjustment of the workforce before month 1 begins.

b. Prepare a production plan using a chase strategy, relying only on hiring and layoffs.

c. Prepare a mixed-strategy production plan that uses only a level workforce and anticipation inventory through month 7 (an adjustment of the workforce may be made before month 1 begins) then switches to a chase strategy for months 8–12.

d. Contrast these three plans on the basis of annual costs.

6. Tax Prep Advisers, Inc., has forecasted the following staffing requirements for tax preparation associates over the next 12 months. Management would like three alternative staffing plans to be developed.

Month	1	2	3	4
Demand	5	8	10	13
Month	5	6	7	8
Demand	18	20	20	14
Month	9	10	11	12
Demand	12	8	2	1

The company currently has 10 associates. No more than 10 new hires can be accommodated in any month because of limited training facilities. No backorders are allowed, and overtime cannot exceed 25 percent of regular time capacity on any month. There is no cost for unused overtime capacity. Regular-time wages are $1,500 per month, and overtime wages are 150 percent of regular time wages. Undertime is paid at the same rate as regular time. The hiring cost is $2,500 per person, and the layoff cost is $2.000 per person.

a. Prepare a staffing plan utilizing a level workforce strategy, minimizing undertime. The plan may call for a one-time adjustment of the workforce before month 1.

b. Using a chase strategy, prepare a plan that is consistent with the constraint on hiring and minimizes use of overtime.

c. Prepare a mixed strategy in which the workforce level is slowly increased by two employees per month through month 5 and is then decreased by two employees per month starting in month 6 and continuing through month 12. Does this plan violate the hiring or overtime constraints set the company?

d. Contrast these three plans on the basis of annual costs.

7. Climate Control, Inc., makes expedition-quality rain gear for outdoor enthusiasts. Management prepared a forecast of sales (in suits) for next year and now must prepare a production plan. The company has traditionally maintained a level workforce strategy. All nine workers are treated like family and have been employed by the company for a number of years. Each employee can produce 2,000 suits per month. At present, finished goods inventory holds 24,000 suits. The demand forecast follows:

Month	1	2	3	4
Demand	25,000	16,000	15,000	19,000
Month	5	6	7	8
Demand	32,000	29,000	27,000	22,000
Month	9	10	11	12
Demand	14,000	15,000	20,000	6,000

a. Management is willing to authorize overtime in periods for which regular production and current levels of anticipation inventory do not satisfy demand. However, overtime must be strictly limited to no more than 20 percent of regular time capacity. Management wants to avoid stockouts and backorders and is not willing to accept a plan that calls for shortages. Is it feasible to hold the workforce constant, assuming that overtime is only used in periods for which shortages would occur?

b. Assume that management is not willing to authorize any overtime. Instead, management is willing to negotiate with customers so that backorders may be used as a supply option. However, management is not willing to carry more than 5,000 suits from one month to the next in backorder. Is it feasible to hold the workforce constant, assuming that a maximum backorder of 5,000 suits may be maintained from month to month?

c. Assume management is willing to authorize the use of overtime over the next four months to build additional anticipation inventory. However, overtime must be strictly limited to no more than 20 percent of regular time capacity. Management wants to avoid stockouts and backorders and is not willing to accept a plan that calls for shortages. Is it feasible to hold the workforce constant, assuming that overtime is only used in months 1–4? If not, in which months would additional overtime be required?

8. Gretchen's Kitchen is a fast-food restaurant located in an ideal spot near the local high school. Gretchen Lowe must prepare an annual staffing plan. The only menu items are hamburgers, chili, soft drinks, shakes, and French fries. A sample of 1,000 customers taken at random revealed that they purchased 2,100 hamburgers, 200 pints of chili, 1,000 soft drinks and shakes, and 1,000 bags of French fries. Thus, for purposes of estimating staffing requirements, Lowe assumes that each customer purchases 2.1 hamburgers, 0.2 pint of chili, 1 soft drink or shake, and 1 bag of French fries. Each hamburger requires 4 minutes of labor, a pint of chili requires 3 minutes, and a soft drink or shake and a bag of fries each take 2 minutes of labor.

The restaurant currently has 10 part-time employees who work 80 hours a month on staggered shifts. Wages are $400 per month for regular time and $7.50 per hour for overtime. Hiring and training costs are $250 per new employee, and layoff costs are $50 per employee.

Lowe realizes that building up seasonal inventories of hamburgers (or any of the products) would not be wise because of shelf-life considerations. Also, any demand not satisfied is a lost sale and must be avoided. Three strategies come to mind.

- Use a level strategy relying on overtime and undertime, with up to 20 percent of regular-time capacity on overtime.

- Maintain a base of 10 employees, hiring and laying off as needed to avoid any overtime.

- Utilize a chase strategy, hiring and laying off employees as demand changes to avoid overtime.

When performing her calculations, Lowe always rounds to the next highest integer for the number of employees. She also follows a policy of not using an employee more than 80 hours per month, except when overtime is needed. The projected demand by month (number of customers) for next year is as follows:

Jan.	3,200	July	4,800
Feb.	2,600	Aug.	4,200
Mar.	3,300	Sept.	3,800
Apr.	3,900	Oct.	3,600
May	3,600	Nov.	3,500
June	4,200	Dec.	3,000

a. Develop the schedule of service requirements (hours per month) for the next year.

b. Which strategy is most effective?

c. Suppose that an arrangement with the high school enables the manager to identify good prospective employees without having to advertise in the local newspaper. This source reduces the hiring cost to $50, which is mainly the cost of charred hamburgers during training. If cost is her only concern, will this method of hiring change Gretchen Lowe's strategy? Considering other objectives that may be appropriate, do you think she should change strategies?

9. The Kool King Company has followed a policy of no layoffs for most of the manufacturer's life, even though the demand for its air conditioners is highly seasonal. Management wants to evaluate the cost-effectiveness of this policy. Competitive pressures are increasing, and ways need to be found to reduce costs. The following demand (expressed in employee–month equivalents) has been forecast for next year:

Jan.	70	May	130	Sept.	110
Feb.	90	June	170	Oct.	60
Mar.	100	July	170	Nov.	20
Apr.	100	Aug.	150	Dec.	40

Additional planning data follow, with costs, inventory, and backorders expressed in employee–month equivalents:

Regular-time production cost	$1,500	Hire cost	$500/ person
Overtime production cost	150% of regular-time production cost	Layoff cost	$2,000/ person
Subcontracting cost	$2,500	Current backorders	10
Inventory holding cost	$100	Current inventory	0
Backorder cost	$1,000	Desired ending inventory	0
Maximum overtime	20% of regular-time capacity	Current employment	130 employees

Hiring costs are lower than layoff costs because the facility is located near a Technical Training School. Undertime is paid at the rate equivalent to regular-time production. Each employee who has been with the company at least one year also received 0.5 months of paid vacation. All 130 employees currently employed qualify for vacations next year, assuming that they remain on the workforce. Answer the following questions using *Sales and Operations Planning with Spreadsheets* Solver in OM Explorer, or an Excel spreadsheet that you developed on your own.

a. Develop an S&OP with the level strategy using overtime, undertime, and vacations as the only supply options. Use the maximum amount of overtime so as to minimize undertime. What is the total cost of this plan, and what are its advantages and disadvantages?

b. Develop an S&OP with the chase strategy. Part of your decision will be when and how many vacation periods to grant. What is the total cost of this plan, and what are its advantages and disadvantages?

c. Develop an S&OP with a lower cost than found with either the level or chase strategy, being open to the full range of supply options (including anticipation inventory). Subcontractors can supply up to 50 employee–month equivalents. What is the total cost of this plan, and what are its advantages and disadvantages?

10. A manager faces peak (weekly) demand for one of her operations, but is not sure how long the peak will last. She can either use overtime from the current workforce, or hire/lay off and just pay regular-time wages. Regular-time pay is $500 per week, overtime is $750 per week, the hiring cost is $2,000, and the layoff cost is $3,000. Assuming that people are available seeking such a short-term arrangement, how many weeks must the surge in demand last to justify a temporary hire? *Hint:* Use break-even analysis (see Supplement A, "Decision Making"). Let w be the number of weeks of the high demand (rather than using Q for the break-even quantity). What is the fixed cost for the regular-time option? Overtime option?

11. Gerald Glynn manages the Michaels Distribution Center. After careful examination of his database information, he has determined the daily requirements for part-time loading dock personnel. The distribution center operates 7 days a week, and the daily part-time staffing requirements are

Day	M	T	W	Th	F	S	Su
Requirements	6	3	5	3	7	2	3

Find the minimum number of workers Glynn must hire. Prepare a workforce schedule for these individuals so that each will have two consecutive days off per week and all staffing requirements will be satisfied. Give preference to the S–Su pair in case of a tie.

12. Cara Ryder manages a ski school in a large resort and is trying to develop a schedule for instructors. The instructors receive little salary and work just enough to earn room and board. They receive free skiing and spend most of their free time tackling the resort's notorious double

black-diamond slopes. Hence, the instructors work only 4 days a week. One of the lesson packages offered at the resort is a 4-day beginner package. Ryder likes to keep the same instructor with a group over the 4-day period, so she schedules the instructors for 4 consecutive days and then 3 days off. Ryder uses years of experience with demand forecasts provided by management to formulate her instructor requirements for the upcoming month.

Day	M	T	W	Th	F	S	Su
Requirements	7	5	4	5	6	9	8

a. Determine how many instructors Ryder needs to employ. Give preference to Saturday and Sunday off. (*Hint:* Look for the group of 3 days with the lowest requirements.)

b. Specify the work schedule for each employee. How much slack does your schedule generate for each day?

13. The mayor of Cambridge, Colorado, wanting to be environmentally progressive, decides to implement a recycling plan. All residents of the city will receive a special three-part bin to separate their glass, plastic, and aluminum, and the city will be responsible for picking up the materials. A young city and regional planning graduate, Michael Duffy, has been hired to manage the recycling program. After carefully studying the city's population density, Duffy decides that the following numbers of recycling collectors will be needed:

Day	M	T	W	Th	F	S	Su
Requirements	12	7	9	9	5	3	6

The requirements are based on the populations of the various housing developments and subdivisions in the city and surrounding communities. To motivate residents of some areas to have their pickups scheduled on weekends, a special tax break will be given.

a. Find the minimum number of recycling collectors required if each employee works 5 days a week and has two consecutive days off. Give preference to the S–Su pair when that pair is involved in a tie.

b. Specify the work schedule for each employee. How much slack does your schedule generate for each day?

c. Suppose that Duffy can smooth the requirements further through greater tax incentives. The requirements then will be eight collectors on Monday and seven on the other days of the week. How many collectors will be needed now? Does smoothing of requirements have capital investment implications? If so, what are they?

14. Little 6, Inc., an accounting firm, forecasts the following weekly workload during the tax season:

				DAY			
	M	T	W	Th	F	S	Su
Personal Tax Returns	24	14	18	18	10	28	16
Corporate Tax Returns	16	10	12	15	24	12	4

Corporate tax returns each require 4 hours of an accountant's time, and personal returns each require 90 minutes. During tax season, each accountant can work up to 10 hours per day. However, error rates increase to unacceptable levels when accountants work more than five consecutive days per week.

Hint: Read Supplement D before doing this problem. Let x_i = number for each working schedule, e.g., x_1 = number for Tuesday through Saturday.

a. Create an effective and efficient work schedule by formulating the problem as a Linear Program and solve using POM for Windows.

b. Assume that management has decided to offer a pay differential to those accountants who are scheduled to work on a weekend day. Normally, accountants earn $1,200 per week, but management will pay a bonus of $100 for Saturday work and $150 for Sunday work. What schedule will cover all demand as well as minimize payroll cost?

c. Assume that Little 6 has three part-time employees available to work Friday, Saturday, and Sunday at a rate of $800. Could these employees be cost effectively utilized?

15. Return to Problem 11 and the workforce schedule for part-time loading dock workers. Suppose that each part-time worker can work only 3 days, but the days must be consecutive. Formulate and solve this workforce scheduling problem as a Linear Program and solve it using POM for Windows. Your objective is to minimize total slack capacity. What is the minimum number of loaders needed now and what are their schedules?

Hint: Read Supplement D before doing this problem. Let x_i = number for each 3-day schedule, e.g., x_1 = number for Tuesday through Thursday.

16. The Hickory Company manufactures wooden desks. Management schedules overtime every weekend to reduce the backlog on the most popular models. The automatic routing machine is used to cut certain types of edges on the desktops. The following orders need to be scheduled for the routing machine:

Order	Time Since Order Arrived (hours ago)	Estimated Machine Time (hours)	Due Date (hours from now)
1	6	10	12
2	5	3	8
3	3	15	18
4	1	9	20
5	0	7	21

The due dates reflect the need for the order to be at its next operation.

a. Develop separate schedules by using the FCFS and EDD rules. Compare the schedules on the basis of average flow time and average past due hours.

b. Comment on the performance of the two rules relative to these measures.

17. Currently a company that designs Web sites has five customers in its backlog. The day when the order arrived, processing time, and promised due dates are given in the following table. The customers are listed in the order of when they arrived. They are ready to be scheduled today, which is the start of day 190.

Customer	Time Since Order Arrived (days ago)	Processing Time (days)	Due Date (days from now)
A	10	20	26
B	8	12	50
C	6	28	66
D	3	24	58
E	2	32	100

a. Develop separate schedules by using the FCFS and EDD rules. Compare the schedules on the basis of average flow time and average days past due.

b. Comment on the performance of the two rules relative to these measures. Which one gives the best schedule? Why?

18. The Mowry Machine Shop still has five jobs to be processed as of 8 A.M. today (day 23) at its bottleneck operation. The day when the order arrived, processing time, and promised due dates are given in the following table. The jobs are listed in the order of arrival.

Job	Time Since Order Arrived (days ago)	Processing Time (days)	Due Date (days from now)
A	11	10	22
B	10	8	13
C	8	4	19
D	6	4	16
E	1	3	30

a. Develop separate schedules by using the FCFS and EDD rules. Compare the schedules on the basis of average flow time and average days past due.

b. Which rule gives the best schedule, in your judgment? Why?

Active Model Exercise

This Active Model appears in MyOMLab. It allows you to evaluate the effects of modifying the size of a constant workforce.

QUESTIONS

1. If we use the same number of workers in each period, what happens as the number of workers increases from 15?

2. If we use the same number of workers in each period, what happens as the number of workers decreases from 15?

3. Suppose the hiring cost is $1,100. What happens as the number of workers increases?

4. Suppose the overtime cost is $3,300. What happens as the number of workers increases?

5. Suppose the undertime cost is the same as the regular-time cost (i.e., paid undertime). What is the best number of workers to have in each month and still meet the demand?

6. If the overtime capacity increases to 30 percent, what is the minimum number of workers that meets the demand in every month?

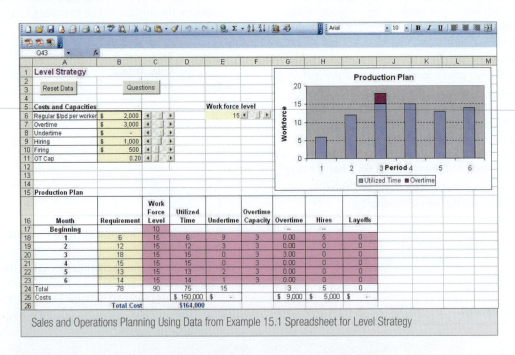

Sales and Operations Planning Using Data from Example 15.1 Spreadsheet for Level Strategy

VIDEO CASE | Sales and Operations Planning at Starwood

Business travel often means staying overnight in a hotel. Upon arrival, you may be greeted by a doorman or valet to assist you with your luggage. Front desk staff awaits your check-in. Behind the scenes, housekeeping, maintenance, and culinary staff prepare for your stay. Making a reservation gives the hotel notice of your plan to stay, but even before your trip is ever conceived, the hotel is staffed and ready. How? Through a process called *sales and operations planning*.

Sales and operations planning is a process every organization performs to some degree. Called a staffing plan (or service resource plan if more detailed) in service organizations, the plan must strike the right level of customer service while maintaining workforce stability and cost control so as to achieve the organization's profit expectations. So where do companies begin? Let us take a look at Starwood Hotels and Resorts to see how it is done.

Starwood operates in more than 750 locations around the globe. At the highest levels, Starwood engages in sales and operations planning on an annual basis, with adjustments made as needed each month by region and by property. Budgeted revenues and other projections come from headquarters; the regions and individual properties then break down the forecasts to meet their expected occupancies. Typically, the director of human resources determines the staffing mix needed across divisions such as food and beverage service, rooms (including housekeeping, spa, and guest services), engineering, Six Sigma (see Chapter 5, "Quality and Performance"), revenue management, and accounting.

At the property level, general managers and their staff must provide input into next year's plan while implementing and monitoring activity in the current year. For most properties, payroll is close to 40 percent of budgeted revenues and represents the largest single expense the hotel incurs. It is also the most controllable expense. Many of Starwood's hotels and most resorts experience patterns of seasonality that affect demand for rooms and services. This seasonality, in turn, significantly affects the organization's staffing plan.

To determine the staffing levels, the company uses a proprietary software program that models occupancy demand based on historical data. The key drivers of staffing are occupied rooms and restaurant meals, called "covers." Starwood knows on a *per room* and *per cover* basis how many staff are required to function properly. When occupancy and covers are entered into the software program, the output models a recommended staffing level for each division. This recommendation is then reviewed by division managers and adjusted, as needed, to be sure staffing is in line with budgeted financial plans. Job fairs to recruit nonmanagement staff are held several times a year so a qualified candidate pool of both part-time and full-time staff is ready when needed. Most hotels maintain a pool of part-time workers who can contract or expand the hours worked if required by property guest levels. Vacations for management are scheduled for the low season. Overtime will be worked as needed, but is less desirable than scheduling the appropriate level of staff in each division.

The program also takes into account both the complexity and positioning of the property within Starwood. For example, a 400-room city hotel that is essentially a high-rise building is not as complex as a 400-room sprawling resort with golf, spa, convention, and other services not offered by the city hotel. Positioning also is important. A five-star resort hotel's customer service

Pearson

A software program that forecasts occupancy based on historical data helps Starwood maintain proper staffing levels at its hotels. Managers know on a per-room, and "per-cover," basis how many hotel employees should be scheduled so that customers get good service.

expectations are much greater than a three-star airport hotel location and requires much higher ratios of staff to guests. Finally, if the hotel is a brand new property, historical data from similar properties is used to model staffing for the first year or two of operation.

Starwood attempts to modify demand and smooth out the peaks and valleys of its demand patterns. Many of the company's hotels experience three seasons: high, mid (called "shoulder"), and low season. Starwood, like its competitors, offers special rates, family packages, and weekend specials to attract different segments of the market during slower business periods. Staff is cross-trained to work in multiple areas, such as front reception and the concierge desk, so additional staff does not have to be added across seasons. Employees may also be temporarily redeployed among Starwood's properties to help out during peak periods. For example, when occupancy is forecast to be high in one region of the country, staff from areas entering their low season will be assigned to cover the demand.

QUESTIONS

1. At what points in the planning process would you expect accounting/finance, marketing, information systems, and operations to play a role? What inputs should these areas provide, and why?

2. Does Starwood employ a chase, level, or mixed strategy? Why is this approach the best choice for the company?

3. How would staffing for the opening of a new hotel or resort differ from that of an existing property? What data might Starwood rely upon to make sure the new property is not over- or understaffed in its first year of operation?

CASE   Memorial Hospital

Memorial Hospital is a 265-bed regional health care facility located in the mountains of western North Carolina. The mission of the hospital is to provide quality health care to the people of Ashe County and the six surrounding counties. To accomplish this mission, Memorial Hospital's CEO has outlined three objectives: (1) maximize customer service to increase customer satisfaction, (2) minimize costs to remain competitive, and (3) minimize fluctuations in workforce levels to help stabilize area employment.

The hospital's operations are segmented into eight major wards for the purposes of planning and scheduling the nursing staff. These wards are listed in Table 15.4, along with the number of beds, targeted patient-to-nurse ratios, and average patient census for each ward. The overall demand for hospital services remained relatively constant over the past few years even though the population of the seven counties served increased. This stable demand can be attributed to increased competition from other hospitals in the area and the rise in alternative health care delivery systems, such as health maintenance organizations (HMOs). However, demand for Memorial Hospital's services does vary considerably by type of ward and time of year. Table 15.5 provides a historical monthly breakdown of the average daily patient census per ward.

The director of nursing for Memorial Hospital is Darlene Fry. Each fall she confronts one of the most challenging aspects of her job: planning the nurse-staffing levels for the next calendar year. Although the average demand for nurses has remained relatively stable over the past couple of years, the staffing plan usually changes because of changing work policies, changing pay structures, and temporary nurse availability and cost. With fall quickly approaching, Fry is collecting information to plan next year's staffing levels.

The nurses at Memorial Hospital work a regular schedule of four 10-hour days per week. The average regular-time pay across all nursing grades is $12.00 per hour. Overtime may be scheduled when necessary. However, because of the intensity of the demands placed on nurses, only a limited amount of overtime is permitted per week. Nurses may be scheduled for as many as 12 hours per day, for a maximum of 5 days per week. Overtime is compensated at a rate of $18.00 per hour. In periods of extremely high demand, temporary part-time nurses may be hired for a limited period of time. Temporary nurses are paid $15.00 per hour. Memorial Hospital's policy limits the proportion of temporary nurses to 15 percent of the total nursing staff.

Finding, hiring, and retaining qualified nurses is an ongoing problem for hospitals. One reason is that various forms of private practice lure many nurses away from hospitals with higher pay and greater flexibility. This situation has caused Memorial to guarantee its full-time staff nurses pay for a minimum of 30 hours per week, regardless of the demand placed on nursing services. In addition, each nurse receives 4 weeks of paid vacation each year. However, vacation scheduling may be somewhat restricted by the projected demand for nurses during particular times of the year.

TABLE 15.4   WARD CAPACITY DATA

Ward	Number of Beds	Patients per Nurse	Patient Census*
Intensive Care	20	2	10
Cardiac	25	4	15
Maternity	30	4	10
Pediatric	40	4	22
Surgery	5	†	†
Post-Op	15	5	8 (T–F daily equivalent)‡
Emergency	10	3	5 (daily equivalent)‡
General	120	8	98

*Yearly average per day

†The hospital employs 20 surgical nurses. Routine surgery is scheduled on Tuesdays and Fridays; five surgeries can be scheduled per day per operating room (bed) on these days. Emergency surgery is scheduled as needed.

‡Daily equivalents are used to schedule nurses because patients flow through these wards in relatively short periods of time. A daily equivalent of 5 indicates that throughout a typical day, an average of five patients are treated in the ward.

At present, the hospital employs 130 nurses, including 20 surgical nurses. The other 110 nurses are assigned to the remaining seven major areas of the hospital. The personnel department informed Fry that the average cost to the hospital for hiring a new full-time nurse is $400 and for laying off or firing a nurse is $150. Although layoffs are an option, Fry is aware of the hospital's objective of maintaining a level workforce.

After looking over the information that she collected, Darlene Fry wants to consider staffing changes in all areas except the surgery ward, which is already correctly staffed.

QUESTIONS

1. Explain the alternatives available to Darlene Fry as she develops a nurse staffing plan for Memorial Hospital. How does each alternative plan meet the objective stated by the CEO?

2. Based on the data presented, develop a nurse staffing plan for Memorial Hospital. Explain your rationale for this plan.

TABLE 15.5 | AVERAGE DAILY PATIENT CENSUS PER MONTH

Ward	MONTH											
	J	F	M	A	M	J	J	A	S	O	N	D
Intensive Care	13	10	8	7	7	6	11	13	9	10	12	14
Cardiac	18	16	15	13	14	12	13	12	13	15	18	20
Maternity	8	8	12	13	10	8	13	13	14	10	8	7
Pediatric	22	23	24	24	25	21	22	20	18	20	21	19
Surgery[*]	20	18	18	17	16	16	22	21	17	18	20	22
Post-Op[†]	10	8	7	7	6	6	10	10	7	8	9	10
Emergency[†]	6	4	4	7	8	5	5	4	4	3	4	6
General	110	108	100	98	95	90	88	92	98	102	107	94

Source: This case was prepared by Dr. Brooke Saladin, Wake Forest University, North Carolina, as a basis for classroom discussion. Copyright © Brooke Saladin. Used with permission.

[*]Average surgeries per day on Tuesday and Thursday.

[†]Daily equivalents

Selected References

Chiang, Wen-Chyuan, Jason C.H. Chen, and Xiaojing Xu. "An Overview of Research on Revenue Management: Current Issues and Future Research." *International Journal of Revenue Management*, vol. 1, no. 1 (2007), pp. 97–128.

Dougherty, John R. "Lessons from the Pros." *APICS Magazine* (November/December, 2007), pp. 31–33.

Dougherty, John R., and Christopher Gray. *Sales & Operations Planning—Best Practices.* Victoria, Canada: Trafford Publishing, 2006.

Esper, Terry L., Alexander E. Ellinger, Theodore P. Stank, Daniel J. Flint, and Mark Moon. "Demand and Supply Integration: A Conceptual Framework of Value Creation through Knowledge Management." *Journal of the Academy of Marketing Science*, vol. 38 (2010), pp. 5–18.

Gray, Christopher. *Sales & Operations Planning—Standard System.* Victoria, Canada: Trafford Publishing, 2007.

Gupta, Jatinder N.D., and Edward Stafford Jr. "Flowshop Scheduling Research after Five Decades." *European Journal of Operational Research*, vol. 169 (2006), pp. 699–711.

Jacobs, F. Robert, William Berry, and D. Clay Whybark. *Manufacturing Planning and Control Systems for Supply Chain Management*, 6th ed. New York: McGraw-Hill/Irwin, 2010.

Kelly, Erin L., and Phyllis Moen. "Rethinking the ClockWork of Work: Why Schedule Control May Pay Off at Work and at Home." *Advances in Developing Human Resources*, vol. 9, no. 4 (Nov 2007), pp. 487–605.

Maher, Kris. "Wal-Mart Seeks New Flexibility in Worker Shifts." *The Wall Street Journal* (January 3, 2007), p. A1.

Muzumdar, Maha, and John Fontanella. "The Secrets to S&OP Success." *Supply Chain Management* (April 2006), pp. 34–41.

Nakano, Mikihisa. "Collaborative Forecasting and Planning in Supply Chains: The Impact on Performance in Japanese Manufacturers." *International Journal of Physical Distribution & Logistics Management*, vol. 39, no. 2, pp. 84–105.

Olhager, Jan, and Erik Selldin. "Manufacturing Planning and Control Approaches: Market Alignment and Performance." *International Journal of Production Research*, vol 45, no. 6 (2007), pp. 1469–1484.

Pinedo, Michael. "Planning and Scheduling in Manufacturing and Services." New York: Springer 2006.

Quadt, Daniel, and Heinrich Kuhn. "A Taxonomy of Flexible Flow Line Scheduling Procedures." *European Journal of Operational Research*, vol. 178 (2007), pp. 686–698.

Rennie, Elizabeth. "All Fired UP! Why Food and Beverage Professionals Must Put S&OP on the Menu." *APICS Magazine* (July/August 2006), pp. 32–35.

Rennie, Elizabeth. "Remote Possibilities: Improved Logistics Management Leads to Promising New Distribution Activities." *APICS Magazine* (July/August 2006), pp. 36–37.

Singhal, Jaya, and Kalyan Singhal. "Holt, Modigliani, Muth, and Simon's Work and Its Role in the Renaissance and Evolution of Operations Management." *Journal of Operations Management*, vol. 25, no. 2 (March 2007), pp. 300–309.

Slone, Reuben, John T. Mentzer, and J. Paul Dittmann. "Are You the Weakest Link in Your Company's Supply Chain?" *Harvard Business Review* (October, 2007), pp.1–11.

Smith, Larry, Joseph C. Andraski, and E. Fawcett. "Integrated Business Planning: A Roadmap to Linking S&OP and CPFR." *Business Forecasting*, vol. 29, no. 4 (Winter 2011), pp. 1–17.

Takey, Flavia, and Marco A. Mesquita. "Aggregate Planning for a Large Food Manufacturer with High Seasonal Demand," *Brazilian Journal of Operations & Production Management*, vol. 3, no. 1 (2006), pp. 5–20.

Trottman, Melanie. "Choices in Stormy Weather: How Airline Employees Make Hundreds of Decisions to Cancel or Reroute Flights." *Wall Street Journal* (February 14, 2006), pp. B1–B3.

Wallace, Thomas F., and Robert A. Stahl. "Sales Forecasting: Improving Cooperation Between the Demand People and the Supply People," *Foresight*. Issue 12 (Winter, 2009), pp. 14–20.

Wallace, Thomas F. *Sales & Operations Planning: The How-To Handbook*, 3rd ed. Cincinnati, OH: T. E. Wallace & Company, 2008.

16

RESOURCE PLANNING

Dow Corning scientists research next generation solar technologies at the company's Solar Application Center

Dow Corning

ow Corning, created as a joint venture between Corning Incorporated and the Dow Chemical Company in 1943, is a global leader that offers over 7,000 innovative products and services using silicon-based technology in diverse industries such as electronics, aviation and aerospace, textile, automotive, and health care, among others. As an example, Dow Corning's Electrical Insulating Compounds are used for creating moisture proof seals for aircraft, automotive, and marine ignition systems. With 11,500 employees, over 25,000 customers, and 45 manufacturing and warehouse locations worldwide, its annual revenues of $6 billion generated a net income of $866 million in 2010. More than half its sales are outside the United States. In order to integrate different business functions and enhance resource planning across the entire firm and its supply chain, Dow Corning turned to SAP, a leading provider of enterprise resource planning (ERP) software solutions. SAP delivers role based access to crucial data, applications, and analytical tools, and has a suite of applications for Business Processes (like Financial Management, Customer Relationship Management, Human Capital Management, and Supply Chain Management), Business Analytics, and Technology. Dow installed the SAP R/3 and mySAP Supply Chain Management solution, with SAP Advanced Planner and Optimizer (APO) at its core.

Prior to the ERP implementation, limited transparency and redundancy within the existing legacy systems made it difficult to access and analyze data needed for effective resource planning, and also hampered decision making and responsiveness. A sequenced implementation of the SAP modules facilitated the linking of key processes from order generation to production planning to warehousing to delivery and final billing. The SAP APO solution enabled the SCOR model of plan, source, make, and deliver (see Chapter 12, "Supply Chain Integration"), which allowed the linkage of shop floor processes and manufacturing operations with the rest of the business. With a transparent view of orders, materials, equipment, product quality, and cost information, Dow Corning can now coordinate plants and processes with greater ease and better match production to market requirements on a global scale. Employee productivity and satisfaction are also up, due mostly to faster response times and accurate on-time deliveries.

Source: End to End Supply Chain Management at Dow Corning, **http://www.sap.com/usa/solutions/business-suite/ erp/operations/pdf/CS_Dow_Corning.pdf**; Dow Corning, Optimizing Operational Performance to Sharpen Competitive Advantage, **www.sap.com**; **http://www.dowcorning.com/content/about/aboutmedia/fastfacts.asp**; May 29, 2011; **http://www12.sap.com/index.epx#/solutions/index.epx**; May 29, 2011.

LEARNING GOALS *After reading this chapter, you should be able to:*

1. Explain how enterprise resource planning (ERP) systems can foster better resource planning.

2. Explain how the concept of dependent demand is fundamental to resource planning.

3. Describe a master production schedule (MPS) and compute available-to-promise quantities.

4. Apply the logic of a material requirements planning (MRP) system to identify production and purchase orders needed for dependent demand items.

5. Apply MRP principles to the provision of services and distribution inventories.

Dow Corning demonstrates that companies can gain a competitive edge by using an effective information system to help with their resource planning. Companies must ensure that all of the resources they need to produce finished services or products are available at the right time. If they are not, a firm risks losing business. For a manufacturer, this task can mean keeping track of thousands of subassemblies, components, and raw materials as well as key equipment capacities. For a service provider, this task can mean keeping track of numerous supplies and carefully scheduling the time and capacity requirements of different employees and types of equipment.

We begin this chapter by describing enterprise resource planning (ERP) systems, which have become a valuable tool for, among other things, resource planning. We then examine a specific approach to resource planning, called material requirements planning (MRP). The concluding section of the chapter illustrates how service providers manage their supplies, human resources, equipment, and financial resources.

resource planning

A process that takes sales and operations plans; processes information in the way of time standards, routings, and other information on how services or products are produced; and then plans the input requirements.

enterprise process

A companywide process that cuts across functional areas, business units, geographical regions, and product lines.

enterprise resource planning (ERP) systems

Large, integrated information systems that support many enterprise processes and data storage needs.

Resource Planning across the Organization

Resource planning lies at the heart of any organization, cutting across all of its different functional areas. It takes sales and operations plans; processes information in the way of time standards, routings, and other information on how services or products are produced; and then plans the input requirements. It also can create reports for managers of the firm's major functional areas, such as human resources, purchasing, sales and marketing, and finance and accounting. In essence, resource planning is a process in and of itself that can be analyzed relative to the firm's competitive priorities.

Enterprise Resource Planning

An **enterprise process** is a companywide process that cuts across functional areas, business units, geographic regions, product lines, suppliers, and customers. **Enterprise resource planning (ERP) systems** are large, integrated information systems that support many enterprise processes and data storage needs. By integrating the firm's functional areas, ERP systems allow an organization to view its operations as a whole rather than having to try to put together the different information pieces produced by its various functions and divisions. Today, ERP systems are being used by traditional brick-and-mortar organizations such as manufacturers, restaurants, hospitals, and hotels, as well as by Internet companies that rely extensively on Web connectivity to link their customers and suppliers.

How ERP Systems Are Designed

ERP revolves around a single comprehensive database that can be made available across the entire organization (or enterprise). Passwords are generally issued to allow certain personnel to access certain areas of the system. Having a single database for all of the firm's information makes it much easier for managers to monitor all of the company's products at all locations and at all times. The database collects data and feeds them into the various modular applications (or suites) of the software system. As new information is entered as a *transaction* in one application, related information is automatically updated in the other applications, including the firm's financial and accounting databases, its human resource and payroll databases, sales, supplier and customer databases, and so forth. In this way, the ERP system streamlines the data flows throughout the organization and supply chain and provides employees with direct access to a wealth of real-time operating information scattered across different functions in the organization. Figure 16.1 shows some of the typical applications with a few subprocesses nested within each one. Some of the applications are for back-office operations such as manufacturing and payroll, while others are for front-office operations such as customer service.

Amazon.com is one company that uses an ERP system. The supply chain application of Amazon's system is particularly important because it allows Amazon.com to link customer orders to warehouse shipments and, ultimately, to supplier replenishment orders. Other applications are more important in other businesses. For example, universities put particular emphasis on the human resources and accounting and finance applications, and manufacturers have an interest in almost every application suite. Not all applications in Figure 16.1 need to be integrated into an ERP system, but those left out will not share their information with the ERP system. Sometimes, however, ERP systems are designed to interface with a firm's existing, older information systems (called "legacy systems").

Designing an ERP system requires that a company carefully analyze its major processes so that appropriate decisions about the coordination of legacy systems and new software can be made. Sometimes, a company's processes that involve redundancies and convoluted information flows must be completely reengineered before the firm can enjoy the benefits of an integrated

Creating Value through Operations Management

Using Operations to Compete
Project Management

Managing Processes

Process Strategy
Process Analysis
Quality and Performance
Capacity Planning
Constraint Management
Lean Systems

Managing Supply Chains

Supply Chain Inventory Management
Supply Chain Design
Supply Chain Location Decisions
Supply Chain Integration
Supply Chain Sustainability and Humanitarian Logistics
Forecasting
Operations Planning and Scheduling
Resource Planning

◀ **FIGURE 16.1**
ERP Application Modules
Source: Based on *Enterprise Resource Planning (ERP)* by Scalle and Cotteleer, Harvard Business School Press. Boston, MA, 1999, No. 9-699-020.

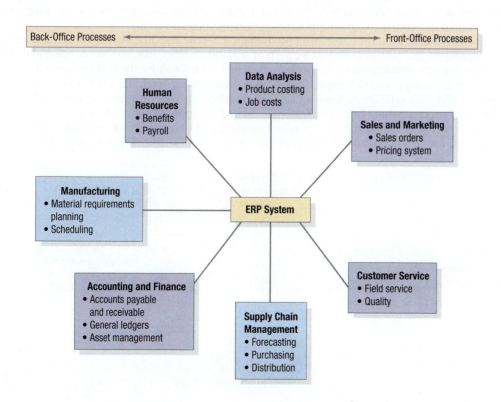

Back-Office Processes ⟷ Front-Office Processes

Data Analysis
• Product costing
• Job costs

Human Resources
• Benefits
• Payroll

Sales and Marketing
• Sales orders
• Pricing system

Manufacturing
• Material requirements planning
• Scheduling

ERP System

Accounting and Finance
• Accounts payable and receivable
• General ledgers
• Asset management

Customer Service
• Field service
• Quality

Supply Chain Management
• Forecasting
• Purchasing
• Distribution

Headquarters of the software company SAP AG in Walldorf, Baden-Wuerttemberg, Germany, Europe

information system. However, a recent study showed that companies reap the greatest rewards when they keep their ERP implementations simple, work with a small number of software vendors, and use standardized systems rather than customizing them extensively. Firms can otherwise end up spending excessive amounts of money on ERP systems that are complex to use and costly to manage. UK confectionary giant Cadbury had to take a 12-million pound hit to their profits due to a build-up of excess inventory of chocolate bars caused by information technology problems arising from the roll out of a new SAP-based ERP system.

Most ERP systems today use a graphical user interface, although the older, keyboard-driven, text-based systems are still popular because of their dependability and technical simplicity. Users navigate through various screens and menus. Training, such as during ERP implementation, focuses on these screens and how users can utilize them to get their jobs done. The biggest suppliers of these off-the-shelf commercial ERP packages are SAP AG, a German company that was also used by Dow Corning in the opening vignette, followed by Oracle Corporation.

Material Requirements Planning

material requirements planning (MRP)

A computerized information system developed specifically to help manufacturers manage dependent demand inventory and schedule replenishment orders.

MRP explosion

A process that converts the requirements of various final products into a material requirements plan that specifies the replenishment schedules of all the subassemblies, components, and raw materials needed to produce final products.

The Manufacturing and Supply Chain Management modules in Figure 16.1 deal with resource planning. Understanding resource planning begins with the concept of *dependent demand*, which sets it apart from the techniques covered in Chapter 9, "Supply Chain Inventory Management." **Material requirements planning (MRP)** is a computerized information system developed specifically to help manufacturers manage dependent demand inventory and schedule replenishment orders. The key inputs of an MRP system are a bill of materials database, a master production schedule, and an inventory record database, as shown in Figure 16.2. Using this information, the MRP system identifies the actions planners must take to stay on schedule, such as releasing new production orders, adjusting order quantities, and expediting late orders.

An MRP system translates the master production schedule and other sources of demand, such as independent demand for replacement parts and maintenance items, into the requirements for all subassemblies, components, and raw materials needed to produce the required parent items. This process is called an **MRP explosion** because it converts the requirements of various final products into a material requirements plan that specifies the replenishment schedules of all the subassemblies, components, and raw materials needed to produce final products.

We first explore the nature of dependent demand and how it differs from independent demand, followed by a discussion of each of the key inputs to the MRP system shown in Figure 16.2.

FIGURE 16.2 ▶
Material Requirements Plan Inputs

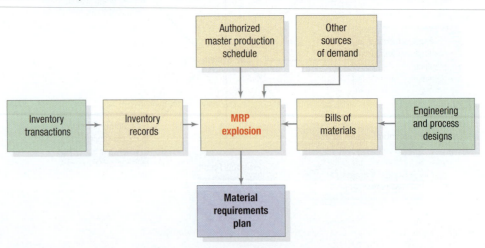

Dependent Demand

For years, many companies tried to manage production and their dependent demand inventories using independent demand systems similar to those discussed in Chapter 9, "Supply Chain Inventory Management," but the outcome was seldom satisfactory because dependent demand is fundamentally different from independent demand. To illustrate the concept of dependent demand, let us consider a Huffy bicycle produced for retail outlets. Demand for a final product, such as a bicycle, is called *independent demand* because it is influenced only by market conditions. In contrast, the demand for spokes going into the bicycle "depends" on the production planned for its wheels. Huffy must forecast this demand using techniques such as those discussed in Chapter 14, "Forecasting." However, Huffy also keeps many other items in inventory—handlebars, pedals, frames, and wheel rims—used to make completed bicycles. Each of these items has a **dependent demand** because the quantity required varies with the production plans for other items held in the firm's inventory—finished bikes, in this case. For example, the demand for frames, pedals, and wheel rims is *dependent* on the production of completed bicycles. Operations can calculate the demand for dependent demand items once the bicycle production levels are laid out in the sales and operations plan. For example, every bicycle needs two wheel rims, so 1,000 completed bicycles need $1,000(2) = 2,000$ rims. Forecasting techniques are not needed for the rims.

The bicycle, or any other product that is manufactured from one or more components, is called a **parent**. The wheel rim is an example of a **component**—an item that goes through one or more operations to be transformed into or become part of one or more parents. A wheel rim, for example, will have several different parents if the rim is used to make more than one style of bicycle. This parent–component relationship can cause erratic dependent demand patterns for components. Suppose that every time inventory falls to 500 units (a reorder point), an order for 1,000 more bicycles is placed, as shown in Figure 16.3(a). The assembly supervisor then authorizes the withdrawal of 2,000 rims from inventory, along with other components for the finished product. The demand for the rim is shown in Figure 16.3(b). So, even though customer demand for the finished bicycle is continuous and reasonably uniform, the production demand for wheel rims is "lumpy"; that is, it occurs sporadically, usually in relatively large quantities. Thus, the production decisions for the assembly of bicycles, which account for the costs of assembling the bicycles and the projected assembly capacities at the time the decisions are made, determine the demand for wheel rims.

Managing dependent demand inventories is complicated because some components may be subject to both dependent and independent demand. For example, the shop floor needs 2,000 wheel rims for the new bicycles, but the company also sells replacement rims for old bicycles directly to retail outlets. This practice places an independent demand on the inventory of wheel rims. Materials requirements planning can be used in complex situations involving components that may have independent demand as well as dependent demand inventories.

Bill of Materials

The replenishment schedule for a component is determined from the production schedules of its parents. Hence, the system needs accurate information on parent–component relationships. A **bill of materials (BOM)** is a record of all the components of an item, the parent–component

dependent demand

The demand for an item that occurs because the quantity required varies with the production plans for other items held in the firm's inventory.

parent

Any product that is manufactured from one or more components.

component

An item that goes through one or more operations to be transformed into or become part of one or more parents.

bill of materials (BOM)

A record of all the components of an item, the parent–component relationships, and the usage quantities derived from engineering and process designs.

◀ **FIGURE 16.3**
Lumpy Dependent Demand Resulting from Continuous Independent Demand

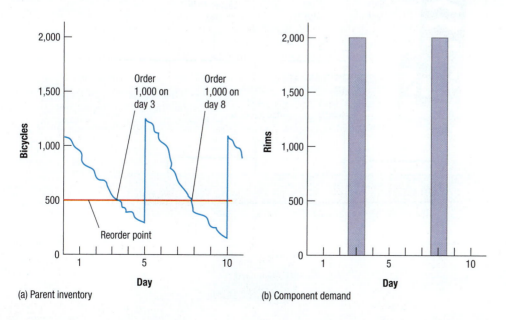

(a) Parent inventory   (b) Component demand

relationships, and the usage quantities derived from engineering and process designs. In Figure 16.4, the BOM of a simple ladder-back chair shows that the chair is made from a ladder-back subassembly, a seat subassembly, front legs, and leg supports. In turn, the ladder-back subassembly is made from back legs and back slats, and the seat subassembly is made from a seat frame and a seat cushion. Finally, the seat frame is made from seat-frame boards. For convenience, we refer to these items by the letters shown in Figure 16.4.

All items except item A are components because they are needed to make a parent. Items A, B, C, and H are parents because they all have at least one component. The BOM also specifies the **usage quantity**, or the number of units of a component that are needed to make one unit of its immediate parent. Figure 16.4 shows usage quantities for each parent–component relationship in parentheses. Note that one chair (item A) is made from one ladder-back subassembly (item B), one seat subassembly (item C), two front legs (item D), and four leg supports (item E). In addition, item B is made from two back legs (item F) and four back slats (item G). Item C needs one seat frame (item H) and one seat cushion (item I). Finally, item H needs four seat-frame boards (item J).

Four terms frequently used to describe inventory items are *end items*, *intermediate items*, *subassemblies*, and *purchased items*. An **end item** typically is the final product sold to the customer; it is a parent but not a component. Item A in Figure 16.4, the completed ladder-back chair, is an end item. Accounting statements classify inventory of end items as either work-in-process (WIP), if work remains to be done, or finished goods. An **intermediate item**, such as item B, C, or H, has at least one parent and at least one component. Some products have several levels of intermediate items; the parent of one intermediate item can also be an intermediate item. Inventory of intermediate items—whether completed or still on the shop floor—is classified as WIP. A **subassembly** is an intermediate item that is assembled (as opposed to being transformed by other means) from more than

usage quantity

The number of units of a component that are needed to make one unit of its immediate parent.

end item

The final product sold to a customer.

intermediate item

An item that has at least one parent and at least one component.

subassembly

An intermediate item that is *assembled* (as opposed to being transformed by other means) from more than one component.

FIGURE 16.4 ▶
BOM for a Ladder-Back Chair

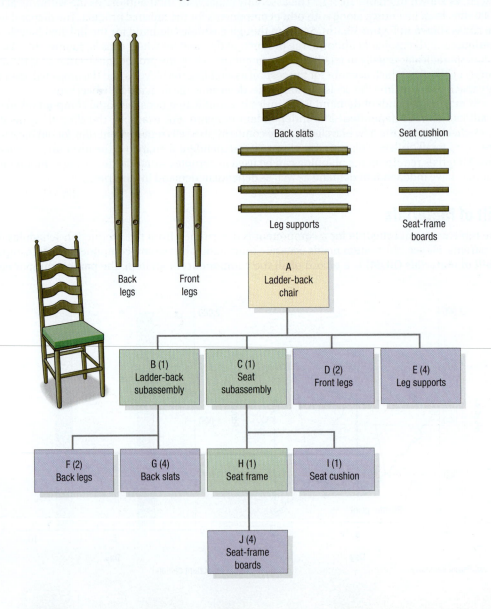

one component. Items B and C are subassemblies. A **purchased item** has no components because it comes from a supplier, but it has one or more parents. Examples are items D, E, F, G, I, and J in Figure 16.4. Inventory of purchased items is treated as raw materials in accounting statements.

A component may have more than one parent. **Part commonality**, sometimes called *standardization of parts* or *modularity*, is the degree to which a component has more than one immediate parent. As a result of commonality, the same item may appear in several places in the bill of materials for a product, or it may appear in the bills of materials for several different products. For example, the seat subassembly in Figure 16.4 is a component of the ladder-back chair and of a kitchen chair that is part of the same family of products. The usage quantity specified in the bill of materials relates to a specific parent–component relationship. The usage quantity for any component can therefore change, depending on the parent item. Part commonality, or using the same part in many parents, increases its volume and repeatability, which provides several process advantages and helps minimize inventory costs.

A careful dissection of the popular iPAD and iPAD 2 products from Apple shows that their bill of materials differ in the enclosure and battery, but are otherwise remarkably similar in their components and design. Same suppliers are used for many components, and costs are very comparable for new revisions of chips found in the previous iPad and iPhones. Standardization of designs and components across different products and generations allows Apple to be very competitive and profitable.

Master Production Scheduling

The second input into a material requirements plan is the **master production schedule (MPS)**, which details how many end items will be produced within specified periods of time. It breaks the sales and operations plan into specific product schedules. Figure 16.5 shows how a sales and operations plan for a family of chairs breaks down into the weekly MPS for each specific chair type (the time period can be hours, days, weeks, or months). The chair example demonstrates the following aspects of master scheduling:

1. The sums of the quantities in the MPS must equal those in the sales and operations plan. This consistency between the plans is desirable because of the economic analysis done to arrive at the sales and operations plan.

2. The production quantities must be allocated efficiently over time. The specific mix of chair types—the number of each type as a percent of the total family's quantity—is based on historic demand and on marketing and promotional considerations. The planner must select lot sizes for each chair type, taking into consideration economic factors such as production setup costs and inventory carrying costs.

purchased item

An item that has one or more parents but no components because it comes from a supplier.

part commonality

The degree to which a component has more than one immediate parent.

master production schedule (MPS)

A part of the material requirements plan that details how many end items will be produced within specified periods of time.

iPad total product costs **$270**		iPad 2 total product costs **$267**
LCD display – 9.7-inch multi-touch screen $59	LCD display – 9.7-inch multi-touch screen $50	
Camera – None $0	Camera – Front & Rear $4.50	
Memory – Samsung $47	Memory – Samsung/Toshiba $30	
Apple Processor – Apple A4 $17	Apple Processor – Apple A5 $25	
Radio Components Infineon/Broadcom $26	Radio Components Infineon/Broadcom or Qualcomm/Broadcom $25	
Sensors – STMicroelectronics $1.00	Sensors – STMicroelectronics $2.50	
Battery $23	Battery $20 – $25	

Bill of Materials for Apple's iPad and iPad2.

FIGURE 16.5 ▶
MPS for a Family of Chairs

	April				May			
	1	2	3	4	5	6	7	8
Ladder-back chair	150					150		
Kitchen chair				120			120	
Desk chair		200	200		200			200
Sales and operations plan for chair family			670				670	

3. Capacity limitations and bottlenecks, such as machine or labor capacity, storage space, or working capital, may determine the timing and size of MPS quantities. The planner must acknowledge these limitations by recognizing that some chair styles require more resources than others and setting the timing and size of the production quantities accordingly.

Figure 16.6 shows the master production scheduling process. Operations must first create a prospective MPS to test whether it meets the schedule with the resources (e.g., machine capacities, workforce, overtime, and subcontractors) provided for in the sales and operations plan. Operations then revises the MPS until a schedule that satisfies all of the resource limitations is developed or until it is determined that no feasible schedule can be developed. In the latter event, the production plan must be revised to adjust production requirements or increase authorized resources. Once a feasible prospective MPS has been accepted by the firm's managers, operations uses the authorized MPS as input to material requirements planning. Operations can then determine specific schedules for component production and assembly. Actual performance data such as inventory levels and shortages are inputs to preparing the prospective MPS for the next period, and so the master production scheduling process is repeated from one period to the next.

FIGURE 16.6 ▶
Master Production Scheduling
Process

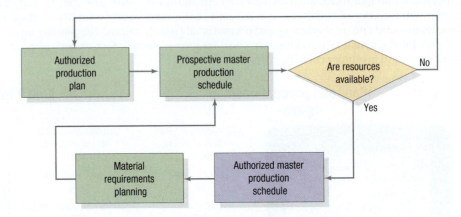

Developing a Master Production Schedule   The process of developing a master production schedule includes (1) calculating the projected on-hand inventory and (2) determining the timing and size of the production quantities of specific products. We use the manufacturer of the ladder-back chair to illustrate the process. For simplicity, we assume that the firm does not utilize safety stocks for end items, even though many firms do. In addition, we use weeks as our planning periods, even though hours, days, or months could be used.

Step 1. *Calculate Projected On-Hand Inventories.* The first step is to calculate the projected on-hand inventory, which is an estimate of the amount of inventory available each week after demand has been satisfied:

$$\begin{pmatrix} \text{Projected on-hand} \\ \text{inventory at end} \\ \text{of this week} \end{pmatrix} = \begin{pmatrix} \text{On-hand} \\ \text{inventory at} \\ \text{end of last week} \end{pmatrix} + \begin{pmatrix} \text{MPS quantity} \\ \text{due at start} \\ \text{of this week} \end{pmatrix} - \begin{pmatrix} \text{Projected} \\ \text{requirements} \\ \text{this week} \end{pmatrix}$$

In some weeks, no MPS quantity for a product may be needed because sufficient inventory already exists. For the projected requirements for this week, the scheduler uses whichever is larger—the forecast or the customer orders booked—recognizing that the forecast is subject to error. If actual booked orders exceed the forecast, the projection will be more accurate if the scheduler uses the booked orders because booked orders are a known quantity. Conversely, if the forecast exceeds booked orders for a week, the forecast will provide a better estimate of the requirements needed for that week because some orders are yet to come in.

The manufacturer of the ladder-back chair produces the chair to stock and needs to develop an MPS for it. Marketing has forecasted a demand of 30 chairs for the first week of April, but actual customer orders booked are for 38 chairs. The current on-hand inventory is 55 chairs. No MPS quantity is due in week 1. Figure 16.7 shows an MPS record with these quantities listed. Because actual orders for week 1 are greater than the forecast, the scheduler uses that figure for actual orders to calculate the projected inventory balance at the end of week 1:

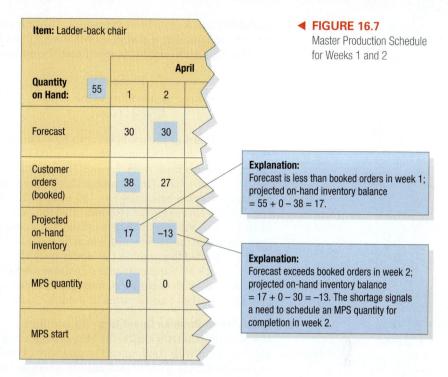

◀ FIGURE 16.7
Master Production Schedule for Weeks 1 and 2

Explanation:
Forecast is less than booked orders in week 1; projected on-hand inventory balance
= 55 + 0 − 38 = 17.

Explanation:
Forecast exceeds booked orders in week 2; projected on-hand inventory balance
= 17 + 0 − 30 = −13. The shortage signals a need to schedule an MPS quantity for completion in week 2.

$$\text{Inventory} = \begin{pmatrix} 55 \text{ chairs} \\ \text{currently} \\ \text{in stock} \end{pmatrix} + \begin{pmatrix} \text{MPS quantity} \\ (0 \text{ for week } 1) \end{pmatrix} - \begin{pmatrix} 38 \text{ chairs already} \\ \text{promised for} \\ \text{delivery in week } 1 \end{pmatrix} = 17 \text{ chairs}$$

▼ FIGURE 16.8
Master Production Schedule for Weeks 1–8

In week 2, the forecasted quantity exceeds actual orders booked, so the projected on-hand inventory for the end of week 2 is 17 + 0 − 30 = −13. The shortage signals the need for more chairs to be produced and available for week 2.

Step 2. Determine the Timing and Size of MPS Quantities. The goal of determining the timing and size of MPS quantities is to maintain a nonnegative projected on-hand inventory balance. As shortages in inventory are detected, MPS quantities should be scheduled to cover them. The first MPS quantity should be scheduled for the week when the projected on-hand inventory reflects a shortage, such as week 2 in Figure 16.7.[1] The scheduler adds the MPS quantity to the projected on-hand inventory and searches for the next period when a shortage occurs. This shortage signals a need for a second MPS quantity, and so on.

Figure 16.8 shows a master production schedule for the ladder-back chair for the next 8 weeks. The order policy requires production lot sizes of 150 units. A shortage of 13 chairs in week 2 will occur unless the scheduler provides for an MPS quantity for that period. Our convention is to show blanks instead of zeroes in all rows, which improves

Item: Ladder-back chair						Order Policy: 150 units Lead Time: 1 week			
			April				May		
Quantity on Hand:	55	1	2	3	4	5	6	7	8
Forecast		30	30	30	30	35	35	35	35
Customer orders (booked)		38	27	24	8	0	0	0	0
Projected on-hand inventory		17	137	107	77	42	7	122	87
MPS quantity			150					150	
MPS start		150					150		

Explanation:
The time needed to assemble 150 chairs is 1 week. The assembly department must start assembling chairs in week 1 to have them ready by week 2.

Explanation:
On-hand inventory balance
= 17 + 150 − 30 = 137. The MPS quantity is needed to avoid a shortage of 30 − 17 = 13 chairs in week 2.

---

[1]In some cases, new orders will be planned before a shortage is encountered. Two such instances occur when safety stocks and anticipation inventories are built up.

readability and is often used in practice. The only exception is in the projected on-hand inventory row, where a number is always shown, even if it is a 0 or negative number.

Once the MPS quantity is scheduled, the updated projected inventory balance for week 2 is

$$\text{Inventory} = \left(\begin{array}{c} 17 \text{ chairs in} \\ \text{inventory at the} \\ \text{end of week 1} \end{array}\right) + \left(\begin{array}{c} \text{MPS quantity} \\ \text{of 150 chairs} \end{array}\right) - \left(\begin{array}{c} \text{Forecast of} \\ 30 \text{ chairs} \end{array}\right) = 137 \text{ chairs}$$

The scheduler proceeds column by column through the MPS record until it reaches the end, filling in the MPS quantities as needed to avoid shortages. The 137 units will satisfy forecasted demands until week 7, when the inventory shortage in the absence of an MPS quantity is $7 + 0 - 35 = -28$. This shortage signals the need for another MPS quantity of 150 units. The updated inventory balance is $7 + 150 - 35 = 122$ chairs for week 7.

The last row in Figure 16.8 indicates the periods in which production of the MPS quantities must *begin* so that they will be available when indicated in the MPS quantity row. In the upper-right portion of the MPS record, a lead time of 1 week is indicated for the ladder-back chair; that is, 1 week is needed to assemble 150 ladder-back chairs, assuming that items B, C, D, and E are available. For each MPS quantity, the scheduler works backward through the lead time to determine when the assembly department must start producing chairs. Consequently, a lot of 150 units must be started in week 1 and another in week 6.

available-to-promise (ATP) inventory

The quantity of end items that marketing can promise to deliver on specified dates.

Available-to-Promise Quantities   In addition to providing manufacturing with the timing and size of production quantities, the MPS provides marketing with information useful for negotiating delivery dates with customers. The quantity of end items that marketing can promise to deliver on specified dates is called **available-to-promise (ATP) inventory**. It is the difference between the customer orders already booked and the quantity that operations is planning to produce. As new customer orders are accepted, the ATP inventory is reduced to reflect the commitment of the firm to ship those quantities, but the actual inventory stays unchanged until the order is removed from inventory and shipped to the customer. An available-to-promise inventory is associated with each MPS quantity because the MPS quantity specifies the timing and size of new stock that can be earmarked to meet future bookings.

Figure 16.9 shows an MPS record with an additional row for the available-to-promise quantities. The ATP in week 2 is the MPS quantity minus booked customer orders until the next MPS quantity, or $150 - (27 + 24 + 8 + 0 + 0) = 91$ units. The ATP indicates to marketing that, of the 150 units scheduled for completion in week 2, 91 units are uncommitted, and total new orders up to that quantity can be promised for delivery as early as week 2. In week 7, the ATP is 150 units because there are no booked orders in week 7 and beyond.

The procedure for calculating available-to-promise information is slightly different for the first (current) week of the schedule than for other weeks because it accounts for the inventory currently in stock. The ATP inventory for the first week equals *current on-hand inventory* plus the MPS quantity for the first week, minus the cumulative total of booked orders up to (but not including) the week in which the next MPS quantity arrives. So, in Figure 16.9, the ATP for the first week is $55 + 0 - 38 = 17$. This information indicates to the sales department that it can promise as many as 17 units this week, 91 more units sometime in weeks 2

▼ **FIGURE 16.9**
MPS Record with an ATP Row

Item: Ladder-back chair					Order Policy: 150 units Lead Time: 1 week			
		April				May		
Quantity on Hand: 55	1	2	3	4	5	6	7	8
Forecast	30	30	30	30	35	35	35	35
Customer orders (booked)	38	27	24	8	0	0	0	0
Projected on-hand inventory	17	137	107	77	42	7	122	87
MPS quantity		150					150	
MPS start	150					150		
Available-to-promise (ATP) inventory	17	91					150	

Explanation:
The total of customer orders booked until the next MPS receipt is 38 units. The ATP = 55 (on-hand) + 0 (MPS quantity) − 38 = 17.

Explanation:
The total of customer orders booked until the next MPS receipt is 27 + 24 + 8 = 59 units. The ATP = 150 (MPS quantity) − 59 = 91 units.

through 6, and 150 more units in week 7 or 8. If customer order requests exceed ATP quantities in those time periods, the MPS must be changed before the customer orders can be booked or the customers must be given a later delivery date—when the next MPS quantity arrives. See Solved Problem 2 at the end of this chapter for an example of decision making using the ATP quantities.

Inventory planners do not create master production plans manually, although they thoroughly understand the logic built into them. Figure 16.10 is typical of the computer support available. It was created with the *Master Production Scheduling* Solver in OM Explorer, and confirms the output shown in Figure 16.9.

Lot Size	150								
Lead Time	1								
Quantity on Hand	55	1	2	3	4	5	6	7	8
Forecast		30	30	30	30	35	35	35	35
Customer Orders (Booked)		38	27	24	8				
Projected On-Hand Inventory		17	137	107	77	42	7	122	87
MPS Quantity			150					150	
MPS Start		150					150		
Available-to-Promise Inv (ATP)		17	91					150	

FIGURE 16.10 ▲

Master Production Scheduling Solver Output Using *OM Explorer*

Freezing the MPS   The master production schedule is the basis of all end item, subassembly, component, and materials schedules. For this reason, changes to the MPS can be costly, particularly if they are made to MPS quantities soon to be completed. Increases in an MPS quantity can result in material shortages, delayed shipments to customers, and excessive expediting costs. Decreases in MPS quantities can result in unused materials or components (at least until another need for them arises) and valuable capacity being used to create products not needed. Similar costs occur when forecasted need dates for MPS quantities are changed. For these reasons, many firms, particularly those with a make-to-stock strategy and a focus on low-cost operations, *freeze*, or disallow changes to, the near-term portion of the MPS.

Reconciling the MPS with Sales and Operations Plans   Because the master production schedule is based on both forecasts as well as actual orders received, it can differ from the sales and operations plan when summed across different periods in a month. For instance, in Figure 16.5, if the sum total MPS quantities of the three models of chairs in the month of April was 725 instead of 670, then either the management must revise the sales and operations plan upwards by authorizing additional resources to match supply with demand, or reduce the quantities of MPS in the month of April to match the sales and operations plan. Master production schedules drive plant and supplier activity, so they must be synchronized with actual customer demands and the sales and operations plans to ensure that the firm's planning decisions are actually being implemented on an ongoing basis.

Inventory Record

Inventory records are a third major input to MRP, and inventory transactions are the basic building blocks of up-to-date records (see Figure 16.2). These transactions include releasing new orders, receiving scheduled receipts, adjusting due dates for scheduled receipts, withdrawing inventory, canceling orders, correcting inventory errors, rejecting shipments, and verifying scrap losses and stock returns. Recording the transactions accurately is essential if the firm's on-hand inventory balances are to be correct and its MRP system is to operate effectively.

The **inventory record** divides the future into time periods called *time buckets*. In our discussion, we use weekly time buckets for consistency with our MPS example, although other time periods could as easily be used. The inventory record shows an item's lot-size policy, lead time, and various time-phased data. The purpose of the inventory record is to keep track of inventory levels and component replenishment needs. The time-phased information contained in the inventory record consists of (1) *gross requirements*, (2) *scheduled receipts*, (3) *projected on-hand inventory*, (4) *planned receipts*, and (5) *planned order releases*.

We illustrate the discussion of inventory records with the seat subassembly, item C that was shown in Figure 16.4. Suppose that it is used in two products: a ladder-back chair and a kitchen chair.

Gross Requirements   The **gross requirements** are the total demand derived from *all* parent production plans. They also include demand not otherwise accounted for, such as demand for replacement parts for units already sold. Figure 16.11 shows an inventory record for item C, the seat subassembly. Item C is produced in lots of 230 units and has a lead time of 2 weeks. The inventory

MyOMLab

Tutor 16.1 in MyOMLab provides a new example to practice completing a master production schedule.

inventory record

A record that shows an item's lot-size policy, lead time, and various time-phased data.

gross requirements

The total demand derived from *all* parent production plans.

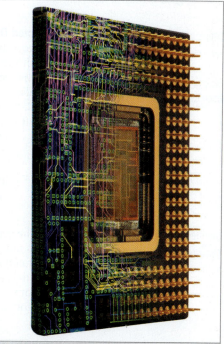

Cypress Semiconductor, a California-based company that manufactures logic devices, USB controllers, general-purpose programmable clocks, memories, and wireless connectivity solutions for consumer and automotive markets, uses commercial software solutions to manage the complexity of its master production scheduling processes.

Item: C Description: Seat subassembly					Lot Size: 230 units Lead Time: 2 weeks				
		Week							
		1	2	3	4	5	6	7	8
Gross requirements		150	0	0	120	0	150	120	0
Scheduled receipts		230	0	0	0	0	0	0	0
Projected on-hand inventory	37	117	117	117	−3	−3	−153	−273	−273
Planned receipts									
Planned order releases									

Explanation:
Gross requirements are the total demand for the two chairs. Projected on-hand inventory in week 1 is 37 + 230 − 150 = 117 units.

FIGURE 16.11 ▲
MRP Record for the Seat Subassembly

projected on-hand inventory

An estimate of the amount of inventory available each week after gross requirements have been satisfied.

record also shows item C's gross requirements for the next 8 weeks, which come from the master production schedule for the ladder-back and kitchen chairs (see Figure 16.5). The MPS start quantities for each parent are added to arrive at each week's gross requirements. The seat subassembly's gross requirements exhibit lumpy demand: Operations will withdraw seat subassemblies from inventory in only 4 of the 8 weeks.

The MRP system works with release dates to schedule production and delivery for components and subassemblies. Its program logic anticipates the removal of all materials required by a parent's production order from inventory at the beginning of the parent item's lead time—when the scheduler first releases the order to the shop.

Scheduled Receipts   Recall that *scheduled receipts* (sometimes called open orders) are orders that have been placed but not yet completed. For a purchased item, the scheduled receipt could be in one of several stages: being processed by a supplier, being transported to the purchaser, or being inspected by the purchaser's receiving department. If the firm is making the item in-house, the order could be on the shop floor being processed, waiting for components, waiting for a machine to become available, or waiting to be moved to its next operation. According to Figure 16.11, one 230-unit order of item C is due in week 1. Given the 2-week lead time, the inventory planner probably released the order 2 weeks ago. Scheduled receipts due in beyond the item's lead time are unusual, caused by events such as a last-minute change in the MPS.

Projected On-Hand Inventory   The **projected on-hand inventory** is an estimate of the amount of inventory available each week after gross requirements have been satisfied. The beginning inventory, shown as the first entry (37) in Figure 16.11, indicates the on-hand inventory available at the time the record was computed. As with scheduled receipts, entries are made for each actual withdrawal and receipt to update the MRP database. Then, when the MRP system produces the revised record, the correct inventory will appear.

Other entries in the row show inventory expected in future weeks. Projected on-hand inventory is calculated as

$$\begin{pmatrix} \text{Projected on-hand} \\ \text{inventory balance} \\ \text{at end of week } t \end{pmatrix} = \begin{pmatrix} \text{Inventory on} \\ \text{hand at end of} \\ \text{week } t-1 \end{pmatrix} + \begin{pmatrix} \text{Scheduled} \\ \text{or planned} \\ \text{reeipts in} \\ \text{week } t \end{pmatrix} - \begin{pmatrix} \text{Gross} \\ \text{requirements} \\ \text{in week } t \end{pmatrix}$$

planned receipts

Orders that are not yet released to the shop or the supplier.

The projected on-hand calculation includes the consideration of **planned receipts**, which are orders not yet released to the shop or the supplier. Planned receipts should not be confused with scheduled receipts. Planned receipts are still at the planning stage and can still change from one week to the next, whereas scheduled receipts are actual orders that are being acted upon by the shop or supplier. In Figure 16.11, the planned receipts are all zero. The on-hand inventory calculations for each week are as follows:

Week 1:	$37 + 230 - 150 = 117$
Weeks 2 and 3:	$117 + 0 - 0 = 117$
Week 4:	$117 + 0 - 120 = -3$
Week 5:	$-3 + 0 - 0 = -3$
Week 6:	$-3 + 0 - 150 = -153$
Week 7:	$-153 + 0 - 120 = -273$
Week 8:	$-273 + 0 - 0 = -273$

In week 4, the balance drops to −3 units, which indicates that a shortage of 3 units will occur unless more seat subassemblies are built. This condition signals the need for a planned receipt to arrive in week 4. In addition, unless more stock is received, the shortage will grow to 273 units in weeks 7 and 8.[2]

Planned Receipts Planning for the receipt of new orders will keep the projected on-hand balance from dropping below zero. The planned receipt row is developed as follows:

1. Weekly on-hand inventory is projected until a shortage appears. Completion of the initial planned receipt is scheduled for the week in which the shortage is projected. The addition of the newly planned receipt should increase the projected on-hand balance so that it equals or exceeds zero. It will exceed zero when the lot size exceeds requirements in the week it is planned to arrive.

2. The projection of on-hand inventory continues until the next shortage occurs. This shortage signals the need for the second planned receipt.

This process is repeated until the end of the planning horizon by proceeding column by column through the MRP record—filling in planned receipts as needed and completing the projected on-hand inventory row. Figure 16.12 shows the planned receipts for the seat subassembly. In week 4, the projected on-hand inventory will drop below zero, so a planned receipt of 230 units is scheduled for week 4. The updated inventory on-hand balance is 117 (inventory at end of week 3) + 230 (planned receipts) − 120 (gross requirements) = 227 units. The projected on-hand inventory remains at 227 for week 5 because no scheduled receipts or gross requirements are anticipated. In week 6, the projected on-hand inventory is 227 (inventory at end of week 5) − 150 (gross requirements) = 77 units. This quantity is greater than zero, so no new planned receipt is needed. In week 7, however, a shortage will occur unless more seat subassemblies are received. With a planned receipt in week 7, the updated inventory balance is 77 (inventory at end of week 6) + 230 (planned receipts) − 120 (gross requirements) = 187 units.

Item: C
Description: Seat subassembly
Lot Size: 230 units
Lead Time: 2 weeks

	Week							
	1	2	3	4	5	6	7	8
Gross requirements	150	0	0	120	0	150	120	0
Scheduled receipts	230	0	0	0	0	0	0	0
Projected on-hand inventory 37	117	117	117	227	227	77	187	187
Planned receipts				230			230	
Planned order releases		230			230			

Explanation:
Without a planned receipt in week 4, a shortage of 3 units will occur: 117 + 0 + 0 − 120 = −3 units. Adding the planned receipt brings the balance to 117 + 0 + 230 − 120 = 227 units. Offsetting for a 2-week lead time puts the corresponding planned order release back to week 2.

Explanation:
The first planned receipt lasts until week 7, when projected inventory would drop to 77 + 0 + 0 − 120 = −43 units. Adding the second planned receipt brings the balance to 77 + 0 + 230 − 120 = 187 units. The corresponding planned order release is for week 5 (or week 7 minus 2 weeks).

▲ **FIGURE 16.12**
Completed Inventory Record for the Seat Subassembly

Planned Order Releases A **planned order release** indicates when an order for a specified quantity of an item is to be issued. We must place the planned order release quantity in the proper time bucket. To do so, we must assume that all inventory flows—scheduled receipts, planned receipts, and gross requirements—occur at the same point of time in a time period. Some firms assume that all flows occur at the beginning of a time period; other firms assume that they occur at the end of a time period or at the middle of the time period. Regardless of when the flows are assumed to occur, we find the release date by subtracting the lead time from the receipt date. For example, the release date for the first planned order release in Figure 16.12 is 4 (planned receipt date) − 2 (lead time) = 2 (planned order release date). Figure 16.12 shows the planned order releases for the seat subassembly. If all goes according to the plan, we will release an order for 230 seat assemblies next week (in week 2). This order release sets off a series of updates to the inventory record. First, the planned order release for the order is removed. Next, the planned receipt for 230 units in week 4 is also removed. Finally, a new scheduled receipt for 230 units will appear in the scheduled receipt row for week 4.

planned order release

An indication of when an order for a specified quantity of an item is to be issued.

_____
[2]There is an exception to the rule of scheduling a planned receipt whenever the projected inventory otherwise becomes negative. When a scheduled receipt is coming in *after* the inventory becomes negative, the first recourse is to expedite the scheduled receipt (giving it an earlier due date), rather than scheduling a new planned receipt.

Planning Factors

The planning factors in a MRP inventory record play an important role in the overall performance of the MRP system. By manipulating these factors, managers can fine-tune inventory operations. In this section, we discuss planning lead time, lot-sizing rules, and safety stock.

Planning Lead Time   Planning lead time is an estimate of the time between placing an order for an item and receiving the item in inventory. Accuracy is important in planning lead time. If an item arrives in inventory sooner than needed, inventory holding costs increase. If an item arrives too late, stockouts, excessive expediting costs, or both may occur.

For purchased items, the planning lead time is the time allowed for receiving a shipment from the supplier after the order has been sent, including the normal time to place the order. Often, the purchasing contract stipulates the delivery date. For items manufactured in-house, a rough-cut estimate of the planning lead time can be obtained by keeping track of the actual lead times for recent orders and computing an average. A more extensive estimating process consists of breaking down each of the following factors:

- Setup time
- Processing time
- Materials handling time between operations
- Waiting time

Each of these times must be estimated for every operation along the item's route. Estimating setup, processing, and materials handling times can be relatively easy, but estimating the waiting time for materials handling equipment or for a workstation to perform a particular operation can be more difficult. In a facility that uses a make-to-order strategy, such as a machine shop, the load on the shop varies considerably over time, causing actual waiting times for a particular order to fluctuate widely. Therefore, being able to accurately estimate the waiting time is especially important when it comes to estimating the planning lead time. However, in a facility that uses a make-to-stock strategy, such as an assembly plant, product routings are more standard and waiting time is more predictable; hence, waiting time generally is a less-troublesome part of planning lead times.

Lot-Sizing Rules   A lot-sizing rule determines the timing and size of order quantities. A lot-sizing rule must be assigned to each item before planned receipts and planned order releases can be computed. The choice of lot-sizing rules is important because they determine the number of setups required and the inventory holding costs for each item. We present three lot-sizing rules: (1) fixed order quantity, (2) periodic order quantity, and (3) lot-for-lot.

fixed order quantity (FOQ)

A rule that maintains the same order quantity each time an order is issued.

Fixed Order Quantity   The **fixed order quantity (FOQ)** rule maintains the same order quantity each time an order is issued.[3] For example, the lot size might be the size dictated by equipment capacity limits, such as when a full lot must be loaded into a furnace at one time. For purchased items, the FOQ could be determined by the quantity discount level, truckload capacity, or minimum purchase quantity. Alternatively, the lot size could be determined by the economic order quantity (EOQ) formula (see Chapter 9, "Supply Chain Inventory Management"). Figure 16.12 illustrated the FOQ rule. However, if an item's gross requirement within a week is particularly large, the FOQ might be insufficient to avoid a shortage. In such unusual cases, the inventory planner must increase the lot size beyond the FOQ, typically to a size large enough to avoid a shortage. Another option is to make the order quantity an integer multiple of the FOQ. This option is appropriate when capacity constraints limit production to FOQ sizes (at most).

periodic order quantity (POQ)

A rule that allows a different order quantity for each order issued but issues the order for predetermined time intervals.

Periodic Order Quantity   The **periodic order quantity (POQ)** rule allows a different order quantity for each order issued but issues the order for predetermined time intervals, such as every two weeks. The order quantity equals the amount of the item needed during the predetermined time between orders and must be large enough to prevent shortages. Specifically, the POQ is

$$\begin{pmatrix} \text{POQ lot size} \\ \text{to arrive in} \\ \text{week } t \end{pmatrix} = \begin{pmatrix} \text{Total gross requirements} \\ \text{for } P \text{ weeks, including} \\ \text{week } t \end{pmatrix} - \begin{pmatrix} \text{Projected on-hand} \\ \text{inventory balance at} \\ \text{end of week } t-1 \end{pmatrix}$$

This amount exactly covers P weeks' worth of gross requirements. That is, the projected on-hand inventory should equal zero at the end of the Pth week.

---

[3]The *kanban* system essentially uses a FOQ rule, except that the order quantity is very small.

Suppose that we want to switch from the FOQ rule used in Figure 16.12 to the POQ rule. Figure 16.13 was created with the *Single-Item MRP* Solver in OM Explorer. It shows the application of the POQ rule, with $P = 3$ weeks, to the seat subassembly inventory. The first order is required in week 4 because it is the first week that projected inventory balance will fall below zero. The first order using $P = 3$ weeks is

$$(\text{POQ lot size}) = \begin{pmatrix} \text{Gross requirements} \\ \text{for weeks} \\ 4, 5, \text{and } 6 \end{pmatrix} - \begin{pmatrix} \text{Inventory at} \\ \text{end of week 3} \end{pmatrix}$$

$$= (120 + 0 + 150) - 117 = 153 \text{ units}$$

The second order must arrive in week 7 with a lot size of $(120 + 0) - 0 = 120$ units. This second order reflects only two weeks' worth of gross requirements—to the end of the planning horizon.

MyOMLab

Tutor 16.2 in MyOMLab provides a new example to practice lot-sizing decisions using FOQ, POQ, and L4L rules.

Periods		8								
Item	Seat Assembly			Period (P) for POQ			3	Lot Size (FOQ)		
Description								Lead Time		2
POQ Rule	▼									
		1	2	3	4	5	6	7	8	
Gross requirements		150			120		150	120		
Scheduled receipts		230								
Projected on-hand inventory	37	117	117	117	150	150				
Planned receipts					153			120		
Planned order releases			153			120				

◀ **FIGURE 16.13**
The POQ ($P = 3$) Rule for the Seat Subassembly

The POQ rule does not mean that the planner must issue a new order every P weeks. Rather, when an order is planned, its lot size must be enough to cover P successive weeks. One way to select a P value is to divide the average lot size desired, such as the EOQ or some other applicable lot size, by the average weekly demand. That is, express the target lot size as the desired weeks of supply (P) and round to the nearest integer.

Lot for Lot   A special case of the POQ rule is the **lot-for-lot (L4L) rule**, under which the lot size ordered covers the gross requirements of a single week. Thus, $P = 1$, and the goal is to minimize inventory levels. This rule ensures that the planned order is just large enough to prevent a shortage in the single week it covers. The L4L lot size is

$$\begin{pmatrix} \text{L4L lot size} \\ \text{to arrive in} \\ \text{week } t \end{pmatrix} = \begin{pmatrix} \text{Gross requirements} \\ \text{for week } t \end{pmatrix} - \begin{pmatrix} \text{Projected on-hand} \\ \text{inventory balance at} \\ \text{end of week } t-1 \end{pmatrix}$$

lot-for-lot (L4L) rule
A rule under which the lot size ordered covers the gross requirements of a single week.

The projected on-hand inventory combined with the new order will equal zero at the end of week t. Following the first planned order, an additional planned order will be used to match each subsequent gross requirement.

This time we want to switch from the FOQ rule to the L4L rule. Figure 16.14 shows the application of the L4L rule to the seat subassembly inventory. As before, the first order is needed in week 4:

$$(\text{L4L lot size}) = \begin{pmatrix} \text{Gross requirements} \\ \text{in week 4} \end{pmatrix} - \begin{pmatrix} \text{Inventory balance} \\ \text{at end of week 3} \end{pmatrix}$$

$$= 120 - 117 = 3$$

Periods		8								
Item	Seat Assembly			Period (P) for POQ			Lot Size (FOQ)			
Description								Lead Time		2
L4L Rule	▼									
		1	2	3	4	5	6	7	8	
Gross requirements		150			120		150	120		
Scheduled receipts		230								
Projected on-hand inventory	37	117	117	117						
Planned receipts					3		150	120		
Planned order releases			3		150	120				

◀ **FIGURE 16.14**
The L4L Rule for the Seat Subassembly

The stockroom must receive additional orders in weeks 6 and 7 to satisfy each of the subsequent gross requirements. The planned receipt for week 6 is 150 and for week 7 is 120.

Comparing Lot-Sizing Rules Choosing a lot-sizing rule can have important implications for inventory management. Lot-sizing rules affect inventory costs and setup and ordering costs. The FOQ, POQ, and L4L rules differ from one another in one or both respects. In our example, each rule took effect in week 4, when the first order was placed. Let us compare the projected on-hand inventory averaged over weeks 4 through 8 of the planning horizon. The data are shown in Figures 16.12, 16.13, and 16.14, respectively.

$$\text{FOQ:} \quad \frac{227 + 227 + 77 + 187 + 187}{5} = 181 \text{ units}$$

$$\text{POQ:} \quad \frac{150 + 150 + 0 + 0 + 0}{5} = 60 \text{ units}$$

$$\text{L4L:} \quad \frac{0 + 0 + 0 + 0 + 0}{5} = 0 \text{ units}$$

The performance of the L4L rule with respect to average inventory levels comes at the expense of an additional planned order and its accompanying setup time and cost. We can draw three conclusions from this comparison:

1. The FOQ rule generates a high level of average inventory because it creates inventory *remnants*. A remnant is inventory carried into a week, but it is too small to prevent a shortage. Remnants occur because the FOQ does not match requirements exactly. For example, according to Figure 16.12, the stockroom must receive a planned order in week 7, even though 77 units are on hand at the beginning of that week. The remnant is the 77 units that the stockroom will carry for 3 weeks, beginning with receipt of the first planned order in week 4. Although they increase average inventory levels, inventory remnants introduce stability into the production process by buffering unexpected scrap losses, capacity bottlenecks, inaccurate inventory records, or unstable gross requirements.

2. The POQ rule reduces the amount of average on-hand inventory because it does a better job of matching order quantity to requirements. It adjusts lot sizes as requirements increase or decrease. Figure 16.13 shows that in week 7, when the POQ rule has fully taken effect, the projected on-hand inventory is zero—no remnants.

3. The L4L rule minimizes inventory investment, but it also maximizes the number of orders placed. This rule is most applicable to expensive items or items with small ordering or setup costs. It is the only rule that can be used for a low-volume item made to order. It can also approximate the small-lot inventory levels of a lean system.

By avoiding remnants, both the POQ and the L4L rule may introduce instability by tying the lot-sizing decision so closely to requirements. If any requirement changes, so must the lot size, which can disrupt component schedules. Last-minute increases in parent orders may be hindered by missing components.

Safety Stock An important managerial decision is the quantity of safety stock to carry. It is more complex for dependent demand items than for independent demand items. Safety stock for dependent demand items with lumpy demand (gross requirements) is helpful only when future gross requirements, the timing or size of scheduled receipts, and the amount of scrap that will be produced are uncertain. As these uncertainties are resolved, safety stock should be reduced and ultimately eliminated. The usual policy is to use safety stock for end items and purchased items to protect against fluctuating customer orders and unreliable suppliers of components but to avoid using it as much as possible for intermediate items. Safety stocks can be incorporated in the MRP logic by using the following rule: Schedule a planned receipt whenever the projected on-hand inventory balance drops below the desired safety stock level (rather than zero, as before). The objective is to keep a minimum level of planned inventories equal to the safety stock quantity. Figure 16.15 shows what happens when the safety stock requirement has just been increased from 0 units to 80 units of safety stock for the seat assembly using an FOQ of 230 units. The beginning projected on-hand quantity is still 37 units when the safety stock policy is introduced, and cannot fall below 80 units in any future period thereafter. Compare the

▼ **FIGURE 16.15**

Inventory Record for the Seat Subassembly Showing the Application of a Safety Stock

FOQ Rule						Lot Size: 230 units			
						Lead Time: 2 weeks			
						Safety Stock: 80 units			
		Week							
		1	2	3	4	5	6	7	8
Gross requirements		150	0	0	120	0	150	120	0
Scheduled receipts		230	0	0	0	0	0	0	0
Projected on-hand inventory	37	117	117	117	227	227	307	187	187
Planned receipts		0	0	0	230	0	230	0	0
Planned order releases		0	230	0	230	0	0	0	0

results in Figure 16.15 to Figure 16.12. The net effect is to move the second planned order release from week 5 to week 4 to avoid dropping below 80 units in week 6.

Outputs from MRP

MRP systems provide many reports, schedules, and notices to help planners control dependent demand inventories, as indicated in Figure 16.16. In this section, we discuss the MRP explosion process, notices that alert planners to items needing attention, resource requirement reports, and performance reports.

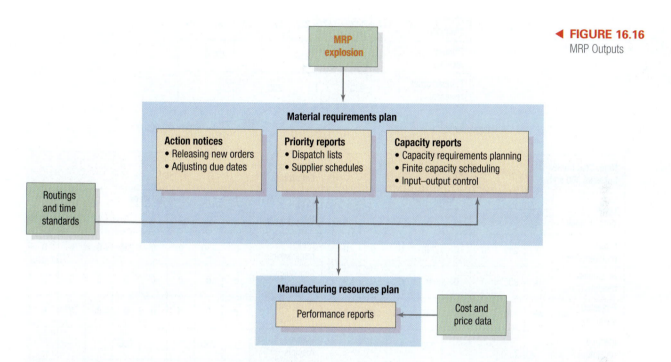

◀ **FIGURE 16.16**
MRP Outputs

MRP Explosion   MRP translates, or *explodes*, the MPS and other sources of demand into the requirements needed for all of the subassemblies, components, and raw materials the firm needs to produce parent items. This process generates the material requirements plan for each component item.

An item's gross requirements are derived from three sources:

1. The MPS for immediate parents that are end items

2. The planned order releases (*not* the gross requirements, scheduled receipts, or planned receipts) for immediate parents below the MPS level

3. Any other requirements not originating in the MPS, such as the demand for replacement parts

Consider the seat subassembly and its inventory record shown in Figure 16.12. The seat subassembly requires a seat cushion and a seat frame, which in turn needs four seat-frame boards. Its BOM is shown in Figure 16.17 (see also Figure 16.4, which shows how the seat subassembly BOM relates to the product as a whole). How many seat cushions should we order from the supplier? How many seat frames should we produce to support the seat subassembly schedule? How many seat-frame boards do we need to make? The answers to these questions depend on the existing inventories of these items and the replenishment orders already in progress. MRP can help answer these questions through the explosion process.

Figure 16.18 shows the MRP records for the seat subassembly and its components. We already showed how to develop the MRP record for the seat subassembly. We now concentrate on the MRP records of its components. The lot-size rules are an FOQ of 300 units for the seat frame, L4L for the seat cushion, and an FOQ of 1,500 for the seat-frame boards. All three components have a 1-week lead time. The key to the explosion process is to determine the proper timing and size of the gross requirements for each component. After we make those determinations, we can derive the planned order release schedule for each component by using the logic already demonstrated.

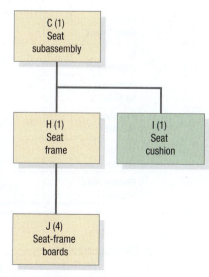

FIGURE 16.17 ▲
BOM for the Seat Subassembly

Item: Seat subassembly
Lot size: 230 units

Lead time: 2 weeks	Week							
	1	2	3	4	5	6	7	8
Gross requirements	150	0	0	120	0	150	120	0
Scheduled receipts	230	0	0	0	0	0	0	0
Projected inventory 37	117	117	117	227	227	77	187	187
Planned receipts				230			230	
Planned order releases		230			230			

Usage quantity: 1   Usage quantity: 1

Item: Seat frames
Lot size: 300 units

Lead time: 1 week	Week							
	1	2	3	4	5	6	7	8
Gross requirements	0	230	0	0	230	0	0	0
Scheduled receipts	0	300	0	0	0	0	0	0
Projected inventory 40	40	110	110	110	180	180	180	180
Planned receipts					300			
Planned order releases				300				

Item: Seat cushion
Lot size: L4L

Lead time: 1 week	Week							
	1	2	3	4	5	6	7	8
Gross requirements	0	230	0	0	230	0	0	0
Scheduled receipts	0	0	0	0	0	0	0	0
Projected inventory 0	0	0	0	0	0	0	0	0
Planned receipts		230			230			
Planned order releases	230			230				

Usage quantity: 4

Item: Seat-frame boards
Lot size: 1,500 units

Lead time: 1 week	Week							
	1	2	3	4	5	6	7	8
Gross requirements	0	0	0	1,200	0	0	0	0
Scheduled receipts	0	0	0	0	0	0	0	0
Projected inventory 200	200	200	200	500	500	500	500	500
Planned receipts				1,500				
Planned order releases			1,500					

FIGURE 16.18 ▲
MRP Explosion of Seat Assembly Components

In our example, the components have no independent demand for replacement parts. Consequently, in Figure 16.18 the gross requirements of a component come from the planned order releases of its parents. The seat frame and the seat cushion get their gross requirements from the planned order release schedule of the seat subassembly. Both components have gross requirements of 230 units in weeks 2 and 5, the same weeks in which we will be releasing orders to make more seat subassemblies. In week 2, for example, the materials handler for the assembly department will withdraw 230 seat frames and 230 seat cushions from inventory so that the assembly department can produce the seat subassemblies in time to avoid a stockout in week 4. The materials plans for the seat frame and the seat cushion must allow for that.

Using the gross requirements in weeks 2 and 5, we can develop the MRP records for the seat frame and the seat cushion, as shown in Figure 16.18. For a scheduled receipt of 300 seat frames in week 2, an on-hand quantity of 40 units, and a lead time of 1 week, we need to release an order of 300 seat frames in week 4 to cover the assembly schedule for the seat subassembly. The seat cushion has no scheduled receipts and no inventory on hand; consequently, we must place orders for 230 units in weeks 1 and 4, using the L4L logic with a lead time of 1 week.

After determining the replenishment schedule for the seat frame, we can calculate the gross requirements for the seat-frame boards. We plan to begin producing 300 seat frames in week 4. Each frame requires 4 boards, so we need to have $300(4) = 1,200$ boards available in week 4. Consequently, the gross requirement for seat-frame boards is 1,200 in week 4. Given no scheduled receipts, 200 boards in stock, a lead time of 1 week, and an FOQ of 1,500 units, we need a planned order release of 1,500 in week 3.

The questions posed earlier can now be answered. We should plan to release the following orders: 300 seat frames in week 4; 230 seat cushions in each of weeks 1 and 4; and 1,500 seat-frame boards in week 3. If MRP plans are updated weekly, only the planned order for week 1 should be released now. Releasing it creates a scheduled receipt of 230 seat cushions that will appear in the updated inventory record. The other orders remain in the planning stage, and even might be revised by the MRP explosion done next week.

Computer Support   In practice a company can have thousands of dependent demand items with an average of six bills of materials levels. Time horizons often stretch out for 30 or more time periods into the future. Doing a MRP explosion by hand (as shown in Figure 16.18) would be impractical. What is needed is massive data processing, the very thing that computers do best, leaving the decision making to the inventory analyst. The *Material Requirements Planning* Solver of OM Explorer in MyOMLab represents a small example of what is done on a much larger scale by commercial packages, can compute requirements for up to two end items. It has the ability to develop inventory records up to 18 items deep with little effort, and can easily re-compute these requirements if there is any change in planning parameters.

Figure 16.19 shows the output from the *Material Requirements Planning* Solver of OM Explorer. It confirms the same results as Figure 16.18. Based on this output, only one thing would be brought to the attention of the inventory planner—the planned order release of 230 units for seat cushions. Its planned order release is now "mature" for release this week. Unless the planner knows of a problem, the planner would place the order with the supplier. At the same time, the planner would input a transaction that automatically eliminates the planned order release in period 1, removes the planned receipt in period 2, and inserts a scheduled receipt for 230 units in period 2. The planner need not look at the records for seat frames or seat-frame boards, because no action is needed for them.

Other Important Reports   Once computed, inventory records for any item appearing in the BOM can be printed in hard copy or displayed on a computer video screen. Inventory planners use a computer-generated memo called an **action notice** to make decisions about releasing new orders and adjusting the due dates of scheduled receipts. These notices are generated every time the system is updated, typically once per week. The action notice alerts planners to only the items that need their attention, such as those items that have a planned order release in the current period or a scheduled receipt that needs its due date revised. Planners can then view the full records for those items and take the necessary actions. An action notice can simply be a list of part numbers for items that need attention; or it can be the full record for such items, with a note at the bottom identifying the action needed.

By itself, the MRP system does not recognize capacity limitations when computing planned orders; that is, it may call for a planned order release that exceeds the amount that can be physically produced. An essential role of planners is to monitor the capacity requirements of material requirements plans, adjusting a plan when it cannot be met. Particular attention is paid to bottlenecks. The planner can apply theory of constraints (TOC) principles (see Chapter 7, "Constraint Management") to keep bottleneck operations fed by adjusting some lot sizing rules or occasionally overriding planned order releases. To facilitate this process, various types of capacity reports can be provided. For example, **capacity requirements planning (CRP)** reports project time-phased

MyOMLab

action notice

A computer-generated memo alerting planners about releasing new orders and adjusting the due dates of scheduled receipts.

capacity requirements planning (CRP)

A technique used for projecting time-phased capacity requirements for work stations; its purpose is to match the materials requirements plan with the capacity of key processes.

FIGURE 16.19 ▶

*Material Requirements Planning
Solver Output of OM Explorer for
Seat Assembly Components*

Master Production Schedule

		1	2	3	4	5	6	7	8	9	10
Item A MPS Start	Descr: Seat		230			230					
Item B MPS Start	Descr:										

☑ Use second finished item

Material Requirements Planning

Item C	Descr: Seat Frames	Period (P) for POQ	Lot Size (FOQ) 300
FOQ Rule			Lead Time 1
			Safety Stock

Usage Quantity for Item: A   1   B

		1	2	3	4	5	6	7	8	9	10
Gross Requirements			230			230					
Scheduled Receipts			300								
Projected On-Hand Inventory	40	40	110	110	110	180	180	180	180	180	180
Planned Receipts						300					
Planned Order Releases					300						

Item D	Descr: Seat Cushion	Period (P) for POQ 1	Lot Size (FOQ)
L4L Rule			Lead Time 1
			Safety Stock

Usage Quantity for Item: A   1   B   C

		1	2	3	4	5	6	7	8	9	10
Gross Requirements			230			230					
Scheduled Receipts											
Projected On-Hand Inventory	0	0	0	0	0	0	0	0	0	0	0
Planned Receipts			230			230					
Planned Order Releases		230			230						

Item E	Descr: Seat-frame boards	Period (P) for POQ	Lot Size (FOQ) 1500
FOQ Rule			Lead Time 1
			Safety Stock

Usage Quantity for Item: A   B   C   4   D

		1	2	3	4	5	6	7	8	9	10
Gross Requirements					1200						
Scheduled Receipts											
Projected On-Hand Inventory	200	200	200	200	500	500	500	500	500	500	500
Planned Receipts					1500						
Planned Order Releases				1500							

capacity requirements for workstations. They calculate workload according to the work required to complete the scheduled receipts already in the shop and to complete the planned order releases not yet released. Bottlenecks are those workstations at which the projected loads exceed station capacities.

Other types of outputs are also possible, such as priority reports on orders already placed to the shop or with suppliers. Priority reports begin with the due dates assigned to scheduled receipts, which planners keep up to date so that they continue to reflect when receipt is really needed. On a broader scale, the information in an MRP system is useful to functional areas other than operations. MRP evolved into **manufacturing resource planning (MRP II)**, a system that ties the basic MRP system to the company's financial system and to other core and supporting processes. For example, management can project the dollar value of shipments, product costs, overhead allocations, inventories, backlogs, and profits by using the MRP plan along with prices and product and activity costs from the accounting system. Also, information from the MPS, scheduled receipts, and planned orders can be converted into cash flow projections, which are broken down by product families. Similar computations are possible for other performance measures of interest to management. In fact, MRP II ultimately evolved into enterprise resource planning (ERP), which was introduced at the beginning of this chapter. Some firms may, however, forego the cost of vendor-delivered MRP and ERP systems because of the huge budgets and company resources involved in their deployment, and instead create their own MRP system implementations in-house. Managerial Practice 16.1 illustrates how Winnebago created and adapted its own homegrown software for the MRP system to achieve business self sufficiency.

manufacturing resource planning (MRP II)

A system that ties the basic MRP system to the company's financial system and to other core and supporting processes.

MANAGERIAL PRACTICE 16.1   Material Requirements Planning at Winnebago Industries

Winnebago Industries, based in Forest City, Iowa, is a leading manufacturer of motor homes and related products and services in the United States since 1958. In a challenging business environment, the firm improves the quality of its products at lowered production costs by emphasizing employee teamwork and involvement. So it is no surprise that Winnebago has for decades kept away from software packages, preferring instead to create and adapt its own applications. It can thereby achieve a closer fit to the company's business needs, an objective that is at times not met by the ERP vendors described at the beginning of this chapter. It can also be more responsive to users of its systems at lowered costs since it can reuse its code as needed.

An example of Winnebago's homegrown approach is its MRP system, which runs on an IBM zSeries mainframe and is used to plan material needs and schedule production orders. When the company added a new model to its fleet of motor homes, it took only a few hundred hours of development, including a new bill of materials, to make the required changes to the MRP system in order to support production of the new vehicle. Such agility in adapting the MRP system to manufacturing and supplier needs can be especially rewarding in an environment where the change in product variety is constant. In 2007 alone, Winnebago offered 20 different models in 86 different floor plans, each of which can be outfitted with a variety of options including colors, wood stains, and drawer pulls. For the 2011 model year, several new floorplans were added. When you are in the business of building customized homes-on-wheels ranging in prices from $60,000 to $285,000, combining cost consciousness with common sense and employee involvement in developing your own MRP, sales order management, and purchasing systems can prove valuable.

CRAIG BORCK KRT/Newscom

Workmen at the Winnebago factory in Forest City, Iowa, lift a completed side of a Winnebago, with interior and exterior surfaces, wiring, framework and foam insulation all in place, onto a cart so it can be hauled to the next step in the assembly process. Since 1995, Winnebago has spent approximately $5 million annually to automate some of its assembly-line processes, up from $2 million per year previously.

Source: "Road Rules: Creating and Adapting Homegrown Software is the Key to Winnebago's Drive for Business Self Sufficiency," October 15, 2006; **http://www .winnebagoind.com/** (May 28, 2011).

MRP, Core Processes, and Supply Chain Linkages

Among the four core processes of an organization that link activities within and across firms in a supply chain, the MRP system interacts with all of them either through its inputs or its outputs. It all begins with customer orders, which consist of orders for end items as well as replacement parts. MRP and resource planning typically reside inside the order fulfillment process. Master Production Scheduling is an integral part of MRP (see Figure 16.2). As shown schematically in Figure 16.20, the MPS drives the feedback between the order fulfillment process and the customer relationship process through confirmation of order receipts and promised due dates. MPS also provides guidance to the sales group within the customer relationship process with respect to when future orders can be promised, and whether the due dates for existing orders can be adjusted in the time frame requested. The new service and product development process provides an updated bill of materials to the MRP system, and makes sure that every component and assembly needed for manufacturing of end items is properly recognized.

FIGURE 16.20 ▶
MRP Related Information
Flows in the Supply Chain

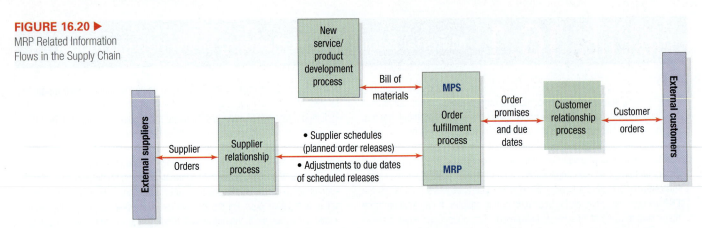

In a similar vein on the inbound side, orders to external suppliers are based on the planned order releases, which come directly from the output of MRP reports. The power of MRP, however, becomes evident when changes to an existing schedule are needed. These changes can be generated, for example, by changes to the MPS because a customer wants to change the timing or size of future orders, by some internal failure such as material shortages or unexpected machine downtime, or by supplier failure. In a supply chain, schedule changes have implications for customers as well as suppliers. Some firms, in partnership with their suppliers, have ERP/MRP systems that can actually "see" into their suppliers' inventory to determine if a particular item is in stock or, if not, when it can be expected. This is an advantage when contemplating a change to the original schedule of order releases. While systems such as this are powerful tools for making changes, care must be taken to avoid unnecessary fluctuations in the timing and size of PORs because of the choice of lot sizing policy. As we have seen, lot sizing rules such as POQ or L4L are susceptible to changes in requirements, and using them indiscriminately can cause instability in replenishment orders. In turn, this instability can be transmitted up the supply chain if the firm's MRP system is electronically linked to the production planning and control systems of its immediate suppliers.

Execution of MRP-based plans using the information flows between core processes as shown in Figure 16.20 properly link a firm with its upstream and downstream supply chain partners. Valid customer and supplier priorities would not be effectively recognized without such an MRP-based framework, which in many firms is actually implemented through an ERP system discussed earlier in the chapter.

MRP and the Environment

Consumer and governmental concern about the deterioration of the natural environment has driven manufacturers to reengineer their processes to become more environmentally friendly. The recycling of base materials is becoming more commonplace, and products are being designed in such a way that they can be remanufactured after their useful lives. Nonetheless, manufacturing processes often produce a number of waste materials that need to be properly disposed of. Wastes come in many forms:

- Effluents, such as carbon monoxide, sulfur dioxide, and hazardous chemicals associated with the processes used to manufacture the product

- Materials, such as metal shavings, oils, and chemicals associated with specific operations

- Packaging materials, such as unusable cardboard and plastics associated with certain products or purchased items

- Scrap associated with unusable products or component defects generated by the manufacturing process

Companies can modify their MRP systems to help them track these wastes and plan for their disposal. The type and amount of waste associated with each item can be entered into its BOM by treating the waste much like you would a component of the item. When the MPS is developed for a product, reports can be generated that project the amount of waste expected during the production process and when it will occur. Although this approach can require that a firm's BOM be modified substantially, the benefits are also substantial. Firms can identify their waste problems in advance to eliminate them in some cases (through process improvement efforts) or plan for their proper disposal in others. It also gives the firm a way to generate any formal documentation required by the government to verify that it has complied with environmental laws and policies.

Resource Planning for Service Providers

We have seen how manufacturing companies can disaggregate an MPS of finished products, which in turn must be translated into the needs for resources, such as staff, equipment, components, and financial assets. The driver for these resource requirements is a material requirements plan. Service providers, of course, must plan their resources just as manufacturers do. However, unlike finished goods, services cannot be inventoried. They must be provided on demand. In terms of resource planning then, service organizations must focus on maintaining the *capacity* to serve their customers. In this section, we will discuss how service providers use the concept of dependent demand and a bill of resources in managing capacity.

Dependent Demand for Services

When we discussed planning and control systems for manufacturers earlier in this chapter, we introduced the concept of *dependent demand*, which is demand for an item that is a function of the production plans for some other item the company produces. For service resource planning, it is useful to define the concept of dependent demand to include demands for resources that are driven by forecasts of customer requests for services or by plans for various activities in support of the services the company provides. Here are some other examples of dependent demands for service providers.

Restaurants   Every time you order from the menu at a restaurant, you initiate the restaurant's need for certain types of goods (uncooked food items, plates, and napkins), staff (chef, servers, and dishwashers), and equipment (stoves, ovens, and cooking utensils). Using a forecast of the demand for each type of meal, the manager of the restaurant can estimate the need for these resources. Many restaurants, for example, feature "specials" on certain days, say, fish on Fridays or prime rib on Saturdays. Specials improve the accuracy of the forecasts managers need to make for different types of meals (and the food products that are required to make them) and typically signal the need for above-average staffing levels. How much of these resources will be needed, however, depends on the number of meals the restaurant ultimately expects to serve. As such, these items—food products and staff members—are dependent demands.

Airlines   Whenever an airline schedules a flight, certain supporting goods are needed (beverages, snacks, and fuel), labor (pilots, flight attendants, and airport services), and equipment (a plane and airport gate). The number of flights and passengers the airline forecasts it will serve determines the amount of these resources needed. Just like a manufacturer, the airline can explode its master schedule of flights to make this determination.

Hospitals   With the exception of the emergency room services, hospitals can use their admission appointments to create a master schedule. The master schedule can be exploded to determine the resources the hospital will need during a certain period. For example, when you schedule a surgical procedure, you generate a need for facilitating goods such as medicines, surgical gowns, linens, staff (a surgeon, nurses, and anesthesiologist), and equipment (an operating room, surgical tools, and recovery bed). As they build their master schedules, hospitals must ensure that certain equipment and personnel do not become overcommitted—that capacity is maintained, in other words. For example, an appointment for a key operation might have to be scheduled in advance at a time a surgeon is available to do it, even though the hospital's other resources—operating room, nurses, and so forth—might be currently be available.

Hotels   A traveler who makes a reservation at a hotel generates demand for facilitating goods (soap and towels), staff (front desk, housekeeping, and concierge), and equipment (fax, television, and exercise bicycle). To determine its dependent resource needs, a hotel adds the number of reservations already booked to the number of "walk-in" customers it forecasts it will have. This figure is used to create the hotel's master schedule. One resource a hotel cannot easily adjust, however, is the number of rooms it has. If the hotel is overbooked, for instance, it cannot simply add more rooms. If it has too few guests, it cannot "downsize" its number of rooms. Given the high capital costs needed

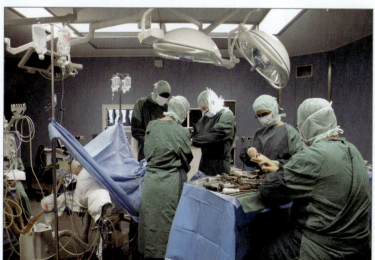

Doctors and nurses in operation room during surgery

for this resource, hotels try to maintain as high a utilization rate as possible by offering group rates or special promotions at certain times of the year. In other words, they try to drive up dependent demand for this particular resource.

Bill of Resources

bill of resources (BOR)

A record of a service firm's parent-component relationships and all of the materials, equipment time, staff, and other resources associated with them, including usage quantities.

The service analogy to the bill of materials in a manufacturing company is the **bill of resources (BOR)**, which is a record of a service firm's parent–component relationships and all of the materials, equipment time, staff, and other resources associated with them, including usage quantities. Once the service firm has completed its master schedule, the BOR can be used to determine what resources the firm will need, how much of them it will need, and when. A BOR for a service provider can be as complex as a BOM for a manufacturer. Consider a hospital that just scheduled treatment of a patient with an aneurysm. As shown in Figure 16.21 (a), the BOR for treatment of an aneurysm has seven levels, starting at the top (end item): (1) discharge, (2) intermediate care, (3) postoperative care (step down), (4) postoperative care (intensive), (5) surgery, (6) preoperative

FIGURE 16.21 ▶

BOR for Treating an Aneurysm

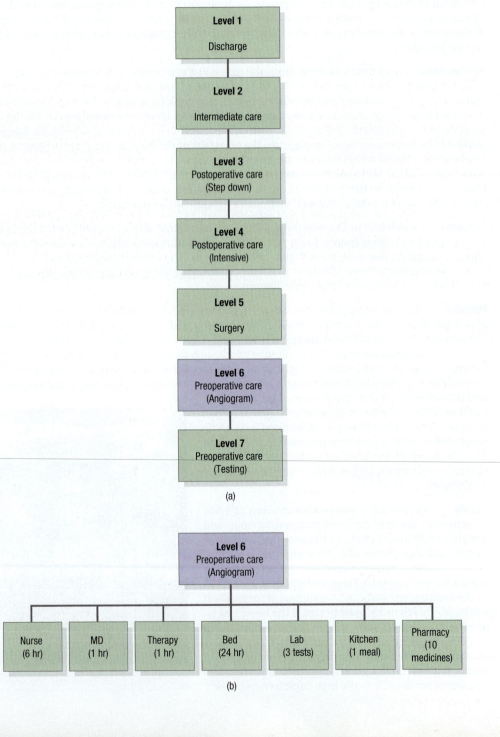

care (angiogram), and (7) preoperative care (testing). Each level of the BOR has a set of material and resource requirements and an associated lead time. For example, at level 6, shown in Figure 16.21(b), the patient needs 6 hours of nurses' time, 1 hour of the primary MD's time, 1 hour of the respiratory therapist's time, 24 hours of bed time, 3 different lab tests, 1 meal, and 10 different medicines from the pharmacy. The lead time for this level is 1 day. The lead time for the entire stay for treatment of the aneurysm is 12.2 days. A master schedule of patient admissions and the BORs for each illness enable the hospital to manage its critical resources. Reports analogous to the MRP II reports we discussed earlier in the chapter can be generated for the people who manage the various functional areas of the hospital.

One resource every service provider needs, however, is cash. Service organizations have to forecast the number of customers they expect to serve so that they have enough cash on hand to purchase materials that support the services—labor and other products. Purchasing these items increases the firm's accounts payable. As services are actually completed for customers, the firm's accounts receivable increases. The firm's master schedule and its accounts receivable and payable help a company predict the amount and timing of its cash flows.

LEARNING GOALS IN REVIEW

1. **Explain how enterprise resource planning (ERP) systems can foster better resource planning.** Review the opening vignette on Dow Corning, and the section on "Enterprise Resource Planning," pp. 544–546. Pay attention to Figure 16.1 to understand how different application modules come together to create functionality and value in the ERP systems.

2. **Explain how the concept of dependent demand is fundamental to resource planning.** See the section on "Dependent Demand," p. 547, which shows how continuous independent demand can lead to lumpy requirements for dependent demand. Then, a separate system, called material requirements planning, is needed to manage dependent demand situations.

3. **Describe a master production schedule (MPS) and compute available-to-promise quantities.** The section "Master Production Scheduling," pp. 549–553, shows you how firms break down a production plan into more detailed schedules. Understand the key relationships between Figures 16.5, 16.6, and 16.8.

4. **Apply the logic of a material requirements planning (MRP) system to identify production and purchase orders needed for dependent demand items.** Using Figure 16.12, page 555, understand how an inventory record is created for a given lot size rule. The section on "Planning Factors," pp. 556–559, shows you how choice of different managerial policies affect material plans. Finally, focus on understanding the MRP explosion process as illustrated in Figure 16.18 on page 560 and Solved Problem 3 on page 570.

5. **Apply MRP principles to the provision of services and distribution inventories.** The section, "Resource Planning for Service Providers," pp. 565–567, illustrates how the Bill of Resources can be used to plan dependent demand for services in settings such as the restaurants, airlines, hospitals, and hotels.

MyOMLab helps you develop analytical skills and assesses your progress with multiple problems on bills of material, parents, intermediate items, components, purchased items, lead times, master production schedules, MRP record, FOQ, L4L, POQ, scheduled receipts, and multi-level material requirements plans.

MyOMLab Resources	Titles	Link to the Book
Video	*Nantucket Nectars: ERP*	Enterprise Resource Planning
Active Model Exercise	16.1 Material Requirements Plan	Bill of Resources
OM Explorer Solvers	Master Production Scheduling Material Requirements Planning Single-Item MRP	Master Production Scheduling; Solved Problem 2 (p. 569) Material Requirements Planning; Solved Problem 3 (p. 570) Material Requirements Planning; Solved Problem 3 (p. 570)
OM Explorer Tutors	16.1 Master Production Scheduling 16.2 FOQ, POQ, and L4L Rules	Master Production Scheduling; Solved Problem 2 (p. 569) Planning Factors; Solved Problem 3 (p. 570)
Tutor Exercises	16.1 Applying Different MRP Lot Sizing Rules	Planning Factors; Solved Problem 3 (p. 570)
Virtual Tours	Vaughn Hockey Equipment Winslow Life Rafts	Bill of Materials Enterprise Resource Planning

MyOMLab Resources	Titles	Link to the Book
MyOMLab Supplements	H. Measuring Output Rates I. Learning Curve Analysis	MyOMLab Supplement H MyOMLab Supplement I
Internet Exercise	Schwinn Hewlett-Packard	Material Requirements Planning Material Requirements Planning
Blank MRP Records		
Key Equations		
Image Library		

Key Terms

action notice 561
available-to-promise (ATP)
　　inventory 552
bill of materials (BOM) 547
bill of resources (BOR) 566
capacity requirements planning
　　(CRP) 561
component 547
dependent demand 547
end item 548
enterprise process 544
enterprise resource planning (ERP)
　　systems 544

fixed order quantity (FOQ) 556
gross requirements 553
intermediate item 548
inventory record 553
lot-for-lot (L4L) rule 557
manufacturing resource planning
　　(MRP II) 562
master production schedule
　　(MPS) 549
material requirements planning
　　(MRP) 546
MRP explosion 546
parent 547

part commonality 549
periodic order quantity (POQ) 556
planned order release 555
planned receipts 554
projected on-hand inventory 554
purchased item 549
resource planning 544
subassembly 548
usage quantity 548

Solved Problem 1

Refer to the bill of materials for product A shown in Figure 16.22.

FIGURE 16.22 ▶
BOM for Product A

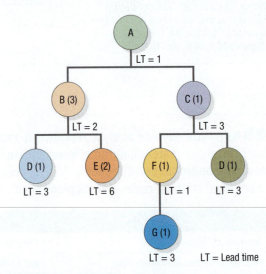

If there is no existing inventory and no scheduled receipts, how many units of items G, E, and D must be purchased to produce 5 units of end item A?

SOLUTION
Five units of item G, 30 units of item E, and 20 units of item D must be purchased to make 5 units of A. The usage quantities shown in Figure 16.22 indicate that 2 units of E are needed to make 1 unit of B and that 3 units of B are needed to make 1 unit of A; therefore, 5 units of A require 30 units of E ($2 \times 3 \times 5 = 30$). One unit of D is consumed to make 1 unit of B, and 3 units of B per unit of A result in 15 units of D ($1 \times 3 \times 5 = 15$); 1 unit of D in each unit of C and 1 unit of C per unit of A result in another 5 units of D ($1 \times 1 \times 5 = 5$). The total requirements to make 5 units of A are 20 units of D ($15 + 5$). The calculation of requirements for G is simply $1 \times 1 \times 1 \times 5 = 5$ units.

Solved Problem 2

The order policy is to produce end item A in lots of 50 units. Using the data shown in Figure 16.23 and the FOQ lot-sizing rule, complete the projected on-hand inventory and MPS quantity rows. Then, complete the MPS start row by offsetting the MPS quantities for the final assembly lead time. Compute the available-to-promise inventory for item A. Finally, assess the following customer requests for new orders. Assume that these orders arrive consecutively and their affect on ATP is cumulative. Which of these orders can be satisfied without altering the MPS Start quantities?

a. Customer A requests 30 units in week 1.

b. Customer B requests 30 units in week 4.

c. Customer C requests 10 units in week 3.

d. Customer D requests 50 units in week 5.

Item: A									Order Policy: 50 units		
									Lead Time: 1 week		
Quantity on Hand:	5	\multicolumn{10}{c} Week									
		1	2	3	4	5	6	7	8	9	10
Forecast		20	10	40	10	0	0	30	20	40	20
Customer orders (booked)		30	20	5	8	0	2	0	0	0	0
Projected on-hand inventory		25									
MPS quantity		50									
MPS start											
Available-to-promise (ATP) inventory											

▲ FIGURE 16.23
MPS Record for End Item A

SOLUTION

The projected on-hand inventory for the second week is

$$\begin{pmatrix} \text{Projected on-hand} \\ \text{inventory at end} \\ \text{of week 2} \end{pmatrix} = \begin{pmatrix} \text{On-hand} \\ \text{inventory in} \\ \text{week 1} \end{pmatrix} + \begin{pmatrix} \text{MPS quantity} \\ \text{due in week 2} \end{pmatrix} - \begin{pmatrix} \text{Requirements} \\ \text{in week 2} \end{pmatrix}$$

$$= 25 + 0 - 20 = 5 \text{ units}$$

where requirements are the larger of the forecast or actual customer orders booked for shipment during this period. No MPS quantity is required.

Without an MPS quantity in the third period, a shortage of item A will occur: $5 + 0 - 40 = -35$. Therefore, an MPS quantity equal to the lot size of 50 must be scheduled for completion in the third period. Then, the projected on-hand inventory for the third week will be $5 + 50 - 40 = 15$.

Figure 16.24 shows the projected on-hand inventories and MPS quantities that would result from completing the MPS calculations. The MPS start row is completed by simply shifting a copy of the MPS quantity row to the left by one column to account for the 1-week final assembly lead time. Also shown are the available-to-promise quantities. In week 1, the ATP is

$$\begin{pmatrix} \text{Available-to-} \\ \text{Promise in} \\ \text{week 1} \end{pmatrix} = \begin{pmatrix} \text{On-hand} \\ \text{quantity in} \\ \text{week 1} \end{pmatrix} + \begin{pmatrix} \text{MPS quantity} \\ \text{in week 1} \end{pmatrix} - \begin{pmatrix} \text{Orderes booked up} \\ \text{to week 3 when the} \\ \text{next MPS arrives} \end{pmatrix}$$

$$= 5 + 50 - (30 + 20) = 5 \text{ units}$$

The ATP for the MPS quantity in week 3 is

$$\begin{pmatrix} \text{Available-to-} \\ \text{Promise in} \\ \text{week 3} \end{pmatrix} = \begin{pmatrix} \text{MPS quantity} \\ \text{in week 3} \end{pmatrix} - \begin{pmatrix} \text{Orderes booked up} \\ \text{to week 7 when the} \\ \text{next MPS arrives} \end{pmatrix}$$

$$= 50 - (5 + 8 + 0 + 2) = 35 \text{ units}$$

The other ATPs equal their respective MPS quantities because no orders are booked for those weeks. As for the new orders, Customer A's request for 30 units in week 1 cannot be accommodated; the earliest it can be shipped is week 3 because the ATP for week 1 is insufficient. Assuming that Customer A's order is rejected, Customer B's request may be satisfied. The ATP for week 1 will

																Lot Size: 50 units Lead Time: 1 week
		Week														
Quantity on Hand:	**5**	1	2	3	4	5	6	7	8	9	10	11	12	13	14	15
Forecast		20	10	40	10			30	20	40	20					
Customer orders (booked)		30	20	5	8		2									
Projected on-hand inventory		25	5	15	5	5	3	23	3	13	43					
MPS quantity		50		50				50		50	50					
MPS start			50				50		50	50						
Available-to- promise (ATP) inventory		5		35				50		50	50					

FIGURE 16.24 ▲
Completed MPS Record for
End Item A

stay at 5 units and the ATP for week 3 will be reduced to 5 units. This acceptance allows the firm the flexibility to immediately satisfy an order for 5 units or less, if one comes in. When the MPS is updated next, the customer orders booked for week 4 will be increased to 38 to reflect the new order's shipping date. Customer C's order for 10 units in week 3 is likewise accepted. The ATP for weeks 1 and 3 will be reduced to 0, and when the MPS is updated, the customer orders booked for week 3 will be increased to 15. Finally, Customer D's order for 50 units in week 5 cannot be satisfied without changing the MPS.

Solved Problem 3

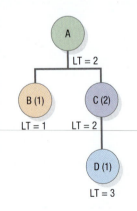

FIGURE 16.25 ▲
BOM for Product A

The MPS start quantities for product A calls for the assembly department to begin final assembly according to the following schedule: 100 units in week 2; 200 units in week 4; 120 units in week 6; 180 units in week 7; and 60 units in week 8. Develop a material requirements plan for the next 8 weeks for items B, C, and D. The BOM for A is shown in Figure 16.25, and data from the inventory records are shown in Table 16.1.

SOLUTION
We begin with items B and C and develop their inventory records, as shown in Figure 16.26. The MPS for product A must be multiplied by 2 to derive the gross requirements for item C because of the usage quantity. Once the planned order releases for item C are found, the gross requirements for item D can be calculated.

MyOMLab

Active Model 16.1 in MyOMLab
provides additional insight
on lotsizing decisions for this
problem.

TABLE 16.1 | INVENTORY RECORD DATA

	ITEM		
Data Category	**B**	**C**	**D**
Lot-sizing rule	POQ (P = 3)	L4L	FOQ = 500 units
Lead time (LT)	1 week	2 weeks	3 weeks
Scheduled receipts	None	200 (week 1)	None
Beginning (on-hand) inventory	20	0	425

◄ **FIGURE 16.26**
Inventory Records for Items B, C, and D

Item: B
Description:
Lot Size: POQ (P = 3)
Lead Time: 1 week

		1	2	3	4	5	6	7	8	9	10
							Week				
Gross requirements			100		200		120	180	60		
Scheduled receipts											
Projected on-hand inventory	20	20	200	200	0	0	240	60	0	0	0
Planned receipts			280				360				
Planned order releases		280			360						

Item: C
Description:
Lot Size: L4L
Lead Time: 2 weeks

		1	2	3	4	5	6	7	8	9	10
							Week				
Gross requirements			200		400		240	360	120		
Scheduled receipts		200									
Projected on-hand inventory	0	200	0	0	0	0	0	0	0	0	0
Planned receipts					400		240	360	120		
Planned order releases			400		240	360	120				

Item: D
Description:
Lot Size: FOQ = 500 units
Lead Time: 3 weeks

		1	2	3	4	5	6	7	8	9	10
							Week				
Gross requirements			400		240	360	120				
Scheduled receipts											
Projected on-hand inventory	425	425	25	25	285	425	305	305	305	305	305
Planned receipts					500	500					
Planned order releases		500	500								

Discussion Questions

1. For an organization of your choice, such as where you previously worked, discuss how an ERP system could be used and whether it would increase effectiveness.

2. Form a group in which each member represents a different functional area of a firm. Provide a priority list of the information that could be generated from an MPS, from the most important to the least important, for each functional area. Rationalize the differences in the lists.

3. Consider the master flight schedule of a major airline, such as Air New Zealand. Discuss the ways in which it is analogous to a master production schedule for a manufacturer.

4. Consider a service provider that is in the delivery business, such as UPS or FedEx. How can the principles of MRP be useful to such a company?

Problems

The OM Explorer and POM for Windows software is available to all students using the 10th edition of this textbook. Go to **www.pearsonhighered.com/krajewski** website to download these computer packages. If you purchased MyOMLab, you also have access to Active Models software and significant help in doing the following problems. Check with your instructor on how best to use these resources. In many cases, the instructor wants you to understand how to do the calculations by hand. At the least, the software provides a check on your calculations. When calculations are particularly complex and the goal is interpreting the results in making decision, the software entirely replaces the manual calculations.

1. Consider the bill of materials (BOM) in Figure 16.27.

    a. How many immediate parents (one level above) does item I have? How many immediate parents does item E have?

    b. How many unique components does product A have at all levels?

    c. Which of the components are purchased items?

    d. How many intermediate items does product A have at all levels?

    e. Given the lead times (LT) in weeks noted on Figure 16.27, how far in advance of shipment must a purchase commitment be for any of the purchased items identified in part (c)?

2. Product A is made from components B, C, and D. Item B is a subassembly that requires 2 units of C and 1 unit of E. Item D also is an intermediate item, made from one unit of F. All other usage quantities are 2. Draw the BOM for product A.

3. What is the lead time (in weeks) to respond to a customer order for product A, based on the BOM shown in Figure 16.28, assuming no existing inventories or scheduled receipts?

4. Product A is made from components B and C. Item B, in turn, is made from D and E. Item C also is an intermediate item, made from F and H. Finally, intermediate item E is made from H and G. Note that item H has two parents. The following are item lead times:

Item	A	B	C	D	E	F	G	H
Lead Time (weeks)	1	2	2	6	5	6	4	3

    a. What lead time (in weeks) is needed to respond to a customer order for product A, assuming no existing inventories or scheduled receipts?

    b. What is the customer response time if all purchased items (i.e., D, F, G, and H) are in inventory?

    c. If you are allowed to keep just one purchased item in stock, which one would you choose?

5. Refer to Figure 16.22 and Solved Problem 1. If inventory consists of 2 units of B, 1 unit of F, and 3 units of G, how many units of G, E, and D must be purchased to produce 5 units of product A?

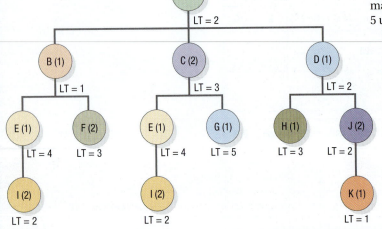

FIGURE 16.27 ▲
BOM for Product A

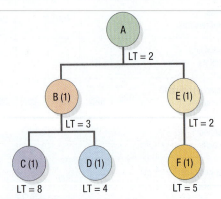

FIGURE 16.28 ▲
BOM for Product A

6. Complete the MPS record in Figure 16.29 for a single item.

◄ **FIGURE 16.29**
MPS Record for Single Item

Item: A						Order Policy: 60 units		
					Lead Time: 1 week			
				Week				
Quantity on Hand: 35	1	2	3	4	5	6	7	8
Forecast	20	18	28	28	23	30	33	38
Customer orders (booked)	15	17	9	14	9	0	7	0
Projected on-hand inventory								
MPS quantity								
MPS start								

7. Complete the MPS record shown in Figure 16.30 for a single item.

◄ **FIGURE 16.30**
MPS Record for Single Item

Item: A						Order Policy: 100 units		
					Lead Time: 1 week			
		January				February		
Quantity on Hand: 75	1	2	3	4	5	6	7	8
Forecast	65	65	65	45	50	50	50	50
Customer orders (booked)	40	10	85	0	35	70	0	0
Projected on-hand inventory								
MPS quantity								
MPS start								

8. An end item's demand forecasts for the next 10 weeks are 30, 20, 35, 50, 25, 25, 0, 40, 0, and 50 units. The current on-hand inventory is 80 units. The order policy is to produce in lots of 100. The booked customer orders for the item, starting with week 1, are 22, 30, 15, 9, 0, 0, 5, 3, 7, and 0 units. At present, no MPS quantities are on-hand for this item. The lead time is 2 weeks. Develop an MPS for this end item.

9. Figure 16.31 shows a partially completed MPS record for ball bearings.

    a. Develop the MPS for ball bearings.

    b. Four customer orders arrived in the following sequence:

Order	Quantity	Week Desired
1	500	4
2	400	5
3	300	1
4	300	7

    Assume that you must commit to the orders in the sequence of arrival and cannot change the desired shipping dates or your MPS. Which orders should you accept?

10. Tabard Industries forecasted the following demand for one of its most profitable products for the next 8 weeks: 120, 120, 120, 100, 100, 100, 80, and 80 units. The booked customer orders for this product, starting in week 1 are: 100, 80, 60, 40, 10, 10, 0, and 0 units. The current on-hand inventory is 150 units, the order quantity is 200 units, and the lead time is one week.

    a. Develop a MPS for this product.

    b. The marketing department revised its forecast. Starting with week 1, the new forecasts are: 120, 120, 120, 150, 150, 150, 100, and 100 units. Assuming that the prospective MPS you developed in part (a) does not change, prepare a revised MPS record. Comment on the situation that Tabard now faces.

    c. Returning to the original forecasted demand level and the MPS record you developed in part (a), assume

that marketing accepted a new customer order for 200 units in week 2 and thereby booked orders in week 2 is now 280 units. Assuming that the prospective MPS you developed in part (a) does not change, prepare a revised MPS record. Comment on the situation that Tabard now faces.

11. Figure 16.32 shows a partially completed MPS record for 2" pneumatic control valves. Suppose that you receive the following orders for the valves (shown in the order of their arrival). As they arrive, you must decide whether to accept or reject them. Which orders would you accept for shipment?

Order	Amount (Units)	Week Requested
1	15	2
2	30	5
3	25	3
4	75	7

12. The forecasted requirements for an electric hand drill for the next 6 weeks are 15, 40, 10, 20, 50, and 30 units. The marketing department has booked orders totaling 20, 25, 10, and 20 units for delivery in the first (current), second, third, and fourth weeks. Currently, 30 hand drills are in stock. The policy is to order in lots of 60 units. Lead time is one week.

    a. Develop the MPS record for the hand drills.

    b. A distributor of the hand drills places an order for 15 units. What is the appropriate shipping date for the entire order?

13. A forecast of 240 units in January, 320 units in February, and 240 units in March has been approved for the

FIGURE 16.31 ▶
MPS Record for Ball Bearings

Item: Ball bearings										Order Policy: 500 units Lead Time: 1 week		

		Week										
Quantity on Hand:	400	1	2	3	4	5	6	7	8	9	10
Forecast		550	300	400	450	300	350	200	300	450	400
Customer orders (booked)		300	350	250	250	200	150	100	100	100	100
Projected on-hand inventory											
MPS quantity	500										
MPS start											
Available-to-promise (ATP) inventory											

Item: 2" Pneumatic control valve					Order Policy: 75 units Lead Time: 1 week				
	Week								
Quantity on Hand: 10	1	2	3	4	5	6	7	8
Forecast	40	40	40	40	30	30	50	50
Customer orders (booked)	60	45	30	35	10	5	5	0
Projected on-hand inventory								
MPS quantity	75	75						
MPS start	75							
Available-to- promise (ATP) inventory								

◀ **FIGURE 16.32**
MPS Record for 2" Pneumatic Control Valve

seismic-sensory product family manufactured at the Rockport facility of Maryland Automated, Inc. Three products, A, B, and C, comprise this family. The product mix ratio for products A, B, and C for the past 2 years has been 35 percent, 40 percent, and 25 percent, respectively. Management believes that the monthly forecast requirements are evenly spread over the 4 weeks of each month. Currently, 10 units of product C are on hand. The company produces product C in lots of 40, and the lead time is 2 weeks. A production quantity of 40 units from the previous period is scheduled to arrive in week 1. The company has accepted orders of 25, 12, 8, 10, 2, and 3 units of product C in weeks 1 through 6, respectively.

Prepare a prospective MPS for product C and calculate the available-to-promise inventory quantities.

14. The partially completed inventory record for the tabletop subassembly in Figure 16.33 shows gross requirements, scheduled receipts, lead time, and current on-hand inventory.

   a. Complete the last three rows of the record for an FOQ of 110 units.

   b. Complete the last three rows of the record by using the L4L lot-sizing rule.

   c. Complete the last three rows of the record by using the POQ lot-sizing rule, with $P = 2$.

Item: M405—X Description: Tabletop subassembly							Lot Size: Lead Time: 2 weeks			
	Week									
	1	2	3	4	5	6	7	8	9	10
Gross requirements	90		85		80		45	90		
Scheduled receipts	110									
Projected on-hand inventory 40										
Planned receipts										
Planned order releases										

◀ **FIGURE 16.33**
Inventory Record for the Tabletop Subassembly

15. The partially completed inventory record for the rotor subassembly in Figure 16.34 shows gross requirements, scheduled receipts, lead time, and current on-hand inventory.

   a. Complete the last three rows of the record for an FOQ of 150 units.

   b. Complete the last three rows of the record by using the L4L lot-sizing rule.

   c. Complete the last three rows of the record by using the POQ lot-sizing rule, with $P = 2$.

FIGURE 16.34 ▶
Inventory Record for the Rotor Subassembly

Item: Rotor subassembly								Lot Size:	
								Lead Time: 2 weeks	
		Week							
		1	2	3	4	5	6	7	8
Gross requirements		65	15	45	40	80	80	80	80
Scheduled receipts		150							
Projected on-hand inventory	20								
Planned receipts									
Planned order releases									

16. The partially completed inventory record for the driveshaft subassembly in Figure 16.35 shows gross requirements, scheduled receipts, lead time, and current on-hand inventory.

   a. Complete the last three rows of the record for an FOQ of 50 units.

   b. Complete the last three rows of the record by using the L4L lot-sizing rule.

   c. Complete the last three rows of the record by using the POQ lot-sizing rule, with $P = 4$.

17. Figure 16.36 shows a partially completed inventory record for the real wheel subassembly. Gross requirements, scheduled receipts, lead time, and current on-hand inventory are shown.

   a. Complete the last three rows of the record for an FOQ of 200 units.

   b. Complete the last three rows of the record by using an FOQ of 100 units.

   c. Complete the last three rows of the record by using the L4L rule.

FIGURE 16.35 ▶
Inventory Record for the Driveshaft Subassembly

Item: Driveshaft subassembly								Lot Size:	
								Lead Time: 3 weeks	
		Week							
		1	2	3	4	5	6	7	8
Gross requirements		35	25	15	20	40	40	50	50
Scheduled receipts		80							
Projected on-hand inventory	10								
Planned receipts									
Planned order releases									

Item: MQ–09 Description: Rear wheel subassembly											Lot Size: Lead Time: 1 week	

◀ **FIGURE 16.36**
Inventory Record for the Rear Wheel Subassembly

Item: MQ–09
Description: Rear wheel subassembly

Lot Size:
Lead Time: 1 week

Week	1	2	3	4	5	6	7	8	9	10
Gross requirements	25	105	110	90		45	110	60		
Scheduled receipts										
Projected on-hand inventory **50**										
Planned receipts										
Planned order releases										

◀ **FIGURE 16.36**
Inventory Record for the Rear Wheel Subassembly

18. A partially completed inventory record for the motor subassembly is shown in Figure 16.37.

   a. Complete the last three rows of the record by using the L4L rule.

   b. Complete the last three rows of the record by using the POQ rule with $P = 2$.

   c. Complete the last three rows of the record by using the POQ rule with $P = 4$.

   d. If it costs the company $1 to hold a unit in inventory from one week to the next, and the cost to release an order is $50, which of the lot sizing rules used above will provide the lowest inventory holding + order release cost?

Item: GF–4
Description: Motor subassembly

Lot Size:
Lead Time: 2 weeks

Week	1	2	3	4	5	6	7	8	9	10
Gross requirements		80	50	35	20	55	15	30	25	10
Scheduled receipts	60									
Projected on-hand inventory **20**										
Planned receipts										
Planned order releases										

▲ **FIGURE 16.37**
Inventory Record for the Motor Subassembly

Advanced Problems

Depending on the time available to cover this chapter, your instructor may not assign these problems. Perhaps, your assignment for the multi-level problems allows the use of the *Material Requirements Planning* Solver of OM Explorer, with main requirement of looking for situations that need the inventory planner's attention (i.e., an action notice).

19. The BOM for product A is shown in Figure 16.38, and data from the inventory records are shown in Table 16.2. In the master production schedule for product A, the MPS start row has 100 units in week 3 and 200 in week 6. Develop the material requirements plan for the next 6 weeks for items B, C, and D.

   a. Develop the material requirements plan for the next 6 weeks for items B, C, and D.

   b. What specific managerial actions are required in week 1?

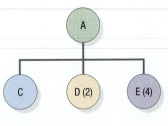

FIGURE 16.38 ▲
BOM for Product A

TABLE 16.2 | **INVENTORY RECORD DATA**

	ITEM		
Data Category	**C**	**D**	**E**
Lot-Sizing Rule	L4L	FOQ = 200	POQ (*P* = 3 weeks)
Lead time	2 weeks	1 week	1 week
Scheduled receipts	None	200 (in week 3)	200 (in week 3)
Beginning inventory	50	200	0

20. The BOMs for products A & B and data from the inventory records are shown in Figure 16.39. Data from the inventory records are shown in Table 16.3. In the master production schedule for product A, the MPS start row has 85 units in week 2 and 200 in week 4 and 50 in week 8. In the master production schedule for product B, the MPS start row has 65 units in week 3 and 50 in week 4 and 50 in week 5 and 75 in week 8.

   a. Develop the material requirements plan for the next 8 weeks for items C, D, and E. Note any difficulties you observe in the inventory records.

   b. Can the difficulties noted in part a be rectified by expediting any Scheduled Receipts?

FIGURE 16.39 ▶
BOMs for Product A and Product B

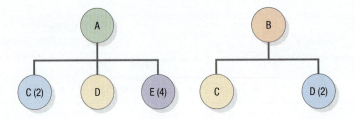

TABLE 16.3 | **INVENTORY RECORD DATA**

	ITEM		
Data Category	**C**	**D**	**E**
Lot-sizing rule	L4L	FOQ = 200	POQ (*P* = 2 weeks)
Lead time	2 weeks	1 week	1 week
Scheduled receipts	200 (in week 3)	0	0
Beginning inventory	0	0	200

21. Figure 16.40 illustrates the BOM for product A. The MPS start row in the master production schedule for product A calls for 50 units in week 2, 65 units in week 5, and 80 units in week 8. Item C is produced to make A and to meet the forecasted demand for replacement parts. Past replacement part demand has been 20 units per week (add 20 units to C's gross requirements). The lead times for items F and C are 1 week, and for the other items the lead time is 2 weeks. No safety stock is required for items B, C, D, E, and F. The L4L lot-sizing rule is used for items B and F; the POQ lot-sizing rule (*P* = 3) is used for C. Item E has an FOQ of 600 units, and D has an FOQ of 250 units. On-hand inventories are 50 units of B, 50 units of C, 120 units of D, 70 units of E, and 250 units of F. Item B has a scheduled receipt of 50 units in week 2. Develop a material requirements plan for the next 8 weeks for items B, C, D, E, and F.

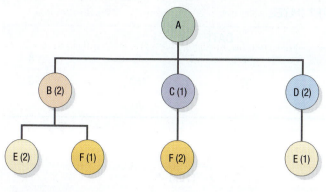

▲ **FIGURE 16.40**
BOM for Product A

22. The following information is available for three MPS items:

Product A	An 80-unit order is to be started in week 3.
	A 55-unit order is to be started in week 6.
Product B	A 125-unit order is to be started in week 5.
Product C	A 60-unit order is to be started in week 4.

Develop the material requirements plan for the next 6 weeks for items D, E, and F. The BOMs are shown in Figure 16.41, and data from the inventory records are shown in Table 16.4. (*Warning:* A safety stock requirement applies to item F. Be sure to plan a receipt for any week in which the projected on-hand inventory becomes less than the safety stock.)

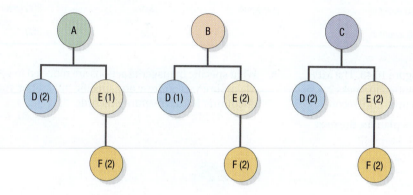

◀ **FIGURE 16.41**
BOMs for Products A, B, and C

TABLE 16.4 | INVENTORY RECORD DATA

| Data Category | ITEM | | |
	D	E	F
Lot-sizing rule	FOQ = 150	L4L	POQ (P = 2)
Lead time	3 weeks	1 week	2 weeks
Safety stock	0	0	30
Scheduled receipts	150 (week 3)	120 (week 2)	None
Beginning inventory	150	0	100

▼ **FIGURE 16.42**
BOMs for Products A and B

23. Figure 16.42 shows the BOMs for two products, A and B. Table 16.5 shows the MPS quantity start date for each one. Table 16.6 contains data from inventory records for items C, D, and E. There are no safety stock requirements for any of the items.

a. Determine the material requirements plan for items C, D, and E for the next 8 weeks.

b. What specific managerial actions are required in week 1?

c. Suppose that a very important customer places an emergency order for a quantity of product A. In order to satisfy this order, a new MPS of 200 units of product A is now required in week 5. Determine the changes to the material requirements plan if this order is accepted and note any problems that you detect.

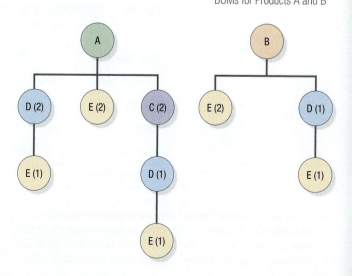

TABLE 16.5 | MPS QUANTITY START DATES

Product	DATE							
	1	2	3	4	5	6	7	8
A		125		95		150		130
B			80			70		

TABLE 16.6 | INVENTORY RECORD DATA

Data Category	ITEM		
	C	D	E
Lot-sizing rule	L4L	POQ ($P = 3$)	FOQ = 800
Lead time	3 weeks	2 weeks	1 week
Scheduled receipts	200 (week 2)	None	800 (week 1)
Beginning inventory	85	625	350

24. The BOM for product A is shown in Figure 16.43. The MPS for product A calls for 120 units to be started in weeks 2, 4, 5, and 8. Table 16.7 shows data from the inventory records.

   a. Develop the material requirements plan for the next 8 weeks for each item.

   b. What specific managerial actions are required in week 1? Make sure you address any specific difficulties you encounter in the inventory records.

FIGURE 16.43 ▶
BOM for Product A

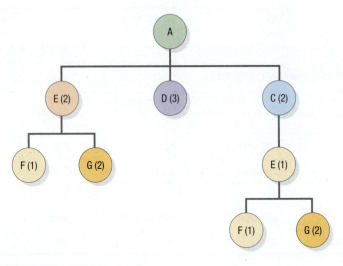

TABLE 16.7 | INVENTORY RECORD DATA

Data Category	ITEM				
	C	D	E	F	G
Lot-sizing rule	L4L	FOQ = 700	FOQ = 700	L4L	L4L
Lead time	3 weeks	3 weeks	4 weeks	2 weeks	1 week
Safety stock	0	0	0	50	0
Scheduled receipts	150 (week 2)	450 (week 2)	700 (week 1)	None	1,400 (week 1)
Beginning inventory	125	0	235	750	0

25. Refer to Solved Problem 1 (Figure 16.22) for the bill of materials and Table 16.8 for component inventory record information. Develop the material requirements plan for all components and intermediate items associated with product A for the next 10 weeks. The MPS for product A calls for 50 units to be started in weeks 2, 6, 8, and 9. (*Warning:* Safety stock requirements apply to items B and C.)

TABLE 16.8 | **INVENTORY RECORD DATA**

Data Category	ITEM					
	B	C	D	E	F	G
Lot-sizing rule	L4L	L4L	POQ (P = 2)	L4L	L4L	FOQ = 100
Lead time	2 weeks	3 weeks	3 weeks	6 weeks	1 week	3 weeks
Safety stock	30	10	0	0	0	0
Scheduled receipts	150 (week 2)	50 (week 2)	None	400 (week 6)	40 (week 3)	None
Beginning inventory	30	20	60	400	0	0

26. An end item's demand forecasts for the next 6 weeks are 30 units, followed by forecasts of 25 units for weeks 7 though 10. The current on-hand inventory is 60 units. The order policy is to produce in lots of 100. The booked customer orders for the item, starting with week 1, are 22, 30, 15, 11, 0, 0, 9, 0, 0, and 0 units. The lead time is 2 weeks.

a. Develop an MPS for this end item.

b. The marketing department has received six orders for this item in the following sequence:

Order 1 is for 40 units to be delivered in period 3

Order 2 is for 60 units to be delivered in period 4

Order 3 is for 70 units to be delivered in period 6

Order 4 is for 40 units to be delivered in period 3

Order 5 is for 20 units to be delivered in period 5

Order 6 is for 115 units to be delivered in period 9

Assuming that the prospective MPS you developed in part (a) does not change, which orders would you be able to accept based on the available to promise (ATP)?

27. An end item's demand forecasts for the next 10 weeks are 30, 30, 30, 30, 20, 20, 30, 30, 30, and 30 units. The current on-hand inventory is 100 units. The order policy is to produce in lots of 75. The booked customer orders for the item, starting with week 1, are 15, 38, 7, 5, 0, 3, 10, 0, 0, and 0 units. The lead time is 2 weeks.

a. Develop an MPS for this end item.

b. The marketing department has received five orders for this item in the following sequence:

Order 1 is for 20 units to be delivered in period 1

Order 2 is for 75 units to be delivered in period 4

Order 3 is for 90 units to be delivered in period 6

Order 4 is for 75 units to be delivered in period 7

Order 5 is for 90 units to be delivered in period 10

Assuming that the prospective MPS you developed in part (a) does not change, which orders would you be able to accept based on the available to promise (ATP)?

28. The bill of materials and the data from the inventory records for product A are shown in Figure 16.44. Assume that the MPS start quantities for A are 100 units in weeks 1, 2, 3, 4, 7, 8, 9, and 10.

Derive an MRP plan for the components going into product A using the data in Table 16.9.

What specific managerial actions are required in week 1? Make sure you address any specific difficulties you encounter in the inventory records.

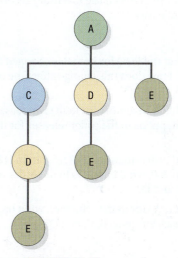

▲ **FIGURE 16.44**
BOM for product A

TABLE 16.9 | **INVENTORY RECORD DATA**

Data Category	ITEM		
	C	D	E
Lot-sizing rule	L4L	L4L	FOQ = 300
Lead time	1 week	2 weeks	2 weeks
Scheduled receipts		100 (in week 2)	300 (in week 2)
Beginning Inventory	225	350	100

29. The bill of materials and the data from the inventory records for product A are shown in Figure 16.45. Assume that the MPS start quantities for A are 50 units in weeks 1, 2 and 3, and 150 units in weeks 6 and 8. Derive an MRP plan for the components going into product A using the data in Table 16.10.

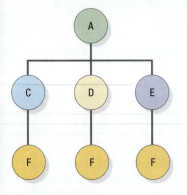

▲ FIGURE 16.45
BOM for Product A

TABLE 16.10 | INVENTORY RECORD DATA

Data Category	ITEM			
	C	**D**	**E**	**F**
Lot-sizing rule	POQ ($P = 2$)	L4L	FOQ = 300	FOQ = 400
Lead time	1 week	1 week	2 weeks	4 weeks
Scheduled receipts		100 (week 2)		400 (week 1)
Beginning inventory	100	0	110	40

Active Model Exercise

This Active Model appears in MyOMLab. It allows you to evaluate the relationship between the inventory record data and the planned order releases.

QUESTIONS

1. Suppose that the POQ for item B is changed from 3 weeks to 2 weeks. How does this change affect the order releases for items B, C, and D?

2. As the on-hand inventory for item C increases from 0 to 200, what happens to the order releases for items B, C, and D?

3. As the fixed order quantity (FOQ) for item D increases from 500 to 750, what happens to the order releases for items B, C, and D?

4. As the lead time for item C changes, what happens to the order releases for items B, C, and D?

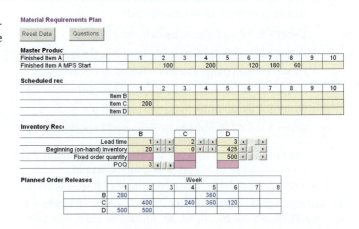

ACTIVE MODEL 16.1 ▲
Material Requirements Planning Using Data from Solved Problem 3 and Table 16.1

CASE   Flashy Flashers, Inc.

Flashy Flashers is a medium-sized firm employing 900 persons and 125 managerial and administrative personnel. The firm produces a line of automotive electrical components. It supplies about 95 auto parts stores and Moonbird Silverstreak car dealers in its region. Johnny Bennett, who serves as the president, founded the company by producing cable assemblies in his garage. By working hard, delivering consistent product quality, and by providing good customer service, he expanded his business to produce a variety of electrical components. Bennett's commitment to customer service is so strong that his company motto, "Love Thy Customers as Thyself," is etched on a big cast-iron plaque under his giant oil portrait in the building's front lobby.

The company's two most profitable products are the automotive front sidelamp and the headlamp. With the rising popularity of Eurosport sedans, such as the Moonbird Silverstreak, Flashy Flashers has enjoyed substantial demand for these two lamp items.

Last year, Kathryn Marley, the vice president of operations and supply chain management, approved the installation of a new MRP system. It is a

first important step toward the eventual goal of a full-fledged ERP system. Marley worked closely with the task force that was created to bring MRP online. She frequently attended the training sessions for selected employees, emphasizing how MRP should help Flashy Flashers secure a better competitive edge.

A year later, the MRP system is working fairly well. However, Marley believes that there is always a better way and seeks to continually improve the company's processes. To get a better sense for potential improvements, she met with the production and inventory control manager, the shop supervisor, and the purchasing manager. Here are some of their observations.

Production and Inventory Control Manager

Inventory records and BOM files are accurate and well maintained. Inventory transactions are faithfully made when inventory is replenished or removed from the stockroom so that current on-hand balances are credible. There is a MRP explosion each week, which gives the company the new material requirements plan. It provides information that helps identify when new orders need to be launched. Information can also be searched to help identify which scheduled receipts need to be expedited and which ones can be delayed by assigning them a later due date, thereby making room for more urgent jobs.

One planner suggested that the MRP outputs should be extended to provide priority and capacity reports, with pointers as to which items need their attention. The original plan was to get the order-launching capability implemented first. However, there is no formal system of priority planning, other than the initial due date assigned to each scheduled receipt when it is released, transforming it from a planned order release into a scheduled receipt. The due dates do not get updated later even when there are unexpected scrap losses, capacity shortages, short shipments, or last-minute changes in the MPS (responding to requests from favorite customers). Jobs are scheduled on the shop floor and by suppliers according to the EDD rule, based on their due dates. If due dates assigned to scheduled receipts were updated, it might help get open orders done when they are really needed. Furthermore, planned order releases in the action bucket are translated into scheduled receipts (using inventory transactions), after checking that its components are available. The current system does not consider possible capacity problems when releasing new orders.

Shop Supervisor

His primary complaint is that the shop workloads are anything but level. One week, they hardly have any work, and the supervisor overproduces (more than called for by the scheduled receipts) just to keep everyone busy. The next week can be just the opposite—so many new orders with short fuses that almost everyone needed to work overtime or else the scheduled receipt quantities are reduced to cover immediate needs. It is feast or famine,

unless they make things work on the shop floor! They do make inventory transactions to report deviations from plan for the scheduled receipts, but these "overrides" make the scheduled receipt information in the MRP records more uncertain for the planners. A particular concern is to make sure that the bottleneck workstations are kept busy.

Purchasing

Buyers are putting out too many fires, leaving little time for creative buying. In such cases, their time is spent following up on orders that are required in the very near future or that are even late. Sometimes, the MRP plan shows planned order releases for purchased items that that are needed almost immediately, not allowing for the planned lead time. In checking the MRP records, the planned lead times are realistic and what the suppliers expect. Last week, things were fine for an item, and this week a rush order needs to be placed. What is the problem?

Marley tried to assimilate all this information. She decided to collect all the required information about the sidelamps and headlamps (shown in Table 16.11 through Table 16.14 and in Figure 16.46) to gain further insight into possible problems and identify areas for improvement.

Your Assignment

Put yourself in Marley's place and prepare a report on your findings. Specifically, you are required to do a manual MRP explosion for the sidelamps and headlamps for the next 6 weeks (beginning with the current week). Assume that it is now the start of week 1. Fill in the planned order releases form provided in Table 16.15. It should show the planned order releases for all items for the next 6 weeks. Include it in your report.

TABLE 16.11 | PART NUMBERS AND DESCRIPTIONS

Part Number	Description
C206P	Screws
C310P	Back rubber gasket
HL200E	Headlamp
HL211A	Head frame subassembly
HL212P	Head lens
HL222P	Headlamp module
HL223F	Head frame
SL100E	Sidelamp
SL111P	Side lens
SL112A	Side frame subassembly
SL113P	Side lens rubber gasket
SL121F	Side frame
SL122A	Side bulb subassembly
SL123A	Flasher bulb subassembly
SL131F	Side cable grommet and receptacle
SL132P	Side bulb
SL133F	Flasher cable grommet and receptacle
SL134P	Flasher bulb

TABLE 16.12 | MASTER PRODUCTION SCHEDULE

Item Description and Part Number	Quantity	MPS Start Date
Headlamp (HL200E)	120	Week 4
	90	Week 5
	75	Week 6
Sidelamp (SL100E)	100	Week 3
	80	Week 5
	110	Week 6

TABLE 16.13 | REPLACEMENT PART DEMAND

Item Description and Part Number	Quantity	Date
Side lens (SL111P)	40	Week 3
	35	Week 6

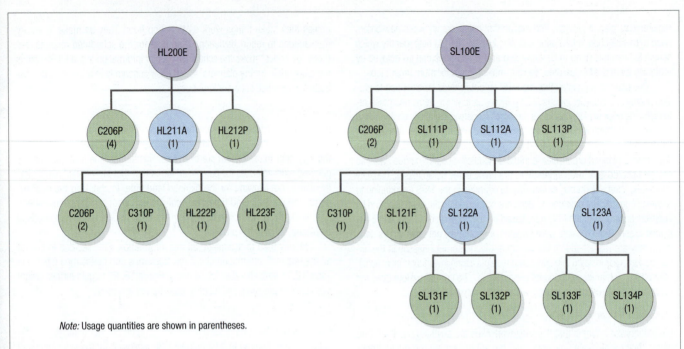

Note: Usage quantities are shown in parentheses.

▲ **FIGURE 16.46**
BOMs for Headlamps and Sidelamps

Supplement your report with worksheets on the manual MRP explosion, and list the actions that planners should consider this week to (1) release new orders, (2) expedite scheduled receipts, and (3) delay a scheduled receipt's due date.

Finally, identify the good and bad points of MRP implementation at Flashy Flashers. Conclude by making suggestions on ways to improve its resource planning process.

TABLE 16.14 | **SELECTED DATA FROM INVENTORY RECORDS**

Part Number	Lead Time (Weeks)	Safety Stock (Units)	Lot-Sizing Rule	On-Hand (Units)	Scheduled Receipt (Units and Due Dates)
C206P	1	30	FOQ = 2,500	270	—
C310P	1	20	FOQ = 180	40	180 (week 1)
HL211A	3	0	L4L	0	
HL212P	2	15	FOQ = 350	15	—
HL222P	3	10	POQ (P = 4 weeks)	10	285 (week 1)
HL223F	1	0	POQ (P = 4 weeks)	0	120 (week 1)
SL111P	2	0	FOQ = 350	15	—
SL112A	3	0	L4L	20	80 (week 2)
SL113P	1	20	FOQ = 100	20	—
SL121F	2	0	L4L	0	80 (week 2)
SL122A	2	0	L4L	0	80 (week 2)
SL123A	2	0	FOQ = 200	0	—
SL131F	2	0	POQ (P = 2 weeks)	0	110 (week 1)
SL132P	1	25	FOQ = 100	35	100 (week 1)
SL133F	2	0	FOQ = 250	0	—
SL134P	1	0	FOQ = 400	100	—

TABLE 16.15 | **PLANNED ORDER RELEASE FORM**

FILL IN THE PLANNED ORDER RELEASES FOR ALL COMPONENTS.

Item Description and Part Number	Week					
	1	2	3	4	5	6
Side lens (SL111P)						
Side lens rubber gasket (SL113P)						
Side frame subassembly (SL112A)						
Side frame (SL121F)						
Side bulb subassembly (SL122A)						
Flasher bulb subassembly (SL123A)						
Side cable grommet and receptacle (SL131F)						
Flasher cable grommet and receptacle (SL133F)						
Side bulb (SL132P)						
Flasher bulb (SL134P)						
Head frame subassembly (HL211A)						
Head lens (HL212P)						
Headlamp module (HL222P)						
Head frame (HL223F)						
Back rubber gasket (C310P)						
Screws (C206P)						

Selected References

Becker, Nathan. "iPad's Bill of Materials Close to first iPad." *Wall Street Journal*, March 14, 2011.

Bendoly, E., and M. Cotteleer. "Understanding Behavioral Sources of Process Variation following Enterprise System Deployment." *Journal of Operations Management*, vol. 26, no. 1 (2008).

Davenport, Thomas H. "Putting the Enterprise into the Enterprise System." *Harvard Business Review* (July–August 1998), pp. 121–131.

Hendricks, Kevin B., Vinod R. Singhal, and Jeff K. Stratman. "The Impact of Enterprise Systems on Corporate Performance: A Study of ERP, SCM and CRM System Implementations." *Journal of Operations Management*, vol. 25, no. 1 (2007), pp. 65–82.

Jacobs, F. Robert, William Berry, and D. Clay Whybark. *Manufacturing Planning and Control Systems for Supply Chain Management*, 6th ed. New York: McGraw-Hill/Irwin, 2010.

Jacobs, F. Robert, and Weston, F.C. (Ted) Jr. "Enterprise Resource Planning (ERP)—A Brief History." *Journal of Operations Management*, vol. 25, no. 2 (2007), pp. 357–363.

Mabert, Vincent A. "The Early Road to Materials Requirements Planning." *Journal of Operations Management*, vol. 25, no. 2 (2007), pp. 346–356.

McAfee, A., and E. Brynjolfsson. "Investing in the IT That Makes a Competitive Advantage." *Harvard Business Review*, vol. 86 (July–August 2008), pp. 98–107.

McCue, Andy. "Too Much Candy: IT Glitch Costs Cadbury." *Business Week*, June 8, 2006.

Scalle, Cedric X., and Mark J. Cotteleer. *Enterprise Resource Planning (ERP)*. Boston, MA: Harvard Business School Publishing, 1999, No. 9-699-020.

Wallace, Thomas F. *Sales & Operations Planning: The How-To Handbook*, 3rd ed. Cincinnati, OH: T. E. Wallace & Company, 2008.

Wallace, Thomas F., and Robert A. Stahl. *Master Scheduling in the 21st Century*. Cincinnati, OH: T. E. Wallace & Company, 2003.

LINEAR PROGRAMMING

In many business situations, resources are limited and demand for them is great. For example, a limited number of vehicles may have to be scheduled to make multiple trips to customers, or a staffing plan may have to be developed to cover expected variable demand with the fewest employees. In this supplement, we describe a technique called **linear programming**, which is useful for allocating scarce resources among competing demands. The resources may be time, money, or materials, and the limitations are known as constraints. Linear programming can help managers find the best allocation solution and provide information about the value of additional resources.

linear programming

A technique that is useful for allocating scarce resources among competing demands.

LEARNING GOALS *After reading this supplement, you should be able to:*

1. Describe formulating models for various problems.
2. Demonstrate graphic analysis and solutions for two-variable problems.
3. Define slack variables, surplus variables, and sensitivity analysis.
4. Interpret computer output of a linear programming solution.
5. Apply the transportation method to Sales and Operations Planning (S&OP) problems.

Basic Concepts

Before we can demonstrate how to solve problems in operations and supply chain management with linear programming, we must first explain several characteristics of all linear programming models and mathematical assumptions that apply to them: (1) objective function, (2) decision variables, (3) constraints, (4) feasible region, (5) parameters, (6) linearity, and (7) nonnegativity.

Linear programming is an *optimization* process. A single **objective function** states mathematically what is being maximized (e.g., profit or present value) or minimized (e.g., cost or scrap). The objective function provides the scorecard on which the attractiveness of different solutions is judged.

Decision variables represent choices that the decision maker can control. Solving the problem yields their optimal values. For example, a decision variable could be the number of units of a product to make next month or the number of units of inventory to hold next month. Linear programming is based on the assumption that decision variables are *continuous*; they can be fractional quantities and need not be whole numbers. Often, this assumption is realistic, as when the decision variable is expressed in dollars, hours, or some other continuous measure. Even when the decision variables represent nondivisible units, such as workers, tables, or trucks, we

objective function

An expression in linear programming models that states mathematically what is being maximized or minimized.

decision variables

Variables that represent the choices the decision maker can control.

sometimes can simply round the linear programming solution up or down to get a reasonable solution that does not violate any constraints, or we can use a more advanced technique, called *integer programming.*

Constraints are limitations that restrict the permissible choices for the decision variables. Each limitation can be expressed mathematically in one of three ways: a less-than-or-equal-to ($\leq$), an equal-to ($=$), or a greater-than-or-equal-to ($\geq$) constraint. A $\leq$ constraint puts an upper limit on some function of decision variables and most often is used with maximization problems. For example, a $\leq$ constraint may specify the maximum number of customers who can be served or the capacity limit of a machine. An $=$ constraint means that the function must equal some value. For example, 100 (not 99 or 101) units of one product must be made. An $=$ constraint often is used for certain mandatory relationships, such as the fact that ending inventory always equals beginning inventory plus production minus sales. A $\geq$ constraint puts a lower limit on some function of decision variables. For example, a $\geq$ constraint may specify that production of a product must exceed or equal demand.

Every linear programming problem must have one or more constraints. Taken together, the constraints define a **feasible region**, which represents all permissible combinations of the decision variables. In some unusual situations, the problem is so tightly constrained that there is only one possible solution—or perhaps none. However, in the usual case, the feasibility region contains infinitely many possible solutions, assuming that the feasible combinations of the decision variables can be fractional values. The goal of the decision maker is to find the best possible solution.

The objective function and constraints are functions of decision variables and parameters. A **parameter**, also known as a *coefficient* or *given constant*, is a value that the decision maker cannot control and that does not change when the solution is implemented. Each parameter is assumed to be known with **certainty**. For example, a computer programmer may know that running a software program will take 30 minutes—no more, no less.

The objective function and constraint equations are assumed to be linear. **Linearity** implies proportionality and additivity—there can be no products (e.g., $10x_1x_2$) or powers (e.g., x_1^3) of decision variables. Suppose that the profit gained by producing two types of products (represented by decision variables x_1 and x_2) is $2x_1 + 3x_2$. Proportionality implies that one unit of x_1 contributes \$2 to profits and two units contribute \$4, regardless of how much of x_1 is produced. Similarly, each unit of x_2 contributes \$3, whether it is the first or the tenth unit produced. Additivity means that the total objective function value equals the profits from x_1 plus the profits from x_2.

Finally, we make an assumption of **nonnegativity**, which means that the decision variables must be positive or zero. A firm that makes spaghetti sauce, for example, cannot produce a negative number of jars. To be formally correct, a linear programming formulation should show a ≥ 0 constraint for each decision variable.

Although the assumptions of linearity, certainty, and continuous variables are restrictive, linear programming can help managers analyze many complex resource allocation problems. The process of building the model forces managers to identify the important decision variables and constraints, which is a useful step in its own right. Identifying the nature and scope of the problem represents a major step toward solving it. In a later section, we show how sensitivity analysis can help the manager deal with uncertainties in the parameters and answer "what-if" questions.

Formulating a Problem

Linear programming applications begin with the formulation of a *model* of the problem with the general characteristics just described. We illustrate the modeling process here with the **product-mix problem**, which is a one-period type of planning problem, the solution of which yields optimal output quantities (or product mix) of a group of services or products subject to resource capacity and market demand constraints. This problem was first introduced in Chapter 7, "Constraint Management," and now we take it up more formally. Formulating a model to represent each unique problem, using the following three-step sequence, is the most creative and perhaps the most difficult part of linear programming.

Step 1. Define the Decision Variables. What must be decided? Define each decision variable specifically, remembering that the definitions used in the objective function must be equally useful in the constraints. The definitions should be as specific as possible. Consider the following two alternative definitions:

x_1 = product 1
x_1 = number of units of product 1 to be produced and sold next month

The second definition is much more specific than the first, making the remaining steps easier.

---

constraints

In linear programming, the limitations that restrict the permissible choices for the decision variables.

feasible region

A region that represents all permissible combinations of the decision variables in a linear programming model.

parameter

A value that the decision maker cannot control and that does not change when the solution is implemented.

certainty

The word that is used to describe that a fact is known without doubt.

linearity

A characteristic of linear programming models that implies proportionality and additivity—there can be no products or powers of decision variables.

nonnegativity

An assumption that the decision variables must be positive or zero.

product-mix problem

A one-period type of planning problem, the solution of which yields optimal output quantities (or product mix) of a group of services or products subject to resource capacity and market demand constraints.

Step 2. Write Out the Objective Function. What is to be maximized or minimized? If it is next month's profits, write out an objective function that makes next month's profits a linear function of the decision variables. Identify parameters to go with each decision variable. For example, if each unit of x_1 sold yields a profit of $7, the total profit from product $x_1 = 7x_1$. If a variable has no impact on the objective function, its objective function coefficient is 0. The objective function often is set equal to Z, and the goal is to maximize or minimize Z.

Step 3. Write Out the Constraints. What limits the values of the decision variables? Identify the constraints and the parameters for each decision variable in them. As with the objective function, the parameter for a variable that has no impact in a constraint is 0. To be formally correct, also write out the nonnegativity constraints.

As a consistency check, make sure that the same unit of measure is being used on both sides of each constraint and in the objective function. For example, suppose that the right-hand side of a constraint is hours of capacity per month. Then, if a decision variable on the left-hand side of the constraint measures the number of units produced per month, the dimensions of the parameter that is multiplied by the decision variable must be hours per unit because

$$\left(\frac{\text{Hours}}{\text{Unit}} \right) \left(\frac{\text{Units}}{\text{Month}} \right) = \left(\frac{\text{Hours}}{\text{Month}} \right)$$

Of course, you can also skip around from one step to another, depending on the part of the problem that has your attention. If you cannot get past step 1, try a new set of definitions for the decision variables. Often the problem can be modeled correctly in more than one way.

EXAMPLE D.1   **Formulating a Linear Programming Model**

The Stratton Company produces two basic types of plastic pipe. Three resources are crucial to the output of pipe: extrusion hours, packaging hours, and a special additive to the plastic raw material. The following data represent next week's situation. All data are expressed in units of 100 feet of pipe.

	PRODUCT		
Resource	Type 1	Type 2	Resource Availability
Extrusion	4 hr	6 hr	48 hr
Packaging	2 hr	2 hr	18 hr
Additive mix	2 lb	1 lb	16 lb

The contribution to profits and overhead per 100 feet of pipe is $34 for type 1 and $40 for type 2. Formulate a linear programming model to determine how much of each type of pipe should be produced to maximize contribution to profits and to overhead, assuming that everything produced can be sold.

SOLUTION

Step 1. To define the decision variables that determine product mix, we let

x_1 = amount of type 1 pipe to be produced and sold next week, measured in 100-foot increments (e.g., $x_1 = 2$ means 200 feet of type 1 pipe)

and

x_2 = amount of type 2 pipe to be produced and sold next week, measured in 100-foot increments

Step 2. Next, we define the objective function. The goal is to maximize the total contribution that the two products make to profits and overhead. Each unit of x_1 yields $34, and each unit of x_2 yields $40. For specific values of x_1 and x_2, we find the total profit by multiplying the number of units of each product produced by the profit per unit and adding them. Thus, our objective function becomes

$$\text{Maximize: } \$34x_1 + \$40x_2 = Z$$

Step 3. The final step is to formulate the constraints. Each unit of x_1 and x_2 produced consumes some of the critical resources. In the extrusion department, a unit of x_1 requires 4 hours and a unit of x_2 requires

6 hours. The total must not exceed the 48 hours of capacity available, so we use the $\leq$ sign. Thus, the first constraint is

$$4x_1 + 6x_2 \leq 48$$

Similarly, we can formulate constraints for packaging and raw materials:

$$2x_1 + 2x_2 \leq 18 \text{ (packaging)}$$
$$2x_1 + x_2 \leq 16 \text{ (additive mix)}$$

These three constraints restrict our choice of values for the decision variable because the values we choose for x_1 and x_2 must satisfy all of the constraints. Negative values for x_1 and x_2 do not make sense, so we add nonnegativity restrictions to the model:

$$x_1 \geq 0 \text{ and } x_2 \geq 0 \text{ (nonnegativity restrictions)}$$

We can now state the entire model, made complete with the definitions of variables.

$$\text{Maximize: } \$34x_1 + \$40x_2 = Z$$

$$\text{Subject to: } 4x_1 + 6x_2 \leq 48$$
$$2x_1 + 2x_2 \leq 18$$
$$2x_1 + x_2 \leq 16$$
$$x_1 \geq 0 \text{ and } x_2 \geq 0$$

where

x_1 = amount of type 1 pipe to be produced and sold next week, measured in 100-foot increments

x_2 = amount of type 2 pipe to be produced and sold next week, measured in 100-foot increments

Graphic Analysis

graphic method of linear programming

A type of graphic analysis that involves the following five steps: plotting the constraints, identifying the feasible region, plotting an objective function line, finding a visual solution, and finding the algebraic solution.

With the model formulated, we now seek the optimal solution. In practice, most linear programming problems are solved with a computer. However, insight into the meaning of the computer output—and linear programming concepts in general—can be gained by analyzing a simple two-variable problem with the **graphic method of linear programming**. Hence, we begin with the graphic method, even though it is not a practical technique for solving problems that have three or more decision variables. The five basic steps are (1) *plot the constraints*, (2) *identify the feasible region*, (3) *plot an objective function line*, (4) *find the visual solution*, and (5) *find the algebraic solution*.

Plot the Constraints

We begin by plotting the constraint equations, disregarding the inequality portion of the constraints ($<$ or $>$). Making each constraint an equality ($=$) transforms it into the equation for a straight line. The line can be drawn as soon as we identify two points on it. Any two points reasonably spread out may be chosen; the easiest ones to find are the *axis intercepts*, where the line intersects each axis. To find the x_1 axis intercept, set x_2 equal to 0 and solve the equation for x_1. For the Stratton Company in Example D.1, the equation of the line for the extrusion process is

$$4x_1 + 6x_2 = 48$$

For the x_1 axis intercept, $x_2 = 0$, so

$$4x_1 + 6(0) = 48$$
$$x_1 = 12$$

To find the x_2 axis intercept, set $x_1 = 0$ and solve for x_2:

$$4(0) + 6x_2 = 48$$
$$x_2 = 8$$

We connect points (0, 8) and (12, 0) with a straight line, as shown in Figure D.1.

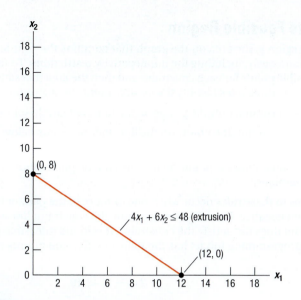

◄ **FIGURE D.1**
Graph of the Extrusion Constraint

| EXAMPLE D.2 | **Plotting the Constraints** |

For the Stratton Company problem, plot the other constraints: one constraint for packaging and one constraint for the additive mix.

SOLUTION
The equation for the packaging process's line is $2x_1 + 2x_2 = 18$. To find the x_1 intercept, set $x_2 = 0$:

$$2x_1 + 2(0) = 18$$
$$x_1 = 9$$

To find the x_2 axis intercept, set $x_1 = 0$:

$$2(0) + 2x_2 = 18$$
$$x_2 = 9$$

The equation for the additive mix's line is $2x_1 + x_2 = 16$. To find the x_1 intercept, set $x_2 = 0$:

$$2x_1 + 0 = 16$$
$$x_1 = 8$$

To find the x_2 axis intercept, set $x_1 = 0$:

$$2(0) + x_2 = 16$$
$$x_2 = 16$$

With a straight line, we connect points (0, 9) and (9, 0) for the packaging constraint and points (0, 16) and (8, 0) for the additive mix constraint. Figure D.2 shows the graph with all three constraints plotted.

MyOMLab

Active Model D.1 in MyOMLab offers many insights on graphic analysis and sensitivity analysis. Use it when studying Examples D.2 through D.4.

MyOMLab

Tutor D.1 in MyOMLab provides a new practice example for plotting the constraints.

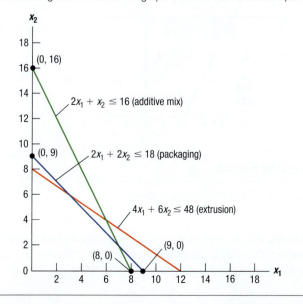

◄ **FIGURE D.2**
Graph of the Three Constraints

Identify the Feasible Region

The feasible region is the area on the graph that contains the solutions that satisfy all the constraints simultaneously, including the nonnegativity restrictions. To find the feasible region, first locate the feasible points for each constraint and then the area that satisfies all constraints. Generally, the following three rules identify the feasible points for a given constraint:

1. For the $=$ constraint, only the points on the line are feasible solutions.

2. For the $\leq$ constraint, the points on the line and the points below or to the left of the line are feasible solutions.

3. For the $\geq$ constraint, the points on the line and the points above or to the right of the line are feasible solutions.

Exceptions to these rules occur when one or more of the parameters on the left-hand side of a constraint are negative. In such cases, we draw the constraint line and test a point on one side of it. If the point does not satisfy the constraint, it is in the infeasible part of the graph. Suppose that a linear programming model has the following five constraints plus the two nonnegativity constraints:

$$2x_1 + x_2 \geq 10$$

$$2x_1 + 3x_2 \geq 18$$

$$x_1 \leq 7$$

$$x_2 \leq 5$$

$$-6x_1 + 5x_2 \leq 5$$

$$x_1, x_2 \geq 0$$

The feasible region is the shaded portion of Figure D.3. The arrows shown on each constraint identify which side of each line is feasible. The rules work for all but the fifth constraint, which has a negative parameter, -6, for x_1. We arbitrarily select $(2, 2)$ as the test point, which Figure D.3 shows is below the line and to the right. At this point, we find $-6(2) + 5(2) = -2$. Because -2 does not exceed 5, the portion of the figure containing $(2, 2)$ is feasible, at least for this fifth constraint.

FIGURE D.3 ▶
Identifying the Feasible
Region

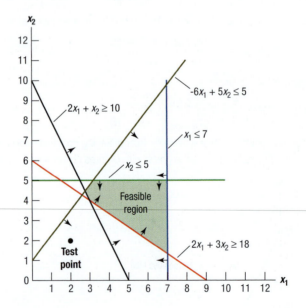

| **EXAMPLE D.3** | **Identifying the Feasible Region** |

Identify the feasible region for the Stratton Company problem.

SOLUTION
Because the problem contains only $\leq$ constraints, and the parameters on the left-hand side of each constraint are not negative, the feasible portions are to the left of and below each constraint. The feasible region, shaded in Figure D.4, satisfies all three constraints simultaneously.

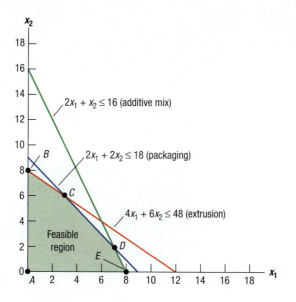

Plot an Objective Function Line

Now we want to find the solution that optimizes the objective function. Even though all the points in the feasible region represent possible solutions, we can limit our search to the corner points. A **corner point** lies at the intersection of two (or possibly more) constraint lines on the boundary of the feasible region. No interior points in the feasible region need be considered because at least one corner point is better than any interior point. Similarly, other points on the boundary of the feasible region can be ignored because a corner point is at least as good as any of them.

In Figure D.4, the five corner points are marked A, B, C, D, and E. Point A is the origin $(0, 0)$ and can be ignored because any other feasible point is a better solution. We could try each of the other corner points in the objective function and select the one that maximizes Z. For example, corner point B lies at $(0, 8)$. If we substitute these values into the objective function, the resulting Z value is 320:

corner point

A point that lies at the intersection of two (or possibly more) constraint lines on the boundary of the feasible region.

$$34x_1 + 40x_2 = Z$$
$$34(0) + 40(8) = 320$$

However, we may not be able to read accurately the values of x_1 and x_2 for some of the points (e.g., C or D) on the graph. Algebraically solving two linear equations for each corner point also is inefficient when there are many constraints and, thus, many corner points.

The best approach is to plot the objective function on the graph of the feasible region for some arbitrary Z values. From these objective function lines, we can spot the best solution visually. If the objective function is profits, each line is called an *iso-profit line* and every point on that line will yield the same profit. If Z measures cost, the line is called an *iso-cost line* and every point on it represents the same cost. We can simplify the search by plotting the first line in the feasible region—somewhere near the optimal solution, we hope. For the Stratton Company example, let us pass a line through point E (8, 0). This point is a corner point. It might even be the optimal solution because it is far from the origin. To draw the line, we first identify its Z value as $34(8) + 40(0) = 272$. Therefore, the equation for the objective function line passing through E is

$$34x_1 + 40x_2 = 272$$

Every point on the line defined by this equation has an objective function Z value of 272. To draw the line, we need to identify a second point on it and then connect the two points. Let us use the x_2 intercept, where $x_1 = 0$:

$$34(0) + 40x_2 = 272$$
$$x_2 = 6.8$$

Figure D.5 shows the iso-profit line that connects points (8, 0) and (0, 6.8). A series of other dashed lines could be drawn parallel to this first line. Each would have its own Z value. Lines above the first line we drew would have higher Z values. Lines below it would have lower Z values.

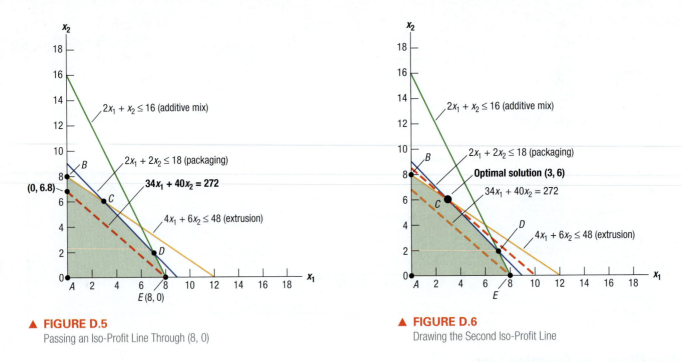

▲ **FIGURE D.5**
Passing an Iso-Profit Line Through (8, 0)

▲ **FIGURE D.6**
Drawing the Second Iso-Profit Line

Find the Visual Solution

We now eliminate corner points A and E from consideration as the optimal solution because better points lie above and to the right of the $Z = 272$ iso-profit line. Our goal is to maximize profits, so the best solution is a point on the iso-profit line *farthest* from the origin, but still touching the feasible region. (For minimization problems, it is a point in the feasible region on the iso-cost line *closest* to the origin.)[1] To identify which of the remaining corner points is optimal (B, C, or D), we draw, parallel to the first line, one or more iso-profit lines that give better Z values (higher for maximization and lower for minimization). The line that just touches the feasible region identifies the optimal solution. For the Stratton Company problem, Figure D.6 shows the second iso-profit line. The optimal solution is the last point touching the feasible region: point C. It appears to be in the vicinity of (3, 6), but the visual solution is not exact.

A linear programming problem can have more than one optimal solution. This situation occurs when the objective function is parallel to one of the faces of the feasible region. Such would be the case if our objective function in the Stratton Company problem were $38x_1 + $38x_2$. Points (3, 6) and (7, 2) would be optimal, as would any other point on the line connecting these two corner points. In such a case, management probably would base a final decision on nonquantifiable factors. It is important to understand, however, that we need to consider only the corner points of the feasible region when optimizing an objective function.

Find the Algebraic Solution

To find an exact solution, we must use algebra. We begin by identifying the pair of constraints that define the corner point at their intersection. We then list the constraints as equations and solve them simultaneously to find the coordinates x_1, x_2 of the corner point. Simultaneous equations can be solved several ways. For small problems, the easiest way is as follows:

Step 1. Develop an equation with just one unknown. Start by multiplying both sides of one equation by a constant so that the coefficient for one of the two decision variables is *identical* in both equations. Then, subtract one equation from the other and solve the resulting equation for its single unknown variable.

Step 2. Insert this decision variable's value into either one of the original constraints and solve for the other decision variable.

---

[1]The statements "farthest from the origin" or "closest to the origin" would no longer be true if there are negative coefficients in the objective function.

| EXAMPLE D.4 | **Finding the Optimal Solution Algebraically** |

Find the optimal solution algebraically for the Stratton Company problem. What is the value of Z when the decision variables have optimal values?

MyOMLab

Tutor D.2 in MyOMLab provides a new practice example for finding the optimal solution.

SOLUTION

Step 1. Figure D.6 showed that the optimal corner point lies at the intersection of the extrusion and packaging constraints. Listing the constraints as equalities, we have

$$4x_1 + 6x_2 = 48 \text{ (extrusion)}$$
$$2x_1 + 2x_2 = 18 \text{ (packaging)}$$

We multiply each term in the packaging constraint by 2. The packaging constraint now is $4x_1 + 4x_2 = 36$. Next, we subtract the packaging constraint from the extrusion constraint. The result will be an equation from which x_1 has dropped out. (Alternatively, we could multiply the second equation by 3 so that x_2 drops out after the subtraction.) Thus,

$$
\begin{array}{r}
4x_1 + 6x_2 = 48 \\
-(4x_1 + 4x_2 = 36) \\
\hline
2x_2 = 12 \\
x_2 = 6
\end{array}
$$

Step 2. Substituting the value of x_2 into the extrusion equation, we get

$$4x_1 + 6(6) = 48$$
$$4x_1 = 12$$
$$x_1 = 3$$

Thus, the optimal point is (3, 6). This solution gives a total profit of $34(3) + 40(6) = \$342$.

DECISION POINT
Management at the Stratton Company decided to produce 300 feet of type 1 pipe and 600 feet of type 2 pipe for the next week for a total profit of $342.

Slack and Surplus Variables

Figure D.6 shows that the optimal product mix will exhaust all the extrusion and packaging resources because at the optimal corner point (3, 6) the two constraints are equalities. Substituting the values of x_1 and x_2 into these constraints shows that the left-hand sides equal the right-hand sides:

$$4(3) + 6(6) = 48 \text{ (extrusion)}$$
$$4(3) + 2(6) = 18 \text{ (packaging)}$$

A constraint (such as the one for extrusion) that helps form the optimal corner point is called a **binding constraint** because it limits the ability to improve the objective function. If a binding constraint is relaxed, or made less restrictive, a better solution is possible. Relaxing a constraint means increasing the right-hand-side parameter for a $\leq$ constraint or decreasing it for a $\geq$ constraint. No improvement is possible from relaxing a constraint that is not binding, such as the additive mix constraint in Figure D.6. If the right-hand side was increased from 16 to 17 and the problem solved again, the optimal solution would not change. In other words, there is already more additive mix than needed.

For nonbinding inequality constraints, knowing how much the left and right sides differ is helpful. Such information tells us how close the constraint is to becoming binding. For a $\leq$ constraint, the amount by which the left-hand side falls short of the right-hand side is called **slack**. For a $\geq$ constraint, the amount by which the left-hand side exceeds the right-hand side is called **surplus**. To find the slack for a $\leq$ constraint algebraically, we *add* a slack variable to the constraint and convert it to an equality. Then, we substitute in the values of the decision variables and solve for the slack. For example, the additive mix constraint in Figure D.6, $2x_1 + x_2 \leq 16$, can be rewritten by adding slack variable s_1:

$$2x_1 + x_2 + s_1 = 16$$

We then find the slack at the optimal solution (3, 6):

$$2(3) + 6 + s_1 = 16$$
$$s_1 = 4$$

binding constraint

A constraint that helps form the optimal corner point; it limits the ability to improve the objective function.

slack

The amount by which the left-hand side of a linear programming constraint falls short of the right-hand side.

surplus

The amount by which the left-hand side of a linear programming constraint exceeds the right-hand side.

The procedure is much the same to find the surplus for a $\geq$ constraint, except that we *subtract* a surplus variable from the left-hand side. Suppose that $x_1 + x_2 \geq 6$ was another constraint in the Stratton Company problem, representing a lower bound on the number of units produced. We would then rewrite the constraint by subtracting a surplus variable s_2:

$$x_1 + x_2 - s_2 = 6$$

The surplus at the optimal solution (3, 6) would be

$$3 + 6 - s_2 = 6$$
$$s_2 = 3$$

Sensitivity Analysis

Rarely are the parameters in the objective function and constraints known with certainty. Often, they are just estimates of actual values. For example, the available packaging and extrusion hours for the Stratton Company are estimates that do not reflect the uncertainties associated with absenteeism or personnel transfers, and the required hours per unit to package and extrude may be time estimates that essentially are averages. Likewise, profit contributions used for the objective function coefficients do not reflect uncertainties in selling prices and such variable costs as wages, raw materials, and shipping.

Despite such uncertainties, initial estimates are needed to solve the problem. Accounting, marketing, and time-standard information systems (see MyOMLab Supplement H) often provide these initial estimates. After solving the problem using these estimated values, the analyst can determine how much the optimal values of the decision variables and the objective function value Z would be affected if certain parameters had different values. This type of postsolution analysis for answering "what-if" questions is called *sensitivity analysis*.

One way of conducting sensitivity analysis for linear programming problems is the brute-force approach of changing one or more parameter values and resolving the entire problem. This approach may be acceptable for small problems, but it is inefficient if the problem involves many parameters. For example, brute-force sensitivity analysis using 3 separate values for each of 20 objective function coefficients requires 3^{20}, or 3,486,784,401, separate solutions! Fortunately, efficient methods are available for getting sensitivity information without resolving the entire problem, and they are routinely used in most linear programming computer software packages. Table D.1 describes the four basic types of sensitivity analysis information provided by linear programming.

TABLE D.1 | SENSITIVITY ANALYSIS INFORMATION PROVIDED BY LINEAR PROGRAMMING

Key Term	Definition
Reduced cost	How much the objective function coefficient of a decision variable must improve (increase for maximization or decrease for minimization) before the optimal solution changes and the decision variable "enters" the solution with some positive number
Shadow price	The marginal improvement in Z (increase for maximization and decrease for minimization) caused by relaxing the constraint by one unit
Range of optimality	The interval (lower and upper bounds) of an objective function coefficient over which the optimal values of the decision variables remain unchanged
Range of feasibility	The interval (lower and upper bounds) over which the right-hand-side parameter of a constraint can vary while its shadow price remains valid

Computer Solution

Most real-world linear programming problems are solved on a computer, so we concentrate here on understanding the use of linear programming and the logic on which it is based. The solution procedure in computer codes is some form of the **simplex method**, which is an iterative algebraic procedure for solving linear programming problems.

Simplex Method

The graphic analysis gives insight into the logic of the simplex method, beginning with the focus on corner points. If there is any feasible solution to a problem, at least one corner point will always be the optimum, even when multiple optimal solutions are available. Thus, the simplex method starts with an initial corner point and then systematically evaluates other corner points in such a way that the objective function improves (or, at worst, stays the same) at each iteration. In the Stratton Company problem, an improvement would be an increase in profits. When no more improvements are possible, the optimal solution has been found.[2] The simplex method also helps generate the sensitivity analysis information that we developed graphically.

Each corner point has no more than m variables that are greater than 0, where m is the number of constraints (not counting the nonnegativity constraints). The m variables include slack and surplus variables, not just the original decision variables. Because of this property, we can find a corner point by simultaneously solving m constraints, where all but m variables are set equal to 0. For example, point B in Figure D.6 has three nonzero variables: x_2, the slack variable for packaging, and the slack variable for the additive mix. Their values can be found by simultaneously solving the three constraints, with x_1 and the slack variable for extrusion equal to 0. After finding this corner point, the simplex method applies information similar to the reduced costs to decide which new corner point to find next that gives an even better Z value. It continues in this way until no better corner point is possible. The final corner point evaluated is the optimal one.

Computer Output

Computer programs dramatically reduce the amount of time required to solve linear programming problems. The capabilities and displays of software packages are not uniform. For example, POM for Windows in MyOMLab can handle small- to medium-sized linear programming problems. Inputs are made easily and nonnegativity constraints need not be entered. Microsoft's *Excel* Solver offers a second option for similar problem sizes. More advanced software for larger problems is available from multiple sources.

Here we show output from POM for Windows when applied to the Stratton Company. Figure D.7 shows the two *data entry* screens. The first screen allows you to enter the problem's

MyOMLab

▼ **FIGURE D.7**
Data Entry Screens

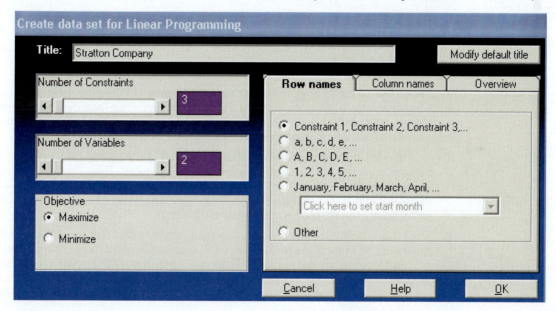

	X1	X2		RHS	Equation form
Maximize	34	40			Max 34X1 + 40X2
Extrusion	4	6	<=	48	4X1 + 6X2 <= 48
Packaging	2	2	<=	18	2X1 + 2X2 <= 18
Additive	3	1	<=	16	3X1 + X2 <= 0

---

[2] For more information on how to perform the simplex method manually, see Render, Barry, Ralph M. Stair, and Michael E. Hanna. *Quantitative Analysis for Management*, 11th ed. or any other current textbook on management science.

name, specify the number of constraints and decision variables, and choose between maximization and minimization. After making these inputs and clicking the OK button, the data table screen is shown. Enter the parameters, give names to each constraint and decision variable (as desired), and specify the type of relationship ($\leq$, $=$, $\geq$) for each constraint. The second screen in Figure D.7 shows the completed data table. The user may customize labels for the decision variables, right-hand-side values, and constraints. Here, the first decision variable is labeled as "X1," the right-hand-side values as "RHS," and the extrusion constraint as "Extrusion." Slack and surplus variables will be added automatically as needed. When all of the inputs are made, click the green arrow labeled "Solve" in the upper-right corner.

Figure D.8 displays the solution with the *Results* screen. All output confirms our earlier calculations and the graphic analysis. Of particular interest is the bottom row that gives the optimal values of the decision variables ($x_1 = 3$ and $x_2 = 6$), and also the optimal value of the objective function ($342). The shadow prices for each constraint are given in the last column.

▼ **FIGURE D.8**
Results Screen

	X1	X2		RHS	Shadow Price
			Stratton Company Solution		
Maximize	34	40			
Constraint 1	4	6	<=	48	3
Constraint 2	2	2	<=	18	11
Constraint 3	2	1	<=	16	0
Solution->	3	6		342	

Click on the Window icon and select the second option in the drop down menu to switch to the *Ranging* screen. It is shown in Figure D.9. The top half deals with the decision variables. Of particular interest are the reduced costs and the lower and upper bounds. Two tips on interpreting the *reduced cost* information are as follows:

1. It is relevant only for a decision variable that is 0 in the optimal solution. If the decision variable is greater than 0, ignore the reduced cost number. Thus, for the Stratton Company problem the reduced cost numbers provide no new insight because they are always 0 when decision variables have positive values in the optimal solution. Look instead at the lower and upper bounds on the objective function coefficients.

2. It tells how much the objective function coefficient of a decision variable that is 0 in the optimal solution must *improve* (increase for maximization problems or decrease for minimization problems) before the optimal solution would change. At that point, the decision variable associated with the coefficient enters the optimal solution at some positive level. To learn the new solution, apply POM for Windows again with a coefficient improved by slightly more than the reduced cost number.

▼ **FIGURE D.9**
Ranging Screen

Variable	Value	Reduced Cost	Original Val	Lower Bound	Upper Bound
X1	3	0	34	26.6667	40
X2	6	0	40	34	51
Constraint	Shadow Price	Slack/Surplus	Original Val	Lower Bound	Upper Bound
Constraint 1	3	0	48	40	54
Constraint 2	11	0	18	16	20
Constraint 3	0	4	16	12	Infinity

The top half of this screen also gives the *range of optimality*, or the lower and upper bound over which the objective function coefficients can range without affecting the optimal values of the decision variables. Note that the objective function coefficient for x_1, which currently has a value of $34, has a range of optimality from $26.6667 to $40. While the objection function's Z value would change with coefficient changes over this range, the optimal values of the decision variables remain the same.

The bottom half of Figure D.9 deals with the constraints, including the slack or surplus variables and the original right-hand-side values. Of particular interest are the shadow prices. Two tips on interpreting a *shadow price* follow:

1. The number is relevant only for a binding constraint, where the slack or surplus variable is 0 in the optimal solution. For a nonbinding constraint, the shadow price is 0.

2. The sign of the shadow price can be a positive or negative. The sign depends on whether the objective function is being maximized or minimized, and whether it is a $\leq$ constraint or $\geq$ constraint. If you simply ignore the sign, interpret the shadow price as the benefit of increasing the right-hand side by one unit for a $\leq$ constraint and reducing it by one unit for a $\geq$ constraint. The benefit is a reduction in the objective function value for minimization problems, and an increase for maximization problems. The shadow price can also be interpreted as the marginal loss (or penalty) in Z caused by making the constraint more restrictive by one unit.

Thus, the Stratton Company problem has 4 pounds of the additive mix slack, so the shadow price is $0. Packaging, on the other hand, is a binding constraint because it has no slack. The shadow price of one more packaging hour is $11.

Finally, Figure D.9 reports the lower and upper bounds for the *range of feasibility*, over which the right-hand-side parameters can range without changing the shadow prices. For example, the $11 shadow price for packaging is valid over the range from 16 to 20 hours.

The number of variables in the optimal solution (counting the decision variables, slack variables, and surplus variables) that are greater than 0 never exceeds the number of constraints. Such is the case for the Stratton Company problem, with its three constraints (not counting the implicit nonnegativity constraints) and three nonzero variables in the optimal solution (x_1, x_2, and the additive mix slack variable). On some rare occasions, the number of nonzero variables in the optimal solution can be less than the number of constraints—a condition called **degeneracy**. When degeneracy occurs, the sensitivity analysis information is suspect. If you want more "what-if" information, simply run your software package again using the new parameter values that you want to investigate.

degeneracy

A condition that occurs when the number of nonzero variables in the optimal solution is less than the number of constraints.

EXAMPLE D.5	**Using Shadow Prices for Decision Making**

The Stratton Company needs answers to three important questions: (1) Would increasing capacities in the extrusion or packaging area pay if it cost an extra $8 per hour over and above the normal costs already reflected in the objective function coefficients? (2) Would increasing packaging capacity pay if it cost an additional $6 per hour? (3) Would buying more raw materials pay?

SOLUTION

Expanding extrusion capacity would cost a premium of $8 per hour, but the shadow price for that capacity is only $3 per hour. However, expanding packaging hours would cost only $6 per hour more than the price reflected in the objective function, and the shadow price is $11 per hour. Finally, buying more raw materials would not pay because a surplus of 4 pounds already exists; the shadow price is $0 for that resource.

DECISION POINT

Management decided to increase its packaging hour's capacity but did not decide to expand extrusion capacity or buy more raw materials.

The Transportation Method

A special case of linear programming is the **transportation problem**, which can be represented as a standard table, sometimes called a *tableau*. Figure D.10 is an example, where the rows are supply sources and the columns are demands. Both the supplies and demands can be broken down into several periods into the future. Typically, the rows of the table are linear constraints that impose capacity limitations, and the columns are linear constraints that require certain demand levels to be met. Each cell in the tableau is a decision variable, and a per-unit cost is shown in the upper-right corner of each cell. Figure D.10 implies 52 decision variables (13 rows $\times$ 4 columns) and five constraints (13 rows + 4 columns).

Transportation problems can be formulated as a conventional linear programming problem and solved as usual. The **transportation method** simplifies data input and is a more efficient solution technique, but does not provide sensitivity analysis as in Figure D.9. Here we show

transportation problem

A special case of linear programming that has linear constraints for capacity limitations and demand requirements.

transportation method

A more efficient solution technique than the simplex method for solving transportation problems.

Source of Supply	Time Period				Capacity
	1	2	3	4	
Period / Initial Inventory					
1 / Type 1					
1 / Type 2					
1 / Type 3					
2 / Type 1					
2 / Type 2					
2 / Type 3					
3 / Type 1					
3 / Type 2					
3 / Type 3					
4 / Type 1					
4 / Type 2					
4 / Type 3					
Demand					

how production planning problems can be formulated as transportation problems. Chapter 11, "Supply Chain Location Decisions" shows an entirely different application of the transportation method: how to solve location problems. We focus on the setup and interpretation of the problem, leaving the rest of the solution process to a computer software package.

Transportation Method for Production Planning

Making sure that demand and supply are in balance is central to sales and operations planning (SOP), so it is no surprise that the transportation method can be applied to it. The *transportation method for production planning* is particularly helpful in determining anticipation inventories. Thus, it relates more to manufacturers' production plans than to service providers' staffing plans. In fact, the workforce levels for each period are inputs to the transportation method rather than outputs from it. Different workforce adjustment plans should be evaluated. Thus, several transportation method solutions may be obtained before a final plan is selected.

Using the transportation method for production planning is based on the assumption that a demand forecast is available for each period, along with a possible workforce adjustment plan. Capacity limits on overtime and the use of subcontractors also are needed for each period. Another assumption is that all costs are linearly related to the amount of goods produced; that is, a change in the amount of goods produced creates a proportionate change in costs.

To develop a sales and operations plan for a manufacturer, we do the following:

1. Obtain the demand forecasts for each period to be covered by the sales and operations plan and identify the initial inventory level currently available that can be used to meet future demand.

2. Select a candidate workforce adjustment plan, using a chase strategy, level strategy, or a mixed strategy. Specify the capacity limits of each production alternative (regular time, overtime, and subcontracting) for each period covered by the plan.

3. Estimate the cost of holding inventory and the cost of possible production alternatives (regular-time production, overtime production, and subcontracting). Identify the cost of undertime, if idle regular-time capacity is paid.

4. Input the information gathered in steps 1–3 into a computer routine that solves the transportation problem. After getting the solution, calculate the anticipation inventory levels and identify high-cost elements of the plan.

5. Repeat the process with other plans for regular-time, overtime, and subcontracting capacities until you find the solution that best balances cost and qualitative considerations. Even though this process involves trial and error, the transportation method yields the best mix of regular time, overtime, and subcontracting for each supply plan.

Example D.6 demonstrates this approach using the *Transportation Method (Production Planning)* module in the POM for Windows package.

EXAMPLE D.6	**Preparing a Production Plan with the Transportation Method**

The Tru-Rainbow Company produces a variety of paint products for both commercial and private use. The demand for paint is highly seasonal, peaking in the third quarter. Initial inventory is 250,000 gallons, and ending inventory should be 300,000 gallons.

Tru-Rainbow's manufacturing manager wants to determine the best production plan using the following demand requirements and capacity plan. Demands and capacities here are expressed in thousands of gallons (rather than employee-period equivalents). The manager knows that the regular-time cost is $1.00 per unit, overtime cost is $1.50 per unit, subcontracting cost is $1.90 per unit, and inventory holding cost is $0.30 per unit per quarter. Undertime is paid and the cost is $0.50 per unit. It is less than the regular-time cost because only labor costs are involved, not materials and variable overhead going into paint production.

	Demand	Regular-time Capacity	Overtime Capacity	Subcontractor Capacity
Quarter 1	300	450	90	200
Quarter 2	850	450	90	200
Quarter 3	1,500	750	150	200
Quarter 4	350	450	90	200
Totals	3,000	2,100	420	800

The following constraints apply:

a. The maximum allowable overtime in any quarter is 20 percent of the regular-time capacity in that quarter.

b. The subcontractor can supply a maximum of 200,000 gallons in any quarter. Production can be subcontracted in one period and the excess held in inventory for a future period to avoid a stockout.

c. No backorders or stockouts are permitted.

SOLUTION

Figure D.11 shows the POM for Windows screen of data inputs. Figure D.12 shows the POM for Windows screen that displays the optimal solution for this particular workforce adjustment plan. It looks much like the table shown previously, but with one exception. The demand for quarter 4 is shown to be 650,000 gallons rather than the demand forecast of only 350,000. The larger number reflects the desire of the manager to have an ending inventory in quarter 4 of 300,000 gallons. Some points to note in Figures D.11 and D.12 include the following:

1. There is a row in Figure D.12 for each supply alternative (instead of the "source of supply" in Figure D.10) on a quarter-by-quarter basis. The last column in each row indicates the maximum amount that can be used

▼ **FIGURE D.11**
POM for Windows Screens for Tru-Rainbow Company

Period	Demand	Regular tm Capacity	Overtime Capacity	Subcontract Capacity		Unit costs	Value
Quarter 1	300	450	90	200		Regular time	1
Quarter 2	850	450	90	200		Overtime	1.5
Quarter 3	1500	750	150	200		Subcontracting	1.9
Quarter 4	650	450	90	200		Holding cost	.3
						Lost sales cost	Not allowed
						Idle RT Capacity	.5
						Initial Inventory	250

Optimal cost = $4,010	Quarter 1	Quarter 2	Quarter 3	Quarter 4	Excess Capacity	Capacity
Init Inventory	230		20			250
Quarter 1 RegTime	50	400				450
Quarter 1 Overtime			90			90
Quarter 1 Subcontract	20				180	200
Quarter 2 RegTime		450				450
Quarter 2 Overtime			90			90
Quarter 2 Subcontract			200			200
Quarter 3 RegTime			750			750
Quarter 3 Overtime			150			150
Quarter 3 Subcontract			200			200
Quarter 4 RegTime				450		450
Quarter 4 Overtime				90		90
Quarter 4 Subcontract				110	90	200
Demand	300	850	1500	650	270	

▲ **FIGURE D.12**
Solution Screen for Prospective Tru-Rainbow Company Production Plan

to meet demand. The first row is the initial inventory available, and the rows that follow are for regular-time, overtime, and subcontracting production in each of the four quarters. The initial inventory can be used to satisfy demand in any of the four quarters. The second row (regular-time production in period 1) can also be used to satisfy demand in any of the four periods the plan will cover, and so on. The numbers in the last column give the maximum capacity made available for the supply alternatives. For example, the regular-time capacity for quarter 3 increases from the usual 450,000 gallons to 750,000 gallons, to help with the peak demand forecasted to be 1,500,000 gallons.

2. A column indicates each future quarter of demand and the last row gives its demand forecast. The demand for the fourth quarter is shown to be 650 units, because it includes the desired amount of ending inventory. The Excess Capacity column shows the cost of unused capacity, and the number in the last row (270 units) is the amount by which total capacity exceeds total demand. The unit costs for unused capacity are 0, except for the $0.50 unit cost of unused regular-time capacity.

3. The numbers in the other cells (excluding the cells in the last row or last column) show the cost of producing a unit in one period and, in some cases, carrying the unit in inventory for sale in a future period. These numbers correspond to the costs in the upper-right corners of the cells in Figure D.10. For example, the cost per unit of regular-time production in quarter 1 is $1.00 per gallon if it is used to meet the demand in quarter 1. This cost is found in row 2 and column 1 of the tableau. However, if it is produced to meet demand in quarter 2, the cost increases to $1.30 (or $1.00 + $0.30) because we must hold the unit in inventory for one quarter. Satisfying a unit of demand in quarter 3 by producing in quarter 1 on regular time and carrying the unit for two quarters costs $1.60, or [$1.00 + (2 × $0.30)], and so on. A similar approach is used for the costs of overtime and subcontracting.

4. The cells in the bottom left portion of the tableau with a cost of $9,999 are associated with backorders (or producing in a period to satisfy demand in a period after it was needed). Here we disallow backorders by making the backorder cost an arbitrarily large number, in this case $9,999 per unit. If backorder costs are so large, the transportation method will try to avoid backorders because it seeks a solution that minimizes total cost. If that is not possible, we increase the staffing plan and the overtime and subcontracting capacities.

5. The least expensive alternatives are those in which the output is produced and sold in the same period. For example, the cost for quarter 2 overtime production is only $1.50 per gallon if it is designated to meet demand in quarter 2 (row 6, column 2). The cost increases to $2.10 if it is designated for quarter 4 demand. However, we may not always be able to avoid alternatives that create inventory because of capacity restrictions.

6. Finally, the per-unit holding cost for the beginning inventory in period 1 is 0 because it is a function of previous production-planning decisions.

The first row in Figure D.12 shows that 230 units of the initial inventory are used to help satisfy the demand in quarter 1. The remaining 20 units in the first row are earmarked for helping supply the demand in quarter 3. The sum of the allocations across row 1 for the four quarters (230 + 0 + 20 + 0) does not exceed the maximum capacity of 250, given in the right column. With the transportation method, this result must occur with each row. Any shortfalls are unused capacity, given in the "Excess Capacity" column. In this case, the undertime cost of $0.50 per unit was sufficiently large that no regular time capacity went unused.

Similarly, the sum of the allocations down each column must equal the total demand for the quarter. For example, the demand for quarter 1 is supplied from 230 units of the initial inventory, 50 units of quarter 1 regular-time production, and 20 units of quarter 1 subcontracting production. Summed together, they equal the forecasted demand of 300 units.

To further interpret the solution, we can convert Figure D.12 into the following table. For example, the total regular-time production in quarter 1 is 450,000 gallons (50,000 gallons to help meet demand in quarter 1 and 400,000 gallons to help satisfy demand in quarter 2).

The anticipation inventory held at the end of each quarter is obtained in the last column. For any quarter, it is the quarter's beginning inventory plus total supply (regular-time and overtime production, plus subcontracting) minus demand. For example, for quarter 1 the beginning inventory (250,000) plus the total from production and subcontracting (560,000) minus quarter 1 demand (300,000) results in an ending inventory of 510,000, which also is the beginning inventory for quarter 2.

Quarter	Regular-time Production	Overtime Production	Subcontracting	Total Supply	Anticipation Inventory
1	450	90	20	560	250 + 560 − 300 = 510
2	450	90	200	740	510 + 740 − 850 = 400
3	750	150	200	1,100	400 + 1,100 − 1,500 = 0
4	450	90	110	650	0 + 650 − 350 = 300
Totals	2,100	420	530	3,050	

Note: Anticipation inventory is the amount at the end of each quarter, where Beginning inventory + Total production − Actual Demand = Ending inventory.

The breakdown of costs can be found by multiplying the allocation in each cell of Figure D.12 by the cost per unit in that cell in Figure D.11 (b). Computing the cost column by column (it can also be done on a row-by-row basis) yields a total cost of $4,010,000, or $4,010 × 1,000.

	Cost Calculations by Column	
Quarter 1	230($0) + 50($1.00) + 20($1.90)	= $ 88
Quarter 2	400($1.30) + 450($1.00)	= 970
Quarter 3	20($0.60) + 90($2.10) + 90($1.80) + 200($2.20) + 750($1.00) + 150($1.50) + 200($1.90)	= 2,158
Quarter 4	450($1.00) + 90($1.50) + 110($1.90)	= 794
		Total = $4,010

DECISION POINT

This plan requires too much overtime and subcontracting, and the anticipation inventory cost is substantial. The manager decided to search for a better capacity plan—with increases in the workforce to boost regular-time production capacity—that could lower production costs, perhaps even low enough to offset the added capacity costs.

Applications

Many problems in operations and supply chain management, and in other functional areas, lend themselves to linear programming and the transportation method. In addition to the examples already introduced, applications also exist in process management, constraint management, shipping assignments, inventory control, and shift scheduling. The review problems at the end of this supplement and at the end of previous chapters illustrate many of these types of problems. Once the decision maker knows how to formulate a problem generally, he or she can then adapt it to the situation at hand.

LEARNING GOALS IN REVIEW

1 **Describe formulating models for various problems**. Review the section "Formulating a Problem," pp. 588–590, and pay particular attention to Example D.1.

2 **Demonstrate graphic analysis and solutions for two-variable problems**. See the section "Graphic Analysis," pp. 590–596, on plotting the constraints, identifying the feasible region, plotting the objective function, and finding the solution.

3 **Define slack variables, surplus variables, and sensitivity analysis**. The sections on "Slack and Surplus Variables" and "Sensitivity Analysis" are found on pp. 595–596.

4 **Interpret computer output of a linear programming solution**. Computer output from POM for Windows is displayed and

interpreted on pp. 597–599. Key information relates to the optimal values of the decision variables, objective function value, slack variables, and surplus variables. The shadow prices, reduced costs, lower bounds, and upper bounds can be valuable information for sensitivity analysis.

5 **Apply the transportation method to Sales and Operations Planning (S&OP) problems**. The section "Transportation Method for Production Planning," pp. 599–603, gives a step-by-step description on how to set up the problem as a transportation problem and then how to interpret the POM for Windows output. Pay particular attention to Figures D.11 and D.12. Example D.6 and Solved Problem 2 are also helpful.

MyOMLab helps you develop analytical skills and assesses your progress with multiple problems on graphic analysis, slack and surplus variables, optimal solution, computer solution, shadow prices, the transportation method, production plans, and shift schedules.

MyOMLab Resources	Titles	Link to the Book
Active Model Exercise	D.1 LP Graph	Graphic Analysis; Solved Problem 1 (p. 605)
OM Explorer Tutors	D.1 Plotting the Constraints	Plot the Constraints; Figure D.1 (p. 591; Example D.2 (pp. 590–591); Figure D.2 (p. 591) Example D.3 (pp. 592–593)
	D.2 Finding the Optimal Solution	Plot the Objective Function Line; Figures D.5 and D.6 (p. 594); Find the Visual Solution; Find the Algebraic Solution; Example D.4 (p. 595)
	D.3 Finding Slack at Optimal Solution	Slack and Surplus Variables; Solved Problem 1 (p. 605)
	D.4 Graphic and Algebraic Solution	Graphic Analysis; Solved Problem 1 (p. 605)
POM for Windows	Linear Programming Results	Computer Solution
	Ranging	Computer Output; Figures D.7, D.8, and D.9 (pp. 597–598)
	Graph	Graphic Analysis
	Transportation Method (Production Planning)	Transportation Method for Production; Example D.6; Figures D.11, D.12, and D.14 (pp. 601, 602, and 606); Solved Problem 2 (pp. 605–606)
Virtual Tours	Virgin Atlantic	Formulating a Problem
	Nike	Formulating a Problem
	Purina	Formulating a Problem
MyOMLab Supplements	F. Financial Analysis	MyOMLab Supplement F
Internet Exercise	Whirlpool	Formulating a Problem
	Marriott	Formulating a Problem
	Starbucks	Formulating a Problem
Key Equations		
Image Library		

Key Terms

binding constraint 595
certainty 588
constraints 588
corner point 593
decision variables 587
degeneracy 599
feasible region 588
graphic method of linear programming 590

linearity 588
linear programming 587
nonnegativity 588
objective function 587
parameter 588
product-mix problem 588
simplex method 596
range of feasibility 596

range of optimality 596
reduced cost 596
shadow price 596
slack 595
surplus 595
transportation method 599
transportation problem 599

Solved Problem 1

O'Connel Airlines is considering air service from its hub of operations in Cicely, Alaska, to Rome, Wisconsin, and Seattle, Washington. O'Connel has one gate at the Cicely Airport, which operates 12 hours per day. Each flight requires 1 hour of gate time. Each flight to Rome consumes 15 hours of pilot crew time and is expected to produce a profit of $2,500. Serving Seattle uses 10 hours of pilot crew time per flight and will result in a profit of $2,000 per flight. Pilot crew labor is limited to 150 hours per day. The market for service to Rome is limited to nine flights per day.

a. Use the graphic method of linear programming to maximize profits for O'Connel Airlines.

b. Identify positive slack and surplus variables, if any.

MyOMLab

Tutor D.4 in MyOMLab provides a practice example for finding the graphic and algebraic solution.

SOLUTION

a. The objective function is to maximize profits, Z:

$$\text{Maximize: } \$2,500x_1 + \$2,000x_2 = Z$$

where

$$x_1 = \text{number of flights per day to Rome, Wisconsin}$$
$$x_2 = \text{number of flights per day to Seattle, Washington}$$

The constraints are

$$x_1 + x_2 \leq 12 \text{ (gate capacity)}$$
$$15x_1 + 10x_2 \leq 150 \text{ (labor)}$$
$$x_1 \leq 9 \text{ (market)}$$
$$x_1 \geq 0 \text{ and } x_1 \geq 0$$

A careful drawing of iso-profit lines parallel to the one shown in Figure D.13 will indicate that point D is the optimal solution. It is at the intersection of the labor and gate capacity constraints. Solving algebraically, we get

$$15x_1 + 10x_2 = 150 \text{ (labor)}$$
$$-10x_1 - 10x_2 = -120 \text{ (gate} \times -10)$$
$$5x_1 + 0x_2 = 30$$
$$x_1 = 6$$
$$6 + x_2 = 12 \text{ (gate)}$$
$$x_2 = 6$$

The maximum profit results from making six flights to Rome and six flights to Seattle:

$$\$2,500(6) + \$2,000(6) = \$27,000$$

b. The market constraint has three units of slack, so the demand for flights to Rome is not fully met:

$$x_1 \leq 9$$
$$x_1 + s_3 = 9$$
$$6 + s_3 = 9$$
$$s_3 = 3$$

▼ **FIGURE D.13**
Graphic Solution for O'Connel Airlines

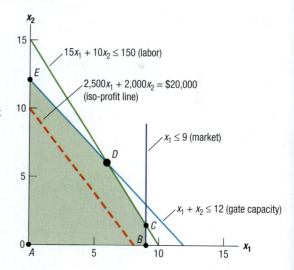

Solved Problem 2

The Arctic Air Company produces residential air conditioners. The manufacturing manager wants to develop a sales and operations plan for the next year based on the following demand and capacity data (in hundreds of product units):

	Demand	Regular-time Capacity	Overtime Capacity	Subcontractor Capacity
Jan–Feb (1)	50	65	13	10
Mar–Apr (2)	60	65	13	10
May–Jun (3)	90	65	13	10
Jul–Aug (4)	120	80	16	10
Sep–Oct (5)	70	80	16	10
Nov–Dec (6)	40	65	13	10
Totals	430	420	84	60

Undertime is unpaid, and no cost is associated with unused overtime or subcontractor capacity. Producing one air conditioning unit on regular time costs $1,000, including $300 for labor. Producing a unit on overtime costs $1,150. A subcontractor can produce a unit to Arctic Air specifications for $1,250. Holding an air conditioner in stock costs $60 for each 2-month period, and 200 air conditioners are currently in stock. The plan calls for 400 units to be in stock at the end of period 6. No backorders are allowed. Use the transportation method to develop a plan that minimizes costs.

SOLUTION

The following tables identify the optimal production and inventory plans. Figure D.14 shows the tableau that corresponds to this solution. An arbitrarily large cost ($99,999 per period) was used for backorders, which effectively ruled them out. Again, all production quantities are in hundreds of units. Note that demand in period 6 is 4,400. That amount is the period 6 demand plus the desired ending inventory of 400. The anticipation inventory is measured as the amount at the end of each period. Cost calculations are based on the assumption that workers are not paid for undertime or are productively put to work elsewhere in the organization whenever they are not needed for this work.

FIGURE D.14 ▶
Tableau for Optimal Production and Inventory Plans

Period	Alternatives	1	2	3	4	5	6	Unused Capacity	Total Capacity
	Initial Inventory	0	60	120	180 **2**	240	300	0	2
1	Regular time	1,000 **50**	1,060 **15**	1,120	1,180	1,240	1,300	0	65
1	Overtime	1,150	1,210	1,270	1,330	1,390	1,450	13	13
1	Subcontract	1,250	1,310	1,370	1,430	1,490	1,550	10	10
2	Regular time	99,999	1,000 **41**	1,060 **12**	1,120 **12**	1,180	1,240	0	65
2	Overtime	99,999	1,150 **4**	1,210	1,270	1,330	1,390	9	13
2	Subcontract	99,999	1,250	1,310	1,370	1,430	1,490	10	10
3	Regular time	99,999	99,999	1,000 **65**	1,060	1,120	1,180	0	65
3	Overtime	99,999	99,999	1,150 **13**	1,210	1,270	1,330	0	13
3	Subcontract	99,999	99,999	1,250	1,310	1,370	1,430	10	10
4	Regular time	99,999	99,999	99,999	1,000 **80**	1,060	1,120	0	80
4	Overtime	99,999	99,999	99,999	1,150 **16**	1,210	1,270	0	16
4	Subcontract	99,999	99,999	99,999	1,250 **10**	1,310	1,370	0	10
5	Regular time	99,999	99,999	99,999	99,999	1,000 **70**	1,060	10	80
5	Overtime	99,999	99,999	99,999	99,999	1,150	1,210	16	16
5	Subcontract	99,999	99,999	99,999	99,999	1,250	1,310	10	10
6	Regular time	99,999	99,999	99,999	99,999	99,999	1,000 **44**	21	65
6	Overtime	99,999	99,999	99,999	99,999	99,999	1,150	13	13
6	Subcontract	99,999	99,999	99,999	99,999	99,999	1,250	10	10
	Demand	50	60	90	120	70	44	132	566

One initially puzzling aspect of this solution is that it allocates the initial inventory of 200 units to meet demand in period 4 rather than in period 1. The explanation is that multiple optimal solutions exist and this solution is only one of them. However, all solutions result in the same production and anticipation inventory plans derived as follows:

PRODUCTION PLAN

Period	Regular-time Production	Overtime Production	Subcontracting	Total
1	6,500	—	—	6,500
2	6,500	400	—	6,900
3	6,500	1,300	—	7,800
4	8,000	1,600	1,000	10,600
5	7,000	—	—	7,000
6	4,400	—	—	4,400

ANTICIPATION INVENTORY

Period	Beginning Inventory Plus Total Production Minus Demand	Anticipation (Ending) Inventory
1	200 + 6,500 − 5,000	1,700
2	1,700 + 6,900 − 6,000	2,600
3	2,600 + 7,800 − 9,000	1,400
4	1,400 + 10,600 − 12,000	0
5	0 + 7,000 − 7,000	0
6	0 + 4,400 − 4,000	400

Discussion Questions

1. A particular linear programming maximization problem has the following less-than-or-equal-to constraints: (1) raw materials, (2) labor hours, and (3) storage space. The optimal solution occurs at the intersection of the raw materials and labor hours constraints, so those constraints are binding. Management is considering whether to authorize overtime. What useful information could the linear programming solution provide to management in making this decision? Suppose a warehouse becomes available for rent at bargain rates. What would management need to know in order to decide whether to rent the warehouse? How could the linear programming model be helpful?

2. Linear programming and the transportation method promise optimal solutions. However, wise managers sometimes, after seeing the optimal results such as in Figures D.8 or D.12, might decide to implement different plans. How do you explain such decision making?

Problems

The OM Explorer and POM for Windows software is available to all students using the 10th edition of this textbook. Go to **www .pearsonhighered.com/krajewski** to download these computer packages. Check with your instructor on how best to use it. For linear programming problems with more than two variables, and for transportation problems, use POM for Windows to solve your model formulations. The emphasis in these cases is on modeling problems and interpreting computer output.

1. The Sports Shoe Company is a manufacturer of basketball and football shoes. The manager of marketing must decide the best way to spend advertising resources. Each football team sponsored requires 120 pairs of shoes. Each basketball team requires 32 pairs of shoes. Football coaches receive $300,000 for shoe sponsorship, and basketball coaches receive $1,000,000. The manager's promotional budget is $30,000,000. The company has a limited supply (4 liters, or 4,000 cubic centimeters) of flubber, a rare and costly compound used in promotional athletic shoes. Each pair of basketball shoes requires 3 cc of flubber, and each pair of football shoes requires 1 cc. The manager wants to sponsor as many basketball and football teams as resources will allow.

   a. Create a set of linear equations to describe the objective function and the constraints.

b. Use graphic analysis to find the visual solution.

c. What is the maximum number of each type of team that the company can sponsor?

2. A business student at Nowledge College must complete a total of 65 courses to graduate. The number of business courses must be greater than or equal to 23. The number of nonbusiness courses must be greater than or equal to 20. The average business course requires a textbook costing $60 and 120 hours of study. Nonbusiness courses require a textbook costing $24 and 200 hours of study. The student has $3,000 to spend on books.

a. Create a set of linear equations to describe the objective function and the constraints.

b. Use graphic analysis to find the visual solution.

c. What combination of business and nonbusiness courses minimizes total hours of study?

d. Identify the slack or surplus variables.

3. In Problem 2, suppose that the objective is to minimize the cost of books and that the student's total study time is limited to 12,600 hours.

a. Use graphic analysis to determine the combination of courses that minimizes the total cost of books.

b. Identify the slack or surplus variables.

4. Mile-High Microbrewery makes a light beer and a dark beer. Mile-High has a limited supply of barley, limited bottling capacity, and a limited market for light beer. Profits are $0.20 per bottle of light beer and $0.50 per bottle of dark beer.

a. The following table shows resource availability of products at the Mile-High Microbrewery. Use the graphic method of linear programming to maximize profits. How many bottles of each product should be produced per month?

	PRODUCT		
Resource	Light Beer (x_1)	Dark Beer (x_2)	Resource Availability (per month)
Barley	0.1 gram	0.6 gram	2,000 grams
Bottling	1 bottle	1 bottle	6,000 bottles
Market	1 bottle	—	4,000 bottles

b. Identify any constraints with slack or surplus.

5. The plant manager of a plastic pipe manufacturer has the opportunity to use two different routings for a particular type of plastic pipe. Routing 1 uses extruder A, and routing 2 uses extruder B. Both routings require the same melting process. The following table shows the time requirements and capacities of these processes:

	TIME REQUIREMENTS (HR/100 FT)		
Process	Routing 1	Routing 2	Capacity (hr)
Melting	1	1	45
Extruder A	3	0	90
Extruder B	0	1	160

Each 100 feet of pipe processed on routing 1 uses 5 pounds of raw material, whereas each 100 feet of pipe processed on routing 2 used only 4 pounds. This difference results from differing scrap rates of the extruding machines. Consequently, the profit per 100 feet of pipe processed on routing 1 is $60 and on routing 2 is $80. A total of 200 pounds of raw material is available.

a. Create a set of linear equations to describe the objective function and the constraints.

b. Use graphic analysis to find the visual solution.

c. What is the maximum profit?

6. A manufacturer of textile dyes can use two different processing routings for a particular type of dye. Routing 1 uses drying press A, and routing 2 uses drying press B. Both routings require the same mixing vat to blend chemicals for the dye before drying. The following table shows the time requirements and capacities of these processes:

	TIME REQUIREMENTS (HR/KG)		
Process	Routing 1	Routing 2	Capacity (hr)
Mixing	2	2	54
Dryer A	6	0	120
Dryer B	0	8	180

Each kilogram of dye processed on routing 1 uses 20 liters of chemicals, whereas each kilogram of dye processed on routing 2 uses only 15 liters. The difference results from differing yield rates of the drying presses. Consequently, the profit per kilogram processed on routing 1 is $50 and on routing 2 is $65. A total of 450 liters of input chemicals is available.

a. Write the constraints and objective function to maximize profits.

b. Use the graphic method of linear programming to find the optimal solution.

c. Identify any constraints with slack or surplus.

7. The Trim-Look Company makes several lines of skirts, dresses, and sport coats. Recently, a consultant suggested that the company reevaluate its South Islander line and allocate its resources to products that would maximize contribution to profits and to overhead. Each product requires the same polyester fabric and must pass through the cutting and sewing departments. The following data were collected for the study:

	PROCESSING TIME (HR)		
Product	Cutting	Sewing	Material (yd)
Skirt	1	1	1
Dress	3	4	1
Sport coat	4	6	4

The cutting department has 100 hours of capacity, sewing has 180 hours of capacity, and 60 yards of material are available. Each skirt contributes $5 to profits and overhead; each dress, $17; and each sport coat, $30.

a. Specify the objective function and constraints for this problem.

b. Use a computer package such as POM for Windows to solve the problem.

8. Consider Problem 7 further.

a. How much would you be willing to pay for an extra hour of cutting time? For an extra hour of sewing time? For an extra yard of material? Explain your response to each question.

b. Determine the range of right-hand-side values over which the shadow price would be valid for the cutting constraint and for the material constraint.

9. Polly Astaire makes fine clothing for big and tall men. A few years ago Astaire entered the sportswear market with the Sunset line of shorts, pants, and shirts. Management wants to make the amount of each product that will maximize profits. Each type of clothing is routed through two departments, A and B. The relevant data for each product are as follows:

	PROCESSING TIME (HR)		
Product	Department A	Department B	Material (yd)
Shirts	2	1	2
Shorts	2	3	1
Pants	3	4	4

Department A has 120 hours of capacity, department B has 160 hours of capacity, and 90 yards of material are available. Each shirt contributes $10 to profits and overhead; each pair of shorts, $10; and each pair of pants, $23.

a. Specify the objective function and constraints for this problem.

b. Use a computer package such as POM for Windows to solve the problem.

c. How much should Astaire be willing to pay for an extra hour of department A capacity? How much for an extra hour of department B capacity? For what range of right-hand values are these shadow prices valid?

10. The Butterfield Company makes a variety of knives. Each knife is processed on four machines. The processing times required are as follows. Machine capacities (in hours) are 1,500 for machine 1; 1,400 for machine 2; 1,600 for machine 3; and 1,500 for machine 4.

	PROCESSING TIME (HR)			
Knife	Machine 1	Machine 2	Machine 3	Machine 4
A	0.05	0.10	0.15	0.05
B	0.15	0.10	0.05	0.05
C	0.20	0.05	0.10	0.20
D	0.15	0.10	0.10	0.10
E	0.05	0.10	0.10	0.05

Each product contains a different amount of two basic raw materials. Raw material 1 costs $0.50 per ounce,

and raw material 2 costs $1.50 per ounce. There are 75,000 ounces of raw material 1 and 100,000 ounces of raw material 2 available.

	REQUIREMENTS (OZ/UNIT)		
Knife	Raw Material 1	Raw Material 2	Selling Price ($/unit)
A	4	2	15.00
B	6	8	25.50
C	1	3	14.00
D	2	5	19.50
E	6	10	27.00

a. If the objective is to maximize profit, specify the objective function and constraints for the problem. Assume that labor costs are negligible.

b. Solve the problem with a computer package such as POM for Windows.

11. The Nutmeg Corporation produces three different products, each in a 1-pound can: Almond-Lovers Mix, Walnut-Lovers Mix, and the Thrifty Mix. Three types of nuts are used in Nutmeg's products: almonds, walnuts, and peanuts. Nutmeg currently has 350 pounds of almonds, 150 pounds of walnuts, and 1000 pounds of peanuts. Each of Nutmeg's products must contain a certain percentage of each type of nut, as shown in the following table. The table also shows the revenue per can as well as the cost per pound to purchase nuts.

	PERCENTAGE REQUIREMENTS PER CAN			
	Almonds	Walnuts	Peanuts	Revenue per can
Almond-Lovers Mix	80%	20%	0%	$8.00
Walnut-Lovers Mix	20%	80%	0%	$10.00
Thrifty Mix	10%	10%	80%	$4.50
Cost per pound	$4.50	$6.00	$3.00	

a. Given Nutmeg's current stock of nuts, how many cans of each product should be produced to maximize revenue?

b. Does the solution you developed in part a change if Nutmeg is interested in maximizing contribution margin (defined as revenue per unit – raw material cost)?

c. If 50 additional pounds of walnuts became available, how would your contribution-margin maximizing solution from part (b) change?

12. A problem often of concern to managers in processing industries is blending. The Nutmeg Corporation, from Problem 11, is considering a new product it intends to sell to active and health-concerned adults. This new product will be a 4-ounce package of nuts that conforms to specific heath requirements. First, the 4 ounce package can contain no more than 720 calories. It must deliver at least 20 grams of protein. Finally, the package must provide at least 15 percent of the adult daily requirement (ADR) of calcium and 20 percent of the ADR of iron. Nutmeg would like to use only almonds, walnuts, and peanuts in this new

product. The following table provides nutritional data on each of these ingredients as well as their cost to Nutmeg.

Ingredients	Calories per ounce	Grams of protein per ounce	Percent ADR of calcium per ounce	Percent ADR of iron per ounce	Cost per ounce
Almonds	180	6	8%	6%	$0.28
Walnuts	190	4	2%	6%	$0.38
Peanuts	170	7	0%	4%	$0.12

a. Use linear programming to find the cost minimizing number of ounces of each ingredient that Nutmeg should use in each 4 ounce package. What is the per package cost of raw materials?

b. The marketing department at Nutmeg insists that each package should contain at least $\frac{1}{2}$ ounce of almonds, at least $\frac{1}{2}$ ounce of walnuts, and no more than 1 ounce of peanuts. Does the solution developed for part (a) satisfy these new constraints? If not, use linear programming to find a solution that includes these marketing requirements. What is the new cost of raw materials?

13. A small fabrication firm makes three basic types of components for use by other companies. Each component is processed on three machines. The processing times follow. Total capacities (in hours) are 1,600 for machine 1; 1,400 for machine 2; and 1,500 for machine 3.

Component	PROCESSING TIME (HR)		
	Machine 1	Machine 2	Machine 3
A	0.25	0.10	0.05
B	0.20	0.15	0.10
C	0.10	0.05	0.15

Each component contains a different amount of two basic raw materials. Raw material 1 costs $0.20 per ounce, and raw material 2 costs $0.35 per ounce. At present, 200,000 ounces of raw material 1 and 85,000 ounces of raw material 2 are available.

	REQUIREMENT (OZ/UNIT)		
Component	Raw Material 1	Raw Material 2	Selling Price ($/unit)
A	32	12	40
B	26	16	28
C	19	9	24

a. Assume that the company must make at least 1,200 units of component B, that labor costs are negligible, and that the objective is to maximize profits. Specify the objective function and constraints for the problem.

b. Use a computer package such as POM for Windows to solve the problem.

14. The following is a linear programming model for analyzing the product mix of Maxine's Hat Company, which produces three hat styles:

Maximize: $\$7x_1 + \$8x_2 + \$6x_3 = Z$

Subject to: $2x_1 + 4x_2 + 2x_3 \leq 120$ (machine A time)

$\qquad\qquad 5x_1 + 3x_2 + 2x_3 \leq 400$ (machine B time)

$\qquad\qquad 2x_1 + 2x_2 + 4x_3 \leq 110$ (machine C time)

$\qquad\qquad x_1 \geq 0 x_2 \geq 0$, and $x_3 \geq 0$

The POM for Windows printout in Figure D.15 shows the optimal solution to the problem. Consider each of the following statements independently, and state whether it is true or false. Explain each answer.

a. If the price of hat 3 were increased to $11.50, it would be part of the optimal product mix. *Hint:* Hat 3 is

(a) Results Screen

Maximize		7	8	6			
machine A time		2	4	2	<=	120	.5
machine B time		5	3	2	<=	400	0
machine C time		2	2	4	<=	110	3
Solution->		50	5	0		390	

(b) Ranging Screen

Variable	Value	Reduced Cost	Original Val	Lower Bound	Upper Bound
X1	50	0	7	4.6667	8
X2	5	0	8	7	14
X3	0	7	6	-Infinity	13
Constraint	Shadow Price	Slack/Surplus	Original Val	Lower Bound	Upper Bound
machine A time	.5	0	120	110	220
machine B time	0	135	400	265	Infinity
machine C time	3	0	110	60	120

▲ **FIGURE D.15**
Solution Screens for Maxine's Hat Company

represented by x_3 and its optimal value currently is 0, which means that it is not to be produced (and not part of the optimal product mix).

b. The capacity of machine B can be reduced to 280 hours without affecting profits.

c. If machine C had a capacity of 115 hours, the production output would remain unchanged.

15. The Washington Chemical Company produces chemicals and solvents for the glue industry. The production process is divided into several "focused factories," each producing a specific set of products. The time has come to prepare the production plan for one of the focused factories. This particular factory produces five products, which must pass through both the reactor and the separator. Each product also requires a certain combination of raw materials. Production data are shown in Table D.2.

The Washington Chemical Company has a long-term contract with a major glue manufacturer that requires annual production of 3,000 pounds of both products 3 and 4. More of these products could be produced because demand currently exceeds production capacity.

a. Determine the annual production quantity of each product that maximizes contribution to profits. Assume the company can sell all it can produce.

b. Specify the lot size for each product.

16. The Warwick Manufacturing Company produces shovels for industrial and home use. Sales of the shovels are seasonal, and Warwick's customers refuse to stockpile them during slack periods. In other words, the customers want to minimize inventory, insist on shipments according to their schedules, and will not accept backorders.

Warwick employs manual, unskilled laborers who require only basic training. Producing 1,000 shovels costs $3,500 on regular time and $3,700 on overtime. These amounts include materials, which account for more than 85 percent of the cost. Overtime is limited to production of 15,000 shovels per quarter. In addition, subcontractors can be hired at $4,200 per thousand shovels, but Warwick's labor contract restricts this type of production to 5,000 shovels per quarter.

The current level of inventory is 30,000 shovels, and management wants to end the year at that level. Holding

1,000 shovels in inventory costs $280 per quarter. The latest annual demand forecast is as follows:

Quarter	Demand
1	70,000
2	150,000
3	320,000
4	100,000
Totals	640,000

Build a linear programming model to determine the *best* regular-time capacity plan. Assume the following:

- The firm has 30 workers now, and management wants to have the same number in quarter 4.
- Each worker can produce 4,000 shovels per quarter.
- Hiring a worker costs $1,000, and laying off a worker costs $600.

17. The management of Warwick Manufacturing Company is willing to give price breaks to its customers as an incentive to purchase shovels in advance of the traditional seasons. Warwick's sales and marketing staff estimates that the demand for shovels resulting from the price breaks would be as follows:

Quarter	Demand	Original Demand
1	120,000	70,000
2	180,000	150,000
3	180,000	320,000
4	160,000	100,000
Totals	640,000	640,000

Calculate the optimal production plan (including the workforce staffing plan) under the new demand schedule. Compare it to the optimal production plan under the original demand schedule. Evaluate the potential effects of demand management.

TABLE D.2 | **PRODUCTION DATA FOR WASHINGTON CHEMICAL**

Resource	PRODUCT					Total Resources Available
	1	2	3	4	5	
Reactor (hr/lb)	0.05	0.10	0.80	0.57	0.15	7,500 hr*
Separator (hr/lb)	0.20	0.02	0.20	0.09	0.30	7,500 hr*
Raw material 1 (lb)	0.20	0.50	0.10	0.40	0.18	10,000 lb
Raw material 2 (lb)	—	0.70	—	0.50	—	6,000 lb
Raw material 3 (lb)	0.10	0.20	0.40	—	—	7,000 lb
Profit contribution ($/lb)	4.00	7.00	3.50	4.00	5.70	

*The total time available has been adjusted to account for setups. The five products have a prescribed sequence owing to the cost of changeovers between products. The company has a 35-day cycle (or 10 changeovers per year per product). Consequently, the time for these changeovers has been deducted from the total time available for these machines.

18. The Bull Grin Company produces a feed supplement for animal foods produced by a number of other companies. Producing 1,000 pounds of supplement costs $810 on regular time and $900 on overtime. These amounts include materials, which account for more than 80 percent of the cost. The plant can produce 400,000 pounds of supplement per quarter using regular time but overtime is limited to the production of 40,000 pounds per quarter. The current level of inventory is 40,000 pounds and management wants to end the year at that level. Holding 1,000 pounds of feed supplement in inventory costs $110 per quarter. The latest annual demand forecast follows:

Quarter	Demand (in Pounds)
1	100,000
2	410,000
3	770,000
4	440,000

a. Formulate this production planning problem as a linear program after defining all decision variables.

b. Solve your formulation using a computer package such as POM for Windows.

c. Assume that subcontractors can be hired at $1,100 per thousand pounds to produce as much supplement as Bull Grin requires. Does this change the cost minimizing solution found in part (b)?

d. If Bull Grin realizes that the current level of inventory is actually 0 pounds, are the resources assumed in part (c) adequate to satisfy all demand and still end the year with 40,000 pounds in ending inventory? If so, how much will the cost of Bull Grin's production plan increase?

19. Supertronics, Inc., would like to know how the firm's profitability is altered by product mix. Currently, product mix is determined by giving priority to the product with the highest per unit contribution margin (defined as the difference between price and material cost). Details on the Supertronics product line, including processing time at each workstation, follow:

	PRODUCT			
	Alpha	**Beta**	**Delta**	**Gamma**
Price	$350.00	$320.00	$400.00	$500.00
Material Cost	$50.00	$40.00	$125.00	$150.00
Weekly Demand in Units	100	60	50	80
Processing Time at Machine 1 in Minutes	20	0	40	10
Processing Time at Machine 2 in Minutes	25	20	0	50
Processing Time at Machine 3 in Minutes	0	20	60	30

a. Assume that Supertronics has 5,500 minutes of capacity available at each workstation each week. Develop a linear program to define the production mix that maximizes contribution margin.

b. Solve your formulation using a computer package such as POM for Windows.

c. Given your solution for part (b), which machine is the bottleneck?

d. How would your formulation and solution in part (b) change if 50 units of each product were already committed to customers and thereby had to be produced?

20. JPMorgan Chase has a scheduling problem. Operators work 8-hour shifts and can begin work at midnight, 4 A.M., 8 A.M., noon, 4 P.M., or 8 P.M. Operators are needed to satisfy the following demand pattern. Formulate a linear programming model to cover the demand requirements with the minimum number of operators.

Time Period	Operators Needed
Midnight to 4 A.M.	4
4 A.M. to 8 A.M.	6
8 A.M. to noon	90
Noon to 4 P.M.	85
4 P.M. to 8 P.M.	55
8 P.M. to 12 midnight	20

21. Revisit Problem 18 on the Bull Grin Company. Some cost and demand parameters have changed. Producing 1,000 pounds of supplement now costs $830 on regular time and $910 on overtime. No additional cost is incurred for unused regular time, overtime, or subcontractor capacity. Overtime is limited to production of a total of 20,000 pounds per quarter. In addition, subcontractors can be hired at $1,000 per thousand pounds, but only 30,000 pounds per quarter can be produced this way.

The current level of inventory is 40,000 pounds, and management wants to end the year at that level. Holding 1,000 pounds of feed supplement in inventory per quarter costs $100. The latest annual forecast is shown in Table D.3.

Use the transportation method of production planning in POM for Windows to find the optimal production plan and calculate its cost, or use the spreadsheet approach to find a good production plan and calculate its cost.

22. The Cut Rite Company is a major producer of industrial lawn mowers. The cost to Cut Rite for hiring a semiskilled worker for its assembly plant is $3,000, and the cost for laying off one is $2,000. The plant averages an output of 36,000 mowers per quarter with its current workforce of 720 employees. Regular-time capacity is directly proportional to the number of employees. Overtime is limited to a maximum of 3,000 mowers per quarter, and subcontracting is limited to 1,000 mowers per quarter. The costs to produce one mower are $2,430 on regular time (including materials), $2,700 on overtime, and $3,300 via subcontracting. Unused regular-time capacity costs $270 per mower. No additional cost is incurred for unused overtime or subcontractor capacity. The current level of

TABLE D.3 | **FORECASTS AND CAPACITIES**

	PERIOD				
	Quarter 1	Quarter 2	Quarter 3	Quarter 4	Total
Demand (pounds)	130,000	400,000	800,000	470,000	1,800,000
Capacities (pounds)					
Regular time	390,000	400,000	460,000	380,000	1,630,000
Overtime	20,000	20,000	20,000	20,000	80,000
Subcontract	30,000	30,000	30,000	30,000	30,000

inventory is 4,000 mowers, and management wants to end the year at that level. Customers do not tolerate backorders, and holding a mower in inventory per quarter costs $300. The demand for mowers this coming year is as follows:

Quarter	1	2	3	4
Demand	10,000	41,000	77,000	44,000

Two workforce plans have been proposed, and management is uncertain as to which one to use. The following table shows the number of employees per quarter under each plan:

Quarter	1	2	3	4
Plan 1	720	780	920	720
Plan 2	860	860	860	860

a. Which plan would you recommend to management? Explain, supporting your recommendation with an analysis using the transportation method of production planning.

b. If management used creative pricing to get customers to buy mowers in nontraditional time periods, the following demand schedule would result:

Quarter	1	2	3	4
Demand	20,000	54,000	54,000	44,000

Which workforce plan would you recommend now?

23. The Holloway Calendar Company produces a variety of printed calendars for both commercial and private use.

The demand for calendars is highly seasonal, peaking in the third quarter. Current inventory is 165,000 calendars, and ending inventory should be 200,000 calendars.

Ann Ritter, Holloway's manufacturing manager, wants to determine the best production plan for the demand requirements and capacity plan shown in the following table. (Here, demand and capacities are expressed as thousands of calendars rather than as employee-period equivalents.) Ritter knows that the regular-time cost is $0.50 per unit, overtime cost is $0.75 per unit, subcontracting cost is $0.90 per unit, and inventory holding cost is $0.10 per calendar per quarter. Unused regular-time capacity is not paid.

	QUARTER				
	1	2	3	4	Total
Demand	250	515	1,200	325	2,290
Capacities					
Regular time	300	300	600	300	1,500
Overtime	75	75	150	75	375
Subcontracting	150	150	150	150	600

a. Recommend a production plan to Ritter, using the Transportation Method (Production Planning) module of POM for Windows. (Do not allow any stockouts or backorders to occur.)

b. Interpret and explain your recommendation.

c. Calculate the total cost of your recommended production plan.

CASE   R.U. Reddie for Location

The R.U. Reddie Corporation, located in Chicago, manufactures clothing specially designed for stuffed cartoon animals such as Snoopy and Wile E. Coyote. Among the popular products are a wedding tuxedo for Snoopy and a flak jacket for Wile E. Coyote. The latter is capable of stopping an Acme rocket at close range . . . sometimes.

For many sales, the company relies upon the help of spoiled children who refuse to leave the toy store until their parents purchase a wardrobe for their stuffed toys. Rhonda Ulysses Reddie, owner of the company, is concerned over the market projections that indicate demand for the product is substantially greater than current plant capacity. The "most likely" projections indicate that the company will be short by 400,000 units next year, and thereafter 700,000 units annually. As such, Rhonda is considering opening a new plant to produce additional units.

Background

The R.U. Reddie Corporation currently has three plants located in Boston, Cleveland, and Chicago. The company's first plant was the Chicago location, but as sales grew in the Midwest and Northeast, the Cleveland and Boston plants were built in short order. As the demand for wardrobes for stuffed animals moved west, warehouse centers were opened in St. Louis and Denver. The capacities of the three plants were increased to accommodate the demand. Each plant has its own warehouse to satisfy demands in its own area. Extra capacity was used to ship the product to St. Louis or Denver.

The new long-term forecasts provided by the sales department contain both good news and bad news. The added revenues will certainly help Rhonda's profitability, but the company would have to buy another plant to realize the added profits. Space is not available at the existing plants, and the benefits of the new technology for manufacturing stuffed animal wardrobes are tantalizing. These factors motivate the search for the best location for a new plant. Rhonda identifies Denver and St. Louis as possible locations for the new plant.

Rhonda's Concerns

A plant addition is a big decision. Rhonda has started to think about the accuracy of the data she was able to obtain. She has market, financial, and operations concerns.

Market

The projected demands for years 2 through 10 show an annual increase of 700,000 units to a total of 2,000,000 units for each year. Rhonda has two concerns here. First, what if the projections for each city are off plus or minus 10 percent equally across the board? That is, total annual demands could be as low as 1,800,000 or as high as 2,200,000, with each city being affected the same as the others. Second, the marketing manager has expressed a concern over a possible market shift from the Midwest and Northeast to the West. Under this scenario, additional demand would reach 50,000 units in St. Louis and 150,000 units in Denver, with the other cities staying at the "most likely" demand projections.

Financial

Rhonda has realized that the net present value (NPV) of each alternative is an important input to the final decision. However, the accuracy of the estimates for the various costs is critical to determining good estimates of cash flows. She wonders whether her decision will change if the COGS (variable production plus transportation costs) for each option are off by ±10 percent. That is, what if the variable production costs and transportation costs of St. Louis are 10 percent higher than estimated while the variable production costs and transportation costs for Denver are 10 percent lower than estimated? Or vice versa? Furthermore, what if the estimate for fixed costs are off by ±10 percent? For example, suppose St. Louis is 10 percent higher while Denver is 10 percent lower, or vice versa. Would the recommendation change under any of these situations?

Operations

The ultimate location of the new plant will determine the distribution assignments and the level of utilization of each plant in the network. Cutting back production in any of the plants will change the distribution assignments of all plants. Because the "most likely" demand projections with a new plant will mean excess capacity in the system, the capacity of the Cleveland plant could be cut in year 2 and beyond. Suppose Cleveland cuts back production by 50(000) units a year from year 2 and beyond. Will this action affect the choice between Denver and St. Louis? What is the impact on the distribution assignments of the plant? Some nonquantifiable concerns also need consideration. First, the availability of a good workforce is much better in Denver than St. Louis because of the recent shutdown of a Beanie Baby factory. The labor market is much tighter in St. Louis and the prognosis is for continued short supply in the foreseeable future. Second, the Denver metropolitan area just instituted strict environmental regulations. Rhonda's new plant would adhere to existing laws, but the area is highly environmentally conscious and more regulations may be coming in the future. It is costly to modify a plant once operations begin. Finally, Denver has a number of good suppliers with the capability to assist in production design (new wardrobe fashions). St. Louis also has suppliers, but they cannot help with product development. Proximity to suppliers with product development capability is a "plus" for this industry.

Data

The following data have been gathered for Rhonda:
1. The per-unit shipping cost based on the average tonmile rates for the most efficient carriers is $0.0005 per mile. The average revenue per outfit is $8.00.
2. The company currently has the following capacity (in thousands) constraints:

	Capacity
Boston	400
Cleveland	400
Chicago	500

3. Data concerning the various locations are found in Table D.4.

4. New plant information:

Alternative	Building and Equipment[2]	Annual Fixed Costs (SGA)[1,3]	Variable Production Costs/Unit	Land[1]
Denver	$12,100	$550	$3.15	$1,200
St. Louis	10,800	750	3.05	800

[1] Figures are given in thousands.

[2] Net book value of plant and equipment with remaining depreciable life of 10 years.

[3] Annual fixed costs do not include depreciation on plant and equipment.

5. The road mileage between the cities is as follows:

	Boston	Cleveland	Chicago	St. Louis	Denver
Boston	—	650	1,000	1,200	2,000
Cleveland		—	350	600	1,400
Chicago			—	300	1,000
St. Louis				—	850
Denver					—

6. Basic assumptions you should follow:

- Terminal value (in 10 years) of the new investment is 50 percent of plant, equipment, and land cost.
- The tax rate is 40 percent.
- Straight-line depreciation is used for all assets over a 10-year life.
- R.U. Reddie is a 100 percent equity company with all equity financing and a weighted average cost of capital (WACC) of 11 percent.
- Capacity of the new plant production for the first year will be 500 (000) units.
- Capacity of the new plant production thereafter will be 900 (000) units.
- Cost of goods sold (COGS) equals variable costs of production plus total transportation costs.
- Costs to ship from a plant to its own warehouse are zero; however, production costs are applicable.

TABLE D.4 | **LOCATION DATA FOR R.U. REDDIE**

City	Most Likely Demand First Year[1]	Most Likely Demand After Years 2–10[1]	Current Costs, Building and Equipment[1,2]	Annual Fixed Costs (SGA)[1,3]	Variable Production Costs/Unit	Land[1]
Boston	80	140	$9,500	$600	$3.80	$500
Cleveland	200	260	7,700	300	3.00	400
Chicago	370	430	8,600	400	3.25	600
St. Louis	440	500				
Denver	610	670				

[1] Figures are given in thousands.

[2] Net book value of plant and equipment with remaining depreciable life of 10 years.

[3] Annual fixed costs do not include depreciation on plant and equipment.

7. R.U. Reddie operations and logistics managers determined the shipping plan and cost of goods sold for the option of *not* building a new plant and simply using the existing capacities to their fullest extent (status quo solution).

Year 1	COGS = $4,692,000
Boston to Boston	80
Boston to St. Louis	320
Cleveland to Chicago	80
Cleveland to Cleveland	200
Cleveland to St. Louis	120
Chicago to Chicago	290
Chicago to Denver	210
Years 2–10	**COGS = $4,554,000**
Boston to Boston	140
Boston to St. Louis	260
Cleveland to Cleveland	260
Cleveland to St. Louis	140
Chicago to Chicago	430
Chicago to St. Louis	70

QUESTIONS

Your team has been asked to determine whether R.U. Reddie should build a new plant and, if so, where it should be located. Your report should consist of six parts.

1. Write a memo from your team to R.U. Reddie indicating your recommendation and a brief overview of the supporting evidence.

2. Model the location decision as a linear programming problem. The objective function should be to minimize the total variable costs (production plus transportation costs). The variables should be the quantity to ship from each of the plants (including one of the alternative new plants) to each of the warehouses. You should have 20 variables (four plants and five warehouses). You should also have nine constraints (four plant capacity constraints and five warehouse demand constraints). See the addendum for hints. You will need two models—one for Denver and one for St. Louis.

3. Use a software package for linear programming, such as with POM for Windows, to solve for the optimal distribution plan for each alternative (i.e., Denver and St. Louis). A linear programming model is preferable for this case because of the sensitivity analysis information that it provides.

4. Compute the NPV of each alternative. Use the results from the linear models for the COGS for each alternative. (*Hint:* Your analysis will be simplified if you think in terms of incremental cash flows.) Create an easy-to-read spreadsheet for each alternative.

5. Do a sensitivity analysis of the quantitative factors mentioned in the case: Forecast errors (across the board and market shift), errors in COGS estimate, and errors in fixed cost estimates. Do each factor independent of the others and use the "most likely" projections as the base case. Summarize the results in one table.

6. Use the analysis in question 5 to identify the key quantitative variables that determine the superiority of one alternative over another. Rationalize your final recommendation in light of all the considerations R.U. Reddie must make.

Addendum

Here are some hints for your model.

1. The capacity constraint for Boston would look like this:

$$1B - B + 1B - CL + 1B - CH + 1B - D + 1B - SL \leq 400$$

The variable B-CH means Boston to Chicago in this example. You will need a total of four capacity constraints, one for each of the existing sites and one for the alternative site you are evaluating. Remember that the new site will have a capacity limit of 500 in the first year, and 900 in the second year.

2. The demand constraint for Boston would look like this:

$$1B - B + 1CL - B + 1CH - B + 1D - B = 140$$

The Denver location alternative is depicted in this example (D-B represents the number of units produced in Denver and shipped to Boston). You will need a total of five demand constraints, one for each warehouse location. Notice that the demand constraints have equal signs to indicate that exactly that quantity must be received at each warehouse.

3. Define your variables to be "thousands of units shipped," and omit the three zeros after each demand and capacity value. Then remember to multiply your final decisions and total variable costs by a thousand after you get your solution from the model.

4. Because the capacity and demand changes from year 1 to year 2, you need to run your model twice for each location to get the necessary data. You will also run the model and spread-sheets multiple times for question 5 of the report.

Selected References

Anderson, D.R., D.J. Sweeney, T.A.Williams, J. D. Camm and K. Martin. *An Introduction to Management Science: A Quantitative Approach to Decision Making*, 13th ed. Cincinnati: South-Western, 2011.

Hillier, Fredrick S., and Mark S. Hillier. *Introduction to Management Science: A Modeling and Case Studies Approach with Spreadsheets*, 4th ed. Burr Ridge, IL: McGraw Hill, 2011.

Ragsdale, Cliff. *Spreadsheet Modeling & Decision Analysis: A Practical Introduction to Management Science*, 6th ed. Cincinnati, OH: South-Western, 2011.

Render, Barry, Ralph M. Stair, and Michael E. Hanna. *Quantitative Analysis for Management*, 11th ed. Upper Saddle River, NJ: Prentice Hall, 2011.

Winston, Wayne L., and S. Christian Albright. *Practical Management Science*, 4th ed. Belmont, CA: Duxbury Press, 2012.

NORMAL DISTRIBUTION

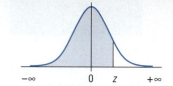

	.00	.01	.02	.03	.04	.05	.06	.07	.08	.09
.0	.5000	.5040	.5080	.5120	.5160	.5199	.5239	.5279	.5319	.5359
.1	.5398	.5438	.5478	.5517	.5557	.5596	.5636	.5675	.5714	.5753
.2	.5793	.5832	.5871	.5910	.5948	.5987	.6026	.6064	.6103	.6141
.3	.6179	.6217	.6255	.6293	.6331	.6368	.6406	.6443	.6480	.6517
.4	.6554	.6591	.6628	.6664	.6700	.6736	.6772	.6808	.6844	.6879
.5	.6915	.6950	.6985	.7019	.7054	.7088	.7123	.7157	.7190	.7224
.6	.7257	.7291	.7324	.7357	.7389	.7422	.7454	.7486	.7517	.7549
.7	.7580	.7611	.7642	.7673	.7704	.7734	.7764	.7794	.7823	.7852
.8	.7881	.7910	.7939	.7967	.7995	.8023	.8051	.8078	.8106	.8133
.9	.8159	.8186	.8212	.8238	.8264	.8289	.8315	.8340	.8365	.8389
1.0	.8413	.8438	.8461	.8485	.8508	.8531	.8554	.8577	.8599	.8621
1.1	.8643	.8665	.8686	.8708	.8729	.8749	.8770	.8790	.8810	.8830
1.2	.8849	.8869	.8888	.8907	.8925	.8944	.8962	.8980	.8997	.9015
1.3	.9032	.9049	.9066	.9082	.9099	.9115	.9131	.9147	.9162	.9177
1.4	.9192	.9207	.9222	.9236	.9251	.9265	.9279	.9292	.9306	.9319
1.5	.9332	.9345	.9357	.9370	.9382	.9394	.9406	.9418	.9429	.9441
1.6	.9452	.9463	.9474	.9484	.9495	.9505	.9515	.9525	.9535	.9545
1.7	.9554	.9564	.9573	.9582	.9591	.9599	.9608	.9616	.9625	.9633
1.8	.9641	.9649	.9656	.9664	.9671	.9678	.9686	.9693	.9699	.9706
1.9	.9713	.9719	.9726	.9732	.9738	.9744	.9750	.9756	.9761	.9767
2.0	.9772	.9778	.9783	.9788	.9793	.9798	.9803	.9808	.9812	.9817
2.1	.9821	.9826	.9830	.9834	.9838	.9842	.9846	.9850	.9854	.9857
2.2	.9861	.9864	.9868	.9871	.9875	.9878	.9881	.9884	.9887	.9890
2.3	.9893	.9896	.9898	.9901	.9904	.9906	.9909	.9911	.9913	.9916
2.4	.9918	.9920	.9922	.9925	.9927	.9929	.9931	.9932	.9934	.9936
2.5	.9938	.9940	.9941	.9943	.9945	.9946	.9948	.9949	.9951	.9952
2.6	.9953	.9955	.9956	.9957	.9959	.9960	.9961	.9962	.9963	.9964
2.7	.9965	.9966	.9967	.9968	.9969	.9970	.9971	.9972	.9973	.9974
2.8	.9974	.9975	.9976	.9977	.9977	.9978	.9979	.9979	.9980	.9981
2.9	.9981	.9982	.9982	.9983	.9984	.9984	.9985	.9985	.9986	.9986
3.0	.9987	.9987	.9987	.9988	.9988	.9989	.9989	.9989	.9990	.9990
3.1	.9990	.9991	.9991	.9991	.9992	.9992	.9992	.9992	.9993	.9993
3.2	.9993	.9993	.9994	.9994	.9994	.9994	.9994	.9995	.9995	.9995
3.3	.9995	.9995	.9995	.9996	.9996	.9996	.9996	.9996	.9996	.9997
3.4	.9997	.9997	.9997	.9997	.9997	.9997	.9997	.9997	.9997	.9998

TABLE OF RANDOM NUMBERS

71509	68310	48213	99928	64650	13229	36921	58732	13459	93487
21949	30920	23287	89514	58502	46185	00368	82613	02668	37444
50639	54968	11409	36148	82090	87298	41396	71111	00076	60029
47837	76716	09653	54466	87987	82362	17933	52793	17641	19502
31735	36901	92295	19293	57582	86043	69502	12601	00535	82697
04174	32342	66532	07875	54445	08795	63563	42295	74646	73120
96980	68728	21154	56181	71843	66134	52396	89723	96435	17871
21823	04027	76402	04655	87276	32593	17097	06913	05136	05115
25922	07122	31485	52166	07645	85122	20945	06369	70254	22806
32530	98882	19105	01769	20276	59401	60426	03316	41438	22012
00159	08461	51810	14650	45119	97920	08063	70819	01832	53295
66574	21384	75357	55888	83429	96916	73977	87883	13249	28870
00995	28829	15048	49573	65277	61493	44031	88719	73057	66010
55114	79226	27929	23392	06432	50200	39054	15528	53483	33972
10614	25190	52647	62580	51183	31338	60008	66595	64357	14985
31359	77469	58126	59192	23371	25190	37841	44386	92420	42965
09736	51873	94595	61367	82091	63835	86858	10677	58209	59820
24709	23224	45788	21426	63353	29874	51058	29958	61220	61199
79957	67598	74102	49824	39305	15069	56327	26905	34453	53964
66616	22137	72805	64420	58711	68435	60301	28620	91919	96080
01413	27281	19397	36231	05010	42003	99865	20924	76151	54089
88238	80731	20777	45725	41480	48277	45704	96457	13918	52375
57457	87883	64273	26236	61095	01309	48632	00431	63730	18917
21614	06412	71007	20255	39890	75336	89451	88091	61011	38072
26466	03735	39891	26361	86816	48193	33492	70484	77322	01016
97314	03944	04509	46143	88908	55261	73433	62538	63187	57352
91207	33555	75942	41668	64650	38741	86189	38197	99112	59694
46791	78974	01999	78891	16177	95746	78076	75001	51309	18791
34161	32258	05345	79267	75607	29916	37005	09213	10991	50451
02376	40372	45077	73705	56076	01853	83512	81567	55951	27156
33994	56809	58377	45976	01581	78389	18268	90057	93382	28494
92588	92024	15048	87841	38008	80689	73098	39201	10907	88092
73767	61534	66197	47147	22994	38197	60844	86962	27595	49907
51517	39870	94094	77092	94595	37904	27553	02229	44993	10468
33910	05156	60844	89012	21154	68937	96477	05867	95809	72827
09444	93069	61764	99301	55826	78849	26131	28201	91417	98172
96896	43769	72890	78682	78243	24061	55449	53587	77574	51580
97523	54633	99656	08503	52563	12099	52479	74374	79581	57143
42568	30794	32613	21802	73809	60237	70087	36650	54487	43718
45453	33136	90246	61953	17724	42421	87611	95369	42108	95369
52814	26445	73516	24897	90622	35018	70087	60112	09025	05324
87318	33345	14546	15445	81588	75461	12246	47858	08983	18205
08063	83575	26294	93027	09988	04487	88364	31087	22200	91019
53400	82078	52103	25650	75315	18916	06809	88217	12245	33053
90789	60614	20862	34475	11744	24437	55198	55219	74730	59820
73684	25859	86858	48946	30941	79017	53776	72534	83638	44680
82007	12183	89326	53713	77782	50368	01748	39033	47042	65758
80208	30920	97774	41417	79038	60531	32990	57770	53441	58732
62434	96122	63019	58439	89702	38657	60049	88761	22785	66093
04718	83199	65863	58857	49886	70275	27511	99426	53985	84077

Glossary

ABC analysis The process of dividing SKUs into three classes, according to their dollar usage, so that managers can focus on items that have the highest dollar value.

acceptable quality level (AQL) A statement of the proportion of defective items (outside of specifications) that the buyer will accept in a shipment.

acceptance sampling The application of statistical techniques to determine whether a quantity of material should be accepted or rejected based on the inspection or test of a sample.

action notice A computer-generated memo alerting planners about releasing new orders and adjusting the due dates of scheduled receipts.

activity The smallest unit of work effort consuming both time and resources that the project manager can schedule and control.

activity-on-node (AON) network An approach used to create a network diagram, in which nodes represent activities and arcs represent the precedence relationships between them.

activity slack The maximum length of time that an activity can be delayed without delaying the entire project, calculated as S = LS – ES or S = LF – EF.

additive seasonal method A method in which seasonal forecasts are generated by adding a constant to the estimate of average demand per season.

advanced planning and scheduling (APS) systems Computer software systems that seek to optimize resources across the supply chain and align daily operations with strategic goals.

aggregate plan See sales and operations plan.

aggregation The act of clustering several similar services or products so that forecasts and plans can be made for whole families.

allowance time The time added to the normal time to adjust for certain factors.

annual plan or financial plan A plan for financial assessment used by a nonprofit service organization.

annuity A series of payments on a fixed amount for a specified number of years.

anticipation inventory Inventory used to absorb uneven rates of demand or supply.

appraisal costs Costs incurred when the firm assess the performance level of its processes.

assemble-to-order strategy A strategy for producing a wide variety of products from relatively few subassemblies and components after the customer orders are received.

assignable causes of variation Any variation-causing factors that can be identified and eliminated.

attributes Service or product characteristics that can be quickly counted for acceptable performance.

auction A marketplace where firms place competitive bids to buy something.

automation A system, process, or piece of equipment that is self-acting and self-regulating.

available-to-promise (ATP) inventory The quantity of end items that marketing can promise to deliver on specified dates.

average aggregate inventory value The total average value of all items held in inventory for a firm.

average outgoing quality (AOQ) The expressed proportion of defects that the plan will allow to pass.

average outgoing quality limit (AOQL) The maximum value of the average outgoing quality over all possible values of the proportion defective.

back office A process with low customer contact and little service customization.

backlog An accumulation of customer orders that a manufacturer has promised for delivery at some future date.

backorder A customer order that cannot be filled when promised or demanded but is filled later.

backorder and stockout Additional costs to expedite past-due orders, the costs of lost sales, and the potential cost of losing a customer to a competitor (sometimes called loss of goodwill).

backward integration A firm's movement upstream toward the sources of raw materials, parts, and services through acquisitions.

balance delay The amount by which efficiency falls short of 100 percent.

Baldrige Performance Excellence Program A program that promotes, recognizes, and publicizes quality strategies and achievements.

bar chart A series of bars representing the frequency of occurrence of data characteristics measured on a yes-or-no basis.

base case The act of doing nothing and losing orders from any demand that exceeds current capacity, or incur costs because capacity is too large.

base-stock system An inventory control system that issues a replenishment order, Q, each time a withdrawal is made, for the same amount of the withdrawal.

batch process A process that differs from the job process with respect to volume, variety, and quantity.

benchmarking A systematic procedure that measures a firm's processes, services, and products against those of industry leaders.

bill of materials (BOM) A record of all the components of an item, the parent–component relationships, and the usage quantities derived from engineering and process designs.

bill of resources (BOR) A record of a service firm's parent-component relationships and all of the materials, equipment time, staff, and other resources associated with them, including usage quantities.

binding constraint A constraint that helps form the optimal corner point; it limits the ability to improve the objective function.

Black Belt An employee who reached the highest level of training in a Six Sigma program and spends all of his or her time teaching and leading teams involved in Six Sigma projects.

block plan A plan that allocates space and indicates placement of each operation.

bottleneck A capacity constraint resource (CCR) whose available capacity limits the organization's ability to meet the product volume, product mix, or demand fluctuation required by the marketplace.

brainstorming Letting a group of people, knowledgeable about the process, propose ideas for change by saying whatever comes to mind.

break-even analysis The use of the break-even quantity; it can be used to compare processes by finding the volume at which two different processes have equal total costs.

break-even quantity The volume at which total revenues equal total costs.

bullwhip effect The phenomenon in supply chains whereby ordering patterns experience increasing variance as you proceed upstream in the chain.

business plan A projected statement of income, costs, and profits.

c-chart A chart used for controlling the number of defects when more than one defect can be present in a service or product.

capacity The maximum rate of output of a process or a system.

capacity cushion The amount of reserve capacity a process uses to handle sudden increases in demand or temporary losses of production capacity; it measures the amount by which the average utilization (in terms of total capacity) falls below 100 percent.

capacity gap Positive or negative difference between projected demand and current capacity.

capacity requirement What a process's capacity should be for some future time period to meet the demand of customers (external or internal), given the firm's desired capacity cushion.

capacity requirements planning (CRP) A technique used for projecting time-phased capacity requirements for workstations; its purpose is to match the material requirements plan with the capacity of key processes.

capital intensity The mix of equipment and human skills in a process.

carbon footprint The total amount of greenhouse gasses produced to support operations, usually expressed in equivalent tons of carbon dioxide (CO_2).

cash flow The difference between the flows of funds into and out of an organization over a period of time, including revenues, costs, and changes in assets and liabilities.

catalog hubs A system whereby suppliers post their catalog of items on the Internet and buyers select what they need and purchase them electronically.

causal methods A quantitative forecasting method that uses historical data on independent variables, such as promotional campaigns, economic conditions, and competitors' actions, to predict demand.

cause-and-effect diagram A diagram that relates a key performance problem to its potential causes.

center of gravity A good starting point to evaluate locations in the target area using the load–distance model.

centralized placement Keeping all the inventory of a product at a single location such as a firm's manufacturing plant or a warehouse and shipping directly to each of its customers.

certainty The word that is used to describe that a fact is known without doubt.

channel One or more facilities required to perform a given service.

channel assembly The process of using members of the distribution channel as if they were assembly stations in the factory.

chase strategy A strategy that involves hiring and laying off employees to match the demand forecast.

checklist A form used to record the frequency of occurrence of certain process failures.

close out An activity that includes writing final reports, completing remaining deliverables, and compiling the team's recommendations for improving the project process.

closed-loop supply chain A supply chain that integrates forward logistics with reverse logistics, thereby focusing on the complete chain of operations from the birth to the death of a product.

closeness matrix A table that gives a measure of the relative importance of each pair of operations being located close together.

collaborative planning, forecasting, and replenishment (CPFR) A nine-step process for supply chain integration that allows a supplier and its customers to collaborate on making the forecast by using the Internet.

combination forecasts Forecasts that are produced by averaging independent forecasts based on different methods, different sources, or different data.

common causes of variation The purely random, unidentifiable sources of variation that are unavoidable with the current process.

competitive capabilities The cost, quality, time, and flexibility dimensions that a process or supply chain actually possesses and is able to deliver.

competitive orientation A supplier relation that views negotiations between buyer and seller as a zero-sum game: Whatever one side loses, the other side gains, and short-term advantages are prized over long-term commitments.

competitive priorities The critical dimensions that a process or supply chain must possess to satisfy its internal or external customers, both now and in the future.

complementary products Services or products that have similar resource requirements but different demand cycles.

component An item that goes through one or more operations to be transformed into or become part of one or more parents.

compounding interest The process by which interest on an investment accumulates and then earns interest itself for the remainder of the investment period.

concurrent engineering A concept that brings product engineers, process engineers, marketers, buyers, information specialists, quality specialists, and suppliers together to design a product and the processes that will meet customer expectations.

consistent quality Producing services or products that meet design specifications on a consistent basis.

constraint Any factor that limits the performance of a system and restricts its output.

constraints In linear programming, the limitations that restrict the permissible choices for the decision variables.

consumer's risk (β) The probability of accepting a lot with LTPD quality (a type II error).

continuous flow process The extreme end of high-volume standardized production and rigid line flows, with production not starting and stopping for long time intervals.

continuous improvement The philosophy of continually seeking ways to improve processes based on a Japanese concept called *kaizen*.

continuous review (Q) system A system designed to track the remaining inventory of a SKU each time a withdrawal is made to determine whether it is time to reorder.

control chart A time-ordered diagram that is used to determine whether observed variations are abnormal.

cooperative orientation A supplier relation in which the buyer and seller are partners, each helping the other as much as possible.

core competencies The unique resources and strengths that an organization's management considers when formulating strategy.

core process A set of activities that delivers value to external customers.

corner point A point that lies at the intersection of two (or possibly more) constraint lines on the boundary of the feasible region.

crash cost (CC) The activity cost associated with the crash time.

crash time (CT) The shortest possible time to complete an activity.

critical mass A situation whereby several competing firms clustered in one location attract more customers than the total number who would shop at the same stores at scattered locations.

critical path The sequence of activities between a project's start and finish that takes the longest time to complete.

critical path method (CPM) A network planning method developed in the 1950s as a means of scheduling maintenance shutdowns at chemical-processing plants.

critical ratio (CR) A ratio that is calculated by dividing the time remaining until a job's due date by the total shop time remaining for the job, which is defined as the setup, processing, move, and expected waiting times of all remaining operations, including the operation being scheduled.

cross-docking The packing of products on incoming shipments so that they can be easily sorted at intermediate warehouses for outgoing shipments based on their final destinations; the items are carried from the incoming-vehicle docking point to the outgoing-vehicle docking point without being stored in inventory at the warehouse.

cumulative sum of forecast errors (CFE) A measurement of the total forecast error that assesses the bias in a forecast.

customer contact The extent to which the customer is present, is actively involved, and receives personal attention during the service process.

customer involvement The ways in which customers become part of the process and the extent of their participation.

customer population An input that generates potential customers.

customer relationship process A process that identifies, attracts, and builds relationships with external customers, and facilitates the placement of orders by customers; sometimes referred to as *customer relationship management*.

customization Satisfying the unique needs of each customer by changing service or product designs.

cycle counting An inventory control method, whereby storeroom personnel physically count a small percentage of the total number of items each day, correcting errors that they find.

cycle inventory The portion of total inventory that varies directly with lot size.

cycle-service level See service level.

cycle time The maximum time allowed for work on a unit at each station.

decision theory A general approach to decision making when the outcomes associated with alternatives are often in doubt.

decision tree A schematic model of alternatives available to the decision maker, along with their possible consequences.

decision variables Variables that represent the choices the decision maker can control.

defect Any instance when a process fails to satisfy its customer.

degeneracy A condition that occurs when the number of nonzero variables in the optimal solution is less than the number of constraints.

delivery speed Quickly filling a customer's order.

Delphi method A process of gaining consensus from a group of experts while maintaining their anonymity.

demand management The process of changing demand patterns using one or more demand options.

dependent demand The demand for an item that occurs because the quantity required varies with the production plans for other items held in the firm's inventory.

dependent variable The variable that one wants to forecast.

design team A group of knowledgeable, team-oriented individuals who work at one or more steps in the process, conduct the process analysis, and make the necessary changes.

development speed Quickly introducing a new service or a product.

discount rate The interest rate used in discounting the future value to its present value.

discounting The process of finding the present value of an investment when the future value and the interest rate are known.

diseconomies of scale Occurs when the average cost per unit increases as the facility's size increases.

distribution center A warehouse or stocking point where goods are stored for subsequent distribution to manufacturers, wholesalers, retailers, and customers.

double-sampling plan A plan in which management specifies two sample sizes and two acceptance numbers; if the quality of the lot is very good or very bad, the consumer can make a decision to accept or reject the lot on the basis of the first sample, which is smaller than in the single-sampling plan.

drum-buffer-rope (DBR) A planning and control system that regulates the flow of work-in-process materials at the bottleneck or the capacity constrained resource (CCR) in a productive system.

earliest due date (EDD) A priority sequencing rule that specifies that the job or customer with the earliest due date is the next job to be processed.

earliest finish time (EF) An activity's earliest start time plus its estimated duration, t, or $EF = ES + t$.

earliest start time (ES) The earliest finish time of the immediately preceding activity.

early supplier involvement A program that includes suppliers in the design phase of a service or product.

economic order quantity (EOQ) The lot size that minimizes total annual inventory holding and ordering costs.

economic production lot size (ELS) The optimal lot size in a situation in which replenishment is not instantaneous.

economies of scale A concept that states that the average unit cost of a service or good can be reduced by increasing its output rate.

economies of scope Economies that reflect the ability to produce multiple products more cheaply in combination than separately.

electronic commerce (e-commerce) The application of information and communication technology anywhere along the supply chain of business processes.

electronic data interchange (EDI) A technology that enables the transmission of routine business documents having a standard format from computer to computer over telephone or direct leased lines.

elemental standard data A database of standards compiled by a firm's analysts for basic elements that they can draw on later to estimate the time required for a particular job, which is most appropriate when products or services are highly customized, job processes prevail, and process divergence is great.

employee empowerment An approach to teamwork that moves responsibility for decisions further down the organizational chart—to the level of the employee actually doing the job.

end item The final product sold to a customer.

enterprise process A companywide process that cuts across functional areas, business units, geographical regions, and product lines.

enterprise resource planning (ERP) systems Large, integrated information systems that support many enterprise processes and data storage needs.

environmental responsibility An element of sustainability that addresses the ecological needs of the planet and the firm's stewardship of the natural resources used in the production of services and products.

Euclidean distance The straight-line distance, or shortest possible path, between two points.

exchange An electronic marketplace where buying firms and selling firms come together to do business.

executive opinion A forecasting method in which the opinions, experience, and technical knowledge of one or more managers are summarized to arrive at a single forecast.

expediting The process of completing a job or finishing with a customer sooner than would otherwise be done.

exponential smoothing method A weighted moving average method that calculates the average of a time series by implicitly giving recent demands more weight than earlier demands.

external customers A customer who is either an end user or an intermediary (e.g., manufacturers, financial institutions, or retailers) buying the firm's finished services or products.

external failure costs Costs that arise when a defect is discovered after the customer receives the service or product.

external suppliers The businesses or individuals who provide the resources, services, products, and materials for the firm's short-term and long-term needs.

facility location The process of determining geographic sites for a firm's operations.

feasible region A region that represents all permissible combinations of the decision variables in a linear programming model.

financial responsibility An element of sustainability that addresses the financial needs of the shareholders, employees, customers, business partners, financial institutions, and any other entity that supplies the capital for the production of services or products or relies on the firm for wages or reimbursements.

finished goods (FG) The items in manufacturing plants, warehouses, and retail outlets that are sold to the firm's customers.

first-come, first served (FCFS) A priority sequencing rule that specifies that the job or customer arriving at the workstation first has the highest priority.

five S (5S) A methodology consisting of five workplace practices—sorting, straightening, shining, standardizing, and sustaining—that are conducive to visual controls and lean production.

fixed automation A manufacturing process that produces one type of part or product in a fixed sequence of simple operations.

fixed cost The portion of the total cost that remains constant regardless of changes in levels of output.

fixed order quantity (FOQ) A rule that maintains the same order quantity each time an order is issued.

fixed schedule A schedule that calls for each employee to work the same days and hours each week.

flexible (or programmable) automation A manufacturing process that can be changed easily to handle various products.

flexible flow The customers, materials, or information move in diverse ways, with the path of one customer or job often crisscrossing the path that the next one takes.

flexible workforce A workforce whose members are capable of doing many tasks, either at their own workstations or as they move from one workstation to another.

flow shop A manufacturer's operation that specializes in medium- to high-volume production and utilizes line or continuous flow processes.

flow time The amount of time a job spends in the service or manufacturing system.

flowchart A diagram that traces the flow of information, customers, equipment, or materials through the various steps of a process.

focus forecasting A method of forecasting that selects the best forecast from a group of forecasts generated by individual techniques.

focused factories The result of a firm's splitting large plants that produced all the company's products into several specialized smaller plants.

forecast A prediction of future events used for planning purposes.

forecast error The difference found by subtracting the forecast from actual demand for a given period.

forward integration Acquiring more channels of distribution, such as distribution centers (warehouses) and retail stores, or even business customers.

forward placement Locating stock closer to customers at a warehouse, DC, wholesaler, or retailer.

front office A process with high customer contact where the service provider interacts directly with the internal or external customer.

future value of an investment The value of an investment at the end of the period over which interest is compounded.

Gantt chart A project schedule, usually created by the project manager using computer software, that superimposes project activities, with their precedence relationships and estimated duration times, on a time line.

geographical information system (GIS) A system of computer software, hardware, and data that the firm's personnel can use to manipulate, analyze, and present information relevant to a location decision.

graphic method of linear programming A type of graphic analysis that involves the following five steps: plotting the constraints, identifying the feasible region, plotting an objective function line, finding a visual solution, and finding the algebraic solution.

graphs Representations of data in a variety of pictorial forms, such as line charts and pie charts.

Green Belt An employee who achieved the first level of training in a Six Sigma program and spends part of his or her time teaching and helping teams with their projects.

green purchasing The process of identifying, assessing, and managing the flow of environmental waste and finding ways to reduce it and minimize its impact on the environment.

gross requirements The total demand derived from *all* parent production plans.

group technology (GT) An option for achieving line-flow layouts with low volume processes; this technique creates cells not limited to just one worker and has a unique way of selecting work to be done by the cell.

heijunka The leveling of production load by both volume and product mix.

hiring and layoff Costs of advertising jobs, interviews, training programs for new employees, scrap caused by the inexperience of new employees, loss of productivity, and initial paperwork. Layoff costs include the costs of exit interviews, severance pay, retaining and retraining remaining workers and managers, and lost productivity.

histogram A summarization of data measured on a continuous scale, showing the frequency distribution of some process failure (in statistical terms, the central tendency and dispersion of the data).

holdout sample Actual demands from the more recent time periods in the time series that are set aside to test different models developed from the earlier time periods.

humanitarian logistics The process of planning, implementing and controlling the efficient, cost-effective flow and storage of goods and materials, as well as related information, from the point of origin to the point of consumption for the purpose of alleviating the suffering of vulnerable people.

hurdle rate The interest rate that is the lowest desired return on an investment; the hurdle over which the investment must pass.

hybrid office A process with moderate levels of customer contact and standard services with some options available.

immediate predecessors Work elements that must be done before the next element can begin.

independent demand items Items for which demand is influenced by market conditions and is not related to the inventory decisions for any other item held in stock or produced.

independent variables Variables that are assumed to affect the dependent variable and thereby "cause" the results observed in the past.

industrial robot Versatile, computer-controlled machine programmed to perform various tasks.

interarrival times The time between customer arrivals.

intermediate item An item that has at least one parent and at least one component.

intermodal shipments Mixing the modes of transportation for a given shipment, such as moving shipping containers or truck trailers on rail cars.

internal customers One or more employees or processes that rely on inputs from other employees or processes in order to perform their work.

internal failure costs Costs resulting from defects that are discovered during the production of a service or product.

internal suppliers The employees or processes that supply important information or materials to a firm's processes.

inventory A stock of materials used to satisfy customer demand or to support the production of services or goods.

inventory holding cost The sum of the cost of capital and the variable costs of keeping items on hand, such as storage and handling, taxes, insurance, and shrinkage.

inventory management The planning and controlling of inventories in order to meet the competitive priorities of the organization.

inventory pooling A reduction in inventory and safety stock because of the merging of variable demands from customers.

inventory position (IP) The measurement of a SKU's ability to satisfy future demand.

inventory record A record that shows an item's lot-size policy, lead time, and various time-phased data.

inventory turnover An inventory measure obtained by dividing annual sales at cost by the average aggregate inventory value maintained during the year.

ISO 9001:2008 A set of standards governing documentation of a quality program.

ISO 14000:2004 Documentation standards that require participating companies to keep track of their raw materials use and their generation, treatment, and disposal of hazardous wastes.

ISO 26000:2010 International guidelines for organizational social responsibility.

jidoka Automatically stopping the process when something is wrong and then fixing the problems on the line itself as they occur.

JIT system A system that organizes the resources, information flows, and decision rules that enable a firm to realize the benefits of JIT principles.

job process A process with the flexibility needed to produce a wide variety of products in significant quantities, with considerable divergence in the steps performed.

job shop A manufacturer's operation that specializes in low- to medium-volume production and utilizes job or batch processes.

Johnson's rule A procedure that minimizes makespan when scheduling a group of jobs on two workstations.

judgment methods A forecasting method that translates the opinions of managers, expert opinions, consumer surveys, and salesforce estimates into quantitative estimates.

judgmental adjustment An adjustment made to forecasts from one or more quantitative models that accounts for recognizing which models are performing particularly well in recent past, or take into account contextual information.

just-in-time (JIT) philosophy The belief that waste can be eliminated by cutting unnecessary capacity or inventory and removing non-value-added activities in operations.

kanban A Japanese word meaning "card" or "visible record" that refers to cards used to control the flow of production through a factory.

labor-limited environment An environment in which the resource constraint is the amount of labor available, not the number of machines or workstations.

latest finish time (LF) The latest start time of the activity that immediately follows.

latest start time (LS) The latest finish time minus its estimated duration, t, or $LS = LF - t$.

layout The physical arrangement of operations created by the various processes.

lead time The elapsed time between the receipt of a customer order and filling it.

lean systems Operations systems that maximize the value added by each of a company's activities by removing waste and delays from them.

learning curve A line that displays the relationship between processing time and the cumulative quantity of a product or service produced.

learning curve analysis A time estimation technique that takes into account the learning that takes place on an ongoing basis, such as where new products or services are introduced.

level strategy A strategy that keeps the workforce constant, but varies its utilization with overtime, undertime, and vacation planning to match the demand forecast.

line balancing The assignment of work to stations in a line process so as to achieve the desired output rate with the smallest number of workstations.

line flow The customers, materials, or information move linearly from one operation to the next, according to a fixed sequence.

line process A process that lies between the batch and continuous processes on the continuum; volumes are high and products are standardized, which allows resources to be organized around particular products.

linear programming A technique that is useful for allocating scarce resources among competing demands.

linear regression A causal method in which one variable (the dependent variable) is related to one or more independent variables by a linear equation.

linearity A characteristic of linear programming models that implies proportionality and additivity—there can be no products or powers of decision variables.

Little's law A fundamental law that relates the number of customers in a waiting-line system to the arrival rate and waiting time of customers.

load–distance method A mathematical model used to evaluate locations based on proximity factors.

lot A quantity of items that are processed together.

lot-for-lot (L4L) rule A rule under which the lot size ordered covers the gross requirements of a single week.

lot size The quantity of an inventory item management either buys from a supplier or manufactures using internal processes.

lot sizing The determination of how frequently and in what quantity to order inventory.

lot tolerance proportion defective (LTPD) The worst level of quality that the consumer can tolerate.

low-cost operation Delivering a service or a product at the lowest possible cost to the satisfaction of external or internal customers of the process or supply chain.

make-or-buy decision A managerial choice between whether to outsource a process or do it in-house.

make-to-order strategy A strategy used by manufacturers that make products to customer specifications in low volumes.

make-to-stock strategy A strategy that involves holding items in stock for immediate delivery, thereby minimizing customer delivery times.

makespan The total amount of time required to complete a group of jobs.

manufacturing resource planning (MRP II) A system that ties the basic MRP system to the company's financial system and to other core and supporting processes.

market research A systematic approach to determine external consumer interest in a service or product by creating and testing hypotheses through data-gathering surveys.

mass customization The strategy that uses highly divergent processes to generate a wide variety of customized products at reasonably low costs.

mass production A term sometimes used in the popular press for a line process that uses the make-to-stock strategy.

Master Black Belt Full-time teachers and mentors to several Black Belts.

master production schedule (MPS) A part of the material requirements plan that details how many end items will be produced within specified periods of time.

material requirements planning (MRP) A computerized information system developed specifically to help manufacturers manage dependent demand inventory and schedule replenishment orders.

mean absolute deviation (MAD) A measurement of the dispersion of forecast errors.

mean absolute percent error (MAPE) A measurement that relates the forecast error to the level of demand and is useful for putting forecast performance in the proper perspective.

mean squared error (MSE) A measurement of the dispersion of forecast errors.

methods time measurement (MTM) A commonly used predetermined data system.

metrics Performance measures that are established for a process and the steps within it.

minimum-cost schedule A schedule determined by starting with the normal time schedule and crashing activities along the critical path, in such a way that the costs of crashing do not exceed the savings in indirect and penalty costs.

mixed-model assembly A type of assembly that produces a mix of models in smaller lots.

mixed-model line A production line that produces several items belonging to the same family.

mixed strategy A strategy that considers the full range of supply options.

Modified Accelerated Cost Recovery System (MACRS) The only acceptable depreciation method for tax purposes that shortens the lives of investments, giving firms larger early tax deductions.

Monte Carlo simulation A simulation process that uses random numbers to generate simulation events.

most likely time (m) The probable time required to perform an activity.

MRP explosion A process that converts the requirements of various final products into a material requirements plan that specifies the replenishment schedules of all the subassemblies, components, and raw materials needed to produce final products.

multiple-dimension rules A set of rules that apply to more than one aspect of a job.

multiplicative seasonal method A method whereby seasonal factors are multiplied by an estimate of average demand to arrive at a seasonal forecast.

nearest neighbor (NN) heuristic A technique that creates a route by deciding the next city to visit on the basis of its proximity.

naïve forecast A time-series method whereby the forecast for the next period equals the demand for the current period, or Forecast = D_t.

nested process The concept of a process within a process.

net present value (NPV) method The method that evaluates an investment by calculating the present values of all after-tax total cash flows and then subtracting the initial investment amount for their total.

network diagram A network planning method, designed to depict the relationships between activities, that consists of nodes (circles) and arcs (arrows).

new service/product development process A process that designs and develops new services or products from inputs received from external customer specifications or from the market in general through the customer relationship process.

nominal value A target for design specifications.

nonnegativity An assumption that the decision variables must be positive or zero.

normal cost (NC) The activity cost associated with the normal time.

normal time (NT) In the context of project management, the time necessary to complete an activity under normal conditions.

normal time (NT) In the context of time study, a measurement found by multiplying the select time ($\bar{t}$), the frequency (F) of the work element per cycle, and the rating factor (RF).

normal time for the cycle (NTC) A measurement found by summing the normal time for each element.

objective function An expression in linear programming models that states mathematically what is being maximized or minimized.

offshoring A supply chain strategy that involves moving processes to another country.

one-worker, multiple-machines (OWMM) cell A one-person cell in which a worker operates several different machines simultaneously to achieve a line flow.

on-time delivery Meeting delivery-time promises.

open orders See scheduled receipts (SR).

operating characteristic (OC) curve A graph that describes how well a sampling plan discriminates between good and bad lots.

operations management The systematic design, direction, and control of processes that transform inputs into services and products for internal, as well as external, customers.

operations planning and scheduling The process of balancing supply with demand, from the aggregate level down to the short-term scheduling level.

operations scheduling A type of scheduling in which jobs are assigned to workstations or employees are assigned to jobs for specified time periods.

operations strategy The means by which operations implements the firm's corporate strategy and helps to build a customer-driven firm.

optimistic time (a) The shortest time in which an activity can be completed, if all goes exceptionally well.

optional replenishment system A system used to review the inventory position at fixed time intervals and, if the position has dropped to (or below) a predetermined level, to place a variable-sized order to cover expected needs.

order fulfillment process A process that includes the activities required to produce and deliver the service or product to the external customer.

order qualifier Minimal level required from a set of crteria for a firm to do business in a particular market segment.

order winner A criterion customers use to differentiate the services or products of one firm from those of another.

ordering cost The cost of preparing a purchase order for a supplier or a production order for manufacturing.

organizational learning The process of gaining experience with products and processes, achieving greater efficiency through automation and other capital investments, and making other improvements in administrative methods or personnel.

outsourcing Paying suppliers and distributors to perform processes and provide needed services and materials.

overtime The time that employees work that is longer than the regular workday or workweek for which they receive additional pay.

***p*-chart** A chart used for controlling the proportion of defective services or products generated by the process.

pacing The movement of product from one station to the next as soon as the cycle time has elapsed.

parameter A value that the decision maker cannot control and that does not change when the solution is implemented.

parent Any product that is manufactured from one or more components.

Pareto chart A bar chart on which factors are plotted along the horizontal axis in decreasing order of frequency.

part commonality The degree to which a component has more than one immediate parent.

past due The amount of time by which a job missed its due date.

path The sequence of activities between a project's start and finish.

payback method A method for evaluating projects that determines how much time will elapse before the total of after-tax flows will equal, or pay back, the initial investment.

payoff table A table that shows the amount for each alternative if each possible event occurs.

performance rating factor (RF) An assessment that describes *how much* above or below average the worker's performance is on each work element.

periodic order quantity (POQ) A rule that allows a different order quantity for each order issued but issues the order for predetermined time intervals.

periodic review (*P*) sytem A system in which an item's inventory position is reviewed periodically rather than continuously.

perpetual inventory system A system of inventory control in which the inventory records are always current.

pessimistic time (*b*) The longest estimated time required to perform an activity.

phase A single step in providing a service.

pipeline inventory Inventory that is created when an order for an item is issued but not yet received.

plan-do-study-act cycle A cycle, also called the Deming Wheel, used by firms actively engaged in continuous improvement to train their work teams in problem solving.

planned order release An indication of when an order for a specified quantity of an item is to be issued.

planned receipts Orders that are not yet released to the shop or the supplier.

planning horizon The set of consecutive time periods considered for planning purposes.

plants within plants (PWPs) Different operations within a facility with individualized competitive priorities, processes, and workforces under the same roof.

poka-yoke Mistake-proofing methods aimed at designing fail-safe systems that minimize human error.

postponement The strategy of delaying final activities in the provision of a product until the orders are received.

precedence diagram A diagram that allows one to visualize immediate predecessors better; work elements are denoted by circles, with the time required to perform the work shown below each circle.

precedence relationship A relationship that determines a sequence for undertaking activities; it specifies that one activity cannot start until a preceding activity has been completed.

predetermined data approach A database approach that divides each work element into a series of micromotions that make up the element. The analyst then consults a published database that contains the normal times for the full array of possible micromotions.

preemptive discipline A rule that allows a customer of higher priority to interrupt the service of another customer.

preference matrix A table that allows the manager to rate an alternative according to several performance criteria.

present value of an investment The amount that must be invested now to accumulate to a certain amount in the future at a specific interest rate.

presourcing A level of supplier involvement in which suppliers are selected early in a product's concept development stage and are given significant, if not total, responsibility for the design of certain components or systems of the product.

prevention costs Costs associated with preventing defects before they happen.

priority rule A rule that selects the next customer to be served by the service facility.

priority sequencing rule A rule that specifies the job or customer processing sequence when several jobs are waiting in line at a workstation.

process Any activity or group of activities that takes one or more inputs, transforms them, and provides one or more outputs for its customers.

process analysis The documentation and detailed understanding of how work is performed and how it can be redesigned.

process capability The ability of the process to meet the design specifications for a service or product.

process capability index, C_{pk} An index that measures the potential for a process to generate defective outputs relative to either upper or lower specifications.

process capability ratio, C_p The tolerance width divided by six standard deviations.

process chart An organized way of documenting all the activities performed by a person or group of people, at a workstation, with a customer, or on materials.

process choice A way of structuring the process by organizing resources around the process or organizing them around the products.

process divergence The extent to which the process is highly customized with considerable latitude as to how its tasks are performed.

process failure Any performance shortfall, such as error, delay, environmental waste, rework, and the like.

process improvement The systematic study of the activities and flows of each process to improve it.

process simulation The act of reproducing the behavior of a process, using a model that describes each step.

process strategy The pattern of decisions made in managing processes so that they will achieve their competitive priorities.

process structure A process decision that determines the process type relative to the kinds of resources needed, how resources are partitioned between them, and their key characteristics.

producer's risk (α) The risk that the sampling plan will fail to verify an acceptable lot's quality and, thus, reject it (a type I error).

product family A group of services or products that have similar demand requirements and common process, labor, and materials requirements.

production plan a manufacturing firm's sales and operations plan that centers on production rates and inventory holdings.

product-mix problem A one-period type of planning problem, the solution of which yields optimal output quantities (or product mix) of a group of services or products subject to resource capacity and market demand constraints.

productivity The value of outputs (services and products) produced divided by the values of input resources (wages, costs of equipment, and so on).

program An interdependent set of projects that have a common strategic purpose.

program evaluation and review technique (PERT) A network planning method created for the U.S. Navy's Polaris missile project in the 1950s, which involved 3,000 separate contractors and suppliers.

project An interrelated set of activities with a definite starting and ending point, which results in a unique outcome for a specific allocation of resources.

project management A systemized, phased approach to defining, organizing, planning, monitoring, and controlling projects.

projected on-hand inventory An estimate of the amount of inventory available each week after gross requirements have been satisfied.

protection interval The period over which safety stock must protect the user from running out of stock.

pull method A method in which customer demand activates production of the service or item.

purchased item An item that has one or more parents but no components because it comes from a supplier.

purchasing The activity that decides which suppliers to use, negotiates contracts, and determines whether to buy locally.

push method A method in which production of the item begins in advance of customer needs.

quality A term used by customers to describe their general satisfaction with a service or product.

quality at the source A philosophy whereby defects are caught and corrected where they were created.

quality circles Another name for problem-solving teams; small groups of supervisors and employees who meet to identify, analyze, and solve process and quality problems.

quality engineering An approach originated by Genichi Taguchi that involves combining engineering and statistical methods to reduce costs and improve quality by optimizing product design and manufacturing processes.

quality loss function The rationale that a service or product that barely conforms to the specifications is more like a defective service or product than a perfect one.

quality of life A factor that considers the availability of good schools, recreational facilities, cultural events, and an attractive lifestyle.

quantity discount A drop in the price per unit when an order is sufficiently large.

R-chart A chart used to monitor process variability.

radio frequency identification (RFID) A method for identifying items through the use of radio signals from a tag attached to an item.

random number A number that has the same probability of being selected as any other number.

range of feasibility The interval (lower and upper bounds) over which the right-hand-side parameter of a constraint can vary while its shadow price remains valid.

range of optimality The interval (lower and upper bounds) of an objective function coefficient over which the optimal values of the decision variables remain unchanged.

raw materials (RM) The inventories needed for the production of services or goods.

rectified inspection The assumption that all defective items in the lot will be replaced with good items if the lot is rejected and that any defective items in the sample will be replaced if the lot is accepted.

rectilinear distance The distance between two points with a series of 90-degree turns, as along city blocks.

reduced cost How much the objective function coefficient of a decision variable must improve (increase for maximization or decrease for minimization) before the optimal solution changes and the decision variable "enters" the solution with some positive number.

reengineering The fundamental rethinking and radical redesign of processes to improve performance dramatically in terms of cost, quality, service, and speed.

regular time Wages paid to employees plus contributions to benefits.

reorder point (R) The predetermined minimum level that an inventory position must reach before a fixed quantity Q of the SKU is ordered.

reorder point (ROP) system See continuous review (Q) system.

repeatability The degree to which the same work can be done again.

resource flexibility The ease with which employees and equipment can handle a wide variety of products, output levels, duties, and functions.

resource plan A plan that determines the requirements for materials and other resources on a more detailed level than the sales and operations plan.

resource planning A process that takes sales and operations plans; processes information in the way of time standards, routings, and other information on how services or products are produced; and then plans the input requirements.

revenue management Varying price at the right time for different customer segments to maximize revenues yielded by existing supply capacity.

reverse logistics The process of planning, implementing, and controlling the efficient, cost effective flow of products, materials, and information from the point of consumption back to the point of origin for returns, repair, remanufacture, or recycling.

risk-management plan A plan that identifies the key risks to a project's success and prescribes ways to circumvent them.

rotating schedule A schedule that rotates employees through a series of workdays or hours.

route planning An activity that seeks to find the shortest route to deliver a service or product.

SA8000:2008 A list of standards covering nine dimensions of ethical workforce management.

safety stock inventory Surplus inventory that a company holds to protect against uncertainties in demand, lead time, and supply changes.

sales and operations plan (S&OP) A plan of future aggregate resource levels so that supply is in balance with demand.

salesforce estimates The forecasts that are compiled from estimates of future demands made periodically by members of a company's salesforce.

salvage value The cash flow from the sale or disposal of plant and equipment at the end of a project's life.

sample size A quantity of randomly selected observations of process outputs.

sampling plan A plan that specifies a sample size, the time between successive samples, and decision rules that determine when action should be taken.

scatter diagram A plot of two variables showing whether they are related.

schedule A detailed plan that allocates resources over short time horizons to accomplish specific tasks.

scheduled receipts (SR) Orders that have been placed but have not yet been received.

SCOR model A framework that focuses on a basic supply chain of plan, source, make, deliver, and return processes, repeated again and again along the supply chain.

select time (t) The average observed time based only on representative times.

self-managed team A small group of employees who work together to produce a major portion, or sometimes all, of a service or product.

sensitivity analysis A technique for systematically changing parameters in a model to determine the effects of such changes.

sequencing Determining the order in which jobs or customers are processed in the waiting line at a workstation.

sequential-sampling plan A plan in which the consumer randomly selects items from the lot and inspects them one by one.

service blueprint A special flowchart of a service process that shows which steps have high customer contact.

service facility A person (or crew), a machine (or group of machines), or both necessary to perform the service for the customer.

service level The desired probability of not running out of stock in any one ordering cycle, which begins at the time an order is placed and ends when it arrives in stock.

service system The number of lines and the arrangement of the facilities.

setup cost The cost involved in changing over a machine or workspace to produce a different item.

setup time The time required to change a process or an operation from making one service or product to making another.

shadow price The marginal improvement in Z (increase for maximization and decrease for minimization) caused by relaxing the constraint by one unit.

shortest processing time (SPT) A priority sequencing rule that specifies that the job requiring the shortest processing time is the next job to be processed.

shortest route problem A problem whose objective is to find the shortest distance between two cities in a network or map.

simple moving average method A time-series method used to estimate the average of a demand time series by averaging the demand for the n most recent time periods.

simplex method An iterative algebraic procedure for solving linear programming problems.

simulation The act of reproducing the behavior of a system using a model that describes the processes of the system.

single-bin system A system of inventory control in which a maximum level is marked on the storage shelf or bin, and the inventor is brought up to the mark periodically.

single-digit setup The goal of having a setup time of less than 10 minutes.

single-dimension rules A set of rules that bases the priority of a job on a single aspect of the job, such as arrival time at the workstation, the due date, or the processing time.

single-sampling plan A decision to accept or reject a lot based on the results of one random sample from the lot.

Six Sigma A comprehensive and flexible system for achieving, sustaining, and maximizing business success by minimizing defects and variability in processes.

slack The amount by which the left-hand side of a linear programming constraint falls short of the right-hand side.

slack per remaining operations (S/RO) A priority sequencing rule that determines priority by dividing the slack by the number of operations that remain, including the one being scheduled.

social responsibility An element of sustainability that addresses the moral, ethical, and philanthropic expectations that society has of an organization.

sole sourcing The awarding of a contract for a service or item to only one supplier.

special-purpose teams Groups that address issues of paramount concern to management, labor, or both.

staffing plan A sales and operations plan for a service firm, which centers on staffing and other human resource-related factors.

standard deviation(σ) for forecasting A measurement of the dispersion of forecast errors.

standard deviation (σ) for statistical quality control The square root of the variance of a distribution.

standard time (ST) A measurement found by incorporating the normal time and allowances; $ST = NTC(1 + A)$, where A equals the proportion of the normal time added for allowances.

statistical process control (SPC) The application of statistical techniques to determine whether a process is delivering what the customer wants.

steady state The state that occurs when the simulation is repeated over enough time that the average results for performance measures remain constant.

stock-keeping unit (SKU) An individual item or product that has an identifying code and is held in inventory somewhere along the supply chain.

stockout An order that cannot be satisfied, resulting in a loss of the sale.

straight-line depreciation method The simplest method of calculating annual depreciation; found by subtracting the estimated salvage value from the amount of investment required at the beginning of the project, and then dividing by the asset's expected economic life.

subassembly An intermediate item that is *assembled* (as opposed to being transformed by other means) from more than one component.

suggestion system A voluntary system by which employees submit their ideas on process improvements.

supplier relationship process A process that selects the suppliers of services, materials, and information and facilitates the timely and efficient flow of these items into the firm.

supply chain An interrelated series of processes within and across firms that produces a service or product to the satisfaction of customers.

supply chain design Designing a firm's supply chain to meet the competitive priorities of the firm's operations strategy.

supply chain processes Business processes that have external customers or suppliers.

supply chain integration The effective coordination of supply chain processes through the seamless flow of information up and down the supply chain.

supply chain management The synchronization of a firm's processes with those of its suppliers and customers to match the flow of materials, services, and information with customer demand.

support process A process that provides vital resources and inputs to the core processes and therefore is essential to the management of the business.

surplus The amount by which the left-hand side of a linear programming constraint exceeds the right-hand side.

sustainability A characteristic of processes that are meeting humanity's needs without harming future generations.

swim lane flowchart A visual representation that groups functional areas responsible for different sub-processes into lanes.

takt time Cycle time needed to match the rate of production to the rate of sales or consumption.

tardiness See past due.

teams Small groups of people who have a common purpose, set their own performance goals and approaches, and hold themselves accountable for success.

technological forecasting An application of executive opinion to keep abreast of the latest advances in technology.

theoretical minimum (TM) A benchmark or goal for the smallest number of stations possible, where the total time required to assemble each unit (the sum of all work-element standard times) is divided by the cycle time.

theory of constraints (TOC) A systematic management approach that focuses on actively managing those constraints that impede a firm's progress toward its goal.

throughput time Total elapsed time from the start to the finish of a job or a customer being processed at one or more workcenters.

time-based competition A strategy that focuses on the competitive priorities of delivery speed and development speed.

time between orders (TBO) The average elapsed time between receiving (or placing) replenishment orders of Q units for a particular lot size.

time compression The feature of simulation models that allows them to obtain operating characteristic estimates in much less time than is required to gather the same operating data from a real system.

time series The repeated observations of demand for a service or product in their order of occurrence.

time-series analysis A statistical approach that relies heavily on historical demand data to project the future size of demand and recognizes trends and seasonal patterns.

time study A work measurement method using a trained analyst to perform four basic steps in setting a time standard for a job or process: selecting the work elements (or nested processes) within the process to be studied, timing the elements, determining the sample size, and setting the final standard.

time value of money The concept that a dollar in hand can be invested to earn a return so that more than one dollar will be available in the future.

tolerance An allowance above or below the nominal value.

top quality Delivering an outstanding service or product.

total inventory The sum of scheduled receipts and on-hand inventories.

total quality management (TQM) A philosophy that stresses three principles for achieving high levels of process performance and quality: (1) customer satisfaction, (2) employee involvement, and (3) continuous improvement in performance.

tracking signal A measure that indicates whether a method of forecasting is accurately predicting actual changes in demand.

transportation method A more efficient solution technique than the simplex method for solving transportation problems.

transportation method for location problems A quantitative approach that can help solve multiple-facility location problems.

transportation problem A special case of linear programming that has linear constraints for capacity limitations and demand requirements.

traveling salesman problem A problem whose objective is to find the shortest possible route that visits each city exactly once and returns to the starting city.

trend projection with regression A forecasting model that is a hybrid between a time-series technique and the causal method.

two-bin system A visual system version of the Q system in which a SKU's inventory is stored at two different locations.

type I error An error that occurs when the employee concludes that the process is out of control based on a sample result that falls outside the control limits, when in fact it was due to pure randomness.

type II error An error that occurs when the employee concludes that the process is in control and only randomness is present, when actually the process is out of statistical control.

uncontrollable variables Random events that the decision maker cannot control.

undertime The situation that occurs when employees do not have enough work for the regular-time workday or workweek.

usage quantity The number of units of a component that are needed to make one unit of its immediate parent.

utilization The degree to which equipment, space, or the workforce is currently being used, and is measured as the ratio of average output rate to maximum capacity (expressed as a percent).

value analysis A systematic effort to reduce the cost or improve the performance of services or products, either purchased or produced.

value stream mapping (VSM) A qualitative lean tool for eliminating waste or *muda* that involves a current state drawing, a future state drawing, and an implementation plan.

variable cost The portion of the total cost that varies directly with volume of output.

variables Service or product characteristics, such as weight, length, volume, or time that can be measured.

variety Handling a wide assortment of services or products efficiently.

vendor-managed inventories (VMI) A system in which the supplier has access to the customer's inventory data and is responsible for maintaining the inventory on the customer's site.

visual system A system that allows employees to place orders when inventory visibly reaches a certain marker.

volume flexibility Accelerating or decelerating the rate of production of services or products quickly to handle large fluctuations in demand.

waiting line One or more "customers" waiting for service.

warranty A written guarantee that the producer will replace or repair defective parts or perform the service to the customer's satisfaction.

weeks of supply An inventory measure obtained by dividing the average aggregate inventory value by sales per week at cost.

weighted-distance method A mathematical model used to evaluate layouts (of facility locations) based on proximity factors.

weighted moving average method A time-series method in which each historical demand in the average can have its own weight; the sum of the weights equals 1.0.

work breakdown structure (WBS) A statement of all work that has to be completed.

work elements The smallest units of work that can be performed independently.

work-in-process (WIP) Items, such as components or assemblies, needed to produce a final product in manufacturing or service operations.

work measurement The process of creating labor standards based on the judgment of skilled observers.

work sampling method A process that estimates the proportion of time spent by people or machines on different activities, based on observations randomized over time.

work standard The time required for a trained worker to perform a task following a prescribed method with normal effort and skill.

workforce scheduling A type of scheduling that determines when employees work.

$\bar{x}$-chart A chart used to see whether the process is generating output, on average, consistent with a target value set by management for the process or whether its current performance, with respect to the average of the performance measure, is consistent with past performance.

Name Index

Adams, Ron, 502
Adelhelm, Mark, 383
Ahire, Sanjay L., 155
Albright, S. Christian, 616
Andersen, Bjørn, 155
Anderson, D. R., 616
Andraski, Joseph C., 542
Ansberry, Clare, 305
Ante, Spencer A., 464
Apte, Uday M., 30, 116
Armstrong, J. Scott, 505
Arndt, Aaron D., 505
Arnold, Tony J. R., 344, 357
Arter, Dennis, 199
Artman, Les B., 462
Attaran, Mohsen, 505
Attaran, Sharmin, 505
Avon, Ravi, 383
Axsäter, Sven, 344, 357

Babbage, Charles, 3
Babbar, Sunil, 199
Bacon, Dawn, 206
Bagchi, Sugato, 383
Baghai, Ramin, 116
Baghel, Amit, 155
Bakke, Nils Arne, 224
Baohong, Liu, 383
Barfield, B., 512
Bastow, B. J., 344, 357
Baumer, David L., 462
Beamon, Benita M., 462
Becker, Nathan, 585
Bendoly, E., 585
Bennett, Johnny, 582
Benton, W. C., 344, 357, 440
Berry, William, 541, 585
Beschorner, Thomas, 462
Besterfield, Dale, 199
Bhuiyan, Nadjia, 155
Block, Dan, 343
Bonini, Sheila, 462
Booth, Alan, 116
Bossert, John M., 326
Bowen, H. Kent, 305
Bower, J. L., 224
Bowersox, Donald, 440
Boylan, John E., 505
Brache, Alan P., 138, 155
Brienzi, Mark, 440
Brink, Harold, 116
Brown, A., 244, 273
Brunswick, Alex, 381–383

Brynjolfsson, E., 585
Bunyaratavej, K., 410
Bylinsky, Gene, 105

Calantone, Roger, 440
Callioni, Gianpaolo, 344, 357
Camm, J. D., 616
Cannon, Alan R., 344, 357
Carey, Susan, 155
Carter, C. R., 462
Cavanagh, Roland R., 30, 199
Cederlund, Jerold P., 464, 505
Chandrasekaran, Anithashree, 155
Chapman, Stephen, 344, 357
Chase, Richard B., 30, 116
Chen, Haozhe, 505
Chen, Jason C. H., 541
Chiang, Wen-Chyuan, 541
Chittum, Ryan, 391, 410
Choi, Thomas Y., 440
Chopra, Sunil, 440
Chowdhury, Santanu, 397, 410
Clemen, Robert T., 48
Clive, Lloyd M., 344, 357
Closs, David J., 462
Coffman, Curt, 155
Collier, David A., 199
Collis, David J., 30
Collis, John, 117
Coltman, T., 462
Connaughton, Patrick, 464
Conner, Martin P., 462
Cook, Robert L., 440
Cooper, Robert B., 242
Cooper, Tim, 462
Corominas, Albert, 273
Cotteleer, M., 585
Cotteleer, Mark J., 585
Cox, J., 273
Crandall, Richard E., 344, 357
Crosby, Philip B., 199
Curkovic, Sime, 462

Daugherty, Patricia J., 505
Davenport, Thomas H., 155, 585
David, E. W., 462
David, Ed, 462
David, Paul, 438
Davis, Donna F., 505
Davis, Mark, 30, 384
Day, J. M., 462
Day, Jamison, 462
De Montgros, Xavier, 344, 357

De Vries, Jan, 117
De Waart, Dick, 383
Decker, Gero, 155
Deeds, David, 410
Deming, W. Edwards, 163, 199
Dhir, Krishna, 19
Dillman, Linda, 307
Dittmann, J. Paul, 384, 542
Donnell, Joe, 343
Dougherty, John R., 541
Drake, Matthew J., 462
Duggasani, A., 512
Duncan, Acheson J., 199
Duray, Rebecca, 30, 383

Edmondson, Amy C., 155
Efrati, Amir, 464
Ellinger, Alexander E., 541
Ellram, Lisa M., 383
Engardio, Pete, 442
Erickson, Gary, 461
Esper, Terry L., 541

Farley, T., 512
Fawcett, E., 542
Feigenbaum, A. V., 199
Ferguson, M., 512
Ferguson, Mark, 462
Field, Joy M., 117
Fiksel, Joseph, 462
Fildes, Robert, 505
Fisher, Anne, 155
Fisher, Marshall L., 116, 383
Fitzsimmons, James A., 30
Fitzsimmons, Mona, 30
Flannery, John, 397
Fleck, Thomas, 440
Fleming, John H., 155
Flint, Daniel J., 541
Flynn, Laurie J., 383
Fontanella, John, 541
Ford, Henry, 4
Foust, Dean, 2
Frei, Frances X., 90
Freund, Brian C., 440
Freund, June M., 440
Fry, Darlene, 540–541
Fry, Timothy D., 273

Gaimon, Cheryl, 30
Gales, Jon, 158
Gallagher, Christi, 308
Gallego, Guillermo, 383

Galuszka, P., 386, 410
Garber, Randy, 383
Gaynor, Elisse, 464
Genchev, Stefan E., 505
Ghosh, Soumen, 584
Gibson, Brian, 440
Gilbert, C. G., 224
Ginsburg, Greg, 461
Glatzel, Christoph, 383
Goel, Ajay K., 383
Goldratt, E. M., 88, 273
Goldratt, Eli, 245
Golicic, Susan L., 505
Goodwin, Paul, 505
Gorner, Steven, 462
Goyal, S. K., 117
Gray, Christopher, 541
Greasley, A., 155
Grey, William, 383
Grosskopf, Alexander, 155
Großpietsch, Jocher, 383
Grout, John R., 305
Grover, Varun, 116
Gryna, Frank, Jr., 199
Gunasekaran, Angappa, 155
Gunther, Marc, 446
Gupta, Jatinder N. D., 541

Hagen, Claus, 117
Hahn, E. D., 410
Hall, Kenji, 276
Hammer, Michael, 30, 117, 155
Hammer Michael, 22
Handfield, Robert, 462
Handfield, Robert B., 440, 462
Hanna, Michael E., 597, 616
Harada, Takashi, 303
Harris, John A., 462
Harrison, L., 512
Harter, James K., 155
Hartly-Urquhart, Roland, 383
Hartvigsen, David, 88, 155, 199, 224, 242, 344, 357, 440, 462
Harv, 384
Hayes, Robert, 117
Hayward, Rob, 462
Hefler, Janet, 158
Heineke, Janelle, 30
Hellberg, Ronald, 224
Hendricks, Kevin B., 585
Higbie, J., 512
Hill, Terry, 30, 117
Hillier, F. S., 242
Hillier, Fredrick S., 616
Hillier, Mark S., 616
Hindo, Brian, 462

Hofman, Debra, 384
Holland, Steven, 341–343
Holloway, Charles, 50
Holt, David, 50
Holweb, Matthias, 305
Horton, Suzanne S., 137
Hoyle, David, 199
Huckman, Robert S., 30

Jack, Eric, 117
Jackson, Marianna, 383
Jacobs, F. Robert, 541, 585
Jain, Rashmi, 155
Jaworski, Bernard J., 117
Jennings, M. M., 462
Jeston, John, 155
Johansson, Pontus, 117
Johnson, Kent A., 155
Jones, Alissa, 462
Jones, Daniel T., 30, 305
Jones, Darren, 88, 254, 273
Jordan, Bob, 223
Junglas, I., 462
Juran, J. M., 199

Kanetkar, Vinay, 155
Kaplan, Robert S., 30
Karetski, Ivan, 154
Karmarkar, Uday, 30, 155
Katircioglu, Kaan, 383
Keating, B., 462
Kekre, Sham, 440
Kelleher, Herb, 222
Kelly, Erin L., 541
Kemper, Steve, 383
Kerwin, Kathleen, 199
Kerzner, Harold, 88
Kincer, Ed, 243–244
King, Neil, Jr., 30
King, Rollin, 222
King-Metters, Kathryn, 117
Klassen, Robert D., 224
Klein, J. A., 305
Kohli, Rajiv, 464, 505
Kopczak, L., 462
Kramer, Mark R., 30
Kratz, Ellen Florian, 442
Krauss, Clifford, 207
Kriz, A., 462
Kroenke, D., 125
Kruger, Daniel, 120
Kuhn, Heinrich, 541
Kulpa, Margaret K., 155
Kulwiec, Ray, 462
Kung, Peter, 117
Kuyumcu, A., 512

La Ferla, Beverly, 155
Lacy, Peter, 462
Lambert, Douglas, 462
Lawrence, Michael, 505
Leaver, Sharyn, 464
Lee, Don, 19
Lee, Hau, 50
Lee, Hau L., 155, 384, 440
Lee, Kay Ree, 384
Lee, Min-Yo, 269
Leonard, Christopher, 207, 224
Levine, Edward H., 116
Lewis, J. P., 88
Lieberman, G. S., 242
Liker, Jeffrey K., 440
Lillywhite, Serena, 462
Lindgreen, Adam, 462
Little, J. D. C., 242
Lovelock, Christopher H., 410
Lucier, Gregory T., 199
Lunsford, J. Lynn, 372

Mabert, Vincent A., 585
MacCurdy, Douglas, 440
Maher, Kris, 541
Malhotra, Manoj K., 116, 117, 155, 273
Malik, Yogesh, 440
Mallick, Debasish, 117
Maloni, M. J., 440
Manikas, Andrew, 357
Mantel, Samuel J., Jr., 88
Maon, Francois, 462
Marley, Kathryn, 582–583
Martha, Joseph, 462
Martin, K., 616
Martin, Karla L., 30
Mascitelli, Ron, 305
McAfee, A., 585
McBride, David, 277, 281, 305
McCarthy, Teresa, 505
McCaskey, Sue, 343
McClain, John O., 273
McCormack, Kevin, 440
McCue, Andy, 585
Meile, Larry, 154, 381
Meindl, Peter, 440
Melnyk, Steven, 440, 462
Melo, M. T., 410
Menor, Larry J., 224
Mentzer, John T., 384, 505, 542
Meredith, Jack R., 88
Mesquita, Marco A., 542
Metters, Richard, 117, 512
Meyer, Christopher, 30
Meyer, Tobias, A., 462
Miller, Alex, 88, 254, 273
Miller, Jamey, 440

Miller, Mason, 115–116
Miller, Tom, 115–116
Millstein, Mitchell, 305
Min, Hokey, 505
Min, Soonhong, 505
Mitra, Amitava, 199
Moen, Phyllis, 541
Mollenkopf, Diane A., 462
Montgomery, David, 483, 505
Moon, Mark, 541
Moore, P. M., 242
Moussavi, Nazgol, 383
Muir, Nancy C., 88
Muller, Martin, 462
Murakami, S., 273
Murphy-Hoye, Mary, 440
Muzumdar, Maha, 541

Naik, Rajan, 116
Nair, Anand, 117
Nakano, Mikihisa, 541
Neilson, Gary L., 30
Nelis, Johan, 155
Netessine, Serguei, 440
Neuberger, Lisa, 462
Neuman, Robert P., 30, 199
Nicanor-Kimball, Char, 501
Nicholas, John M., 88
Nickel, S., 410
Niemeyer, Alex, 440
Nikolopoulos, Konstantinos, 505
Norton, David P., 30

O'Connor, Leo, 162
O'Connor, Marcus, 505
Ohno, Taiichi, 4, 277, 284
Olhager, Jan, 541
Olhger, Jan, 117
Onkal, Dilek, 505

Pande, Peter, 12
Pande, Peter S., 30, 199
Pastor, Rafael, 273
Pearson, David, 19
Philipoom, Patrick R., 273
Pinedo, Michael, 541
Place, Sharon, 502
Plambeck, Erica L., 462
Plans, Joan, 273
Porter, Michael, 30
Porter, Michael E., 30
Powell, Bill, 30
Powers, Elizabeth, 30
Prajogo, Daniel, 117
Pulaski, Bradley, 383
Pullman, Madeleine, 117

Quadt, Daniel, 541
Quazi, A., 462
Queenan, C., 512

Ragsdale, Cliff, 48, 616
Randall, Taylor, 440
Rayport, Jeffrey F., 117
Reddie, Rhonda Ulysses, 614
Reeve, James M., 384
Reilly, Terence, 48
Render, Barry, 597, 616
Rennie, Elizabeth, 372, 541
Rice, James B., Jr., 440
Richey, R. Glenn, 505
Ritzman, Larry P., 30, 117, 224
Roath, Anthony S., 505
Roberts, Alan, 502
Roberts, Dexter, 384
Roberts, Vicky, 87
Roos, Daniel, 30
Roth, Aleda, 30
Roth, Michael, 65
Rother, Mike, 305
Rubinstein, Ed, 410
Rudi, Nils, 440
Rukstad, Michael G., 30
Rummler, Geary A., 138, 155
Russell, J. P., 199
Ruwadi, Brian, 440
Ryan, Oliver, 446

Saaty, T. L., 242
Saffo, Paul, 505
Safizadeh, M. Hossein, 30, 117, 224
Sahay, B. S., 117
Saladin, Brooke, 30, 116, 305, 439, 541
Saldanha-da-Gama, F., 410
Sandor, J., 462
Sarkar, Suman, 383
Scalle, Cedric X., 585
Schaller, Jeff, 305
Schlachter, John Teepen, 462
Schonberger, Richard J., 305
Schwager, Andre, 30
Schwarz, Anne, 199
Scott, Bradley S., 155
Sehgal, Sanjay, 117
Selldin, Erik, 541
Serwer, Andy, 162
Seshadri, Sridhar, 199
Sester, Dennis, 199
Seybold, Dave, 383
Shafer, Scott M., 88
Shah, Rachna, 305
Share, Hugh M., 462
Sharma, Amol, 158

Sharma, Deven, 30
Sherer, Susan A., 464, 505
Shi, Dailun, 383
Shook, John, 305
Siferd, Sue P., 88
Silva, Ildefonso, 383
Silva, L., 462
Singh, Jitendra V., 383
Singhal, Jaya, 542
Singhal, Kalyan, 542
Singhal, Vinod R., 585
Skinner, Wickham, 30, 117
Slagmulder, Regine, 344, 357
Sloan, Alfred, 4
Slobig, Zachary, 446
Slone, Reuben, 542
Slone, Reuben E., 384
Smaros, Johanna, 505
Smith, Larry, 505, 542
Sodhi, ManMohan S., 440
Spear, Steven, 305
Spear, Steven J., 305
Speckman, R. E., 462
Spriggs, Ed, 343
Srikanth, Mokshagundam L., 273
Srinivasan, Mandyam, 88, 254, 273, 384
Srivastan, Vats N., 383
Sroufe, Robert, 440, 462
Stahl, Robert A., 485, 542, 585
Stair, Ralph M., 597, 616
Stank, Theodore P., 541
Stanton, Phil, 502
Stavrulaki, Euthemia, 384
Steele, Daniel C., 273
Stefanis, Stavros, 383
Stein, Herman, 88
Stern, Sam, 65
Stewart, Douglas M., 305
Stratman, Jeff K., 585
Subbakrishna, Sunil, 462
Sullivan, Laurie, 308, 440
Sutaria, Saumya S., 116
Sutton, Margaret M., 88
Svensson, Peter, 19, 30
Sweeney, D. J., 616
Swink, Morgan, 117
Syntetos, Aris A., 505
Taguchi, Genichi, 179–180
Takey, Flavia, 542
Taylor, Alex, 10
Taylor, Frederick, 3
Tenhiala, A., 224
Thomas, A., 443, 462
Thomas, Chad, 29
Thomas, L. Joseph, 273
Tiede, Tom, 384

Timme, Stephen G., 344, 357
Tonkin, Lea, 305
Toyoda, Eiji, 284
Trent, Robert J., 440
Trent, William, 202
Trottman, Melanie, 542

Umble, E., 273
Umble, Michael, 273

Van Wassenhove, L., 462
Van Wassenhove, L.N., 442
Van Wassenhove, Luk N., 344, 357
VanHamme, Joelle, 462
Voss, Chris, 30

Wagner, Richard H., 440
Wallace, Thomas F., 485, 542, 585

Walters, Donald, 344, 357
Walton, S., 462
Ward, Peter T., 30, 305
Ward, S., 512
Werner, Ben, 17
Weske, Mathias, 155
Weston, F. C., Jr., 585
Whitcomb, Gus, 308
Whybark, D. C., 462
Whybark, D. Clay, 541, 585
Wilcock, Anne E., 155
Williams, Janet, 304–305
Williams, Sean P., 326
Williams, T. A., 616
Williams, Wayne, 304–305
Williams-Timme, Christine, 344, 357

Wilson, Jim, 155
Winston, Wayne L., 616
Wiseman, Paul, 16, 30
Wolf, John, 439
Womack, James P., 30, 305
Wood, Craig, 30
Wright, Linda, 344, 357

Xu, Xiaojing, 541

Yannick, Julliard, 199
Yao, Yuliang, 464, 505
Yip, George S., 410
Yu, Wen-Bin Vincent, 505

Zimmerman, Richard S., 137
Zinner, Darren E., 30
Zomerdijk, Leonieke G., 117

Subject Index

Page numbers followed by f have figures. Page numbers followed by t have tables.

3PL (third-party logistics provider), 426

5S (Five S) practices, 282–283, 283f, 283t

99 Restaurants and Pubs, 391

Abasco, 244

ABB, 101f, 102f

ABC analysis, 314–315, 314f

Accenture Institute for High Performance, 443

Acceptance quality level (AQL), 165

Acceptance sampling, 165–166, 166f

Action notice, 561

Activity, definition of, 53

Activity ownership, 53

Activity relationships, 54f

Activity slack, 58f, 59

Activity time, 55, 56

Activity-on-node (AON) network, 55

Additive seasonal method, 481

Administrative costs, 418

Advanced planning and scheduling (APS) systems, 528

AFC Enterprises, 391

Aggregate plan, 509t

Aggregation, 466, 509

Agility, 453

Air New Zealand, 507–508

Aircraft construction, 57f

Airline industry, 13–14, 206, 565

Algebraic solution, in linear programming, 594–595

Alternatives, 36, 210–211

Amazon.com, 1, 12t, 395f, 545

Analysis. *See also* Process analysis
  ABC analysis, 314–315, 314f
  break-even analysis, 31–35, 33f, 369–370, 395–396, 396f
  of cost–time trade-offs, 60–64, 61f
  data analysis tools, 132–135
  error analysis, 467f
  graphic analysis, 590–596
  learning curve analysis, 129, 129f
  market analysis, 10–11
  multiple regression analysis, 472
  new service/product development process, 417
  planning and, 66–69
  sensitivity analysis, 33, 319, 319t, 596, 596t

total cost analysis, for supplier selection, 418–419
  value analysis, 421

Andon, 280

Annual plan, 510

Anticipation inventory, 311, 312, 314, 513t, 514

AON (activity-on-node) network, 55

Appraisal costs, 159

APS (advanced planning and scheduling) systems, 528

AQL (acceptance quality level), 165

ArcInfo, 390

Arrival distribution, 229–230

Arrival rates, waiting lines and, 237

Assemble-to-order (ATO) strategy, 96–97, 368, 368f, 373, 415

Assembly line, 4

Assignable causes of variation, 168–169, 169f

Astro Studios, 49

ATI, 49

ATO (assemble-to-order) strategy, 96–97, 368, 368f, 373, 415

ATP (available-to-promise) inventory, 552–553

Attributes, 65, 167, 174–177

Auctions, 423

Autodaq Private Auction service, 423

Autoliv, 303–304

Automation, 4
  fixed, 103
  flexible, 103, 105
  in lean systems, 282
  of manufacturing processes, 103, 104f
  of service processes, 103–104

Available-to-promise (ATP) inventory, 552–553, 552f

Average aggregate inventory value, 364

Average payoff, 38

B. Manischewitz Company, 511

Backlogs, 513, 513t

Back-office processes, 94

Backorder, 310, 513, 513t

Backorder costs, 516t

Backward integration, 370

Bal Seal Engineering, 246f

Balance delay, 256–257

Baldrige Performance Excellence Program, 182

Baptist Memorial Hospital, 137

Bar, Pareto, and Line Charts Solver, 133

Bar charts, 132, 133f, 136f

Barns & Noble, 1

Base case, 210

Base-stock system, 329

Batch processes, 96, 249–250

Bavarian Motor Works (BMW), 385–387

Baxter Planning Systems, 411

Behavior factors, in line processes, 258–259

Benchmarking, 137–138, 139f

Benchmarking Partners, Inc., 485f

Beta distribution, 66f

Bias error, 467

Bill of materials (BOM), 547–549, 548f, 549f, 559f

Bill of resources (BOR), 566–567, 566f

Billing and payments process, 15, 15t

Binding constraint, 595

BIOTA, 421f

Black Belts, 165

Block plan, 98, 98f, 100f

BMW (Bavarian Motor Works), 385–387

Boeing, 57f, 204, 372, 388

BOM (Bill of materials), 547–549, 548f, 549f, 559f

BOR (Bill of resources), 566–567, 566f

Bose Corporation, 278

Boston's Big Dig project, 65

Bottleneck method, 250

Bottlenecks, 247–254
  in batch process, 249–250
  definition of, 244
  Drum-Buffer-Rope (DBR) systems, 252–254, 253f
  exploitation of, 246
  floating, 248
  identification of, 246, 248–250
  increase capacity of, 247
  in manufacturing processes, 248–254
  product mix decisions, 250–252
  relieving, 250
  in service processes, 247–248, 247f

Brainstorming, 135–137
Break-even analysis, 31–35, 33f, 369–370, 395–396, 396f
Break-Even Analysis Solver, 34f, 35
Break-even quantity, 31–33
British Petroleum oil spill, 243–244
Brunswick Distribution, Inc., 381–383
BTS (build-to-stock), 373
Buffer, 252
Buffer management, 253
Build-to-stock (BTS), 373
Bullwhip effect, 413, 413f
Business plan, 509–510
Business-to-business systems, 429
Business-to-consumer systems, 429
Buyer-supplier relationships, 455–456
Buying, 422–424

CAD (computer aided design), 4
Calculated values, 519
California Academy of Family Physicians, 159
Call centers, 161f, 525
CAM (computer aided manufacturing), 4
Capacity. *See also* Capacity planning; Long-term capacity planning
    definition of, 202
    input measures of, 203, 208–210
    logistics, 426
    output measures of, 203, 208
    perishable, 512
    total, 288–290
Capacity constrained resource (CCR), 244, 252. *See also* Bottlenecks
Capacity cushions, 205–206, 374t
Capacity gaps, 210
Capacity planning, 201–213
    across the organization, 202
    expansion, timing and sizing, 206–207
    at Fitness Plus, 223–224
    introduction to, 201–202
    long-term, 202–205
    other decisions and, 207–208
    at Sharp Corporation, 201–202
    simulation, 213
    at Southwest Airlines, 222–223
    systematic approach, 208–211
    timing and sizing strategies, 205–208
    tools for, 212–213
Capacity requirements, 208–210
Capacity requirements planning (CRP), 561–562

Capital, cost of, 309
Capital intensity, 92, 102–104, 105, 107
Carbon footprint, 447
Cargo security, 454
Carrying cost, 309
Cash flow, 211, 366
Cash-to-cash, 366
Catalog hubs, 422–423
Causal methods, 467, 470–472
Cause-and-effect diagrams, 133–134, 134f, 136f
CC (crash cost), 61
c-charts, 176–177, 177f
CCR (capacity constrained resource), 244, 252. *See also* Bottlenecks
CDI (Clinical Design Initiative), 294
CDP (customer demand planning), 425
Celanese, 326
Celestica, 50
Cell phones, 157–158
Center of gravity, 393–394
Centralized placement, 367
Certainty, 36–37, 588
CFE (cumulative sum of forecast errors), 467
Chad's Creative Concepts, 29–30
Chance events, 36
Change, strategies for, 108–109
Changeover, 279
Channel, 227–228
Channel assembly, 369
Chase strategy, 515, 519–521, 520f
Checklists, 132, 136f
Chief operations officer (COO), 3
Child labor, 455
China, 18
Church's Chicken, 391
Clif Bar & Company, 461
Clinical Design Initiative (CDI), 294
Close out, 70
Close supplier ties, 278–279
Closed-loop supply chain, 444–445, 445f
Closeness matrix, 98
Coin catapult, 198–199
Collaborative activities, 430
Collaborative effort, 9–10
Collaborative planning, forecasting, and replenishment (CPFR), 463–464, 485, 485f
Collective bargaining, 455
Combination forecasts, 482, 483
Command and control, in international disasters, 454
Common causes of variation, 168

Communication technology, globalization and, 18
Communicator, project manager as, 52
Community programs for recycling, 447
Comparative cost advantages, 18
Comparative labor costs, 370–371
Compensation, 455
Competitive benchmarking, 138
Competitive capabilities, 11–14, 15, 15t
Competitive orientation, 421–422
Competitive priorities, 7, 11–14, 14t, 373t
Competitors, location of, 390
Complementary products, 511, 513t
Component, 547
Computer aided design (CAD), 4
Computer aided manufacturing (CAM), 4
Computer simulation, 65–66
Computer support, material requirements planning, 561
Computerized decision-making tools, 21
Concurrent engineering, 417
Conformance to specifications, 161
Consistent quality, 12t, 14, 14t, 15, 15t
Constraint management, 243–259
    across the organization, 245
    bottlenecks, identification and management, 247–254
    British Petroleum oil spill, 243–244
    introduction to, 243–244
    in lean systems, 319
    in a line process, 254–259
    at Min-Yo Garment Company, 269–272
    at Southwest Airlines, 272–273
    theory of constraints, 245–247, 246t
Constraints, 244, 588, 589, 590–591, 591f
Construction costs, 204
Container system, 293
Containerless system, 293
Contextual knowledge, 470
Continuous flow process, 96
Continuous improvement, 163–164, 179, 277–278, 278f, 285f
Continuous review (*Q*) system, 320–325, 328–329
Contribution margin, 250–252, 366, 373t

Control charts
    for attributes, 174–177
    introduction to, 169–170, 169f, 170f
    tracking signals and, 482, 482f
    for variables, 170–174
COO (chief operations officer), 3
Cooperative orientation, 422
Copper Kettle Catering (CKC), 304–305
Coral Princess, 428
Core competencies, 9
Core processes, 7, 9, 563–564
Corner points, 593
Corporate strategy, 7, 8–10, 8f
Cost of goods sold, 364, 366
Cost to crash, 61
Costco, 12t
Costs
    administrative, 418
    appraisal, 159
    backorder, 516t
    of capital, 309
    carrying, 309
    construction, 204
    direct, 60, 61, 61f
    external failure, 159
    fixed, 32, 204
    freight, 418
    of goods sold, 364, 366
    handling, 309
    hiring and layoff, 516t
    indirect, 60
    internal failure, 159
    inventory, 418
    inventory holding, 309, 315, 516t
    logistics, 371
    of lot-sizing policy, 317
    materials, 204, 418
    minimization of, 61–64
    ordering, 310, 315
    overtime, 516t
    prevention, 159
    of purchased materials, 204
    of quality, 159
    for real estate, 389
    regular time, 516t
    setup, 310, 315
    stockout, 516t
    storage, 309
    total project, 60
    transportation, 310, 389–390
    types of with sales and operations planning, 516t
    variable, 32
Cost–time trade-offs, 60–64, 61f
Cousins Subs, 391

CPFR (collaborative planning, forecasting, and replenishment), 463–464, 485, 485f
CPM (critical path method), 54
Crash cost (CC), 61
Crash time (CT), 61
Crashing an activity, 60, 61
Critical mass, 390
Critical path, 57, 59
Critical path method (CPM), 54
CRM (customer relationship management), 428
Cross-docking, 91t, 427–428
CRP (capacity requirements planning), 561–562
CT (crash time), 61
Cultural change, total quality management, 161–162
Cumulative sum of forecast errors (CFE), 467
Current state drawing, 287, 289f
Custom Molds, Inc., 115–116
Customer contact, 6, 92–93, 93t, 106f
Customer demand planning (CDP), 425
Customer involvement
    decision patterns, 105, 106
    in process strategy, 92, 100–101
Customer population, 226–227, 226f
Customer relationship management (CRM), 428
Customer relationship process, 7, 428–430
Customer satisfaction, 160–161
Customer service, 91t, 310, 361, 429–430
Customer support, 161
Customer-contact matrix, 93, 93f
Customers
    external, 4, 161
    internal, 4, 161
    proximity to, 389
Customization, 12t, 14t, 97, 367–369
Cycle counting, 315
Cycle inventory, 311–313, 315, 317f, 346
Cycle time, 256, 259
Cycle-service level, 322
Cyclical demand patterns, 465, 466f
Cypress Semiconductor, 553

Daiichi Nuclear Reactors, 19
Data analysis tools
    bar charts, 132, 133f, 136f
    cause-and-effect diagrams, 133–134, 134f, 136f
    checklists, 132, 136f

graphs, 134–135
histograms, 132
Pareto charts, 132–133, 133f, 136f, 314
performance evaluation, 131f, 132–135
scatter diagrams, 133
Data entry screens, 597–598, 597f
Data sharing, 430
Data snooping, 135
DBR (Drum-Buffer-Rope) systems, 252–254, 253f
DCs (distribution centers), 367, 387
Deceptive business practices, 160
Decision maker, project manager as, 52
Decision making, 31–41
    break-even analysis, 31–35, 33f, 369–370, 395–396, 396f
    under certainty, 36–37
    decision theory, 36–39
    decision trees, 39–41, 39f, 213, 213f
    preference matrix, 35–36, 35f, 392–393, 420
    process strategy and, 91–92, 92f
    under risk, 38–39
    under uncertainty, 37–38
Decision patterns
    for manufacturing processes, 106–107, 106f, 107f
    for service processes, 105–106, 106f
Decision rule, 36
Decision theory, 36–39
Decision trees, 39–41, 39f, 213, 213f
Decision variables, 587, 588
Decision-making tools, computerized, 21
Deckers Outdoor Corporation, 500–502
Dedication, project manager and, 52
Deepwater Horizon, 243–244
Defects, 159, 277t
Degeneracy, 599
Delay, on process charts, 129
Delivery speed, 12t, 14t, 15, 15t
Delivery system needs, 10
Dell Computers, 12t, 211, 211f
Delphi method, 470
Demand
    distribution of during lead time, 322, 322f
    management of, 511–513
    patterns of, 465–466, 466f
    reorder point and, 320–322, 320f, 321f, 323

supply chain design and, 373t
target inventory level and, 327–328
Demand During the Protection Interval Simulator, 328
Demand options, 511–513, 513t
Deming Wheel, 163–164, 164f
Dependent demand, 547, 547f, 565–566
Dependent demand items, 319–320
Dependent variables, 470–471
Deposit fee, 447
Derived values, 519
Design collaboration, 421
Design stage, new service/product development process, 416–417
Design team, 122
Design-to-order (DTO), 373–374
Destinations, 397
Deterioration, 310
Development, of new products and services, 417
Development speed, 12t, 14t
Direct costs, 60, 61, 61f
Disaster, 453
Disaster relief supply chains, 444t, 453–455, 453f
Discipline, 455
Discrimination, 455
Diseconomies of scale, 204–205, 205f
Disney MGM Studios, 231f
Disney World, 104, 225, 231f
Distribution centers (DCs), 367, 387
DMAIC process. *See* Six Sigma Improvement Model
Documentation of processes, 122
    flowcharts, 123–125, 124f, 165, 237f
    service blueprints, 126–127, 126f
    swim lane flowcharts, 125–126, 125f
    work measurement techniques, 127–131
Domino's Pizza, 391
Donor independence, 455
Dow Corning, 543–544
Downstream, in supply chain integration, 413
Dreamliner airplane, 372, 388
Drum-Buffer-Rope (DBR) systems, 252–254, 253f
DTO (design-to-order), 373–374
Dummy plants or warehouses, 398
Dynamic sales volume, 361

E&B Marine, 484f
Earliest finish time (EF), 57
Earliest promised due date (EDD), 229, 526–527

Earliest start time (ES), 57
Early supplier involvement, 421
Earthquake, in Japan, 19
Eastman Kodak, 411–412
EBay, 89–90
E-commerce, 428–429
Economic dependency, 421
Economic order quantity (EOQ)
    assumptions of, 345
    calculation of, 315–318
    managerial insights from, 319
    sensitivity analysis, 319, 319t
Economic production lot size (ELS), 346–347, 347f
Economic Production Lot Size Solver, 347f
Economies of scale, 204, 205f
Economies of scope, 104
EDD (earliest promised due date), 229, 526–527
EDI (electronic data interchange), 422
EDLP (every day low pricing), 430
EF (earliest finish time), 57
Efficiency, 256–257, 444t
Efficiency curve, for supply chains, 361, 361f
Elastec, 244
Electronic commerce, 428–429
Electronic data interchange (EDI), 422
Elemental standard data approach, 127
ELS (economic production lot size), 346–347, 347f
Employee empowerment, 163
Employee involvement, in total quality management, 161–163
Employee-period evaluation, 519
Employees. *See* Workforce
End item, 548
Energy efficiency, 447–452
    freight density, 448–451, 450t
    transportation distance, 447–448
    transportation mode, 452
    transportation technology, 452
Enterprise process, 544
Enterprise Resource Planning (ERP) Systems, 4, 544–546, 545f
Environmental impact, 388
Environmental issues, 19–20
Environmental management system, 181
Environmental protection, 443
Environmental responsibility, 443–452, 444t
    definition of, 442
    energy efficiency, 447–452

material requirements planning, 564
    reverse logistics, 444–447
Environmental scanning, 8
Equipment, 102
Equipment utilization, 310
ERP (Enterprise Resource Planning) Systems, 4, 544–546, 545f
Error analysis, 467f
Error measures, 468–469, 468f
ES (earliest start time), 57
Estimating the average, in forecasting, 473–475
Ethanol industry, 207
Ethical issues, 19–20
Ethics
    quality and, 159–160
    supply chain and, 455–457
EU (European Union), 18
Euclidean distance, 99
European Union (EU), 18
Events, in decision theory, 36
Every day low pricing (EDLP), 430
Exchange, 423
Execution phase, 70
Executive opinion, 470
Expansion, timing and sizing, 206–207
Expansionist strategy, 206–207, 206f
Expected payoff, 40
Expected value, 38
Expected value decision rule, 426, 427
Expected value of an alternative, 426
Expediting, 513t, 526
Exponential distribution, 230
Exponential smoothing method, 474–475
External customers, 4, 161
External failure costs, 159
External suppliers, 4
Extrusion constraint, 591f

Facilitator, project manager as, 52
Facilitators, in process analysis, 122
Facilities, 9
Facility location, 387, 396–400, 426, 456. *See also* Supply chain location
Fast food industry, 391
FCFS (first-come, first-served) rule, 229, 526–527
Feasibility, range of, 596t, 599
Feasible region, 588, 592–593, 592–593f

Federal Express, 4, 441–442, 453
FG (finished goods), 311
Finance, integration of in a business, 3, 3f
Financial institutions, regulation and, 18
Financial measures, 245t, 365–367
Financial plan, 510
Financial responsibility, 442, 444t
Finish times, 57–59, 58f
Finished goods (FG), 311
Finite-source model, 235–236
First-come, first-served (FCFS) rule, 229, 526–527
Fishbone diagram, 134
Fiskars Brands, Inc., 483
Fitness for use, 161
Fitness Plus, 223–224
Five S (5S) practices, 282–283, 283f, 283t
Fixed automation, 103
Fixed costs, 32, 204
Fixed interval reorder system, 325
Fixed order quantity (FOQ), 556
Fixed order-quantity system, 320
Fixed schedule, 523
Flashy Flashers, Inc., 582–585
Flexible automation, 103, 105
Flexible flow, 93
Flexible workforce, 102, 282
Flextronics, 50
Floating bottlenecks, 248
Flow time, 526
Flowcharts, 123–125, 124f, 165, 237f
Flowers-on-Demand, 362
Focus, strategic fit, 107–108
Focus forecasting, 483
Focused factories, 108, 108f
Focused service operations, 107–108
Follow the leader strategy, 207
FOQ (fixed order quantity), 556
ForAgentsOnly.com, 22
Forced labor, 455
Ford, Henry, 3f
Ford Motor Company, 3f
Forecast error, 467–469, 467f, 468f
Forecast Pro, 484
Forecasting, 463–485
    across the organization, 465
    causal methods, 470–472
    collaboration and, 485
    combination forecasts, 482, 483
    computer support, 470
    at Deckers Outdoor Corporation, 500–502
    definition of, 464
    demand patterns, 465–466, 466f

errors in, 467–469, 467f, 468f
estimating the average, 473–475
focus, 483
with holdout sample, 481, 503–504
introduction to, 463–465
judgment methods, 466–467, 470
key decisions in, 466–470
linear regression, 470–472, 471f
at Motorola, 463–464
multiple techniques, using, 482–483
as a nested process, 485
as a process, 483–485, 485t
quantitative forecasting method, 481–482
technique selection, 466–467
time-series methods, 467, 472–481
a vital energy statistic, 504–505
what to forecast, 466
at Yankee Fork and Hoe Company, 502–503
Forward integration, 370
Forward placement, 367
Forward Response Centers, 453
Forward stock locations (FSLs), 411
Freedom of association, 455
Freight costs, 418
Freight density, 448–451, 450t
Front-office processes, 94
FSLs (forward stock locations), 411
FUJIFILM Imaging Colorants, 518
Full launch, of new services/products, 417
Functional benchmarking, 138
Functional organizational structure, 52
Future state drawing, 287

Gantt chart, 59–60, 60f, 521–522
Gantt progress charts, 522, 522f
Gantt workstation charts, 522, 522f
The Gap, Inc., 10
General Electric (GE), 164, 396, 397
General Motors (GM), 4, 10, 19, 206, 206f, 294, 371f
Geographical information systems (GIS), 390–391, 397
Gillette, 103, 375f
GIS (geographical information systems), 390–391, 397
Global competition, 17–19
Global strategies, 9–10
Globalization, 17–19
GM (General Motors), 4, 10, 19, 206, 206f, 294, 371f

Goods and services, demand for imports, 18
Goods sold, costs of, 364, 366
Google, 464
Graphic analysis, 590–596
    find the algebraic solution, 594–595
    find the visual solution, 594
    identify the feasible region, 588, 592–593, 592–593f
    plot an objective function line, 587, 589, 593–594
    plot the constraints, 590–591, 591f
    sensitivity analysis, 596, 596t
    slack and surplus variables, 595–596
Graphic method of linear programming, 590–596
Graphs, 134–135
Green Belts, 165
Green purchasing, 420
Gross requirements, 553–554
Group technology (GT), 286–287, 286f
Gulf of Mexico oil spill, 243–244

Hallmark, 515
Handling costs, 309
Handoffs, 126
Harrah's Cherokee Casino & Hotel, 512
Harry Potter and the Deathly Hallows (Rowling), 1
Harry Potter series, 1–2
Health and safety, 455
Heart to Heart International, 453
Heijunka, 281, 285, 285f, 303
Heuristic decision rules, 258t
Hewlett-Packard, 446
Hiring and layoff costs, 516t
Histograms, 132
History file, 469, 484
Holdout sample, 481, 503–504
Honda, 279
Honda Civic, 13
Horizontal demand patterns, 465, 466f
Hospitals, 565
Hotels industry, 565–566. *See also* Starwood Hotels and Resorts
House of Toyota, 284–285, 285f
Humanitarian logistics, 452.
    See also Logistics
    definition of, 442
    disaster relief supply chains, 453–455, 453f
    supply chain ethics, 455–457

Hurricane Katrina, 441
Hybrid-office processes, 94

IBM, 49
Idle time, 256–257
ILOG Solver, 528
Imaging Products, 359
Immediate predecessors, 254
Import quotas, 18
Imports, demand for, 18
Inappropriate processing, 277t
Independent demand, 547
Independent demand items, 319
Independent variables, 471
India, 18, 397
Indirect costs, 60
Industrial robot, 103
Industry Week, 390
Information exchange, 424
Information flows in the supply
    chain, 564f
Information gathering, 97–98
Information inputs, 514, 514f
Information technology, 4
Initial tableau, 398, 398f
Innovation
    operational, 22
    at Progressive Insurance, 22
    supply chains and, 22
    sustainability and, 443
In-plant representative, 278
Input measures of capacity, 203,
    208–210
Inputs, 5
Inspection, on process charts, 129
Insurance, on inventory, 309–310
Integer programming, 588
Intel, 203f
Interarrival times, 230
Intermediate item, 548
Intermodal shipments, 452
Internal benchmarking, 138
Internal customers, 4, 161
Internal failure costs, 159
Internal suppliers, 4
International Federation of Red
    Cross and Red Crescent
    Societies, 453
International Organization for
    Standardization, 181
International quality documenta-
    tion standards, 181–182
Internet, 371
Invacare Corporation, 389
Inventory (I)
    anticipation, 312, 314, 513t, 514
    Anticipation inventory, 311

available-to-promise, 552–553, 552f
creation of, 309f
cycle, 311–313, 315, 317f, 346
definition of, 309
estimation of, 312–313
financial measures and, 245t
insurance on, 309–310
large inventories, pressures for,
    310–311
in lean systems, 295
pipeline, 311, 312, 314
reduction tactics, 313–314
safety stock, 312, 313–314, 323,
    323f, 558–559, 558f
small inventories, pressures for,
    309–310
of successive stocking points,
    311f
types of, 311–313
waste in, 277t
Inventory control systems, 319–329,
    345–352
    comparative advantages of *Q* and
        P systems, 328–329
    continuous review (*Q*) system,
        320–325, 328–329
    hybrid systems, 329
    noninstantaneous replenish-
        ment, 345–347, 346f
    one-period decisions, 350–352,
        351f
    periodic review (*P*) system,
        325–329, 325f, 457
    quantity discounts, 311, 348–350,
        348f, 350f
Inventory costs, 418
Inventory holding cost, 309, 315,
    516t
Inventory investment, 374t
Inventory management, 307–329
    ABC analysis, 314–315, 314f
    across the organization, 308
    at Celanese, 326
    control systems, 319–329
    economic order quantity,
        315–319
    introduction to, 307–308
    supply chain ethics, 444t,
        456–457
    supply chains and, 309–314
    at Walmart, 307–308
Inventory measures, 364–365, 365f
Inventory placement, 367
Inventory pooling, 367
Inventory position (IP), 320
Inventory record, 553–555, 555f, 558f
Inventory remnants, 558

Inventory strategies, 96–97
Inventory turnover, 364
IP (inventory position), 320
IPad, 549f
ISO 9001:2008 documentation
    standards, 181
ISO 14000:2004 environmental
    management system, 181
ISO 26000:2010 social responsibility
    guidelines, 181–182
Iso-cost line, 593
Iso-profit line, 593, 594f

Japan, earthquake in, 19
JB Hunt Transport Services, 1
JCPenney, 320f
Jidoka, 280, 285, 285f, 303
JIT (just-in-time) philosophy, 277,
    285, 285f, 456
JIT II system, 278–279
JIT system, 277
Job and customer sequences, 513t
Job process, 95
Joint venture, 10
Jollibee Foods Corporation, 9f, 10
José's Authentic Mexican
    Restaurant, 154
Judgment methods, in forecasting,
    466–467, 470
Judgmental adjustments, 483
Just Born, Inc., 104f
Just-in-time (JIT) philosophy, 277,
    285, 285f, 456

Kaizen, 163, 277–278
Kanban system, 290–293, 291f, 556
Keppel Port, 373f
KitchenHelper Corp., 382
Kvichak Marine, 244

L4L (lot-for-lot) rule, 557, 557f
Labor classification, 294–295
Labor climate, 388
Labor laws, 371
Labor productivity, 16
Labor unions, 371
Labor utilization, 310
Laplace decision rule, 37
Latest finish time (LF), 57
Latest start time (LS), 57–58
Layout
    design of, 100
    information gathering, 97–98
    for lean systems, 285–287
    in process strategy, 92, 97–100
    weighted-distance method,
        98–100

Layout Solver, 100, 100f
LCL (lower control limit), 169
Lead time
  definition of, 9
  demand, distribution of, 322, 322f
  in line balancing, 256
  reorder point and, 320–323, 321f
  in resource planning, 556
  in supply chains, 374t, 430
  target inventory level and, 327–328
Lean systems, 275–297
  across the organization, 276
  at Autoliv, 303–304
  automation, 282
  continuous improvement, 277–278, 278f, 285f
  definition of, 276
  Five S (5S) practices, 282–283, 283f, 283t
  flexible workforce, 282
  human costs of, 293–294
  introduction to, 275–276
  inventory and scheduling, 295
  Kanban system, 290–293, 291f
  layout design, 285–287
  operational benefits and implementation issues, 293–295
  at Panasonic Corporation, 275–276
  process considerations, 279–283
  pull method of work flow, 279–280
  quality at the source, 280
  standardization, 282, 283t, 285f
  supply chain considerations, 278–279
  Total Preventive Maintenance (TPM), 283, 285f
  Toyota production system, 284–285
  uniform workstation loads, 281–282
  value stream mapping (VSM), 287–290, 287f
Learning curve analysis, 129, 129f
Lenovo, 19
Level of aggregation, 466
Level strategy, 515, 519–521, 521f
LF (latest finish time), 57
Life cycle, 70, 70f
Life-cycle assessment, 181
Lincolnway Energy, 207f
Line balancing, 254–258
  cycle time, 256, 259
  desired output rate, 255

idle time, efficiency, and balance delay, 256–257
  precedence diagram, 255
  solution, 258
  theoretical minimum, 256
Line charts, 134–135
Line flow, 94
Line process, 96, 254–259
Linear programming, 587–603
  basic concepts, 587–590
  computer output, 597–599
  computer solution, 596–599
  definition of, 587
  graphic analysis, 590–596
  problem formulation, 588–590
  at R.U. Reddie Corporation, 614–616
  simplex method, 596–597
  transportation method, 599–603, 600f
Linear regression, 470–472, 471f
Linearity, 588
Little's law, 234–235, 292
Load-distance method, 99, 393–394
Locating abroad, 10, 17
Location. *See* Supply chain location
Locus of control, 423–424
Logistics. *See also* Humanitarian logistics; Reverse logistics
  capacity and, 426
  costs, 371
  in lean systems, 295
  order fulfillment process, 425–428
  supply chain processes, 91t
Long-term capacity planning, 202
  diseconomies of scale, 204–205, 205f
  economies of scale, 204, 205f
  measurement of capacity and utilization, 203–204
  systematic approach, 208–211
Lot sizes, 279, 308
Lot sizing, 311, 317, 346f, 430, 556, 558
Lot-for-lot (L4L) rule, 557, 557f
Lots, 279
Low-cost operations, 12t, 14, 14t, 15, 15t
Lower control limit (LCL), 169
Lowe's, 514
LS (latest start time), 57–58

Machine productivity, 16
MAD (mean absolute deviation), 468
Make-or-buy decisions, 33–35, 369

Make-to-order (MTO) strategy, 96, 105, 373, 415
Make-to-stock (MTS) strategy, 97
Malcolm Baldrige National Quality Improvement Act, 182
Management, role of, 17
Management systems, 455
Manufacturing
  process structure in, 94–97
  productivity, 16
  supply chain design, 362–363, 363f
  supply chain location in, 388–389
Manufacturing firms, 510, 518f, 519
Manufacturing processes, 5–6
  automation, 103, 104f
  bottlenecks, management of, 248–254
  characteristics of, 5f
  decision patterns for, 106–107, 106f, 107f
  vs. service processes, 5–6
Manufacturing resource planning (MRP II), 562
Manugistics, 484
MAPE (mean absolute percent error), 468
MapInfo, 390, 391
MapPoint, 390
Marco's Franchising, 391
Market analysis, 10–11
Market effect, of offshoring, 371
Market research, 470
Market segmentation, 10
Marketing, 3, 3f, 428–429
Markets, proximity to, 388, 389–390
Mass customization, 97, 367–369
Mass production, 4, 97
Master Black Belts, 165
Master production schedule (MPS), 549
  available-to-promise quantities, 552–553, 552f
  development of, 550–552
  freezing, 553
  illustration of, 551f, 553f, 554f
  process of, 550f
  reconciling with sales and operations plans, 553
Master Production Scheduling Solver, 553f
Material requirements planning (MRP), 510, 546–564
  bill of materials, 547–549, 548f, 549f, 559f
  core processes and supply chain linkages, 563–564

dependent demand, 547, 547f
information flows in the supply chain, 564f
inputs, 546f
inventory record, 553–555, 555f, 558f
master production scheduling, 549–553
outputs, 559–562, 559f, 562f
planning factors, 556–559
at Winnebago Industries, 563
Material Requirements Planning Solver, 562f
Materials costs, 204, 418
Materials handling, 129
Matrix organizational structure, 52
Matsushita Corporation, 275
Maximax, in decision theory, 37
Maximin, in decision theory, 37
McDonald's, 10, 12t, 119–120, 123f, 293
McDonnell Douglas, 204
Mean absolute deviation (MAD), 468
Mean absolute percent error (MAPE), 468
Mean bias, 467
Mean squared error (MSE), 468
Memorial Hospital, 540–541
Metrics, 122, 131f, 139f. *See also* Benchmarking
Micrografx, 124
Microsoft Corporation, 49–50
Microsoft Project, 60
Microsoft Visio, 124
Minimax regret, in decision theory, 37
Minimum-cost schedule, 62–64
Min-Yo Garment Company, 269–272
Mixed arrangement, 228, 228f
Mixed strategy, 516, 518f
Mixed-model assembly, 281
Mixed-model line, 259
Mode selection, 426
Modular design, 369
Modularity, 549
Moe's Southwest Grill, 396f
Moment of truth, 93
Most likely time, 66
Motion, 277t
Motorola, 164, 463–464
MPS (master production schedule). *See* Master production schedule (MPS)
MRP explosion, 546, 560f
MRP II (manufacturing resource planning), 562

MSE (mean squared error), 468
MTO (make-to-order) strategy, 96, 105, 373, 415
MTS (make-to-stock) strategy, 97
Muda, 277, 287
Multifactor productivity, 16
Multiple regression analysis, 472
Multiple-channel, multiple-phase arrangement, 228, 228f, 236f
Multiple-channel, single-phase arrangement, 228, 228f, 235f
Multiple-server model, 233–234, 234f
Multiplicative seasonal method, 479–480
Mustang restoration, 87–88

NAFTA (North American Free Trade Agreement), 18
Naïve forecast, 473
National Family Opinion, 197
National Institute of Standards and Technology (NIST), 182
National Labor Relations Board (NLRB), 388
NC (normal cost), 61
Near-critical paths, 68–69
Nearest neighborhood (NN) heuristic, 448, 449–450
Needs assessment, 10–11
Negotiation, 421–422
Nested processes, 5, 485
Net income, 366
Network diagram, 54–56, 58f
New Balance, 293f
New service/product development process, 7, 416–417, 416f
New service/product introduction, 373t
Newsboy problem, 350
Nikon, 359–361
Nintendo, 19
Nissan, 528
NIST (National Institute of Standards and Technology), 182
NLRB (National Labor Relations Board), 388
NN (nearest neighborhood) heuristic, 448, 449–450
Nominal value, 177
Noninstantaneous replenishment, 345–347, 346f
Nonnegativity, 588
Normal cost (NC), 61
Normal distribution, 66f
Normal time (NT), 61
North American Free Trade Agreement (NAFTA), 18

NT (normal time), 61
Number of Containers Solver, 292f

Objective function, 587, 589, 593–594
Obsolescence, 309
OE (operating expense), 245t, 366
Offshoring, 370–371
Olympic Stadium, 53f
OM Explorer
  Bar, Pareto, and Line Charts Solver, 133
  break-even analysis, 395
  Break-Even Analysis Solver, 34f, 35
  common units of measurement, 519
  Demand During the Protection Interval Simulator, 328
  Economic Production Lot Size Solver, 347f
  exponential smoothing, 475
  forecasting routines in, 484
  introduction to, 21
  Layout Solver, 100, 100f
  Master Production Scheduling Solver, 553f
  Material Requirements Planning Solver, 562f
  Number of Containers Solver, 292f
  One-Period Inventory Decisions Solver, 351–352f
  Preference Matrix Solver, 35
  Process Solver, 130
  Quantity Discounts Solver, 350f
  Sales and Operations Planning with Spreadsheets Solver, 518
  Sales and Operations Planning with Spreadsheets Solver, 518
  Seasonal Forecasting Solver, 480, 480f
  sensitivity analysis, 352
  Single-Item MRP Solver, 557
  Time Series Forecasting Solver, 476, 482
  Trend Projection with Regression Solver, 476, 480
  Waiting-Lines Solver, 232, 234, 236
Omgeo, 138f
One-period decisions, 350–352, 351f
One-Period Inventory Decisions Solver, 351–352f
One-worker, multiple-machines (OWMM) cell, 285–286, 285f
On-time delivery, 12t, 14, 14t
Open issues, 69

Open order file, 270f
Open orders, 320
Operating expense (OE), 245t, 366
Operation, definition of, 2, 97
Operation strategy, 374t
Operational benefits, of lean
    systems, 293–295
Operational innovation, 22
Operational measures, financial
    measures, relationship with, 245t
Operations, 1–3, 3f, 65, 129
Operations management, 1–22
    at Chad's Creative Concepts, 29–30
    challenges in, 21–22
    competitive priorities and capa-
        bilities, 11–14
    historical evolution of, 3–4
    introduction to, 2–3
    as pattern of decisions, 15
    process view, 4–6
    as set of decisions, 20–21
    supply chain view, 6–11
    trends in, 16–20
    value creation through, 21
Operations planning and schedul-
    ing, 507–528
    across the organization, 508–509
    aggregation, 509
    Air New Zealand, 507–508
    definition of, 508
    demand management, 511–513
    demand options, 511–513, 513t
    introduction to, 507–508
    Memorial Hospital, 540–541
    relationship to other plans,
        509–511, 510f
    sales and operations plans,
        513–521
    scheduling, 521–528
    stages in, 509–511
    at Starwood Hotels and Resorts,
        539
    supply options, 513t, 514–515
Operations strategy, 7–8
Operations Terminal Information
    System (OTIS), 223
Opportunities, identification of,
    121–122
Optimality, range of, 596t, 598
Optimistic time, 66
Optimization, 587
Optional replenishment system, 329
Order fulfillment process, 7
    customer demand planning, 425
    logistics, 425–428
    production, 425
    supply planning, 425

Order placement, 429
Order qualifiers, 13, 13f
Order winners, 11–13, 13f
Ordering cost, 310, 315
Organizational structure, 52
OTIS (Operations Terminal
    Information System), 223
Output measures of capacity, 203,
    208
Output rate, 255
Outputs, 5
    desired rate of, 255
    material requirements planning,
        559–562, 559f, 562f
    nature of in service vs. manufac-
        turing, 5
    variation of, 166–169
Outsourcing, 91t, 369–370
Outsourcing processes, 369–372
Overproduction, 277t
Overtime, 514–515, 517
Overtime costs, 516t
OWMM (one-worker, multiple-
    machines) cell, 285–286, 285f
Ownership, 426

Pacing, 258
Paid undertime, 515
Panama Canal project, 60f
Panasonic Corporation, 275–276, 282
Parameters, 588
Parent, 547
Parent company facilities, proxim-
    ity to, 389
Pareto charts, 132–133, 133f, 136f,
    314
Part commonality, 549
Parts Emporium, Inc., 343–344
Part-time workers, 513t, 515
Passenger security process,
    236–237, 237f
Past due, 526
Path, 57
Payload ratio, 452
Payments to suppliers, 311
Payoff, in decision theory, 36
Payoff table, 36
PCA (percentage of calls answered),
    525
p-chart, 174–176, 176f
Percentage of calls answered (PCA),
    525
Performance, 158. See also Quality
Performance evaluation, 122–123,
    131–135, 131f
Performance measurements, 167,
    363–367, 431, 526

Periodic order quantity (POQ),
    556–558, 557f
Periodic reorder system, 325
Periodic review (P) system, 325–329,
    325f, 457
Perishable capacity, 512
Perpetual inventory system, 328
PERT (project evaluation and
    review technique), 54
Pessimistic time, 66
Phase, 227–228
Phoenician, 86–87, 152–153
Pie charts, 134
Pilferage, 309–310
Pinnacle Strategies, 243–244, 245
Pipeline inventory, 311, 312, 314
Plan-do-study-act cycle, 163–164,
    164f
Planned order release, 555
Planned receipts, 554–555
Planning, 53–69. See also
    Operations planning and
    scheduling
    analysis, 66–69
    cost–time trade-offs, analysis of,
        60–64
    network diagram, 54–56, 58f
    risk assessment, 64–66
    schedule development, 57–60
    work breakdown structure,
        53–54, 54f
Planning horizon, 208
Planning strategies, 515–516
Plants within plants (PWPs), 107
PlayStation, 49–50
Poka-yoke, 280, 281f
POM for Windows, 469, 472, 476,
    482, 484, 597
Popeye's Chicken, 391
POQ (periodic order quantity),
    556–558, 557f
Postponement, 96–97, 369
Power of Innovation program,
    152–153
PowerPoint, 40, 124
Precedence diagram, 255, 255f, 258f
Precedence relationships, 54
Precision Equipment, 359
Predetermined data approach, 128
Preemptive discipline, 229
Preference matrix, 35–36, 35f,
    392–393, 420
Preference Matrix Solver, 35
Preschedules appointments, 511,
    513t
Presourcing, 421
Prestige Products, 243–244

Prevention costs, 159
Priority rule, 226, 226f, 238
Priority sequencing rules, 526–527
Private carrier, 426
Probabilities, 36, 68
Probability distributions, 229–230
Problem-solving teams, 163
Process advantages, 204
Process analysis, 119–140
  across the organization, 121
  definition of, 121
  documentation, 122, 123–131
  illustration of, 121f
  introduction to, 119–121
  at José's Authentic Mexican
    Restaurant, 154
  management and implementa-
    tion, 123, 138–140
  at McDonald's, 119–120
  opportunities, identification of,
    121–122
  performance evaluation,
    122–123, 131–135, 131f
  redesign, 123, 135–138
  scope, definition of, 122
  at Starwood Hotels and Resorts,
    152–153
  systematic approach, 121–123
Process capability, 177–180, 178f
Process capability index, 178–179
Process capability ratio, 179
Process charts, 129–131, 130f, 165
Process choice, 95
Process costs, product volume and,
  102f
Process distribution, 167, 168f
Process divergence, 93
Process failure, 132, 135–137
Process improvement, 109, 250
Process integration, 371
Process reengineering, 108, 108t,
  250
Process segments, 107
Process simulation, 135
Process Solver, 130
Process strategy, 89–109
  across the organization, 90–91
  capital intensity, 102–104, 105,
    107
  at Custom Molds, Inc., 115–116
  customer involvement, 92,
    100–101
  decisions, 91–92, 92f
  definition of, 90
  at eBay, 89–90
  introduction to, 89–90
  layout, 92, 97–100

process structure in manufacturing, 94–97
process structure in services,
  92–94
resource flexibility, 102, 105, 107
strategic fit, 104–108
strategies for change, 108–109
supply chains and, 90–91
Process structure in manufacturing
  manufacturing process structur-
    ing, 95–96
  production and inventory strate-
    gies, 96–97
  product-process matrix, 94–95,
    95f
Process structure in services
  customer contact, 92–93, 93t
  customer-contact matrix, 93
  decision patterns for manufac-
    turing processes, 106
  service process structuring, 93–94
  strategic fit, 105
Processes, 4–6
  definition of, 2
  evaluation of, 33–35
  forecasting as, 483–485, 485t
  illustration of, 4f
  in lean systems, 279–283
  management of, 21–22, 21f
  nested, 5, 485
  outside of operations, 91t
  service and manufacturing, 5–6
  supply chain, 91, 91t
  working of, 4–5
Product families, 509
Product mix decisions, 250–252
Product returns, 371
Product variety, 373t
Product volume, process costs and,
  102f
Production, order fulfillment pro-
  cess, 425
Production plan, 509t
Production planning, transporta-
  tion method for, 600–603
Production schedule, 271f
Production strategies, 96–97
Productivity, 16–17
Productivity improvement, 16–17,
  443
Product-mix problem, 588–590
Product-process matrix, 94–95, 95f
Profit and loss schedule, 271f
Program, definition of, 51
Programmable automation, 103
Progressive Insurance, 22
Project, definition of, 50–51

Project evaluation and review technique (PERT), 54
Project management, 49–70
  definition and organization of
    projects, 51–52
  definition of, 51
  introduction to, 49–51
  monitoring and controlling,
    69–70
  at the Phoenician, 86–87
  planning, 53–69
  at Roberts Auto Sales and Service,
    87–88
  Xbox, 49–50
Project manager, selection of, 52
Project objective statement, 51–52
Project resources, monitoring, 70
Project schedule, development of,
  57–60
Project team capability, 65
Projected on-hand inventory,
  554–555
Promotional pricing, 511, 513t
Propulsion systems, 452
Protection interval, 322
Psychological impressions, 161
Publishing industry, 1–2, 105
Pull method, 279–280
Purchased items, 549
Purchasing, 295, 417
Pure Project organizational struc-
  ture, 52
Push method, 279–280
PWPs (plants within plants), 107

Qualitative factors, 392
Quality, 157–182
  acceptance sampling, 165–166,
    166f
  across the organization, 158
  Baldrige Performance Excellence
    Program, 182
  costs of, 159
  definition of, 160
  ethics and, 159–160
  international quality documenta-
    tion standards, 181–182
  introduction to, 157–158
  process capability, 177–180, 178f
  Six Sigma, 164–165, 164f, 178,
    197–198
  at Starwood Hotels and Resorts,
    197–198
  statistical process control,
    166–170
  statistical process control meth-
    ods, 170–177

total quality management, 160–164, 160f
at Verizon Wireless, 157–158
Quality at the source, 162, 280
Quality circles, 163
Quality engineering, 179–180
Quality loss function, 180, 181f
Quality management standards, 181
Quality of life, 389
Quantitative factors, 392
Quantitative forecasting method, 481–482
Quantity discount, 311, 348–350, 348f, 350f
Quantity Discounts Solver, 350f
Quebecor World, 1
QVC, 158

Radio-frequency identification (RFID), 308, 424
Ramp-up, 417
Random demand patterns, 465
Range, 168
Ranging screen, 598, 598f
Raving Brands, 396f
Raw materials (RM), 311
R-chart, 170–173, 171t
Real estate costs, 389
Rectilinear distance, 99
Recycling, 445–447
Redesign of processes, 123, 135–138, 139f
Reduction tactics, for inventory, 313–314
Reengineering, 108, 108t
Regular time wages, 516t
Relative drag, 452
Reliance Power, Ltd., 397
Reorder point (R)
    for constant demand and lead time, 320–321, 320f
    safety stock and, 323, 323f
    for variable demand and constant lead time, 321–322, 321f, 323
    for variable demand and lead time, 323–324
Reorder point (ROP) system, 320
Repeatability, 313
Reservations, 512, 513t
Resource acquisition, 70
Resource allocation, 70
Resource flexibility, 92, 102, 105, 107
Resource leveling, 70
Resource plan, 509t

Resource planning, 510, 543–567
    across the organization, 544
    bill of resources, 566–567, 566f
    definition of, 544
    dependent demand for services, 565–566
    at Dow Corning, 543–544
    enterprise resource planning, 544–546, 545f
    Flashy Flashers, Inc., 582–585
    introduction to, 543–544
    material requirements planning, 546–564
    for service providers, 565–567
Resources, proximity to, 389
Restaurants, 565
Results screen, 598, 598f
Return on assets (ROA), 366f, 367
Returns processor, 445
Revenue management, 512, 513t
Reverse auction, 423
Reverse logistics, 444–447, 444t. *See also* Logistics
Reward systems, 294–295
Rework, 159, 371
RFID (radio-frequency identification), 308, 424
Ridge Golf Course, 395f
Risk
    assessment of, 64–66
    decision making under, 38–39
    definition of, 64
    simulation, 65–66
    statistical analysis, 66, 66f
Risk minimization, 443
Risk-management plans, 64–65
Ritz Carolton resort, 11f, 12t
RM (raw materials), 311
ROA (return on assets), 366f, 367
Roberts Auto Sales and Service, 87–88
Rolex, 12t
ROP (reorder point) system, 320
Rope, 252
Rotating schedule, 523
Route planning, 447–448
Rowling, J. K., 1
R.R. Donnelly & Sons, 1, 105
R.U. Reddie Corporation, 614–616

SA8000:2008, 455
Safety stock inventory, 312–314, 323, 323f, 558–559, 558f
SAIC (Shanghai Automotive Industry Corporation), 10, 371f
Sales and Operations Planning with Spreadsheets Solver, 518

Sales and operations plans, 513–521
    constraints and costs, 516
    costs, types of, 516t
    definition of, 509t
    information inputs, 514
    for make-to-stock product, 517f
    manufacturer's plan using spreadsheet and mixed strategy, 518f
    planning strategies, 515–516
    as a process, 516–518
    spreadsheets, use of, 518–521
    supply options, 513t, 514–515
Salesforce estimates, 470
Sample coefficient of determination, 471
Sample correlation coefficient, 471
Sample mean, 167–168, 168f
Sample size, 167
Sampling, 167
Sampling distributions, 167–168, 169f
Sampling plan, 167
Sam's Club, 307
SAP, 543–544, 546
SAS/GIS, 390, 484
Scatter diagrams, 133
Schedule
    development of, 57–60
    fixed, 523
    stability of, 295
    status, 69–70
    workforce, 510, 513t, 523–525
Scheduled receipts (SR), 320, 554
Scheduling, 521–528. *See also* Operations planning and scheduling
    definition of, 509t
    of employees, 522–525
    Gantt charts, 521–522
    in lean systems, 295
    planning, 510
    sequencing jobs at a workstation, 525–527
    software support, 528
Scholastic, 1–2
Scope creep, 52
Scope of a project, 51–52, 122
SCOR model, 415–416, 416f
Scrap, 159
Seaport Access Tunnel, 65
Seasonal Forecasting Solver, 480, 480f
Seasonal index, 479
Seasonal patterns, 465, 466f, 479–481
Self-managed teams, 163
Sensitivity, project manager and, 52

Sensitivity analysis, 33, 319, 319t, 596, 596t
Sequencing, 525
Sequencing jobs at a workstation, 525–527
Service blueprints, 126–127, 126f
Service encounter, 93
Service facility, 226, 226f, 227–229, 228f, 237
Service level, 322
Service objective (SO), 525
Service or product needs, 10
Service processes, 5–6, 93f
  automation of, 103–104
  bottlenecks, management of, 247–248, 247f
  characteristics of, 5f
  decision patterns for, 105–106, 106f
  vs. manufacturing processes, 5–6
  structuring, 93–94
Service providers, 510, 519, 565–567
Service system, 226, 226f
Service time distribution, 230
Service/product attributes, 65
Service/product proliferation, 361–362
Services
  evaluation of, 32–33
  process structure in, 92–94
  productivity and, 16
  supply chain design, 362–363, 363f
  supply chain location in, 389–390
Setup, 279
Setup cost, 310, 315
Setup time, 209, 248
Shadow price, 596t, 599
Shanghai Automotive Industry Corporation (SAIC), 10, 371f
Sharp Corporation, 201–202
Shine, 283t
Shin-Etsu Chemical Company, 19
Shortest expected processing time (SPT), 229
Shortest route problem, 448
Shrinkage, 309–310
Simple moving average method, 473–474
Simplex method, 596–597
SimQuick, 236–237
Simulation
  in capacity planning, 213
  process simulation, 135
  risk and, 65–66
  waiting lines, 236–237, 237f
Single facility, location decision, 391–396

Single-bin system, 328
Single-channel, multiple-phase arrangement, 228, 228f
Single-channel, single-phase arrangement, 228, 228f, 232, 232f
Single-digit setup, 279
Single-Item MRP Solver, 557
Single-server model, 231–233
Site comparison, for supply chain location, 392–393
Six Sigma, 164–165, 164f, 178, 197–198
Six Sigma Improvement Model, 164–165, 165f
Skoda Automotive, 257f
SKU (stock-keeping unit), 314, 414–415, 466, 501
Slack variables, 595–596
Small lot sizes, 279
SmartDraw, 124
SO (service objective), 525
Social responsibility, 442, 444t, 452–457
Software support for scheduling systems, 528
Sole sourcing, 422
Sonic Distributors, 380–381
Sonoco, 17, 17f
Sort, 283t
Sources of supply, 397
Sourcing, 91t, 418–420, 438
Southwest Airlines, 222–223, 272–273
SPC (statistical process control). See Statistical process control (SPC)
Special inventory models
  noninstantaneous replenishment, 345–347, 346f
  one-period decisions, 350–352, 351f
  quantity discounts, 348–350, 348f, 350f
Special-purpose teams, 163
Spirit AeroSystems, 282f
Sports Optics, 359
Spreadsheets
  for chase strategy, 520f
  for level strategy, 521f
  for manufacturing firm, 519
  in sales and operations planning, 518–521
  for service providers, 519
SPT (shortest expected processing time), 229
SR (scheduled receipts), 320, 554
Staffing plan, 509t

Standard deviation, 168, 174
Standard deviation of the errors (σ), 468
Standard error of the estimate, 471
Standardization, 283t, 285f
Standardization of parts, 549
Standardized components, 282
Standardized work methods, 282
Starbucks, 100–101, 101f, 388f
Start times, 57–59, 58f
Starwood Hotels and Resorts, 29
  operations planning and sales, 539
  process analysis, 152–153
  process performance and quality, 197–198
  quality, 197–198
  sourcing, 438
  supply chain integration, 438
States of nature, 36
Statistical analysis, of risk, 66, 66f
Statistical criteria, 481
Statistical process control (SPC)
  coin catapult, 198–199
  control charts, 169–170, 169f, 170f
  control charts for attributes, 174–177
  control charts for variables, 170–174
  lean systems and, 293
  methods, 163, 170–177
  outputs, variation of, 166–169
Steering team, 122
Steinway & Sons, 162
Stock-keeping unit (SKU), 314, 414–415, 466, 501
Stockout, 310, 513, 513t
Stockout costs, 516t
Storage, on process charts, 130
Storage costs, 309
Straighten, 283t
Strategic alliances, 9–10
Strategic fit, 64, 90, 104–108, 106f
Subassembly, 548–549
Subcontractors, 513t, 515
Successive stocking points, 311f
Suggestion system, 122
Sunbelt Pool Company, 398
Supplier certification and evaluation, 420
Supplier relationship process, 7, 417–424
  buying, 422–424
  design collaboration, 421
  information exchange, 424
  negotiation, 421–422
  sourcing, 418–420, 438

Supplier selection, 374t, 418–420
Suppliers, 278–279, 389
Supply chain design, 359–375
  across the organization, 361–362
  for the Boeing Dreamliner airplane, 372
  for Brunswick Distribution, Inc., 381–383
  definition of, 361
  efficiency and responsiveness, 372–375, 373t, 374t, 375f
  financial measures, 365–367
  introduction to, 359–361
  inventory measures, 364–365, 365f
  inventory placement, 367
  for manufacturing, 362–363, 363f
  mass customization, 367–369
  Nikon, 359–361
  outsourcing processes, 369–372
  performance measurement, 363–367
  reverse logistics, 444–445
  for services, 362–363, 363f
  for Sonic Distributors, 380–381
  strategic implications, 372–375
Supply chain ethics, 444t, 456–457
Supply chain integration, 411–431
  across the organization, 412–413
  customer relationship process, 428–430
  definition of, 412
  dynamics of, 413–416, 413f
  Eastman Kodak, 411–412
  external causes of disruptions, 414, 415f
  illustration of, 412f
  implications of disruptions, 414–415
  internal causes of disruptions, 414
  introduction to, 411–412
  new service/product development process, 416–417, 416f
  order fulfillment process, 425–428
  process measures, 431t
  at Starwood Hotels and Resorts, 438
  supplier relationship process, 417–424
  supply chain performance, 430–431
  at Wolf Motors, 439
Supply chain linkages, 563–564
Supply chain location, 385–400
  across the organization, 387
  Bavarian Motor Works (BMW), 385–387

break-even analysis, 395–396, 396f
for a facility within a supply chain network, 396–400
factors affecting, 387–390
for General Electric (GE), 396, 397
geographical information systems, 390–391, 397
GIS method for multiple facilities, 397
introduction to, 385–387
load-distance method, 393–394
in manufacturing, 388–389
in services, 389–390
for a single facility, 391–396
site comparison, 392–393
transportation method, 397–400
Supply chain operations reference model, 415
Supply chain processes, 91, 91t
Supply chain sustainability, 441–457
  across the organization, 443
  at Clif Bar & Company, 461
  environmental responsibility, 443–452, 444t
  examples of, 444t
  Federal Express, 441–442, 453
  humanitarian logistics, 452–457, 453f
  illustration of, 443f
  introduction to, 441–443
  management of, 457
  social responsibility, 442, 444t, 452–457
Supply chains, 6–11
  at Celanese, 326
  close supplier ties, 278–279
  closed-loop, 444–445, 445f
  corporate strategy, 7, 8–10, 8f
  at Deckers Outdoor Corporation, 500–502
  definition of, 2, 360, 361
  disaster relief, 444t, 453–455, 453f
  efficiency curve, 361, 361f
  in House of Toyota, 284–285, 285f
  illustration of, 6f, 360f
  innovation, adding value with, 22
  inventory and scheduling, 309–314
  Japanese earthquake, 19
  lead time in, 374t, 430
  in lean systems, 278–279
  management of, 21–22, 21f

market analysis, 10–11
operations strategy, 7–8
performance measurement, 363–367, 430–431
process strategy and, 90–91
processes in, 6–11
small lot sizes, 279
Supply options, 513t, 514–515
Supply planning, 425
Supply Pro, 244
Support, customer, 161
Support processes, 7
Surplus variables, 595–596
Sustain, 283t
Sustainability, 442–443. *See also* Supply chain sustainability
Swift Electronic Supply, Inc., 341–343
Swim lane flowcharts, 125–126, 125f

Tableau, 398, 398f, 599, 600f
Taguchi's quality loss function, 180, 181f
Take back, 447
Takt time, 281, 288–290
Tardiness, 526
Target, 395f
Target inventory level, 327–328
Tariffs, 371
Taxes, 309–310, 371, 389
TBO (time between orders), 318
Team, selection of, 52
Teams, total quality management, 162–163
Technical competence, project manager and, 52
Technological forecasting, 470
Technology, advances in, 4
Technology licensing, 10
Technology transfer, 371
Texas Instruments, 280
Theoretical minimum (TM), 256
Theory of constraints (TOC), 245–247, 246t
Third-party logistics provider (3PL), 426
Throughput (T), 245t
Throughput time, 247
Timberland, 20, 20f
Time, aggregation, 509
Time between orders (TBO), 318
Time series, 465
Time Series Forecasting Solver, 476, 482
Time study method, 127–128
Time-based competition, 11

Time-series methods, of forecasting, 467, 472–481
  estimating the average, 473–475
  naïve forecast, 473
  seasonal patterns, 479–481
  selection criteria, 481
  trend projection with regression, 467, 476–478
TM (theoretical minimum), 256
TMMI (Toyota Motor Manufacturing, Indiana), 281f
TNT, 453
TOC (theory of constraints), 245–247, 246t
Tolerance, 177
Top quality, 12t, 14, 14t
Total cost analysis, for supplier selection, 418–419
Total Preventive Maintenance (TPM), 283, 285f
Total project costs, 60
Total quality management (TQM)
  continuous improvement, 163–164
  customer satisfaction, 160–161, 160f
  employee involvement, 161–163
Total revenue, 365
Toyota Corolla, 13
Toyota Corporation, 277, 278, 281
Toyota Motor Manufacturing, Indiana (TMMI), 281f
Toyota Production System (TPS), 4, 284–285, 303
TPM (Total Preventive Maintenance), 283, 285f
TPS (Toyota Production System), 4, 284–285, 303
Tracking signals, 481–482, 482f
Trade barriers, 18
Trade-in, 447
Trading blocks, 18
Traditional method, 250
Trans Alaska Pipeline System, 426f
Transocean Ltd, 243
Transportation, 277t
  costs of, 310, 389–390
  globalization and, 18
  on process charts, 129
Transportation distance, 447–448
Transportation method, 397–400
  in linear programming, 599–603, 600f
  for location problems, 397–400
  for production planning, 600–603
  for supply chain location, 397–400

Transportation mode, 452
Transportation problem, 599
Transportation technology, 452
Traveling salesman problem, 448, 448f
Trend, 476
Trend demand patterns, 465, 466f
Trend projection with regression, 467, 476–478
Trend Projection with Regression Solver, 476, 480
Triveni Engineering and Industries, Ltd., 397
Two-bin system, 324–325
Type I error, 170
Type II error, 170

UCL (upper control limit), 169
UN Development Program, 453
Uncertainty, decision making under, 37–38
Undertime, 515
Uniform workstation loads, 281–282
Unilever, 474f
United Nations, 453
United Parcel Service (UPS), 4, 12t, 282, 359–360, 411–412, 453
United States Postal Service (USPS), 12t
Units of measurement, 466
University of Pittsburgh Medical Center Shadyside, 294
Upper control limit (UCL), 169
UPS (United Parcel Service), 4, 12t, 282, 359–360, 411–412, 453
Upstream, in supply chain integration, 413
U.S. Marine Corps Maintenance Center, 253–254
Usage quantity, 548
USPS (United States Postal Service), 12t
Utilities, 389
Utilization, 204
Utilization (U), 245t
Utilized time, 519

Vacation schedules, 513t, 515
Value, 161
Value analysis, 421
Value stream mapping (VSM), 287–290, 287f
Variable cost, 32
Variables, 167, 170–174
Variety, 12t, 14t
Vendor managed inventories (VMI), 424

Verizon Wireless, 157–158
Vertical integration, 370
Visual solution, in linear programming, 594
Visual system, 324
VMI (vendor managed inventories), 424
Volkswagen (VW), 10, 103f
Volume flexibility, 12t, 14t, 15, 15t
Volume needs, 10
VSM (value stream mapping), 287–290, 287f
VW (Volkswagen), 10, 103f

WACC (weighted average cost of capital), 309
Wait-and-see strategy, 206–207, 206f
Waiting, 277t
Waiting lines, 225–238
  arrangement of, 227–229, 228f
  arrival rates, 237
  decision area, 237–238
  definition of, 225
  formation of, 225–226
  probability distributions, 229–230
  simulation, 236–237, 237f
  single vs. multiple, 227
  structure of problems with, 226–229
Waiting-line models, 212, 212f, 226f
  finite-source model, 235–236
  Little's law, 234–235, 292
  multiple-server model, 233–234, 234f
  operations analysis using, 230–236
  single-server model, 231–233
Waiting-line theory, 226
Waiting-Lines Solver, 232, 234, 236
Walmart, 4, 307–308, 446, 452, 485f
Walt Disney World, 225, 231f
Warehousing, 91t
Warner-Lambert, 485f
Warranty, 159
Waste, 277–278, 277t, 287
Watt, James, 3
WBS (work breakdown structure), 53–54, 54f
Weeks of supply, 364
Weighted average cost of capital (WACC), 309
Weighted moving average method, 474
Weighted payoff, 37
Weighted-distance method, 98–100
West Marine, 484f

Whirlpool, 514
Whitney, Eli, 3
Wii game system, 18f, 19
Winnebago Industries, 563
WIP (work-in-process), 234–235, 311
Wipro, 397
Wistron, 50
Wolf Motors, 439
Work breakdown structure (WBS), 53–54, 54f
Work elements, 254
Work flow, pull method of, 279–280
Work measurement techniques, 127–131
    elemental standard data approach, 127
    learning curve analysis, 129, 129f

    predetermined data approach, 128
    process charts, 129–131, 130f
    time study method, 127–128
    work sampling method, 128, 128f
Work sampling method, 128, 128f
Workforce
    aggregation, 509
    core competencies, 9
    flexibility of, 102, 282
    scheduling, 522–525
    underutilization of, 277t
Workforce adjustment, 513t, 514
Workforce diversity, 19–20
Workforce schedule, 510, 513t, 523–525
Workforce utilization, 513t, 514–515

Working capital, 366
Working hours, 455
Work-in-process (WIP), 234–235, 311
Workstation loads, 281–282
World Food Program, 453

Xbox, 49–50
$\bar{x}$-Chart, 171–174, 171t

Yankee Fork and Hoe Company, 502–503
Yellow Transportation, 1
Yield management, 512

Zara, 12t
Zastava Corporation, 13
ZTE Corporation, 19